Adaptive Catchment Management and Reservoir Operation

Adaptive Catchment Management and Reservoir Operation

Special Issue Editors

Guangtao Fu
Guangheng Ni
Chi Zhang

MDPI • Basel • Beijing • Wuhan • Barcelona • Belgrade

MDPI

Special Issue Editors
Guangtao Fu
University of Exeter
UK

Guangheng Ni
Tsinghua University
China

Chi Zhang
Dalian University of Technology
China

Editorial Office
MDPI
St. Alban-Anlage 66
4052 Basel, Switzerland

This is a reprint of articles from the Special Issue published online in the open access journal *Water* (ISSN 2073-4441) from 2017 to 2019 (available at: https://www.mdpi.com/journal/water/special_issues/Adaptive_Catchment_Reservoir_Operation)

For citation purposes, cite each article independently as indicated on the article page online and as indicated below:

LastName, A.A.; LastName, B.B.; LastName, C.C. Article Title. *Journal Name* **Year**, *Article Number*, Page Range.

ISBN 978-3-03897-738-4 (Pbk)
ISBN 978-3-03897-739-1 (PDF)

Contents

About the Special Issue Editors

Guangtao Fu, Professor of Water Intelligence at the University of Exeter, has a research focus on developing and applying new computer models, data analytics and artificial intelligence tools to tackle urban water challenges such as water supply resilience, network leakage, urban flooding and urban wastewater management. He is a Royal Society Industry Fellow and a Turing Fellow at the Alan Turing Institute. He has authored 130 papers in international peer-reviewed journals and conference papers, and received several awards including the 2014 'Quentin Martin Best Practice Oriented Paper' and the 2018 'Best Research-Oriented Paper' awards from the American Society of Civil Engineers.

Guangheng Ni is Professor of Hydrology and Water Resource in the Department of Hydraulic Engineering at Tsinghua University. He has worked on a variety of state and international projects including reasonable use of transboundary waters, impacts of changing environments on hydrology and water resources, ecological regulation and remediation for cascade hydropower development, urban flood forecasting system with coupling of hydrologic and atmospheric processes. He has published over 100 research papers and is the developer of several hydrological models. He received several awards from the Ministry of Education, Ministry of Water Resources.

Chi Zhang, Professor at Dalian University of Technology, has a research focus on water resources modelling and analysis, such as watershed hydrology simulation under changing environments, optimal operation of multi-reservoir systems, flood control and decision support. He is a Yangtze River Distinguished scholar awarded by the Ministry of Education in China and is supported by the 'New Century Excellent Talent of the Ministry of Education Plan'. He has published 90 papers, 40 of which were indexed by SCI and published in leading journals in his field, and received several awards including the first prize for 'Science and Technology Progress, Ministry of Education' (ranking No. 1), and 2018 'Best Research-Oriented Paper' awards from the American Society of Civil Engineers.

water

MDPI

Editorial

Recent Advances in Adaptive Catchment Management and Reservoir Operation

Guangtao Fu [1,*], Guangheng Ni [2] and Chi Zhang [3]

[1] Centre for Water Systems, College of Engineering, Mathematics and Physical Sciences, University of Exeter, North Park Road, Exeter, Devon EX4 4QF, UK

[2] Department of Hydraulic Engineering, Tsinghua University, No. 1 Qinghuayuan, Haidian, Beijing 100084, China; ghni@tsinghua.edu.cn

[3] School of Hydraulic Engineering, Dalian University of Technology, Dalian 116023, China; czhang@dlut.edu.cn

* Correspondence: G.Fu@exeter.ac.uk; Tel.: +44-1392-723692

Received: 9 February 2019; Accepted: 25 February 2019; Published: 27 February 2019

Abstract: This editorial introduces the latest research advances in the special issue on catchment management and reservoir operations. River catchments and reservoirs play a central role in water security, community wellbeing and social-economic prosperity, but their operators and managers are under increasing pressures to meet the challenges from population growth, economic activities and changing climates in many parts of the world. This challenge is tackled from various aspects in the 27 papers included in this special issue. A synthesis of these papers is provided, focusing on four themes: reservoir dynamics and impacts, optimal reservoir operation, climate change impacts, and integrated modelling and management. The contributions are discussed in the broader context of the field and future research directions are identified to achieve sustainable and resilient catchment management.

Keywords: adaptive management; catchment modelling; integrated management; reservoir operation

1. Introduction

Catchment management and reservoir operation play a central role in water security, community wellbeing and social-economic prosperity. Building reservoirs, which can store water during wet periods and release it during dry periods, is an ancient approach to supply water to meet the ever growing municipal, industrial and agricultural demands and to protect communities and cities from flooding. Reservoirs are estimated to contribute directly to 12–16% of global food production as they provide irrigation for 30–40% of a total of 268 million hectares of irrigated lands worldwide [1]. Nowadays, however, many reservoirs have to meet new demands in hydropower generation and environmental flow regulation. For example, hydropower from reservoirs is the main source of renewable, clean energy, and it accounts for about 19% of the world's electricity supply and 97% of all electricity generated from renewable sources [1,2]. Reservoir operation and management have to be considered in the context of a system of systems to address the different, often conflicting, needs of various stakeholders and their interdependency in the catchment and beyond.

The management of river catchments and reservoirs is now under increasing pressure from population growth, economic activities and changing climate means and extremes in many parts of the world. By the year 2050, the world population is expected to increase to nine billion and agricultural production will need to increase by 70% to cope with the population increase and rising food consumption [3]. This poses a huge challenge for land expansion and water withdrawals for irrigation from reservoirs. Further, the total world energy consumption has been projected to rise from 549 quadrillion British thermal units (Btu) in 2012 to 815 quadrillion Btu in 2040, an increase of 48% [4], of which an increasing proportion will be generated from renewable sources, including hydropower. According to the United Nations estimates, climate change could lead to an increase of 20% in water scarcity in the coming decades [5]. The above-mentioned factors place an increasing pressure on the effective management of surface water resources and environments. Adaptive management of river catchments and reservoirs is crucial to guarantee sustainability in the water-energy-food-environment nexus, which may become a major problem for sustainable development by 2050 [6,7].

Adaptive management of river catchments and reservoirs requires an in-depth understanding of the various hydrological processes and the impacts of future uncertainties, and then the development of robust, sustainable solutions to meet the needs of various stakeholders and the environment. Research shows that small perturbations in precipitation frequency and/or quantity can result in significant impacts on the discharge [8], and modest changes in natural inflows result in large changes in reservoir storage [9]. Further, the changes in the hydrologic cycle will affect energy production and water management adaptation strategies should be developed [10]. Climate change may confound water resources planning because of the deep uncertainty in the local effects [11] and the system robustness and resilience need to be fully understood [12]. Under deep uncertainty, the adaptive operational approach may prove a reliable and sustainable overall management strategy [13].

To tackle the huge challenges in moving towards adaptive catchment management, a special issue on adaptive catchment management and reservoir operation was proposed to review the latest developments in cutting-edge knowledge, novel methodologies, innovative management options and case studies in the field of water resources and catchment management. The main research of the special issues focuses on the following four themes: reservoir dynamics and their impacts on the sediment concentration in the reservoir and river, optimal operation of reservoirs, climate change impacts and integrated catchment modelling and management. These themes are covered by the 27 papers included in this special issue, as introduced in Section 2. This special issue will help researchers and practical engineers understand the current challenges in catchment and reservoir management and the current state-of-the-art knowledge and technologies employed to tackle these challenges. It will encourage managers and operators to use advanced tools for better planning and management of catchments and reservoirs, and thus improve the sustainability and resilience of water resources systems.

2. Overview of The Special Issue

2.1. Reservoir Dynamics and Impacts

The construction of dams interrupts the natural continuity of rivers; this not only alters river hydrology, hydraulics and aquatic ecology in the catchment, but also makes the reservoir itself a complex system in which various processes need to be better understood. The studies in this special issue provide an enhanced understanding of the processes within a reservoir and at the catchment scale that could be used to improve catchment and reservoir management.

The sediment deposition within reservoirs has been a key issue that affects reservoir capacity during the design life time. In China, 8 billion m^3 of storage capacity of 20 large reservoirs has been lost due to sedimentation, which is 66% of the total reservoir capacity of these reservoirs [14]. The research topics in this special issue range from the loads of the sediments and the distribution of sediments in the reservoir to the sediment flushing efficiency of the reservoir. Ezz-Aldeen et al. [15] assessed the annual runoff and sediment loads of the Dokan Dam watershed using the Soil and Water Assessment Tool (SWAT) and identified the basins with a high sediment load per unit area. Chen and Tsai [16] proposed a two-dimensional bed evolution model to estimate the sediment distribution, bed evolution within a reservoir. He [17] and He et al. [18] quantified the effects of near-bed concentration on sediment flux after the construction of the reservoir. To reduce the sediments, Esmaeili et al. [19] studied the effects of water and discharge manipulation and the construction of an auxiliary channel on sediment flushing efficiency with a three-dimensional numerical analysis.

The construction of dams poses risks to the deterioration of upstream and downstream riverine and riparian ecosystems as they can affect the flow regimes, sediment transport, biogeochemical cycle, and downstream water temperature. Marcinkowski et al. [20] quantified the long-term downstream effects of the Siemianówka Reservoir on the river's flow regime, including the flow duration and recurrence of floods and droughts, and concluded that the upstream dam is the main driver inducing the deterioration of the anastomosing stretch of downstream. Jiang et al. [21] investigated the effects of the impoundment and the operation of the Jinghong Reservoir on downstream thermal regimes through a three-dimensional hydro- and thermodynamic model. Yang et al. [22] evaluated the impacts of water transfers on nitrogen (N) and phosphorus (P) uptake in the inner protection zone of the receiving reservoir of the largest inter-basin water transfer project in China, i.e., the South-to-North Water Transfer Project in China.

Recent research has confirmed that reservoirs emit a significant amount of greenhouse gas emissions, but one of the challenges is how to accurately quantify greenhouse gas emissions from individual reservoirs. Chen et al. [23] used two artificial neutral networks to estimate the total carbon dioxide emissions from the world's reservoirs and concluded that the models can be used to predict CO_2 emissions from new reservoirs.

2.2. Optimal Reservoir Operation

The optimal design and operation of reservoirs has long been studied, but challenges remain in many areas, such as improving the search efficiency, balancing objectives and increasing system resilience, which are addressed in this special issue.

In addition to reliability and risk, there is a need to consider the performance of reservoirs from other aspects, such as vulnerability and resilience [24]. Paseka et al. [25] considered the resilience and robustness of the reservoir as key criteria to address the uncertainties from a range of future climate scenarios, and demonstrated an optimal design approach using a multipurpose reservoir with a number of objectives, including downstream environmental flow, water supply and hydropower generation. Chen et al. [26] suggested that vulnerability, which is quantified as the expected violation of the generation yield, should be considered in the optimal scheduling of hydropower generation.

In order to restore the natural stream flows and reduce the negative impacts of reservoirs, the optimal operation of the reservoirs should consider social-economic and ecological objectives. Liu et al. [27] developed the hedging rules to consider economic and ecologic objectives during reservoir operation. Zhou et al. [28] demonstrated how the joint operation of several reservoirs can effectively reduce the flood damage in several areas downstream.

New optimisation algorithms have been developed to improve the search efficiency and solution quality when optimising complex, large water resources systems. Wen et al. [29] proposed an improved differential evolution algorithm to solve the optimal operation model of the long-term scheduling of large-scale cascade hydropower stations. Uysal et al. [30] used probabilistic streamflow forecasts with a lead time of 48 hours to improve real-time flood control solutions. In the short-term operation of hydropower plants, Ji et al. [31] proposed a new progressive optimality algorithm to consider the interactions between two cascaded reservoirs. Bhatia et al. [32] used time-varying hedging policies to improve the reservoir performance, which significantly reduced the water shortage ratio and vulnerability in the case of Hemavathy Reservoir in Southern India.

2.3. Climate Change Impacts

Understanding the river runoff uncertainty is essential for the better adaptive management of water resources under changing environments. For the changes of historical runoff, Ye et al. [33] proposed two methods to study the quantitative relationship between daily and monthly flow duration curves. Kinouchi et al. [34] quantified the basin-scale seasonal rainfall and elucidated the quantitative relationship with existing climate indices. For the change of runoff in the future,

Zhu et al. [35] investigated the variations of future climate and water resources availability in the Biliu River basin in the northeast China based on the downscaled climate data.

The impacts of climate change on water systems have gained a lot of attention in the past decades [36]. Abera et al. [37] assessed existing and future hydropower operation at the Tekeze Reservoir in the face of climate change. Jiang et al. [38] built a system dynamics model to simulate the evolution of the land and water resource systems in Heilongjiang Province under different climate, economic and policy scenarios.

2.4. Integrated Modelling and Management

Integrated catchment management has long been promoted for sustainable resource management. It recognises the complex relationships between hydrological, ecological and socio-economic systems within a catchment, and seeks to integrate different systems, models and stakeholders for water management. Zhao et al. [39] integrated a 1D water quality model and an environmental fluid dynamics model to assess the environmental capacity in the Huangshi Reservoir basin, which helped to determine the reduction targets to achieve the water quality requirements in the reservoir. An integrated model was also developed to investigate the flood risk of a key water infrastructure—the South-to-North Water Diversion project in China—and key model parameters were identified by Jin et al. [40]. Tian et al. [41] revealed that the joint operation of surface water and groundwater reservoirs is key to achieve balance among the agricultural water demand, ecological water demand and groundwater sustainability.

The impact of flooding has to be considered from an integrated perspective. Choi et al. [42] conducted a multi-scale analysis to investigate the relationships among the bitterling and mussel communities, lentic habitat structures and channel characteristics, and provided new insights into flood and sediment management at the catchment scale. The integration is also required at the stakeholder level. Indeed, effective cooperation among stakeholders was demonstrated to have a significant impact on water resource allocation in the Hanjiang River Basin through a game theory-based bi-level optimisation model [43].

3. Conclusions

The research articles included in this special issue addressed the challenges in catchment and reservoir management and proposed new methods, models and tools for a wide range of contemporary issues in the following themes: reservoir dynamics and impact analysis of dam construction, optimal reservoir operation, climate change impacts on hydrological processes and water management, and integrated catchment management.

With a better understanding of the interdependency and complexity of various processes and systems in a catchment, the utilization of water resources must be considered from an integrated perspective, including the integration of physical, chemical and ecological processes; integration of information and communications technology (ICT) and infrastructure [44]; and cooperation between institutions and stakeholders. Meanwhile, growing populations and economic activities increase the demands on food, energy and water, and their nexus needs to be addressed in the context of deep uncertainty arising from climate change [7,45]. To achieve sustainable and resilient catchment management, significant efforts are required from the research and practical communities to develop integrated models, new artificial intelligence tools, and robust and adaptive management options to meet the needs of various stakeholders and the environment.

funding: This work is funded by the UK Royal Society through an international exchanges project (Ref: IEC\NSFC\170249) and an Industry Fellowship to the first author (Ref: IF160108).

Acknowledgments: We thank all authors for their notable contributions to this special issue and the Water editorial team for their great support during the review of the submitted manuscripts.

Conflicts of Interest: The authors declare no conflict of interest.

References

1. The World Commission on Dams. *Dams and Development: A Framework for Decision Making*; Earthscan: London, UK, 2000.
2. Demirbas, A. Focus on the World: Status and Future of Hydropower. *Energy Sources Part B Econ. Plan. Policy* **2007**, *2*, 237–242. [CrossRef]
3. Bruinsma, J. *The Resource Outlook to 2050: By How Much Do Land, Water and Crop Yields Need to Increase by 2050?* Food and Agriculture Organization of the United Nations: Rome, Italy, 2009.
4. The US Energy Information Administration. *International Energy Outlook 2016 with Projections to 2040*; Energy Information Administration: Washington, DC, USA, 2016.
5. The United Nations. *The United Nations World Water Development Report 2: Water—A Shared Responsibility*; UN: New York, NY, USA, 2006.
6. Chen, J.; Shi, H.; Sivakumar, B.; Peart, M.R. Population, water, food, energy and dams. *Renew. Sustain. Energy Rev.* **2016**, *56*, 18–28. [CrossRef]
7. Cai, X.; Wallington, K.; Shafiee-Jood, M.; Marston, L. Understanding and managing the food-energy-water nexus—Opportunities for water resources research. *Adv. Water Resour.* **2018**, *111*, 259–273. [CrossRef]
8. Risbey, J.S.; Entekhabi, D. Observed Sacramento Basin streamflow response to precipitation and temperature changes and its relevance to climate impact studies. *J. Hydrol.* **1996**, *184*, 209–223. [CrossRef]
9. Christensen, N.S.; Wood, A.W.; Voisin, N.; Lettenmaier, D.P.; Palmer, R.N. The effects of climate change on the hydrology and water resources of the Colorado River basin. *Clim. Chang.* **2004**, *62*, 337–363. [CrossRef]
10. Xu, C.; Singh, V.P. Review on regional water resources assessment models under stationary and changing climate. *Water Resour. Manag.* **2004**, *18*, 591–612. [CrossRef]
11. Minville, M.; Brissette, F.; Leconte, R. Uncertainty of the impact of climate change on the hydrology of a nordic watershed. *J. Hydrol.* **2008**, *358*, 70–83. [CrossRef]
12. Casal-Campos, A.; Sadr, S.M.K.; Fu, G.; Butler, D. Reliable, Resilient and Sustainable Urban Drainage Systems: An Analysis of Robustness under Deep Uncertainty. *Environ. Sci. Technol.* **2018**, *52*, 9008–9021. [CrossRef] [PubMed]
13. Ajami, N.K.; Hornberger, G.M.; Sunding, D.L. Sustainable water resource management under hydrological uncertainty. *Water Resour. Res.* **2008**, *44*. [CrossRef]
14. Wang, Z.; Hu, C. Strategies for managing reservoir sedimentation. *Int. J. Sediment Res.* **2009**, *24*, 369–384. [CrossRef]
15. Ezz-Aldeen, M.; Hassan, R.; Ali, A.; Al-Ansari, N.; Knutsson, S. Watershed Sediment and Its Effect on Storage Capacity: Case Study of Dokan Dam Reservoir. *Water* **2018**, *10*, 858. [CrossRef]
16. Chen, C.; Tsai, C. Estimating Sediment Flushing Efficiency of a Shaft Spillway Pipe and Bed Evolution in a Reservoir. *Water* **2017**, *9*, 924. [CrossRef]
17. He, L. Quantifying the Effects of Near-Bed Concentration on the Sediment Flux after the Operation of the Three Gorges Dam, Yangtze River. *Water* **2017**, *9*, 986. [CrossRef]
18. He, L.; Chen, D.; Zhang, S.; Liu, M.; Duan, G. Evaluating Regime Change of Sediment Transport in the Jingjiang River Reach, Yangtze River, China. *Water* **2018**, *10*, 329. [CrossRef]
19. Esmaeili, T.; Sumi, T.; Kantoush, S.; Kubota, Y.; Haun, S.; Rüther, N. Three-Dimensional Numerical Study of Free-Flow Sediment Flushing to Increase the Flushing Efficiency: A Case-Study Reservoir in Japan. *Water* **2017**, *9*, 900. [CrossRef]
20. Marcinkowski, P.; Grygoruk, M. Long-Term Downstream Effects of a Dam on a Lowland River Flow Regime: Case Study of the Upper Narew. *Water* **2017**, *9*, 783. [CrossRef]
21. Jiang, B.; Wang, F.; Ni, G. Heating Impact of a Tropical Reservoir on Downstream Water Temperature: A Case Study of the Jinghong Dam on the Lancang River. *Water* **2018**, *10*, 951. [CrossRef]
22. Yang, S.; Bai, J.; Zhao, C.; Lou, H.; Wang, Z.; Guan, Y.; Zhang, Y.; Zhang, C.; Yu, X. Decline of N and P Uptake in the Inner Protection Zone of a Terminal Reservoir during Inter-Basin Water Transfers. *Water* **2018**, *10*, 178. [CrossRef]
23. Chen, Z.; Ye, X.; Huang, P. Estimating Carbon Dioxide (CO_2) Emissions from Reservoirs Using Artificial Neural Networks. *Water* **2018**, *10*, 26. [CrossRef]
24. Zhang, C.; Xu, B.; Li, Y.; Fu, G. Exploring the relationships among reliability, resilience, and vulnerability of water supply using many-objective analysis. *J. Water Resour. Plan. Manag.* **2017**, *143*, 04017044. [CrossRef]

25. Paseka, S.; Kapelan, Z.; Marton, D. Multi-Objective Optimization of Resilient Design of the Multipurpose Reservoir in Conditions of Uncertain Climate Change. *Water* **2018**, *10*, 1110. [CrossRef]
26. Chen, C.; Kang, C.; Wang, J. Stochastic Linear Programming for Reservoir Operation with Constraints on Reliability and Vulnerability. *Water* **2018**, *10*, 175. [CrossRef]
27. Liu, Y.; Zhao, J.; Zheng, H. Piecewise-Linear Hedging Rules for Reservoir Operation with Economic and Ecologic Objectives. *Water* **2018**, *10*, 865. [CrossRef]
28. Zhou, C.; Sun, N.; Chen, L.; Ding, Y.; Zhou, J.; Zha, G.; Luo, G.; Dai, L.; Yang, X. Optimal Operation of Cascade Reservoirs for Flood Control of Multiple Areas Downstream: A Case Study in the Upper Yangtze River Basin. *Water* **2018**, *10*, 1250. [CrossRef]
29. Wen, X.; Zhou, J.; He, Z.; Wang, C. Long-Term Scheduling of Large-Scale Cascade Hydropower Stations Using Improved Differential Evolution Algorithm. *Water* **2018**, *10*, 383. [CrossRef]
30. Uysal, G.; Alvarado-Montero, R.; Schwanenberg, D.; Şensoy, A. Real-Time Flood Control by Tree-Based Model Predictive Control Including Forecast Uncertainty: A Case Study Reservoir in Turkey. *Water* **2018**, *10*, 340. [CrossRef]
31. Ji, C.; Yu, H.; Wu, J.; Yan, X.; Li, R. Research on Cascade Reservoirs' Short-Term Optimal Operation under the Effect of Reverse Regulation. *Water* **2018**, *10*, 808. [CrossRef]
32. Bhatia, N.; Srivastav, R.; Srinivasan, K. Season-Dependent Hedging Policies for Reservoir Operation—A Comparison Study. *Water* **2018**, *10*, 1311. [CrossRef]
33. Ye, L.; Ding, W.; Zeng, X.; Xin, Z.; Wu, J.; Zhang, C. Inherent Relationship between Flow Duration Curves at Different Time Scales: A Perspective on Monthly Flow Data Utilization in Daily Flow Duration Curve Estimation. *Water* **2018**, *10*, 1008. [CrossRef]
34. Kinouchi, T.; Yamamoto, G.; Komsai, A.; Liengcharernsit, W. Quantification of Seasonal Precipitation over the upper Chao Phraya River Basin in the Past Fifty Years Based on Monsoon and El Niño/Southern Oscillation Related Climate Indices. *Water* **2018**, *10*, 800. [CrossRef]
35. Zhu, X.; Zhang, C.; Qi, W.; Cai, W.; Zhao, X.; Wang, X. Multiple Climate Change Scenarios and Runoff Response in Biliu River. *Water* **2018**, *10*, 126. [CrossRef]
36. Zhang, C.; Zhu, X.; Fu, G.; Zhou, H.; Wang, H. The impacts of climate change on water diversion strategies for a water deficit reservoir. *J. Hydroinform.* **2014**, *16*, 872–889. [CrossRef]
37. Abera, F.F.; Asfaw, D.H.; Engida, A.N.; Melesse, A.M. Optimal Operation of Hydropower Reservoirs under Climate Change: The Case of Tekeze Reservoir, Eastern Nile. *Water* **2018**, *10*, 273. [CrossRef]
38. Jiang, Q.; Zhao, Y.; Wang, Z.; Fu, Q.; Wang, T.; Zhou, Z.; Dong, Y. Simulating the Evolution of the Land and Water Resource System under Different Climates in Heilongjiang Province, China. *Water* **2018**, *10*, 868. [CrossRef]
39. Zhao, F.; Li, C.; Chen, L.; Zhang, Y. An Integrated Method for Accounting for Water Environmental Capacity of the River–Reservoir Combination System. *Water* **2018**, *10*, 483. [CrossRef]
40. Jin, S.; Liu, H.; Ding, W.; Shang, H.; Wang, G. Sensitivity Analysis for the Inverted Siphon in a Long Distance Water Transfer Project: An Integrated System Modeling Perspective. *Water* **2018**, *10*, 292. [CrossRef]
41. Tian, Y.; Xiong, J.; He, X.; Pi, X.; Jiang, S.; Han, F.; Zheng, Y. Joint Operation of Surface Water and Groundwater Reservoirs to Address Water Conflicts in Arid Regions: An Integrated Modeling Study. *Water* **2018**, *10*, 1105. [CrossRef]
42. Choi, M.; Takemon, Y.; Ikeda, K.; Jung, K. Relationships Among Animal Communities, Lentic Habitats, and Channel Characteristics for Ecological Sediment Management. *Water* **2018**, *10*, 1479. [CrossRef]
43. Han, Q.; Tan, G.; Fu, X.; Mei, Y.; Yang, Z. Water Resource Optimal Allocation Based on Multi-Agent Game Theory of HanJiang River Basin. *Water* **2018**, *10*, 1184. [CrossRef]

44. Meng, F.; Fu, G.; Butler, D. Cost-effective River Water Quality Management using Integrated Real-Time Control Technology. *Environ. Sci. Technol.* **2017**, *51*, 9876–9886. [CrossRef] [PubMed]

45. Zhang, C.; Chen, X.; Li, Y.; Ding, W.; Fu, G. Water-energy-food nexus: Concepts, questions and methodologies. *J. Clean. Prod.* **2018**, *195*, 625–639. [CrossRef]

water

Article

Watershed Sediment and Its Effect on Storage Capacity: Case Study of Dokan Dam Reservoir

Mohammad Ezz-Aldeen [1], Rebwar Hassan [1], Ammar Ali [2], Nadhir Al-Ansari [1,*] and Sven Knutsson [1]

[1] Department of Civil, Environmental and Natural Resources Engineering, Lulea University of Technology, 971 87 Lulea, Sweden; mohezz@ltu.se (M.E.-A.); rebwar.hassan@ltu.se (R.H.); Sven.Knutsson@ltu.se (S.K.)

[2] Department of Water Resources Engineering, Baghdad University, Baghdad 10071, Iraq; ammali_75@yahoo.com

* Correspondence: nadhir.alansari@ltu.se; Tel.: +46-920-491-858

Received: 28 May 2018; Accepted: 26 June 2018; Published: 28 June 2018

Abstract: Dokan is a multipurpose dam located on the Lesser Zab River in the Iraq/Kurdistan region. The dam has operated since 1959, and it drains an area of 11,690 km^2. All reservoirs in the world suffer from sediment deposition. It is one of the main problems for reservoir life sustainability. Sustainable reservoir sediment-management practices enable the reservoir to function for a longer period of time by reducing reservoir sedimentation. This study aims to assess the annual runoff and sediment loads of the Dokan Dam watershed using the soil and water assessment tool (SWAT) model to evaluate the relative contributions in comparison with the total values delivered from both watershed and Lesser Zab River and to identify the basins with a high sediment load per unit area. These help in the process of developing a plan and strategy to manage sediment inflow and deposition. The SUFI-2 program was applied for a model calibrated based on the available field measurements of the adjacent Derbendekhan Dam watershed, which has similar geological formations, characteristics and weather. For the calibration period (1961–1968), the considered statistical criteria of determination coefficients and Nash–Sutcliffe model efficiency were 0.75 and 0.64 for runoff while the coefficients were 0.65 and 0.63 for sediment load, respectively. The regionalization technique for parameter transformation from Derbendekhan to Dokan watershed was applied. Furthermore, the model was validated based on transformed parameters and the available observed flow at the Dokan watershed for the period (1961–1964); they gave reasonable results for the determination coefficients and Nash–Sutcliffe model efficiency, which were 0.68 and 0.64, respectively. The results of SWAT project simulation for Dokan watershed for the period (1959–2014) indicated that the average annual runoff volume which entered the reservoir was about 2100 million cubic meters (MCM). The total sediment delivered to the reservoir was about 72 MCM over the 56 years of dam life, which is equivalent to 10% of the reservoir dead storage. Two regression formulas were presented to correlate the annual runoff volume and sediment load with annual rain depth for the studied area. In addition, a spatial distribution of average annual sediment load was constructed to identify the sub basin of the high contribution of sediment load.

Keywords: Dokan Dam; runoff; sediment load; SWAT

1. Introduction

Most of the dams, storage schemes, and different hydraulic structures around the world suffer from sedimentation problems. For dams and reservoirs, this effect is mainly concerned with the design capacity and operation schedule. The main source of reservoir sediments is the main river flow in addition to the runoff water, carrying the sediment load from watersheds and valleys surrounding the reservoir.

After a long period of dam operation, it is usually necessary to evaluate the current storage capacity of the reservoir relative to that in the design stage. The runoff and sediment load delivered to the reservoir could be estimated based on measured values of continuous river flow. Schleiss et al. [1] highlights and discusses the main matters concerning reservoir sedimentation. The reservoir sedimentation problem should be considered from the early stages of planning design and operation. In addition, the sedimentation process can create problems downstream from the dam which should also be considered in the planning and design stages.

Physically based models are usually used in cases where runoff and sediment load data records are not available. These models are of two types, which are referred to as single storm models and continuous simulation models. The Areal Non-Point Source Watershed Environment Response Simulation (ANSWERS) is developed by Beasley et al. [2]; and the European Soil Erosion Model (EUROSEM) was improved by Morgan et al. [3]; all of those models mentioned are examples of the former models. Examples of the latter are the spatially distributed erosion and sediment yield component, chemicals, runoff and erosion from agricultural management systems CREAMS (Science and Education Administration; Department of Agriculture, Washington, WA, USA) which are proposed by Knisel [4]; the SHESED model (the hydrologic and sediment transportation model of hydrological modeling system (SHE)) which is proposed by Wicks and Bathurst [5]; and the soil and water assessment model (SWAT) that was developed by Arnold et al. [6]. The SWAT model is the most commonly used model and, for this reason, a number of researchers have modified this model for different purposes (see [7,8]). Durao et al. [9] estimates the transported nutrient load in the Ardila River watershed in Spain by applying the SWAT model in order to identify the contribution of this load to the whole watershed. The model is applied to simulate long-period data; the real daily precipitation data is considered for the period 1930–2000. The considered flow data for model calibration and validation extend from 1950 to 2000, and nutrient data stretch from 1981 to 1999. The results indicate that the main source of diffusion prolusion comes from the main tributaries of Spain. Wang et al. [10] tests the possible conservation practices within a rangeland watershed using the agricultural policy/environmental extender (APEX). The model is calibrated and validated for both flow and sediment yield for the Cowhouse Creek watershed in north Texas. The analysis of the scenario extends from 1951 to 2008. It shows that a significant reduction reaches to 58.8% of overland sediment losses from the area covered by a range brush to range grass. This reduction is due to the replacement of shrub species with herbaceous species within the subareas. Samaras and Koutitas [11] evaluates the effect of land-use changes in a watershed on coastal erosion for a selected area in north Greece. They apply both the SWAT model and a shoreline evolution model, a shoreline evolution model, for this purpose. The simulation is applied before and after land use change using three formulas of sediment transportation. The result indicates that a reduction in crop/land use cover from 23.3% to 5.1% leads to a reduction in both watershed sediment yield and sediment discharge at the outlet by (56.4%) and (26.4 to 12.8%), respectively. This study can be considered as a suitable tool and guide for future work in the same field. Samaras and Koutitas [12] studies the effect of climate change on sediment transport and morphology. The study is applied to a selected sandy coast area and its watershed in North Greece. Both SWAT models are implemented for the watershed and PELNCON-M is implemented for the coastal area to achieve the study aims. Two scenarios are employed; the first one is considered to be an extreme rise in the precipitation depth on the watershed, and the second one is considered to be an extreme rise in waves in the coastal area. Results of the first scenario shows a significant effect on erosion, sediment transport, sediment yield and discharge at the watershed outlet, while the second scenario indicates a lower effect on the coastline variation. Arnold [13] developed the SWAT + CUP model (SWAT Calibration and Uncertainty Programs, Swiss Federal Institute of Aquatic Science and Technology, Zurich, Switzerland), which provides a semi-automatic tool for decision-making for the SWAT model by applying both manual and automated calibration and incorporating both sensitivity and uncertainty analysis. A number of previous studies [14–17] were applied using the SWAT model to estimate runoff, sediment yield and/or other soluble materials for ungagged watersheds based on neighboring or similar property watersheds. Another technique is used for flow and erosion, and sediment transport is the distributed mode. Juez et al. [18] simulates the hydro-sedimentary response of the Western Mediterranean catchment to a representative rainfall storm. The simulation combines the distributed flow surface model with the empirical model for infiltration, the Soil Conservation Services Model (SCS), and the erosion model, which is the Hillslope Erosion Model (HEM), considering water depth and flow as a 2D model. The present model is a tool for analyzing the hydro-sedimentary process at a temporal and special scale.

Most countries in the Middle East suffer from water shortage problems, where the annual allocation per capita does not exceed 500 m^3 [19]. For this reason, water is essential to life, socioeconomic development, and political stability in this region. Future prospects are negative; therefore, this problem is expected to be more chronic and severe in future [20]. Iraq used to be considered to be a relatively rich country in terms of its water resources, until the mid-1970s, because of the presence of the Tigris and Euphrates Rivers [21]. Due to regional and internal problems in Iraq, the estimation of the overall water required is about 75–81 billion cubic meters (BCM) [22], while the available quantity is 59–75 BCM and will drop to 17.61 BCM in 2025 [23]. In view of this situation, it is very important to know the actual storage capacity of the reservoirs—which are unknown now—so that prudent water resources planning can be done. The sedimentation rate of several reservoirs was recently investigated in Mosul and Dohuk. This is the third reservoir to be dealt with in Iraq. The bed of the Dokan Dam reservoir (located in the northeast of Iraq) is surveyed by Hassan et al. [24], and this studied the bed sediment using 32-bed samples distributed spatially over the reservoir. The results indicate that the bed sediments of the reservoir are composed of silt (48%), clay (23%), gravel (15%) and sand (14%).

All reservoirs in the world suffer from sediment deposition. This is one of the main problems for reservoir life sustainability. Sustainable reservoir sediment-management practices enable continued reservoir functioning for a longer period by reducing reservoir sedimentation. Iraq suffers from water shortage problems, especially after the construction of a series of storage reservoirs in source countries (Turkey, Syria and Iran), so the evaluation of the actual live storage capacity of dams is important for the prudent management of the operation schedule. The aim of this study is to assess the annual runoff and sediment loads of the Dokan Dam watershed (ungagged area) using a SWAT model set-up based on the parameter transformation technique of the modeling-gauged Derbendekhan watershed to learn the hydrological behavior of the area and to assess its contribution to the total values pouring into the reservoir. Moreover, the set-up model helps us to find the spatial distribution of erosion and annual sediment yield for the sub basins. This will help us to find the sub basins with a high sediment yield and evaluate effective factors for them. These assessments help in the process of developing a plan and a strategy for managing the sediment inflow and deposition.

2. Study Area

2.1. Location and Topography

The considered study area is the watershed of the Dokan Dam reservoir, situated in the northeast of Iraq (Figure 1). Dokan dam is a concrete arch dam located in the Lower Zab River, about 65 km southeast of Sulaimaniyah city and 295 km north of Baghdad, the capital city of Iraq. The dam height is about 116 m at maximum river depth, having a total storage capacity of 6.87 × 10^9 m^3 (6.14 × 10^9 m^3 live storage and 0.73 × 10^9 m^3 dead storage) at normal operation level of 511 m.a.s.l. [22]. The dam has been built to serve irrigation, power generation, water supply and flood control needs. Due to the limited observed data of flow at the Dokan Dam watershed and the unavailability of sediments load data, the second watershed considered for this research is the Derbendekhan Dam watershed; it is the nearest watershed to the study area. The properties of the Dokan and Derbendikhan watersheds are shown in Table 1. The digital elevation model (DEM) with 30 m resolution is considered to identify the watershed boundary, classification of overland and channel flow, slopes and other properties.

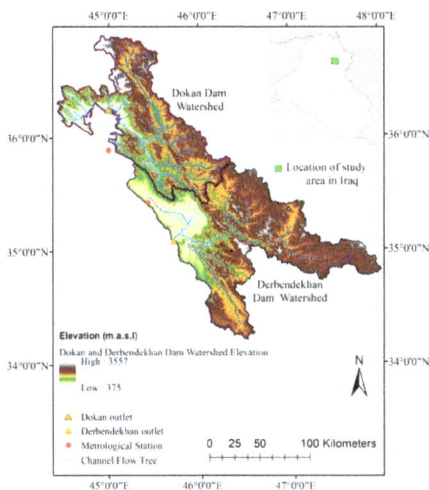

Figure 1. Topographic map of the watershed areas of the Dokan and Derbendekhan Dams and their locations in Iraq.

Table 1. Properties of the Dokan and Derbendekhan watersheds.

Watershed	Area (km^2)	Max. Elevation (m.a.s.l)	Min. Elevation (m.a.s.l)	Average Slope (%)	Maximum Annual Rain (mm)	Minimum Annual Rain (mm)
Dokan Dam	11,690	3557	489	26.5	1125	182
Derbendekhan Dam	15,280	3332	375	23.3	970	174

2.2. Soil Type and Land Use

The exposed rocks at the Dokan and Derbendekhan watershed areas are mainly limestone and minor exposures of dolomitic limestone, dolomite, and Quaternary alluvial deposits [25,26]. Based on the Reconnaissance Soil Map of the three Northern Governorates, Iraq [26] and the Food and Agriculture Organization of the United Nations (FAO) soil map [27], both watersheds are located on a common extended type of soil classification. Samples for different soil classes were taken depending on the soil map of the study area. A map of soil types is prepared for this study as a shape file for each watershed to be used in the SWAT model (see below for model details). The soil samples analysis includes grain size distribution in different types, organic matter content and hydraulic conductivity. The analysis of soil samples indicates that the area generally consists of four major soil types. Most of the area (85.6%) is covered by gravelly sandy mud; 6.9% is gravelly mud; and 7.5% of two types of muddy gravel (the main differences between the two types are the percent of gravel, which is 74% and 56% for types 1 and 2, respectively). Figure 2 shows the shape file of soil type considered in the SWAT model for the Dokan watershed.

The land use map for the years (1976–1979) [28] and available satellite image (NASA's Landsat GeoCover, 2007, with a spatial resolution of 14.25 m) indicates that the winter plants and pastures represent the main part of the land use map of the studied area. This depends on rainfall as a main source of irrigation water. The other parts are forests, vegetables and urban areas (villages). The land use change is limited (see Table 2). This is mainly due to the geological nature of both watersheds. In addition, the topography of the area does not enhance any changes. It is noteworthy to mention that rain is the main source of irrigation. For these reasons, the land cover did not change widely through the study period since the operation of the dam from the year 1959 to the year 2014. The Dam and the studied watershed are located away from the main cities, so changes in the urban and rural areas are limited. Table 2 shows the percentage of different land use cover for the two periods of the available land use map.

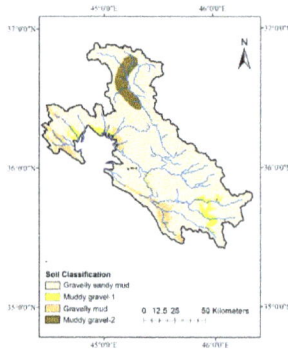

Figure 2. Soil type classification of the Dokan Reservoir watershed.

Table 2. Percentages of different land use types for two years of the studied period.

Year	Winter Plant and Pasture	Forest	Vegetables	Urban Area (Village)
1976–1979	77.3	22.0	0.5	0.2
2007	82.7	15.6	1.6	0.1

Due to the small difference between the percentage of land cover for the two available years, a map of land use for the study area is prepared as a shape file (Figure 3). The area consists of four types of land use/cover. Winter plants (pasture) and forests of different types of trees cover the main part of the study area, while the remaining small area is planted with vegetables near the reservoir boundary and/or in urban areas (villages).

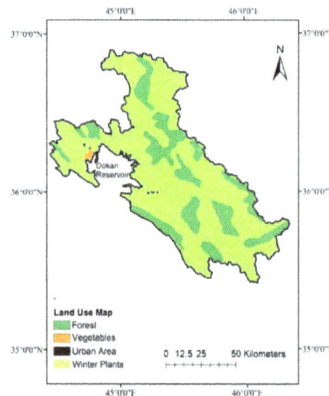

Figure 3. Land use and land cover classification of the Dokan Reservoir watershed.

3. Applied Model

The soil and water assessment tool (SWAT) is a physically based continuous simulation model for short or long times that can be applied to large river basins and complex watersheds. It was developed by the US Department of Agriculture, Agricultural Research Service [6]. The model is an efficient tool to estimate the flow and sediment load in addition to different chemical and nutrient materials. The model divides the watershed into sub basins based on their DEM date and hydrological response units (HRU); each unit has the same soil type, land use and land slope.

The hydrological simulation is based on the topographical terrain, soil type, land use and hydrological data of daily precipitation, maximum and minimum temperature, wind speed, relative humidity and solar radiation. The flow can be estimated based on a water balance equation; this equation is simulated in the SWAT model by different modular: the land phase and routing phase [14]. For the land phase, the soil water balance calculation is based on the following form [29]:

$$SW_t = SW_0 + \sum_{i=1}^{t} (R_{day} - Q_{sur} - E_a - w_{seep} - Q_{qw})$$ (1)

where,

SW_t: Water content of the soil (mm);
SW_0: Initial water content (mm);
R_{day}: Depth of precipitation (mm);
Q_{sur}: Equivalent depth of surface runoff (mm);
E_a: Evapotranspiration depth (mm);
w_{seep}: Depth of water seepage out of considered surface profile (mm);
Q_{qw}: Equivalent depth of return flow (mm).

The Penman–Monteith method is considered for potential evapotranspiration estimation. The required input data to estimate the potential evapotranspiration (PET) using the Penman–Monteith method are daily solar radiation, air temperature, relative humidity and wind speed. The formula of this method considers three effective factors for evapotranspiration, which are the required energy to sustain evaporation, the required strength to remove the water vapor and the aerodynamic in addition to resistance of the surface. The Penman–Monteith method is in the following form [29]:

$$\lambda E = \frac{\Delta(H_{net} - G) + \rho_{air} \cdot c_p \cdot [e_z^o - e_z]/r_a}{\Delta + \gamma \cdot \left(1 + \frac{r_c}{r_a}\right)}$$ (2)

where,

λE: Latent heat flux density (MJ/m^2/day);
E: Evaporation rate (mm/day);
Δ: Saturation vapor pressure-temperature slope (de/dt) (kPa/Co);
H_{net}: Net radiation (MJ/m^2/day);
G: Density of heat flux to the ground (MJ/m^2/day);
ρ_{air}: Density of the air (kg/m^3);
c_p: Specific heat at constant pressure (MJ/kg/day);
e_z^o: Saturated vapor pressure of air at height z, (kPa);
e_z: Water vapor pressure of air at height z (kPa);
λ: Psychrometric constant (kPa/Co);
r_c: Resistance of plant canopy (s/m);
r_a: Diffusion resistance of the air layer (s/m).

Also, the different parameters of land management are recognized based on soil type, land use and land cover. The soil water content and soil infiltration can be estimated by two methods based on the available data, either by the Green–Ampt infiltration equation or curve number methods. The Green–Ampt equation requires rainfall data of a sub daily interval, which is not available in Iraqi weather stations, so the curve number method is utilized throughout this work using the following form [29]:

$$Q_{surf} = \frac{\left(R_{day} - 0.2S\right)^2}{\left(R_{day} + 0.8S\right)} \tag{3}$$

where,

Q_{surf}: Equivalent depth of surface runoff (mm);
R_{day}: Rainfall depth of the considered day (mm);
S: Retention parameter (mm).

The value of S can be estimated by the following equation [29]:

$$S = 25.4\left(\frac{1000}{CN} - 10\right) \tag{4}$$

where, CN is the curve number of that considered day.

The second process includes the estimation of soil erosion from the overland due to rainfall detachment and surface runoff in addition to channel erosion and deposition. The sediments, routing in both the overland and channel flow, are estimated based on rainfall data, soil properties, land use/land cover and topography. The maps of soil type and land use are required with the digital elevation model (DEM) data to identify the topography of the watershed and to classify it into overland and channel sediment flow. The modified universal soil loss equation (MUSLE) is considered in the following form [29]:

$$sed = 11.8 \times \left(Q_{surf} \cdot q_{peak} \cdot are_{hru}\right)^{0.56} K_{USLE} \cdot C_{USLE} \cdot P_{USLE} \cdot LS_{USLE} \cdot CFRG \tag{5}$$

where,

sed: Yield of sediment for the considered storm or day (ton.);
Q_{surf}: Volume of surface runoff (mm/ha);
q_{peak}: Greatest surface runoff rate (m^3/s);
are_{hru}: Hydrologic response unit area (ha);
K_{USLE}: Soil erodibility factor of Universal Soil Loss Equation (USLE);
C_{USLE}: Cover and management factor of USLE;
P_{USLE}: Soli practice factor of USLE;
LS_{USLE}: Topographic factor of USLE;
$CFRG$: Factor of coarse fragment.

4. Model Calibration and Validation

4.1. Runoff and Sediment Load Calibration

Although the mathematical and conceptual models are considered widely in hydrological studies to simulate different events, such as runoff flow, sediment and both suspended and dissolved material transport, they still require calibration with measured values to ensure the accuracy of the model outputs.

Two types of dataset are prepared to be applied to SWAT model. The metrological (climate) data include daily precipitation, maximum and minimum temperature, wind speed, relative humidity and solar radiation; hydrometric data are also present. The second dataset is the topography data, which includes the DEM map.

In view of the limited available measurements of flow and unavailable sediment records of the Dokan Dam watershed, the available data of the Derbendekhan Dam watershed (the adjacent watershed to Dokan, Figure 1), were considered for both flow and sediment model calibration. Due to the similarity in the geological formation, soil type, land use, topographical and watershed characteristics (Table 1), the parameters of the calibrated watershed can be transformed to an ungauged watershed model [17]. To calibrate the results of the SWAT model for both runoff and sediment, the SWAT-CUP software is applied. It is an efficient tool to adjust different parameters of the SWAT model to obtain optimal local results and create an uncertainty analysis of SWAT model parameters to provide an easy and quick method of calibration and standardized calibration [30]. The considered software for the model is the Sequential Uncertainly Fitting version 2 (SUFI-2, Swiss Federal Institute of Aquatic Science and Technology, Zurich, Switzerland). In this program, all the uncertainty parameters can be used in the model calibration, including uncertainly in driving variables, parameters of the conceptual model and considered data [30]. Different statistical criteria can be considered in the model objective function to evaluate model performance, such as the determination coefficient, Nash–Sutcliff model efficiency, root mean square error and Chi-square. The determination coefficient is considered to be effective criteria to obtain the optimal values of flow and sediment concentration between the observed and measured data.

The SWAT project for the Derbendekhan watershed is set-up including the required DEM data, soil type land use, as shown in Figures 1–3, respectively, and meteorological data based on the nearest stations to the area as shown in Figure 1. The monthly average flow rate data at Derbendekhan station are considered for model calibration. To obtain an enhanced calibration of the model and for more understanding of the model's performance, the monthly recorded flow data are separated into base flow and surface runoff [14]. The recursive digital filter technique [31] is used to obtain a monthly separation based on the original daily separation technique. The separated monthly runoff from the total flow as measured values is applied in SUFI-2 to calibrate the model parameters. The statistical criterion of the determination coefficient is used as an objective function criterion. Besides this, the Nash–Sutcliffe model is employed to evaluate the model performance. For monthly runoff flow, the highest obtained values are 0.75 and 0.64 for the coefficient of determination and Nash–Sutcliffe model, respectively. The model uncertainty was measured using two factors: the P-factor reflects the percentage of measured data bracketed by 95% prediction uncertainty (95PPU). This means that one minus the P-factor represents the presence of poorly simulated values. The R-factor is another measurement of model uncertainly equal to the average thickness of the 95PPU band divided by the standard deviation of measured data. For the runoff calibration period, the P-factor is 0.78 and the R-factor is 1.27. Figure 4 shows the observed and simulated runoff at the Derbendekhan watershed outlet and the uncertainty band (95PPU) for the period (1961–1968).

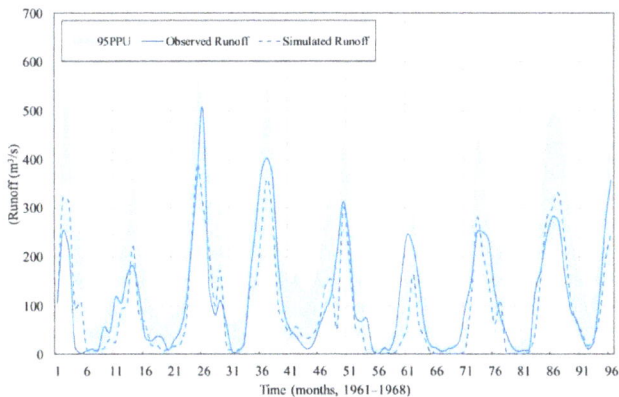

Figure 4. Monthly observed and simulated runoff at the Derbendekhan outlet and 95PPU for the period 1961–1968.

The available measured sediment concentration data of the Diyala River at Derbendekhan station, as presented by Assad [32], were utilized to calibrate the SWAT model parameters for the same period of runoff flow calibration. Figure 5 shows the measured and optimal simulated values of sediment load concentration at Derbendekhan outlet. The same statistical criteria are implemented, and the optimal resultant values are 0.65 and 0.63 for determination coefficient and Nash–Sutcliffe model efficiency, respectively, while the P-factor and R-factor equal to 0.68 and 2.27 respectively.

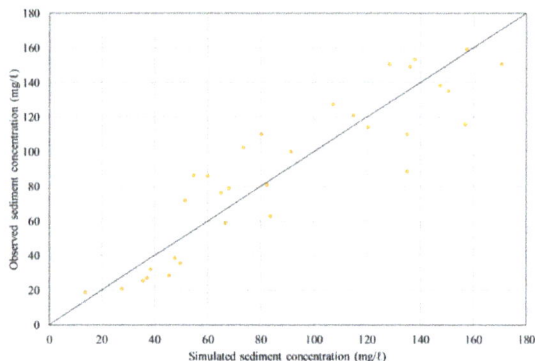

Figure 5. Observed and simulated sediment concentration for period (1961–1968) at the Derbendekhan watershed.

The most effective parameters are selected for the runoff and sediment model calibration as proposed by [28] in addition to other parameters. The resultant best-fitted values (optimal values) of the parameters and the considered range are shown in Table 3. Parameters listed in Table 3 are also mentioned in other previous literature [14,33].

Table 3. The range, optimal (fitted) values and sensitivity analysis of considered parameters.

Parameter	Description	Min_Value	Max_Value	Fitted_Value
V_GWQMN.gw	Threshold depth in shallow aquifer (mm).	0	2	1.698
V_CH_COV1.rte	Channel erodibility factor.	0.05	0.6	0.38715
R_USLE_K(..).sol	USLE, equation soil erodibility (k) factor.	−0.8	0.8	−0.1168
R_LAT_SED.hru	Sediment conc. In lateral and ground flow.	0	5000	2525
R_SPCON.bsn	Linear par. for calculating max. amount of sediment that can be re-entrained during channel sediment routing.	0.0001	0.01	0.008743
V_SPEXP.bsn	Exponential Par. for calculating max. amount of sediment that can be re-entrained during channel sediment routing.	1	2	1.549
V_GW_REVAP.gw	Groundwater "revap" coefficient.	0.02	0.2	0.07166
V_ALPHA_BF.gw	Base flow alpha factor (days).	0	1	0.555
V__CH_COV2.rte	Channel cover factor.	0.001	1	0.281719
V_GW_DELAY.gw	Groundwater delay (days).	0.02	0.2	0.07166
R_SOL_BD(..).sol	Moist bulk density.	−0.5	1	−0.3755
R_USLE_C(..)plant.dat	Min value of USLE-C factor for land cover/plant.	−0.5	0.5	0.151000
R_CH_N2.rte	Manning's "n" value for the main channel.	−0.5	0.5	−0.279
R_USLE_P.mgt	USLE, support practice parameter.	0	1	0.941
R_SOL_K(..).sol	Saturated hydraulic conductivity.	−0.5	0.5	0.177
R_CN2.mgt	SCS runoff curve number.	−0.2	0.2	0.198
R_SOL_AWC(..).sol	Available water content of the soil layer.	−0.5	1	0.5005

Note: **R**: Relative, **V**: Replace. (..) for different soil or plant type.

4.2. Runoff Validation

A SWAT model project is set-up for the Dokan Dam watershed, the main part of the study area. The limited recorded data of the monthly flow rate at the outlet of the Dokan watershed and the absence of any sediment load measurements leads to utilizing the regionalization technique to transfer the effective parameters from the adjacent gauged (Derbendekhan) watershed. Due to its similarity in geological formation, soil type, land use, watershed characteristics and weather data, the effective hydrological parameters obtained from the Derbendekhan (gauged) watershed can be transformed to the Dokan (ungauged) watershed. The process of parameter transformation is called regionalization. There are a number of presented methods for the regionalization of the watershed hydrological parameters: Kokkonen [34] applies the regression approach, while Parajka et al. [35] employs kringing and a similarity approach and Heuvelmans et al. [36] investigates the application of artificial neural nets and other methods. Since the Derbendekhan watershed is adjacent to the Dokan watershed (Figure 1), and the physical, topographical properties and rainfall are similar (Table 1) along with the geological formation, land use/land cover, and soil type, both watersheds have similar flow and sediment parameters. In this case, the effective parameters can be transformed from a donor watershed to an ungauged watershed. The fitted values of the Derbendekhan parameters calibrated by the SUFI-2 program are transferred to the Dokan watershed SWAT project. The SUFI-2 program is implemented for the calibration, uncertainly analysis and regionalization of the considered parameters of the SWAT model for the Debendekhan watershed for runoff and sediment load of concentration. Here, the SUFI-2 algorithm is used for the calibration, validation and measurement of the uncertainty for input data, model and sensitive parameters. The degree of uncertainty is measured by two values: P-factor and R-factor. The percent of measured values bracketed by 95% prediction uncertainly represent (95PPU), which is the P-factor while the ratio of 95PPU thickness divided by standard deviation of measured values is equal to the R-factor. When the simulated values are exactly the measured ones, the value of P-factor equals 1 and the value of R-factor equals zero [37].

Based on the transformed parameters, the simulated runoff flows are compared with measured values at the Dokan watershed outlet after the separation of the base flow for the period 1961–1964 (Figure 6) to evaluate the effectiveness of the transformation of the hydrological parameters. For the validation period of runoff, the obtained values of the determination coefficient and Nash–Sutcliffe model efficiency are 0.68 and 0.64, respectively, indicating that the transformation process is successful. The P-factor and R-factor for the model uncertainty also indicate a reasonable model performance: the P-factor is 0.71 and R-factor is 0.97.

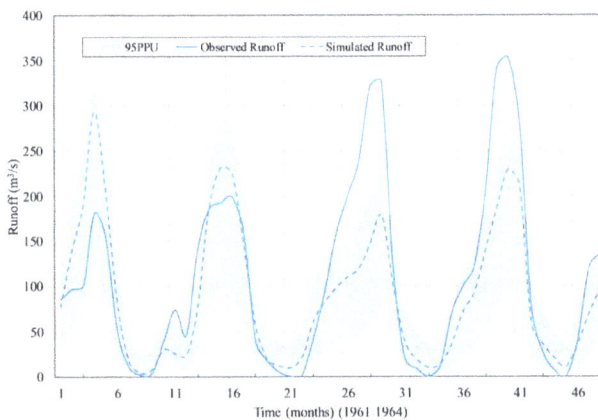

Figure 6. Monthly observed and simulated runoff at the Dokan outlet and 95PPU for the period 1962–1965.

5. Results and Discussion

After Durbendekhan SWAT project calibration for runoff and sediment data, the best-fitted values of the hydrological watershed parameters are transformed by the regionalization technique to the

Dokan SWAT project, the main project of the study. The sensitivity analysis is also studied for the most effective parameters on both the runoff and sediment load. The sensitivities are accomplished to identify the effective parameters on runoff and sediment load values for the watershed. The parameter sensitivity is estimated in the SUFI-2 model based on the multiple regression system presented by Abbaspour [38] to evaluate the effect of the considered parameter value (b_i) on the objective function (g); its sensitivity is in the following form:

$$g = \alpha + \sum_{i=1}^{m} \beta_i b_i \tag{6}$$

This formula calculates the average changes in the objective function due to the change in a given parameter while other parameters are changing. The comparative significance and sensitivity of each parameter are estimated based on the statistical criteria of the *t*-stat and *p*-value. The *t*-stat value is obtained from the coefficient of a parameter in the multiple regression analysis divided by its standard error. If the coefficient value is large in comparison to the standard errors, this mean that the parameter is sensitive. The *p*-value can be obtained by comparing the *t*-stat value with the student's distribution table [37]. The *p*-value of each term test is the null hypothesis, in which the coefficient is not affected. If the *p*-value is less than 0.05, it indicates that the null hypothesis can be rejected. The *t*-stat and *p*-values of different effective parameters are shown in Table 4. The parameters are arranged from low to high sensitivity, i.e., from low *t*-stat value or high *p*-value. The result of the test indicates that the soil water content, soil curve number at normal conditions (CN2) and the soil saturated hydraulic conductivity are the most effective parameters while threshold depth in a shallow aquifer, the channel erodibility factor and soil erodibility (k) factor in USLE have the lowest effect on runoff and sediment load simulation.

Table 4. Effective parameters arranged from low to high sensitivity based on *t*-stat and *p*-value.

Parameter	Absolute *t*-Stat	*p*-Value
GWQMN.gw	0.61	0.54
CH_COV1.rte	0.62	0.54
USLE_K(..).sol	0.77	0.44
LAT_SED.hru	0.78	0.43
SPCON.bsn	0.92	0.36
SPEXP.bsn	0.97	0.33
GW_REVAP.gw	0.98	0.33
ALPHA_BF.gw	1.20	0.23
CH_COV2.rte	1.25	0.21
GW_DELAY.gw	1.25	0.21
SOL_BD(..).sol	1.34	0.18
USLE_C(..)plant.dat	1.96	0.05
CH_N2.rte	2.08	0.04
USLE_P.mgt	3.69	0.03
SOL_K(..).sol	9.34	0.02
CN2.mgt	11.57	0.01
SOL_AWC(..).sol	16.05	0.00

Note: (..) for different soil or plant type.

The model is applied for the study period to estimate the runoff and sediment load reaching the Dokan Reservoir. The considered years of simulation have begun since the operation of the dam in 1959 to the year of the bathometry survey (2014) carried out by [39]. The resultant annual runoff volume that enters the Dukan Reservoir from the HRUs ranges from 300 to 4600 MCM (Figure 7), depending on rainfall intensity, depth and distribution through the rainy season. The runoff average volume from the watershed represents 35% of the live storage capacity of the dam, indicating that watershed runoff makes a significant contribution to reservoir inflow.

The SWAT model is an efficient tool to estimate the runoff hydrograph, but, in some hydrological studies, such as scheduling reservoir operations to supply the demand rate, the assessment of water resource income only is required. A regression formula is determined based upon the input–output data of SWAT model of the study area. This is a simple and quick tool to correlate the annual runoff depth with annual rainfall with good correlation ($R^2 = 0.9$) results without the need for more detailed input data as required in SWAT projects. The relationship used is in the following form:

$$Run_{Ann} = 0.075 \times R_{Ann}^{1.21} - 1.92, \ R^2 = 0.90 \ (for \ R_{Ann} > 16 \ mm) \tag{7}$$

where

Run_{Ann}: Annual runoff depth mm);
R_{Ann}: Annual rainfall depth (mm).

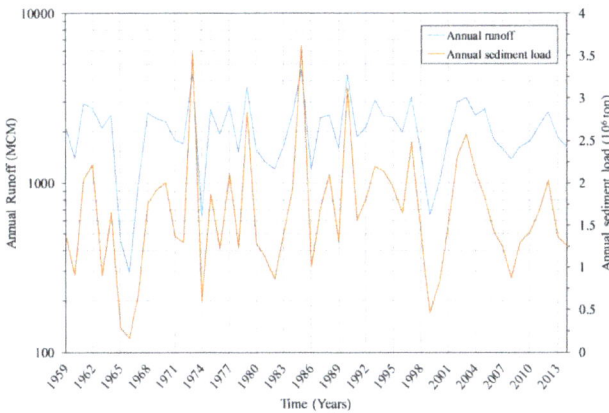

Figure 7. Annual runoff volume and sediment load delivered to the Dokan Reservoir for the period 1959–2014.

The formula is suitable when the annual runoff depth is greater than 15 mm to avoid negative runoff; this value is already much lower than the minimum historical recorded value.

The sediment load delivered to the Dokan reservoir is also estimated based on MUSLE programmed into the SWAT model for each single storm. The results are presented here annually. The average annual sediment load concentration is 650 mg/ℓ. This concentration can be considered relatively low in comparison with other locations or measurements in the region as proposed by [40,41] as well as the worldwide rate [42]. This is due to the nature of the rocks of the area and the effect of plant cover, such as winter pasture and plants and some forest trees throughout the region which reduce the detachment of soil particles transported with runoff flow. The estimated annual sediment load delivered to the Dokan reservoir from the watershed ranges from 3.6 to 0.16 × 10^6 ton for studied period, Figure 7. The average annual value is 1.63 × 10^6 ton.

The sediment trap efficiency of the reservoir is estimated based on the method presented by Garg V. and Jothiprakash V. [43]. This depends on reservoir storage capacity and annual inflow. The trap efficiency of the Dokan reservoir changes through the study period from 1 to 0.985.

Based on the results obtained from the simulation model, the estimated sediment load volume deposited in the reservoir for the considered period is about 10% of the dead storage capacity. This value is for the watershed only, which is considered a reasonable value and does not affect the designed project life. However, the Lesser Zab load should be also considered to evaluate the total amount of sediment load delivered and deposited within the reservoir. The total amount of sediment deposited in the Dokan reservoir for the period (1959–2014) is 209 MCM [38]. This means that the sediment load delivered into the reservoir from the watershed based on simulated results is about 34% of the total sediments deposited within the reservoir, which mean that the watershed sediment contribution is an effective value.

Due to the huge amount of data required, including topographical, metrological and hydrological data, different maps of soil, and land use to estimate the sediment load based on the applied model, a simple regression formula is used. It is based on simulated values to correlate the sediment load per unit area of the Dokan watershed with annual runoff depth in the following form:

$$Sed_{Ann} = 0.056 \cdot R_{Ann}^{1.195} - 7.33, \ R^2 = 0.97 \ (for \ R_{Ann} > 60mm) \tag{8}$$

where

Sed_{Ann}: Annual sediment load per unit area, (ton/km^2);
R_{Ann}: Annual rainfall depth (mm).

The formula is suitable when the annual runoff depth is greater than 60 mm; this value is already lower than the minimum historical recorded value in Dokan area.

A special distribution map of average annual sediment yield per unit area of sub basins is also prepared (Figure 8a). It can be noticed that the annual sediment load contribution ranges from 13 to 950 ton/km^2 approximately. The rate of erosion and sediment yield depends on a number of factors: topography, soil type, land cover and rainfall intensity. Comparing the sub basins of different soil types and land uses, in both the annual sediment yield map (Figure 8a) and the sub basin slope map (Figure 8b), it can be noticed that the most effective factor for the sediment yield is the land slope rather than other factors. This can be clearly noticed in that some basins have the same soil type and land use but a higher slope gives higher sediment yield. Sub basins having an average slope between 25 to 45% represent the area of high sediment yields from 400 to 950 t/km^2; however, when the sub basin slope is less than 20%, the sediment yield per unit area is reduced to about 40 t/km^2.

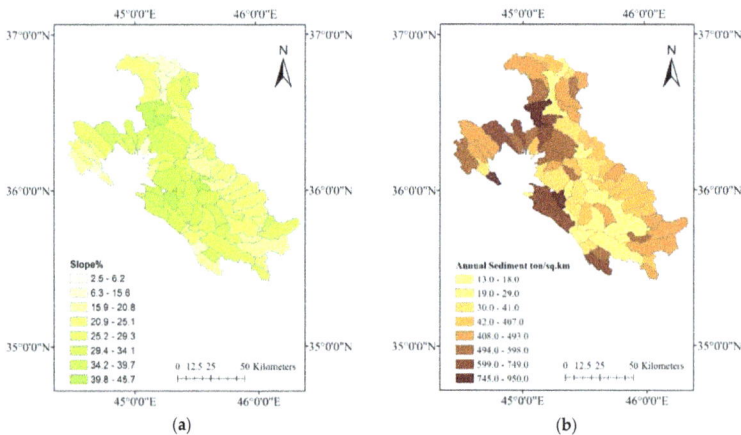

Figure 8. (a) Spatial distribution of average annual sediment load for the Dokan watershed sub basins; (b) average slope of sub basins.

This map is a tool that can enable decision-makers to apply a suitable method to reduce the erosion load, especially from high erosion rate areas. Depending on the selected area, the treatment may include practicing strip planting, terracing, or contour forming to reduce the effect of slope on surface runoff flow velocity, erosion and sediment transport capacity.

6. Conclusions

The soil and water assessment tool (SWAT) model is applied to assess the runoff and sediment delivered from the Dokan Dam watershed. Due to the limited recorded flow data and the absence of sediment measurements data at a station near the inlet of Dokan Reservoir, the model is calibrated for both runoff and sediment load for the Derbendekhan watershed adjacent to the Dokan watershed. The regionalization technique is employed to transfer the calibrated parameters of the SWAT project from a gauged (Derbendekhan) to an ungauged (Dokan) watershed. The resultant monthly runoff flow for the Dokan SWAT project is based on transformed parameters which were compared with measured values to evaluate the regionalization technique and model performance. The determination coefficient (R^2) and Nash–Sutcliffe model efficiency (Eff.) are 0.68 and 0.64, respectively, indicating a reasonable model performance with this technique. The average watershed contribution for annual runoff represents 35% of the dam life storage; this percentage is considered effective in the dam operation schedule. The total sediment load delivered to the Dokan reservoir from the watershed for the studied period is about 72 MCM. This load forms about 10% of the dead storage capacity of the reservoir. Generally, the total sediment load delivered and deposited in the reservoir for the period of dam operation is considered acceptable within the allowed limits. The map of special distribution of annual sediment load yield per unit area of each sub basins is presented; the average slopes map reflects a good agreement with the map of annual sediment load yield in comparison to other effective factors to be considered. This indicates that the land slope is the most effective factor on erosion and sediment transport. This can be used for soil conservation treatment to reduce the erosion rate.

Author Contributions: M.E.-A. and R.H. did the field, methodology and modelling, A.A. helped in the modelling, N.A.-A. and S.K. did the supervision.

funding: This research received no external funding.

Acknowledgments: The authors would like to express their thanks to John McManus (University of St. Andrews, St. Andrews, UK) and Ian Foster (Northampton University, Northampton, UK) for reading the manuscript and for their fruitful discussions and suggestions, and to Lulea University of Technology for some of the financial support for this research.

Conflicts of Interest: The authors declare no conflict of interest.

References

1. Schleiss, A.J.; Franca, M.J.; Juez, C.; De Cesare, G. Reservoir Sedimentation—Vision Paper. *J. Hydraul. Res.* **2016**, *54*, 595–614. [CrossRef]
2. Beasley, D.B.; Huggins, L.F.; Monke, E.J. ANSWERS: A model for watershed planning. *Trans. ASAE* **1980**, *23*, 938–944. [CrossRef]
3. Morgan, R.P.C.; Quinton, J.N.; Smith, R.E.; Govers, G.; Poesen, J.W.; Auerswald, K.; Chisci, G.; Torri, D.; Styczen, M.E.; Folly, A.J. The European soil erosion model (EUROSEM): A dynamic approach for predicting sediment transport from fields and small catchments. *Earth Surface Process. Landf.* **1998**, *23*, 527–544. [CrossRef]
4. Knisel, W.G. *CREAMS: A Field-Scale Model for Chemical, Runoff and Erosion from Agricultural Management Systems*; USDA Conservation Research Report; Conservation Model Development: Washington, WA, USA, 1980.
5. Wicks, J.M.; Bathurst, J.C. SHESED: A physically based, distributed erosion and sediment yield component for the SHE hydrological modelling system. *J. Hydrol.* **1996**, *175*, 213–238. [CrossRef]
6. Arnold, J.G.; Srinisvan, R.; Muttiah, R.S. Large area hydrologic modeling and assessment. Part I: Model development. *Am. Water Resour. Assoc.* **1998**, *34*, 73–89. [CrossRef]
7. Easton, Z.M.; Fukal, D.R.; White, E.D.; Collick, A.S.; Ashagre, B.B.; McCartney, M.; Awulachew, S.B.; Ahmed, A.A.; Steenhuis, T.S. A multi basin SWAT model analysis of runoff and sedimentation in the Blue Nile, Ethiopia. *Hydrol. Earth Syst. Sci.* **2010**, *14*, 827–1841. [CrossRef]
8. Swami, V.A.; Kulkarni, S.S. Simulation of runoff and sediment yield for a Kaneri Watershed Using SWAT Model. *J. Geosci. Environ. Protect.* **2016**, *4*, 1–15. [CrossRef]

9. Durao, A.; Leitao, P.C.; Brito, D.; Fernandes, R.; Neves, R.; Morais, M.M. Estimation of Transported Pollutant Load in Ardila Catchment Using the SWAT Model. In Proceedings of the International SWAT Conference, Toledo, Spain, 15–17 June 2011.

10. Wang, X.; Amonett, C.; Williams, J.R.; Fox, W.E.; Tu, C. Assessing Impacts of Rangeland Conservation Practices Prior to Implementation: A Simulation Case Study using APEX. In Proceedings of the International SWAT Conference, Toledo, Spain, 15–17 June 2011.

11. Samaras, A.G.; Koutitas, C.G. The impact of watershed management on coastal morphology: A case study using an integrated approach and numerical modeling. *Geomorphology* **2014**, *211*, 52–63. [CrossRef]

12. Samaras, A.G.; Koutitas, C.G. Modeling the impact of climate change on sediment transport and morphology in coupled watershed-coast systems: A case study using an integrated approach. *Int. J. Sediment Res.* **2014**, *29*, 304–315. [CrossRef]

13. Arnold, J.G.; Youssef, M.A.; Yen, H.; White, A.Y.; Sheshukov, A.M.; Sadeghi, D.N.; Moriasi, D.N.; Steiner, J.L.; Amatya, D.M.; Haney, E.B.; et al. Hydrological processes and model representation: Impact of soft data on calibration. *Trans. ASABE* **2015**, *58*, 1637–1660.

14. Emam, A.R.; Kappas, M.; Linh, N.H.K.; Renchin, T. Hydrological Modeling and Runoff Mitigation in an Ungauged Basin of Central Vietnam Using SWAT Model. *Hydrology* **2017**, *4*, 1–17.

15. Emam, A.R.; Kappas, M.; Linh, N.H.K.; Renchin, T. Hydrological Modeling in an Ungauged Basin of Central Vietnam. *Hydrol. Earth Syst. Sci. Discuss.* **2016**, 1–33. [CrossRef]

16. Prabhanjan, A.; Rao, E.P.; Eldho, T.I. Application of SWAT Model and Geospatial Techniques for Sediment-Yield Modeling in Ungauged Watersheds. *J. Hydrol. Eng.* **2014**, *20*, C6014005. [CrossRef]

17. B'ardossy, A. Calibration of hydrological model parameters for ungauged catchments. *Hydrol. Earth Syst. Sci.* **2007**, *11*, 703–710. [CrossRef]

18. Juez, C.; Tena, A.; Fernández-Pato, J.; Batalla, R.J.; Garcia-Navarro, P. Application of A Distributed 2d Overland Flow Model for Rainfall/Runoff and Erosion Simulation in A Mediterranean Watershed. *Cuadernos de Investigación Geográfica* **2018**. (In Spanish) [CrossRef]

19. Al-Ansari, N. Hydropolitics of the Tigris and Euphrates Basins. *Engineering* **2016**, *8*, 140–172. [CrossRef]

20. Bazzaz, F. Global climatic changes and its consequences for water availability in the Arab world. In *Water in the Arab Word: Perspectives and Prognoses*; Roger, R., Lydon, P., Eds.; Harvard University: Cambridge, MA, USA, 1993; pp. 243–252.

21. Al-Ansari, N.; Knutsson, S. Toward Prudent management of Water Resources in Iraq. *Adv. Sci. Eng. Res.* **2011**, *1*, 53–67.

22. World Bank. *Iraq. Country Water Resources, Assistance Strategy: Addressing Major Threats to People's Livelihoods*; Water, Environment, Social and Rural Development Department: Washington, DC, USA, 2006.

23. UNESCO. *Iraq's Water in the International Press*; UNESCO Office for Iraq: Erbil, Iraq, 2010.

24. Hassan, R.; Al-Ansari, N.; Ali, S.A.; Ali, A.A.; Abdullah, T.; Knutsson, S. Dukan Dam Reservoir Bed Sediment, Kurdistan Region, Iraq. *Engineering* **2016**, *8*, 582–596. [CrossRef]

25. Karim, K.H.; Al-Hakari, S.H.S.; Kharajiany, S.O.A.; Khanaqa, A.P. Surface Analysis and Critical Review of the Darbandikhan (Khanaqin) Fault, Kurdistan Region, Northeast Iraq. *Kurdistan Acad.* **2016**, *12*, 61–75.

26. Berding, F. *Reconnaissance Soil Map of the Three Northern Governorates*; FAO Erbil Coordination Office: Erbil, Iraq, 2001.

27. Food and Agriculture Organization of the United Nations. *The Digital Soil Map of the World and Derived Soil Properties*; Version 3.6; FAO: Rome, Italy, 2003.

28. Ministry of Agricultural and Irrigation, Republic of Iraq. *Map of Soil and Land Use*; Ministry of Agricultural and Irrigation: Baghdad, Iraq, 1990.

29. Neitsch, S.L.; Arnold, J.G.; Kiniry, J.R.; Williams, J.R. *Soil and Water Assessment Tool*; Theoretical Documentation Version 2009; Texas Water Resources Institute: College Station, TX, USA, 2011.

30. Abbaspour, K.C.; Vejdani, M.; Haghighat, S. SWAT-CUP Calibration and Uncertainty Programs for SWAT. In Proceedings of the International Congress on Modelling and Simulation (MODSIM 2007), Melbourne, Australia, 10 December 2007.

31. Smakhtin, V.U. Estimating continuous monthly baseflow time series and their possible applications in the context of the ecological reserve. *Water SA* **2001**, *27*, 213–218. [CrossRef]

32. Assad, M.N. Sediments and Sediment Discharge in the River Diyala. Master's Thesis, Baghdad University, Baghdad, Iraq, 1978.

33. Salimi, T.E.; Nohegar, A.; Malekian, A.; Hosseini, M.; Holisaz, A. Runoff simulation using SWAT model and SUFI-2 algorithm. *Casp. J. Environ. Sci.* **2016**, *14*, 69–80.

34. Kokkonen, T.; Jakeman, A.; Young, P.; Koivusalo, H. Predicting daily flows in ungauged catchments: Model regionalization from catchment descriptors at the Coweeta Hydrologic Laboratory, North Carolina. *Hydrol. Process.* **2003**, *11*, 2219–2238. [CrossRef]

35. Parajka, J.; Merz, R.; Bloschl, G. A comparison of regionalisation methods for catchment model parameters. *Hydrol. Earth Syst. Sci.* **2005**, *9*, 157–171. [CrossRef]

36. Heuvelmans, G.; Muys, B.; Feyen, J. Regionalisation of the parameters of a hydrological model: Comparison of linear regression models with artificial neural nets. *J. Hydrol.* **2006**, *319*, 245–265. [CrossRef]

37. Tang, F.F.; Xu, H.S.; Xu, Z.Z. Model calibration and uncertainty analysis for runoff in the Chao River Basin using sequential uncertainty fitting. *Procedia Environ. Sci.* **2012**, *13*, 1760–1770. [CrossRef]

38. Abbaspour, K.C. *SWAT-CUP: SWAT Calibration and Uncertainty Programs—A User Manual*; Swiss Federal Institute of Aquatic Science and Technology, Eawag: Dubendorf, Switzerland, 2015.

39. Hassan, R.; Al-Ansari1, N.; Ali, A.; Ali, S.; Knutsson, S. Bathymetry and siltation rate for Dokan Reservoir, Iraq. *J. Lakes Reserv. Res. Manag.* **2017**, *20*, 1–11. [CrossRef]

40. Walling, D.E. *The Impact of Global Change on Erosion and Sediment Transport by Rivers: Current Progress and Future Challenges*; United Nations Educational, Scientific and Cultural Organization: Paris, France, 2009.

41. Basson, G. *Sedimentation and Sustainable Use of Reservoirs and River Systems*; International Commission on Large Dams: Paris, France, 2009.

42. Mahmood, R. *Reservoir Sedimentation: Impact, Extent, Mitigation*; World Bank: Washington, DC, USA, 1987.

43. Garg, V.; Jothiprakash, V. Trap Efficiency Estimation of a Large Reservoir. *J. Hydraul. Eng.* **2008**, *14*, 88–101. [CrossRef]

water | MDPI

Article

Estimating Sediment Flushing Efficiency of a Shaft Spillway Pipe and Bed Evolution in a Reservoir

Ching-Nuo Chen [1,*] and Chih-Heng Tsai [2]

[1] International Master Program of Soil and Water Engineering, National Pingtung University of Science and Technology, Pingtung 91201, Taiwan
[2] Department of Recreation and Healthcare Management, Chia Nan University of Pharmacy and Science, Tainan 71710, Taiwan; jht581212@gmail.com
* Correspondence: ginrochen@mail.npust.edu.tw

Received: 9 October 2017; Accepted: 22 November 2017; Published: 28 November 2017

Abstract: Control of reservoir sedimentation in order to ensure their sustainable use has drawn attention among water engineers and water resource managers. Several methods have been proposed, but most of the developed methodologies are incapable of modelling bed evolutions, while at the same time, compute sediment flushing efficiency. In this study a two-dimensional bed evolution model is proposed to estimate sediment distribution, bed evolution and sediment flushing efficiency of reservoirs. A-Gong-Dian reservoir, in southern Taiwan, is used as an illustrative example. Typhoon events were used to verify the proposed model. Simulations were conducted for one and two-day storm events under return periods, 2, 5, 10, 25, 50, 100, and 200-year. The results indicated that the average sediment flushing efficiency of the shaft spillway under one and two-day storms were close, 58.50% and 59.39%, respectively. These results were similar to observed laboratory tests experiments, where an efficiency of 65.34% was obtained. This study suggests that the applied model could be adopted to ensure the sustainable use of reservoirs, and also to find an optimal area for the location of a shaft spillway pipe. Therefore, the proposed model could serve as a reference to the reservoir management personnel.

Keywords: two-dimensional bed evolution model; sediment flushing of empty storage; shaft spillway pipe; sediment flushing efficiency

---

1. Introduction

Reservoirs are often affected by accelerated sediment deposition rates and has shortened the life of reservoirs by more than 65% in China alone [1]. As a result, the economic value of such projects has severely declined. Not only do they influence the life of reservoirs, they also pose safety hazards, as illustrated by [2]. Their sustainability is strongly dependent on how well the rate of sediment deposition is reduced and on the techniques of managing the reservoirs. Several techniques are available for their management, amongst which are mechanical excavation, dredging (conventional dredging, dry excavation), and hydraulic desilting. For an exhaustive review of the different techniques, the reader is referred to [3], who explored sustainable sediment management in reservoirs based on experience from five continents. Mechanical excavation and dredging boats, however, are associated with higher costs when compared to hydraulic desilting, and are often plagued by subsequent disposal problems. Hydraulic desilting employs stream power and hydraulics to cut down sediment deposits downstream. Flushing out sediments in reservoirs has been shown to cut costs [4], despite the large amount consumed by the flushing operations. Emamgholizadeh and Samadi [5] classify flushing into two, complete (also termed empty) and partial drawdown flushing. These, in turn, include hydro-suction, sediment sluicing, sediment bypass, density current venting, and hydraulic flushing through the reservoir, used independently or in combination [6]. The efficiency of sediment flushing depends on the geometry of the reservoir, sediment particle size, characteristics of sediments deposited, flow discharge, and flow depth. Several authors have argued that the efficiency of sediment flushing is influenced by the ratio of storage volume to incoming runoff [7], which should be less than 0.05 [8] for this technique to work. Moreover, we did not evaluate this threshold in this study, since the reservoir under investigation already applied hydraulic sediment flushing. Madadi, et al. [9] managed to improve flushing efficiencies by up to 280% through reconfigured reservoir bottom outlets in laboratory experiments.

Effective management of reservoirs system require a model that can predict future behaviour and response to perturbation [10], and all of the models are developed through experiments and depending on the status quo, they may grow in complexity to include conceptual frameworks, computer calculations, numerical simulations, and physical scale modelling [11]. Physical models have been applied to study the process and efficiency of sediment flushing in a reservoir. Although they have been successfully used to understand and reproduce to some extent complex physical processes that occur in nature, and have contributed significantly in hydraulic construction designs, they are relatively costly and time consuming [12]. More recently photogrammetry-based surveys using unmanned aerial systems have been used to evaluate flushed sediments [13–15]. Moreover, such techniques can only compute the amount of flushed sediments only when the reservoir is dry (i.e., empty), and subsequent images are necessary to compute the Digital Elevation Models (DEMs) of difference, from which the flushing efficiencies may be computed. Network-based programming techniques have also been employed in multi-reservoir systems [16], though their core emphasis is on determining empty flushing of sediments.

Given the recent advances in computational power, multi-dimensional models have increased the capability of assessing sedimentation problems and the multi-dimensional models have been extensively adopted in engineering application and analysis. For models to be adopted, they should reflect the physical characteristics of the reservoir and complexity in question. Numerical sediment transport models are available in one, two, and three dimensions. The widely used models, however, are one- (1D) and two-dimensional (2D) models when compared to the high computer intensive three-dimensional (3D) models. Examples of 1D sediment transport models are HEC-6 [17], HEC-RAS [18], and FLUVIAL-12 [19,20]. Castillo, Carrillo, and Álvarez [7] employed four complementary methods, which included 1D model, to determine sedimentation and flushing in a reservoir These models are capable of simulating longitudinal flows in rivers, moreover, they run short in the simulation of sediment transport and bed evolution in reservoirs. To be applied in sediment flushing, several assumptions should be made, thus, compromising the accuracy and efficiency in reservoir management. In such situations, reservoirs are narrow in shape, flow highly channelized, while closely following the thalweg [12]. However, most reservoir pools are wide and have no single clear flow direction, and they often constitute complex topography and geometry. As a result, multi-dimensional models are used. Olsen [21] used a depth-averaged 2D model to study the flushing process in a water reservoir in Nepal. Besides two-dimensional models, three-dimensional models have also been applied to study sediments in reservoirs. Olsen and Skoglund [22] applied a 3D model to calculate the sediment deposition in a hydropower reservoir, and also in a sand trap. Fang and Rodi [23] used a 3D model to simulate flow and sediment transport in the Three Gorges Project (TGP) reservoir in Yangtze River. Khosronejad, et al. [24] used a three-dimensional finite volume model to study the effects of various parameters on the quantity of sediment that was released from a reservoir in the reservoir flushing process.

Although the above models could estimate sediment erosion and deposition, bed evolution in a reservoir, and the efficiency of flushing, we have not found a model that is capable of combining all of these key reservoir management strategies in a single package. In addition, the above stated models require suspended sediment concentration, and sediment yield hydrograph into the reservoir, which are not easily obtained. Consequently, rating curves of discharge and suspended sediment transport rate are used, and these are associated with high errors [25]. Incorrect estimation of sediment inflow into reservoirs especially during flood events, will eventually lead to inefficient flushing of sediments and to misleading bed evolution in the reservoirs. It is therefore imperative to develop models that are highly efficient in estimating inflow hydrographs and sedigraphs, in turn, correctly estimating the amount of sediments to be flushed, while estimating the resultant bed evolution. A two-dimensional bed evolution model having these capabilities is developed and applied in this study. The upstream boundary condition hydrographs of inflow discharge and suspended sediment concentration for the 2D dimensional bed evolution model were calculated by the Physiographic Soil Erosion and Deposition (PSED) [26]. The PSED model can accurately estimate discharge hydrographs, concentration of suspended sediment hydrograph and suspended sediment transport rate from a watershed.

2. Numerical Model

The depth-averaged two-dimensional bed evolution model is divided into three parts: (1) water flow calculations; (2) sediment transport calculations; and, (3) bed elevation variation calculations.

2.1. Governing Equations for Water

The depth-averaged continuity and momentum equations are given below:

$$\frac{\partial h}{\partial t} + \frac{\partial (uh)}{\partial x} + \frac{\partial (vh)}{\partial y} = 0 \tag{1}$$

$$\frac{\partial (uh)}{\partial t} + \frac{\partial (uuh)}{\partial x} + \frac{\partial (uvh)}{\partial y} = \frac{\partial}{\partial x}\left(2\varepsilon h\frac{\partial u}{\partial x}\right) + \frac{\partial}{\partial y}\left[\varepsilon h\left(\frac{\partial u}{\partial y} + \frac{\partial v}{\partial x}\right)\right] - \frac{\tau_{bx}}{\rho} - gh\frac{\partial H}{\partial x} \tag{2}$$

$$\frac{\partial (vh)}{\partial t} + \frac{\partial (vuh)}{\partial x} + \frac{\partial (vvh)}{\partial y} = \frac{\partial}{\partial x}\left[\varepsilon h\left(\frac{\partial u}{\partial y} + \frac{\partial v}{\partial x}\right)\right] + \frac{\partial}{\partial y}\left(2\varepsilon h\frac{\partial v}{\partial y}\right) - \frac{\tau_{by}}{\rho} - gh\frac{\partial H}{\partial y} \tag{3}$$

in which t is time; x and y are horizontal Cartesian coordinates; h is the depth; u and v are depth-average flow velocities in x and y directions; H is water surface elevation; g is gravitational acceleration; ρ is density of flow; ε is the depth-average kinematic eddy viscosities of water; and, τ_{bx} and τ_{by} are bed shear stresses τ_b in x and y directions, $\tau_b = \rho g h S_f$, S_f is the friction slope.

The depth-average kinematic eddy viscosities of water can be approximated and expressed as [27]:

$$\varepsilon = \frac{\kappa}{6}u_* h \tag{4}$$

in which κ is the von Karman constant, and $\kappa = 0.4$ is chosen in this study. u_* is the shear velocity and $u_* = \sqrt{ghS_f}$.

2.2. Governing Equations for Sediment Transport with Source Terms

The convective-diffusive equation of suspended sediment, can be expressed as [28]:

$$\frac{\partial (Ch)}{\partial t} + \frac{\partial (uCh)}{\partial x} + \frac{\partial (vCh)}{\partial y} = \frac{\partial}{\partial x}\left(\varepsilon\frac{\partial C}{\partial x}\right) + \frac{\partial}{\partial y}\left(\varepsilon\frac{\partial C}{\partial y}\right) + q_{se} - q_{sd} \tag{5}$$

where C is the depth-averaged volumetric concentration of suspended sediment; q_{se} and q_{sd} are the entrainment and deposition terms of river bed, respectively. According to Itakura and Kishi [29], the entrained rate of channel bed can be expressed as:

$$q_{se} = 0.008\sqrt{sgd}\left[0.14\frac{\rho}{\rho_s}\left(14\sqrt{\tau_*} - \frac{0.9}{\sqrt{\tau_*}}\right) - \frac{\omega_s}{\sqrt{sgd}}\right] \tag{6}$$

in which $s = (\rho_s - \rho)/\rho$ is the submerged specific gravity of the sediment; ρ_s is density of the sediment; d is diameter of the sediment; ω_s is fall velocity of the sediment; and, τ_* is non-dimensional bed shear stress, $\tau_* = u_*^2/sgd$.

The deposition rate of suspended sediment can be expressed as:

$$q_{sd} = \omega_s C_a \tag{7}$$

where, C_a is the concentration of sediment near the channel bed. C_a can be estimated from the volumetric concentration of suspended sediment obtained at 0.05 depth from the channel bed. Using the exponential law, the volumetric concentration of suspended sediment may be expressed as [30]:

$$C_a = \frac{P_e}{[1 - \exp(-P_e)]} C \tag{8}$$

where P_e is the Peclet's number, which may be expressed as $\omega_s h / \varepsilon$.

An extensively used bed load transport formula is the Meyer Peter and Muller formula (MPM) [31]. Moreover, Wong and Parker [32] amended the MPM formula, and more accurate estimates of bed load transport rate were obtained. Nonetheless, since the original MPM formula is relatively easy to apply when establishing a numerical model, bed load transport was calculated using the original MPM formula in this study.

$$\left(\frac{k_n}{k'}\right)^{\frac{3}{2}} \gamma h S_f = 0.047(\gamma_s - \gamma)d + 0.25\left(\frac{\gamma}{g}\right)^{1/3}\left(\frac{\gamma_s - \gamma}{\gamma}\right)^{2/3} q_b^{2/3} \tag{9}$$

where γ and γ_s are the specific weight of water and the specific weight of sediment, respectively; q_b is the bed load transport rate per unit width of bed; k_n is Strickler's roughness coefficient, which can be represented as the reciprocal of Manning's roughness coefficient; and, $k' = 26/d_{90}^{1/6}$, d_{90} is the size of sediment in the unit of meter for which 90% of the material is finer.

2.3. Governing Equations for Bed Variation

The bed evolution due to sediment transport rate is not equal throughout an alluvial river. The continuity equation for bed elevation variations can be written as [33,34]:

$$\frac{\partial z}{\partial t} + \frac{1}{1 - \lambda}\left[\frac{\partial q_{bx}}{\partial x} + \frac{\partial q_{by}}{\partial y} + (q_{se} - q_{sd})\right] = 0 \tag{10}$$

where z is the channel bed elevation; λ is the porosity, $\lambda = 0.245 + 0.0864/d_m^{0.21}$, d_m is the mean sediment diameter [35]; and, q_{bx} and q_{by} are the components of q_b in x and y directions, respectively.

2.4. Numerical Scheme

MacCormack explicit finite-difference method [36] is adopted and is divided into predictor and corrector steps. The forward finite-difference is used to discretize the predictor step while the backward finite-difference is used to discretize the corrector step.

The forward finite-difference is used to compute the water depth h by the continuity equation (Equation (1)) and the u and v are computed by the momentum equations (Equations (2) and (3)). The predicted value may be written as:

$$h_{i,j}^* = h_{i,j}^n + \Omega_{hf}' \tag{11}$$

$$u_{i,j}^* = u_{i,j}^n \frac{h_{i,j}^n}{h_{i,j}^{n+1}} + \Omega_{uf}' \tag{12}$$

$$v_{i,j}^* = v_{i,j}^n \frac{h_{i,j}^n}{h_{i,j}^{n+1}} + \Omega_{vf}'$$

(13)

The superscript * denotes the predicted value; the superscript n and $n + 1$ refers to the variables at the known and unknown time levels; the subscript i and j denote the grid in x- and y-directions; and, Ω_{hf}', Ω_{uf}' and Ω_{vf}' are the functions of the known value of variables h, u, v at the time level n.

The backward finite-difference is used to calculate the water depth h by the continuity equation (Equation (1)) and the u and v are computed by the momentum equations (Equations (2) and (3)). The corrected value may be written as:

$$h_{i,j}^{**} = h_{i,j}^n + \Omega_{hb}'$$

(14)

$$u_{i,j}^{**} = u_{i,j}^n \frac{h_{i,j}^n}{h_{i,j}^{n+1}} + \Omega_{ub}'$$

(15)

$$v_{i,j}^{**} = v_{i,j}^n \frac{h_{i,j}^n}{h_{i,j}^{n+1}} + \Omega_{vb}'$$

(16)

The superscript ** denotes the corrected value; Ω_{hb}', Ω_{ub}', and Ω_{vb}' are the functions of the known value of variables h, u, v at the time level n.

The value of the variables at the unknown time level could be calculated by the predicted and corrected values that may be written as:

$$h_{i,j}^{n+1} = \frac{1}{2}\left(h_{i,j}^* + h_{i,j}^{**}\right)$$

(17)

$$u_{i,j}^{n+1} = \frac{1}{2}\left(u_{i,j}^* + u_{i,j}^{**}\right)$$

(18)

$$v_{i,j}^{n+1} = \frac{1}{2}\left(v_{i,j}^* + v_{i,j}^{**}\right)$$

(19)

Bed Evolution Model—MacCormack Explicit Finite—Difference Method

The explicit finite-difference method is used to discretize the suspended sediment concentration convection-diffusion equation (Equation (5)) and the continuity equation for bed elevation variations (Equation (10)) to calculate the volumetric concentration of suspended sediments and bed elevation. The volumetric concentration of suspended sediments C is calculated by:

$$C_{i,j}^{n+1} = \frac{C_{i,j}^n h_{i,j}^n}{h_{i,j}^{n+1}} + \Omega_c'$$

(20)

where Ω_c' is composed of the known value of variables h, u, v, C at the time level n.

The bed elevation may be written as:

$$z_{i,j}^{n+1} = z_{i,j}^n + \Omega_z'$$

(21)

where Ω_z' is composed of the known value of variables h, u, v, C at the time level n.

3. Study Area

A-Gong-Dian Reservoir (Figure 1) is used as an illustrative example in this study. This reservoir, located in Kaohsiung City, (southern Taiwan), collects water from Joushui River and Wanglai River. The total watershed area is 29.58 with 12.81 km^2 (43%) from Joushui River watershed and 16.77 km^2 (57%) from Wanglai River watershed. The length of the dam is 2.38 km, making it the longest dam in Taiwan. Its major purpose is flood control, while other uses, such as irrigation, water supply, and tourism benefit.

The elevation of the dam top, design water level, and maximum water level are 42, 37, and 40 m, respectively. The reservoir was completed in 1953. However, since its completion, large amounts of green-grey clay and yellow silty clay from the upstream watersheds of Joushui and Wanglai rivers have been washed into the reservoir and severe sedimentation has been observed [37]. The effective reservoir capacity was slashed from 20.5 to 5.9 million cubic meters in 1996. In order to revamp the reservoir, the A-Gong-Dian reservoir improvement project was implemented in 1997. Large-scale sediment flushing of the reservoir was executed and 11.6 million cubic meters of sediments were dredged, and the reservoir reached empty storage. The reservoir improvement project involved dam improvement, conduit spillway reconstruction, water intake tower reconstruction, trans-basin waterway, etc. It was finally completed in 2005, and re-opened in June 2006. The shaft spillway pipe (Figure 2) has been operated ever since, for the period 1 June to 10 September annually, which corresponds to the wet season in this region.

Although its design capacity storage is 20.5 million cubic meters, currently, the effective storage capacity is 16.69 million cubic meters, and the total water storage is 45 million cubic meters. The reservoir adopts a shaft spillway pipe having a 2.8 m diameter to reduce the pipe top elevation to 27 m. Based on hydraulic model test, flow discharge of the shaft spillway pipe can be expressed by Equations (22) and (23) [37]:

$$\text{Free overfall } Q_{\text{out}} = 34.12(H_s - 27)^{1.5}, \ H_s \leq 28.57 \text{ m} \tag{22}$$

$$\text{Pipe flow } Q_{\text{out}} = 17.63(H_s - 14)^{0.5}, \ H_s > 28.57 \text{ m} \tag{23}$$

where H_s is water level in the reservoir, Q_{out} is the releasing discharge. The maximum releasing discharge of pipe flow for the shaft spillway pipe is 89.90 m^3/s when the water level reaches the 40-m design maximum flood retention level.

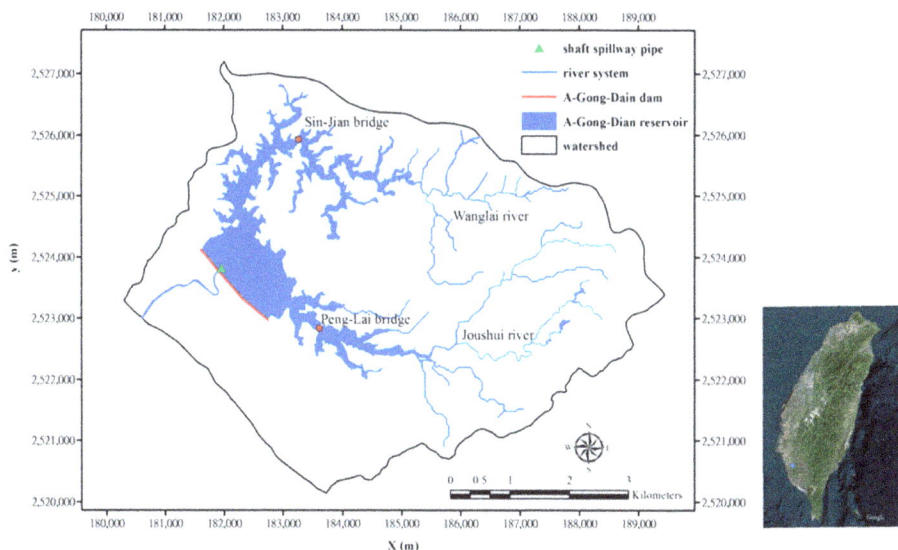

Figure 1. A-Gong-Dian Reservoir watershed and its river system.

Figure 2. Shaft spillway pipe.

4. Methodology

In order to understand the sedimentation pattern, distribution of sediments into the A-Gong-Dian reservoir, bed evolution within the reservoir, and the sediment flushing efficiency of an empty storage operation of the shaft spillway pipe during a flood season, a depth-averaged two-dimensional bed evolution model was developed and applied. The upstream boundary condition hydrographs of inflow discharge and suspended sediment concentration for the two-dimensional bed evolution model were calculated by the PSED model [26,38].

The PSED model can accurately estimate discharge hydrographs, sediment hydrograph, suspended sediment transport rate, and sediment yield from a watershed. In this model, GIS is used to partition the river catchment into computed river cells and land cells, according to the spatial distribution of the physiographical characteristics, such as topography, landform, and vegetation distribution in the watershed. Due to the complex nature of these earth features, a large amount of data is often generated. With the assistance of GIS, the PSED model can handle enormous hydrologic and physiographic datasets, simulating the physical erosion process without the need for simplification.

Computations were done for the storage region between Sin-Jian Bridge of the Wanglai river and the Peng-Lai Bridge of the Joushui river, which is about 5 km long. Cross-sections were measured by the Southern Water Resources Bureau in 2008, and the measured elevations are shown in Figure 3. The study area was discretised into 26,251, squared, $\Delta x = \Delta y = 10$ m, computational cells and the time step was 0.05 s ($\Delta t = 0.05$). Conducted field experiments have shown the average sediment particle size to be 0.015 mm, while the Manning's roughness coefficient was 0.028. The initial water depth condition was 0.1 m, while velocity and sediment concentration were 0 for each computed grid.

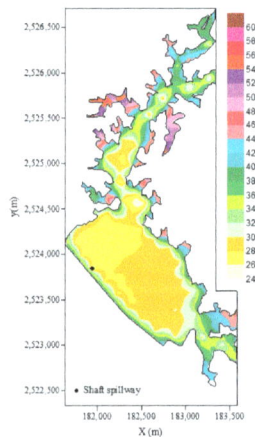

Figure 3. Bed elevation in storage area of A-Gong-Dian reservoir.

The hydrograph boundary conditions of inflow discharge and suspended sediment concentration for the typhoon event and design rainfall of one- and two-day storms under various return periods (2, 5, 10, 25, 50, 100, 200-year) for the depth-averaged two-dimensional bed evolution model were simulated and calculated by the PSED model.

5. Results and Discussions

5.1. Analysis of the Sediment Flushing Efficiency of an Empty Storage

Reducing sediment deposition to increase flood control volume is crucial for Agongdian reservoir. Therefore, sediment flushing of empty storage by a shaft spillway pipe is operated during the wet period to decrease the sediment deposition within the reservoir bed. To understand the sediment flushing efficiency of empty storage by a shaft spillway pipe under one-day and two-day storms of various return periods, the operations for sediment flushing of the empty storage were simulated using the depth-averaged two-dimensional bed evolution model, and the results are shown in Tables 1 and 2.

The results indicate that the average sediment flushing efficiency for one-day and two-day storms are close, 58.50% and 59.39%, respectively. For the simulated storms, the lowest efficiencies are 58.16%, 58.62% (200-year return period), and the highest efficiencies are 58.98%, 60.19% (10-year, 2-year return period). The simulation results are close to 65.34% [39], which agree well with the efficiency obtained from the hydraulic model test in the laboratory. The results show that the two-dimensional bed evolution model can reasonably simulate sediment flushing of empty storage by a shaft spillway pipe, as sediment flushing efficiency reached up to 60%.

Table 1. Empty flushing efficiency in one-day storms under different return periods.

Return Period (Year)	Sediment Yield of the Watershed (m^3)	Sediment Deposited in the Reservoir (m^3)	Sediment Flushing by Shaft Spillway (m^3)	Sediment Flushing Efficiency (%)
2	201,341	83,154	118,187	58.70
5	280,522	116,875	163,647	58.34
10	331,857	136,121	195,736	58.98
25	375,159	156,963	218,196	58.16
50	427,718	176,549	251,169	58.72
100	464,325	193,141	271,184	58.40
200	498,867	208,575	290,292	58.16

Table 2. Empty flushing efficiency in two-day storms under different return periods.

Return Period (Year)	Sediment Yield of the Watershed (m^3)	Sediment Deposited in the Reservoir (m^3)	Sediment Flushing by Shaft Spillway (m^3)	Sediment Flushing Efficiency (%)
2	284,238	113,155	171,083	60.19
5	397,096	159,791	237,305	59.76
10	471,082	189,846	281,236	59.70
25	541,196	220,050	321,146	59.34
50	625,933	255,506	370,427	59.18
100	701,485	287,889	413,596	58.96
200	775,718	320,992	454,726	58.62

5.2. Analysis of Sediment Delivery Behaviour in Reservoirs

Sediment flushing efficiency is related to the life and function of the reservoir. Therefore, this topic of great importance for reservoir management. In addition, it is necessary to carry out the analysis from sediment transport into the reservoir to bed evolution during the process of flooding and sediment flushing. Management of reservoirs does not only entail improving the flushing efficiencies, but in depth investigations are needed to understand the whole process from catchment erosion, to sediment transport and bed evolution, and finally to the removal of the sediments. To understand the whole complex process of sediments transport during heavy storms or floods in a reservoir, the severe storm event typhoon Morakot, which hit Taiwan in 6 to 10 August 2009, was numerically simulated to estimate the variations of water depth, suspended sediment concentration, sediment delivery, and bed evolution. Results are shown in Figures 4–6. Figure 4 show water depth variation in the reservoir at 36, 60, 72, 84, 96, and 120-h, while Figure 5 show distribution of suspended sediment concentration, and Figure 6, bed evolution.

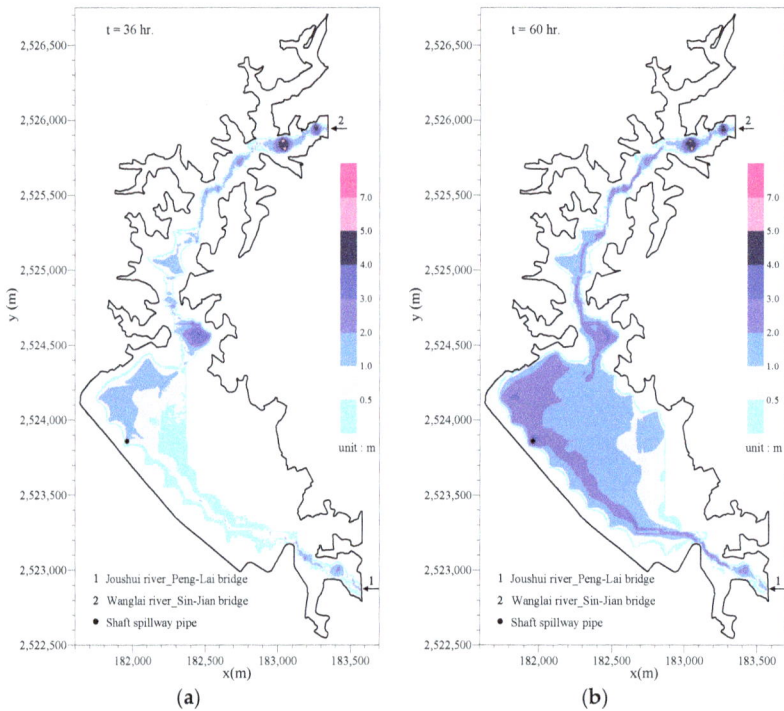

Figure 4. *Cont.*

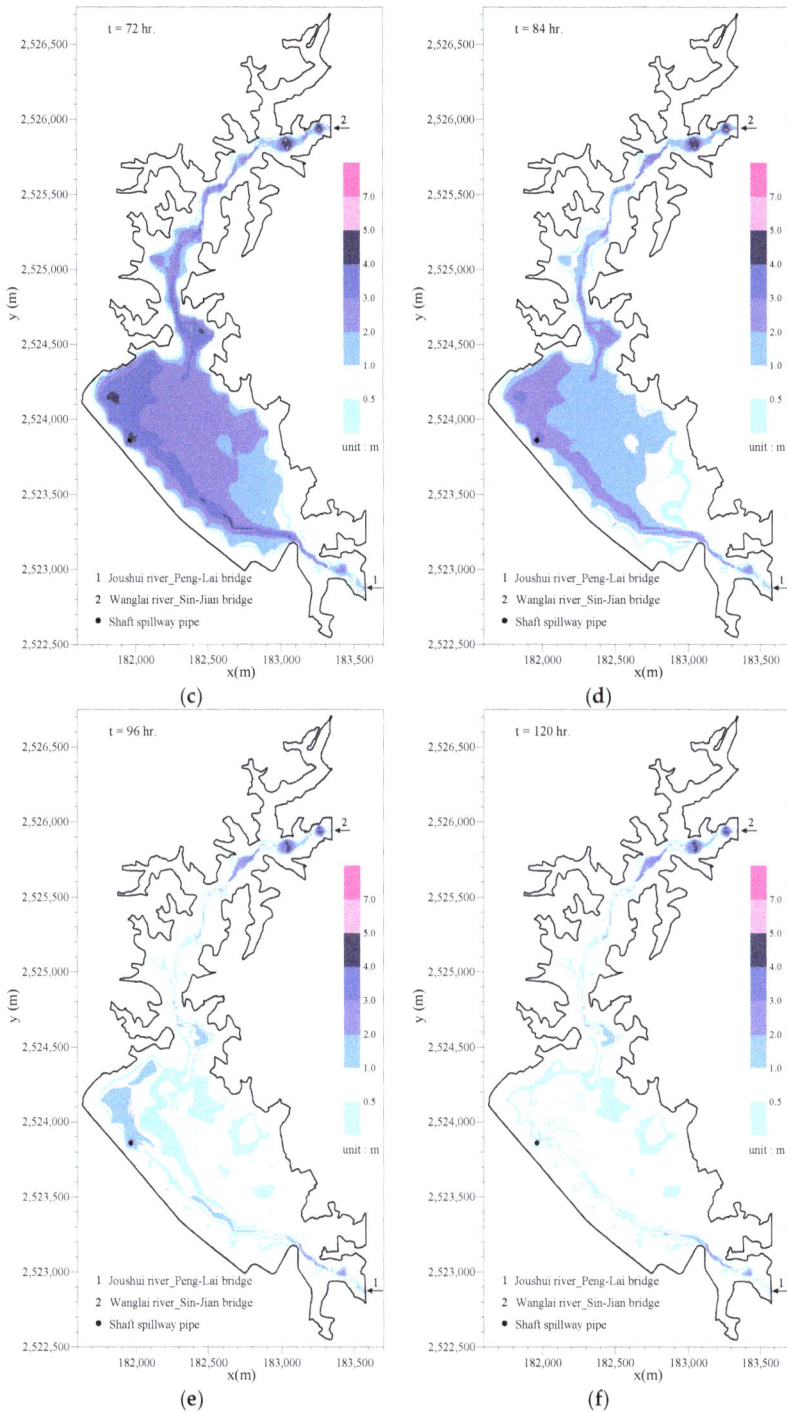

Figure 4. (**a**) Water depth in the reservoir at *t* = 36 h; (**b**) Water depth in the reservoir at *t* = 60 h; (**c**) Water depth in the reservoir at *t* = 72 h; (**d**) Water depth in the reservoir at *t* = 84 h; (**e**) Water depth in the reservoir at *t* = 96 h; and, (**f**) Water depth in the reservoir at *t* = 120 h.

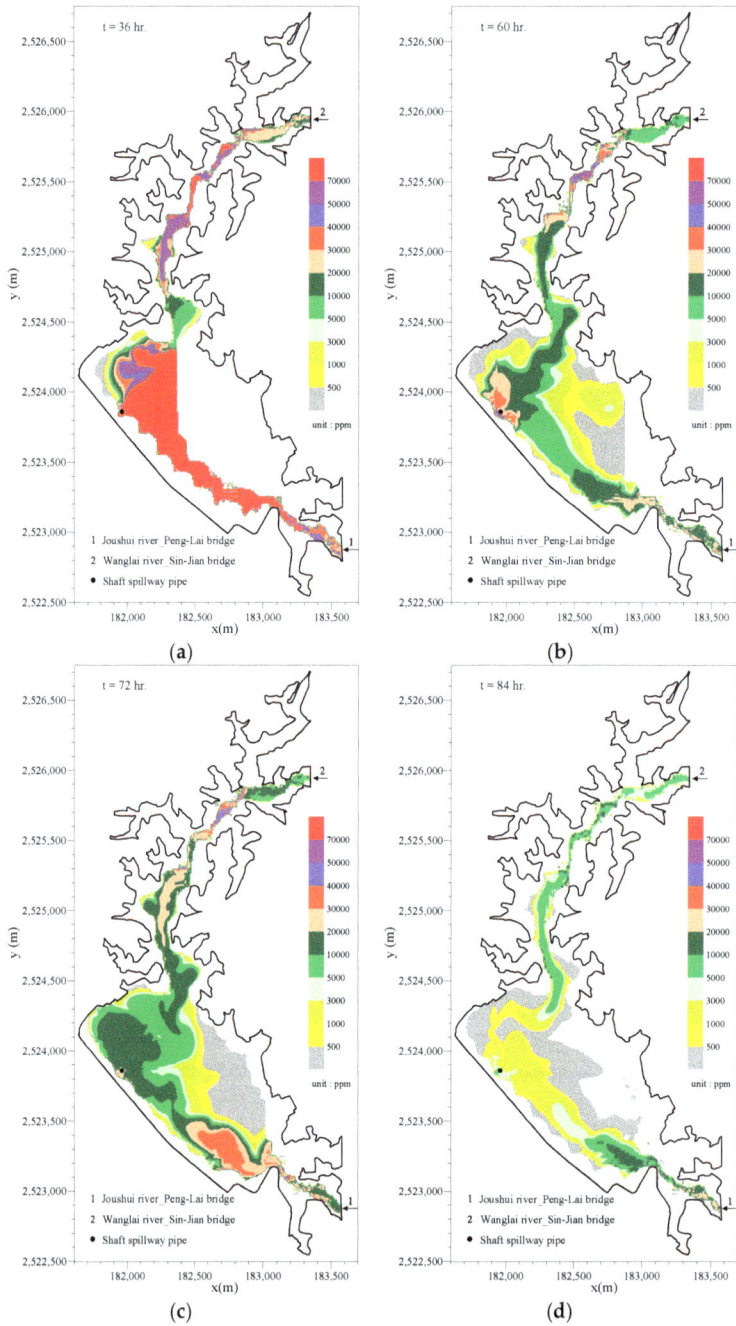

Figure 5. *Cont.*

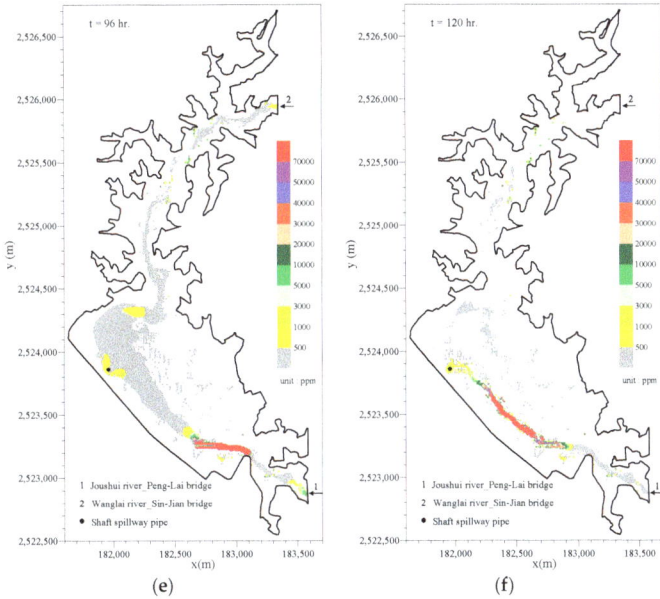

Figure 5. (**a**) Distribution of suspended sediment concentration at *t* = 36 h; (**b**) Distribution of suspended sediment concentration at *t* = 60 h; (**c**) Distribution of suspended sediment concentration at *t* = 72 h; (**d**) Distribution of suspended sediment concentration at *t* = 84 h; (**e**) Distribution of suspended sediment concentration at *t* = 96 h; and, (**f**) Distribution of suspended sediment concentration at *t* = 120 h.

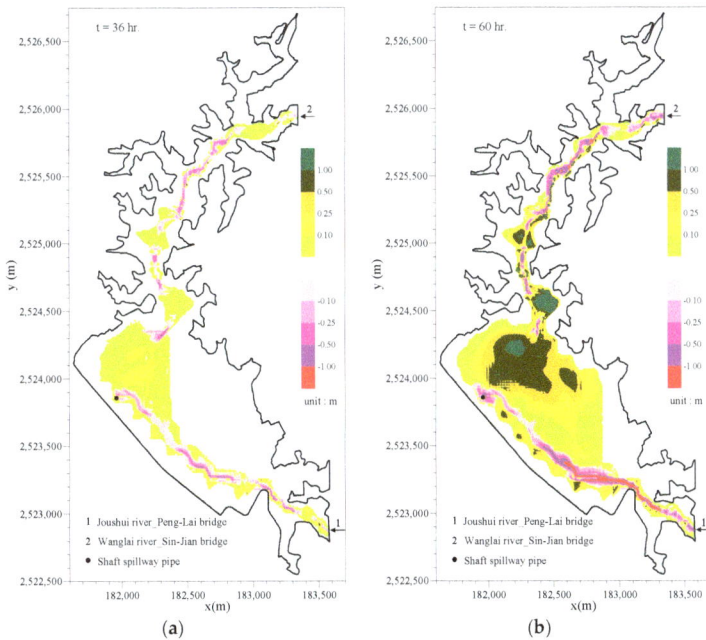

Figure 6. *Cont.*

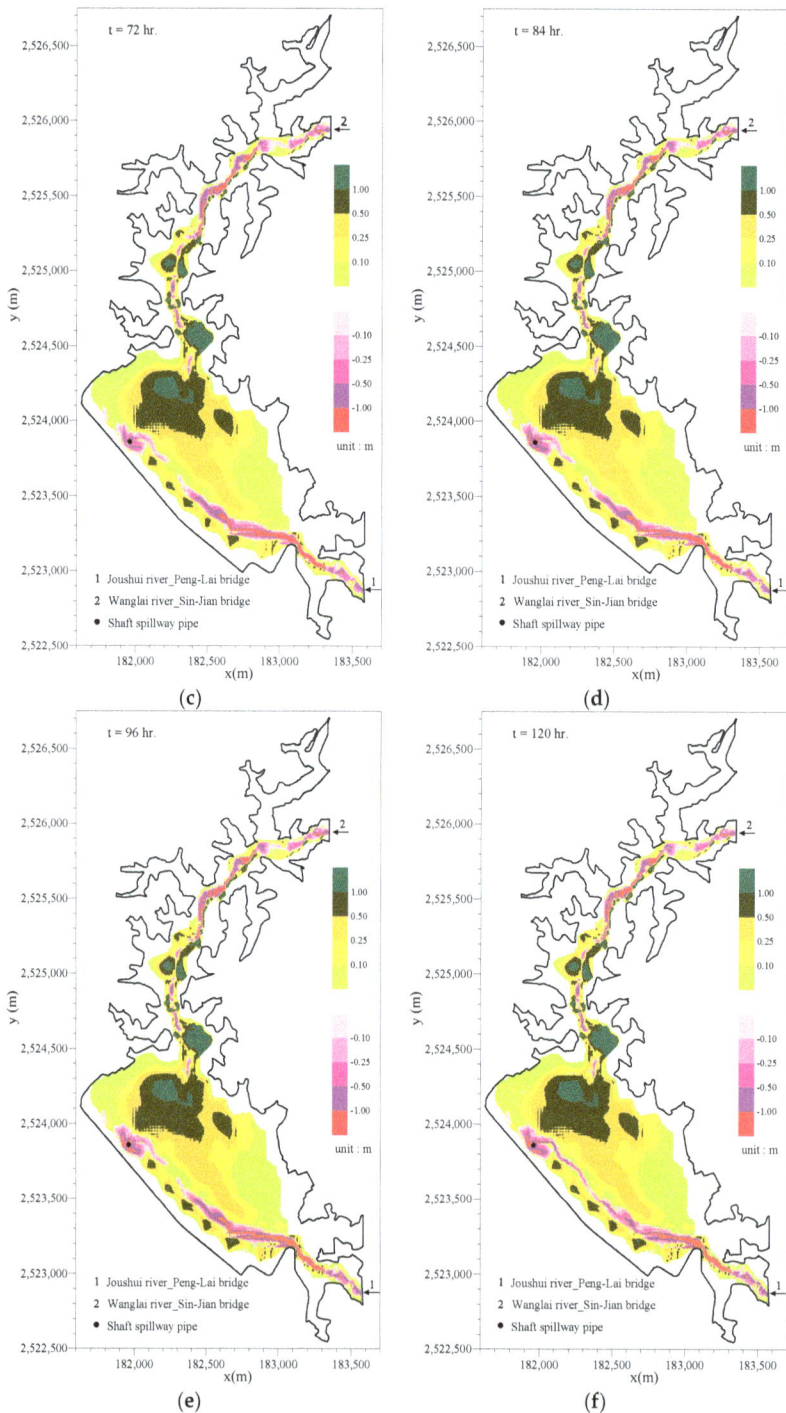

Figure 6. (**a**) Bed evolution in the reservoir at t = 36 h; (**b**) Bed evolution in the reservoir at t = 60 h; (**c**) Bed evolution in the reservoir at t = 72 h; (**d**) Bed evolution in the reservoir at t = 84 h; (**e**) Bed evolution in the reservoir at t = 96 h; and, (**f**) Bed evolution in the reservoir at t = 120 h.

Water depth is shown to initially increase in Figure 4 while flow entered the reservoir, and after *t* = 72-h it began to gradually decrease as the inflow discharge became less than the releasing discharge of the shaft spillway. Figure 5 shows the distribution of suspended sediment concentration. The results show muddy water into the reservoir and convection-diffusion of sediment in the reservoir. In addition, the high bed shear stress that was caused by the increased velocity entrained more sediments into the flow, thus the high turbid flow in the reservoir. Sediment concentration reached more than 70,000 ppm and the high turbid flow was released by the shaft spillway pipe, and the siltation in the reservoir was due to the unreleased turbid flow.

Figure 6 shows both the erosion and the deposition depth during typhoon Morakot. Erosion is seen to be more dominant at the thalweg (>0.5 m), and is more apparent at the erosion ditch that is generated between Joushui river and the shaft spillway. This could be attributed to the high flow velocity, resulting from the shaft spillway pipe, and the steep slopes around this area. Moreover, a not significant erosion ditch is seen between the Wanglai river and the shaft spillway pipe. Deposition is prevalent from the 60th hour, especially downstream of the Wanglai river, greater than 0.5 m, and is not severe from the Joushui side as it hardly reaches 0.25 m. It is worth mentioning that the downstream of Wanglai river where it joins the reservoir is wider (Figure 2) when compared to the Joushui river, hence the flow velocity is greatly reduced and it increases the rate of sediment deposition. Part of the sediments deposited in the reservoir could not be flushed out by the shaft spillway pipe, resulting in erosion-deposition interplay.

Final bed evolution under the different return periods of one and two-day storms, with a total simulation time of 120 h is shown in Figures 7–13. Similar patterns of erosion and deposition are seen with the different return periods; moreover, these intensify with increasing the return period. There is a significant erosion ditch (>1 m) from the Joushui river down to all around the shaft spillway pipe. The maximum deposition area in the north-eastern of the shaft spillway pipe is mainly due to the non-significant erosion ditch that is seen between the Wanglai river and the shaft spillway pipe. Hence, if the shaft spillway pipe were shifted to the applicable distance in the north-eastern direction, it will improve the probability of forming an erosion ditch between Wanglai river and the shaft spillway. There would be two significant erosion ditches formed to enhance flushing.

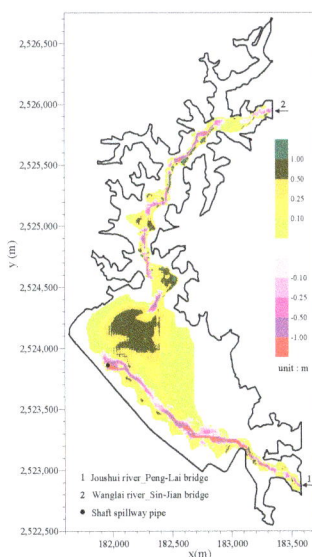

Figure 7. Bed evolution in the reservoir during a two-year return period, *t* = 120 h.

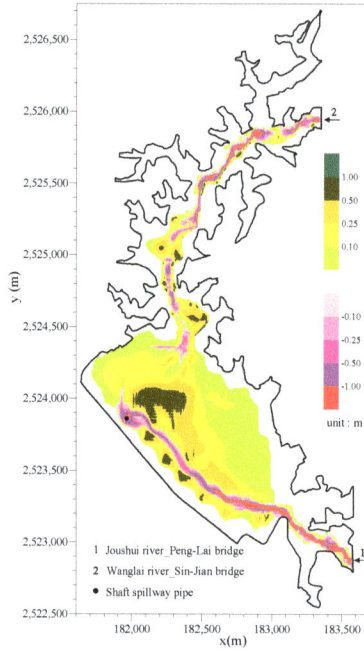

Figure 8. Bed evolution in the reservoir during a five-year return period, *t* = 120 h.

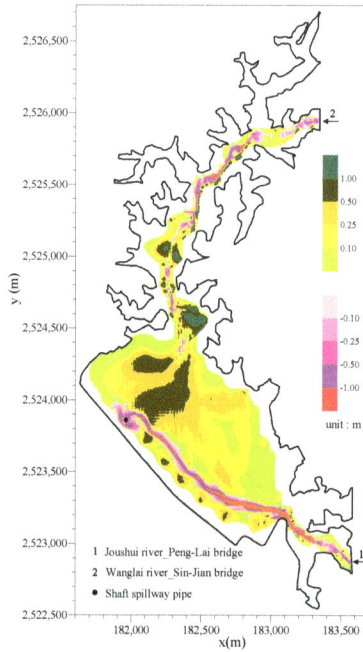

Figure 9. Bed evolution in the reservoir during a 10-year return period, *t* = 120 h.

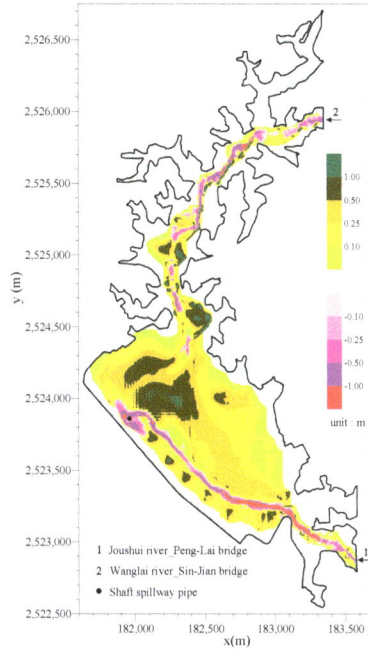

Figure 10. Bed evolution in the reservoir during a 25-year return period, *t* = 120 h.

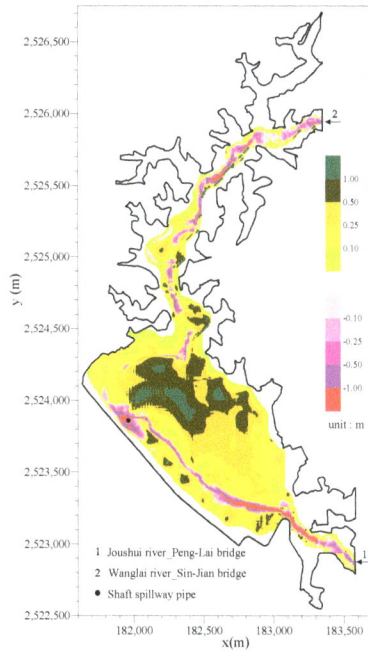

Figure 11. Bed evolution in the reservoir during a 50-year return period, *t* = 120 h.

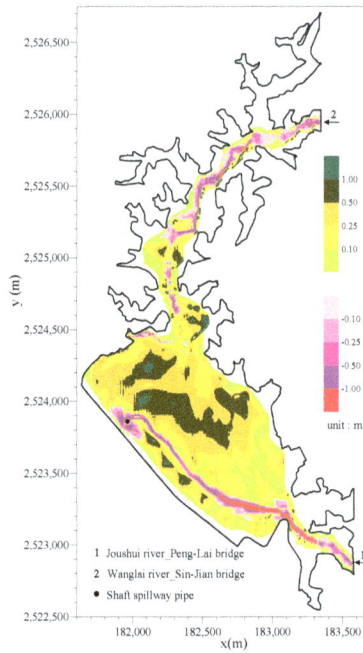

Figure 12. Bed evolution in the reservoir during a 100-year return period, t = 120 h.

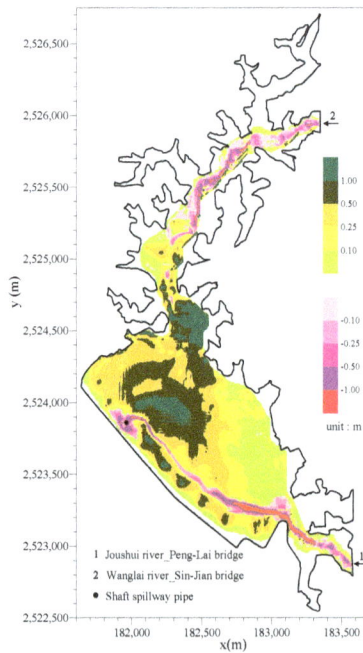

Figure 13. Bed evolution in the reservoir during a 200-year return period, t = 120 h.

6. Conclusions

A model is developed and applied to the simulation of sediment erosion/deposition and sediment distribution within a reservoir, and to simulate the flushing efficiency of a shaft spillway pipe. Hydrograph boundary conditions of inflow and suspended sediment concentration of one-day and two-day storms of different return periods, (2, 5, 10, 25, 50, 100, 200-year), were computed by the 1D, PSED model. Based on the results, the following conclusions may be drawn.

The average efficiency of the empty flushing by a shaft spillway pipe under one- and two-day storms of various return periods were almost similar, 58.49% and 59.39%, respectively. These results were found to be adequate as the efficiency was almost 60%. The overall simulation results are close to 65.34%, which is the efficiency that is obtained from a hydraulic model test in a laboratory.

Bed evolution in the reservoir was significantly driven by flow velocity under empty flushing by the shaft spillway pipe after the rainfall had stopped. A significant erosion ditch was generated after 96 h of simulation time between Joushui river and the shaft spillway pipe. At the end of the simulation time, no erosion ditch was developed from the Wanglai river due to the relatively wider cross section and low flow velocity. The shaft spillway could not completely flush out the sediments that were deposited; as a result, bed evolution was a mixture of erosion and deposition. Similar patterns of erosion and deposition were observed for the selected storm, typhoon Morakot, and the design rainfall events under different return periods.

A common observation was the significant erosion ditch that formed between the Wanglai river and the shaft spillway pipe that enhanced flushing. From our findings, we propose a relocation of the shaft spillway pipe towards the north-eastern direction. Shifting the pipe to this direction could improve the probability of forming the erosion ditch between Wanglai river and the shaft spillway. Two significant erosion ditches would improve the empty flushing efficiency even further.

Acknowledgments: Observed and hydraulic model test data for this study was provided by A-Gong-Dian Reservoir Management Centre, Southern Region Water Resources office, Water Resources Agency, Ministry of Economic Affairs. The authors gratefully acknowledge their support.

Author Contributions: Ching-Nuo Chen conceived and designed the study and further wrote the manuscript while Chih-Heng Tsai contributed significantly with data analysis.

Conflicts of Interest: The authors declare no conflict of interest.

References

1. Wang, Z.-Y.; Hu, C. Strategies for Managing Reservoir Sedimentation. *Int. J. Sediment Res.* **2009**, *24*, 369–384. [CrossRef]
2. Schleiss, A.J.; Boes, R.M. *Dams and Reservoirs under Changing Challenges*; CRC Press: Boca Raton, FL, USA, 2011.
3. Kondolf, G.M.; Gao, Y.; Annandale, G.W.; Morris, G.L.; Jiang, E.; Zhang, J.; Cao, Y.; Carling, P.; Fu, K.; Guo, Q. Sustainable Sediment Management in Reservoirs and Regulated Rivers: Experiences from Five Continents. *Earth Future* **2014**, *2*, 256–280. [CrossRef]
4. Chang, F.-J.; Lai, J.-S.; Kao, L.-S. Optimization of Operation Rule Curves and Flushing Schedule in a Reservoir. *Hydrol. Process.* **2003**, *17*, 1623–1640. [CrossRef]
5. Emamgholizadeh, S.; Samadi, H. Desilting of Deposited Sediment at the Upstream of the Dez Reservoir in Iran. *J. Appl. Sci. Environ. Sanit.* **2008**, *3*, 25–35.
6. Chaudhry, M.A.; Habib-ur-Rehman, M.; Akhtar, M.N.; Hashmi, H.N. Modeling Sediment Deposition and Sediment Flushing through Reservoirs Using 1-D Numerical Model. *Arab. J. Sci. Eng.* **2014**, *39*, 647–658. [CrossRef]
7. Castillo, L.G.; Carrillo, J.M.; Álvarez, M.A. Complementary Methods for Determining the Sedimentation and Flushing in a Reservoir. *J. Hydraul. Eng.* **2015**, *141*, 05015004. [CrossRef]
8. Basson, G.R.; Rooseboom, A. Sediment Pass-through Operations in Reservoirs. In Proceedings of the International Conference on Reservoir Sedimentation, Fort Collins, CL, USA, 9–13 September 1996; pp. 1107–1130.

9.   Madadi, M.R.; Rahimpour, M.; Qaderi, K. Improving the Pressurized Flushing Efficiency in Reservoirs: An Experimental Study. *Water Resour. Manag.* **2017**, *31*, 4633–4647. [CrossRef]

10. Morris, G.L.; Fan, J. *Reservoir Sedimentation Handbook: Design and Management of Dams, Reservoirs, and Watersheds for Sustainable Use*; McGraw-Hill: New York, NY, USA, 1998.

11. United States Society on Dams. *Modeling Sediment Movement in Reservoirs*; United States Society on Dams: Denver, CO, USA, 2015.

12. Basson, G.R.; Rooseboom, A. *Dealing with Reservoir Sedimentation: Guidelines and Case Studies*; International Commission on Large Dams: Paris, France, 1996.

13. Pagliari, D.; Rossi, L.; Passoni, D.; Pinto, L.; De Michele, C.; Avanzi, F. Measuring the Volume of Flushed Sediments in a Reservoir Using Multi-Temporal Images Acquired with Uas. *Geomat. Nat. Hazards Risk* **2017**, *8*, 150–166. [CrossRef]

14. Su, T.-C.; Chou, H.-T. Application of Multispectral Sensors Carried on Unmanned Aerial Vehicle (Uav) to Trophic State Mapping of Small Reservoirs: A Case Study of Tain-Pu Reservoir in Kinmen, Taiwan. *Remote Sens.* **2015**, *7*, 10078–10097. [CrossRef]

15. Guertault, L.; Camenen, B.; Peteuil, C.; Paquier, A. Long Term Evolution of a Dam Reservoir Subjected to Regular Flushing Events. *Adv. Geosci.* **2014**, *39*, 89–94. [CrossRef]

16. Chou, F.N.F.; Wu, C.W. Assessment of Optimal Empty Flushing Strategies in a Multi-Reservoir System. *Hydrol. Earth Syst. Sci. Discuss* **2016**, *2016*, 1–49. [CrossRef]

17. U.S. Army Corps of Engineers. *Hec-6 Scour and Deposition in Rivers and Reservoir, User's Manual*; Hydrologic Engineering Centre: Davis, CA, USA, 1991.

18. U.S. Army Corps of Engineers. *Hec-Ras (Hydrologic Engineering Center River Analysis System) Manual*; Hydrologic Engineering Centre: Davis, CA, USA, 2005.

19. Chang, H.H. *Fluvial Processes in River Engineering*; Wiley: Hoboken, NJ, USA, 1988.

20. Chang, H.H.; Harrison, L.L.; Lee, W.; Tu, S. Numerical Modeling for Sediment-Pass-through Reservoirs. *J. Hydraul. Eng.* **1996**, *122*, 381–388. [CrossRef]

21. Olsen, N.R.B. Two-Dimensional Numerical Modelling of Flushing Processes in Water Reservoirs. *J. Hydraul. Res.* **1999**, *37*, 3–16. [CrossRef]

22. Olsen, N.R.B.; Skoglund, M. Three-Dimensional Numerical Modeling of Water and Sediment Flow in a Sand Trap. *J. Hydraul. Res.* **1994**, *32*, 833–844. [CrossRef]

23. Fang, H.-W.; Rodi, W. Three-Dimensional Calculations of Flow and Suspended Sediment Transport in the Neighborhood of the Dam for the Three Gorges Project (Tgp) Reservoir in the Yangtze River. *J. Hydraul. Res.* **2003**, *41*, 379–394. [CrossRef]

24. Khosronejad, A.; Rennie, C.D.; Neyshabouri, A.A.S.; Gholami, I. Three-Dimensional Numerical Modeling of Reservoir Sediment Release. *J. Hydraul. Res.* **2008**, *46*, 209–223. [CrossRef]

25. Wang, Y.M.; Tfwala, S.S.; Chan, H.C.; Lin, Y.C. The Effects of Sporadic Torrential Rainfall Events on Suspended Sediments. *Arch. Sci. J.* **2013**, *66*, 211–224.

26. Chen, C.-N.; Tsai, C.-H.; Tsai, C.-T. Simulation of Sediment Yield from Watershed by Physiographic Soil Erosion–Deposition Model. *J. Hydrol.* **2006**, *327*, 293–303. [CrossRef]

27. Tsai, C.H.; Tsai, C.T. Simulation of Two-Dimensional Gradually-Varied Unsteady Flow in Curved Channels. *J. Chin. Inst. Civ. Hydraul. Eng.* **1995**, *7*, 461–473.

28. Tsai, C.H. Numerical Simulation of Flow and Bed Evolution in Alluvial River with Levees. Ph.D. Thesis, National Cheng-Kung University, Tainan, Taiwan, 2000.

29. Itakura, T.; Kishi, T. Open Channel Flow with Suspended Sediments. *J. Hydraul. Div.* **1980**, *106*, 1325–1343.

30. Shimizu, Y.; Yamaguchi, H.; Itakura, T. Three-Dimensional Computation of Flow and Bed Deformation. *J. Hydraul. Eng.* **1990**, *116*, 1090–1108. [CrossRef]

31. Meyer-Peter, P.E.; Muller, R. Formulas for Bedload Transport. In Proceedings of the 2nd International Association for Hydraulic Research, Stockholm, Sweden, 7–9 June 1948; pp. 39–64.

32. Wong, M.; Parker, G. Reanalysis and Correction of Bed-Load Relation of Meyer-Peter and Müller Using Their Own Database. *J. Hydraul. Eng.* **2006**, *132*, 1159–1168. [CrossRef]

33. Paola, C.; Voller, V.R. A Generalized Exner Equation for Sediment Mass Balance. *J. Geophys. Res. Earth Surf.* **2005**, *110*, 2156–2202. [CrossRef]

34. Parker, G.; Paola, C.; Leclair, S. Probabilistic Exner Sediment Continuity Equation for Mixtures with No Active Layer. *J. Hydraul. Eng.* **2000**, *126*, 818–826. [CrossRef]

35. Komura, S.; Simons, D.B. River-Bed Degradation Below Dams. *J. Hydraul. Div.* **1967**, *93*, 1–14.

36. Bhallamudi, M.S.; Chaudhry, H.M. Computation of Flows in Open-Channel Transitions. *J. Hydraul. Res.* **1992**, *30*, 77–93. [CrossRef]

37. Southern Water Resources Bureau. *Effect Assessment of Empty Storage Operation for Sediment Prevention Observation Program of Agongdian Reservoir*; Water Resources Agency, MOEA: Taichung, Taiwan, 2010.

38. Chen, C.-N.; Tsai, C.-H. The Impact of Climate Change on Inundation Potential. *Int. J. Environ. Sci. Dev.* **2013**, *4*, 496–500. [CrossRef]

39. Water Resources Planning Institute. *Hydraulic Model Studies on the Functions and Operations of Silting Prevention in a-Kung-Tien Reservior*; Water Resources Agency, Ministry of Economic Affairs: Taichung, Taiwan, 2003.

water

MDPI

Article

Quantifying the Effects of Near-Bed Concentration on the Sediment Flux after the Operation of the Three Gorges Dam, Yangtze River

Li He

Key Laboratory of Water Cycle and Related Land Surface Processes, Institute of Geographic Sciences and Natural Resources Research, Chinese Academy of Sciences, Beijing 100101, China; heli@igsnrr.ac.cn; Tel.: +86-10-6488-8151

Received: 12 November 2017; Accepted: 15 December 2017; Published: 18 December 2017

Abstract: The regime of sediment transport in the Jingjiang Reach has significantly changed from quasi-equilibrium to sub-saturation since the impoundment of the Three Gorges Dam (TGD), and vertical profiles of suspended sediment concentration (SSC) have changed accordingly. Vertical profiles of SSC data measured at three hydrological stations in the Jingjiang Reach (Zhicheng, Shaishi, and Jianli), before and after the impoundment of TGD, were collected and analyzed. Analytic results indicate a remarkably large concentration in the near-bed zone (within 10% of water depth from the river-bed) in a sub-saturated channel. The maximum measured concentration was up to 15 times the vertical average concentration, while the ratio in quasi-equilibrium channel was less than four times that. Concentrations normalized with reference concentration at the same height, and may decrease with increasing values of suspension index (settling velocity over shear velocity). In addition, concentration near the water surface may be larger than concentration in the near-bed region when the suspension index is smaller than 0.01. Sediment flux transported in the near-bed zone may be up to 35% of the total sediment flux in unsaturated flows. The relationship between deviations of estimating sediment flux when ignoring the near-bed concentration and discharge in flood season and non-flood season are different in unsaturated and quasi-equilibrium channels. Analysis indicates that, in the quasi-equilibrium channel, more attention should be paid to near-bed concentration during non-flood season, the same as measurements during flood season with larger discharge.

Keywords: sediment regime; suspended sediment concentration; vertical profiles of concentration; Jingjiang River Reach; Yangtze River

---

1. Introduction

The majority (>90%) of river-borne flux is closely associated with sediment [1]. A significant proportion of sediments are transported in suspension, as the bed load at river mouths is often less than 1% of the total solid transport [2].

Various vertical profiles of suspended sediment concentration (SSC) have been observed by field observation. For example, these profiles include linear type including linear or quasi-linear (with different gradients), parabolic curve, and mixed linear type [3,4]. For different linear types, the difference between concentration on the water surface and concentration near the bed varies. For linear type, the difference of concentration in the vertical direction increases with increasing slope, and the rate of variation in a vertical direction is constant. Furthermore, the linear types can be observed in estuaries with smaller concentrations [3]. For the parabolic type, the rate of variation in a vertical direction varies, and exists in a relatively larger concentration in the near-bed region. For the parabolic type, the concentration in the near-bed region is a kind of tailing phenomenon, and this kind of vertical profile can be observed in tide water [3]. Various theories, i.e., gravity, diffusion, mixing, energy dissipation, and stochastic models have been applied to simulate the vertical distribution of SSC [5,6]. And various efforts have been focused on the temporal and spatial variation of SSC by field survey, especially the near-surface and near-bed concentration [7,8]. Zuo et al. [3] pointed out that larger concentrations in the near-bed region may be caused by flocculate, saline water and tidal wave. Based on observation in the Jingjiang River Reach after Three Gorges Dam's impoundment, vertical profiles of SSC in channels with a changed sediment regime have revealed a remarkably large concentration in the near-bed region [9]. This indicates that, a larger concentration in the near-bed region can also be observed in a reservoir down-channel [9]. Erosion downstream is a living topic coping with sediment trapping in reservoirs. Varied sediment transport regimes and sediment-related problems in the middle reach (Jingjiang Reach, China) have a profound morphological impact on the lower reach, that is, navigation, pollutant and deposition in channels and ports, water and sediment management in stem channels, social and economic problems and so on [10]. Various studies focus on the changed sediment regime and its influences on the channel downstream (i.e., [11–13]). However, the characteristics of the distinct larger near-bed concentration and its effects on the river reach have not been widely analyzed.

Thus, the study aims to analyze the changed vertical profiles of SSC with changed sediment regimes. Based on vertical profiles of concentration, detailed characteristics of the tailing phenomena are analyzed. Then, the effects of high concentration in the near-bed region on sediment flux are estimated. Finally, the vertical distribution of suspended sediment concentration and its effects after dam operation are compared with data before operation.

2. Materials and Methods

2.1. Study Area and Data

The Yangtze River (YR) is the largest and longest river in China, and the third largest river in the world. The length of the YR is approximately 6.3×10^3 km, and the drainage area is approximately 1.8×10^6 km^2.

The Three Gorges Dam (TGD) is located at the exit of the upper YR, Yichang in Hubei province [14]. The dam is 185 m high and the storage capacity of the reservoir is 3.9×10^8 m^3. The main purposes of the project are flood control, power generation and navigation. It started to impound water in 2003. After 2003, bedload and suspended load from the upper drainage area of the YR are trapped in the reservoir of the TGD. For suspended sediment load (SSL), more than 70% may be trapped during the first 10 years of operation, and approximately half of the SSL may be trapped during the operation over 40–100 years; the ratio may decrease to 15% and 10% after 80 years and 100 years, respectively [15]. Therefore, the SSL entering the middle and lower channel decreases, which leads to erosion of river reach down the dam [16]. During the operation of the TGD, the downstream channel erosion was extensive, and riverbed incision was accelerated [17].

The Jingjiang River is the river reach between Zhicheng and Chenglingji stations, with another two controlling stations of Shashi and Jianli (Figure 1). The length of the Jingjiang River Reach is approximately 348 km, and it is 64 km downstream from the TGD. The length of the upper and the lower part of the Jingjiang River Reach are approximately 172 km and 176 km, respectively [18,19]. During the operation of the TGD, the flow regime and sediment regime of the Jingjiang River Reach have changed [14].

Data measured at these three gauges in the Jingjiang Reach before and after the operation of TGD are collected (Table 1). Observed data are the vertical profiles of concentrations, discharges, wetted area, water depth, temperature, water stages, velocities and corresponding gradations. Data are measured by the Jingjiang Hydrology and Water Resources Surveying Bureau (JHWRSB), and published by the Yangtze River Water Resources Commission (YRWRC).

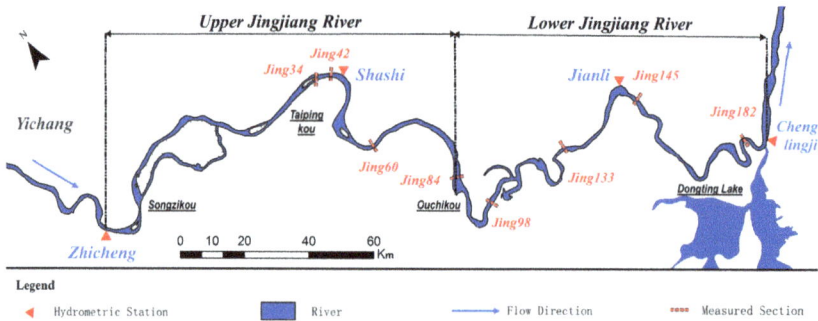

Figure 1. Sketch of the Yangtze River and Jingjiang River Reach.

Table 1. Statics of data measured at the Jingjiang River Reach.

Stations	Measured Year	Groups
Zhicheng	1996, 1998, 2002	55
Shashi	1996, 1998, 2010, 2011, 2012, 2013	130
Jianli	1986, 1998, 2002, 2010, 2011, 2012, 2013	180

Vertical profiles measured before and after 2010 are different. For data measured before dam operation, there are only five measuring points in each vertical line. Typical vertical profiles of data measured before dam operation may be described with relative heights (y/H) of 0.94, 0.8, 0.6, 0.2, and 0.04 (data measured 24 September 2002, at Zhicheng station with distance left of 700 m), where y is the distance of each measured point from the riverbed (m), and H is the averaged water depth (m) of this vertical line. This means that the near-bed region (less than 10% depth) has one measuring point.

For data measured after dam operation, there are seven points in each vertical profile, with typical relative heights (y/H) of 0.98, 0.8, 0.4, 0.2, 0.1, 0.0307, 0.0061 (data measured 11 July 2011, at Jianli station with distance left of 1170 m). The two near-bed points are measured with distance of 0.5 m and 0.1 m from the bed, respectively.

2.2. Equations to Estimate the Effects of Near-Bed Concentration on Sediment Flux

The sediment load by the two near-bed points are compared with the total sediment load of the local vertical profile,

$$R_0 = \frac{Q_{sv(7)} - Q_{sv(5)}}{Q_{sv(7)}} \times 100\% \tag{1}$$

where $Q_{sv(7)}$ and $Q_{sv(5)}$ are vertical sediment flux by seven points and upper five points, respectively.

The result of ignoring these two points can also be estimated. The deviation ratio of sediment transport rate by omitting the two near-bed points can be calculated by forms as:

$$R = \frac{Q_{s(7)} - Q_{s(5)}}{Q_{s(7)}} \times 100\% \tag{2}$$

where $Q_{s(7)}$ and $Q_{s(5)}$ are sediment flux by seven points and five points, respectively. $Q_{s(5)}$ is estimated with two measured near-bed points being omitted artificially. For the cross-sectional estimation, there are only several vertical profiles. The delta-shaped area between the left bank and the first vertical line from the left bank is also considered. The delta-shaped area between the right bank and the last vertical line from the left bank is not considered due to the difficulty of identifying the right bank.

The specific values of data measured during pre- and post-operation may vary in different vertical lines, but the differences are limited. These vertical profiles measured before dam operation missed the near-bed region due to what has been done after operation. They are interpolated and extended to the near-bed region, with minimum relative height (y/H) of 0.006.

3. Results

3.1. Distinct Non-Uniform Vertical Distribution

3.1.1. Comparing with Vertical-Averaged Concentration

In order to describe the vertical profiles of SSC, the vertical profiles are compared with vertical-averaged concentration. The relative height of the vertical coordinate is expressed as y/H, in which y is the distance of each measured point from the riverbed (m), and H is the averaged water depth (m) of this vertical line. The relative concentration in the horizontal coordinate is expressed as S_i/S_{avg}, where S_i is the measured concentration of the vertical profile (kg/m^3), and S_{avg} represents the vertical averaged concentration (kg/m^3). The minimum distance of measuring points from the riverbed can be named as the reference height, and the corresponding concentration is the reference concentration.

Figure 2 compares the vertical distribution of sediment concentrations in the Jingjiang Reach before and after the TGD impounded water in 2003. Figure 2 shows that, after the dam operation, measured vertical profiles of suspended sediment at these two gauges show a distinctly high concentration in the near-bed zone.

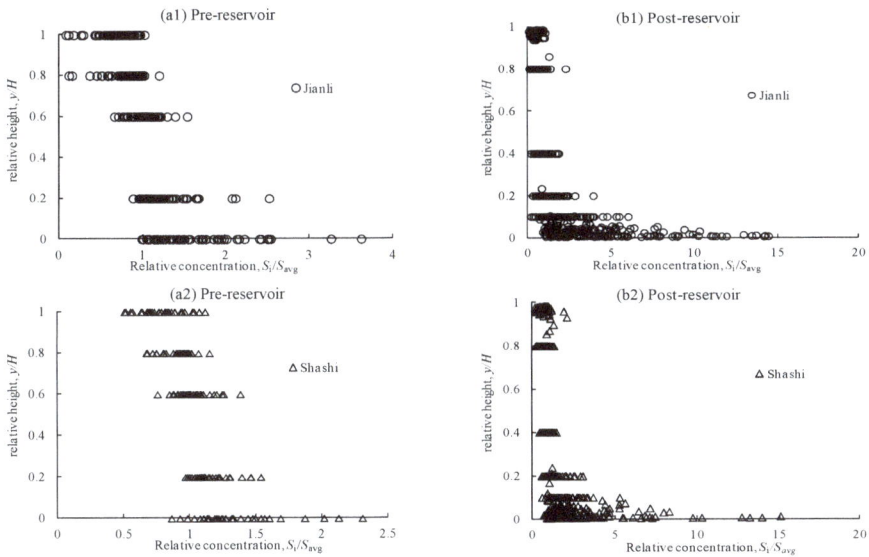

Figure 2. Vertical distributions of sediment concentrations at Shashi and Jianli hydrological stations; data measured before (**a**) and after (**b**) the Three Gorges Dam (TGD) operation in 2003. S_i is the measured concentration of the vertical profile (kg/m^3), and S_{avg} represents the vertical averaged concentration (kg/m^3).

For data measured after operation, the minimum relative height of measured points from the riverbed (y/H) is approximately 0.0051. The maximum relative concentrations (S_i/S_{avg}) are approximately 14.45 (Jianli) and 15.05 (Shashi). This kind of distinctly high concentration in the near-bed zone can be named as "tailing phenomena". It indicates that more sediment is transported in areas near the riverbed; the measured two near-bed points make it much worse. According to the data, average concentrations of these two near-bed points may account for up to 1.86 times of vertical averaged concentration at Shashi hydrological station, and 2.73 times at Jianli hydrological station.

For data measured before operation, the minimum relative height of measured points from the riverbed (y/H) is approximately 0.014 (0.3 m from the riverbed). The maximum relative concentrations (S_i/S_{avg}) are approximately 3.62 (Jianli) and 2.30 (Shashi). The maximum relative concentrations (S_i/S_{avg}) for Zhicheng station is approximately 2.76. For Zhicheng station, there are no data measured after the dam operation. Thus, data measured at Zhicheng are not included in Figure 2. Data measured before dam operation can be viewed as short-tail tailing phenomena.

Relatively high concentrations near the bed region have also been pronounced. Relatively larger near-bed concentrations can also be observed in the estuary of Yangtze River, Qiantangjiang River and Taizhou sea area [20,21]. Figure 3 illustrates relative concentrations (S_i/S_{avg}) measured at several regions, which also demonstrated a kind of tailing phenomena. Relative concentration (S_i/S_{avg}) is estimated with measured concentration over vertical averaged concentration, as in Figure 2. Data measured at Yangtze Estuary are drawn from Zhang [20], data measured at Wuhan and Jiujiang River Reach are drawn from Tang et al. [22], and data at Oujiangkou and Taizhou sea areas are drawn from Dong et al. [21]. The maximum relative concentration (S_i/S_{avg}) is approximately 4.3, and the minimum relative height of measured points from the riverbed (y/H) is approximately 0.03. Therefore, this kind of vertical profile can be viewed as short-tail tailing phenomena. The tailing phenomena which is much apparent in the Middle Jingjiang River Reach can be viewed as long-tail tailing phenomena. The near-bed concentration of the S-type vertical distribution is also relatively larger [23]. The tailing phenomena in the estuary area can be observed during tide periods and non-flood seasons [20].

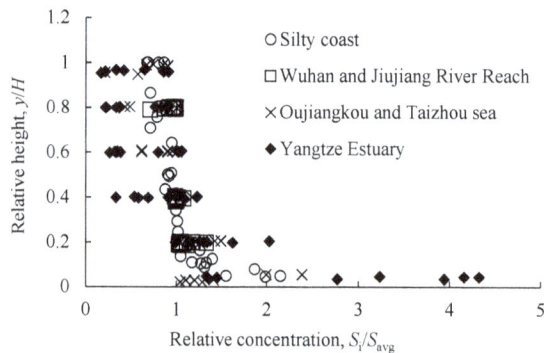

Figure 3. Vertical distribution of sediment concentration measured at different regions. Data at Yangtze Estuary are drawn from Zhang [20], data at Wuhan and Jiujiang River Reach are drawn from Tang et al. [22], data at Oujiangkou and Taizhou sea area are drawn from Dong et al. [21], and data at silty coast are drawn from Xia et al. [24]. S_i is the measured concentration of the vertical profile (kg/m^3), and S_{avg} represents the vertical averaged concentration (kg/m^3).

3.1.2. Comparing with Near-Bed Point

In order to highlight the concentration near the bed, the vertical profiles of suspended sediment load are compared with concentration of the near-bed point. The horizontal coordinate is S_i/S_{ia}, where S_{ia} represents the measured concentration of the near-bed point (kg/m^3). Only data measured after the operation are analyzed.

According to S_i/S_{ia}, the concentration near the bed is not absolutely larger than that of upper points, as shown in Figure 4. Vertical profiles can be grouped into two kinds: the first one is profiles in which concentration of the near-bed point is the largest (Type I), and the second one is profiles in which maximum concentration occurred in the near-bed region (Type II). All the vertical profiles observed in Zhang [20], Tang et al. [22], and Dong et al. [21] can be grouped as Type I with short-tail.

Figure 4 illustrates the vertical profiles measured after dam operation at Shashi and Jianli stations. For data measured after operation, approximately 27% of vertical profiles can be grouped as Type II (Jianli), and the ratio for Shashi station is approximately 48% (Table 2). The suspension index (ω/u_*) in Table 2 is estimated with settling velocity (ω) over shear velocity (u_*). According to Table 2, vertical profiles of Type I have a relatively larger chance of occurring in flood season, while vertical profiles of Type II have a relatively larger chance of occurring in non-flood season.

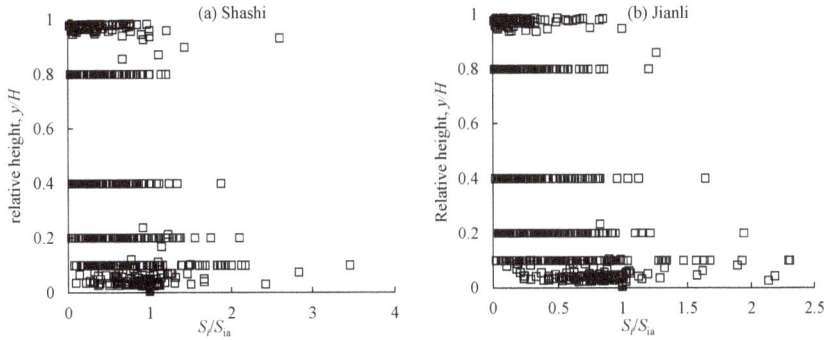

Figure 4. Vertical concentrations comparing with near-bed point concentration at Shashi (**a**) and Jianli (**b**) hydrological stations (data measured after dam operation). S_i = measured concentration of the vertical profile (kg/m^3), and S_{ia} = measured concentration of the near-bed point (kg/m^3).

Table 2. Characteristics of vertical profiles of Type I and Type II measured at Shashi and Jianli hydrological stations (after dam operation).

Stations	Type I						Type II					
	Flood Season			Non-Flood Season			Flood Season			Non-Flood Season		
	Amount of Group	Range of (ω/u_*)	D_{50} (mm)	Amount of Group	Range of (ω/u_*)	D_{50} (mm)	Amount of Group	Range of (ω/u_*)	D_{50} (mm)	Amount of Group	Range of (ω/u_*)	D_{50} (mm)
Shashi	33	0.16–95.38	0.01	17	2.09–108.89	0.06	20	0.16–1.37	0.03	25	1.84–95.67	0.07
Jianli	47	0.14–76.15	0.06	29	2.38–61.64	0.10	10	0.13–32.58	0.06	19	1.38–63.92	0.10

For Type I, five typical profiles are shown in Figure 5a, and five typical profiles of Type II are shown in Figure 5b. This shows that the concentrations of the near-bed region of Type I vary with different coefficients (ω/u_*). For type II, the vertical distribution of concentration is much more complex, while concentration with relative height less than 0.04 is relatively large.

Figure 6 illustrates the vertical distribution of S_i/S_{ia} with different (ω/u_*). In order to make it more readable, the coefficient in Figure 6 is (ω/u_*). It shows that, the concentration of S_i/S_{ia} at the same relative height decreases with increasing coefficients (Figure 6a). When the coefficient decreases to less than 0.01, vertical profiles of Type I change to Type II (Figure 6b). One thing that should be pointed out is that lines in Figure 6 are estimated with data measured in flood seasons with Type I at Shashi station. The lines may vary with different data adopted, but the rules are the same: the concentration of S_i/S_{ia} at the same relative height (y/H) decreases with increasing coefficient (ω/u_*), and Type I may change to Type II with a coefficient less than 0.01.

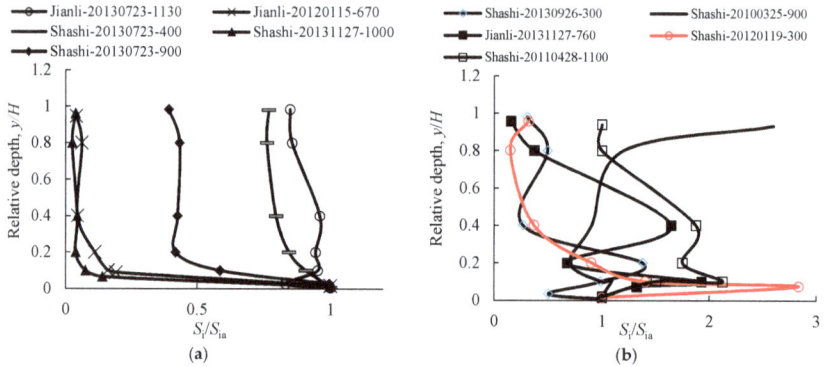

Figure 5. Typical vertical profiles of type I (**a**) and type II (**b**) with tailing phenomena. The legends are composed of station name, date of measurement, and lateral distance of vertical line from left bank. S_i = measured concentration of the vertical profile (kg/m^3), and S_{ia} = measured concentration of the near-bed point (kg/m^3).

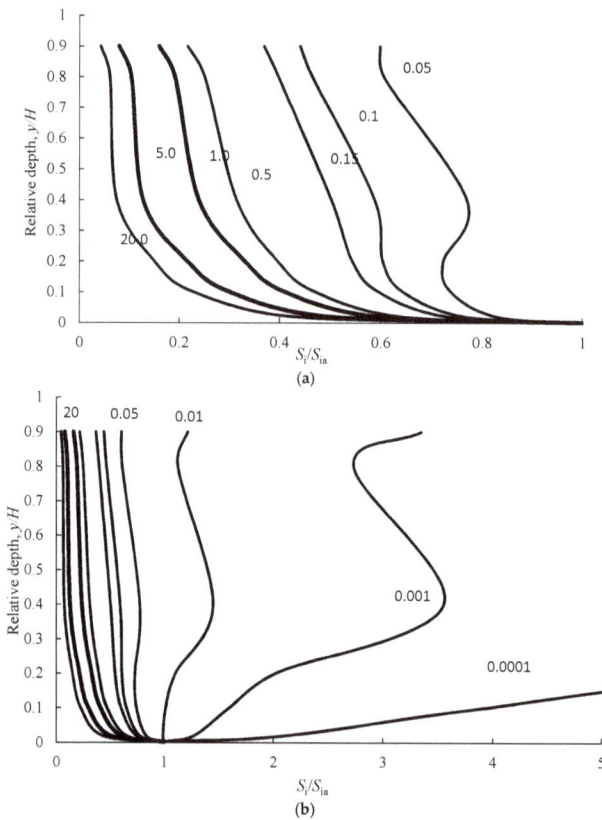

Figure 6. Vertical distribution of S_i/S_{ia} with different ω/u_*, (**a**) for $(\omega/u_*) > 0.05$ and (**b**) for $(\omega/u_*) < 0.0001$. ω = settling velocity, u_* = over shear velocity, S_i = measured concentration of the vertical profile (kg/m^3), and S_{ia} = measured concentration of the near-bed point (kg/m^3).

3.2. Effects on Sediment Flux

Suspended sediment transport is the most significant factor influencing estuaries in emorpho-dynamics, yet it is often one of the largest unknowns. As to sediment trapped by the TGD and changed sediment regime, the vertical distribution of suspended sediment varies in the Jingjiang River. The estimation of sediment flux of the Jingjiang River may also be influenced, which is important to navigation, channel management of the whole river system, and so on.

The tailing phenomena means that, sediment concentration within 10% of water depth from the river-bed cannot be ignored, otherwise it may lead to different results when estimating erosion or deposition by sediment-transport balance methods and volume methods. Thus, the effects of remarkably larger concentration in the near-bed region are analyzed.

3.2.1. Vertical Sediment Flux

In Figure 7, the B_i is the lateral distance of measured vertical lines from the left bank, and B_{avg} is the maximum width of the cross section. The maximum width of the cross section is assumed to be the maximum lateral distance from the left bank, namely the distance of the last vertical profiles from the left bank. It shows that these two kinds of vertical profiles may occur in the whole lateral cross section. The maximum R_0 value is larger than 47%. The average ratios for these two, type profiles at Shashi station are 18% (Type I) and 13% (Type II). The average ratios for these two, type profiles at Jianli station are 22% (Type I) and 18% (Type II). Table 3 indicates that the R_0-values for non-flood season are relatively larger than those of flood season.

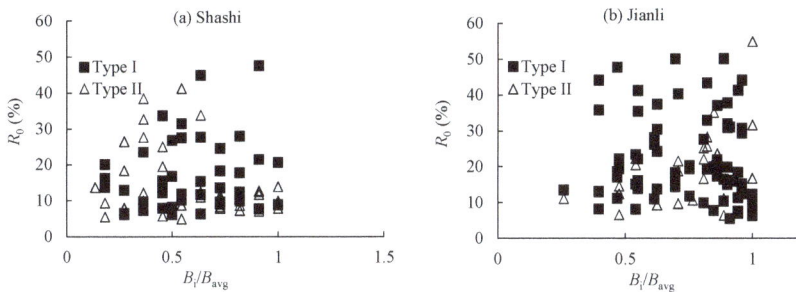

Figure 7. Lateral distribution of contribution ratios of the two near-bed points, (**a**) for Shashi hydrological station, and (**b**) for Jianli hydrological station. R_0 = contribution ratio of the two near-bed points, B_i = lateral distance of measured vertical lines from the left bank, and B_{avg} = maximum width of the cross section.

Table 3. Vertical averaged R_0-values of data measured after TGD dam operation.

Stations	Type I		Type II	
	Flood Season	Non-Flood Season	Flood Season	Non-Flood Season
Shashi	15.67	21.28	12.51	14.10
Jianli	19.57	27.70	16.94	19.61

Figure 8 indicates that the R_0 value may increase with increasing particle size and ω/u_*.

Zhang [20] also pointed out that the near-bed concentration may affect the estimated sediment load when $A = \omega/(\beta k u_*) > 0.15$, where ω is the settling velocity of uniformed particles, k is the Karmen coefficient, u_* is the shear velocity of approaching flow, and β is a coefficient for non-uniform sediment. When suspension index $A = \omega/(\beta k u_*)$ equals two, suspended sediment may concentrate in an area with a distance of 0.2 times the water depth from the riverbed, and the ratio of near-bed transportation over the total transportation (R) is approximately 30% [20]. Tang et al. [22] pointed out that, the near-bed concentration may be more apparent when $\omega/(ku*) > 5$. The values of $\omega/u*$ are 0.06, 0.8 and 2 (with $k = 0.4$ and $\beta = 1$) by Zhang [20] and Tang et al. [22], respectively. For data illustrated in Figure 8, the values of $\omega/u*$ range from 0.001 to 0.201. Zhang [20] also pointed out that the tailing phenomena is more apparent with larger values of relative particle size (D_i/D_m, particle-size over averaged particle size) of non-uniform sediment. The ratio (R) is approximately 25% when $D_i/D_m > 3$ [20].

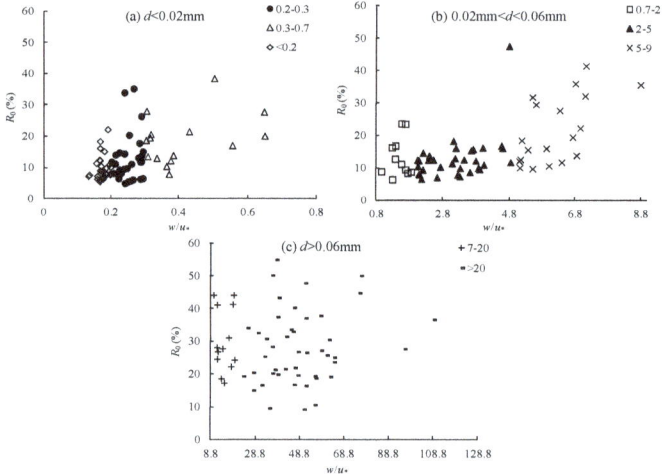

Figure 8. Relationship between R_0, $\omega/u*$ and particle diameters, (**a**) $d < 0.02$ mm, (**b**) 0.02 mm $< d <$ 0.06 mm, and (**c**) $d > 0.06$ mm. The legends are the values of $\omega/u*$. R_0 = contribution ratio of the two near-bed points, ω = settling velocity, and $u*$ = shear velocity.

In total, initiation of motion, suspension threshold, and the vertical distribution of sand concentration in the water column all need to be determined for an accurate estimate of sediment transport.

3.2.2. Sectional Sediment Flux

Due to the larger contribution ratio of the two near-bed points (R_0) in Figure 7, the contributions on sectional sediment flux can also be estimated. Figure 9 shows that, if the two near-bed points are not included, the sediment flux may be underestimated by approximately 40%. During flood season, the R-value may increase with increasing discharge, while the largest R-value may occur by measures in non-flood season. In total, the larger average R-value for flood season at Shashi station may be caused by larger discharge (Table 4).

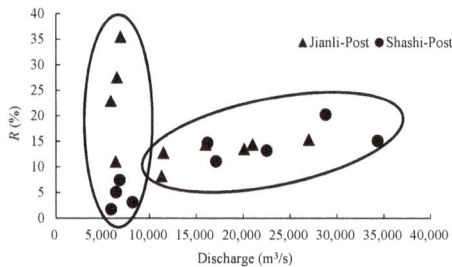

Figure 9. Deviation ratio of sediment transport rate by data measured after TGD dam operation. The two ellipses indicate non-flood season and flood season, respectively. R = contribution ratio of the two near-bed points on vertical sediment flux.

Table 4. Sectional averaged *R*-values of data measured after TGD dam operation.

Stations	Total	Flood Season	Non-Flood Season
Shashi	10.21	14.88	4.36
Jianli	17.56	13.10	24.24

This described situation may result in the different erosion amounts by sediment-transport balance method and volume method [25]. Yuan et al. [26] showed that, the contribution ratio by the near-bed concentration varies by different hydrological years and river reach. For a certain hydrological station, the contribution ratios may also vary with the inflow condition. Therefore, the impact of high concentration near the bed surface must be given more attention in the future, especially during non-flood season and flood season with larger discharge.

The influence of near-bed concentration on estimating the annual deposition by the sediment-balance method has also been analyzed by various studies [26,27]. Based on hydrology data and geometry data measured between 2002 and 2008, contributions of factors that may cause differences were distinguished. For the Yichang–Zhicheng River Reach, there are three kinds of factors that may contribute to deposition estimation with sediment-balance method, namely, sand dredging in the channel, the near-bed concentration, and non-balanced water runoff. For the Zhicheng–Shashi River Reach and Shashi–Jianli River Reach, there are four kinds of contributing factors, namely, sand dredging in the channel, the near-bed concentration, non-balanced water runoff, and simplified measurement.

Figure 10 illustrates the contribution ratios of different factors. The two ratios in Figure 10 are the contribution ratio by near-bed concentration over the total influencing factors, and contribution ratio by total influencing factors on the total deposition estimation. It shows that, for estimating the deposition with sediment-balance method, the contribution by near-bed concentration cannot be ignored, with ratios of 12%, 9% and 18%, respectively. As to its large magnitude of total deposition, the net deposition caused by near-bed concentration is 1.267×10^7 m^3, 9.80×10^9 m^3 and 2.303×10^7 m^3, respectively.

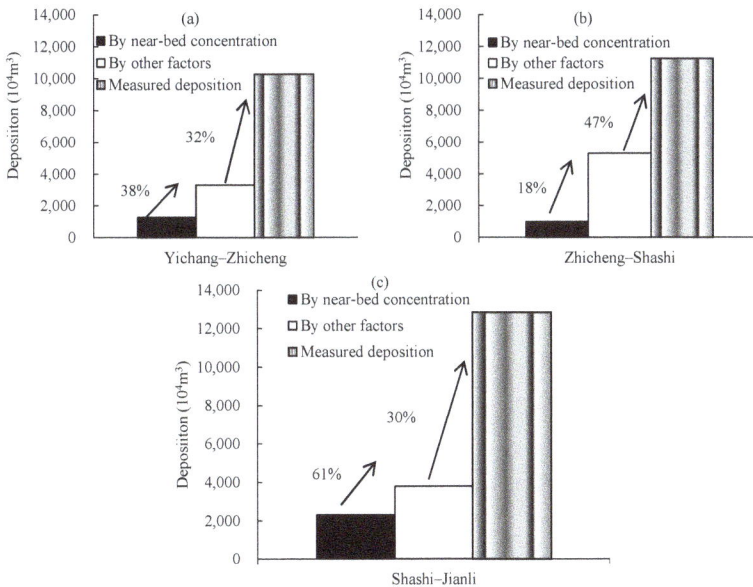

Figure 10. Contributing ratios of near-bed sediment concentration on total deposition estimation, at three sites, (**a**) the Yichang–Zhicheng, (**b**) the Zhicheng–Shashi, and (**c**) the Sahshi–Jianli reaches, based on data measured during 2002–2008 (from Yuan et al. [26]).

Dam construction has primarily contributed to sediment zonation (with different patterns of size distribution) weakening in the Jingjiang reaches (from Yichang to Chenglingji) over the last half-century [28]. The percentage of finer grouped particles (diameter less than 0.01 mm) decreased from 60% (2002) to approximately 40% (2008), and the medium sediment diameter increased from 0.052 mm (2002) to 0.081 mm (2008) [29]. The operation of the TGD and the ensuing decline of SSC was the main reason for the coarsening of the bed sediment in 1977–2003, especially at Shashi station (located 173 km downstream from Yichang station) [28]. It indicates that the coarser particulates had settled in the reservoir, and the river channel between Yichang and Shashi stations was badly eroded by clean water from the reservoir [28]. The coarser eroded sediment gradually settled along the channel between Shashi and Hankou stations due to the decrease in slope and current [28].

4. Discussion

Based on data measured after dam operation, the vertical profiles of SSC in the starved channel show a distinct tailing phenomena. Furthermore, it also influences the estimation of sediment flux. Data measured before dam operation are also collected to make a comparison, between both the vertical distribution and effects on sediment flux estimation.

4.1. Comparing with Data Measured Before Dam Operation

4.1.1. Vertical Profiles of SSC

Tailing phenomena can also be pronounced, however, the tailing phenomena is much more limited (relative concentration of 3.6 for Jianli and 2.3 for Shashi, as shown in Figure 5), the same as verticals by Zhang [20], Tang et al. [22], and Dong et al. [21]. Thus, vertical profiles before operation can be termed as short-tailing.

For data measured before operation, more data are measured in flood season. Approximately 20% (Jianli, Shashi, and Zhicheng stations) of vertical profiles can be grouped as Type II. In total, the occurring frequency of Type II increased after the operation of the TGD, especially at Shashi station (Table 5). In addition, the value of (ω/u_*) increases dramatically.

Table 5. Characteristics of vertical profiles measured at Shashi and Jianli hydrological stations before TGD dam operation.

Stations	Flood Season		Non-Flood Season	
	Amount of Group	Range of (ω/u_*)	Amount of Group	Range of (ω/u_*)
Shashi	27	0.08–0.46	9	0.23–1.34
Jianli	53	0.02–0.78	22	0.05–12.12

Based on theoretical analysis, Zhang and Tan [30] stated that, if the concentration of coarser particles (eight times the finer particles) is larger than 33.5 kg/m^3, the vertical distribution of concentration for finer particles may be different from that of the distribution of uniform particles, with $dS_f/dy > 0$ (Figure 11). It also stated that this kind of distribution of particles finer than 0.01 mm can be observed in Tongguan and Huayuankou stations, Yellow River.

Figures 12 and 13 illustrate the coarseness of suspended sediment and bed load before and after dam operation. For Shashi station, the percentage of particles coarser than 0.008 mm increases after the dam operation (Figure 12). For Jianli station, the percentage of particles coarser than 0.062 mm increases after the dam operation (Figure 12). The coarsening of bed materials after operation is apparent, especially at Shashi station (Figure 13). The coarsened suspended sediment and bed materials lead to the occurrence of remarkably large concentration in the near-bed region.

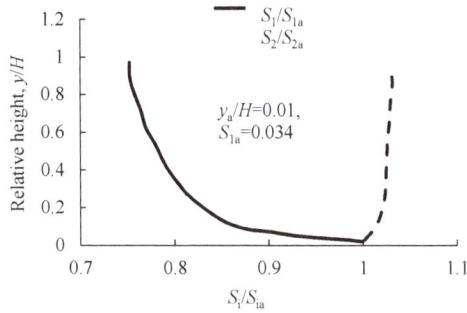

Figure 11. Calculated vertical sediment concentrations comparing with near-bed point concentration (After Zhang and Tan [30]). S_i = measured concentration of the vertical profile (kg/m^3), and S_{ia} = measured concentration of the near-bed point (kg/m^3).

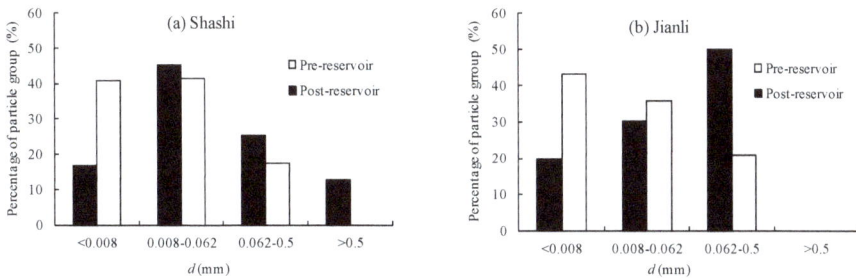

Figure 12. Percentages of particle groups before and after TGD dam operation at Shashi (**a**) and Jianli (**b**) hydrological stations. Data for pre-reservoir are average values between 1991 and 2002, and data for post-reservoir are average values between 2003 and 2016.

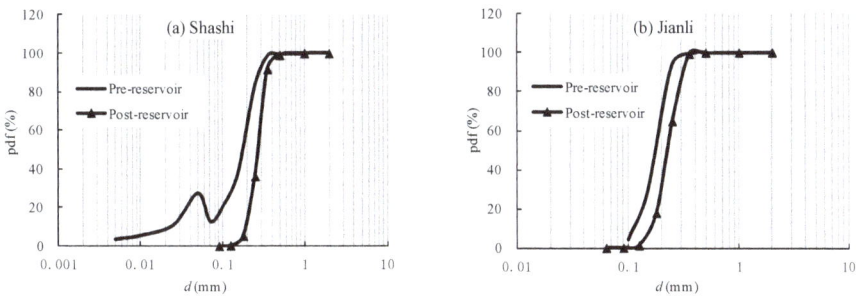

Figure 13. Bed load of Shashi (**a**) and Jianli (**b**) hydrological stations. Data in (**a**) were measured 9 July 1986 at Xinchang hydrological station, and 6 June 2013 at Shashi hydrological station; data in (**b**) were measured 26 May 1986 and 13 March 2013 at Jianli hydrological station.

4.1.2. Sediment Flux

The interpolation made for data measured before dam operation has extended five point profiles to seven point profiles (with almost the same relative height of data after operation). Then, the deviations of flux estimation can also be calculated during the cross-sectional estimation. The maximum ratio may be nearly 35% (Figure 14). The maximum R-values may have occurred during non-flood season, and more attention should be paid on measuring during non-flood season (Table 6). The variances of R-value before and after dam operation are almost the same. However, the R-values during flood season may remain constant with increasing discharge for data measured before operation.

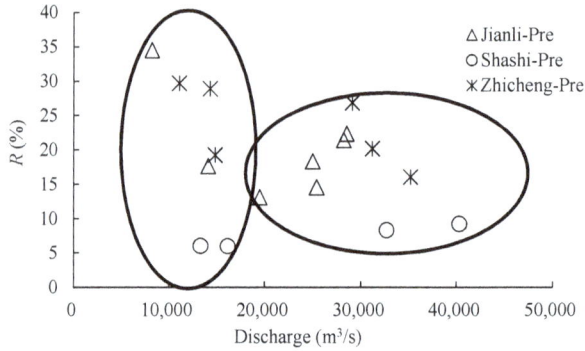

Figure 14. Deviation ratio of sediment transport rate by data measured before TGD dam operation. The two ellipses indicate non-flood season and flood season, respectively. R = contribution ratio of the two near-bed points on vertical sediment flux.

Table 6. Sectional averaged R-values of data measured before TGD dam operation.

Stations	Total	Flood Season	Non-Flood Season
Zhicheng	23.50	21.04	25.86
Shashi	7.43	8.81	6.04
Jianli	32.95	17.96	45.93

During the estimation of sectional averaged sediment flux, there is an assumption that may lead to uncertainties of estimation. For data measured before operation, the extension and interpolation may also lead to uncertainties. This kind of discrepancy can be verified by comparison between estimated and measured sectional discharge and concentration, as shown in Figure 15.

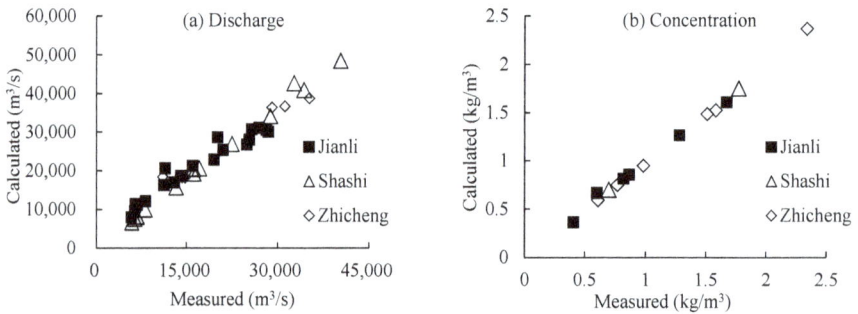

Figure 15. Comparison between measured and calculated sectional discharge (**a**) and concentration (**b**).

4.2. Contribution of Large Floods

The characteristics of suspended sediment concentration and yield at the event scale have been widely analyzed in various environments worldwide [31–33]. Floods are relevant for most of the suspended sediment load [34]. While sediment transport by floods correlated with flood regime, different flood regimes lead to different sediment transportation. For instance, analysis of a typical agro-catchment of the Loess Plateau shows that, the contribution of accumulative total sediment yield by the different flood regimes to the summed sediment yield of all of the examined 158 events are 4%, 13%, 6%, 21%, and 56% for Regimes A to E, respectively [35]. Thus, the flood regimes are analyzed and shown in Figure 16.

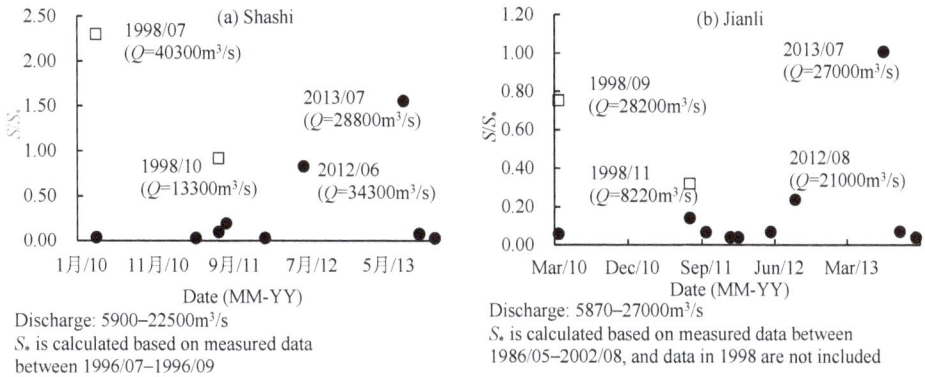

Discharge: 5900–22500m³/s
S_* is calculated based on measured data between 1996/07–1996/09

Discharge: 5870–27000m³/s
S_* is calculated based on measured data between 1986/05–2002/08, and data in 1998 are not included

Figure 16. Sediment regimes of floods occurring before and after the operation of the TGD, the Jingjiang River Reach. (**a**) for Shashi Hydrological Station, and (**b**) for Jianli Hydrological Station, S = average concentration, and S_* = average carrying capacity.

Measured data during flood period, such as the floods in 1998, 2013 and 2012, show an over-saturated state, and after the peak flood, the regime may change to pre-equilibrium or starved (Figure 16, Table 7). Figure 16 shows that, the relationships between discharge and concentration measured in floods with high discharges are different from those of floods with low discharges.

Shifts in the discharge-rating curves after three extreme flood events have also been observed by Whitaker et al. [36]. This indicates that the Jingjiang Reach of the YR has changed its sediment regime from pre-equilibrium to starved of sediment because of the operation of the TGD, and the flood regime may be different from the normal sediment regime. Shifts in the discharge-rating curves caused a period of channel instability [36]. Higher suspended sediment concentration may reduce the channel capacity and lead to channel erosion [37]. Finally, the fluvial erosion intensity during flood seasons is highly correlated with a corresponding incoming sediment coefficient [19].

The 1998 flood is the largest flood in the YR's history since China's economy entered the high-speed development period at the beginning of the 1980s, and it caused severe damage to the basin, causing losses exceeding 2×10^{10} RMB (Chinese currency) [38]. The higher water levels also led to overbank flows in the middle Yangtze River [39]. The sediment flux during this flood has been estimated [40]. The water stage in the 1998 flood was higher than that in the 1954 flood, despite the smaller total discharge during the 1998 flood.

Table 7. Characteristics of major floods in China in 1998 (Data adapted from Zhou and Chen [41]).

Stations	Flood Volumes 10^9 m³		Maximum Peak Flow Discharge		Maximum Flood Stage	
	In 30 Days	In 60 Days	Discharge (m³/s)	Date	Stage (m)	Date
Yichang	137.9	254.5	63,600	16/8	54.49	16/8
Shashi	–	–	53,700	17/8	45.22	17/8
Jianli	–	–	45,200	17/8	38.31	17/8
Luoshan	–	–	68,600	27/7	34.95	20/8
Hankou	175.4	336.5	72,300	19/8–20/8	29.43	20/8
Jiujiang	–	–	–	–	23.03	2/8
Datong	202.7	395.1	82,300	–	16.32	2/8
Nanjing	–	–	–	–	10.14	29/7

The variability in the relationship between sediment concentration and water discharge (namely hysteretic patterns) has been used to explain the variability of sediment sources from one flood to another [42]. Whitaker et al. [36] compared the estimated sediment yield for an unsampled flood event, a single-event surrogate SSC–Q rating curve, and a long-term SSC–Q rating curve. Analysis revealed that, the long-term SSC–Q rating curve is estimated to be approximately 37% greater than the single-event SSC–Q estimates, which indicates a high degree of uncertainty. Analysis shows a high variability in the rating curve between similar floods [36]. Thus, large floods may also contribute to the occurrence of remarkably large concentration in the near-bed region (Type I), and the deviation of estimated sediment flux with sediment-balance method.

4.3. Connection with Dongting Lake

Reservoir sedimentation would disrupt previous flow and sediment delivery systems [43], especially channels down the dam. The TGD is the world's biggest dam built on the largest river in China, the YR [14]. With impoundment of the TGD, channel reach downstream of the reservoir has adjusted to a changed flow and sediment situation, especially in the Jingjiang reach [29,44].

The Dongting Lake is connected with the Jingjiang River Reach with three distributaries' channels and one outlet. It receives water from the main stream of the Yangtze River via three distributaries' channels, and it discharges into the main stream of the Yangtze River at Chenglingji (the outlet of Dongting lake, Hunan Province, China). The annual runoff received from the Yangtze River is approximately 92.3 km^3 through the Sankou distributary channels [45]. Li et al. [46] pointed out that, the channel changes serve as the primary factor in facilitating the decrease in the discharge diversion ratio, but not the main factor for the decreased amount of the discharge diversion. The occurring frequency of Type II increases after the operation of the TGD, especially at Shashi station, while two of the three distributary channels (Songzikou, Jingzhou City, Hubei Province, China and Taipingkou, Jingzhou City, Hubei Province, China) are located in the reach of upper Shashi station. Thus, vertical profiles of Type II may contribute to this relationship with much concentration transported into the lake.

4.4. Uncertainties

Turbidity monitoring is developed and applied attempting to better predict the continuous variability of suspended sediment concentration during a flood event, and in turn the total sediment yield [47,48]. The accurate in situ quantification of suspended sediment concentration is essential for model calibration and to identify sediment pathways and sediment flux [49].

The two near-bed points are measured with distances of 0.5 m and 0.1 m from the bed, respectively. As to the near-bed measuring, the small distance from the riverbed may be doubtful, as it is difficult to distinguish the riverbed. Equipment used in the field survey includes real-time kinematic Global Navigation Satellite System (RTK–GNSS), Total Station, vessel-mounted Acoustic Doppler Current Profilers (ADCP), Digital Level, Laser Particle Size Analyzer (LPSA), and so on. All the equipment and criteria have been strictly handled and executed by professional surveyors in JHWRSB. Several standards from the technical manual by JHWRSB [25] are listed in Table 8.

The main advantage of a RTK–GNSS survey is that it is faster, cheaper and easier to use during the surveying phase than classical topographic surveying [50]. Pagliari et al. [50] compared the differences between coordinates of the ground-controlled points estimated using RTK–GNSS, and those computed from the classical topographic measurements, during processing photogrammetric block. It pointed out that a RTK–GNSS survey may be sufficient to determine the requested tolerance [50]. Thus, in recent years, miniaturized components (GNSS receivers) have been adopted in field surveys [51].

Table 8. Instruments, methods and criteria adopted in the field survey.

Measuring	Instruments or Methods	Criteria or Error
Positioning of instrument	Total station Digital Level	Positioning error: ±0.3 m (vertical), ±1.5 m (vertical lines)
Vertical lines for measuring velocity	RTK–GNSS (antenna and receiver)	Position errors of antenna: 0.5° RMS (orientation) and 0.5–3 m RMS (position). Position errors of receiver: ±10 mm + 1 ppm (horizontal), ±20 mm + 1 ppm (vertical), where ppm means additional error per km of baseline.
Velocity	Acoustic Doppler Current Profilers (ADCP)	The deviation between each measured discharge and the averaged discharge should be less than ±5%, otherwise, data should be re-measured.
Water depth	ADCP	Verified with fish lead. Frequency: 600 KHz Sounding range: 0.7–75 m Velocity measurement: ±20 m/s Resolution: 0.1 cm/s
Sampling Suspended sediment	Bottom-touched automatic-closing sampler	The dropping speed of the sampler is reduced when approaching the riverbed. The sampler may close automatically when it is brought into contact with the riverbed.
Suspended sediment gradation	Sieving method	Field (2 mm, 5 mm, 10 mm, 25 mm, 50 mm, 75 mm, 100 mm, 150 mm, 200 mm, 250 mm, and 300 mm) Lab (0.002 mm, 0.004 mm, 0.008 mm, 0.016 mm, 0.032 mm, 0.062 mm, 0.125 mm, 0.25 mm, 0.5 mm, 1.0 mm, and 2.0 mm)
Near-bed concentration	Double-checked after sampling and grain-size analysis	The measured concentration is reliable when the vertical profiles of concentration of particles with d_{50} < 0.062 mm has no changing point in near-bed region. Otherwise, related measurements should be omitted.

ADCP is relatively easy to use, enables gathering various data types at once, can be attached to a RC platform, and has rapid three-dimensional flow structure measuring [52,53]. However, disadvantages of ADCP include snap-shot measurement of flow-change to errors, being time-consuming over large areas, and not being able to measure flow or depth in depths under 0.2 m [53]. As it is rather expensive, ADCP is relatively adopted to measure flow structure, flow discharge, bathymetry, bed load discharge, suspended load and channel change in field surveys [53]. Zhang et al. [54] pointed out that the error of ADCP in measuring cross-sections in the middle and lower YR channel is ±1% and the random uncertainty is 2–5%, and the integrated system (GPS and ASCP) delivered positioning with <15 cm accuracy vertically and <30 cm horizontally.

For sediment grain-size analysis, the results from LPSA and sieving are comparable when particles have a spherical shape [55]. Using a sieving method for coarser particles is more acceptable than for finer ones [55]. A sieving method is used in field surveys by JHWRSB [25]. According to the backscatter intensity of ADCP beams, suspended sediment concentration can also be estimated with inversion [56].

A major focus of research has been the development and application of turbidity monitoring, in an attempt to better predict the continuous variability of suspended sediment concentration during a flood event, and in turn the total sediment yield [47]. The concentration by sampling can be validated by echo intensity of the ADCP for future analysis.

5. Conclusions

The sediment regime of the Jingjiang Reach of the Yangtze River has changed dramatically because of the operation of the TGD. Based on vertical distribution data at the three controlling stations (Zhicheng, Shashi, and Jianli) on the Jingjiang Reach, the characteristics of remarkably large concentration in near-bed region and its implications are analyzed in this study. Our conclusions can be summarized as:

(1) In sub-saturated channels, vertical distribution of suspended sediment concentration (SSC) is characterized by a remarkably large concentration in the near-bed zone (within 10% of water depth from the river-bed). The maximum measured concentration may be up to 15 times of vertical average concentration, while the ratio in quasi-equilibrium channels is less than four.

(2) Concentrations normalized with reference concentration at the same height may decrease with increasing values of suspension index (ω/u_*). Additionally, concentration near water surface may be larger than concentration in near-bed region when suspension index is smaller than 0.01.

(3) After the dam operation, ignoring the near-bed concentration may cause up to 35% deviation when applied to estimate the sediment flux in the unsaturated flows. Deviations may increase with increasing discharge in flood season, while the maximum value may occur in non-flood season. Deviations in quasi-equilibrium channel may not increase with increasing discharge in flood season, while the maximum value may also occur in non-flood season. Deviations may increase with increasing particle size and suspension index.

Analytic results indicate that, in sub-saturated channels, more attention should be paid to near-bed concentration during non-flood season, the same as measurements during flood season with larger discharge.

Acknowledgments: This work was supported by the National Program on Key Basic Research Project (No. 2016YFC0402303); and the National Natural Science Foundation of China (Nos. 51579230, 51109198). We wish to express our deep gratitude to the Jingjiang Hydrology and Water Resources Surveying Bureau, Yangtze River Water Resources Commission for permission to access the hydrological data.

Conflicts of Interest: The author declares no conflict of interest.

References

1. Ludwig, W.; AmiotteSuchet, P.; Probst, J.L. River discharge of carbon to the world's oceans: Determining local inputs of alkalinity and of dissolved and particulate organic carbon. *Comptes Rendus de L Academie des Sciences Serie II Fascicule A-Sciences de la Terre et des Planetes* **1996**, *323*, 1007–1014.

2. Asselman, N.E.M. Fitting and interpretation of sediment rating curves. *J. Hydrol.* **2000**, *234*, 228–248. [CrossRef]

3. Zuo, S.H.; Li, J.F.; Wan, X.N.; Shen, H.T.; Fu, G. Characteristics of temporal and spatial variation of suspended sediment concentration in the Changjiang Estuary. *J. Sediment Res.* **2006**, *3*, 68–75. (In Chinese with English Abstract)

4. Pal, D.; Ghoshal, K. Hydrodynamic interaction in suspended sediment distribution of open channel turbulent flow. *Appl. Math. Modell.* **2017**, *49*, 630–646. [CrossRef]

5. Higgins, A.; Restrepo, J.C.; Otero, L.J.; Ortiz, J.C.; Conde, M. Vertical distribution of suspended sediment in the mouth area of the Magdalena River, Colombia. *Lat. Am. J. Aquat. Res.* **2017**, *45*, 724–736. [CrossRef]

6. Nir, S.Q.; Sun, H.G.; Zhang, Y.; Chen, D.; Chen, W.; Chen, L.; Schaefer, S. Vertical distribution of suspended sediment under steady flows: Existing theories and fractional derivative model. *Discret. Dyn. Nat. Soc.* **2017**, *2017*, 1–11.

7. Sahin, C.; Verney, R.; Sheremet, A.; Voulgaris, G. Acoustic backscatter by suspended cohesive sediment: Field observations, Seine Estuary, France. *Cont. Shelf Res.* **2017**, *134*, 39–51. [CrossRef]

8. Armijos, E.; Crave, A.; Espinoza, R.; Fraizy, P.; Dos Santos, A.L.M.R.; Sampaio, F.; De Oliveria, E.; Santini, W.; Martinez, J.M.; Autin, P.; et al. Measuring and modelling vertical gradients in suspended sediments in the Solimões/Amazon River. *Hydrol. Process.* **2017**, *31*, 654–667. [CrossRef]

9. He, L.; Chen, D.; Duan, G.L.; Zhu, Z.H.; Zhang, S.Y. Sediment suspension in the "starving" Jingjiang reach, downstream from the Three Gorges Dam, China. In Proceedings of the World Environmental and Water Resources Congress, West Palm Beach, FL, USA, 22–26 May 2016; pp. 303–313.

10. Fang, H.Y.; Cai, Q.G.; Chen, H.; Li, Q.Y. Temporal changes in suspended sediment transport in a gullied loess basin: The lower Chabagou Creek on the Loess Plateau in China. *Earth Surf. Process. Landf.* **2008**, *33*, 1977–1992. [CrossRef]

11. Zhang, R.; Zhang, S.H.; Xu, W.; Wang, B.D.; Wang, H. Flow regime of the three outlets on the south bank of Jingjiang River, China: An impact assessment of the Three Gorges Reservoir for 2003–2010. *Stoch. Environ. Res. Risk Assess.* **2015**, *29*, 2047–2060. [CrossRef]

12. Fan, P.; Li, J.C.; Liu, Q.Q.; Singh, V.P. Case Study: Influence of morphological changes on flooding in Jingjiang River. *J. Hydraul. Eng.* **2008**, *134*, 1757–1766. [CrossRef]

13. Yang, C.; Cai, X.B.; Wang, X.L.; Yan, R.R.; Zhang, T.; Zhang, Q.; Lu, X.R. Remotely sensed trajectory analysis of channel migration in lower Jingjiang Reach during the period of 1983–2013. *Remote Sens.* **2015**, 7, 16241–16256. [CrossRef]

14. Gao, B.; Yang, D.W.; Yang, H.B. Impact of the three Gorges Dam on flow regime in the middle and lower Yangtze River. *Quat. Int.* **2013**, 304, 43–50. [CrossRef]

15. Yang, H.; Tang, R. *Study on the Evolution of Jingjiang Reach of Middle Yangtze River*; China Water Power Press: Beijing, China, 1999. (In Chinese)

16. Dai, Z.J.; Fagherazzi, S.; Mei, X.F.; Gao, J.J. Decline in suspended sediment concentration delivered by the Changjiang (Yangtze) River into the East China Sea between 1956 and 2013. *Geomorphology* **2016**, 268, 123–132. [CrossRef]

17. Xu, K.H.; Milliman, J.D. Seasonal variations of sediment discharge from the Yangtze River before and after impoundment of the Three Gorges Dam. *Geomorphology* **2009**, 104, 276–283. [CrossRef]

18. Xia, J.Q.; Zong, Q.L.; Deng, S.S.; Xu, Q.X.; Lu, J.Y. Seasonal variations in composite riverbank stability in the Lower Jingjiang Reach, China. *J. Hydrol.* **2014**, 519, 3664–3673. [CrossRef]

19. Xia, J.Q.; Zong, Q.L.; Zhang, Y.; Xu, Q.X.; Li, X.J. Prediction of recent bank retreat processes at typical sections in the Jingjiang Reach. *Sci. China* **2014**, 57, 1490–1499. [CrossRef]

20. Zhang, C.F. Vertical Profile of Non-Uniform Suspended Sediment Concentration in Yangtze Estuary. Master's Thesis, Zhejiang University, Hangzhou, China, 2016. (In Chinese with English Abstract)

21. Dong, X.T.; Li, R.J.; Fu, G.C.; Zhang, H.C. Relationship between vertical distribution of velocity and sediment concentration of sediment-laden flow. *J. Hohai Univ. Natl. Sci.* **2015**, 43, 371–376. (In Chinese with English Abstract)

22. Tang, M.L.; Cao, H.Q.; Li, Q.Y.; Zhai, W.L. Vertical distribution of suspended sediment in middle reaches of Yangtze River. *Yangtze River* **2017**, 48, 6–11. (In Chinese with English Abstract)

23. Feng, Q.; Xiao, Q.L. A new S-type vertical distribution of suspended sediment concentration. *J. Sediment Res.* **2015**, 1, 19–24. (In Chinese with English Abstract)

24. Xia, Y.F.; Xu, H.; Chen, Z.; Wu, D.W.; Zhang, S.Z. Experimental study on suspended sediment concentration and its vertical distribution under spilling breaking wave actions in silty coast. *China Ocean Eng.* **2011**, 25, 565–575. [CrossRef]

25. JHWRSB (Jingjiang Hydrology & Water Resources Surveying Bureau, Hydrology Bureau). *Technical Report: Observation of Unbalanced Sediment Transport in the Lower Reaches of the Three Gorges Reservoir*; Changjiang Water Resources Commission: Wuhan, Hubei, China, 2016. (In Chinese)

26. Yuan, Y.; Zhang, X.F.; Duan, G.L. Modifying the channel erosion and deposition amount of Yichang-Jianli reach calculated by sediment discharge method. *J. Hydroelectr. Eng.* **2014**, 33, 163–169. (In Chinese with English Abstract)

27. Duan, G.L.; Peng, Y.B.; Guo, M.J. Comparative analysis on riverbed erosion and deposition amount calculated by different methods. *J. Yangtze River Sci. Res. Inst.* **2014**, 31, 108–118. (In Chinese with English Abstract)

28. Dai, S.B.; Lu, X.X. Sediment load change in the Yangtze River (Changjiang): A review. *Geomorphology* **2014**, 215, 60–70. [CrossRef]

29. Zhao, G.S.; Lu, J.Y.; Visser, P.J. Fluvial river regime in disturbed river systems: A case study of evolution of the Middle Yangtze River in post-TGD (Three Gorges Dam), China. *J. Geol. Geophys.* **2015**, 4, 6. [CrossRef]

30. Zhang, X.F.; Tan, G.M. Characteristics of vertical concentration distribution of non-uniform particles. *J. Hydraul. Eng.* **1992**, 10, 48–52. (In Chinese with English Abstract)

31. Lenzi, M.A.; Mao, L.; Comiti, F. Interannual variation of suspended sediment load and sediment yield in an alpine catchment. *Hydrol. Sci. J.* **2003**, 48, 899–915. [CrossRef]

32. Zabaleta, A.; Martinez, M.; Uriarte, J.A.; Antiguedad, I. Factors controlling suspended sediment yield during runoff events in small headwater catchments of the Basque Country. *Catena* **2007**, 71, 179–190. [CrossRef]

33. Langlois, J.L.; Johnson, D.W.; Mehuys, G.R. Suspended sediment dynamics associated with snowmelt runoff in a small mountain stream of Lake Tahoe (Nevada). *Hydrol. Process.* **2005**, 19, 3569–3580. [CrossRef]

34. Tena, T.; Batalla, R.J.; Vericat, D. Reach-scale suspended sediment balance downstream from dams in a large Mediterranean River. *Hydrol. Sci. J.* **2012**, 57, 831–849. [CrossRef]

35. Zhang, W.; Yuan, J.; Han, J.Q.; Huang, C.T.; Li, M. Impact of the Three Gorges Dam on sediment deposition and erosion in the middle Yangtze River: A case study of the Shashi Reach. *Hydrol. Res.* **2016**, 175–186. [CrossRef]

36. Whitaker, A.C.; Sato, H.; Sugiyama, H. Changing suspended sediment dynamics due to extreme flood events in a small pluvial-nival system in northern Japan. In *Sediment Dynamics in Changing Environments*; IAHS-AISH Publication: Christchurch, New Zealand, 2008; Volume 325, pp. 192–199.

37. Li, L.Q.; Lu, X.X.; Chen, Z.Y. River channel change during the last 50 years in the middle Yangtze River, the Jianli reach. *Geomorphology* **2007**, *85*, 185–196. [CrossRef]

38. Zhu, Y.H.; Fan, B.L.; Yao, S.M.; Sun, G.Z.; Li, F.Z. Propagation features of the 1998 big floods in the Jinagjiang Reach of the Yangtze River. In *Advances in Water Resources and Hydraulic Engineering*; Springer: Berlin/Heidelberg, Germany, 2009; pp. 1032–1037.

39. Zong, Y.Q.; Chen, X.Q. The 1998 flood on the Yangtze, China. *Nat. Hazards* **2000**, *22*, 165–184. [CrossRef]

40. Xu, K.Q.; Chen, Z.Y.; Zhao, Y.W.; Wang, Z.H.; Zhang, J.Q.; Hayashi, S.J.; Murakami, S.; Watanabe, M. Simulated sediment flux during 1998 big-flood of the Yangtze (Changjiang) River, China. *J. Hydrol.* **2005**, *313*, 221–223. [CrossRef]

41. Zhou, W.H.; Chen, Y.H. The 1998 flood of the Yangtze River. *Int. J. Sediment Res.* **1999**, *14*, 61–66.

42. Seeger, M.; Errea, M.-P.; Begueria, S.; Arnaez, J.; Marti, C.; Garcia-Ruiz, J.M. Catchment soil moisture and rainfall characteristics as determinant factors for discharge/suspended sediment hysteretic loops in a small headwater catchment in the Spanish Pyrenees. *J. Hydrol.* **2004**, *288*, 299–311. [CrossRef]

43. Ran, L.; Lu, X.; Xin, Z.; Yang, X. Cumulative sediment trapping by reservoirs in large river basins: A case study of the Yellow River basin. *Glob. Planet. Chang.* **2013**, *100*, 308–319. [CrossRef]

44. Xia, J.Q.; Deng, S.S.; Lu, J.Y.; Xu, Q.X.; Zong, Q.L.; Tan, G.M. Dynamic channel adjustments in the Jingjiang reach of the middle Yangtze River. *Sci. Rep.* **2016**, *6*, 22802. [CrossRef] [PubMed]

45. Dou, H.; Jiang, J. *Dongting Lake*; Press of University of Science and Technology of China: Hefei, China, 2000.

46. Li, N.; Wang, L.C.; Zeng, C.F.; Wang, D.; Liu, D.F.; Wu, X.T. Variations of runoff and sediment load in the Middle and Lower Reaches of the Yangtze River, China (1950–2013). *PLoS ONE* **2016**, *11*, e160154. [CrossRef] [PubMed]

47. Orwin, J.F.; Smart, C.C. An inexpensive turbidimeter for monitoring suspended sediment. *Geomorphology* **2005**, *68*, 3–15. [CrossRef]

48. Pfannkuche, J.; Schmidt, A. Determination of suspended particulate matter concentration from turbidity measurements: Particle size effects and calibration procedures. *Hydrol. Process.* **2003**, *17*, 1951–1963. [CrossRef]

49. Stutter, M.; Dawson, J.J.C.; Glendell, M.; Napier, F.; Potts, J.M.; Sample, J.; Vinten, A.; Watson, H. Evaluating the use of in-situ turbidity measurements to quantify fluvial sediment and phosphorus concentration and flux in agricultural streams. *Sci. Total Environ.* **2017**, *607–608*, 391–402. [CrossRef] [PubMed]

50. Pagliari, D.; Rossi, L.; Passoni, D.; Pinto, L.; De Michele, C.; Avanzi, F. Measuring the volume of flushed sediments in a reservoir using multi-temporal images acquired with UAS. *Geomat. Nat. Hazards Risk* **2017**, *8*, 150–166. [CrossRef]

51. Bandini, F.; Jakobsen, J.; Olesen, D.; Reyna-Gutierrez, J.A.; Bauer-Gottwein, P. Measuring water level in rivers and lakes from lightweight unmanned aerial vehicles. *J. Hydrol.* **2017**, *548*, 237–250. [CrossRef]

52. Togneri, M.; Lewis, M.; Neill, S.; Masters, I. Comparison of ADCP observations and 3D model simulations of turbulence at a tidal energy site. *Renew. Energy* **2017**, *114*, 273–282. [CrossRef]

53. Kasvi, E.; Hooke, J.; Kurkela, M.; Vaaja, M.T.; Virtanen, J.P.; Hyyppa, H.; Alho, P. Modern empirical and modelling study approaches in fluvial geomorphology to elucidate sub-bend-scale meander dynamics. *Prog. Phys. Geogr.* **2017**, *41*, 533–569. [CrossRef]

54. Zhang, Q.; Shi, Y.F.; Xiong, M. Geometric properties of river cross sections and associated hydrodynamic implications in Wuhan-Jiujiang river reach, the Yangtze River. *J. Geogr. Sci.* **2009**, *19*, 58–66. [CrossRef]

55. Li, W.K.; Wu, Y.X.; Huang, Z.M.; Fan, R.; Lv, J.F. Measurement results comparison between Laser Particle Analyzer and Sieving method in particle size distribution. *Zhongguo Fenti Jishu* **2007**, *5*, 10–13. (In Chinese with English Abstract)

56. Jin, W.F.; Liang, C.J.; Zhou, B.F.; Li, J.D. Measurement and analysis of high suspended sediment concentration in Jintang channel with vessel-mounted ADCP. *J. Mar. Sci.* **2009**, *27*, 31–39. (In Chinese with English Abstract)

water

MDPI

Article

Evaluating Regime Change of Sediment Transport in the Jingjiang River Reach, Yangtze River, China

Li He [1,*], Dong Chen [1,2], Shiyan Zhang [1], Meng Liu [1,2] and Guanglei Duan [3]

[1]   Key Laboratory of Water Cycle and Related Land Surface Processes, Institute of Geographic Sciences and Natural Resources Research, Chinese Academy of Sciences, Beijing 100101, China; dchen@igsnrr.ac.cn (D.C.); zhangshy@igsnrr.ac.cn (S.Z.); lium.17b@igsnrr.ac.cn (M.L.)
[2]   College of Resources and Environment, University of Chinese Academy of Sciences, Beijing 100049, China
[3]   Jingjiang Hydrology and Water Resources Surveying Bureau, Yangtze River Water Resources Commission, Wuhan 434020, China; jjduangl@cjh.com.cn
*   Correspondence: heli@igsnrr.ac.cn; Tel.: +86-10-6488-8151

Received: 6 February 2018; Accepted: 14 March 2018; Published: 15 March 2018

Abstract: The sediment regime in the Jingjiang river reach of the middle Yangtze River has been significantly changed from quasi-equilibrium to unsaturated since the impoundment of the Three Gorges Dam (TGD). Vertical profiles of suspended sediment concentration (SSC) and sediment flux can be adopted to evaluate the sediment regime at the local and reach scale, respectively. However, the connection between the vertical concentration profiles and the hydrologic conditions of the sub-saturated channel has rarely been examined based on field data. Thus, vertical concentration data at three hydrological stations in the reach (Zhicheng, Shashi, and Jianli) are collected. Analyses show that the near-bed concentration (within 10% of water depth from the riverbed) may reach up to 15 times that of the vertical average concentration. By comparing the fractions of the suspended sediment and bed material before and after TGD operation, the geomorphic condition under which the distinct large near-bed concentrations occur have been examined. Based on daily discharge-sediment hydrographs, the reach scale sediment regime and availability of sediment sources are analyzed. In total, remarkable large near-bed concentrations may respond to the combination of wide grading suspended particles and bed material. Finally, several future challenges caused by the anomalous vertical concentration profiles in the unsaturated reach are discussed. This indicates that more detailed measurements or new measuring technologies may help us to provide accurate measurements, while a fractional dispersion equation may help us in describing. The present study aims to gain new insights into regime change of sediment suspension in the river reaches downstream of a very large reservoir.

Keywords: sediment regime; suspended sediment concentration; vertical profiles of concentration; the Jingjiang River Reach; the Yangtze River

---

1. Introduction

Characteristics of the sediment regime are one of the most important factors that control the processes in the transportation and sedimentation zones of a fluvial system [1]. Changes in the sediment regime cause not only varied fluvial evolution processes but also the inner physical processes of sediment suspension. Using a reach scale, the sediment regime of a river reach can be expressed as the suspended sediment load (sediment flux). Regarding the physical processes of suspension on a local scale, they may be related to the vertical profiles of suspended sediment concentration (SSC), which are essential for estimating sediment flux, and also numerical simulation [2]. Thus, a large number of researchers have paid, and are paying, attention to the vertical concentration profiles.

During the early stage of physical or analytical analyses, many studies have investigated steady and equilibrium sediment transport [3]. For equilibrium sediment transport, traditional advection-diffusion equation (ADE) theory and the improvements (including sediment turbulent diffusion coefficient, vertical flow velocity distribution, and suspension index) are widely used [4,5]. For calculating profiles of SSC, numerical models produce good results for uniform sediments in equilibrium situations [6,7]. However, this is not the case for non-uniform sediments in equilibrium situations, not to mention non-uniform sediments in non-equilibrium situations [6,7]. For instance, Nicholas et al. [8] compared observed flow widths with results from a theoretical model developed for non-equilibrium (aggradation) conditions. Analysis results illustrated that models derived for equilibrium conditions may have limited utility in non-equilibrium situations, despite their widespread use to date [8].

In the non-equilibrium condition, the deviations of the vertical profiles of SSC from the local equilibrium state are different from those of the equilibrium state [9,10]. For hydraulic engineers, the vertical concentration profiles are essential for estimating erosion and deposition using the sediment balanced method, and also for estimating sediment-associated contaminants [11,12]. For geomorphologies, non-equilibrium suspended sediment transport (SST) is one of the key processes driving the channel evolution by erosion or deposition [13]. Thus, various approaches have been employed in non-uniform sediments in non-equilibrium sediment transport. However, most of these analyses are based on flume data or theoretical analysis [14,15]. The connection between the hydrologic condition of the non-equilibrium channel and the vertical concentration profiles has rarely been examined based on field data [16].

The regime of sediment transport in the Jingjiang river reach (JJRR, about 102 km downstream of the Three Gorges Dam (TGD)) has been significantly changed from quasi-equilibrium to sub-saturation, as the impoundment of the TGD. More and more researchers are focusing on the changed sediment regime of the sub-saturated channel in the reach scale. However, field observation has revealed frequent occurrences of remarkable large concentrations in the near-bed zone by extending the vertical measuring to the near-bed zone (i.e., with a distance of 0.1 m from the riverbed). Thus, it is also necessary to analyze the changed sediment regime in the local scale. The contribution of the remarkable, large, near-bed concentration on the sediment flux in the unsaturated channel has been analyzed [17]. Here, this paper aims to evaluate the changed sediment regime of the JJRR from both a local-scale view (vertical concentration profiles) and a reach-scale view (sediment flux). It also helps one to understand the connection between the concentration profiles (local-scale) and the hydrologic conditions of the sub-saturated channel.

Firstly, the vertical profiles of SSC are analyzed by their characteristics, hydrodynamic conditions, and geomorphic conditions to exhibit the changed sediment regime at the local scale. Secondly, the temporal variation of sediment flux is analyzed with the cumulative anomaly method to exhibit the changed sediment regime at the reach scale. Thirdly, the relationship between SSC and discharge is analyzed with the method of the sediment rating curve (SRC), which is used to investigate the conditions for the occurrence of the remarkable large near-bed concentration. Finally, the challenges that may be caused by the remarkable, large, near-bed concentration are discussed.

2. Study Reach, Data, and Methodology

2.1. The Study Reach

The length of the Yangtze River (YR) is 6.3×10^3 km, and the drainage area is 1.8×10^6 km^2 (Figure 1). The TGD, located at the end of the upper reach of the YR, controls a drainage area of 1.0×10^6 km^2. The dam is 185 m high, and storage capacity of the reservoir is 3.93×10^4 Hm^3. The main purposes of the project are flood control, power generation, and navigation.

The JJRR (between Zhicheng and Chenglingji stations) is about 60 km downstream from the Yichang hydrological station. There are five hydrological stations in the JJRR: Zhicheng, Shashi (Jing 45), Xinchang (Jing 84), Jianli, and Chenglingji (Figure 1b). Moreover, there are three distributary channels on the reach through which YR delivers water and sediment into Dongting Lake, Songzikou, Taipingkou, and Ouchikou. The river reach can be divided by Ouchikou into the upper and lower sub-reaches (Figure 1b). The lengths of these two sub-reaches are 172 km and 176 km, respectively.

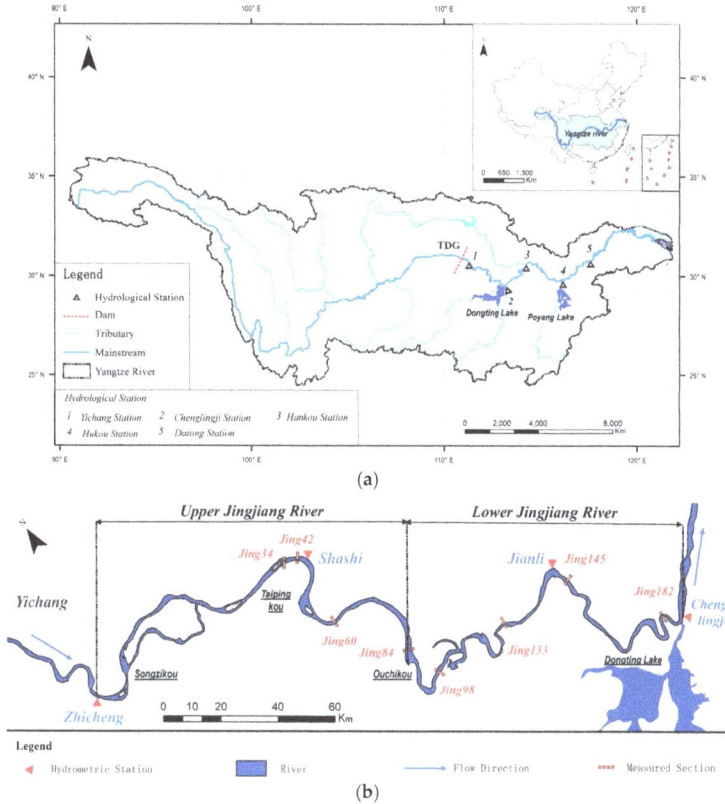

Figure 1. Sketch of the YR (**a**) and JJRR (**b**); (**b**) is drawn from [17].

2.2. Data

Both daily data and field surveyed data were used in the present study.

Daily discharge and suspended sediment load (SSL) are measured and published by the Yangtze River Water Resources Commission (YRWRC). Data at three stations (Zhicheng, Shashi, and Jianli) in 1987–2014 were obtained from yearbooks. Distances of these three stations downstream from the Yichang station are 60 km, 157 km, and 300 km, respectively.

Field surveyed data were provided by the Jingjiang Hydrology and Water Resources Surveying Bureau (JHWRSB), YRWRC. Data include vertical concentration, sediment gradation, and corresponding velocities. Table 1 manifests the years of measurement and the numbers of vertical profiles. Profiles measured before and after 2010 are different. Before 2010, there are five measuring points in each vertical profile. The normalized depths are approximately $1.0H$, $0.8H$, $0.4H$, $0.2H$, and $0.1H$ (H is the water depth of the vertical line, in meters), respectively. After 2010, there are seven measuring points in each vertical profile. The normalized depths of the upper five points are approximately $1.0H$, $0.8H$, $0.4H$, $0.2H$, and $0.1H$, respectively. The other two points were measured in the near-bed region with constant distances from the riverbed of 0.5 m and 0.1 m.

For profiles measured at Shashi station, measured flow depths range from 2.09 m to 19.7 m, and the normalized depths of these two near-bed points are approximately $(0.02–0.2)H$. For profiles measured at Jianli station, measured flow depths range from 2.13 m to 21.2 m, and the normalized depths of these two near-bed points are approximately $(0.0047–0.047)H$.

Table 1. Statistics of data measured at the JJRR (following [17]).

Stations	Observation Year	No. of Verticals
Zhicheng	1996, 1998, 2002	55
Shashi	1996, 1998, 2010, 2011, 2012, 2013	130
Jianli	1986, 1998, 2002, 2010, 2011, 2012, 2013	180

Equipment adopted in the field survey includes global navigation satellite system (GNSS, antenna, and receiver), total station, vessel-mounted acoustic Doppler current profilers (ADCP), digital level, GPS, and laser particle size analyzer (LPSA) [17]. All of this equipment has been strictly examined by professional surveyors. According to the technical manual by JHWRSB, several technical standards and criteria are illustrated, as follows:

(1) Water depth is measured with ADCP, and verified with fish lead.
(2) ADCP is applied to measure the flow twice. The deviation between each measured discharge and the average discharge should be less than ±5%. Otherwise, the data should be re-measured.
(3) Vertical lines for measuring the velocity are positioned with a real-time kinematic (RTK) GNSS.
(4) Suspended sediments are sampled with samplers in the field, and its grain-size distributions are analyzed by the sieving method. All the sampling procedures should be finished within dozens of minutes, and sieving analyses should be finished following the manual operation. For all the five points measured before 2010 and the upper five points measured after 2010, the sampler shown in Figure 2a is adopted. For the two near-bed points measured after 2010, bottom-touched automatic-closing samplers shown in Figure 2b,c are adopted. With this kind of sampler (Figure 2b,c), the distances of the two near-bed points from the riverbed can be ensured. However, the potential disturbance on the riverbed when settling the sampler may still be questionable. Therefore, except a slowed down settling velocity of the sampler, the near-bed concentrations may be double-checked after sampling and grain-size analyses. The assumption is that the measured concentration is reliable only if the vertical concentration profiles of particles finer than 0.062 mm have no abrupt changing point in the near-bed region [17]. Thus, bed materials can be distinguished from suspended sediments, which may be caused by disturbance during the sampling processes.

(a)

Figure 2. *Cont.*

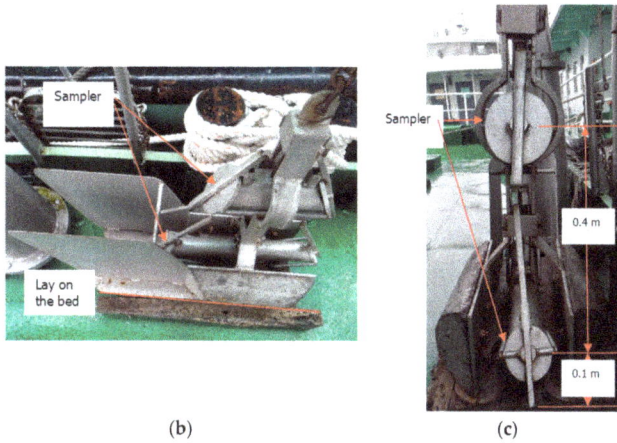

(b)   (c)

Figure 2. Samplers for suspended sediment in the field survey; (**a**) is the sampler with fish lead, and (**b,c**) are the samplers for sampling suspended sediment at the two near-bed points (different camera angles).

2.3. Methodology

Methods of cumulative anomaly and SRCs are applied to investigate the changed sediment regime of un-saturated channel in reach scales. ADEs and the sediment-balance method are adopted to evaluate new challenges for the non-saturated channel.

2.3.1. Cumulative Anomaly

The cumulative anomaly has been widely applied to detect the trends and stages of time series [18]. For time series, the cumulative anomaly (X_t) for data point x_i can be expressed as:

$$X_t = \sum_{i=1}^{t}(x_i - x_m) \quad 1 \leq t \leq n \quad x_m = \frac{1}{n}\sum_{i=1}^{n} x_i \tag{1}$$

in which x_i is the value of the time series; x_m is the mean value of the series x_i; and n is the length of the time series.

The mass curve of cumulative anomaly values describes the change process of a particular parameter by comparing it with the mean value of that parameter over the whole study period [18]. For instance, if the observed data of a given year is greater than the overall mean, the anomaly values will be positive and the mass curve will rise.

2.3.2. SRCs

The SRC is defined as the statistical relationship between SSC ($kg \cdot m^{-3}$) and discharge. SRCs are normally used to describe the flow-sediment relationships in river systems for various purposes [19]. The method is often applied to estimate the SSC or SSL in ungauged regions [20].

The power function of SRC is usually expressed as one of the following two formats:

$$\text{SSC} = eQ^f \tag{2}$$

$$\log(\text{SSC}) = \log(e) + f\log(Q) \tag{3}$$

in which Q is the discharge ($m^3 \cdot s^{-1}$), and e and f are the sediment rating coefficient and exponent, respectively. Linear regression, i.e., Equation (3), is often used to derive the values of the rating coefficient e and exponent f from empirical data (e.g., [21]). The e-parameter contains information on the conversion of Q into SSC, namely, the erosion severity index [22]. A high value of the e-parameter indicates easily-erodible materials and high loads of transported materials [20]. The exponent f-coefficient corresponds to the erosive power of the river and the influence on the sediment supplement. High values of the f-coefficient indicate a slight increase in discharge would significantly enhance the erosive power of the river [20]. Moreover, the f-coefficient is also related to climate, channel morphology, the grain-size distribution of sediments, and erodibility within the river basin [22].

2.3.3. ADEs

Three equations are adopted to evaluate the challenge in simulating the vertical concentration profiles in the unsaturated channel: the Rouse Equation, the Han Equation, and the fADE. These three equations may represent typical concentration profiles in equilibrium transport, non-equilibrium transport, and non-Fickion suspension in the non-equilibrium condition, respectively.

By assuming Fick's first law for sediment diffusion in turbulence, the diffusion theory of sediment suspension can be expressed with traditional ADE model. The equation for equilibrium sediment transport can be expressed as:

$$\omega S + \varepsilon_{sy}\frac{\partial S}{\partial y} = 0 \tag{4}$$

in which S is the sediment concentration (W·L^{-3}), y is the vertical coordinate (L), ω is the sediment settling velocity (L·T^{-1}), and ε_{sy} is the sediment turbulent diffusion coefficient in the y direction (L·T^{-1}).

By replacing ε_{sy} with the fluid eddy viscosity and assuming $\varepsilon_m = \kappa u_*(1 - y/H)y$ based on the Karman-Prandtl logarithmic velocity profile, Rouse et al. [4] obtained the analytical vertical concentration profile:

$$\frac{S}{S_a} = \left[\frac{\frac{H}{y} - 1}{\frac{H}{a} - 1}\right]^{\frac{\omega}{\kappa u_*}} \tag{5}$$

in which S_a is a reference concentration at a given height above the river bed (W·L^{-3}), a is the reference height (L), H is the flow depth (L), κ is von Karman's constant, and u_* is the shear velocity (L·T^{-1}). The frictional flow velocity can be calculated as $u_* = \sqrt{\tau_b/\rho}$, in which τ_b is the near-bed shear stress and ρ is the water density. The Rouse equation, Equation (5), has been widely used for decades. However, its drawbacks are obvious: the concentration is calculated as zero at the water surface and infinity at the river bed. These drawbacks are caused by implicit assumptions, equilibrium sediment transport, and Fick's first law. Extensive efforts have been put forward to improve the equation [10,23–25].

The assumption of equilibrium sediment transport limits its application in the field in which the sediment regime is far from equilibrium [26]. Among others, Han et al. [9] have modified Equation (4) to meet the non-equilibrium condition:

$$\omega S + \varepsilon_{sy}\frac{\partial S}{\partial y} = q_s \tag{6}$$

in which q_s is the net flux due to the imbalance between downward sediment settling and upward turbulent dispersion. Han et al. [9] obtained the solution of Equation (6):

$$\frac{S}{S_a} = \left[\frac{\frac{H}{y} - 1}{\frac{H}{a} - 1}\right]^{\frac{c\omega}{\kappa u_*}} \tag{7}$$

in which c is non-equilibrium coefficient defined as a function of the degree of saturation $c = 1 - f\left(\frac{\bar{S}}{\bar{S}_*}\right)$. $\bar{S}$ is the depth-averaged sediment concentration (kg·m^{-3}), and $\bar{S}_*$ is the depth-averaged sediment capacity. If the saturation degree $\frac{\bar{S}}{\bar{S}_*} = 1$, $c = 1$; if $\frac{\bar{S}}{\bar{S}_*} < 1$, $c > 1$, which means the flow is unsaturated with sediment; and if $\frac{S}{S_*} > 1$, $c < 1$, which indicates an super-saturated flow. The von Karman's constant κ is usually given a value of 0.408 for an unstratified flow.

According to Fick's First Law, particles may make "local" jumps, as most particle jumps induced by turbulence are constrained to a small distance (Δy) in a given Δt according to the Central Limit Theorem (CLT), and Δy is characterized by the length of Representative Elementary Volume (REV) [27,28]. However, modern observations of coherent structures in turbulence have proven that high-speed currents may intermittently sweep the bed, carry much sediment, and eject directly into the upper part of the water column [27,29]. The non-Fickion suspension indicates that particles may make many "nonlocal" jumps in burst-like suspension events [27]. The non-Fickion suspension in non-equilibrium sediment suspension can be described by a fractional advection-dispersion equation (fADE) [27]:

$$\omega S + \bar{\varepsilon}_{sy} \frac{\partial^a S}{\partial^a y} = q_s \tag{8}$$

in which α is the order of the fractional derivative ($0 < \alpha \leq 1$); $\bar{\varepsilon}_{sy}$ ($L^\alpha \cdot T^{-1}$) is the depth-averaged diffusivity expressed as follows:

$$\bar{\varepsilon}_{sy} = \frac{\int_{0.01H}^{H} \kappa u_* (1 - y/h) y \, dy}{0.99H} = 0.175 \kappa u_* H \tag{9}$$

We assume:

$$q_s = -c_0 s \tag{10}$$

in which c_0 is called the non-equilibrium coefficient, and c_0 is a function of (S/S_*):

$$c_0 = f(S - S_*) \tag{11}$$

in which S_* is the sediment transport capacity of flow; if $S = S_*$, then $c_0 = q_s = 0$, which means the flow reaches equilibrium; if $S < S_*$, then $c_0 > 0$ and $q_s < 0$, which means the flow is unsaturated with sediment; and if $S > S_*$, then $c_0 < 0$ and $q_s > 0$, which means flow is supper-saturated with sediment.

The analytical solution of Equation (8) is obtained:

$$\frac{S}{S_a} = E_a \left[-\frac{\omega + c_0}{\varepsilon_{sy}} (y - a)^\alpha \right] \tag{12}$$

in which $E_a()$ is called Mittag-Leffler function ($y \leq a \leq H$), which can be calculated by series expansion or the MATLAB open-source code.

2.3.4. Contribution of Near-Bed Concentrations on SST Rate

The SST rate can be estimated with measured vertical concentration profiles, and the deposition/erosion amount of the channel can be estimated by the sediment balance method. Changed vertical concentration profiles in the unsaturated channel lead to new challenges in field surveying and estimating the SST rate. The field survey should be conducted according to its new characteristic (as extended measuring of near-bed zone by JHWRSB). The effect of the changed vertical concentration profiles on the SST rate can be estimated by evaluating the contribution of two near-bed concentrations on the SST rate:

$$R = \frac{Q_{s(7)} - Q_{s(5)}}{Q_{s(5)}} \times 100\% \tag{13}$$

in which $Q_{s(7)}$ and $Q_{s(5)}$ are cross-sectional SST rates by seven points and, artificially, by five points, respectively. The cross-sectional sediment flux can be estimated with all measured vertical profiles. The delta-shaped area between the left bank and the first vertical line from the left bank is also considered [17]. The delta-shaped area between the right bank and the last vertical line from the left bank is not considered due to the difficulty of identifying the right bank [17].

3. Change of the Sediment Regime

Vertical concentration profiles and SST rates are analyzed to evaluate the sediment regime of local and reach scales, respectively. Then, sediment rating curves are analyzed to investigate the availability of sediment for suspension from the riverbed.

3.1. Vertical Profiles of SSC

3.1.1. Remarkable Large Concentrations in the Near-Bed Zone

e depth-averaged concentration S_{avg} (kg·m^{-3}) is calculated as $S_{avg} = \frac{1}{H} \int S_i dy$, in which S_i is the measured concentration at each point in a vertical line (kg·m^{-3}) and y is the vertical axis of the Cartesian coordinate system. The coordinate origin is located on the riverbed. To present the deviation of the near-bed concentration from the average, the relative concentration is defined as S_a/S_{avg}, in which S_a is the measured concentration at the reference height. The reference height is the minimum distance of measuring points from the river bed. The reference height a equals 0.4–0.5 m for profiles measured before 2010, while the value is 0.1 m for profiles after 2010.

Figure 3 shows the typical vertical profiles measured at Shashi station before and after the impoundment of TGD. The relative concentrations (S_a/S_{avg}) of these four vertical profiles are 2.47, 4.1, 2.10, and 1.52, respectively. Thus, remarkable large concentrations can be observed in the near-bed zone. The remarkable large near-bed concentration is defined by comparison to the vertical average concentration. This can be observed in vertical concentration profiles with large or small measured near-bed concentrations (3.01 kg·m^{-3} for Figure 3a and 0.041 kg·m^{-3} for Figure 3b).

The statistic characteristics of hundreds of vertical profiles are summarized in Table 2. Before TGD operation, the maximum relative concentrations (S_a/S_{avg}) are 2.76 (Zhicheng), 2.30 (Shashi), and 3.62 (Jianli), respectively. After TGD operation, the maximum relative concentrations (S_a/S_{avg}) are 15.05 (Shashi), and 14.45 (Jianli), respectively. The average relative concentrations before TGD are 1.35 (Shashi) and 1.57 (Jianli), respectively. These values jump to 3.22 (Shashi) and 4.89 (Jianli) after TGD, respectively. Thus, both the average and maximum values of relative concentrations (S_a/S_{avg}) increase significantly after the impoundment of the TGD.

Relatively high concentrations near the bed region have also been pronounced in the estuary of YR, Qiantangjiang River, and the Taizhou sea area [30–33]. The near-bed concentration of the S-type vertical distribution is also relatively large [34]. However, the relative concentrations are relatively small (shown in Table 2). The maximum relative concentration (S_a/S_{avg}) is approximately 4.3, and the minimum relative height of measured points from river bed is approximately 0.02. In total, the tailing phenomena in the unsaturated channel are much more apparent.

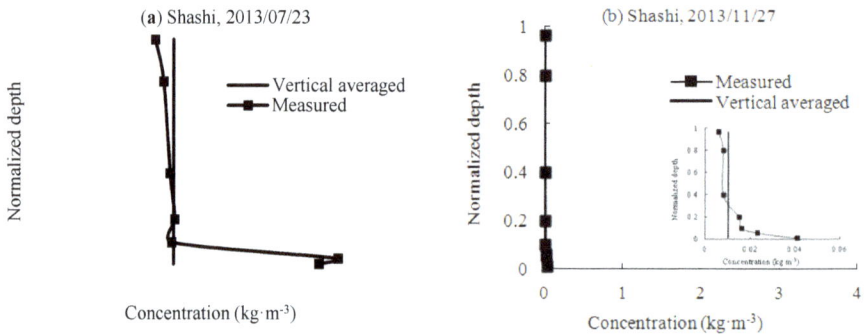

Figure 3. *Cont.*

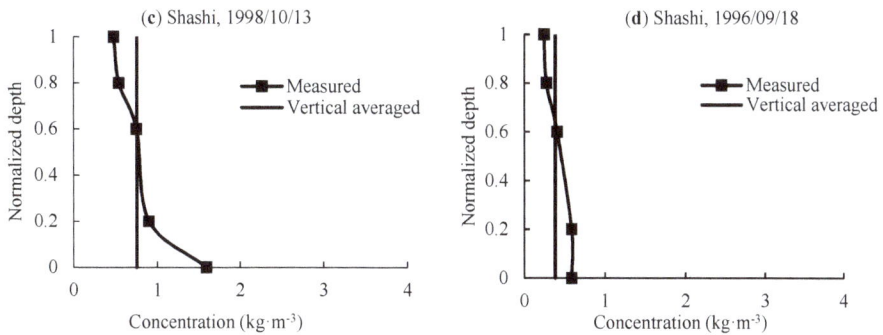

Figure 3. Typical measured vertical profiles of SSC and vertical averaged concentration in the JJRR, measured at Shashi hydrological station before (**c,d**) and after 2010 (**a,b**).

The coarsening of suspended sediment can be revealed by mean diameters of vertical profiles (shown in Table 2). For vertical lines measured at Jianli station, the average mean diameters after TGD range from 0.0087 mm to about 0.22 mm, while the values before TGD may range from 0.0037 mm to 0.07 mm. For vertical lines measured at Shashi station, the average mean diameters after TGD change from 0.0058–0.027 mm to 0.0098–0.224 mm. Thus, the coarsening process of suspended sediment after TGD operation is apparent, especially at Shashi station.

Both of the near-bed SSC and sediment flux are connected with hydrodynamic conditions [35]. Thus, hydrodynamic parameters (depth, velocity, and vertical average concentration) are also summarized in Table 2. The values of ω/u_* vary dramatically (shown in Table 2). Remarkable large values of ω/u_* indicate that relatively coarser particles may be suspended with a small bed shear velocity. Tang et al. [32] pointed out that the near-bed concentration may be more apparent when $(\omega/\kappa u_*) > 5$. Therefore, larger values of $\log(\omega/u_*)$ in non-equilibrium channels contribute to remarkable large concentrations in the near-bed zone.

Normally, concentration at the water surface should be smaller than that at the riverbed. This means that the concentration gradient (estimated with $d(S_i - S_{i-1})/d(y_i - y_{i-1})$) should be negative. However, the occurrence of vertical concentration profiles with positive gradients increases (Table 3). For data measured before TGD, about 20% (Zhicheng, Shashi, and Jianli stations) vertical concentration profiles are characterized with positive gradients. For data measured after TGD, about 28–47% (Shashi and Jianli) of vertical concentration profiles are characterized with positive gradients (Table 3). The differential equation for vertical SSC distribution of non-uniform particles indicates that the concentration of finer particles group near the surface may be greater than that near the bed with wide grading non-uniformed suspended sediment [36]. Moreover, concentration gradients may be positive when the diameter of coarser particles is about eight times that of finer particles [36]. This indicates wide grading sediment particles in the non-equilibrium channel contribute to the remarkable large near-bed concentration.

Table 2. Statistical characteristics of vertical profiles.

River Reach or Hydrological Station		S_a/S_{avg} Maximum Value	S_a/S_{avg} Average Value	Minimum Relative Depth (Reference Height a)	Water Depth (m)	(ω/u_*)	Mean Diameter (mm)	Vertical Averaged Concentration (kg·m^{-3})	Velocity (m·s^{-1})	Source
Post-reservoir	Shashi	15.05	3.22	0.0051 (0.1 m from river bed)	2.09|19.70	0.16|108.89	0.0098|0.224	0.0052|1.29	0.42|2.12	Field survey
	Jianli	14.45	4.89	0.0047 (0.1 m from river bed)	2.13|21.2	0.13|76.15	0.0087|0.220	0.015|1.26	0.27|2.69	
	Zhicheng	2.76	1.33	0.04	1.69|52.88	—	0.0052|0.0134	0.55|1.76	—	
Pre-reservoir	Shashi	2.30	1.35	0.018 (0.4 m from river bed)	4.3|22.6	0.08|1.34	0.0058|0.0265	0.188|1.99	0.08|1.34	
	Jianli	3.62	1.57	0.019 (0.4 m from river bed)	3|21.3	0.024|12.12	0.0037|0.0699	0.26|1.71	0.17|2.62	
Changjiang Estuary		4.33		0.044	—	—	0.009	0.11|0.83	0.24|2.04	[30]
Oujiangkou and Taizhou sea area		2.38		0.02	—	—	—	—	—	[31]
Wuhan and Jiujiang River Reach		1.32		0.19	10|25	0.007|0.036	0.003|0.01	0.01|0.1	—	[32]
Silty Coast		2.15		0.045 (0.012 m from river bed)	0.14|1.95	—	0.12|0.33	1.50|5.30	—	[33]

Table 3. Characteristics of vertical profiles with positive and negative gradations at Shashi and Jianli hydrological stations.

Stations	Post-Reservoir				Pre-Reservoir	
	Negative		Positive		Negative	Positive
	No. of Verticals	D_{50} (mm)	No. of Verticals	D_{50} (mm)	No. of Verticals	No. of Verticals
Shashi	50	0.01–0.06	45	0.03–0.07	29	7
Jianli	76	0.06–0.10	29	0.06–0.10	60	15

3.1.2. Coarsening of Suspended and Bed Materials

Relatively large concentrations in the near-bed zone in estuary areas can be observed during tide periods and non-flood seasons [30]. Thus, the hydrologic condition (suspended sediment and bed materials) under which the remarkable large concentrations in the near-bed zone may occur is also analyzed.

With the operation of TGD, both the suspended and bed materials become much coarser, except for the dramatic decrease of the annual suspended sediment load (SSL) [37]. The temporal variation of suspended sediment size in the downstream channel may exhibit some complex processes [38]. This means that the saturation recovery process of fine sediments that responds to changing hydraulic conditions is different from that of the coarser one. Additionally, under non-uniform and non-equilibrium conditions, suspended sediment dynamics require a relatively large distance or time to approach equilibrium [13]. Thus, the variation trend of non-uniform sediment varies in different river reaches.

The mean diameter of SSC measured at Yichang station decreases from 0.009 mm (before impoundment) to 0.005 mm (2003–2008) [37]. Size distributions of suspended sediment and bed materials in the JJRR are shown in Figures 4 and 5. For Zhicheng station, the proportion of grouped particles with diameters of 0.008–0.062 mm increases, while the proportions of finer (<0.008 mm) and coarser (0.062–0.5 mm) groups decreases. For Shashi station, the non-uniform sediment becomes a kind of wide grading (as shown in Figure 4b), as the proportion of grouped particles with diameters finer than 0.008 mm decreases and the proportion of grouped particles with diameters coarser than 0.062 mm increases. The mean diameter measured at Jianli station decreases from 0.009 mm (before impoundment) to 0.045 mm (2003–2008) [37]. This indicates that the coarser particles (>0.062 mm) around Shashi station have recovered by bed erosion. When flow transports the reach around Jianli station, the change of coarser particles (>0.062 mm) is much more apparent. Thus, wide grading suspended sediment at Shashi and Jianli stations may contribute to the remarkable large concentration in the near-bed zone.

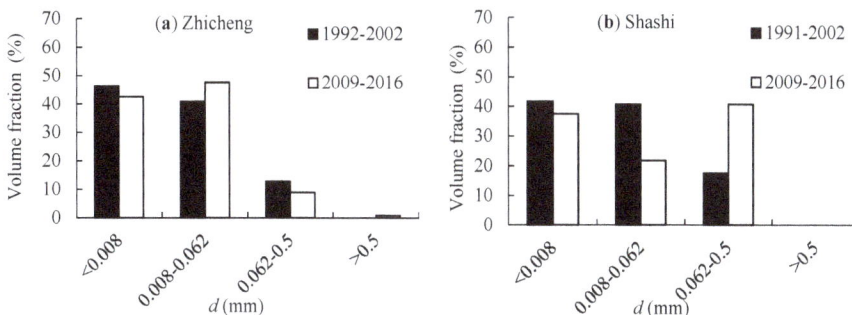

Figure 4. *Cont.*

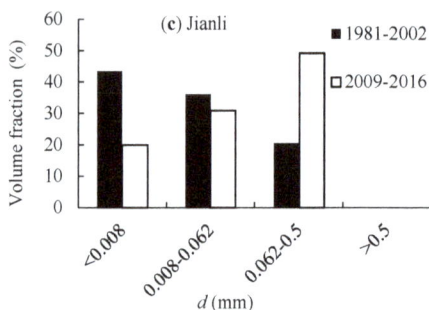

Figure 4. Volume fractions of suspended sediment measured at Zhicheng (**a**), Shashi (**b**), and Jianli (**c**) stations.

The variations of volume fractions of bed materials indicate a remarkable coarsening process in the JJRR, especially the river reach of Zhicheng–Shashi (Figure 5). Zhang [30] pointed out that concentrations in the near-bed zone are more apparent with larger values of relative particle size (D_i/D_m, particle-size over averaged particle size) of non-uniform sediment. Thus, the lifting and suspension of wide grading bed materials may also contribute to the remarkable large concentration in the near-bed region of the non-equilibrium channel.

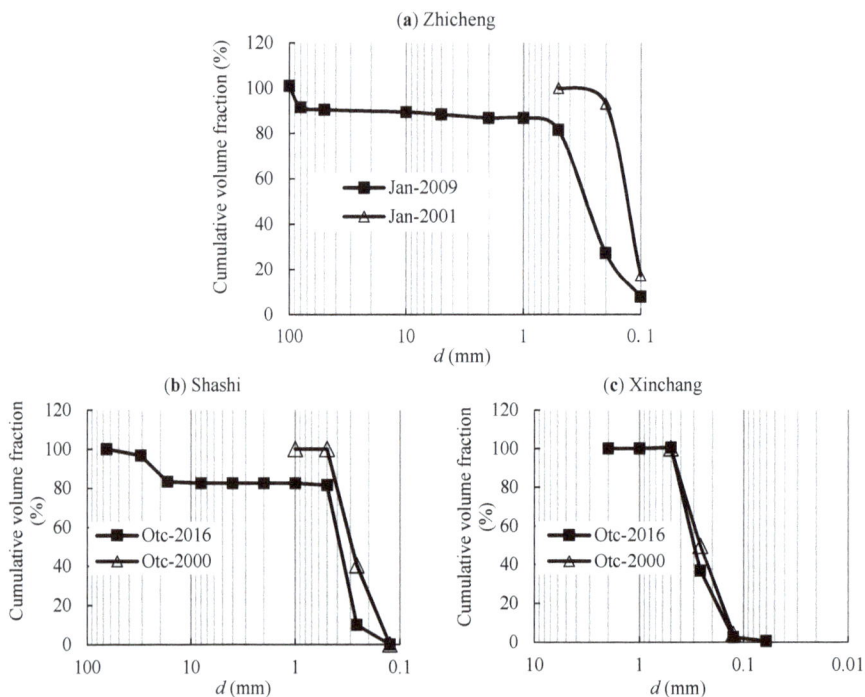

Figure 5. Cumulative volume fractions of bed materials measured at Zhichegn (**a**), Shashi (**b**), and Xinchang (**c**) stations.

3.2. Temporal Change of SSL

At the reach scale, the changed sediment regime can be described with temporal and spatial variations of SSL. The temporal and spatial variations of water runoff have also been analyzed to make a comparison. The temporal variation of water runoff and SSL show a decreasing trend, especially the SSL (Figure 6).

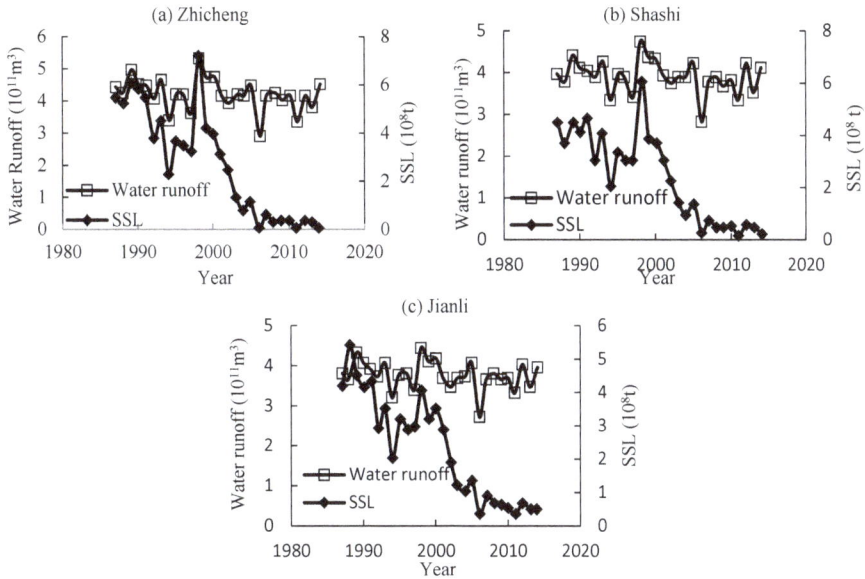

Figure 6. Temporal variation of water runoff and SSL in the JJRR.

Figure 7 shows the mass curves of cumulative anomaly of annual runoff and SSL at the three stations (1987–2014). It shows that the water runoff and SSL show an increasing trend before 2000–2001 and then a decreasing trend after 2000–2001. Three Gorges Reservoir (TGR) was launched to construct in 1994, and TGD was completed and firstly impounded water in June 2003. The TGR started to operate regularly in June, 2006 and became fully operation in 2008. The other changing points of water runoff are 1993, 2005, and 2010, which are in accordance with the operation of TDR. The changing point of 1997 may be caused by the great 1998 flood, which was the highest in recorded history.

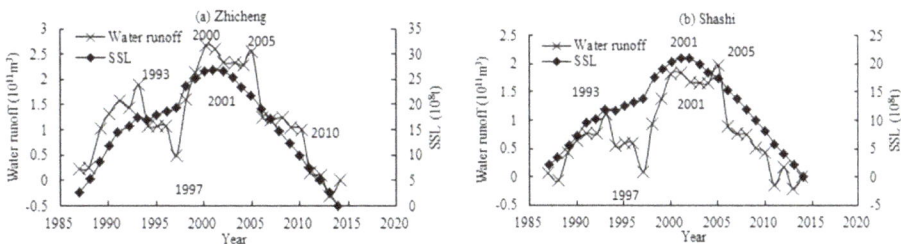

Figure 7. *Cont.*

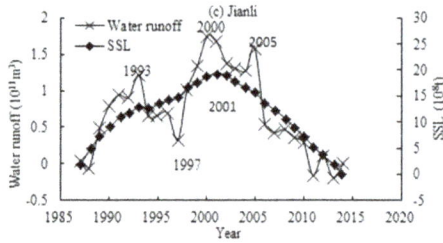

Figure 7. Mass curves of anomalies of runoff and SSL at Zhicheng (**a**), Shashi (**b**), and Jianli (**c**) hydrological stations (1987–2014).

Thus, after 2005, both the runoff and SSL are less than the long-term average value.

3.3. Availability of Sediment Sources

The availability of weathered sediments in the unsaturated channel can be revealed by rating parameters. Thus, the SSC–Q rating curves with data measured before and after TGD operation at the JJRR is adopted to investigate the hydrological condition.

According to data measured in 2013, values of the f-parameter are 2.27 (Shashi) and 1.67 (Jianli), respectively, and values of the log(e)-parameter are −7.70 (Shashi) and −10.70 (Jianli), respectively (Figure 8). The long-term average values by Yang et al. [39] are illustrated as data before dam operation (Figure 8). The long-term average values of the log(e)-parameter and the f-parameter for Xinchang station are −1.92 and 1.17, respectively, and the values for Jianli station are −0.97 and 0.96, respectively [39]. Historical data adopted in the analysis of Yang et al. [39] were measured at four stations: Yichang (1950–1985), Xinchang (1956–1984), Jianli (1954–1986), and Chenglingji (1951–1982). According to Figures 6 and 7, parameters by Yang et al. [39] may represent the hydrological conditions during the period before the changing point. Thus, with the operation of TGD, the values of the f-parameter increase while the values of the log(e)-parameter decrease.

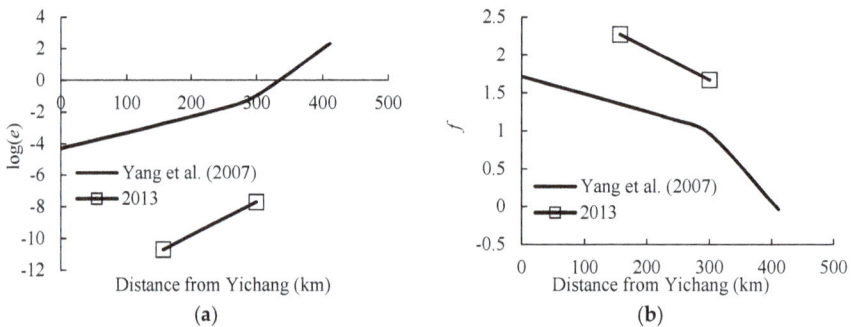

Figure 8. Coefficients and exponents of sediment rating curves (**a**) for log(e) and (**b**) for f; solid lines are data drawn from Yang et al. [39].

Instantaneous SST rates are not only a function of the transport capacity of a river, but also of sediment availability [22]. For sediment rating parameters estimated through regression analysis, the e-coefficient is an erosion severity index in the river channel and is associated with the availability of weathered sediments in the basin area [22]. A high e-value indicates that this area is characterized with easily erodible materials and high loads of materials transported by runoffs [20]. The exponent f-coefficient corresponds to the erosive power of the river and the influence on the sediment supply from the entire basin surface. High values indicate a considerable increase in erosive power and sediment-carrying capacity with an increase in river discharge [20]. Thus, a slight increase in discharge enhances the erosive power of the river significantly. For the JJRR, the increased value of the *f*-parameter indicates a considerable increase of erosive power and sediment-carrying capacity with an increase in river discharge. A decreased value of the log(*e*)-parameter indicates a decreased availability of suspended sediments in the river channel.

The saturation recovery process in the JJRR can be described with Figures 4 and 5, while the changed interactions between suspended sediment and bed material can be explained by the changed values of rating curves' coefficients (Figure 8).

For river reaches around Zhicheng station, coarsening of bed material is obvious. As to its limited distance from the TGD dam (60 km), the saturation recovery process of suspended sediment is limited.

For river reaches around Shashi station, coarsening of bed material is also obvious. Finer-grouped suspended sediments (particles with diameter < 0.008 mm) are transported from the upper river reach instead of local bed materials. Coarser suspended particles (with diameter of 0.062–0.05 mm) have recovered, which are eroded from local bed materials as an increased value of the sediment rating exponent (*f*-parameter).

For river reaches around Jianli station, the coarsening of the bed material is limited. The coarsening of suspended sediment is obvious, especially after flooding (Figure 9). Coarser suspended particles (with diameter of 0.062–0.05 mm) may come from the erosion of the local river bed and transportation from upper reaches by an increased value of the *f*-parameter.

The variability in the relationship between the sediment concentration and water discharge (namely hysteretic patterns) has also been used to explain the variability of sediment sources from one flood to another [40]. Whitaker et al. [41] compared the estimated sediment yield for an un-sampled flood event, a single-event surrogate SSC–Q rating curve, or a long-term SSC–Q rating curve. Analysis revealed that the long-term SSC–Q rating curve estimates approximately 37% higher than the single-event SSC–Q estimates, which indicates a high degree of uncertainty. Analysis shows a high variability in the rating curve between similar floods [41]. This is the reason for erosion during the flood season.

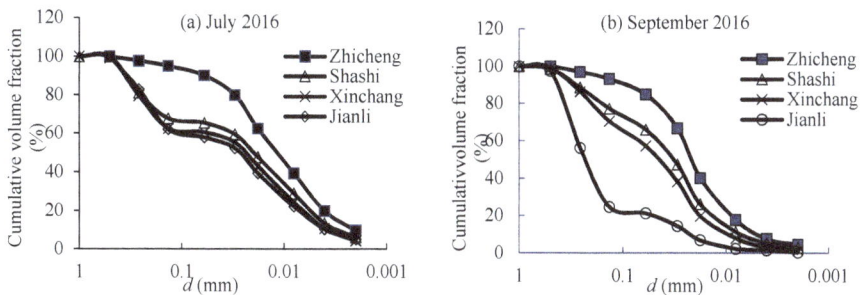

Figure 9. Cumulative volume fractions of suspended sediment at 2016: (**a**) measured in July 2016 and (**b**) data measured in September 2016.

4. Future Challenges

4.1. Accurate Measuring in Field Surveys

Accurate measurement of vertical concentration profiles is important for investigating the vertical distribution of SSC and the accurate estimation of sediment transport. For the field survey in the JJRR, the two near-bed points are measured with constant distances of 0.5 m and 0.1 m from the bed, respectively. However, the small distances may be doubtful due to difficulties in distinguishing the bed material and suspended sediment, especially when the bed is disturbed during settling sampling.

Equipment and criteria adopted in the field survey have been described in detail in the Methodology section, and their availability and methods of improvement with regard to accuracy in the future are discussed.

The availabilities of ADCP and GNSS have been certified by field surveys conducted by other researchers. The error of hydrologic measurement with ADCP is ±1%, and the random uncertainty is 2–5% for measuring cross-sections in the middle and lower YR channel [42]. The integrated system (GPS and ADCP) delivers positioning with <15 cm accuracy vertically and <30 cm horizontally [42]. For GNSS, the differences between coordinates of the ground control points estimated using RTK–GNSS, and those computed from the classical topographic measurements during processing of the photogrammetric blocks have been analyzed by Pagliari et al. [43]. It was pointed out that an RTK–GNSS survey may be sufficient to reach the requested tolerance [43]. GNSS receivers have been adopted in field surveys in recent years, as their main advantages are of being faster, cheaper, and easier to use during the surveying phase than classical topographic survey [43,44]. For sediment grain-size analysis, the results by LPSA and sieving methods are comparable when particles are spherical [45]. Sieving methods for coarser particles are more acceptable than for finer ones [45]. Thus, the sieving method adopted by JHWRSB for fine suspended particles needs further verification.

In total, various efforts have being applied to measure the SSC profiles in field surveying, including methods and instrumentation [46–48]. Some new field instrumentation, including magnetic tracers, drone-based or plane-based topography surveys, and video analysis by spectral cameras can also be used to measure sediment transport and morphology evolution in future field surveys [49]. Except for these new instruments, already existing instruments can also be used with more details. For instance, ADCPs are increasingly used to measure the three-dimensional velocity distribution and sediment transport by measuring the Doppler shift in the backscattered signals from an array of acoustic beams [49,50]. According to the backscatter intensity of ADCP beams, SSC can also be estimated with inversion [51].

4.2. Describing Vertical Concentration Profiles

The vertical distribution of SSC needs to be described accurately for an accurate simulation of channel migration, i.e., a one-dimensional simulation for the transport of sediment mixture in non-equilibrium conditions [52]. Thus, finding a proper way to describe the vertical concentration profiles of the unsaturated channel is one of the key problems for hydraulic researchers [53].

Measured sediment profiles are compared with calculations by three equations, i.e., the Rouse Equation (Equation (5)), the Han Equation (Equation (7)), and the new fADE (Equation (12)). Figure 10 indicates the Rouse Equation is not applicable in the non-equilibrium reach. Compared to the Rouse Equation, the Han Equation for non-equilibrium, sediment-laden flow performs better in the upper part of a vertical but underestimates the concentration near the river bed, i.e., within 10% of the water depth. The fADE is able to characterize the dynamics of the non-equilibrium sediment suspension in "starving" river reaches.

No doubt, Equation (7) for non-equilibrium conditions can offer a better description of sediment profiles in river reaches with unsaturated or oversaturated flows. The inaccurate prediction is partly due to our poor understanding of the mechanism of sediment suspension, i.e., the spatially-stochastic transport behavior in sediment dispersion. For the changed sediment regime of the JJRR, more efforts are required to improve the description of vertical profiles of SSC, especially the physical processes. Additionally, "nonlocal" jumps in burst-like suspension events indicate that bursting eddies may sweep particles from the water bottom and directly eject them to the upper part of water bodies, which may lead to ecological-related problems. The ecology consequences include water quality of the riverine ecosystems, as nutrients and pollutants are often closely connected with sediments [49]. The three outlets (Songzikou, Taipingkou, and Ouchikou) have a special setting of the surface water system between YR and Dongting Lake, and the three diversion channels could interact directly with the lake. Thus, it may also have a profound impact on the biological connection between the main stream and the downstream lakes of Poyang and Dongting.

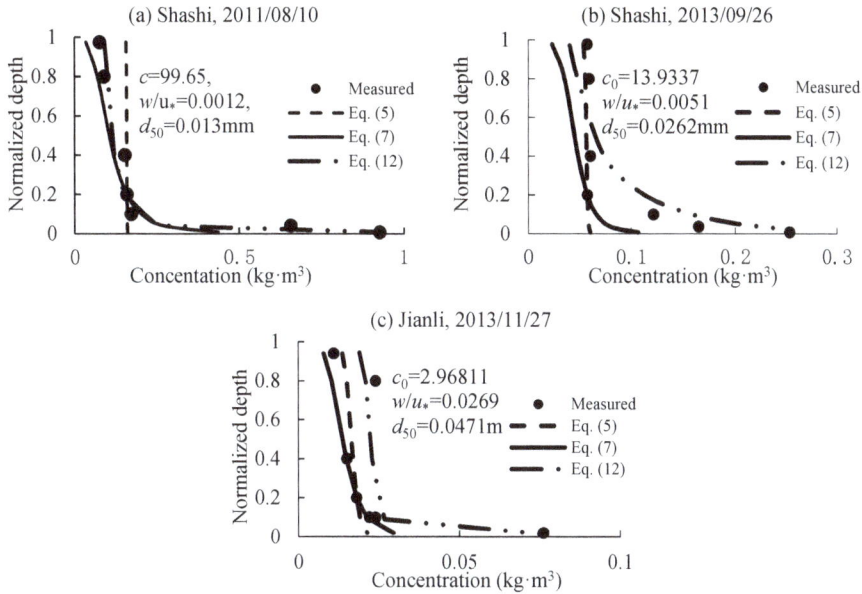

Figure 10. Comparison between measured and calculated vertical profiles of suspended sediment concentration. Data in (**a**) were measured on 10 August 2011 at Shashi station, data in (**b**) were measured on 26 September 2013 at Shashi station, and data in (**c**) were measured on 27 November 2017 at Jianli station.

4.3. Estimating SST Rate

Measuring and describing the vertical SSC profiles accurately is also imperative to the quantification of the SST rate [2]. If near-bed concentrations are not measured or simulated correctly, the estimation of the SST rate may also be influenced.

Figure 11 shows the estimated contributions by two near-bed points. Both of these two extremely large ratios (larger than 60%) occurred at Jianli station in 1996. For these two sections, the estimated sectional average velocities are different from measured data with a large deviation. This indicates that the assumption for Equation (13) is not adoptable in these two cross-sections. Thus, these two sections are excluded from discussion. Figure 11 shows that, if the near-bed concentrations are not included, the sediment flux may be underestimated by approximately 23.5% (Zhicheng), 9.35% (Shashi), and 18.68% (Jianli), respectively. Thus, for the unsaturated JJRR, sediment concentration within 10% of the water depth from the riverbed cannot be ignored. Otherwise, it will lead to underestimated results when calculating the erosion/deposition of the JJRR by the sediment-transport balance method and volume method. Additionally, an underestimated SST rate at Zhicheng station may also lead to overestimated sedimentation in the TGR, which is important for the water storage capacity of the reservoir. The reservoir's functions of water supply, energy production, navigation, flood control, and increased maintenance costs are correlated with the reservoir's water storage capacity [49]. Thus, underestimated sedimentation at Zhicheng station may also affect the reservoir's waterway systems, hydraulic schemes, intakes and outlets, and flow regulation [49].

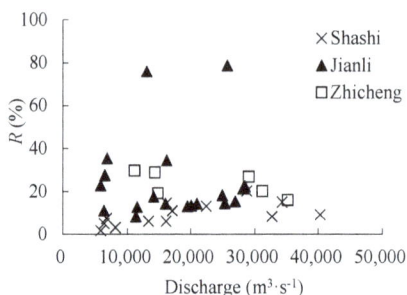

Figure 11. Contribution of the near-bed load on sediment flux.

Fortunately, the influence of near-bed concentration on estimating sediment flux can also be verified by comparing the estimated deposition with the sediment-balance method and the geometric method.

Based on the hydrology data and geometry data measured in 2002–2008, contributing factors that may cause deviation during estimating deposition were distinguished [54,55]. Yuan et al. [53] pronounced that the contribution ratio by near-bed concretion varies in different hydrology years and river reaches (Table 4). For a certain hydrological station, the contribution ratios may also vary with the inflow condition. In total, they partially verified the reliability of measured vertical concentration profiles by comparing contribution ratios.

When suspension index $A = \omega/(\beta k u_*) = 2.0$, suspended sediment may concentrate in areas with a distance of 0.2 times of the water depth from the riverbed, in which β is a coefficient for non-uniform sediment [30]. The near-bed concentration may affect the estimated sediment load only if $A > 0.15$ [30]. If $A = 2.0$, the ratio of near-bed transportation over the total transportation is approximately 30% [30]. The values of ω/u_* with data measured after TGD in Table 2 are much larger than the criteria by Zhang [30], and a large contribution of the near-bed transportation over the total transportation in Figure 11 can be certified. The ratio of the near-bed transportation over the total transportation is approximately 25% when $D_i/D_m > 3$, in which D_i/D_m is the particle size the over average size [30]. The wide grading distribution of SSC after TGD (as shown in Figures 4 and 5) also indicates a large contribution to the total transportation. Thus, the potential doubt that refers to difficulties in distinguishing bed material and suspended sediment can partially be certified. Moreover, the impact of a remarkable large near-bed concentration must be given more attention in the future, especially at river reaches around Shashi station.

Table 4. Contribution by various factors on sediment flux [1].

River Reach	Total Sediment Flux [2] (10⁴ m³)	Contribution by Near-Bed Concentration		Contribution by Other Factors	
		Sediment Flux (10⁴ m³)	Ratio (%)	Sediment Flux (10⁴ m³)	Ratio (%)
Yichang–Zhicheng	10,275	1267	12.33	2047	19.92
Zhicheng–Shashi	11,242	980	8.72	4317	38.40
Shashi–Jianli	12,840	2303	17.94	1494	11.64

[1] Data drawn from Yuan et al. [53] and Duan et al. [54]; [2] total sediment flux is estimated with the geometry method.

In total, fluvial and dynamic characteristics of distinct, large concentrations in near-bed regions may affect sediment flux, erosion/deposition estimation, sediment rating curve, and the channel geometry followed. This brings river engineers a lot of new problems, i.e., degradation of navigation, and dramatic bank erosion [56]. Additionally, the estimation of erosion in the JJRR has a profound impact on the hydrologic connection between the middle YR and Dongting Lake [57].

5. Conclusions

Sediment regime of the JJRR has changed dramatically with the operation of TGD. Based on daily and vertical concentration data at the three hydrological stations (Zhicheng, Shashi, and Jianli) in the JJRR, the sediment regime and its potential challenges are analyzed in this study. The conclusions can be summarized as follows:

(1) In the unsaturated JJRR, measurements have revealed anomalous vertical profiles of SSC, as the near-bed concentrations normalized with the vertical average concentration are dramatically larger than that of the pre-equilibrium channel. The near-bed concentration (within 10% of the water depth from the river bed) may reach up to 15 times that of the vertical average concentration in the non-equilibrium channel.

(2) In the unsaturated JJRR, the combination of wide grading suspended sediment and coarsened bed materials in non-equilibrium channel contribute to a remarkably large concentration in the near-bed zone. For the river reach around Shashi station, the remarkably large near-bed concentration is more apparent.

(3) More detailed measurements or new measuring technologies may assist in the accurate measurement of vertical concentration profiles. A fractional dispersion equation may help to provide accurate descriptions. The outcomes can provide useful information for modeling the morphologic change of the JJRR.

Acknowledgments: The financial support from National Key R&D Program of China (no. 2016YFC0402303); and the National Natural Science Foundation of China (Nos. 51579230, 51779242, 51509234, 41330751 and 51479179).

Author Contributions: L.H. and D.C. conceived and designed the paper, S.Z. analyzed the data, M.L. contributed to figures, G.D. provided materials of vertical concentration, and L.H. wrote the paper.

Conflicts of Interest: The authors declare no conflict of interest.

References

1. Schumm, S.A. *The Fluvial System*; Wiley: New York, NY, USA, 1977.
2. McLean, S.R. Depth-integrated suspended load calculations. *J. Hydraul. Eng.* **1991**, *117*, 1440–1458. [CrossRef]
3. Papanicolaou, A.N.; Elhakeem, M.; Krallis, G.; Prakash, S.; Edinger, J. Sediment transport modelling review-current and future developments. *J. Hydraul. Eng. ASCE* **2008**, *134*, 1–14. [CrossRef]
4. Rouse, H. Modern conceptions of the mechanics of fluid turbulence. *Trans. Am. Soc. Civ. Eng.* **1938**, *102*, 463–543.
5. Hunt, J.N. On the turbulent transport of sediment in open channel. *Proc. R. Soc. Lond. A* **1954**, *224*, 1158. [CrossRef]
6. Aragon, J.A.G.; Salazar, S.S.; Reyes, P.M.; Delgado, C.D. Concentration profiles of suspended sediments calculation in non-uniform sediments, nonequilibrium situations. *Ing. Hidraul. Mex.* **2000**, *15*, 29–35.
7. Maleki, F.S.; Khan, A.A. 1-D coupled non-equilibrium sediment transport modelling for unsteady flows in the discontinuous Galerkin framework. *J. Hydrodyn.* **2016**, *28*, 534–543. [CrossRef]
8. Nicholad, A.P.; Clarke, L.; Quine, T.A. A numerical modelling and experimental study of flow width dynamics on alluvial fans. *Earth Surf. Process. Landf.* **2009**, *34*, 1985–1993. [CrossRef]
9. Han, Q.W.; Chen, X.J.; Xue, X.C. On the vertical distribution of sediment concentration in non-equilibrium transportation. *Adv. Water Sci.* **2010**, *21*, 512–523.
10. Zhao, L.J.; Wu, G.Y.; Wang, J.Y. Review and prospect of vertical distribution by non-equilibrium sediment concentration theory. *J. Hydroelectr. Eng.* **2015**, *34*, 63–69.
11. Regüés, D.; Nadal-Romero, E. Uncertainty in the evaluation of sediment yield form badland areas: Suspended sediment transport estimated in the Araguás catchment (central Spanish Pyrenees). *Catena* **2013**, *106*, 93–100. [CrossRef]
12. Zhu, H.W.; Cheng, P.D.; Li, W.; Chen, J.H.; Pang, Y.; Zhang, D.W. Empirical model for estimating vertical concentration profiles of re-suspended, sediment-associated contaminants. *Acta Mech. Sin.* **2017**, *33*, 846–854. [CrossRef]

13. Cheng, C.; Huang, H.M.; Liu, C.Y.; Jiang, W.M. Challenges to the representation of suspended sediment transfer using a depth-averaged flux. *Earth Surf. Process. Landf.* **2016**, *41*, 1337–1357. [CrossRef]

14. Wu, W. Depth-averaged two-dimensional numerical modelling of unsteady flow and nonuniform sediment transport in open channels. *J. Hydraul. Eng.* **2004**, *130*, 1013–1024. [CrossRef]

15. Tayfur, G.; Singh, V.P. Transport capacity models for unsteady and non-equilibrium sediment transport in alluvial channels. *Comput. Electron. Agric.* **2012**, *86*, 26–33. [CrossRef]

16. Singh, A.K.; Kothyari, U.C.; Raju, K.G.R. Rapidly varying transient flows in alluvial rivers. *J. Hydraul. Res.* **2004**, *42*, 473–486. [CrossRef]

17. He, L. Quantifying the effects of near-bed concentration on the sediment flux after the operation of the Three Gorges Dam, Yangtze River. *Water* **2017**, *9*, 986. [CrossRef]

18. He, Y.; Wang, F.; Tian, P.; Mu, X.M.; Gao, P.; Zhao, G.J.; Wu, Y.P. Impact assessment of human activities on runoff and sediment of Beiluo River in the Yellow River Based on Paired Years of Similar Climate. *Pol. J. Environ. Stud.* **2016**, *25*, 121–135. [CrossRef]

19. Wang, H.J.; Yang, Z.S.; Wang, Y.; Saito, Y.; Liu, J.P. Reconstruction of sediment flux from the Changjiang (Yangtze River) to the sea since the 1860s. *J. Hydrol.* **2008**, *349*, 318–332. [CrossRef]

20. Iadanza, C.; Napolitano, F. Sediment transport time series in the Tiber River. *Phys. Chem. Earth* **2006**, *31*, 1212–1227. [CrossRef]

21. Desilets, S.L.E.; Nijssen, B.; Ekwurzel, B.; Ferre, T.P.A. Post-wildfire changes in suspended sediment rating curves: Sabino Canyon, Arizona. *Hydrol. Process.* **2007**, *21*, 1413–1423. [CrossRef]

22. Asselman, N.E.M. Fitting and interpretation of sediment rating curves. *J. Hydrol.* **2000**, *234*, 228–248. [CrossRef]

23. Huang, S.H.; Sun, Z.L.; Xu, D.; Xia, S.S. Vertical distribution of sediment concentration. *J. Zhejiang Univ. Sci. A* **2008**, *9*, 1560–1566. [CrossRef]

24. Wilson, K.C. Rapid increase in suspended load at high bed shear. *J. Hydraul. Eng.* **2005**, *131*, 46–51. [CrossRef]

25. Brown, G.L. Approximate profile for nonequilibrium suspended sediment. *J. Hydraul. Eng.* **2008**, *134*, 1010–1014. [CrossRef]

26. Jobson, H.E.; Serge, W.W. Predicting concentration profiles in open channels. *J. Hydraul. Div.* **1970**, *96*, 1983–1996.

27. Chen, D.; Sun, H.; Zhang, Y. Fractional dispersion equation for sediment suspension. *J. Hydrol.* **2013**, *491*, 13–22. [CrossRef]

28. Nie, S.Q.; Sun, H.G.; Zhang, Y.; Chen, D.; Chen, W.; Chen, L.; Schaefer, S. Vertical distribution of suspended sediment under steady flow: Existing theories and fractional derivative model. *Discret. Dyn. Nat. Soc.* 2017. [CrossRef]

29. Noguchi, K.; Nezu, I. Particle-turbulence interaction and local particle concentration in sediment-laden open-channel flows. *J. Hydro-Environ. Res.* **2009**, *3*, 54–68. [CrossRef]

30. Sun, Z.L.; Zhang, C.F.; Du, L.H.; Xu, D. Transport rate of nonuniform suspended load. *Shuili Xuebao* **2016**, *47*, 501–508.

31. Dong, X.T.; Li, R.J.; Fu, G.C.; Zhang, H.C. Relationship between vertical distribution of velocity and sediment concentration of sediment-laden flow. *J. Hohai Univ. Natl. Sci.* **2015**, *43*, 371–376. [CrossRef]

32. Tang, M.L.; Cao, H.Q.; Li, Q.Y.; Zhai, W.L. Vertical distribution of suspended sediment in middle reaches of Yangtze River. *Yangtze River* **2017**, *48*, 6–11.

33. Xia, Y.F.; Xu, H.; Chen, Z.; Wu, D.W.; Zhang, S.Z. Experimental study on suspended sediment concentration and its vertical distribution under spilling breaking wave actions in silty coast. *China Ocean Eng.* **2011**, *25*, 565–575. [CrossRef]

34. Feng, Q.; Xiao, Q.L. An new S-type vertical distribution of suspended sediment concentration. *J. Sedim. Res.* **2015**, *1*, 19–24.

35. Parsons, A.J.; Cooper, J.; Wainwright, J. What is suspended sediment? *Earth Surf. Process. Landf.* **2015**, *40*, 1417–1420. [CrossRef]

36. Zhang, X.F.; Tan, G.M. Characteristics of vertical concentration distribution of non-uniform particles. *Shuli Xuebao* **1992**, *10*, 48–52.

37. Han, J.Q.; Sun, Z.H.; Li, Y.T.; Tang, J.W. Changed and caused of lower water level in Yichang–Chenglingji reach after impounding of Three Gorges Reservoir. *Eng. J. Wuhan Univ.* **2011**, *44*, 685–695.

38. Xu, J.X. Complex behaviour of suspended sediment grain size downstream from a reservoir: An example from the Hanjiang River, China. *Hydrol. Sci. J.* **1996**, *41*, 837–849.

39. Yang, G.F.; Chen, Z.Y.; Yu, F.; Wang, Z.; Zhao, Y.; Wang, Z. Sediment rating parameters and their implications: Yangtze River, China. *Geomorphology* **2007**, *85*, 166–175. [CrossRef]

40. Seeger, M.; Errea, M.-P.; Begueria, S.; Arnaez, J.; Marti, C.; Garcia-Ruiz, J.M. Catchment soil moisture and rainfall characteristics as determinant factors for discharge/suspended sediment hysteretic loops in a small headwater catchment in the Spanish Pyrenees. *J. Hydrol.* **2004**, *288*, 299–311. [CrossRef]

41. Whitaker, A.C.; Sato, H.; Sugiyama, H. Changing suspended sediment dynamics due to extreme flood events in a small pluvial-nival system in northern Japan. *Sedim. Dyn. Chang. Environ.* **2008**, *325*, 192–199.

42. Zhang, Q.; Shi, Y.F.; Xiong, M. Geometric properties of river cross sections and associated hydrodynamic implications in Wuhan-Jiujiang river reach, the Yangtze River. *J. Geogr. Sci.* **2009**, *19*, 58–66. [CrossRef]

43. Pagliari, D.; Rossi, L.; Passoni, D.; Pinto, L.; De Michele, C.; Avanzi, F. Measuring the volume of flushed sediments in a reservoir using multi-temporal images acquired with UAS, Geomatics. *Nat. Hazards Risk* **2017**, *8*, 150–166. [CrossRef]

44. Bandini, F.; Jakobsen, J.; Olesen, D.; Reyna-Gutierrez, J.A.; Bauer-Gottwein, P. Measuring water level in rivers and lakes from lightweight unmanned aerial vehicles. *J. Hydrol.* **2017**, *548*, 237–250. [CrossRef]

45. Li, W.K.; Wu, Y.X.; Huang, Z.M.; Fan, R.; Lv, J.F. Measurement results comparison between Laser Particle Analyzer and Sieving method in particle size distribution. *Zhongguo Fenti Jishu* **2007**, *13*, 10–13.

46. Pedocchi, F.; Garcia, M.H. Acoustic measurement of suspended sediment concentration profiles in an oscillatory boundary layer. *Cont. Shelf Res.* **2012**, *46*, 87–95. [CrossRef]

47. Yu, Q.; Flemming, B.W.; Gao, S. Tide-induced vertical suspended sediment concentration profiles: Phase lag and amplitude attenuation. *Ocean Dyn.* **2011**, *61*, 403–410. [CrossRef]

48. Pitarch, J.; Odermatt, D.; Kawak, M.; Wuest, A. Retrieval of vertical particle concentration profiles by optical remote sensing: A model study. *Opt. Express* **2014**, *22*, 947–959. [CrossRef] [PubMed]

49. Schleiss, A.; Franca, M.J.; Juez, C.; Cesare, G. Reservoir sedimentation. *J. Hydraul. Res.* **2016**, *54*, 595–614. [CrossRef]

50. Togneri, M.; Lewis, M.; Neill, S.; Masters, I. Comparison of ADCP observations and 3D model simulations of turbulence at a tidal energy site. *Renew. Energy* **2017**, *114*, 273–282. [CrossRef]

51. Jin, W.F.; Liang, C.J.; Zhou, B.F.; Li, J.D. Measurement and analysis of high suspended sediment concentration in Jintang channel with vessel-mounted ADCP. *J. Mar. Sci.* **2009**, *27*, 31–39.

52. Merkhali, S.P.; Ehteshami, M.; Sadrnejad, S.A. Assessment quality of a nonuniform suspended sediment transport model under unsteady flow condition (case study: Aras River). *Water Environ. J. Promot. Sustain. Solut.* **2015**, *29*, 189–498. [CrossRef]

53. Eggenhuisen, J.T.; McCaffrey, W.D. The vertical turbulence structure of experimental turbidity currents encountering basal obstructions: Implications for vertical suspended sediment distribution in non-equilibrium currents. *Sedimentology* **2012**, *59*, 1101–1120. [CrossRef]

54. Yuan, Y.; Zhang, X.F.; Duan, G.L. Modifying the channel erosion and deposition amount of Yichang–Jianli reach calculated by sediment discharge method. *J. Hydroelectr. Eng.* **2014**, *33*, 163–169.

55. Duan, G.L.; Peng, Y.B.; Guo, M.J. Comparative analysis on riverbed erosion and deposition amount calculated by different methods. *J. Yangtze River Sci. Res. Inst.* **2014**, *31*, 108–118.

56. Xia, J.Q.; Deng, S.S.; Lu, J.Y.; Xu, Q.X.; Zong, Q.L.; Tan, G.M. Dynamic channel adjustments in the Jingjiang reach of the middle Yangtze River. *Sci. Rep.* **2016**, *6*, 22802. [CrossRef] [PubMed]

57. Li, Y.Y.; Yang, G.S.; Li, B.; Wan, R.R.; Duan, W.L.; He, Z. Quantifying the effects of channel change on the discharge diversion of Jingjiang Three outlets after the operation of the Three Gorges Dam. *Hydrol. Res.* **2016**, *47*, 161–174. [CrossRef]

water

MDPI

Article

Three-Dimensional Numerical Study of Free-Flow Sediment Flushing to Increase the Flushing Efficiency: A Case-Study Reservoir in Japan

Taymaz Esmaeili [1,*], Tetsuya Sumi [2], Sameh A. Kantoush [2], Yoji Kubota [3], Stefan Haun [4] and Nils Rüther [5]

[1] Department of Civil Engineering, Gorgan Branch, Islamic Azad University, Gorgan 49147-39975, Iran
[2] Disaster Prevention Research Institute (DPRI), Kyoto University, Uji, Kyoto 611-0011, Japan; sumi.tetsuya.2s@kyoto-u.ac.jp (T.S.); kantoush.samehahmed.2n@kyoto-u.ac.jp (S.A.K.)
[3] Hydro-Soft Technology Institute Co., Ltd., Nakanoshima, Osaka 530-6126, Japan; kubotayj@hydro-soft.co.jp
[4] Institute for Modelling Hydraulic and Environmental Systems, University of Stuttgart, 70569 Stuttgart, Germany; stefan.haun@iws.uni-stuttgart.de
[5] Department of Civil and Environmental Engineering, Norwegian University of Science and Technology, S.P. Andersens vei 5, 7491 Trondheim, Norway; nils.ruther@ntnu.no
* Correspondence: t.esmaeili@gorganiau.ac.ir or taymaz.esmaeili@gmail.com; Tel.: +98-17-325-37682

Received: 12 October 2017; Accepted: 14 November 2017; Published: 19 November 2017

Abstract: The catchment of the Dashidaira reservoir located on the Kurobe River has high sediment yield. Because of the sufficient available amount of water in the catchment during flood events, the free-flow sediment flushing operation with full water-level drawdown is employed every year to preserve the effective storage capacity of the Dashidaira reservoir. This paper focuses first on the numerical simulation of a previously conducted free-flow flushing operation in the Dashidaira reservoir using the available in situ obtained data. Afterwards, to improve the flushing efficiency, the effects of water and discharge manipulation and the construction of an auxiliary channel on the total volume of the flushed sediment were studied. A fully 3D numerical model using the finite volume approach in combination with a wetting/drying algorithm was utilized to reproduce the flow velocity field and simulate the movable bed variations. The outcomes revealed that increasing the average free-flow discharge during the free-flow stage by approximately 56%, in the form of multiple discharge pulses, can enhance the flushing efficiency by up to 13%, and the construction of an auxiliary channel in the wide midstream of the reservoir can locally increase the sediment erosion from this area.

Keywords: reservoir flushing; numerical simulation; flushing efficiency; Kurobe River

1. Introduction

Dams interrupt the natural continuity of sediment transport through rivers, which results in sediment deposition in the reservoir behind the dam [1]. In a global context, sediment deposition is a challenging issue for the long-term utilization of dam reservoirs [2]. It is estimated that 0.5% of total storage volume of the reservoirs is lost annually in the world because of the sedimentation [3]. The loss of effective reservoir storage volume due to sediment deposition reduces the effective lifespan of dams and decreases reservoir functionality for flood control purposes, hydropower generation, irrigation and water supply, thereby generating a substantial economic loss [1,4,5]. Diverse measures, including sediment dredging, density current venting, bypassing, flushing, sluicing and upstream sediment trapping, have been used to control progressive sedimentation in reservoirs to prolong their life [1,6]. Among these measures, drawdown flushing (i.e., free-flow sediment flushing) plays a major role in reservoir storage capacity restoration and conservation since it is an efficient hydraulic technique for sediment removal [7]. Free-flow sediment flushing involves a complete lowering of the water level by opening low-level outlets to temporarily establish riverine flow (i.e., free flow) through the outlets. The accelerated flow erodes a channel through the deposits and flushes out both fine and coarse sediments [4,5,7]. However, only few experimental studies have focused on flushing channel formation [8], and detailed explanations of flushing channel formation and evolution in prototype reservoirs are scarce [9].

In Japan, the sediment yield coming from the catchments is generally high due to the geologically young mountains, steep slopes, flashy flow regimes, and frequent landslides, especially in the mountainous areas. Thus, a large amount of sediment is transported in Japanese rivers, and the sediment inputs in the reservoirs fed by these rivers are high. Kurobe River, located in the Toyama prefecture, is one of the most important rivers in Japan because of the cascade reservoir system along this river and the considerable amount of electricity generated by water from these reservoirs. Therefore, the reservoir owners are interested in implementing the applicable measures for increasing the sediment removal from these reservoirs especially from the specific areas, which encounters the excess deposition problems.

To optimize the sediment management strategies in reservoirs that feature conditions that change from shallow to deep areas and that also contain sandbars, numerical models that are more complicated than the simple one-dimensional (i.e., 1D) models should be used [10]. Advanced two-dimensional (i.e., 2D) numerical models have been employed to solve practical problems in rivers (e.g., simulation of morphological bed changes in river meanders) [11]. Nevertheless, 2D and quasi three-dimensional (i.e., 3D) numerical models are not able to directly simulate a complex 3D flow field that includes secondary currents. However, these phenomena contribute significantly to the natural sediment transport processes [12]. Thus, the application of 3D numerical models is essential if the velocity variation over the flow depth, i.e., helical flow, plays a major role in sediment transport (e.g., in channel bends). 3D numerical models were used by various researchers for simulating the flow field and sediment transportation in the dam reservoirs [13,14]. As for sediment flushing from reservoirs, a 3D numerical model has been shown to perform better than a 2D one in simulating the bed deformation, notably in channel bends [15]. Recently 3D numerical models were applied to investigate the sediment dynamics inside the reservoirs during a flushing operation at the both lab scale (e.g., [15,16]) and the full scale (e.g., [17–19]).

A Computational Fluid Dynamic (CFD) software package called SSIIM (Simulation of Sediment Movements In water Intakes with Multiblock Option) was utilized in this study. The SSIIM program implements a 3D numerical model of the flow field by solving the mass and momentum conservation equation in three dimensions using different turbulence closure approaches [20]. SSIIM was successfully applied to model the 3D flow field under various hydraulic and geometric boundary conditions [21]. SSIIM was also used for coupled computation of the flow and sediment field in physical model and prototype-scale studies by various researchers [22–28]. Recently, this CFD program was used with enhanced grid generation features to simulate the sedimentation and flushing channel evolution in reservoirs [19,29,30]. Nevertheless, it should be noted that simulation of a sediment-flushing event in a steep full-scale reservoir, such as the Dashidaira reservoir, is complex owing to the rapid variation of hydraulic boundary conditions and riverbed materials even in short distances. In addition, numerical studies focusing on potential measures for increasing the sediment flushing efficiency in a full-scale reservoir are scarce.

In the present study, the SSIIM model was employed to simulate the 2012 free-flow sediment flushing operation in the Dashidaira reservoir using the field-measured data. To do this, the sensitivity analysis of computed Total Volume of Flushed Sediments (i.e., TVFS) as a result of changes in the selected empirical parameters together with assessment of morphological bed changes was performed and subsequently model was calibrated. Because of the interest to increase the Flushing Efficiency (i.e., FE), when it is defined as the ratio of the flushed sediment volume to the used water volume, various scenarios, such as using additional artificial discharge during the free-flow state, increasing the water-level drawdown speed, and the construction of an auxiliary longitudinal channel, were numerically modeled to investigate how they affect the morphological bed changes in specific zones of the reservoir and whether they increase the Flushing Efficiency. To provide a more detailed and quantitative insight on morphological bed changes, the Bed Changes Index (i.e., BCI) is introduced and subsequently calculated bed changes were compared for each scenario with a reference case (e.g., calibrated case) to figure out the performance of each scenario. Moreover, the optimum relationship between the flushing efficiency and the amount of water used for flushing was established under the mentioned scenario conditions.

2. Study Case Description

2.1. Site Background

The Kurobe River originates from Washiba Mountain with an elevation of 3000 m, located in the northern Japanese Alps, and flows into Toyama bay in the Japan Sea (Figure 1a) [31]. The catchment area of the river is 682 km^2, and the length of the river is 85 km. The bed slope is steep and varies between 1% and 20%. In the catchment area of the Kurobe River, the average rainfall and total sediment yield are 4000 mm and 1.4 × 10^6 m^3/year, respectively, which are both among the highest in Japan [32]. Dashidaira dam, with a height of 76.7 m, was constructed in 1985 by Kansai Electric Power Company in the Kurobe River and has a power of 124 MW. The average bed slope of the reservoir is about 2.2% and also the gross and effective storage capacities are 9.01 and 1.66 million m^3, respectively [7,32]. The mean annual sediment load and the ratio of the total storage to the mean annual runoff in the Dashidaira reservoir are 0.62 × 10^6 m^3 and 0.00674, respectively [33]. The Dashidaira dam is one of the first dams in Japan constructed with sediment-flushing facilities. The first flushing operation in the Dashidaira reservoir was performed 6 years after the dam construction in 1991. Subsequently, the accumulated distorted sediments within 6 years were diffused to the downstream and estuary zone with many negative environmental implications. After that, the flushing operation is performed every year during the first major flood event in the rainy season to reduce negative environmental impacts on the downstream areas of the dam since aquatic animals have been adapted to the perturbation caused by floods. A flood not only provides enough energy to transport the flood-born sediments through the reservoir, but also scour the previously deposited sediments from the reservoir.

2.2. Field Data Organization

The bathymetric survey for measuring the bed levels is performed regularly by the reservoir owners before and after the annual flushing operation in the Dashidaira reservoir. The measured bed levels before the flushing operation in June 2012 and the locations of cross-sections A-A to L-L, which were used for further assessment of bed variations caused by the flushing, are shown in Figure 1b. The approximately 2-km length of the reservoir has been divided into three areas, namely, areas I, II and III, to analyze the study outcomes based on the generally similar types of bed materials in each area. In area I, the bed material is coarse (e.g., gravel), while in downstream half of area II and the area III, the bed material transitions to finer sediments (e.g., fine sand and even mud). Besides, temporal variation of the water level and discharge magnitudes are recorded at the dam site during the flushing. The inflow discharge and water level fluctuations during the 2012 flushing operation, in which the major sediment inflow was a wash load (i.e., transported without deposition in the reservoir), are illustrated in Figure 1c. After the start of the operation, the preliminary drawdown occurred from 8 to 24 h, and the free-flow state occurred from 24 to 38 h. The water level drawdown from the beginning of the operation to the free-flow state was approximately 25 m. Furthermore, based on the available onsite samples, seven sediment sizes ranging between 316 and 0.37 mm were considered to be representative grain sizes for the bed of the reservoir. Table 1 shows the average sediment size distribution in different cross-sections prior to the flushing operation. The large sediment sizes in cross-section L-L despite the size reduction in sections immediately upstream can be attributed to the last flushing operation process, in which closing the bottom outlets during the water level recovery stage causes the transported large sediments to deposit nearby the dam. During the 2012 flushing operation of the Dashidaira reservoir, the total volume of the eroded sediment from the reservoir was equal to 408,700 m^3.

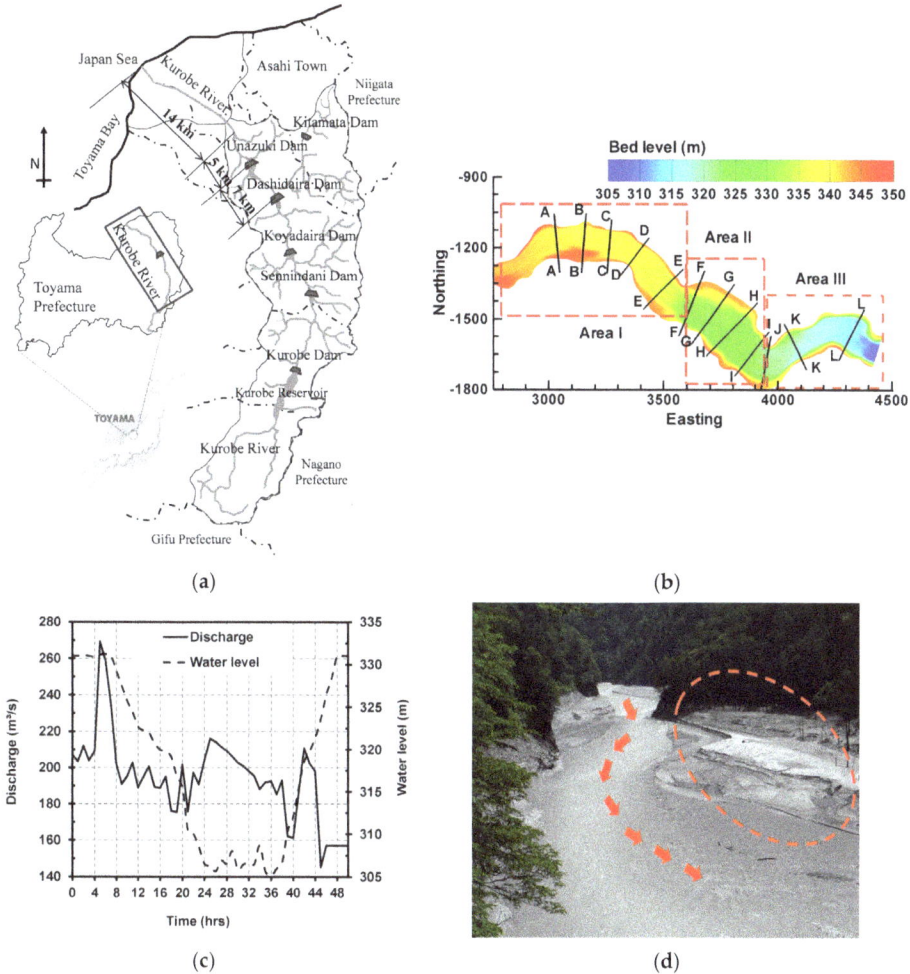

(a)

(b)

(c)

(d)

Figure 1. (**a**) Site map of the Kurobe River showing the location of the Dashidaira reservoir [31]; (**b**) Measured bed topography of the Dashidaira reservoir before the flushing operation in June 2012 and locations of the cross-sections A-L for further assessment of the bed variations in the upstream (i.e., area I), midstream (i.e., area II), and downstream (i.e., area III) portions of the reservoir; (**c**) Water level and discharge rates during the flushing operation in June 2012; (**d**) Onsite view of the dead zone area during the 2012 flushing operation (the dead zone area has been shown with ellipse dashed line and arrows show the flow direction).

The thalweg of the river channel exists close to the right bank of the wide middle area (i.e., area II) and subsequently fine sediments deposited along the left bank, indicated with an ellipse dashed line in Figure 1d, were not removed effectively during the flushing operation. Therefore, fine sediments have been accumulated and consolidated that may contain the degraded organic matters. The area highlighted by the ellipse dashed line is hereafter called the dead zone. Dam owners are interested in removing the accumulated sediments from this zone by annual flushing operation to prevent the formation of consolidated and distorted fine sediments and also to keep the effective storage capacity as much as possible.

Table 1. Average sediment size distribution in the specified cross-sections shown in Figure 1b. Cs. is an abbreviation for Cross-section.

Sediment Size (mm)	Cs. A-A (%)	Cs. B-B (%)	Cs. C-C (%)	Cs. D-D (%)	Cs. E-E (%)	Cs. F-F (%)	Cs. G-G (%)	Cs. H-H (%)	Cs. I-I (%)	Cs. J-J (%)	Cs. K-K (%)	Cs. L-L (%)
316	2	0	0	0	0	0	0	0	0	0	0	0
118.3	74	0	0	0	0	0	0	0	0	0	0	40
37.4	6	73	75	70	69	4	1	0	0	0	0	30
11.8	4	7	6	8	13	14	5	0	0	0	0	16
3.7	3	14	11	14	12	25	18	0	0	0	0	3
1.2	5	4	1	3	2	23	21	0	6	13	0	1
0.37	6	2	7	5	4	35	55	100	94	87	100	10

3. Numerical Model

3.1. Flow Field Modeling

The fully 3D numerical model SSIIM solves the continuity equation together with Reynolds-averaged Navier-Stokes equations in three dimensions and within non-orthogonal coordinates to compute the water motion for turbulent flow [20].

The finite-volume approach is applied as a discretization method to transform the partial differential equations into algebraic equations. The convection term in the Navier-Stokes equation is solved using the second-order upwind scheme [15,20]. The Reynolds stress term is modeled using the standard k-ε turbulence model with constant empirical values [34].

The unknown pressure field in the Reynolds-averaged Navier-Stokes equations is calculated by employing the semi-implicit method for pressure-linked equations [35]. An implicit free-water surface algorithm was used in the computations and was based on the pressure gradient between a cell and the neighbor cells. Then, the free-water surface is computed using the local elevation difference between a cell and the neighboring cells [19].

$$\frac{\partial p}{\partial x_i} = \rho g \frac{\partial z}{\partial x_i} \tag{1}$$

where p is the pressure, x_i is the special geometrical scale, ρ is the water density, g is the acceleration due to gravity and z is the water-level elevation.

Because the water level in all cells is unknown, an iterative method was used. Additionally, because a number of neighbor cells were used to compute the water elevation differences for each cell, different values appeared for the water elevation difference depending on the number of neighboring cells used in the computation. Therefore, a weighted average of these values was applied. The weighted average coefficient for each neighboring cell (a_i) is a function of the Froude number, the flow direction and the location of neighboring cells:

$$a_i = \begin{cases} min(2 - Fr; 1.0) & for \ w > -0.1 \ and \ Fr < 2.0 \\ w^2(Fr - 1.0) & for \ w < -0.5 \ and \ Fr > 2.0 \\ 0.0 & \end{cases} \tag{2}$$

with:

$$w = \frac{\vec{r} \times \vec{u}}{\left|\vec{r}\right| \times \left|\vec{u}\right|} \tag{3}$$

where Fr is the Froude number, w is the dot product of $\vec{r}$ and $\vec{u}$, is $\vec{r}$ the direction vector pointing from the center of a cell to the center of the neighbor cell aimed to take into account the upstream/downstream effect and $\vec{u}$ is the velocity vector of the cell. This coefficient is then used for discretizing the following equation [19]:

$$\sum_{i=1}^{8} a_i z_p = \sum_{i=1}^{8} a_i \left(z_i + \frac{1}{\rho g}(p_p - p_i) \right) \tag{4}$$

where z_p is the water level elevation in the cell, z_i is the water level elevation in the ith neighbor cell, p_p is the pressure in the cell and p_i is the pressure in the ith neighbor cell.

This implicit and iterative approach is a robust and stable method that can also be used in connection with estimates of the sediment transport and morphological bed changes under unsteady flow conditions. The use of an implicit discretization scheme allows large time-step sizes to be employed in the model [23].

In the SSIIM model, the grid is non-orthogonal, unstructured and adaptive, and it moves vertically with changes in the bed and free-water surface elevation. During the computations, only the water body is modeled. The water surface is recomputed after each time step, and the employed wetting/drying algorithm enables the model to have a varying number of grid cells (e.g., in the vertical and lateral directions) with respect to the water depth in the computational domain using Equation (5) [36]. The wetting/drying algorithm causes the cells to dry up and disappear from the computational domain if the water depth is smaller than a user-defined lower boundary. If the water depth becomes greater than the lower boundary in the dry area, this algorithm regenerates a cell. This approach makes it possible to have a dynamic grid that can move in the lateral direction, allowing changes in the bed and water level to be accurately modeled.

$$n = n_{max} \left(\frac{depth}{depth_{max}} \right)^p \tag{5}$$

where n is the number of grid cells in the vertical direction, n_{max} is the maximum number of grid cells in the vertical direction, and p is a user-defined parameter for the number of grid cells.

The Dirichlet boundary condition (logarithmic velocity distribution) was used for the water inflow, whereas the zero-gradient boundary condition was specified for the water outflow and the sediment concentration calculation. For the boundary condition at the bed and walls, where there is no water flux, the empirical wall law introduced by Schlichting (1979) was utilized [20,37]. Also, bed roughness in the form of dunes and ripples is taken into account using an empirical formula introduced by Van Rijn, which employs the characteristics of sediment size distribution and bed form height within the computational domain [20,38].

3.2. Sediment Transport Modeling

The sediment transport computation for simulating the morphological changes is divided into suspended sediment and bed load transport. The suspended sediment transport is calculated by solving the transient convection-diffusion equation.

$$\frac{\partial c}{\partial t} + U_j \frac{\partial c}{\partial x_j} + w \frac{\partial c}{\partial z} = \frac{\partial}{\partial x_j} \left(\Gamma_T \frac{\partial c}{\partial x_j} \right) \tag{6}$$

where U_j is the water velocity, w is the fall velocity of the sediments, c is the sediment concentration over time t and spatial geometries (i.e., x and z), and Γ_T is the turbulent diffusivity. To compute the equilibrium suspended sediment concentration, used as the boundary condition in the cells close to the bed, an empirical formula developed by Van Rijn is used [39]:

$$C_{bed} = 0.015 \frac{d_i}{a} \frac{\left[\frac{\tau - \tau_{c,i}}{\tau_{c,i}}\right]^{1.5}}{\left(d_i \left[\frac{(\rho_s - \rho_w)g}{\rho_w v^2}\right]^{1/3}\right)^{0.3}} \tag{7}$$

where a is a reference level set equal to the roughness height, d_i is the diameter of the i-th fraction, v is the kinematic viscosity, τ is the shear stress, $\tau_{c,i}$ is the critical shear stress for d_i, which was calculated from the Shield's curve, and ρ_w and ρ_s are the density of the water and sediment, respectively.

In the SSIIM model, the bed load can be simulated using the Van Rijn formula or alternatively by the Meyer-Peter-Müller (MPM) formula [40,41]. The MPM formula is appropriate for steep rivers, which mainly transport coarse sediments close to the bed:

$$q_{b,i} = \frac{1}{g} \left[\frac{\rho_w g r I - 0.047 g (\rho_s - \rho_w) d_{50}}{0.25 \rho_w^{\frac{1}{3}} \left(\frac{\rho_s - \rho_w}{\rho_s}\right)^{\frac{2}{3}}} \right]^{\frac{3}{2}} \tag{8}$$

where $q_{b,i}$ is the sediment transport rate for the i-th fraction of the bed load per unit width, d_{50} is the characteristic sediment size (median sediment size), r is the hydraulic radius, and I is the slope of the energy line. The Van Rijn formula has been used to simulate a wide variety of sediment transport issues in both physical model and prototype scales:

$$\frac{q_{b,i}}{d_i^{1.5} \sqrt{\frac{(\rho_s - \rho_w)g}{\rho_w}}} = 0.053 \frac{\left[\frac{\tau - \tau_{c,i}}{\tau_{c,i}}\right]^{1.5}}{d_i^{0.3} \left[\left(\frac{(\rho_s - \rho_w)g}{\rho_w v^2}\right)\right]^{0.1}} \tag{9}$$

The model accounts for side slope effects by utilizing a reduction function of the critical shear stress for incipient motion by introducing the formula of Brooks together with a sand slide algorithm [20,42]. The sand slide algorithm corrects the bed slope if it exceeds a defined critical angle of repose of the sediments during excessive erosion and thereby accounts for side bank erosion. In fact, the implemented sand slide algorithm acts as a limiter when erosion continues and the bed slope increases [26].

4. Numerical Simulations

4.1. Model Setup and Calibration

The computational grid was constructed based on the bathymetric surveys before the flushing (Figure 1b). The mesh cell sizes in the streamwise and transversal directions were 10–20 m and 5–10 m, respectively. The bed material density was assumed to be 2650 kg/m^3. The water levels and inflow discharge fluctuations, which were employed as the hydrodynamic boundary conditions in the simulations, are shown in Figure 1c. In addition, a non-uniform bed material size distribution with spatially varying fractions was introduced to the model by using the seven representative sediment sizes shown in Table 1. More specifically, the computational domain was divided into a number of small segments, and each segment had its own non-uniform grain size distribution. Because the wash load, by its definition, is transported without deposition through the reservoir, its effect on the simulation of the flushing process was neglected in the computations.

For calibration purposes, a reference case was first established, assuming general values for the empirical parameters in the SSIIM model. Due to the good performance of the MPM bed load sediment transport formula for flushing simulations of Alpine reservoirs revealed by Haun et al. with conditions almost similar to those of the Dashidaira reservoir (e.g., steep slopes and a wide variety of sediment size distributions) [29], this sediment transport formula was selected for the reference case and subsequently for the sensitivity analysis and model calibration. Then, the sensitivity of the computed TVFS to changes in the empirical parameters was investigated. The computed TVFS was compared with the measured TVFS (i.e., 408.7×10^3 m^3), and if it was larger than 75% compared to the measured one, it was considered a model case for further assessment. Then, the final simulated bed topography pattern was compared, both qualitatively and quantitatively, with the measurements. A qualitative assessment was performed to check whether the erosion in area I, where the coarser sediments exist, could be captured, and a quantitative assessment was conducted to measure the deviation of the simulated bed levels from the measured ones. Therefore, Mean Absolute Error (MAE) of the simulated bed levels after the flushing operation was calculated. Table 2 summarizes the results of the TVFS sensitivity analysis for the major selected empirical parameters. Also, variations in the MAE of the simulated bed levels for each case have been calculated. The TVFS increased with increasing the active layer thickness and water content but decreased with increasing the critical angle of repose. When a larger amount of bed materials can be eroded during one time step (i.e., a thicker active layer), a higher volume of erosion from the deposits is expected. A higher water content in the sediment deposits (e.g., 50%), which decreases the submerged density of the bed material, is assumed to lead to greater sediment entrainment. A higher critical angle of repose retains a steeper side bank of the flushing channel after each time step, resulting in further deepening of the channel, which is not an efficient approach for increasing the TVFS in the numerical model. In contrast, the lateral development of the flushing channel (i.e., channel widening), which is favorable for increasing the TVFS, can be achieved using a lower critical angle of repose. The roughness, active layer thickness, water content of the bed material, and critical angle of repose were after the calibration set to 0.5 m, 1 m, 40% and 32 degrees, respectively, because the application of these values can result in a reasonable TVFS and accuracy (i.e., 313.0×10^3 m^3, and 1.8 m) and can also satisfy the mentioned qualitative criteria.

Table 2. Sensitivity analysis of the TVFS in reference case with respect to the selected empirical parameters along with variation of the MAE of simulated bed levels for each case.

Parameter	Active Layer Thickness (m)			Water Content of the Bed Material(%)			Critical Angle of Repose (Degree)		
	0.3	0.45	0.85	50	43	38	33	34	35
TVFS ($\times 10^{-3}$ m^3)	261.6	290.5	299.8	369.1	306.2	316.8	311.0	302.7	284.5
MAE (m)	2.17	1.73	1.95	2.25	2.10	1.75	1.98	1.54	1.98

4.2. Evaluation of the Flow Field and Morphological Bed Changes in the Reservoir

Figure 2 shows the computational grid adjustment at the beginning (i.e., $t = 10$ h; Figure 2(a1)) and during the free-flow conditions with a low water head in the reservoir (i.e., $t = 32$ h; Figure 2(a2)), and corresponding surface water velocity field (i.e., Figure 2(b1,b2)). The cells with a smaller water head than a specified value were removed from the computational domain due to the employed wetting/drying algorithm. The grid adjustment also reveals that the flow was deflected to the right-hand side of area II during the free-flow conditions; consequently, the flushing channel location was close to the right bank. Figure 2(b1) shows that a complex flow field with a strong reverse flow pattern and water stagnant zone develops in the lower half of the reservoir. This can be attributed to the complex geometry of the computational domain, the variation in flow depths from shallow conditions in the upstream areas to deep conditions in the downstream areas, and the existing bed roughness of the computational domain. In addition, Figure 2(b2) shows the water flow concentration in the flushing channel when the water level is lowest in the reservoir. During the free-flow condition, as shown in Figure 2(b2), the velocities rise to approximately 4.5 m/s, and supercritical flows are likely to develop in several zones of the flushing channel. However, the 3D numerical model can capture the complex characteristics of the flow field in channel bends (i.e., erosion along the outside of the bend and deposition along the inside) and reproduced the non-symmetrical velocity profile over the width and the tilting the lateral water surface at the apex of the channel bend [43]. As can be observed from Figure 2(b1,b2), the wet area of the computational domain contains the water body and the corresponding surface velocity vectors so that it is distinguished from the dry area without surface velocity vectors. An approximately 2-m-thick deposit along the left bank of the dead zone, shown in Figure 1d, can be attributed to the reverse flows that developed beside the bank of the reservoir during the drawdown stage as shown in Figure 2(b1) and that transported suspended fine materials into the stagnant water zone.

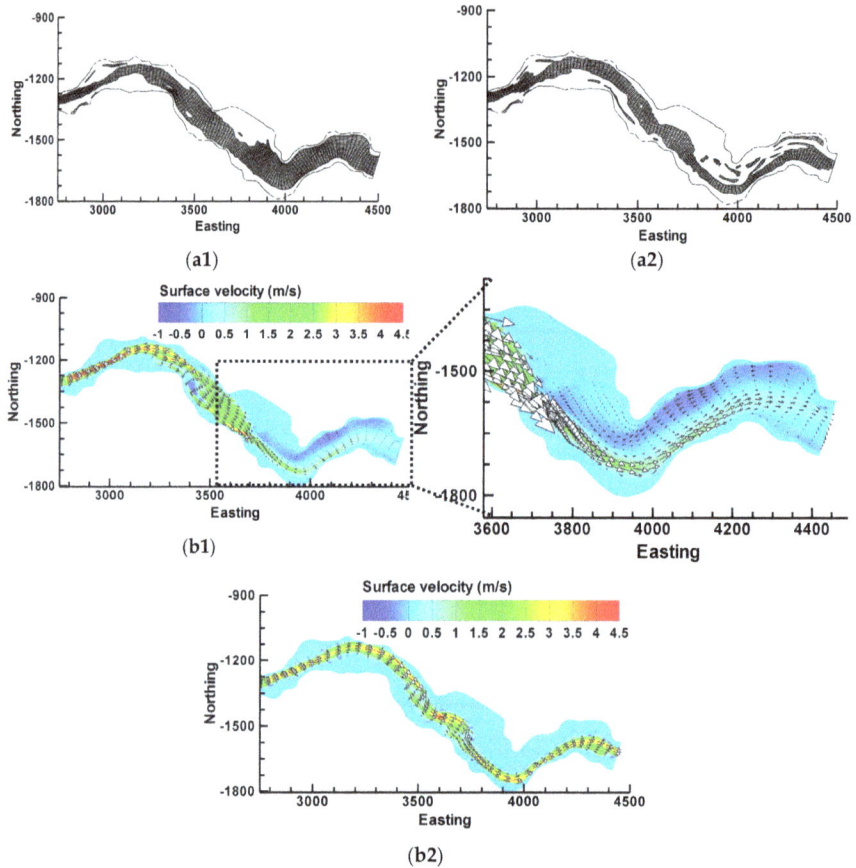

Figure 2. Computational grid (**a1**) at the beginning of the drawdown stage (t = 10 h) and (**a2**) during the free-flow condition in the Dashidaira reservoir (t = 32 h); corresponding surface velocity fields: (**b1**) at the beginning of the drawdown stage (t = 10 h) and (**b2**) during the free-flow condition (t = 32 h). The illustration on the right in (**b1**) shows the reverse flow domain and stagnant water zone.

Figure 3a illustrates the measured bed topography after the 2012 flushing operation in comparison with the simulated final bed topography after the flushing operation using the MPM and Van Rijn formulas (Figure 3b,c, respectively). To provide more quantitative insights into the simulated final bed topography after the flushing operation the BCI is defined as follows:

$$\text{BCI} = \frac{\sum\limits_{i=1}^{n} \left(z_{i\_ms} - z_{i\_reference} \right)}{n} \tag{10}$$

where z_{i_ms} is the measured or simulated bed level after the flushing operation at each grid node and $z_{i\_reference}$ is the measured or simulated reference bed level at the corresponding node, which is used for comparison purposes to provide information about erosion or deposition over a specific zone of the computational grid. Furthermore, n is the number of grid nodes considered for comparison purposes. Positive and negative values of BCI represent depositional and erosional conditions, respectively. In other words, BCI reveals the average change in the bed level of each target zone and readily indicates the dominant morphological process (i.e., erosion or deposition) in the zone compared to the reference case.

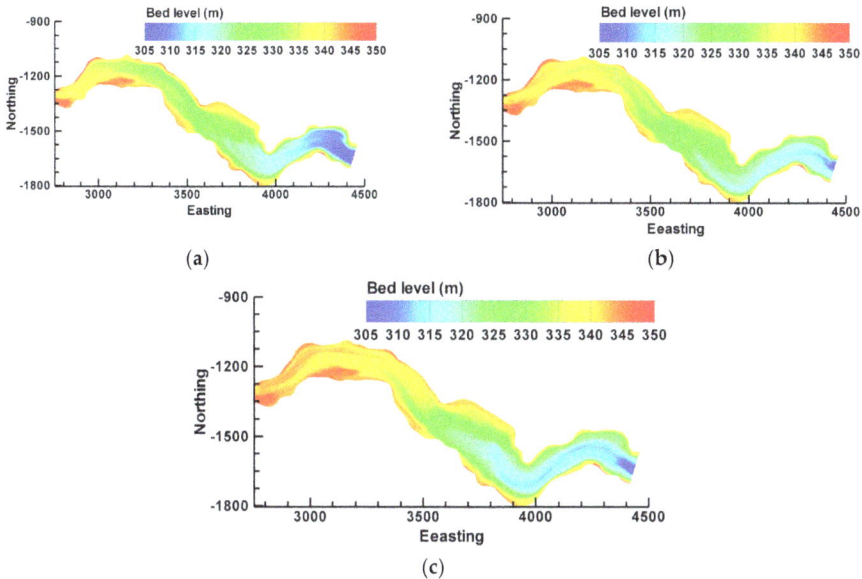

Figure 3. (a) Plan view of the measured bed topography after the 2012 flushing operation; Plan view of the simulated bed topography using (b) the MPM formula and (c) the Van Rijn formula.

BCI parameter was extracted in different areas of the reservoir using the measured bed levels after flushing and for calculations based on the MPM and Van Rijn bed load sediment transport formula when the measured bed levels before flushing (Figure 1b) are considered as reference case. In the upstream area I, the measurements reveal an average erosion value of 2.03 and simulated bed levels via MPM and Van Rijn formula in the calibrated case show average erosion value of 1.61 and 0.15 m, respectively. In this area, the performance of the MPM formula was better than Van Rijn formula in prediction of the morphological bed changes. In the wider midstream segment (i.e., area II), the measurements using the BCI parameter reveal an average erosion value of 0.67 m whereas simulated bed levels via MPM and Van Rijn formula show average erosion value of 1.59 and 3.11 m, respectively. The predicted morphological bed changes in this area obtained using the MPM formula were less overestimated than those obtained using the Van Rijn formula. In the area close to the dam (i.e., area III), the results of the simulations showed a narrower flushing channel compared to the measurements, regardless of the sediment transport formula used [43].

5. Discussion

5.1. Hydrodynamic Scenarios and Their Impacts on the Bed Morphology and Flushing Efficiency

5.1.1. Discharge Scenarios

One of the feasible scenarios in the Dashidaira reservoir is the introduction of an Additional inflow Discharge during the Free-flow stage of the flushing operation (i.e., the ADF scenario).

The additional discharge could be supplied from reservoirs located upstream of the Dashidaira reservoir. This additional inflow could enhance sediment entrainment by increasing the average flow velocity and bed shear stress in the Dashidaira reservoir. The ADF scenario is currently applicable, and preliminary tests for this scenario are being implemented in the Dashidaira reservoir.

Figure 4a shows the additional inflow that could be supplied in different ADF scenarios, and Figure 4b shows the discharge rates and original water level (i.e., Q and h, respectively) together with modified discharge rates under different ADF scenarios in the Dashidaira reservoir. Under the different ADF scenarios, the original boundary conditions of the water level (i.e., recorded during the 2012 flushing operation) have been retained, but the original boundary conditions of the discharge during the free-flow stage of the 2012 flushing operation change depending on which ADF scenario is used. For instance, ADF 110 indicates that 110 m^3/s of additional inflow discharge has been added to the original discharges during the free-flow stage of the 2012 flushing operation. With an original TVFS value equals to 313.0×10^3 m^3 in the reference case (i.e., resulting bed topography after simulation of 2012 flushing), the TVFS increased to 356.0×10^3 in the ADF 75 scenario, 396.1×10^3 in the ADF 110 scenario, and 425.0×10^3 m^3 in the ADF 170 scenario. The effects of introducing constant additional discharges under various ADF scenarios on the FE and the TVFS are illustrated in Figure 4c. The horizontal axis shows the ratio of average discharge during the free-flow stage using different ADF scenarios (i.e., Q_2) to the average discharge during the free-flow stage when no additional discharge is introduced in the reference case (i.e., Q_1). FE$_2$ and FE$_1$ are the flushing efficiencies when an ADF scenario is employed and when no additional discharge is employed in the reference case, respectively. The TVFS increases when the discharge increases during the free-flow stage. The FE values reached approximately −6.5% when the ADF 60 scenario was used. In this case, the increase in the flushed sediment volume was smaller than the increase in the used water volume according to the FE definition. Under the ADF 75, 90, and 110 scenarios, both the FE and TVFS increased with increasing average discharge during the free-flow stage. Then, the FE variation trended downward until a stable level was reached for the ADF 150 and ADF 170 scenarios. In contrast, the TVFS continued to increase. Under the given conditions, increasing the discharge magnitude during the free-flow condition can increase the TVFS, but this increase is not proportional to the discharge increase that causes the decrease in the FE for some cases. According to the diagram shown in Figure 4c, when the average discharge during the free-flow stage increased by approximately 56% under the ADF 110 scenario, (i.e., $Q_2/Q_1 = 1.56$), the FE increased approximately by 5%. Under these conditions, the total used water volume for the flushing operation increased by approximately 21%. In Table 3 the average bed level changes compared to the reference case using the BCI parameter in the upstream, midstream and downstream areas have been revealed for the ADF 75, ADF 110, and ADF 170 scenarios. As shown in Figure 4c and Table 3, introducing additional discharge increases the erosion in all areas by between 3% and 36% depending on the additional discharge, but the effect in the areas close to the dam (i.e., areas II and III) is more pronounced. Instead of adding a constant discharge to the original discharge values during the free-flow condition in the reference case, another scenario (i.e., the PDF scenario) using the same additional water volume over a shorter duration and in the form of discharge pulses was tested. The PDF scenario was introduced to determine whether changing the characteristics of the additional inflowing water (e.g., the discharge intensity) markedly affects the quantity of flushed sediments and the bed changes in specific zones of the reservoir. The concept of the PDF scenario has been illustrated schematically in Figure 4d. It should be noted that before the flushing operation, area III was mainly covered with fine materials. During the simulation of the flushing operation, eroded coarser materials from area I were deposited in the lower parts of area III due to the reduced bed shear stress. Thus, introducing an additional discharge during the free-flow condition can contribute to flushing the deposited sediments out of this area.

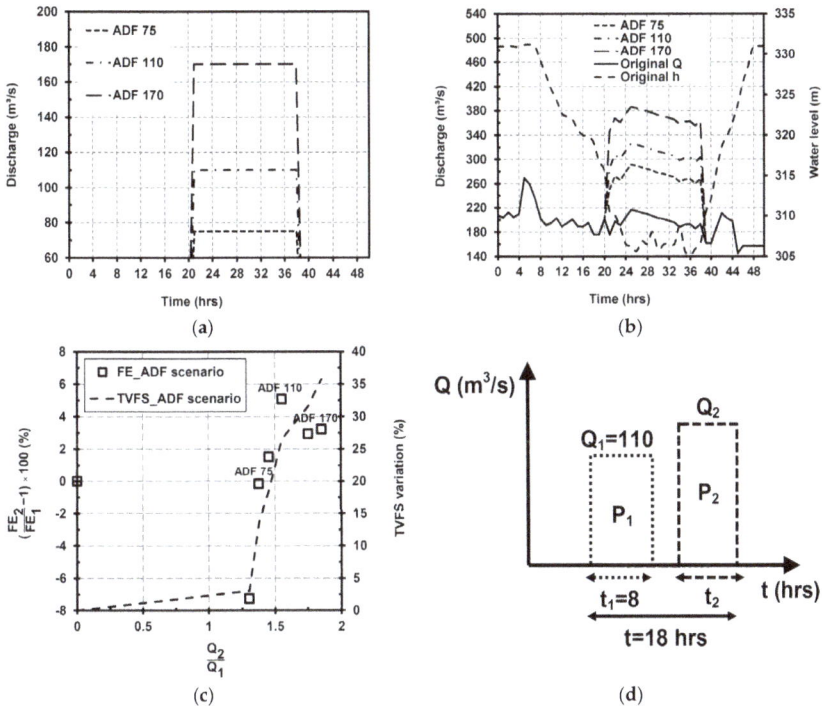

Figure 4. (**a**) Constant additional discharge rates used for different ADF scenarios in the Dashidaira reservoir. These additional rates are added to the original discharge rates shown in Figure 1c; (**b**) Modified hydrodynamic boundary conditions for different ADF scenarios in the Dashidaira reservoir; (**c**) Non-dimensional curves showing the relationships among TVFS, FE, and the water discharge used under different ADF scenarios compared to the reference case; (**d**) Schematic figure illustrating the PDF scenario with a variable discharge pulse in the second half of the free-flow stage.

Because the best FE correlation was found for the ADF 110 scenario as shown in Figure 4c, the total volume of the additional water used during the 18 h of the free-flow stage was calculated, and this additional water volume was introduced into the model in the form of two discharge pulses, the first one in the first half of the free-flow stage (i.e., P1) and the second one (i.e., P2) in the second half. More specifically, in the PDF scenario, a constant 110 m^3/s discharge pulse within 8 h in the first half of the free-flow flushing (i.e., P1 110 8) and a second discharge pulse with variable magnitude and duration in the second half (i.e., P2 Q$_2$ t$_2$) were introduced for further assessments.

In Figure 5, the bed changes with and without the introduction of the additional discharge (i.e., the reference case, ADF 110, and ADF 170) have been plotted at the cross-sections A-A, E-E, F-F, H-H, K-K, and L-L. As can be observed from Figure 5(f1,f2), instead of erosion, deposition occurs in cross section L-L. At the end of free-flow stage during the flushing process, bottom outlets are closed and water level starts to increase beside the dam while it is still low in the upstream portions of the reservoir and free-flow condition exists. In such a condition, the induced bed shear stress is reduced noticeably in areas close to the dam. Therefore, the eroded large size sediments are still transported towards downstream but they are deposited close to the dam without chance to flush. Table 3 reveals that the introduction of an additional discharge during the free-flow condition, in ADF scenarios, has a marginal effect on the erosion of coarser sediments in area I. This pattern can be attributed to the major role of the water-level drawdown stage in the initial development and evolution of the flushing channel, as revealed by experimental and numerical model studies by Esmaeili et al. and Kantoush and Schleiss [30,44]. Notably, from the reservoir entrance up to the cross-section A-A, the BCI values showed 0.04, 0.06, and 0.10 m of erosion under ADF 75, 110, and 170 scenarios compared to the simulated bed levels after 2012 flushing operation. Because the BCI values indicate overall deposition in area I, the eroded coarser bed materials from the area upstream of section A-A are likely deposited again after passing the cross-section A-A as a result of the increasing flow depth and the consequent velocity reduction. This process prevents the smaller grain size bed materials underneath from being eroded. However, if the deposition of coarser materials occurs in the main flushing channel, the erosion will increase, even with a smaller additional discharge (e.g., in the ADF 75 scenario) due to an increase in the bed shear stresses in the channel. In addition, the inflowing discharge during the free-flow condition is concentrated in the central flushing channel, which mainly contributes to a slow widening and deepening of the existing channel. This widening and deepening process is more effective in areas covered by finer sediments (e.g., areas II and III). Thus, use of the 170 m^3/s of additional inflow during the free-flow stage may result in a further increase in the flushing channel width and depth in the lower part of area II and throughout area III.

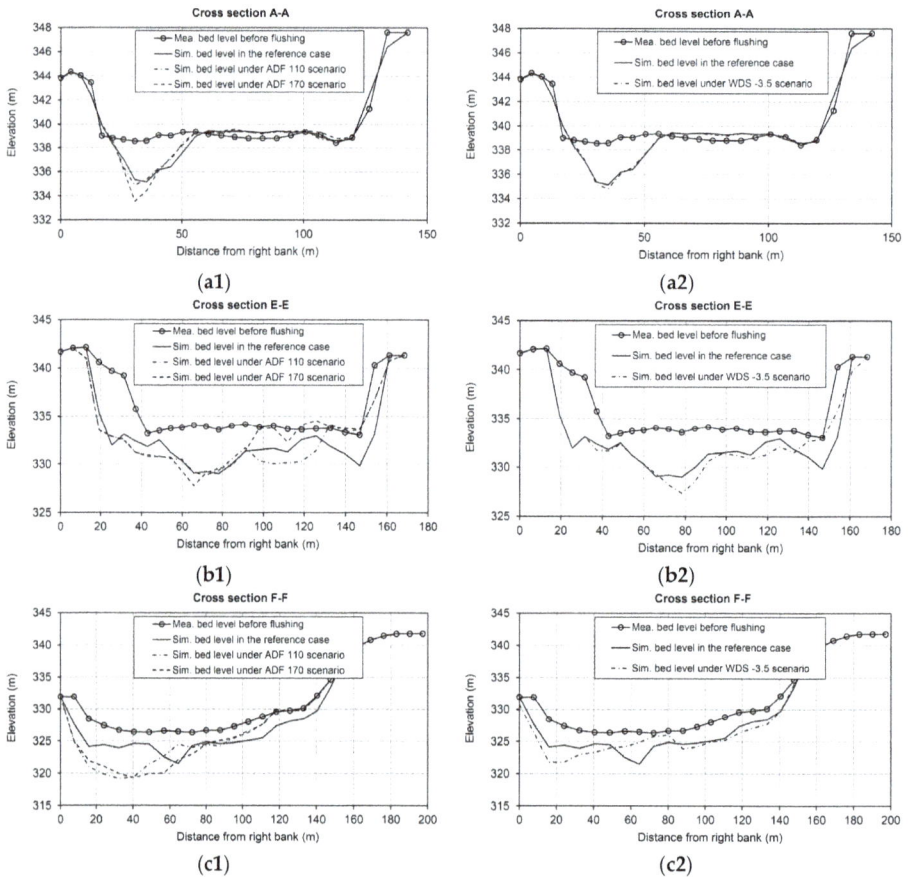

Figure 5. *Cont.*

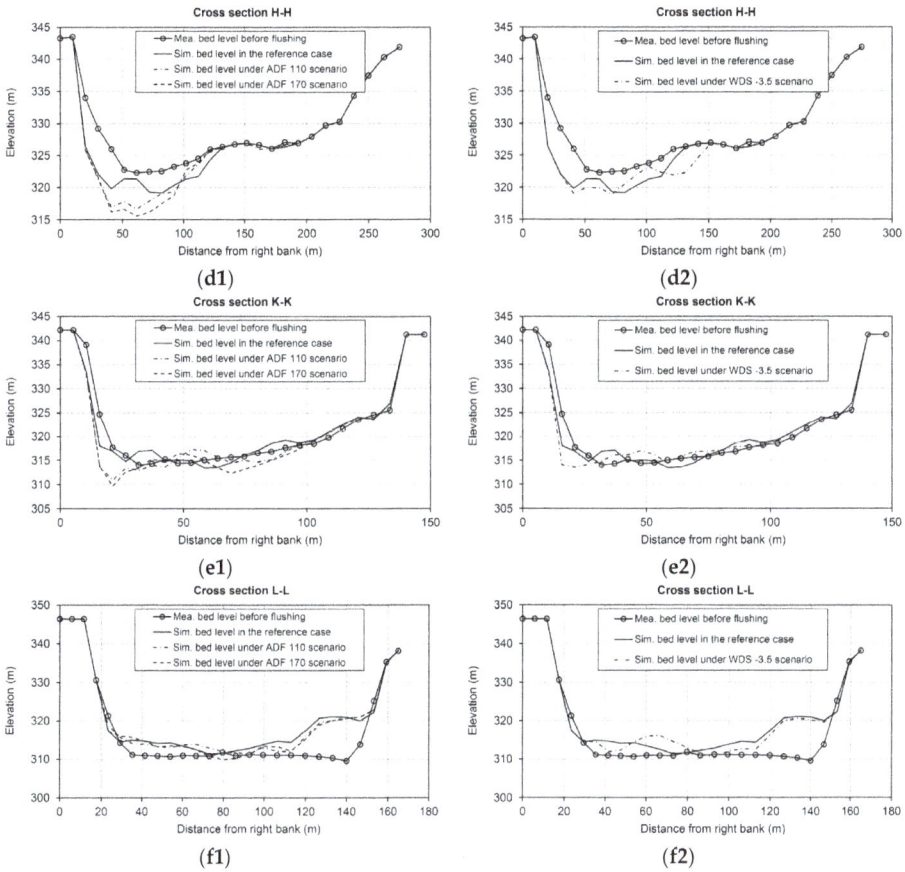

Figure 5. Measured bed levels before flushing along with the simulated bed levels after flushing in the reference case and under the ADF 110, ADF 170 and WDS −3.5 scenarios at (**a1,a2**) cross section A-A; (**b1,b2**) cross section E-E; (**c1,c2**) cross section F-F; (**d1,d2**) cross section H-H; (**e1,e2**) cross section K-K; and (**f1,f2**) cross-section L-L. Locations of the cross sections can be found in Figure 1b. Mea. and Sim. are abbreviations for Simulated and Measured, respectively.

Also, the resulting BCI parameter for the PDF scenario has been shown in Table 3. Using the PDF scenario not only enhanced the TVFS by increasing the erosion from areas II and III but also increased the erosion of coarser material from area I when the second discharge pulse was sufficiently high (i.e., PDF P1 110 8-P2 183.5 6). These increases can be attributed to the transport of already eroded coarser materials from area I to farther downstream areas due to a higher induced bed shear stress. Increasing the TVFS under the PDF scenarios, can increase the FE about 13% compared to the reference case. Nevertheless, due to the further deposition of sediments eroded from the head of the reservoir in the dam vicinity, the bed level changes appear to be marginal compared to the ADF 110 scenario when the second discharge pulse is high. Figure 6 shows the bed changes in different cross-sections in area I under the PDF P1 110 8-P2 183.5 6 scenario compared to the bed changes using the ADF 110 scenario to quantitatively show the advantageous performance of the PDF scenario in eroding the coarser bed materials in area I. However, the use of ADF or PDF scenario does not affect the main flushing channel location close to the right bank in area II. Consequently, the deposited bed materials close to the left bank in area II (i.e., the dead zone) still cannot be effectively removed.

Table 3. Average bed level changes in different areas of the reservoir under different ADF, PDF, and WDS scenarios. For ADF and WDS scenarios, the simulated bed levels after 2012 flushing operation were used as the reference case to extract the BCI parameter, whereas for PDF scenario the simulated bed levels under the ADF 110 scenario were used as the reference case.

Scenario	ADF 75			ADF 110			ADF 170		
Area	I	II	III	I	II	III	I	II	III
BCI (m)	0.32	−0.47	−0.54	0.06	−0.49	−0.55	0.20	−0.76	−0.90
TVFS ($\times 10^{-3}$ m^3)	356.0			396.1			425.0		
Scenario	PDF P1 110 8-P2 137.5 8			PDF P1 110 8-P2 157 7			PDF P1 110 8-P2 183.5 6		
Area	I	II	III	I	II	III	I	II	III
BCI (m)	−0.01	−0.09	−0.21	0.02	−0.04	−0.20	−0.09	0.01	0.04
TVFS ($\times 10^{-3}$ m^3)	426.2			417.3			410.9		
Scenario	WDS −0.5			WDS −2.5			WDS −3.5		
Area	I	II	III	I	II	III	I	II	III
BCI (m)	0.03	−0.14	0.02	0.10	−0.13	−0.23	−0.08	−0.33	−0.22
TVFS ($\times 10^{-3}$ m^3)	322.8			331.9			378.9		

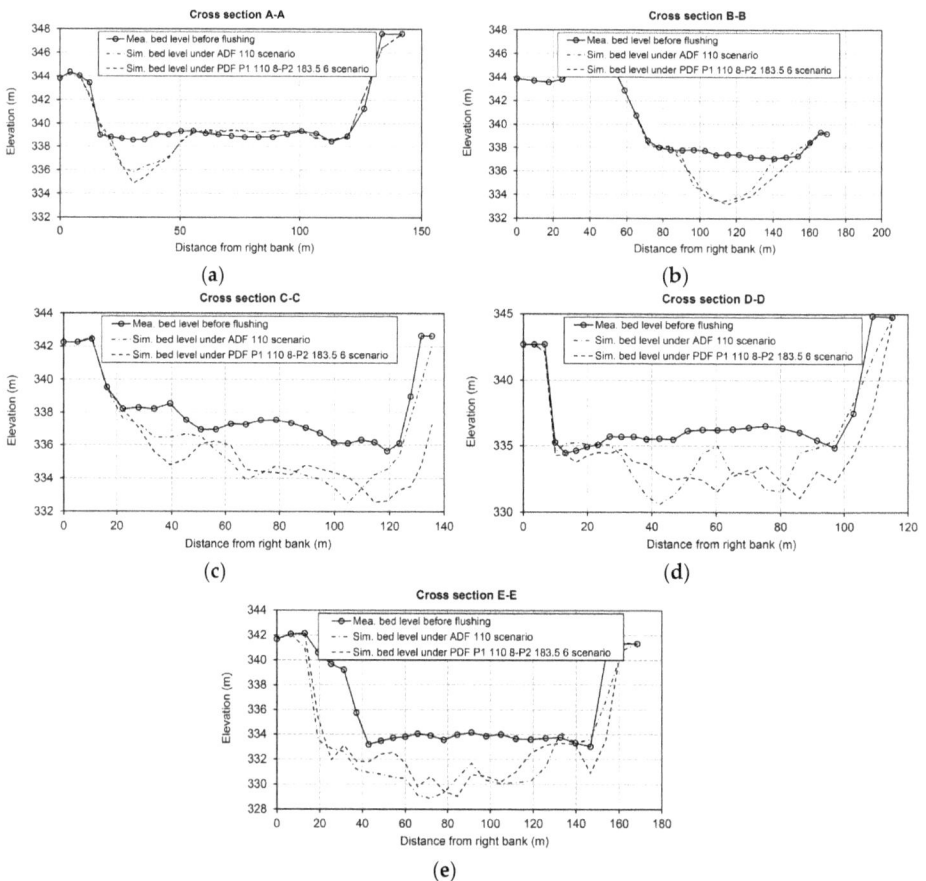

Figure 6. Measured bed levels before flushing and the simulated bed levels after flushing under the ADF 110 and PDF P1 110 8-P2 183.5 6 scenarios at (**a**) cross section A-A; (**b**) cross section B-B; (**c**) cross section C-C; (**d**) cross section D-D; and (**e**) cross section E-E. Locations of the sections can be found in Figure 1b.

5.1.2. Water Level Scenarios

Dominant role of the water-level drawdown stage in scouring the flushing channel has been already highlighted in physical model studies [30,44]. Therefore, another feasible scenario for the Dashidaira reservoir is increasing the Water-level Drawdown Speed (i.e., the WDS scenario). A target limb during the second half of the drawdown stage (i.e., between t = 12 and 20 h after starting the flushing) in the water-level variation curve is selected. Then, the original water-level drawdown rate is increased significantly for 1 hour at the beginning of the target limb (i.e., between t = 12 and 13 h after starting the flushing). Afterwards, the original drawdown rates remain unchanged during the remaining time (i.e., from t = 13 until t = 20 h). As a result, introducing an extra drop in the water level (e.g., 0.5, 2.5, and 3.5 m) for one hour while the original discharge rates remain unchanged causes the water velocity to increase abruptly, which acts as a strong driving force over the bed materials. For example, the WDS -0.5 scenario represents an extra 0.5-m drop in the water level at the beginning of the relevant limb (i.e., between t = 12 and 20 h). This scenario can be created via gate operations during a flushing event (e.g., a temporary increase in the gate opening speed) or with modification of the bottom outlet geometry to increase the discharge capacity of the bottom outlets, which can contribute to a faster drawdown process. Figure 7a illustrates the original water level and discharge rates, together with water-level modifications, during the target limb for three different WDS scenarios in the Dashidaira reservoir.

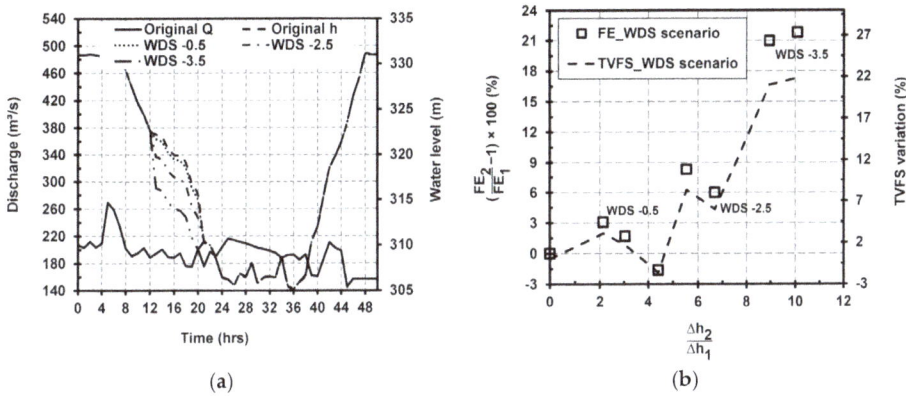

Figure 7. (a) Utilized water levels and discharge rates for different WDS scenarios; (b) Non-dimensional curves showing the relationship among TVFS, FE, and Δh under different WDS scenarios compared to the reference case.

The effects of introducing a faster drawdown of the water level on the FE and the TVFS under various WDS scenarios are shown in Figure 7b. The horizontal axis shows the ratio of the extra imposed water level drop (i.e., Δh_2) to the original one (i.e., Δh_1) at the beginning of the target limb during the drawdown stage. On the left vertical axis, FE_2 represents the flushing efficiency when a WDS scenario is employed, and FE_1 represents the flushing efficiency when the original water level of the 2012 flushing operation is applied. All calculated and presented values are relative to the reference case (i.e., the 2012 flushing operation). As shown in Figure 7b, the FE variations are overall directly related to the variations in $\Delta h_2/\Delta h_1$. However, removing the coarser material from the far upstream area of the reservoir requires a high extra drop in the water level (i.e., a high $\Delta h_2/\Delta h_1$). Moreover, in some cases, increasing $\Delta h_2/\Delta h_1$ results in lower TVFS and FE values because coarser eroded material moves from upstream areas to downstream areas and is deposited on finer materials in the deeper areas. If the driving forces produced by the extra drop in the water level are not strong enough to remobilize the newly deposited coarser sediments overlying finer sediments, the erosion of finer fractions may be lower, resulting in lower TVFS values and consequently lower FE_2 values. Table 3 also shows the average BCI values in the upstream, midstream and downstream areas of the Dashidaira reservoir after the application of the WDS −0.5, WDS −2.5, and WDS −3.5 scenarios. As shown in Table 3, compared to the reference case with the TVFS of 313.0×10^3 m^3, the TVFS increases slightly when the magnitude of the extra drop in the water level is small (e.g., 0.5 m in the WDS −0.5 scenario). When the extra drop in the water level is larger (e.g., 2.5 m), the TVFS increases but not remarkably. Although the water-level decrease can enhance the relative roughness (i.e., the ratio of the roughness height to the water depth), this increased roughness is not high enough to lead to the erosion of the coarser materials in the upstream areas. Thus, the main effect is limited to the finer materials in the WDS −0.5 and WDS −2.5 scenarios. Due to the larger extra water level drop in the WDS −3.5 scenario, higher bed erosion occurs over the entire reservoir, including the upstream areas covered with coarser materials. In Figure 5, the bed changes in different cross-sections located in areas I, II, and III under the WDS −3.5 scenario are compared to the bed levels in the reference simulations.

Although a marked drop in the water level enhances the sediment erosion from the entire reservoir area and increases the FE more than 20% as shown in Figure 7b, the flushing channel location remains close to the right bank in area II, and the deposits along the left bank (i.e., the dead zone) are not effectively removed. However, constraints imposed by the gate facilities for safe and quick opening, and existing risks regarding the side bank failures are necessary to be assessed carefully.

5.2. Auxiliary Channel Scenario

According to the literature, construction of longitudinal channels in the wide areas of the reservoir has been proposed for affecting the scouring pattern [4]. Thus, a potential scenario for improving the erosion in the dead zone area of the Dashidaira reservoir is the application of a longitudinal auxiliary (i.e., secondary) channel in the dead zone to act as a side channel beside the main channel. In this scenario, the flushing flow is partly diverted from the main channel into the auxiliary channel and enters the main channel again at a confluence further downstream of the diversion point. Thus, the morphological processes occur along two flushing channels in area II. In contrast to the discharge and water-level scenarios, which affect the hydrodynamic characteristics within the entire reservoir during the flushing operation, auxiliary channel scenario influences the hydrodynamic characteristics in a specific segment of the reservoir. Thus, the auxiliary channel scenario may be more useful for resolving the issue of ineffective sediment scouring within the dead zone area in the reservoir.

The presence of two flushing channels in area II is better than just one in terms of several aspects. First, erosion occurs in two flushing channels instead of one, which means that flushing occurs over a longer channel (i.e., larger area). Consequently, higher TVFS and FE values are expected. When a longitudinal auxiliary channel is not used, the flushing channel merely develops along the river thalweg (i.e., pre-existing flushing channel). The flushing channel attracts the entire flow, and the channel mainly deepens, which is not an effective approach for increasing the TVFS and FE. Second, a remarkable amount of previously unerodible sediments can be eroded from the dead zone following the construction of a longitudinal auxiliary channel.

The schematic illustration showing the location of the applied auxiliary channel and also the surface velocity vectors in different stages of the flushing using an auxiliary channel are shown in Figure 8. The location of the entrance of auxiliary channel was selected in a place where a portion of the flushing flow can be easily diverted into the channel (Figure 8a). In addition, in the numerical model, side banks of the auxiliary channel were implemented along the existing grid lines in the selected area of the dead zone. The depth and length of the channel were set to be 2.5 and 403 m, respectively. As shown in Figure 8b, the auxiliary channel deflects a portion of the water flow towards the dead zone, and the remaining flow continues its original path close to the right bank of area II along the original thalweg of the main flushing channel. At the beginning of the drawdown stage (i.e., t = 10 h) in Figure 8c, the incoming flow is mainly deflected towards the auxiliary channel. At the middle of the drawdown stage (i.e., t = 16 h), the flow bifurcates completely, with a major portion of the incoming flow being deflected towards the main channel and a smaller portion being deflected towards the auxiliary channel, as shown in Figure 8d. After erosion of the existing sandbar, between the main and auxiliary channel, and the consequent water exchange between the channels due to the small water level fluctuations during the free-flow stage, the water flow is distributed over a wider flushing channel. These conditions continue during the free-flow stage (i.e., t = 32 h), as shown in Figure 8e. More specifically, using a longitudinal auxiliary channel along the dead zone develops another thalweg away from the main flushing channel thalweg by attracting a portion of the incoming flow and preventing the entire flow from heading towards the right bank of area II.

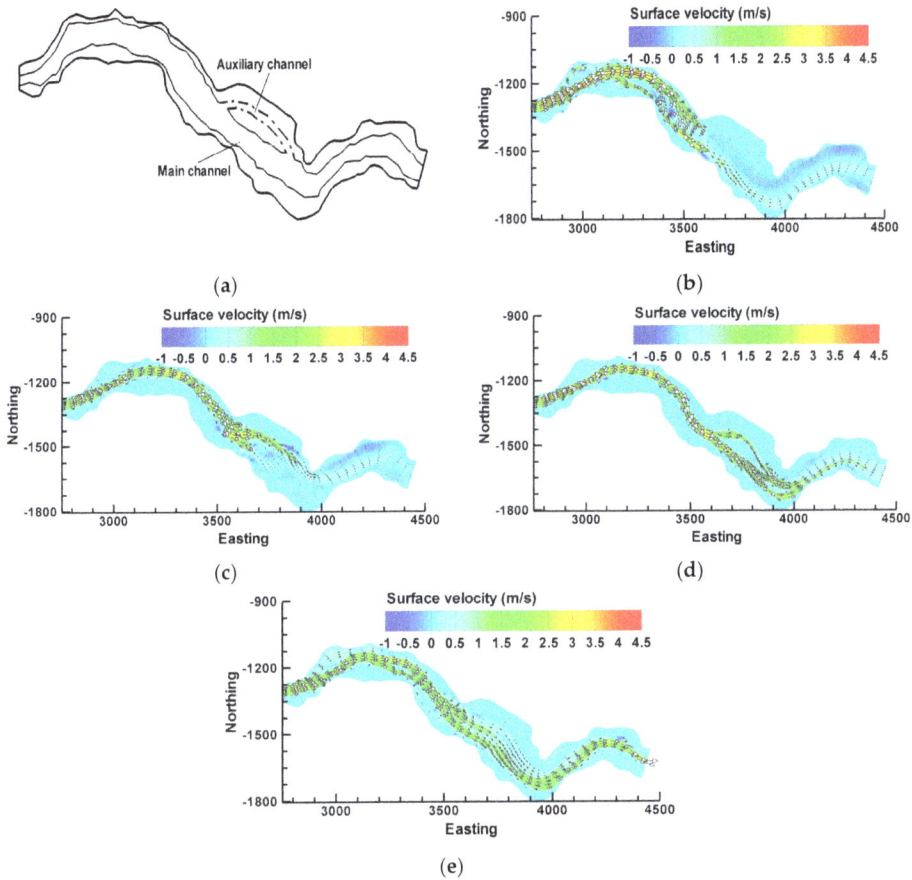

(a)

(b)

(c)

(d)

(e)

Figure 8. (**a**) Schematic illustration of the concept of longitudinal auxiliary flushing channel in the dead zone area of the Dashidaira reservoir; The surface velocity field in different stages using an auxiliary flushing channel: (**b**) before onset of the drawdown stage (i.e., t = 5 h); (**c**) at the beginning of the drawdown stage (i.e., t = 10 h); (**d**) at the middle of the drawdown stage (i.e., t = 16 h); and (**e**) during the free-flow condition (i.e., t = 32 h).

Figure 9a illustrates the plan view of the final bed morphology in the scenario with an auxiliary longitudinal channel in the dead zone of area II. Utilizing this scenario increases the TVFS to 340.0×10^3 m^3, compared to the 313.0×10^3 m^3 of the reference case. In Figure 9b–d, the final bed levels in different cross-sections of area II are shown after the flushing operation and are compared to the bed levels in the reference case. One can clearly see that the bed levels decreased noticeably along the thalweg of the auxiliary flushing channel and meanwhile the channel widened. Both deepening and widening process of the auxiliary channel contributes in increasing the TVFS and subsequently the flushing efficiency.

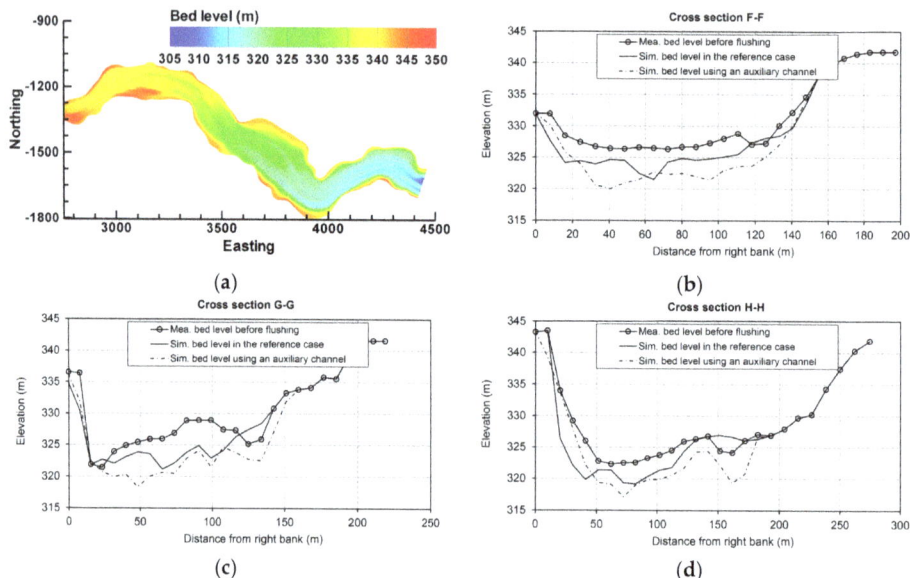

Figure 9. (**a**) Plan view of the final simulated bed levels using an auxiliary flushing channel along the dead zone of area II. Measured bed levels before flushing and simulated ones after flushing in the reference case along with simulated ones using an auxiliary flushing channel, at the location of (**b**) cross section F-F; (**c**) cross section G-G; (**d**) cross section H-H. Location of the cross sections can be found in Figure 1b.

6. Conclusions

The following results were obtained from the present work:

- Both the MPM and Van Rijn formulas yielded satisfactory performances in the simulation of bed changes in specific segments of the reservoir during the flushing operation (e.g., MPM formula in the upper half of the reservoir and Van Rijn formula in the vicinity of the Dam). These sediment transport formulas have been developed empirically to calculate the sediment transport for a given set of sediment sizes and hydrodynamic boundary conditions. However, the bed sediment size distribution, bed roughness, and hydrodynamic boundary conditions change dynamically during the free-flow flushing process. Such significant changes cannot be handled by empirical sediment transport formulas due to their inherent limitations. Nevertheless, the MPM bed load sediment transport formula qualitatively and quantitatively performed better than the Van Rijn formula for the entire reservoir. The MPM formula was able to achieve TVFS values that were more than 75% that of the measured TVFS values. Due to the application of the empirical formulas, the alluvial roughness also could not be estimated appropriately, which further magnifies the mentioned inability of the sediment transport formulas to accurately represent the morphological bed changes.
- For the Dashidaira reservoir, introducing an artificial additional discharge during the free-flow stage is practically feasible since this discharge can be supplied from upstream reservoirs. In addition, because this additional discharge is introduced when the flushing gates are fully

opened and the water level is low, this discharge can be passed through the bottom outlets if its value is less than the maximum capacity of the outlets. Additional discharge has two major effects: first, it increases the induced bed shear stress and bed erosion and supplies an additional driving force to transport eroded sediments farther downstream in the reservoir and flush them out from the reservoir; second, it causes the water level to increase in the downstream river channel, which can be beneficial from an environmental point of view because it washes away fine materials from the downstream channel terraces (thereby preventing river channel clogging). However, it was found that introducing the extra discharge in the form of two discharge pulses with a larger discharge pulse in the second half of the free-flow stage more efficiently increases the FE ratio and the bed degradation in the upstream areas covered with coarser materials. The numerical outcomes showed that introducing approximately 21% more water from upstream reservoirs (i.e., an approximately 56% increase in the average free-flow discharge) can enhance the FE by approximately 5−13% compared to the reference case (i.e., the 2012 flushing operation), depending on how this additional discharge is delivered.

- The construction of an auxiliary longitudinal flushing channel in the dead zone area of the Dashidaira reservoir causes a portion of the flushing flow to deviate from the main channel into the auxiliary channel and to enter the main channel again via a confluence downstream of the diversion point. The non-diverted flow continues along its original path along the thalweg of the main flushing channel and the diverted flow towards the auxiliary channel scour the deposited sediments from the targeted dead zone in the reservoir. The flushing processes associated with the auxiliary longitudinal channel result in a flushing channel that is overall longer and wider. Hence, the FE is higher by as much as approximately 9% compared to the reference case.

Acknowledgments: The first author was funded by the Ministry of Education, Culture, Sports, Science, and Technology of Japan (MEXT) through the HSE program for Asian Megacities. The support of the Disaster Prevention Research Institute (DPRI) of Kyoto University for open access publication is acknowledged. Authors would like to give their thanks to Kansai Electric Power Co., Inc., NEWJEC Inc., and also Shuhei Minami for providing the field data used in this study framework.

Author Contributions: Taymaz Esmaeili analyzed and processed the data, performed the numerical model tests and prepared the first and final draft of this manuscript. Tetsuya Sumi, Sameh A. Kantoush, Stefan Haun, and Nils Rüther provided their theoretical and practical insights and expertise on the outcomes and also contributed to preparation of the manuscript with reading and corrections. Yoji Kubota contributed to preparation and interpretation of a part of input and output data in the numerical model development stage.

Conflicts of Interest: The authors declare no conflict of interest. The funding sponsors had no role in the design of the study; in the collection, analyses, or interpretation of data; in the writing of the manuscript, and in the decision to publish the results.

References

1. Kondolf, G.M.; Gao, Y.; Annandale, G.W.; Morris, G.L.; Jiang, E.; Zhang, J.; Cao, Y.; Carling, P.; Fu, K.; Guo, Q.; et al. Sustainable sediment management in reservoirs and regulated rivers: Experiences from five continents. *Earth's Future* **2014**, *2*, 256–280. [CrossRef]
2. Morris, G.L. Chapter 5: Sediment management and sustainable use of reservoirs. In *Modern Water Resources Engineering*; Wang, L.K., Yang, C.T., Eds.; Humana Press: Totowa, NJ, USA, 2014; Volume 15, pp. 279–337.
3. White, R. *Evacuation of Sediments from Reservoir*; Thomas Telford: London, UK, 2001.
4. Morris, G.L.; Fan, J. *Reservoir Sedimentation Handbook: Design and Management of Dams, Reservoirs and Watersheds for Sustainable Use*; McGraw-Hill: New York, NY, USA, 1998.
5. Shen, H.W. Flushing sediment through reservoirs. *J. Hydraul. Res.* **1999**, *37*, 743–757. [CrossRef]
6. Sumi, T.; Kantoush, S.A. Integrated management of reservoir sediment routing by flushing, replenishing, and bypassing sediments in Japanese river basins. In Proceedings of the 8th International Symposium on Ecohydraulics, Seoul, Korea, 12–16 September 2010.
7. Liu, J.; Minami, S.; Otsuki, H.; Liu, B.; Ashida, K. Prediction of concerted sediment flushing. *J. Hydraul. Eng.* **2004**, *130*, 1089–1096. [CrossRef]

8. Kantoush, S.A. Experimental Study on the Influence of the Geometry of Shallow Reservoirs on Flow Patterns and Sedimentation by Suspended Sediments. Ph.D. Thesis, École Polytechnique Fédérale de Lausanne, Lausanne, Switzerland, 2008.

9. Sloff, C.J.; Jagers, H.R.A.; Kitamura, Y. Study on the channel development in a wide reservoir. In Proceedings of the 2nd River Flow conference, Napoli, Italy, 23–25 June 2004.

10. Fukuoka, S.; Sumi, T.; Horiuchi, S. Sediment management on the arase dam removal project. In Proceedings of the 12th International Symposium on River Sedimentation, Kyoto, Japan, 2–5 September 2013.

11. Asahi, K.; Shimizu, Y.; Nelson, J.; Parker, G. Numerical simulation of river meandering with self-evolving banks. *J. Geophysic. Res.* **2013**, *118*, 2208–2229. [CrossRef]

12. Fukuoka, S.; Uchida, T. Toward integrated multi-scale simulations of flow and sediment transport in rivers. *J. Jpn. Soc. Civ. Eng.* **2013**, *69*, II_1–II_10. [CrossRef]

13. Fang, H.; Rodi, W. Three-dimensional calculation of flow and suspended sediment transport in the neighbourhoods of the dam for the Three Gorges Project (TGP) reservoir in the Yangtze River. *J. Hydraul. Res.* **2003**, *41*, 379–394. [CrossRef]

14. Lu, Y.J.; Wang, Z.Y. 3D numerical simulation for water flows and sediment deposition in dam areas of the Three Gorges project. *J. Hydraul. Eng.* **2009**, *135*, 755–769. [CrossRef]

15. Haun, S.; Olsen, N.R.B. Three-dimensional numerical modelling of the flushing process of the Kali Gandaki hydropower reservoir. *Lakes Reserv.* **2012**, *17*, 25–33. [CrossRef]

16. Khosronejad, A.; Rennie, C.D.; Salehi Neyshabouri, A.A.; Gholam, I. Three-dimensional numerical modeling of reservoir sediment release. *J. Hydraul. Res.* **2008**, *46*, 209–223. [CrossRef]

17. Harb, G.; Dorfmann, C.; Schneider, J.; Badura, H. Numerical analysis of sediment transport processes during a flushing event of an Alpine reservoir. In Proceedings of the 7th River Flow conference, Lausanne, Switzerland, 3–5 September 2014.

18. Gallerano, F.; Cannata, G. Compatibility of Reservoir Sediment Flushing and River Protection. *J. Hydraul. Eng.* **2011**, *137*, 1111–1125. [CrossRef]

19. Haun, S.; Olsen, N.R.B. Three-dimensional numerical modelling of reservoir flushing in a prototype scale. *Int. J. River Basin Manag.* **2012**, *10*, 341–349. [CrossRef]

20. Olsen, N.R.B. *A Three Dimensional Numerical Model for Simulation of Sediment Movement in Water Intakes with Multiblock Option*; Department of Hydraulic and Environmental Engineering, The Norwegian University of Science and Technology: Trondheim, Norway, 2014.

21. Esmaeili, T.; Sumi, T.; Kantoush, S.A.; Haun, S.; Rüther, N. Three-dimensional numerical modelling of flow field in shallow reservoirs. *Proc. ICE-Water Manag.* **2015**, *169*, 229–244. [CrossRef]

22. Dehghani, A.A.; Esmaeili, T.; Chang, W.Y.; Dehghani, N. 3D numerical simulation of local scouring under hydrographs. *Proc. ICE-Water Manag.* **2013**, *166*, 120–131. [CrossRef]

23. Haun, S.; Kjærås, H.; Løvfall, S.; Olsen, N.R.B. Three-dimensional measurements and numerical modelling of suspended sediments in a hydropower reservoir. *J. Hydrol.* **2013**, *479*, 180–188. [CrossRef]

24. Esmaeili, T.; Dehghani, A.A.; Zahiri, A.R.; Suzuki, K. 3D Numerical simulation of scouring around bridge piers (Case Study: Bridge 524 crosses the Tanana River). *World Acad. Sci. Eng. Technol.* **2009**, *58*, 1028–1032.

25. Fischer-Antze, T.; Olsen, N.R.B.; Gutknecht, D. Three-dimensional CFD modeling of morphological bed changes in the Danube River. *Water Resour. Res.* **2008**, *44*. [CrossRef]

26. Rüther, N.; Olsen, N.R.B. Modelling free-forming meander evolution in a laboratory channel using three-dimensional computational fluid dynamics. *Geomorphology* **2007**, *89*, 308–319. [CrossRef]

27. Ruether, N.; Singh, J.M.; Olsen, N.R.B.; Atkinson, E. 3-D computation of sediment transport at water intakes. *Proc. ICE-Water Manag.* **2005**, *158*, 1–7. [CrossRef]

28. Olsen, N.R.B.; Kjellesvig, H.M. Three-dimensional numerical modelling of bed changes in a sand trap. *J. Hydraul. Res.* **1999**, *37*, 189–198. [CrossRef]

29. Haun, S.; Dorfmann, C.; Harb, G.; Olsen, N.R.B. Numerical modelling of the reservoir flushing of the bodendorf reservoir, Austria. In Proceedings of the 2nd IAHR European Congress, Munich, Germany, 27–29 June 2012.

30. Esmaeili, T.; Sumi, T.; Kantoush, S.A. Experimental and numerical study of flushing channel formation in shallow reservoirs. *J. Jpn. Soc. Civ. Eng.* **2014**, *70*, I_19–I_24. [CrossRef]

31. Kantoush, S.A.; Sumi, T.; Suzuki, T.; Murasaki, M. Impacts of sediment flushing on channel evolution and morphological processes: Case study of the Kurobe River, Japan. In Proceedings of the 5th River Flow Conference, Braunschweig, Germany, 8–10 June 2010.

32. Minami, S.; Noguchi, K.; Otsuki, H.; Fukuri, H.; Shimahara, N.; Mizuta, J.; Takeuchi, M. Coordinated sediment flushing and effect verification of fine sediment discharge operation in Kurobe River. In Proceedings of the ICOLD Conference, Kyoto, Japan, 6–8 June 2012.

33. Sumi, T. Evaluation of efficiency of reservoir sediment flushing in Kurobe River. In Proceedings of the 4th International Conference on Scour and Erosion, Tokyo, Japan, 5–7 November 2008.

34. Launder, B.E.; Spalding, D.B. *Lectures in Mathematical Models of Turbulence*; Academic Press: London, UK, 1972.

35. Patankar, S. *Numerical Heat Transfer and Fluid Flow*; Taylor & Francis: New York, NY, USA, 1980.

36. Olsen, N.R.B.; Haun, S. Free surface algorithms for 3-D numerical modeling of flushing. In Proceedings of the 5th River Flow Conference, Braunschweig, Germany, 8–10 June 2010.

37. Schlichting, H. *Boundary Layer Theory*; McGraw-Hill: New York, NY, USA, 1979.

38. Van Rijn, L.C. Sediment transport, part III: Bed forms and alluvial roughness. *J. Hydraul. Eng.* **1984**, *110*, 1733–1754. [CrossRef]

39. Van Rijn, L.C. Sediment transport, part II: Suspended load Transport. *J. Hydraul. Eng.* **1984**, *110*, 1613–1641. [CrossRef]

40. Van Rijn, L.C. Sediment transport, part I: Bed load transport. *J. Hydraul. Eng.* **1984**, *110*, 1431–1456. [CrossRef]

41. Meyer-Peter, E.; Müller, R. Formulas for bed load transport. In Proceedings of the 2nd Meeting of the International Association for Hydraulic Structures Research, IAHR, Stockholm, Sweden, 7–9 June 1948.

42. Brooks, H. Boundary shear stress in curved trapezoidal channels. *J. Hydraul. Div. Am. Soc. Civ. Eng.* **1963**, *89*, 327–333.

43. Esmaeili, T.; Sumi, T.; Kantoush, S.A.; Kubota, Y.; Haun, S. Numerical study on flushing channel evolution, case study of Dashidaira reservoir, Kurobe river. *J. Jpn. Soc. Civ. Eng.* **2015**, *71*, I_115–I_120. [CrossRef]

44. Kantoush, S.A.; Schleiss, A.J. Channel formation in large shallow reservoirs with different geometries during flushing. *J. Environ. Technol.* **2009**, *30*, 855–863. [CrossRef] [PubMed]

water

MDPI

Article

Long-Term Downstream Effects of a Dam on a Lowland River Flow Regime: Case Study of the Upper Narew

Paweł Marcinkowski * and Mateusz Grygoruk

Department of Hydraulic Engineering, Faculty of Civil and Environmental Engineering, Warsaw University of Life Sciences, Nowoursynowska 166 Street, 02-787 Warsaw, Poland; m.grygoruk@levis.sggw.pl
* Correspondence: p.marcinkowski@levis.sggw.pl; Tel.: +48-22-593-52-68

Received: 7 September 2017; Accepted: 10 October 2017; Published: 12 October 2017

Abstract: Most European riverine ecosystems suffer from the negative influence of impoundments on flow regime. Downstream effects of dams lead to a number of environmental and socioeconomic risks and, therefore, should be thoroughly examined in specific contexts. Our study aims to quantify the downstream effects of the Siemianówka Reservoir (Upper Narew, Poland), using statistical analysis of key elements of the river's flow regime, such as the flow duration and recurrence of floods and droughts. In a comparative study on control catchments not influenced by impoundments (the Supraśl and Narewka Rivers), we revealed the following downstream effects of the analyzed dam: significant shortening of spring floods, reduction of the duration and depth of summer droughts, decrease of the maximum discharge, and homogenization of the discharge hydrographs. Although we determined a significant decrease in the duration of summer floods in the "before" and "after" dam function periods, we showed that this issue is regional, climate-related, and replicated in control catchments, rather than an evident downstream effect of the dam. We conclude that significant hydrological downstream effects of the Siemianówka dam–reservoir system could have been the main driver inducing the deterioration of the anastomosing stretch of the Narew River downstream of the dam.

Keywords: Siemianówka; hydrology; Narew River; dam; reservoir; discharge; flow regime

---

1. Introduction

The vast majority of rivers in the northern hemisphere have been segmented by dams, which poses the risk of significant deterioration of downstream riverine and riparian ecosystems [1]. The influence of dams on river systems is reportedly several times greater than, for example, the influence of climate change [2]; therefore, the analysis of downstream effects of dams (reservoirs) remains an important issue in international scientific literature [3–5]. Based on the modification of river discharge on an hourly and daily basis, seasonal changes of flow regimes [6,7], and water temperature modification [8] in impounded rivers, dams can be considered a key element affecting the hydrology of downstream reaches of rivers.

Dams negatively influence all elements of riverine ecosystems. First and foremost, downstream effects of dams induce sediment transport and river channel sedimentation/erosion balance [9]. Changes of the river channel caused by sediment trapping by dams and reservoirs affect the distribution of flow velocities and induce incisions in river channels [10]. Changed water exchange and in-stream biogeochemical cycles affect the water quality of impounded rivers [11]. The changes in the longitudinal connectivity of river reaches and unstable flow regime and water temperature affect fish communities [12,13]. The alteration of the river baseflow due to the operation of the dam negatively affects macroinvertebrate communities by changing their trophic structure [14]. The changed flow regime, including modified inundation periods and seasons, also affects the riparian vegetation [15]. Complex interactions of riverine ecohydrological processes affected by dams need to be addressed in river management strategies [16] because observed and documented downstream effects of dams influence the broad spectrum of the anthroposphere, challenging livelihoods and economies [17].

Large dams have significant downstream effects [3,4,13,16,18–20] and are thus more frequently studied than small ones [8,21]. However, small- and medium-sized dams and associated reservoirs influence the vast majority of the world's rivers [1]. The role of these dams in shaping the global environment remains widespread and locally critical.

In Europe, north-east Poland is known for its high environmental awareness; the area is home to four national parks, five landscape parks, and multiple environmental reserves. Nearly 56% of the area of the Podlaskie Voivodeship province—the highest rate among all regions of the country—is covered under the EU environmental conservation program Natura 2000. The cultural landscape, consisting of a mixture of agricultural land, forests, wetlands, and settlements, is called the "Green Lungs" of Poland. The core of this unique region is the Narew River, which remains the backbone of local agricultural and environmental issues. Part of the middle reach of the Narew River formed the continent's unique anastomosing system of river branches [22,23], which was the main reason for establishing a national park in this region. The uppermost reach of the Narew River was dammed in 1990 when the Siemianówka Reservoir (SR) was constructed. Since then, specific studies have revealed changes in discharge and water-level dynamics [24–28] and biodiversity [29–32] of the riverine ecosystem. These changes have been attributed to the SR. Some observed biogeochemical changes in the downstream anastomosing river system have also been attributed to the dam [33]. The most recent negative hydrological and geomorphological changes of the anastomosing stretch of the Narew River are thought to be accelerated by the downstream effects of the SR [23]. However, no comprehensive long term data-supported hydrological research on the downstream effects of the SR has been conducted so far.

A dataset of more than 20 years of hydrological discharge data for the Narew River after the SR was established is available. Therefore, we aim to conduct a data-based and statistically supported assessment of the river's flow regime based on comparative before–after control–impact comparisons. We investigated the impact of the reservoir construction on the flow regime of downstream reaches of the Narew River. Statistical analyses of the precipitation pattern, daily discharge records (minimum, maximum, 1st, 2nd, and 3rd quartile), and occurrence and recurrence of floods and droughts were conducted for two different time periods: 1951–1989 (pre-dam) and 1990–2012 (post-dam). Our study provides hydrological evidence of the influence of the SR downstream of the Narew River; thus, we open new avenues for the interpretation of biological, ecological, hydrological, and geomorphological research.

2. Materials and Methods

2.1. Study Area

The study area, the Upper Narew Catchment (NE Poland), is the sub-catchment of the largest Polish river basin, namely, the Vistula (Figure 1). The analyzed catchment covers an area of 6656 km^2 (of which 17% belongs to Belarus). The region was formed by glacial erosion and accumulation in the Pleistocene. It is characterized by a flat relief with an average elevation of 152 m above sea level.

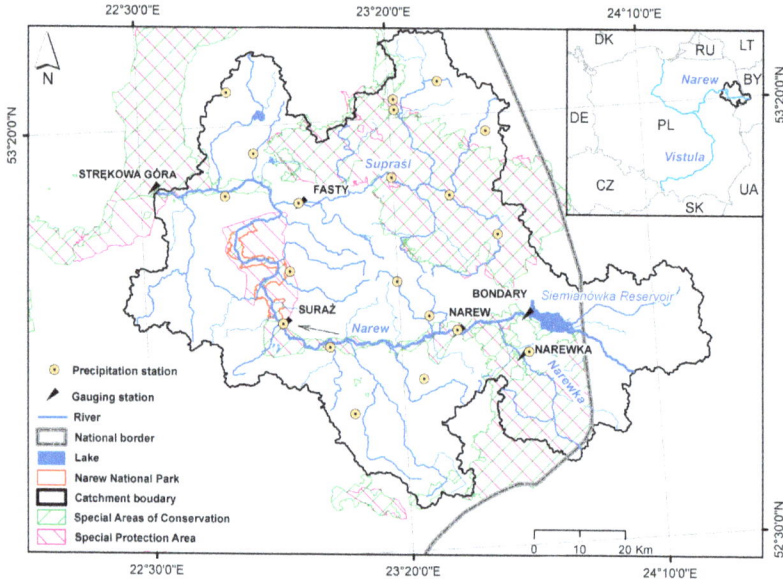

Figure 1. Case study location: Siemianówka Reservoir (SR) and Upper Narew River Catchment, and location of water gauges used to record the data used for the analysis.

The climate of the Upper Narew Catchment is continental; it is influenced by cold polar air masses originating in Russia and Scandinavia. This is reflected in the mean annual air temperature, which equals 7.1 °C, and annual magnitude of air temperature changes that can reach up to 55 °C according to the data of the Institute of Meteorology and Water Management-National Research Institute (IMGW-PIB). Mean annual sums of precipitation oscillate at about 600 mm. Land cover in the Upper Narew Catchment is dominated by agriculture (53%) with arable lands (39%) and pastures (14%) [34]. Urban and settled areas cover less than 3% of the catchment. Dominant types of soil are pure and loamy sands; very heavy impermeable soils (clay, clay loam, and silt loam) are rare [34]. The topographical setting of the catchment determines the typical lowland character of the river valley, which in primeval conditions was supposed to be exposed to regular and long-lasting inundation, most frequently during the early spring snowmelt season. The rivers of this region of Europe are subject to thaw floods, which are considered to be the most important regular extreme phenomenon shaping riparian wetlands, including the anastomosing channel of the Narew downstream of the SR [22]. Substantially lower rain-event driven floods occur occasionally in the summer season.

The Upper Narew Catchment, being a part of the Narew Basin, is surrounded by multiple environmental protection sites, including national parks and Natura 2000 sites (Birds and Habitat Directive; Figure 1). Among the protected areas in the catchment, the Narew National Park (NNP), Upper Narew Valley Refuge Special Area of Conservation (SAC), and Upper Narew Valley Special Protection Area (SPA) are the most important due to their location in an area downstream of the SR. All three protected areas embrace unique wetland habitats shaped by and dependent on regular inundations.

2.2. Siemianówka Reservoir

The SR has a total capacity of 79.5 million m³ and was created through dam construction in the course of the River Narew, (432 km, measured from the outlet). The first construction started in 1977, whereas the complete filling dates back to 1992 when the SR became entirely operational [35].

The dimensions of the reservoir are as follows: 11 km length, 0.8 to 4.5 km width, 2.5 m mean depth, 7 m maximum depth, and 1050 km^2 catchment area. The reservoir and dam were created for multiple purposes including the enhancement of local tourism and recreation, energy production, flood protection, fishing, and irrigation of agricultural lands [36]. According to reservoir management instructions [35], one of the main objectives is to increase the low flow during the summer season in the Narew to maintain biological life, and to mitigate high peak flow to reduce flood risk in the valley during the spring season. Reference catchments used as control sites in our research, that is, from the Narew River down to the Narewka gauge station, and from the Supraśl River down to the Fasty gauge station (Figure 1; Table 1), are located in the same region. Their physiographic features are similar to the analyzed catchment of the Narew River.

Table 1. Gauge stations in the Upper Narew Catchment. The asterisk '*'denotes catchments of major tributaries of the Narew River, which are used as references in this study.

Gauge Station	River	Flow Data Availability	Catchment Area (km^2)
Narewka *	Narewka	1951–2013	590.4
Bondary	Narew	1963–2013	1094.6
Narew	Narew	1951–2013	1978.0
Suraż	Narew	1951–2013	3376.5
Fasty *	Supraśl	1975–2013	1818.0
Strękowa Góra	Narew	1951–2012	6656.0

2.3. Hydrometeorological Analysis

This study used daily precipitation records from 20 meteorological stations for the period 1951–2012 for the analysis. Daily flow records for six gauge stations covering different time periods (Table 1) were subjected to a statistical analysis. Both meteorological and hydrological data were acquired from the IMGW-PIB. Whilst four flow gauge stations (Bondary, Narew, Suraża, and Strękowa Góra) are located on the Narew River and are used for direct dam impact assessment, the remainig two (Narewka and Fasty, located on the Narew tributaries and not influenced by the dam) constitute control stations to eliminate potentially wrong inferences due to the climate impact on the flow regime. To assess the significance level of the precipitation trend, the non-parametric Pettitt's test [37] was applied, with a significance level of $p \leq 0.05$. This test provides the assessment of the null hypothesis H$_0$, implying that the data are homogeneous throughout the period of observation. Pettitt's test was reported in the literature to be sufficient for detecting break points in a set of long-term observations [38,39]. In addition, flow-duration analysis was conducted, yielding the annual minimum; maximum; and 25, 50, and 75 percentile discharge for all gauge stations. Statistical analysis of flow data homogeneity based on Pettitt's test was conducted to indicate significant breakpoints and timescale trends to assess the impact of dam construction on the flow regime.

Additionally, flow-duration curves (FDCs) have been created for every gauge station to express the overall regime change in the analyzed time frame. The FDC is a plot that shows the percentage of time during that discharge in a stream is likely to equal or exceed some specified value of interest. In this study, the flow datasets for each gauge station were divided into two subsets (pre-dam, until 1988; and post-dam, since 1989) for which FDCs were created separately. Finally, temporal patterns of floods and droughts were analyzed. We applied standard thresholds of river discharge assessment, such as the median of the highest annual discharges (MHQ, proxy for a bankfull discharge) and average lowest discharge (ALQ), both from multi-year records, for flood and drought analysis, respectively. The study calculated the number of days with discharge higher than MHQ and lower than ALQ for selected gauge stations to reveal whether the occurrence and recurrence of floods and droughts, that is, extreme hydrological phenomena critical for the functioning and preservation of anastomosing river branches and related biocenoses, have maintained trends similar to pre-establishment of the dam or have changed. For a comprehensive view on the aspect of floods and drought distribution and frequency, we analyzed these phenomena in for all years considered (1951–2013) and separately for summer (May–October) and winter (November–April). In addition, we analyzed floods and droughts of the Narew River for the period from 1975 to 2013 because reference data for the SR were available. This approach allowed us to draw conclusions on the importance of the length of the river discharge dataset with respect to its usefulness for before–after control–impact studies of downstream effects of dams, which has already proven to be a critical issue in downstream effect analysis [40].

3. Results

3.1. Precipitation

The rainfall statistics (Figure 2) based on Pettitt's test clearly indicate a moderate increasing trend over the last 60 years with some wetter and drier periods. Over the period analysed, no statistically significant differences in monthly sums of precipitation were detected between the results from 20 stations analysed (statistics available in the Supplementary material, Section 1). Average annual precipitation sum within the Upper Narew catchment in the years 1951–2012 equalled 598 mm. The total annual precipitation was slightly higher in 1970–1980 than in 1951–1969 or 1981–2012. This pattern was significantly broken in 2010 due to an extremely wet year with flooding across Poland; however, the pattern appears to be continuing in recent years. This characteristic precipitation pattern is a crucial result, and in addition to dam construction, it needs to be taken into account as a potential key controlling factor with respect to further interpretation of flow regime changes.

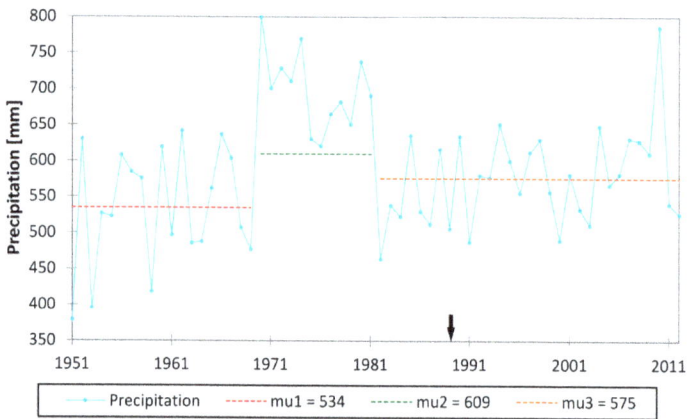

Figure 2. Total annual precipitation for the time period 1951–2012 in the Upper Narew Catchment. "mu" stands for the average value at each time period. The black arrow indicates the year the SR started operation.

3.2. Discharge

3.2.1. Minimum Discharge

The analysis of the annual minimum discharge pattern based on Pettitt's test indicated diverse results for each gauge station. No significant trend was detected in the analyzed period for the Fasty (Figure 3A) and Narewka (Figure 3B) gauge stations located on the tributaries of the Narew River, where the influence of the SR on the flow regime was not an issue. In contrast, characteristic trends and breakpoints were noted for gauge stations located along the Narew River. However, due to different occurrence times, the detected trends were most likely driven by varied factors. For the Bondary gauge station (Figure 3C), located immediately downstream of the SR, one significant break point occurred in 1994, indicating an average increase of 60% between the separated time periods. Moving on further downstream of the Narew gauge station (Figure 3D), two significant breakpoints were detected

(1971 and 1994). During the first time period (1951–1971), the annual minimum flow was $1.6 \text{ m}^3 \cdot \text{s}^{-1}$, increasing from 1972 to 1993 by 87%, and increasing yet again in 1994–2013 by 17%. The Suraż gauge station (Figure 3E) recorded trends identical to those at Narew but of lower magnitude in the corresponding time periods (an increase of 71% from 1972 to 1993 and of 14% from 1994 to 2013). Whilst the first increase of the annual minimum flow is most likely related to the precipitation increase, as the occurrence of both coincide in time, the latter is most likely due to reservoir operation. The Strękowa Góra gauge station (Figure 3F), which is located in the most downstream part of the catchment seems to reflect only precipitation-driven changes of the minimum flow pattern, which increased from 1972 to 1987 by 55% and decreased by 11% from 1988 to 2012. This indicates that the dam impact subsides at some point of the Narew River between the Suraż and Strękowa Góra gauge stations.

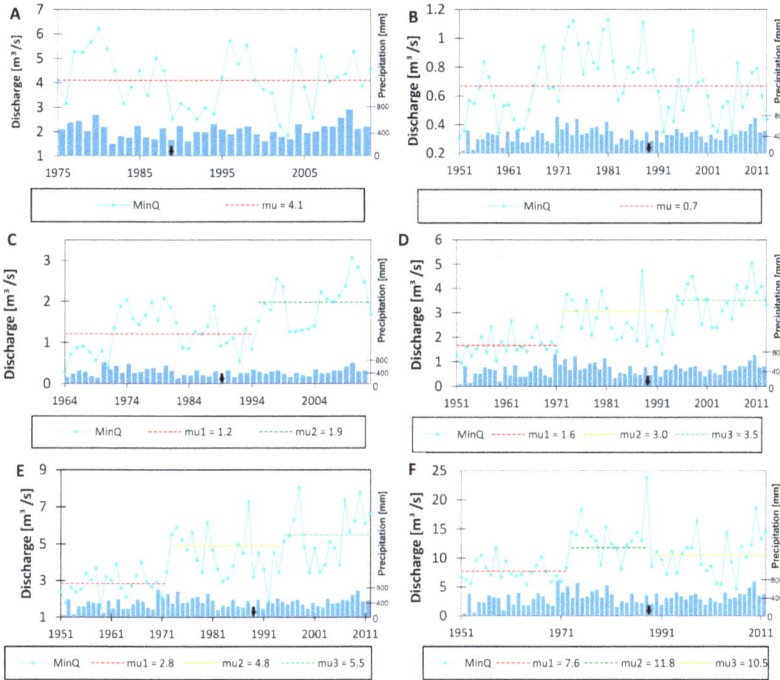

Figure 3. Annual minimum discharge for selected gauge stations of the Upper Narew Catchment: (**A**) Fasty, (**B**) Narewka, (**C**) Bondary, (**D**) Narew, (**E**) Suraż, and (**F**) Strękowa Góra. "mu" stands for the average value of each time period. The black arrow indicates the year the SR started operation.

3.2.2. First Quartile Discharge

The analysis of the first quartile flow, which represents the median of the lower half of the dataset indicates moderate changes. At the three gauge stations (Fasty, Narewka, and Bondary; Figure 4A–C, respectively), no significant trend was recognized during the analyzed period. In the case of minimum flow, a similar response was observed at the Narew (Figure 4D) and Suraż gauge stations (Figure 4E). An increase occurred in 1970, with rates reaching 71% and 60% at the Narew and Suraż gauge stations, respectively. Taking into account the time of appearance, this correlates with the break point of the precipitation pattern shift observed in 1970. It is similar to the response of the minimum flow pattern, with one major exception: it did not subside after 1980 when the precipitation decreased. Considering changes detected at the Strękowa Góra gauge station (Figure 4F), the flow trends precisely follow the precipitation pattern, increasing at first by 50% in 1972–1987, and finally decreasing by 11%.

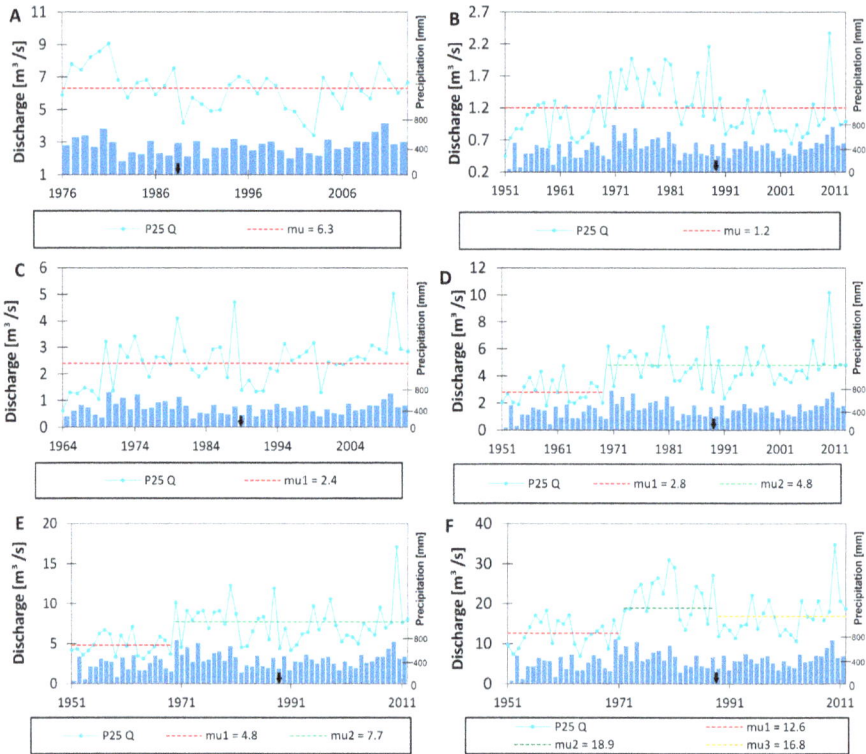

Figure 4. Annual first quartile discharge for selected gauge stations in the Upper Narew Catchment: (**A**) Fasty, (**B**) Narewka, (**C**) Bondary, (**D**) Narew, (**E**) Suraż, and (**F**) Strękowa Góra. "mu" stands for the average value of each time period. The black arrow indicates the year the SR started operation.

3.2.3. Second Quartile Discharge

The analysis of the second quartile flow, which simply represents the median of the observed dataset, indicated no significant trends for the two gauge stations (Fasty and Bondary; Figure 5A,C). Two characteristic patterns were observed at the remaining gauge stations. First, at the Narewka (Figure 5B) and Strękowa Góra (Figure 5F) gauge stations, we observed two significant breakpoints of the trend reflecting variation in precipitation from 1970 to 1980. Second, at the Narew (Figure 5D) and Suraż (Figure 5E) gauge stations, we observed one significant trend shift in 1970. Regarding the rates of increase, the highest rate was noted at the Narew and Suraż gauge stations (64% and 60%, respectively) and a significantly lower one was detected at the Fasty and Strękowa Góra gauge stations (42% and 34%, respectively). The decreasing trends observed at the latter two gauge stations reached 10%.

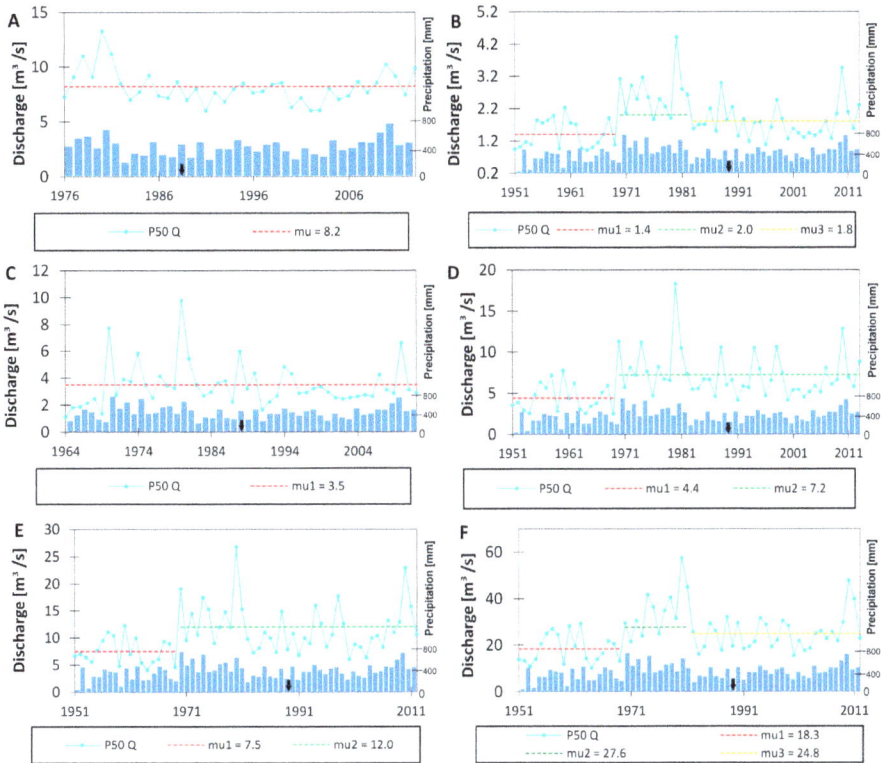

Figure 5. Annual second quartile discharge for selected gauge stations of the Upper Narew Catchment: (**A**) Fasty, (**B**) Narewka, (**C**) Bondary, (**D**) Narew, (**E**) Suraż, and (**F**) Strękowa Góra. "mu" stands for the average value of each time period. The black arrow indicates the year the SR started operation.

3.2.4. Third Quartile Discharge

The analysis of the third quartile flow, which represents the median of the higher half of the dataset, indicates diverse changes. No significant alteration of the trend during the analyzed period was detected at the three gauge stations (Fasty, Bondary, and Strękowa Góra; Figure 6A,C and F). Identical changes were observed at the Narewka (Figure 6B) and Narew (Figure 6D) gauge stations, accurately reflecting the precipitation pattern, which increases in 1970 (by 42%) and decreases in 1983 (by 15%). The flow pattern at the Suraż gauge station (Figure 6E) seems to have partial fluctuations in precipitation, because the increase date coincides with the time in both cases. However, there is a disagreement in the subsidence seen in total precipitation, which is not visible in the flow alteration noticed around 1980.

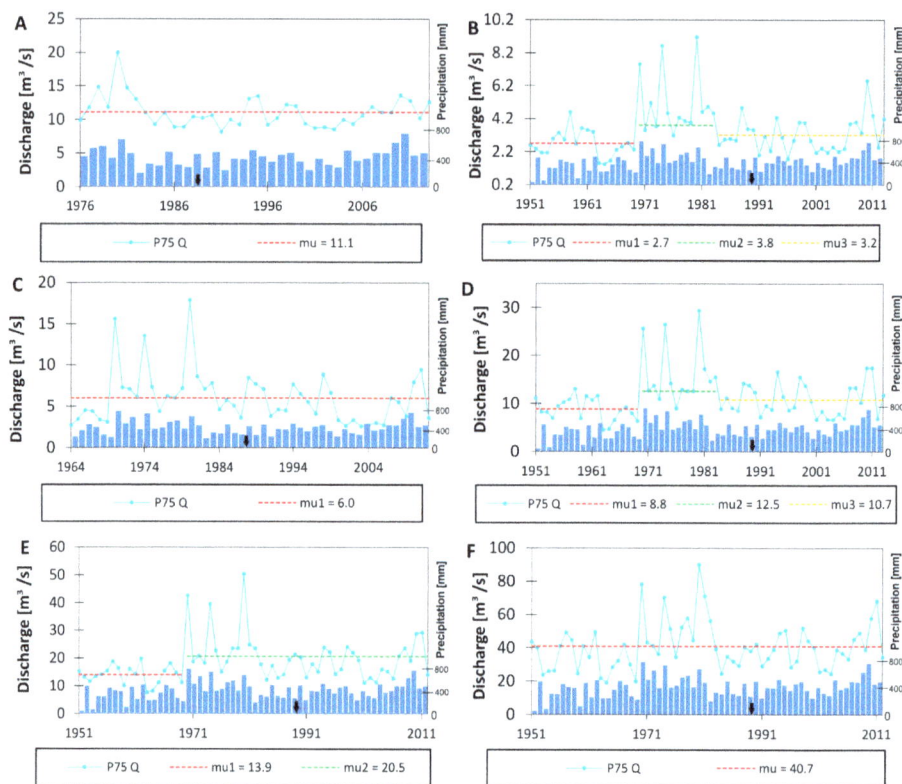

Figure 6. Annual third quartile discharge for selected gauge stations of the Upper Narew Catchment: (**A**) Fasty, (**B**) Narewka, (**C**) Bondary, (**D**) Narew, (**E**) Suraż, and (**F**) Strękowa Góra. "mu" stands for the average value of each time period. The black arrow indicates the year the SR started operation.

3.2.5. Maximum Discharge

The changes in the annual maximum flow trends in the analyzed period seem to be the most conspicuous. Among the investigated gauge stations, two indicated no significant trend changes (Fasty and Narewka; Figure 7A,B). However, an explicit decrease was noted at the remaining four stations. Although the direction of change is consistent, the time of occurrence differs. This proves that the driving force of such change is not homogenous at all stations. In particular, it is most probable that the shifts occurring in 1988 at the Bondary (Figure 7C), Narew (Figure 7D), and Suraż (Figure 7E) gauge stations are caused by the dam construction. A substantial decrease was detected in all cases, reaching 67%, 52% and 33% at the aforementioned gauge stations, respectively. Although the decrease (by 43%) was also noted at the Strękowa Góra gauge station (Figure 7F), the time of the breakpoint occurrence (1983) suggests that it is most likely driven by precipitation decrease.

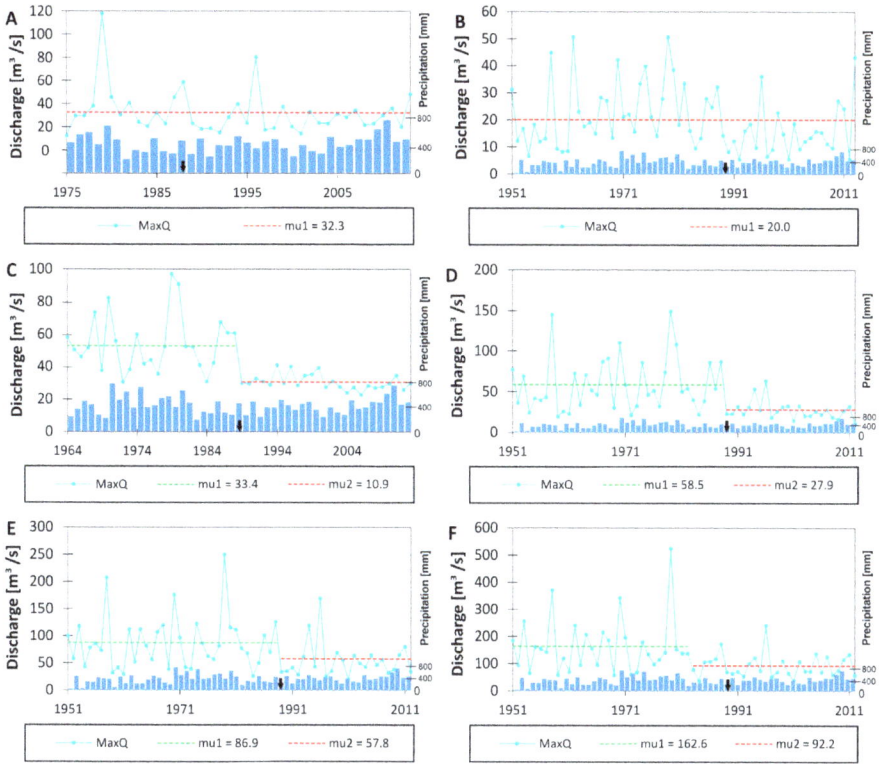

Figure 7. Annual maximum discharge for selected gauge stations of the Upper Narew Catchment: (**A**) Fasty, (**B**) Narewka, (**C**) Bondary, (**D**) Narew, (**E**) Suraż, and (**F**) Strękowa Góra. "mu" stands for the average value of each time period. The black arrow indicates the year the SR started operation.

3.3. Flow-Duration Curve

The results indicating the change in the FDC prove the undeniable impact of reservoir construction on the flow regime of the Narew River. The alterations at the Fasty (Figure 8A) and Narewka (Figure 8B) gauge stations exclusively reflect climate-driven changes because both are located at the Narew tributaries, which are not impacted by the dam construction. Figure 8A,B show that the shape of the curve remains similar during pre- and post-dam periods; only a regular shift (overall decrease) in magnitude is detected in all exceedance probability intervals. In contrast, such a shift does not occur for gauge stations located directly on the Narew River (Figure 8C–F). Instead, a flattening of the curve is observed at all gauge stations during the post-dam period, leading to varying intersections of the curves at different exceedence probability points. The response is homogenous at all gauges, indicating that the daily streamflow decreases in the lower percentiles (i.e., higher flow) and increases in the higher percentiles (i.e., lower flow). Moving from upstream (Bondary) to downstream (Strękowa Góra), the magnitude of change seems to decline gradually, whereas the curve intersections shift to lower percentiles.

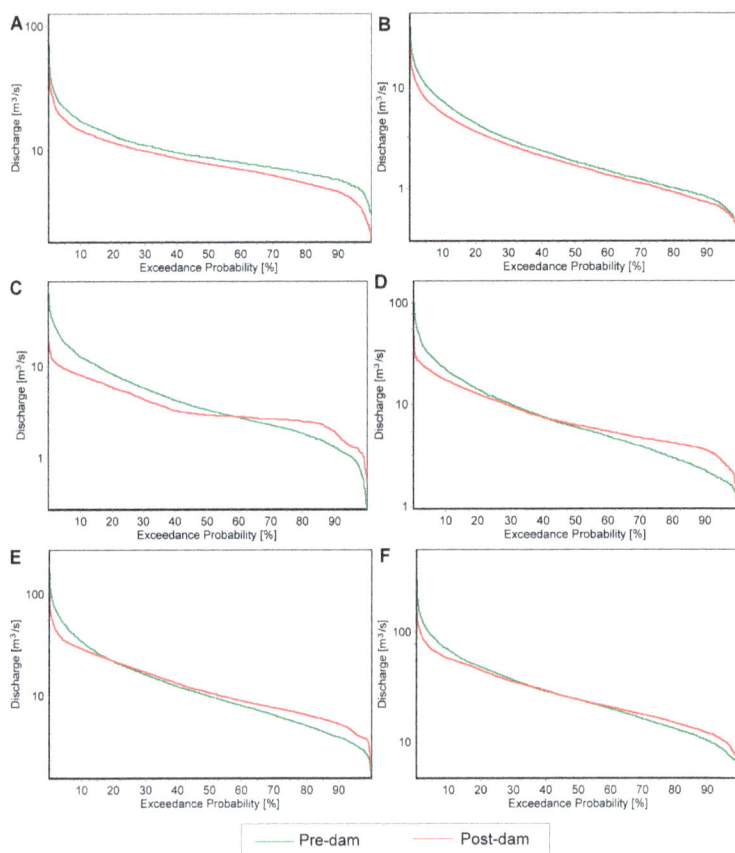

Figure 8. The changes in FDCs at selected gauge stations of the Upper Narew Catchment: (**A**) Fasty, (**B**) Narewka, (**C**) Bondary, (**D**) Narew, (**E**) Suraż, and (**F**) Strękowa Góra. The discharge is presented on a logarithmic scale.

3.4. Ocurrence and Recurrence of Floods and Droughts in Different Seasons

The analysis of temporal trends of floods and droughts revealed significant changes in the ocurrence and recurrence of these extreme phenomena (Figures 9 and 10). In the period "before" the SR was established, the frequency of years without flooding (daily average discharge higher than MHQ) reached only 0.1 (floods did not occur in 4 out of 39 analyzed years). In the period "after" the SR was established, the frequency of years without flooding and higher than the applied threshold reached 0.3 (floods did not occur in 7 out of the 23 analyzed years). The longer inundation periods observed in the Narew River Valley in the past, mainly in March and April, which were assessed using records from the Suraż gauge station, are most likely related to the unmanaged flow regime, which is affected by snowmelts and thaws. Although the annual recurrence of floods tends to decrease on the regional scale (a statistically significant decrease was recorded both for Narew in Suraż and the reference catchment Narewka in Narewka; Figure 10G,T), the decrease of the spring thaw flood recurrence and duration is statistically significant only for Narew (Figure 10I,L vs. Figure 10V,Z). This observation leads to the conclusion that the SR has a significant and considerably high influence on the recurrence and duration of floods for the Narew River: (1) the contemporary recurrence of Narew floods on the annual basis is 34% lower than before the SR was established; and (2) the contemporary duration of spring thaw floods declines by 56%, from an average of 26 days to 15 days before and after the SR started to operate, respectively.

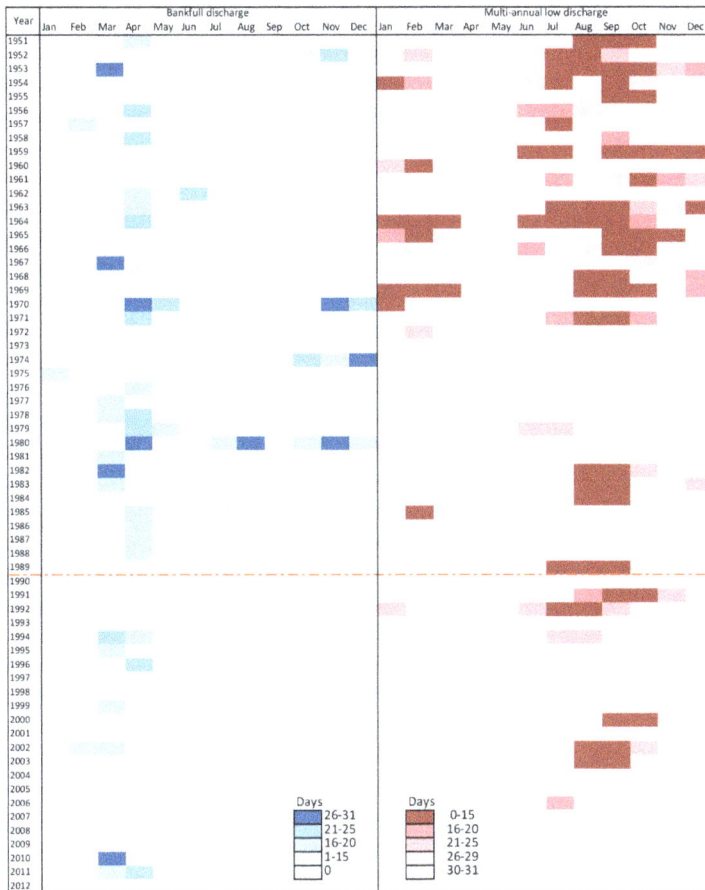

Figure 9. Matrix presenting the occurrence and duration of floods (blue) and droughts (red) of the Narew River at the Suraż gauge station.

The detection of a significant difference of the flood duration (decline) in Narewka (reference catchment) between the whole year (Figure 10T) and summer (V-X; Figure 10U) leads to the conclusion that the operation of the SR does not significantly reduce summer flooding. In terms of flooding, the flow regime of the other reference catchment (Supraśl River, Fasty gauge station) does not show significant changes throughout the analyzed years (Figure 10Y–Z). Despite the fact that the recurrence and duration of flooding generally decreases with respect to the whole year from 1951–2013, for both summer–autumn and winter–spring phenomena, the before–after comparison is not statistically significant (p values range from 0.188 to 0.608).

The SR has a completely different influence on the low flow of the Narew River. On the regional scale, the duration and recurrence of droughts lower than the applied threshold tends to significantly increase (e.g., in the reference catchments of Narewka and Supraśl; Figure 10M–O and P–S, respectively). In contrast, both the short- (1975–2013) and long-term (1951–2013) horizon recurrence and duration of droughts decrease in the case of the Narew downstream SR, similar to short-term observations presented by Romanowicz and Osuch [28]. A statistically significant decline of the drought frequency was recorded in summer (May–October; Figure 10B,E), winter, and in the long term (whole year; Figure 10A,C), which is an important downstream effect of the SR. Although no statistical significance was revealed in the case of Narewka when comparing the durations of droughts before and after the establishment of the SR (Figure 10M–O), we observed a significant increase in the recurrence and duration of droughts in the case of the Supraśl River (Figure 10P,R). Discrepancies in the statistical significance of the before–after regime of floods and droughts between short-term (Figure 10D,J) and long-term analyses (Figure 10A,G) prove that using short sets of data to reveal flow regime changes due to dams may lead to wrong interpretations, as also shown by Huh et al. [40]. Different conclusions drawn by Mioduszewski et al. [27] and Cygan et al. [24] confirmed the hypothesis of Huh et al. [40] who stated that the use of too short a series of discharge records may lead to wrong conclusions about the influence of dams on the flow regime of a particular river. Considering the presented data, one can conclude that the reduction of the recurrence and duration of droughts remains the most notable and unilateral downstream effect of the SR on the flow regime of the Narew River. In combination with the interpretation of floods, we conclude that the SR supports the significant homogenization of the Narew River's discharge, which likely has important consequences for downstream riverine and riparian ecosystems.

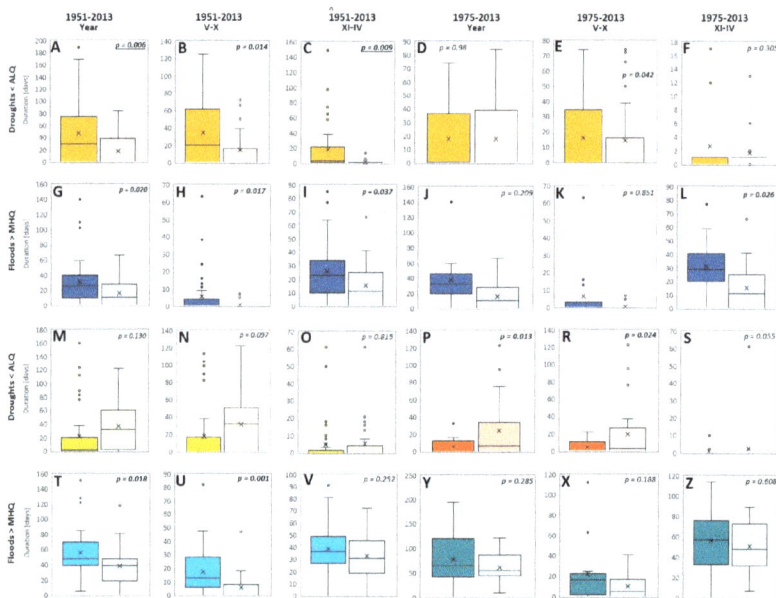

Figure 10. Comparison of the "before" (1951–1988/1975–1988) and "after" (1989–2013) durations of droughts (Q < ALQ; orange/yellow) and floods (Q > MHQ; blue/dark blue) for Narew–Suraż (**A–L**), Narewka–Narewka (**MNO–TUV**), and Supraśl–Fasty (**PRS–YXZ**). (**A–C,G–I,M–O,T–V**) represent data from 1951 to 2013. (**D–F,J–L,P–S,Y–Z**) reflect data from 1975 to 2013.

4. Discussion

4.1. Hydrological Aspects

The results show that the SR altered the flow regime of the Narew River. Although the influence of the SR operation on the average and median flows is not significant compared with studies of Mioduszewski et al. [27] and Cygan et al. [24], and as presented by Kiczko et al. [26]—strongly depends on the river stretch, we revealed the influence of the SR on the extreme flow of the Narew River. Marcinkowski et al. [23] investigated seasonal changes in water flow and indicated that average flows are lower in April but higher in February and March, which is determined by dam operations, in which water is released from the reservoir in advance of the spring thaw to prevent flooding in municipalities close to the reservoir (see Supplementary material, Section 2).

A temporal precipitation increase recorded in the past (1970–1980) was taken into account in addition to the construction of the SR, as one of the most crucial factors altering the river flow regime, which was also observed by Mioduszewski et al. [27]. The impact of precipitation fluctuations is clearly visible in Figures 5B,F and 6B,D, which show that the temporal increase of the median flow precisely follows the precipitation pattern.

Marcinkowski et al. [41], by means of a hydrological model and nine General Circulation Model–Regional Climate Model (GCM–RCM) runs, evaluated pure effect of climate on water resources for the Upper Narew catchment for the 2020–2050 time horizon. They postulated that the median of projected changes in water yield, i.e., the portion of precipitation that reaches the stream, indicate an average annual increase of 9%. Notably, they observed the most pronounced increase in winter, and a substantially lower increase in other seasons. Given the fact, that the SR controls the release of water, storing most of it during the spring season, the increased flow caused by climate change might be significantly suppressed.

The differences observed for the recurrence and duration of droughts between the Narew River and reference catchments of Narewka and Supraśl may result from land-use changes [28]. However, the aspects of land use in Narew and the analyzed reference catchments are controlled by similar drivers originating from environmental policies (e.g., the Common Agricultural Policy of the EU, responsible for subsidising grassland farming in valuable riparian wetlands). Hence, completely antagonistic trends of the recurrence and duration of droughts of the Narew River compared with control catchments, also reported by Romanowicz and Osuch [28], tend to be a clear flow-regime change attributed to the SR's operation. This downstream effect of the SR can be considered positive, both from environmental and socioeconomic perspectives. Higher water levels during droughts are likely to prevent desiccation, which may reduce CO_2 emissions from drained wetlands in some areas of the valley. However, so far, no links have been found between increasing river water levels during droughts and rising groundwater levels in the valley [27,42], indicating that it may be difficult to mitigate general groundwater decline in the area with an appropriate water spill control from the SR. Hence, the role of the SR in improving downstream habitat quality by decreasing the frequency of deep droughts and increasing the water levels in the river appears to be negligible. On the other hand, the lack of long droughts in recent years could also be considered as less stressful for the riverine ecosystem. One might suspect that if this stress factor occurs repeatedly over time, biota would have to adapt to this phenomenon, and species with higher resistance to this stress factor would become more abundant in the ecosystem. However, so far, not much is known about the role of droughts as an ecosystem stress factor influencing the specific and unique biodiversity of river systems.

The proven and evident reduction of the spring thaw flood duration and frequency that can be attributed to the operation of the SR in the analyzed years along with recent milder winters and less snow accumulation than before [42] can contradict the environmental effects of wetland restoration. Bush encroachment in open areas of Narew Valley wetlands, which was reported to be successfully mitigated by birch removal [43], can continue because lowered and shortened inundation allows the re-establishment of young birch stands and feedbacks with evapotranspiration [44]. Hence, shrub removal, although so far successful and important, appears in this case to be a measure tackling the result of the process rather than affecting the process of bush encroachment itself.

4.2. Hydromorphological Aspects

Results of the reconnection of abandoned side arms and anabranches of the Narew River in the NNP area appear to be contradicted by the operation of the SR as well. Although sidearm reconnection projects implemented in the Narew Valley were reported to have positive environmental effects [45], one could suspect that facing a continuing trend of high flow reduction attributed to the function of the SR, managing appropriate levels of sidearm connectivity to the main river channel in the future will remain a challenging task. In addition to negative responses of biota to river maintenance work performed in the Narew River System [46], flow-regime changes remain an important issue to be addressed in future river management strategies and also within the NNP.

The NNP, located downstream of the SR, has been facing the loss of anabranches in recent decades [23]. It has been recognized that the highly variable flood-prone flow regime characterized by the occurrence of seasonal high discharges is one of the most crucial factors for global anastomosing rivers to persist [47,48]. Frequent or high-magnitude flooding is assumed as a precondition for avulsion and the eventual formation of new channels. The SR-induced decrease of the magnitude and duration of floods can, therefore, be deemed as one of the factors, among others recognized by Marcinkowski et al. [23], speeding up the gradual extinction of the valuable anastomosing character of the Narew River. It is critical to address this issue in any amendments to contemporary water management instructions governing the flow regime of the Narew River.

4.3. Ecological Aspects

Another crucial factor responsible for anabranch loss was reported by Marcinkowski et al. [23] and relates to the uncontrolled expansion of Common Reed (*Phragmites australis*) in the NNP in recent decades. As reported by Próchnicki [49], the analysis of aerial imagery revealed a two-fold increase of the reed share in the valley vegetation between 1987 and 1997. It was justified by the overall shift in management strategies and cessation of floodplain mowing, which was very popular until 1980 in this region. However, as recognized in other studies, it could also be determined based on water-level changes. These water-level changes could reflect intensification of drainage and land reclamation directly within the river valley, but—noteworthy—could also result from the SR operation. Stromberg et al. [50] stated that aquatic plant species have been observed to increase in association with riverine alterations such as river channelization, stabilized water levels, reduced frequency of inundation, and altered timing of water and sediment flow. Moreover, Galatowitsch et al. [51] showed that the intensive invasion of *Phragmites australis* in the Platte River (Nebraska, USA) was caused by changes of the natural flow regime after the catchment has been highly altered by water development for irrigation and hydropower production. They stated that rapid expansion of reed followed the significant decrease of the flood flow. Therefore, it is highly probable that the invasion of Common Reed in the NNP, which coincides with the construction of the SR, might be caused by SR-induced flow regime changes, in addition to positive correlations of reed expansion and the cessation of mowing. Reporting downstream effects of the SR on the flow regime of the Narew River, and considering a number of environmental issues examined in the Narew Valley in recent years, similarly to Romanowicz and Osuch [28] we stress that further research is critically needed to reveal the influence of the SR-induced flow regime change on biota and biocoenosis of this area in a site-specific context. Based on the knowledge of the response of riparian environments to reservoir-influenced flow regime changes [1,3,4,6,7,50,51], we stress that keeping the status quo with respect to the regulation of river discharge by the SR can eventually result in irreversible changes of the downstream environment of the Narew River and its valley, such as the deterioration of the main objects of environmental conservation of the NNP related to a functioning anastomosing river channel. Considering the results of Piniewski et al. [52], and given the findings presented in this paper, we also stress that the SR-induced flow regime changes of Narew are likely to be more unilateral and stronger than the ones resulting from prospective climatic changes.

Now that 25 years have passed since the first spill of water was released from the fully filled SR, and given that data-based quantification of the influence of the SR on the Narew River flow regime allows drawing comprehensive and statistically relevant conclusions, the environmental management of the area should anticipate SR-induced flow regime changes of the Narew River.

Although not many studies present acceptable solutions allowing functioning dam-reservoir systems and sustaining resilient anastomosing rivers downstream, we hope that our findings will allow other researchers to view environmental factors of the Narew River and its valley with a different perspective. This perspective should emphasize flow-regime alteration as potentially a critical aspect, followed by the response of ecosystems.

4.4. Implications for Management

In view of our results, all considerations concerning instructions for water release and water management in lowland reservoirs, in particular those relating to the SR, gain new meaning and should be revisited. The observations of Kiczko and Napiórkowski [53] and Kiczko et al. [54], addressing the need for anticipation of environmental issues in water management measures implemented at the reservoir, should be used to provide answers to new management questions, such as 'How should water release from the reservoir be managed in order to keep the level of habitat degradation risk as low as possible?' and 'Is optimization of water release possible when limitations originate from both environmental and agricultural requirements?'.

Finally, a possible negative influence of the SR on downstream environments should be revisited in terms of gains and losses. Although the reservoir itself provides new habitats for birds, remains an important local attraction for recreational fishing, and provides some hydropower energy, it is necessary to calculate environmental and socioeconomic benefits and trade-offs of the SR. This calculation should anticipate actions related to environmental restoration in the Narew Valley and declining areas of valuable habitats downstream, because they might result from the negative influence of the downstream effects of the SR. Revealing the full set of benefits and losses associated with the function of the dam is expected to support (economically) decisions on the maintenance of the dam or on its removal in the future. Observing quick and positive responses of river systems to dam removals [5] fosters hope that there is still a chance to preserve Europe's unique anastomosing system before its complete degradation, which appears to result partly from the operation of the SR, by managing the landscape of the Narew Valley.

5. Conclusions

(1) The SR impact on the flow regime differs among gauge stations and indicators. The most significant and direct change was observed for extreme flow (minimum and maximum) of three subsequent gauges (Bondary, Narew, and Suraż). In each of them, a substantial decrease of high flow and an increase of low flow were detected. The quartile discharge alteration was more blurred and seems to be rather climate change-affected, reflecting precipitation pattern fluctuations.

(2) The SR has a significant and considerable influence on the recurrence and duration of floods of the Narew River: (1) the contemporary recurrence of Narew floods on an annual basis is 34% lower than before the SR was established; and (2) the contemporary duration of spring thaw floods has declined by 56%, from an average of 26 days before to 15 days after the SR has started to operate.

(3) An overall and significant increase in low flow was noted along the course of the Narew River downstream of the SR (in particular, at the Suraż gauge station), while a significant increase in the frequency and duration of droughts was detected in reference catchments of Narewka and Supraśl. However, the role of the SR in improving the quality of downstream riparian habitats tends to be insignificant, because the continuously reported trend of groundwater-level decline in the valley seems not to be reversed once the SR has been established.

(4) The significant change in floods and the flood–drought balance revealed in this study and attributed to the function of the SR poses a great threat with respect to the extinction of anabranches depending on the regular occurrence of high flow. This conclusion needs to be accounted for in the future management of the most valuable anastomosing stretch of the river in the NNP.

(5) The flow regime of rivers in control catchments (Narewka and Supraśl) reflects exclusively temporal changes of precipitation. The flow regime of the Narew River downstream of the SR from

1951 to 2013 presents different dynamics. Flow regime alterations of the Narew River observed for neighboring gauges—that is, the decrease of the duration and extent of spring flooding, increase of low flow, and altered flow durations—can be attributed to the operation of the SR. This observation is contradictory to several previous hydrological studies [24,27], which might have drawn wrong conclusions on the small (or even lack of) influence of the SR on the flow regime of the Narew based on the analysis of too short a data series. However, our observations on the influence of the SR on the flow regime of the Narew River confirm some of the previous suspicions and hypotheses of hydrologists working with limited sets of data [28] and environmentalists tackling biocoenosis of the area [42,55,56].

(6) With respect to the presented conclusions, some changes of riparian and riverine ecosystems of the Narew River and its valley attributed to climate change, land-use modification, or a switch in the role of riparian vegetation in water consumption, and reported in numerous scientific studies, require revision in terms of considering the SR as the dominant factor inducing the flow regime of the Upper Narew River.

Supplementary Materials: The following are available online at www.mdpi.com/2073-4441/9/10/783/s1, Figure S1: Monthly precipitation sum for 20 gauging stations in the Upper Narew catchment, Figure S2: Flow regime in Suraż gauging station 1951–2012. A—1st quartile monthly discharge, B—median monthly discharge, C—3rd quartile monthly discharge, Table S1: T test results of statistical significance of monthly precipitation sum between meteorological stations in the Upper Narew catchment.

Acknowledgments: We would like to thank three anonymous reviewers for their comments and suggestions, which helped to improve the manuscript. The study presented in this paper was financed by the National Science Centre, Poland, under grant number 2015/19/N/ST10/01629.

Author Contributions: Both authors equally contributed to all steps of the data analysis and manuscript preparation.

Conflicts of Interest: The authors declare no conflict of interest.

References

1. Dynesius, M.; Nilsson, C. Fragmentation and flow regulation of river systems in the northern third of the world. *Science* **1994**, *266*, 753–762. [CrossRef] [PubMed]
2. Graf, W.L. Dam nation: A geographic census of American dams and their large-scale hydrologic impacts. *Water Resour. Res.* **1999**, *35*, 1305–1311. [CrossRef]
3. Graf, W.L. Downstream hydrologic and geomorphic effects of large dams on American rivers. *Geomorphology* **2006**, *79*, 336–360. [CrossRef]
4. Magilligan, F.J.; Nislow, K.H. Changes in hydrologic regime by dams. *Geomorphology* **2005**, *71*, 61–78. [CrossRef]
5. O'Connor, J.E.; Duda, J.J.; Grant, G.E. 1000 dams down and counting. *Science* **2015**, *348*, 496–497. [CrossRef] [PubMed]
6. Bejarano, M.D.; Sordo-Ward, A.; Alonso, C.; Nilsson, C. Characterizing effects of hydropower plants on sub-daily flow regimes. *J. Hydrol.* **2017**, *550*, 186–200. [CrossRef]
7. Kibler, K.M.; Alipour, M. Flow alteration signatures of diversion hydropower: An analysis of 32 rivers in Southwestern China. *Ecohydrology* **2017**, *10*, e1842. [CrossRef]
8. Maheu, A.; St-Hilaire, A.; Caissie, D.; El-Jabi, N. Understanding the thermal regime of rivers influenced by small and medium size dams in Eastern Canada. *River Res. Appl.* **2016**, *32*, 2032–2044. [CrossRef]
9. Sanyal, J. Predicting possible effects of dams on downstream river bed changes of a Himalayan river with morphodynamic modelling. *Quat. Int.* **2017**, *453*, 48–62. [CrossRef]
10. Williams, G.P.; Wolman, M.G. *Downstream Effects of Dams on Alluvial Rivers*; Geological Survey Professional Paper 1286; USGS: Washington, DC, USA, 1984.
11. Ahn, J.M.; Jung, K.Y.; Shin, D. Effects of coordinated operation of weirs and reservoirs on the water quality of the Geum River. *Water* **2017**, *9*, 423.
12. Cooper, A.R.; Infante, D.M.; Wehrly, K.E.; Wang, L.; Brenden, T.O. Identifying indicators and quantifying large-scale effects of dams on fishes. *Ecol. Indic.* **2016**, *61*, 646–657. [CrossRef]

13. Song, Y.; Cheng, F.; Murphy, B.R.; Xie, S. Downstream effects of the Three Gorges Dam on larval dispersal, spatial distribution and growth of the four major Chinese carps call for reprioritizing conservation measures. *Can. J. Fish. Aquat. Sci.* **2017**. [CrossRef]

14. Valentin, S.; Wasson, J.G.; Philippe, M. Effects of hydropower peaking on epilithon and invertebrate community trophic structure. *River Res. Appl.* **1995**, *10*, 105–119. [CrossRef]

15. Chen, Q.; Zhang, X.; Chen, Y.; Qiongfang, L.; Qiu, L.; Liu, M. Downstream effects of a hydropeaking dam on ecohydrological conditions at subdaily to monthly time scales. *Ecol. Eng.* **2015**, *77*, 40–50. [CrossRef]

16. Han, J.; Zhang, W.; Fan, Y.; Yu, M. Interacting effects of multiple factors on the morphological evolution of the meandering reaches downstream the Three Gorges Dam. *J. Geogr. Sci.* **2017**, *27*, 1268–1278. [CrossRef]

17. Owusu, K.; Obour, P.B.; Nkansah, M.A. Downstream effects of dams on livelihoods of river-dependent communities: The case of Ghana's Kpong Dam. *Geogr. Tidsskr. Dan. J. Geogr.* **2016**, *117*, 1–10. [CrossRef]

18. Jiang, L.; Ban, X.; Wang, X.; Cai, X. Assessment of hydrologic alteration caused by the Three Gorges Dam in the middle and lower reaches of Yangtze River, China. *Water* **2014**, *6*, 1419–1434. [CrossRef]

19. Lee, J.E.; Heo, J.-H.; Lee, J.; Kim, N.W. Assessment of flood frequency alteration by dam construction via SWAT Simulation. *Water* **2017**, *9*, 264. [CrossRef]

20. Zhang, X.; Dong, Z.; Gupta, H.; Wu, G.; Li, D. Impact of the Three Gorges Dam on the hydrology and ecology of the Yangtze River. *Water* **2016**, *8*, 590. [CrossRef]

21. Mbaka, J.G.; Mwaniki, M.W. A global review of the downstream effect of small impoundments on stream habitat conditions and macroinvertebrates. *Environ. Rev.* **2015**, *23*, 257–263. [CrossRef]

22. Gradziński, R.; Baryła, J.; Doktor, M.; Gmur, D.; Gradziński, M.; Kędzior, A.; Paszkowski, M.; Soja, R.; Zieliński, T.; Żurek, S. Vegetation-controlled modern anastamosing system of the upper Narew River (NE Poland) and its sediments. *Sediment. Geol.* **2003**, *157*, 253–276. [CrossRef]

23. Marcinkowski, P.; Grabowski, R.C.; Okruszko, T. Controls on anastomosis in lowland river systems: Towards process-based solutions to habitat conservation. *Sci. Total Environ.* **2017**, *609*, 1544–1555. [CrossRef] [PubMed]

24. Cygan, B.; Niedbała, J.; Piekarski, M.K. Wpływ Zbiornika Siemianówka na kształtowanie się charakterystyk hydrologicznych rzeki Narwi. In Proceedings of the Conference Materials Zagospodarowanie Zlewni Bugu i Narwi w Ramach Zrównoważonego Rozwoju 2003, Popowo, Poland, 23–24 May 2003. (In Polish)

25. Jekatierynczuk-Rudczyk, E.; Górniak, A. Influence of Siemianówka Reservoir on Narew River below dam. In *Ecosystem of Siemianówka Reservoir in 1990–2004 and its Restoration*; Górniak, A., Ed.; Department of Hydrobiology, University of Białystok: Białystok, Poland, 2006; pp. 193–199. (In Polish)

26. Kiczko, A.; Romanowicz, R.J.; Osuch, M. Impact of water management policy on flow conditions in wetland areas. *Phys. Chem. Earth* **2011**, *36*, 638–645. [CrossRef]

27. Mioduszewski, W.; Gajewski, G.; Biesiada, M. Zróżnicowanie stosunków wodnych w dolinie Narwi w granicach Narwiańskiego Parku Narodowego. *Water Environ. Rural Areas* **2004**, *11*, 39–50. (In Polish)

28. Romanowicz, R.J.; Osuch, M. Assessment of land use and water management induced changes in flow regime of the Upper Narew. *Phys. Chem. Earth* **2011**, *36*, 662–672. [CrossRef]

29. Grabowska, M.; Ejsmont-Karabin, J.; Karpowicz, M. Reservoir-river relationships in lowland, shallow, eutrophic systems: An impact of zooplankton from hypertrophic reservoir on river zooplankton. *Pol. J. Ecol.* **2013**, *61*, 759–768.

30. Grabowska, M.; Mazur-Marzec, H. The effect of cyanobacterial blooms in the Siemianówka Dam Reservoir on the phytoplankton structure in the Narew River. *Oceanol. Hydrobiol. Stud.* **2011**, *40*, 19–26. [CrossRef]

31. Karpowicz, M. Influence of eutrophic lowland reservoir on Crustacean zooplankton assemblages in river valley oxbow lakes. *Pol. J. Environ. Stud.* **2014**, *23*, 2055–2061.

32. Karpowicz, M. Microcrustacean (*Cladocera, Copepoda*) source-sink dynamics in a lowland river ecosystem with a dam reservoir. *Oceanol. Hydrobiol. Stud.* **2016**, *45*, 297–303. [CrossRef]

33. Banaszuk, P.; Wysocka-Czubaszek, A. Phosphorus dynamics and fluxes in a lowland river: The Narew anastomosing river system, NE Poland. *Ecol. Eng.* **2005**, *25*, 429–441. [CrossRef]

34. Banaszuk, H.; Banaszuk, P.; Gradziński, R.; Kamocki, A.K.; Mioduszewski, W.; Okruszko, T.; Próchnicki, P.; Szewczyk, M. *Przyroda Podlasia: Narwiański Park Narodowy*; Ekonomia Podlasia: Podlasie, Poland, 2004; ISBN 83-87231-07-X. (In Polish)

35. BIPROMEL. *Siemianówka Reservoir–Water Management Rules*; Technical Report; Bipromel: Warszawa, Poland, 1999. (In Polish)

36. Sokołowski, J. *Monografia Zbiornika Wodnego Siemianówka*; WZMIUW: Warszawa, Poland, 1999. (In Polish)

37. Pettitt, A.N. A non-parametric approach to the change-point problem. *Appl. Stat.* **1979**, *28*, 126–135. [CrossRef]
38. Rybski, D.; Neumann, J. A Review on the Pettitt Test. In *In Extremis*; Kropp, J., Schellnhuber, H.J., Eds.; Springer: Berlin/Heidelberg, Germany, 2011; pp. 202–213, ISBN 978-3-642-14863-7.
39. Tan, X.; Gan, T.Y.; Shao, D. Effects of persistence and large-scale climate anomalies on trends and change points in extreme precipitation of Canada. *J. Hydrol.* **2017**, *550*, 453–465. [CrossRef]
40. Huh, S.; Dickey, D.A.; Meador, M.R.; Ruhl, K.E. Temporal analysis of frequency and duration of low and high streamflow: Years of record needed to characterise streamflow variability. *J. Hydrol.* **2005**, *310*, 78–94. [CrossRef]
41. Marcinkowski, P.; Piniewski, M.; Kardel, I.; Szcześniak, M.; Benestad, M.; Srinivasan, R.; Ignar, S.; Okruszko, T. Effect of Climate Change on Hydrology, Sediment and Nutrient Losses in Two Lowland Catchments in Poland. *Water* **2017**, *9*, 156. [CrossRef]
42. Banaszuk, P.; Kamocki, A. Effects of climatic fluctuations and land-use changes on the hydrology of temperate fluviogenous mire. *Ecol. Eng.* **2008**, *32*, 133–146. [CrossRef]
43. Kamocki, A.; Kołos, A.; Banaszuk, P. Can we effectively stop the expansion of trees on wetlands? Results of a birch removal experiment. *Wetl. Ecol. Manag.* **2016**, *25*, 359–367. [CrossRef]
44. Grygoruk, M.; Batelaan, O.; Mirosław-Świątek, D.; Szatyłowicz, J.; Okruszko, T. Evapotranspiration of bush encroachments on a temperate mire meadow—A nonlinear function of landscape composition and groundwater flow. *Ecol. Eng.* **2014**, *73*, 598–609. [CrossRef]
45. Deoniziak, K.; Hermaniuk, A.; Wereszczuk, A. Effects of wetland restoration on the amphibian community in the Narew River Valley (Northeast Poland). *Salamandra* **2017**, *53*, 50–58.
46. Grygoruk, M.; Frąk, M.; Chmielewski, A. Agricultural rivers at risk: Dredging results in a loss of macroinvertebrates. Preliminary observation from the Narew Catchment, Poland. *Water* **2015**, *7*, 4511–4522. [CrossRef]
47. Nanson, G.C.; Knighton, A.D. Anabranching rivers: Their cause. character and classification. *Earth Surf. Process. Landf.* **1996**, *21*, 217–239. [CrossRef]
48. Schumann, R.R. Morphology of Red Creek, Wyoming, an arid-region anastomosing channel system. *Earth Surf. Process. Landf.* **1989**, *14*, 277–288. [CrossRef]
49. Próchnicki, P. The expansion of common reed (phragmites australis (cav.) trin. ex steud.) in the anastomosing river valley after cessation of agriculture use (Narew River valley, NE Poland). *Pol. J. Ecol.* **2005**, *53*, 353–364.
50. Stromberg, J.C.; Lite, S.J.; Marler, R.; Paradzick, C.; Shafroth, P.B.; Shorrock, D.; White, J.M.; White, M.S. Altered stream-flow regimes and invasive plant species: The Tamarix case. *Glob. Ecol. Biogeogr.* **2007**, *16*, 381–393. [CrossRef]
51. Galatowitsch, S.M.; Larson, D.L.; Larson, J.L. Factors affecting post-control reinvasion by seed of an invasive species, Phragmites australis, in the central Platte River, Nebraska. *Biol. Invasions* **2016**, *18*, 2505–2516. [CrossRef]
52. Piniewski, M.; Laize, C.L.R.; Acreman, M.; Okruszko, T.; Schneider, C. Effects of climate change on environmental flow indicators in the Narew Basin, Poland. *J. Environ. Qual.* **2011**, *43*, 155–167. [CrossRef] [PubMed]
53. Kiczko, A.; Napiórkowski, J. Aspiration-Reservation Decision Support System fo Siemianówka Reservoir. In *Modelling of Hydrological Processes in the Narew Catchment*; Springer: Berlin, Germany, 2011; pp. 111–121.
54. Kiczko, A.; Romanowicz, R.; Napiórkowski, J.; Piotrowski, A. Integration of reservoir management and flow routing model—Upper Narew Case Study. *Publ. Inst. Geophys. Pol. Acad. Sci.* **2008**, *E-9*, 41–56.
55. Pugacewicz, E. Zmiany w awifaunie lęgowej doliny Górnej Narwi w latach 1986–2007. *Dubelt* **2012**, *4*, 1–41. (In Polish)
56. Szewczyk, M.; Dembek, W.; Kamocki, A. Response of Riparian vegetation to the decrease of flooding: Narew National Park, Poland. In Proceedings of the International Conference 'Towards Natural Flood Reduction Strategies', Warsaw, Poland, 6–13 September 2003. Available online: http://www.academia.edu/download/36018017/3_9l.pdf (accessed on 2 August 2017).

Article

Heating Impact of a Tropical Reservoir on Downstream Water Temperature: A Case Study of the Jinghong Dam on the Lancang River

Bo Jiang, Fushan Wang * and Guangheng Ni

Department of Hydraulic Engineering, Tsinghua University, Beijing 100084, China; jiangbo117@gmail.com (B.J.); ghni@tsinghua.edu.cn (G.N.)
* Correspondence: wfs14@mails.tsinghua.edu.cn; Tel.: +86-178-888-34703

Received: 1 June 2018; Accepted: 11 July 2018; Published: 17 July 2018

Abstract: Reservoirs change downstream thermal regimes by releasing water of different temperatures to that under natural conditions, which may then alter downstream biodiversity and ecological processes. The hydropower exploitation in the mainstream Lancang-Mekong River has triggered concern for its potential effects on downstream countries, especially the impact of the released cold water on local fishery production. However, it was observed recently that the annual water temperature downstream of the Jinghong Reservoir (near the Chinese border) has increased by 3.0 °C compared to its historical average (1997–2004). In this study, a three-dimensional (3D) model of the Jinghong Reservoir was established to simulate its hydro- and thermodynamics. Results show that: (1) the impoundment of the Jinghong Reservoir contributed about 1.3 °C to the increment of the water temperature; (2) the solar radiation played a much more important role in comparison with atmosphere-water heat exchange in changing water temperatures; and (3) the outflow rate also imposed a significant influence on the water temperature by regulating the residence time. After impoundment, the residence time increased from 3 days to 11 days, which means that the duration that the water body can absorb solar radiation has been prolonged. The results explain the heating mechanism of the Jinghong Reservoir brought to downstream water temperatures.

Keywords: tropical reservoir; heating impact; Langcang-Mekong River

1. Introduction

The riverine environment can be influenced by reservoirs and dams, as well as their operations in many forms, including the changing of riverine thermal regimes and downstream water temperatures [1–3]. The health, distribution, and functions of aquatic creatures can be influenced by water temperatures [4–7], so the extensive construction of dams worldwide has long drawn attention to potential effects of damming on downstream thermal regimes [2,8].

Generally, the impact degree of a dam on downstream thermal regimes is decided by its mode of operation and specific mechanism of water release [9]. Many large dams release water from deep portals which are located under the thermocline, namely the hypolimnetic layer of a reservoir. As a result, the cold water is released to the downstream thermal regimes [2,10–12]. This case was rarely reported, but some small dams release water from above the thermocline, namely the epilimnetic layer, so temperatures of the downstream water increase [7].

In addition to annual water temperature, reservoirs also influence seasonal thermal patterns of downstream water. In general, in large reservoirs, the moderate temperatures of downstream water in spring and summer are lower than those in winter when the seasonal fluctuations decrease, while they also display a delay of maxima in comparison with natural rivers [2]. Such phenomena were observed at many dams and reservoirs across the world, such as the Dartmouth Dam on the Mitta Mitta River and Burrendong Dam on the Macquarie River, located in Australia [13,14], the regulated Lyon River located in Scotland [15], and the Hills Creek Dam on the Willamette River, located in America [16].

As the upstream part of the Lancang-Mekong River, the Lancang River is the largest international river in Southern Asia [17]. Since the 1950s, to fully exploit the river's resources, the Chinese government has come up with proposals concerning damming of mainstreams on the Lancang River [18]. At present, six large dams have been put into operation along the mainstream of the Lancang River. These dams include Nuozhadu, Manwan, Gongguoqiao, Dachaoshan, Xiaowan, and Jianghong dams (ranging from upstream to downstream), wherein the Nuozhadu Reservoir was recently put into operation in 2015 [19]. The cascade of dams is constructed with the primary purpose of hydropower generation. According to the hydropower development plan by Huaneng Lancang River Hydropower Company, the hydropower installed capacity will reach 30.0 GW in the Lancang mainstream by 2020 [19]. Nevertheless, the potential hydrological and environmental effects of reservoir impoundment and water release have drawn more attention to downstream countries [17,20–22]. Among these potential effects, altered downstream water temperature has become an important focus as it plays a significant role in influencing the combination of structure, growth, reproduction, distribution, and stream productivity of aquatic organisms [23], in addition to which fish in freshwater serve as the major protein source for local animals which live in downstream countries [19]. In view of those deep and large dams located along the Lancang River, the low-temperature outflow has drawn major concern. In addition, the Nuozhadu Dam is equipped with a multi-level stop log door so as to moderate the downstream water temperature to a pre-dam level [24]. Nevertheless, after the impoundment of six large reservoirs, the water temperatures downstream of the Jinghong Reservoir (the most downstream one of the cascade dams) increased, rather than decreased, in all seasons, which is against what has been observed in other large dams. In other words, the upstream multi-level intake structures could probably be unnecessary. As numerical modeling is a powerful and useful tool, we applied it in investigating the reasons why the Jinghong Reservoir had a unique heating impact on downstream water.

Some one-dimensional (1D) models are applicable to simulate the vertical distribution of the water temperature and chemical/biological materials in a lake or reservoir through time [25]. In general, these models are established under the one-dimensional assumption. Specifically speaking, variations in the lateral directions are relatively small in comparison with those in the vertical directions [26] and are widely used because they are fast enough to facilitate long-term simulation and have performed well in simulating the seasonal and inter-annual variations of lake water temperatures [27]. These models include the bulk model of Kraus and Turner [28], DYRESM [29], PROBE [30], the Hostetler Model [31,32], SEEMOD [33], LIMNMOD [34], MASAS and CHEMSEE [35], Minlake [36], GOTM [37], SIMSTRAT [38], LAKE [39,40], CLM4-LISSS [41], and WRF-Lake [42].

Although 1D lake models are more efficient in computation, under certain conditions (such as long, deep reservoirs) water mass exchange in both vertical and longitudinal directions and temperature gradients may be important [25]. Thus, many two-dimensional (2D) models have been developed through the integration of the reservoir, with the aim to reach the motion equations with lateral averaging [43]. Laterally averaged 2D models are applicable to modeling of long and relatively narrow reservoirs with no or negligible lateral inflow or outflows. These models include the model of Box Exchange Transport Temperature and Ecology of Reservoirs (BETTER) [44]; the Computation of Reservoir Stratification (COORS) model [45]; the Laterally Averaged Reservoir Model (LARM) [46]; the model of Generalized Longitudinal-Vertical Hydrodynamics and Transport (GLVHT) [47], which was developed from LARM; the CE-QUAL-W2 developed through GLVHT expansion to include water quality constituents [48]; the Laterally-Averaged Hydrodynamics Model (LAHM) [43]; and the MIKE21 Flow Model, developed by Danish Hydraulic Institute (DHI) Water and Environment [49–52], which is widely applied. In addition to the laterally-averaged models, vertically-averaged 2D models are also applied when vertical flow variations are not as important as the lateral case, e.g., when the water is shallow [53–55]. These models include the North Sea Model [56], the Tokyo Bay Model [57], the Haringvliet Model, and the depth-averaged mathematical model developed by McGuirk and Rodi [55].

Despite the extensive use of 1D and 2D models, the limit in one-dimensional or vertical and lateral averaging has long been realized and the needs for 3D modeling of lakes and reservoirs have raised more discussion [58–61]. In the past decades, the complex water quality models and 3D hydrodynamic models have been developed thanks to the high-performing super computers at a reasonable cost, as well as the efficient numerical algorithms [62]. The mentioned models include the Curvilinear-grid Hydrodynamics model in three dimensions (CH3D) [63], the Estuarine, Coastal, and Ocean Model (ECOM) [64]; the Environmental Fluid Dynamics Code (EFDC) [65]; the CE-QUAL-ICM model [66]; the model of Water Quality Analysis Simulation Program (WASP) [67]; the Delft3D-FLOW model [68]; the Row-Column AESOP (RCA) model [69]; and the MIKE 3 model by DHI Water and Environment [70].

In order to analyze the complex 3D hydrodynamic and thermodynamic processes more deeply, it is necessary to establish 3D models. Nevertheless, researchers have so far rarely established 3D models for simulation of the environmental and thermal results of damming along the Lancang River [19] in view of the high cost of computation and the lack of observational data. Nowadays, with computational resources and enough pre-dam and post-dam measurement data, we are capable of conducting 3D simulations at the Jinghong Reservoir. The Delft3D-FLOW model was selected for the simulations as it has been widely used around the world and evidence proves that it can simulate sediment transport, flows, water quality, waves, morphological developments, and ecological courses in coastal regions, rivers, and lakes [71–78]. After model calibration, the 3D model of the Jinghong Reservoir was used to simulate the thermodynamic and hydrodynamic situations in the reservoir, as well as the related outflow. The research also carried out some numerical experiments to analyze the heating impact brought by the Jinghong Reservoir to downstream water.

2. Study Area

The Jinghong Reservoir system is located in the lower reaches of the Lancang River, which is, in general, a south-flowing river in southwestern China, and called the Mekong River when entering Laos (Figure 1). This system is defined by the Nuozhadu Dam (22°38′ N, 100°26′ E), the Jinghong Dam (22°03′ N, 100°46′ E), and the 105-km long water body between the two (i.e., the Jinghong Reservoir). The Jinghong Dam went into service in 2009 with a height of 108 m, while the Nuozhadu Dam was put into operation in 2015 with a height of 262 m. The Jinghong Reservoir is a long and narrow channel-shaped reservoir with a length of 105 km and an average width of 312 m. It has a normal water level of 602 m above sea level, a maximum depth of 70 m and a total reservoir capacity of 1.14 billion m^3. Since the operation of the reservoir, the water surface area has increased from 12.7 km^2 to 32.8 km^2. The mean annual inflow is about 1674 m^3 s^{-1}. A tropical reservoir, the Jinghong Reservoir, has a perennial mean air temperature reaching 22.2 °C at the Jinghong Dam.

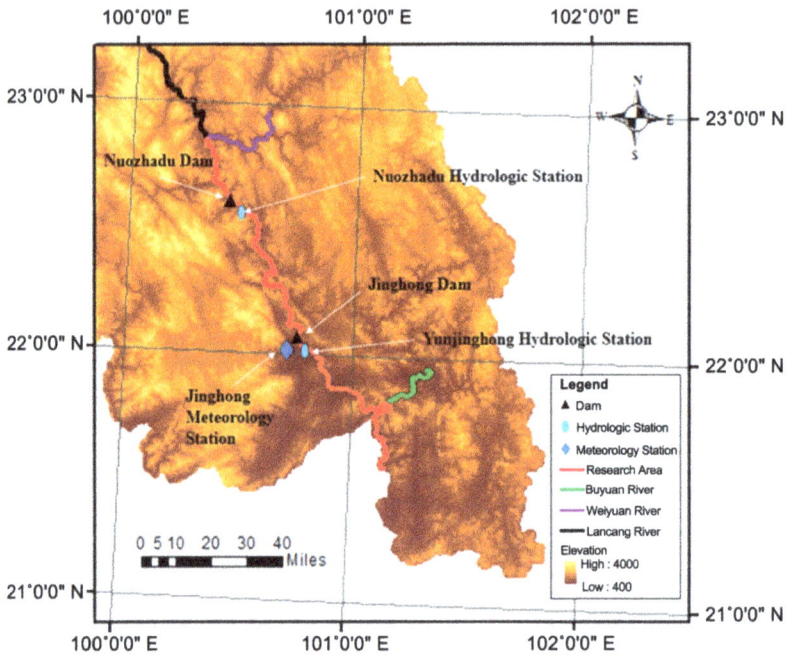

Figure 1. Location and topography of the Jinghong Reservoir system.

2.1. Observation Data

The two hydrological stations involved in the study area are the Nuozhadu Hydrological Station (NHS) and the Yunjinghong Hydrological Station (YHS), both located on the right bank of the river, with the former 5 km downstream of the Nuozhadu Reservoir, and the latter 3 km downstream of the Jinghong Reservoir. Both NHS and YHS are located at the riverine section of the water body, and the fully mixing assumption can be applied. Accordingly, point-measured water temperature can be used as an indicator of the cross-sectional water temperature. To measure the water temperature, water pressure, and air pressure, a Hobo Onset U20-001-02 water level data logger (U20 hereafter) is adopted at each hydrological station on an hourly basis throughout the study period. The water level at each station is computed by:

$$P_{water} = P_{total} - P_{air} = \rho_{water} g H_{U20}, \tag{1}$$

where P_{total} (Pa) is the U20-measured pressure underwater (water pressure and air pressure), P_{air} (Pa) is the U20-measured atmospheric pressure by the river bank, P_{water} (Pa) is the pressure of pure water, and H_{U20} is the depth of the U20, from which water levels are computed. The discharge at YHS is further computed using its rating curve obtained from the Water Resources Department of Yunnan Province, China [79].

2.2. Observation Data Analysis

2.2.1. Comparison with Historical Water Temperature

To illustrate the annual variations of water temperature at NHS and the Jinghong Reservoir, in situ observations were conducted (Figure 2). Water temperatures almost changed synchronously at the two sites throughout the year, with NHS always displaying a lower temperature (a mean difference of 2.2 °C) than the other. This systematic temperature difference reflected the latitude difference between the two sites as water flowed south, a phenomenon not observed at reservoirs flowing in other directions. According to a study on the Cougar Reservoir (an eastern-flowing reservoir in the U.S.), the peak of outflow water temperature lagged about 2 months behind that of inflow, but the inflow and outflow temperatures shared the same maximum and minimum values [80], indicating no large amount of energy was taken in, or lost, as the water body flowed into the reservoir.

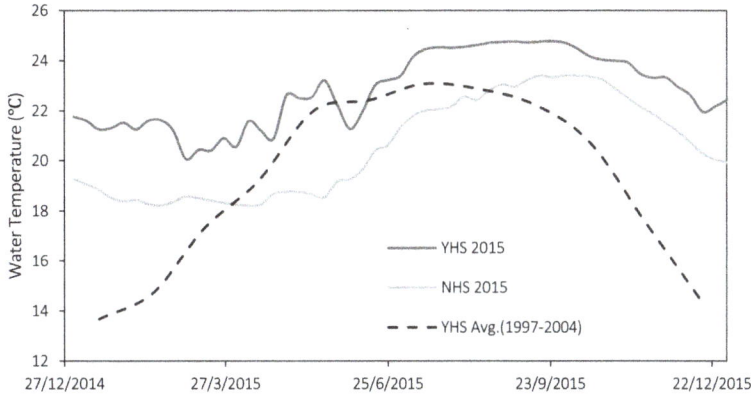

Figure 2. Comparison of observed water temperature data and historical averaged data at YHS. The temporal resolution of the observation data is daily.

Obtained and averaged from 1997 to 2004 at YHS, historical water temperature observations (Figure 2) were shown to range between 13.7 and 23.1 °C annually. However, after the impoundment of the Jinghong Reservoir, the water temperature range was changed to between 19.8 and 25.0 °C, with the annual variation reduced from 9.5 °C to 5.2 °C. This reduction in annual variation mainly resulted from a 6.1 °C increase in the minimum temperature. Overall, the annual average water temperature rose from 19.2 °C to 22.8 °C after impoundment.

2.2.2. Discharge and Water Temperature

Observed water temperatures at YHS were compared with outflow rates from the reservoir in Figure 3. As they were inversely correlated, the discharge rate of the Jinghong Dam could also be a potential factor behind the water temperature alteration.

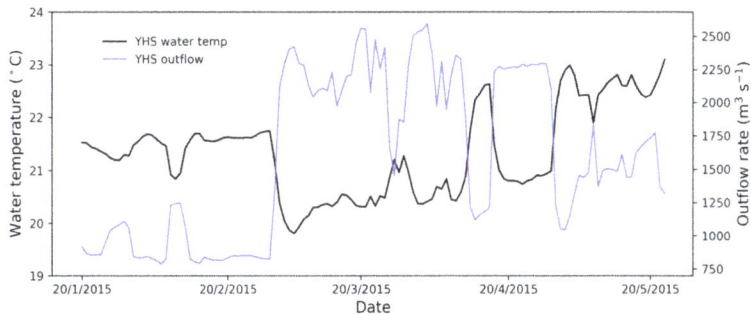

Figure 3. Comparison of the observed YHS water temperature and the YHS discharge rate.

3. Methods

3.1. The Delft3D-FLOW Model

The numerical hydrodynamic modeling system—the Delft3D-FLOW model—was developed by Deltares [68], a Dutch-based research institute. The Delft3D-FLOW software used for this study is based on Fortran and acquired from Delft3D website (https://oss.deltares.nl/web/delft3d/). With Delft3D-FLOW, the 2D (based on averaged depth) and 3D shallow water equations that were unsteady could be solved. The equation system comprises momentum equations in the horizontal plane, the continuity equation and the transport equations of conservative elements [68]. The basic equations are retained in their original form and are not changed in this study, which are introduced briefly as follows.

As for the hydrodynamic equations, Delft3D-FLOW is used for the solution of the Navier-Stokes equations with regard to a fluid which is incompressible under the assumption of Boussinesq and the shallow water environment [81]. In the equation of vertical momentum, the accelerations in the vertical direction are omitted, leading to the equation of hydrostatic pressure. Vertical speeds are computed according to the continuity equation in 3D models. The set of partial differential equations, together with a proper set of boundary and initial situations is solved via a mesh with finite differences. Delft3D-FLOW uses orthogonal curvilinear Cartesian coordinates (ξ, η) in the horizontal direction. Two different vertical grid systems are offered by Delft3D-FLOW, vertically: the Cartesian Z coordinate system (Z-model) and the σ coordinate system (σ-model). The σ coordinate system was first introduced by Phillips [82] and it is applied in this research.

To obtain the continuity equation with averaged depth, the continuity equation shall be integrated for incompressible fluids ($\nabla \bullet \vec{u} = 0$) based on the total depth, wherein the conditions of the kinematic boundary on the bed level and at the water surface are taken into account. It is expressed as follows:

$$\frac{\partial \zeta}{\partial t} + \frac{1}{\sqrt{G_{\xi\xi}}\sqrt{G_{\eta\eta}}}\frac{\partial\big((d+\zeta)U\sqrt{G_{\eta\eta}}\big)}{\partial \xi} + \frac{1}{\sqrt{G_{\xi\xi}}\sqrt{G_{\eta\eta}}}\frac{\partial\big((d+\zeta)V\sqrt{G_{\xi\xi}}\big)}{\partial \eta} = (d+\zeta)Q, \qquad (2)$$

where ζ (m) is the water level on a number of horizontal planes of reference (datum), the coefficients $\sqrt{G_{\xi\xi}}$ and $\sqrt{G_{\eta\eta}}$ (m) are used for transformation from curvilinear coordinates to rectangular coordinates, t (s) is the time, d (m) is the depth under a number of horizontal planes of reference (datum), U (m s^{-1}) is the depth-averaged speed in ξ-direction, V (m s^{-1}) is the depth-averaged speed in η-direction, and Q (m s^{-1}) is the contributions made for each unit area because of the withdrawal or release of water, evaporation and precipitation.

The momentum equations in ξ-direction and η-direction are expressed as follows:

$$\frac{\partial u}{\partial t} + \frac{u}{\sqrt{G_{\xi\xi}}}\frac{\partial u}{\partial \xi} + \frac{v}{\sqrt{G_{\eta\eta}}}\frac{\partial u}{\partial \eta} + \frac{\omega}{d+\zeta}\frac{\partial u}{\partial \sigma} - \frac{v^2}{\sqrt{G_{\xi\xi}}\sqrt{G_{\eta\eta}}}\frac{\partial\sqrt{G_{\eta\eta}}}{\partial \xi} + \frac{uv}{\sqrt{G_{\xi\xi}}\sqrt{G_{\eta\eta}}}$$
$$\frac{\partial\sqrt{G_{\xi\xi}}}{\partial \eta} - fv = -\frac{1}{\rho_0\sqrt{G_{\xi\xi}}}P_\xi + F_\xi + \frac{1}{(d+\zeta)^2}\frac{\partial}{\partial \sigma}\left(v_V\frac{\partial u}{\partial \sigma}\right) + M_\xi, \qquad (3)$$

$$\frac{\partial v}{\partial t} + \frac{u}{\sqrt{G_{\xi\xi}}}\frac{\partial v}{\partial \xi} + \frac{v}{\sqrt{G_{\eta\eta}}}\frac{\partial v}{\partial \eta} + \frac{\omega}{d+\zeta}\frac{\partial v}{\partial \sigma} - \frac{uv}{\sqrt{G_{\xi\xi}}\sqrt{G_{\eta\eta}}}\frac{\partial\sqrt{G_{\eta\eta}}}{\partial \xi} + \frac{u^2}{\sqrt{G_{\xi\xi}}\sqrt{G_{\eta\eta}}}$$
$$\frac{\partial\sqrt{G_{\xi\xi}}}{\partial \eta} - fu = -\frac{1}{\rho_0\sqrt{G_{\eta\eta}}}P_\eta + F_\eta + \frac{1}{(d+\zeta)^2}\frac{\partial}{\partial \sigma}\left(v_V\frac{\partial v}{\partial \sigma}\right) + M_\eta, \qquad (4)$$

where u, v, and ω (m s^{-1}) are speeds in ξ, η, and σ-directions. f (s^{-1}) is the Coriolis parameter (inertial frequency), ρ_0 (kg m^{-3}) is the referential density of water, P_ξ and P_η (kg m^{-2} s^{-2}) are the gradient-based hydrostatic pressure in ξ-direction and η-direction, F_ξ and F_η (m s^{-2}) are the imbalances existing in horizontal Reynold's stresses in the ξ-direction and η-direction, v_V (m^2 s^{-1}) is the vertical eddy viscosity, M_ξ and M_η (m s^{-2}) are source or sink of momentum in the ξ-direction and η-direction. Reynold's stresses F_ξ and F_η are simulated on the basis of the concept of eddy viscosity [83]. The eddy viscosities in vertical and horizontal directions are as follows:

$$v_H = v_{3D} + v_H^{back}; v_V = v_{mol} + \max\left(v_{3D}, v_V^{back}\right), \tag{5}$$

where v_{3D} is the viscosity computed with the 3D k-ε closure scheme, v_H^{back} is the user-defined background horizontal viscosity, v_V^{back} is the user-defined background vertical viscosity, v_{mol} (m^2 s^{-1}) is the kinematic viscosity (molecular) coefficient of water.

Transport of heat and matter is simulated by an equation of advection diffusion in three coordinate directions, and is expressed as follows:

$$
\begin{aligned}
\frac{\partial(d+\zeta)c}{\partial t} &+ \frac{1}{\sqrt{G_{\xi\xi}}\sqrt{G_{\eta\eta}}}\left\{\frac{\partial\left[\sqrt{G_{\eta\eta}}(d+\zeta)uc\right]}{\partial\xi} + \frac{\partial\left[\sqrt{G_{\xi\xi}}(d+\zeta)vc\right]}{\partial\eta}\right\} + \frac{\partial\omega c}{\partial\sigma} \\
&= \frac{d+\zeta}{\sqrt{G_{\xi\xi}}\sqrt{G_{\eta\eta}}}\left\{\frac{\partial}{\partial\xi}\left(D_H\frac{\sqrt{G_{\eta\eta}}}{\sqrt{G_{\xi\xi}}}\frac{\partial c}{\partial\xi}\right) + \frac{\partial}{\partial\eta}\left(D_H\frac{\sqrt{G_{\xi\xi}}}{\sqrt{G_{\eta\eta}}}\frac{\partial c}{\partial\eta}\right)\right\} + \frac{1}{d+\zeta}\frac{\partial}{\partial\sigma}\left(D_V\frac{\partial c}{\partial\sigma}\right) \\
&- \lambda_d(d+\zeta)c + S,
\end{aligned}
\tag{6}
$$

where c (kg m^{-3}) is the concentration of mass, D_H (m^2 s^{-1}) is the total diffusion coefficient in the horizontal direction, D_V (m^2 s^{-1}) is the diffusion coefficient in the vertical direction, λ_d (s^{-1}) is the first order decay course and S (kg m^{-2} s^{-1}) is the source and sink terms in the unit area generated from the withdrawal or release of water and/or heat exchange on the free surface. The diffusion coefficient in the horizontal direction is user-specified, whereas the coefficient in the vertical direction is computed as follows:

$$D_H = D_{SGS} + D_V + D_H^{back}; D_V = \frac{v_{mol}}{\sigma_{mol}} + \max\left(D_{3D}, D_V^{back}\right), \tag{7}$$

where D_{SGS} is the 2D part of diffusion based on the turbulence model of sub-grid scale, D_H^{back} is the user-defined eddy diffusivity in the horizontal direction, D_V^{back} is the eddy diffusivity in the vertical direction, σ_{mol} is the Prandtl-Schmidt number for molecular mixing, and D_{3D} is the diffusion due to turbulent eddy viscosity [68].

3.2. Model Set-Up

3.2.1. Grid Coverage

Along the river flow direction, the upper boundary of the model is set to the Nuozhadu Hydrological Station, whereas the lower boundary of the model is set to the Jinghong Dam. With a longitudinal length of about 95 km, the model covers most parts of the Jinghong Reservoir. The computational grid of the Jinghong Reservoir hydrodynamic model is shown in Figure 4. Grid density varies with topography, i.e., in order to increase the computation accuracy, finer horizontal grids have been placed at the part of the reservoir close to the Jinghong Dam, resulting in a total of about 11,000 horizontal grid cells per layer. Meanwhile, for the vertical direction, the σ coordinate system (σ-model) is applied, with the reservoir discretized into 40 vertical layers, with each layer less than 2 m in thickness.

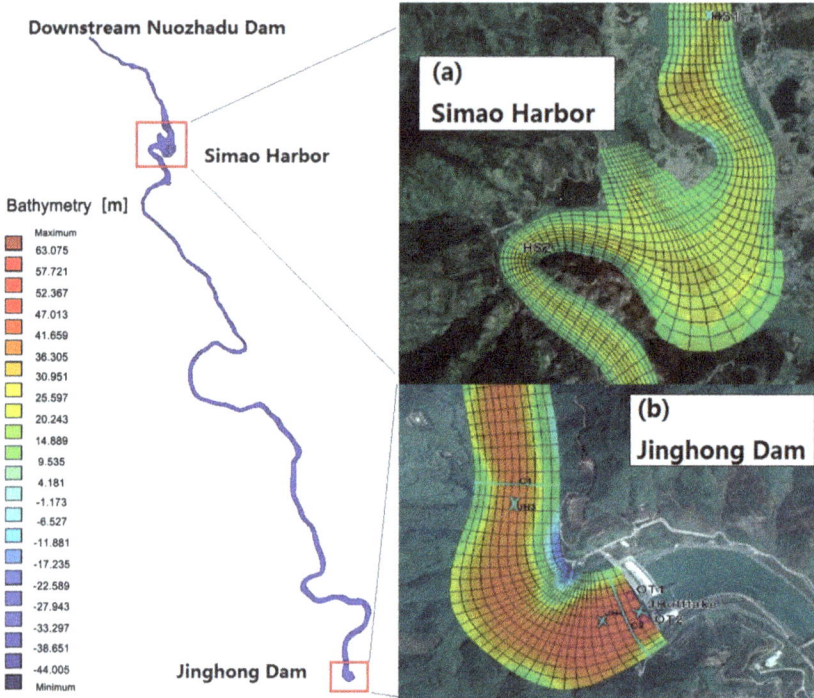

Figure 4. Horizontal discretization (about 11,000 horizontal grid points) and bathymetry of the Jinghong Reservoir with local enlargements at the site of (**a**) the Simao Harbor, and (**b**) the Jinghong Dam. We set 590 m above sea level as the datum (0 m) in the model so as to keep the water level a positive value through simulation period. Grid points with the riverbed higher than 590 m have a negative bathymetry value.

3.2.2. Boundary Conditions

The upper boundary is set at the water level at NHS and the lower boundary is set at the discharge rate downstream of the Jinghong Dam. The terrain data is extracted from a digital elevation model (DEM) with a spatial resolution of 25 m × 25 m.

3.2.3. Meteorological Data and Model Flow Chart

Restrained by the poor data availability, the simulation only covers a period from 20 December 2014 to 22 May 2015. The computational time step is set to 1 minute to meet the demand of numerical stability, as well as computational efficiency.

Meteorological data for this study, which is obtained from the Jinghong National Meteorological Station (Figure 5), is not readily available. As a result, such data is only gathered for a relatively short period, from 20 December 2014 to 20 May 2015, which also defines the study period for this research. These data include local air temperature data, relative humidity data, and solar radiation data with a temporal resolution of 1 day. We acknowledge that it would be better if a full year simulation could be conducted. However, due to solar radiation data availability, 5-month simulation is the best we can practice at the current stage. Similar simulations with temporal coverage of less than 1 year can also be found in the literature [84–86].

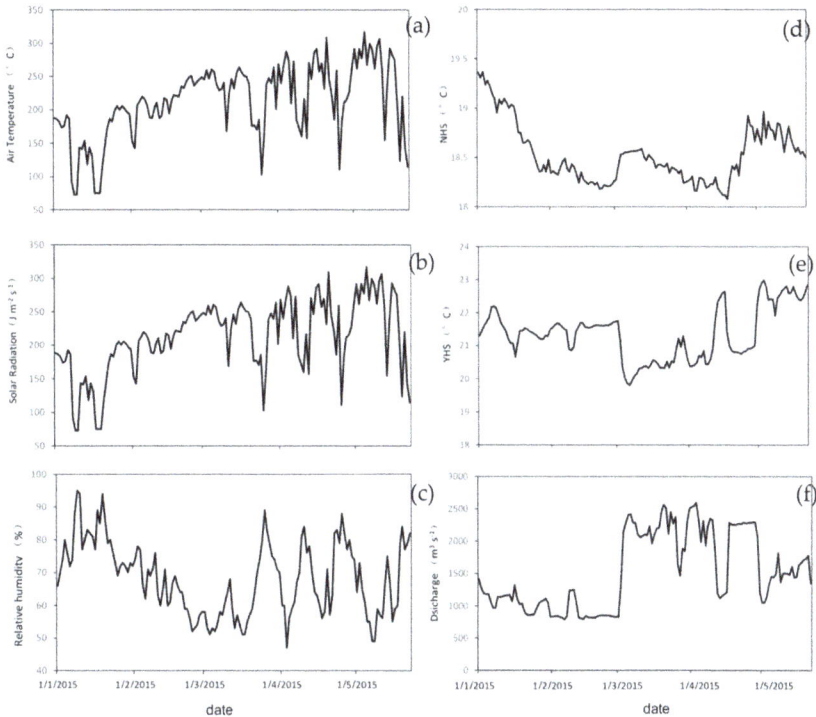

Figure 5. Daily meteorological data (1 January 2015 to 20 May 2015) measured at the Jinghong National Meteorological Station: (**a**) air temperature, (**b**) solar radiation, and (**c**) relative humidity. Water temperature data is measured at the NHS and YHS station: (**d**) water temperature at NHS, (**e**) water temperature at YHS, and (**f**) discharge at YHS.

A flowchart illustrating the Jinghong Reservoir hydrodynamic model is shown in Figure 6.

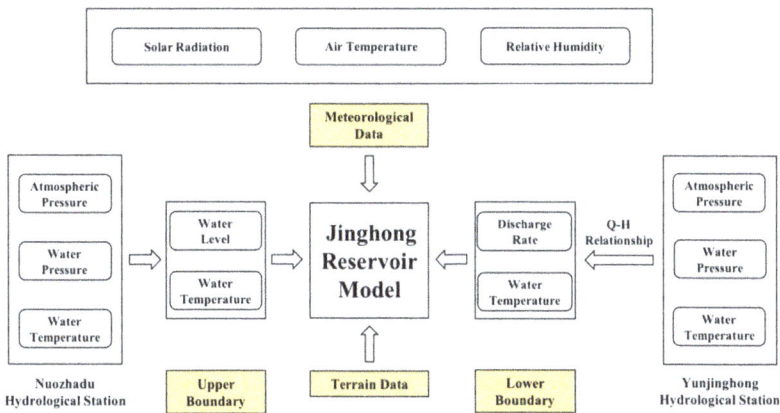

Figure 6. Flowchart of the Jinghong Reservoir model set-up.

3.3. Model Calibration and Validation

The warm-up period of this study occurs from 22 December 2014 to 31 January 2015.

For the purpose of verifying the hydro-dynamic condition of this simulation, the simulated water levels were firstly calibrated against measurements. As the model output temporal resolution was 10 minutes, the hourly average for simulated water levels was calculated for comparison. Data on 2 days' on-site observed water levels at Simao Harbor were provided by field experiment. The best match of the model output and measured data (Figure 7) were achieved by applying critical parameters, as shown in Table 1.

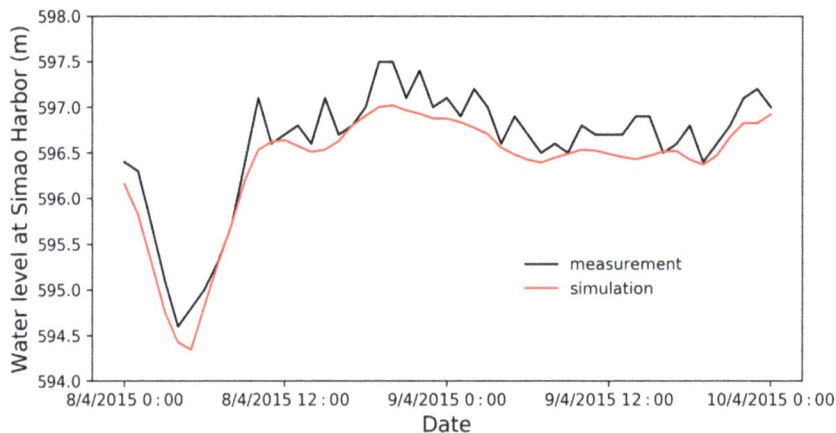

Figure 7. Comparison between simulated and measured water levels at Simao Harbor. Simulated water levels are plotted on an hourly average.

Five main parameters were calibrated by Chanudet et al. (2012): the Chezy's coefficient, which represents the roughness at the water bottom, background horizontal and vertical eddy viscosity, and background horizontal and vertical eddy diffusivity. The studied range of Chezy's coefficient can be referred to Chow (1959), while the range of the background eddy viscosity and diffusivity can also be referred to in previous studies [87–89].

Table 1. Parameters studied for the calibration of the Jinghong Reservoir model and their chosen value.

Parameter	Studied Range	Chosen Value
Chezy's coefficient	$10\text{–}100 \text{ m}^{\frac{1}{2}} \text{ s}^{-1}$	$65 \text{ m}^{\frac{1}{2}} \text{ s}^{-1}$
Background horizontal eddy viscosity	$1 \times 10^{-6}\text{–}1 \text{ m}^2 \text{ s}^{-1}$	$5 \times 10^{-4} \text{ m}^2 \text{ s}^{-1}$
Background horizontal eddy diffusivity	$1 \times 10^{-6}\text{–}1 \text{ m}^2 \text{ s}^{-1}$	$1 \times 10^{-4} \text{ m}^2 \text{ s}^{-1}$
Background vertical eddy viscosity	$1 \times 10^{-10}\text{–}1 \text{ m}^2 \text{ s}^{-1}$	$1 \times 10^{-6} \text{ m}^2 \text{ s}^{-1}$
Background vertical eddy diffusivity	$1 \times 10^{-10}\text{–}1 \text{ m}^2 \text{ s}^{-1}$	$1 \times 10^{-6} \text{ m}^2 \text{ s}^{-1}$

The focus of the calibration then shifted to outlet water temperatures downstream the Jinghong Dam. As YHS was within 3 km downstream of the Jinghong Reservoir in the riverine section of the water body, the simulated outlet water temperature was compared with the measured water temperature at YHS for calibration, with the difference between the two shown in Figure 8. The simulated results were in good agreement with measurements, with the absolute error limited to 1.5 °C.

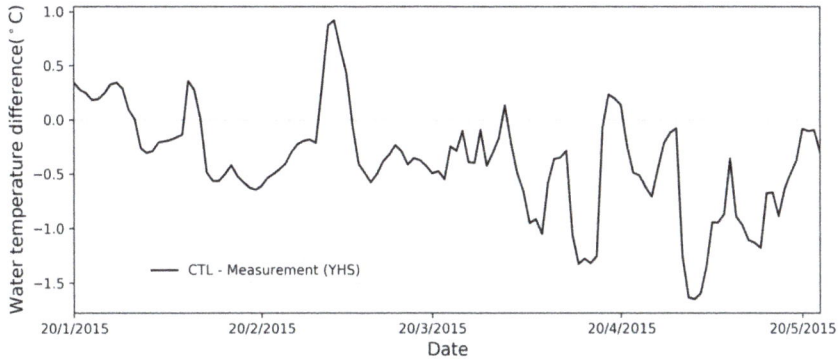

Figure 8. Outlet water temperature difference between the simulation results (CTL) and measurements at YHS.

Compared with measurements, the Jinghong Reservoir model produced good simulations of water temperatures. Nevertheless, since insufficient observations of the water velocity fields were available, no comparison concerning water velocity was carried out either for calibration or for validation.

3.4. Numerical Scenario Setups

As presented in Section 2.2, three main issues are ought to be found out through numerical modeling: the impact of impoundment of Jinghong Reservoir; the contributing factor to water temperature increment; and the relation between outflow rate and outflow temperature from Jinghong Dam. The wind speed and relative humidity were precluded as the driving forces and several numerical experiments were performed (Table 2).

● **Impact of Impoundment of the Jinghong Reservoir**

A control run (CTL) was set up as described in Sections 3.2 and 3.3.

S1 (i.e., scenario 1) was set up to examine the impact of the impoundment of the Jinghong Reservoir, in which the Jinghong Dam was removed and the water body returned to the natural river conditions. A depth-averaged two-dimensional flow model with the same horizontal grid discretization was set up, with the upper boundary condition as described in Section 3.2.2, the Q-H rating-curve adapted as the downstream boundary condition and the atmospheric data as for CTL.

● **Contributing Factors**

To examine the contribution of solar energy to the increase of water temperatures, S2 was imposed with no solar radiation. To further evaluate the impact of heat exchange with the overlying atmosphere on water temperature, atmosphere-water heat conduction, together with solar radiation was turned down in S3.

● **Outflow Rate and Outflow Temperature**

Different scenarios were further established so as to investigate the relationship between the outflow rate and the outflow water temperature at the Jinghong Dam. The water temperature measured at YHS was constantly higher than the one measured at NHS, suggesting that the water parcel experienced an energy absorption process travelling from NHS to YHS. In addition to the rate of energy transfer, the amount of energy absorbed by the water parcel was also determined by the duration of the absorption process. The outflow rate directly affected the time during which the water parcel absorbed energy. In order to measure the duration of the energy absorption process, the tracer function of the model was enabled. The duration of energy absorption is represented by the resident time of the tracer. Different scenarios were established to assess the relationship among the outflow rate, resident time, and inflow-outflow temperature difference. In addition, for the sheer size difference between the Nuozhadu Reservoir (upstream) and the Jinghong Reservoir (downstream), the Jinghong Reservoir's outflow rate was determined by the outflow rate of the Nuozhadu Reservoir. To further simplify the simulation, an assumption was made that the Jinghong Reservoir's inflow rate is equal its outflow rate. The inflow water temperature at the upper boundary was set to the same as for CTL. The simulation duration remained the same as CTL.

The real outflow rate for the Jinghong Reservoir ranged from 536 to 3398 $m^3 s^{-1}$ and with an annual average of 1674 $m^3 s^{-1}$ in 2015. Eight scenarios were established to examine the effect of outflow rates of the Jinghong Dam, with outflow rates prescribed from 500 to 4000 $m^3 s^{-1}$, with 500 $m^3 s^{-1}$ as intervals. The eight scenarios had been denoted as F1–F8 in the study.

Table 2. An overview of numerical experiments.

Experiments	The Jinghong Reservoir	Solar Radiation	Atmosphere-Water Heat Exchange	Outflow Rate ($m^3 s^{-1}$)	Tracer
CTL	ON	ON	ON	Real	OFF
S1	OFF	ON	ON	Real	OFF
S2	ON	OFF	ON	Real	OFF
S3	ON	OFF	OFF	Real	OFF
F1	ON	ON	ON	500	ON
F2	ON	ON	ON	1000	ON
F3	ON	ON	ON	1500	ON
F4	ON	ON	ON	2000	ON
F5	ON	ON	ON	2500	ON
F6	ON	ON	ON	3000	ON
F7	ON	ON	ON	3500	ON
F8	ON	ON	ON	4000	ON

4. Results and Discussion

4.1. Thermal Structure of the Jinghong Reservoir

Covering a distance of 105 km, the Jinghong Reservoir has a water depth roughly increasing along the direction of the river flow. The thermal structure along the reservoir from 10 km down the Nuozhadu Dam to the Jinghong Dam (about 90 km) is shown in Figure 9 (3 days corresponding to different months were selected for demonstration). Generally, the water temperature was almost homogeneous during the cool season from December 2014 to February 2015. As the air temperature increased from February on, a warmer water layer appeared at the surface, which eventually developed into a weakly stratified thermocline at the depth of approximately 5 m by the end of May 2015. Unlike many natural lakes, which experienced overturn in early spring, no overturn was observed during the study period at the Jinghong Reservoir, as it was a tropical one and the surface water temperature never went below 4 °C.

During the study period, only weak stratification was observed in the Jinghong Reservoir in May with the surface-bottom temperature difference reaching only 3 °C. In comparison, much stronger stratifications were reported at other deep reservoirs along the Lancang River after impoundment. For example, the Nuozhadu Reservoir (upstream reservoir to the Jinghong Reservoir) experienced a surface-bottom temperature difference of 10 °C in summer and the largest temperature difference in the Xiaowan Reservoir even reached 14 °C [90]. One main reason is the depth difference between these reservoirs. The Jinghong Reservoir is 70 m deep, while the Nuozhadu and Xiaowan Reservoirs are much deeper, each more than 200 m. By preventing solar radiation and wind-driven eddy penetration, large water bodies in these two deep reservoirs retard heat transfer and are difficult to reach a homogeneous temperature. Another reason is the water exchange index, or the α index, which is a widely used approach [91] in identifying reservoir thermal structures, defined by:

$$\alpha = \frac{w}{v} \, ,\tag{8}$$

where w (m^3) is average annual inflow and v (m^3) is the total capacity of the reservoir. When $\alpha \leq 10$, the reservoir is stably stratified; when $10 < \alpha < 20$, the reservoir is unstably stratified; when $\alpha \geq 20$, the reservoir is mixed. The α index of the Jinghong Reservoir is much higher compared with that of the other two (Table 3). The large quantity of annual inflow (~25 times that of reservoir capacity) serves as a major disturbance, leaving the water body well mixed.

Table 3. The α index of three major reservoirs along the middle and lower reaches of the Lancang River and their thermal structure type according to the index.

Reservoir Name	Average Annual Inflow (billion m^3)	Total Capacity (billion m^3)	α Index	Type
Xiaowan	38.2	15.1	2.5	Stably stratified
Nuozhadu	55.5	23.7	2.3	Stably stratified
Jinghong	58.0	1.1	25.4	Mixed

Vertically, the water columns along the reservoir showed almost no stratification regardless of their location and depth. The water temperature difference between surface and bottom was less than 1 °C in the first 40 km along the reservoir in all simulated time, while its maximum of 3 °C was reached at the Jinghong Dam in May. Longitudinally, water surface temperature rose as the river flowed southward, with its value down the Nuozhadu Dam increasing from around 18.5 °C all the way down the river, to approximately 22 °C before the Jinghong Dam. Temporally, water temperature changed in accordance with air temperature. When the air temperature was lower than the water temperature in January and February, the water temperature experienced a drop, while when the air temperature was higher than the water temperature from March to May, the water temperature rose. However, the water temperature variation also increased longitudinally, with the variation down the Nuozhadu Dam being only 0.5 °C and reaching 3 °C at the Jinghong Dam.

(a)

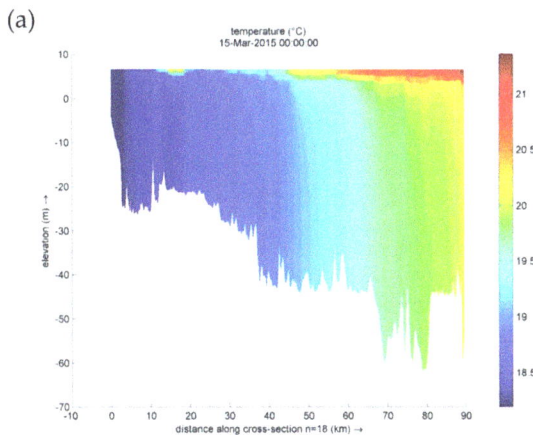

Figure 9. *Cont.*

(b)

temperature (°C)
15-Apr-2015 00:00:00

(c)

temperature (°C)
15-May-2015 00:00:00

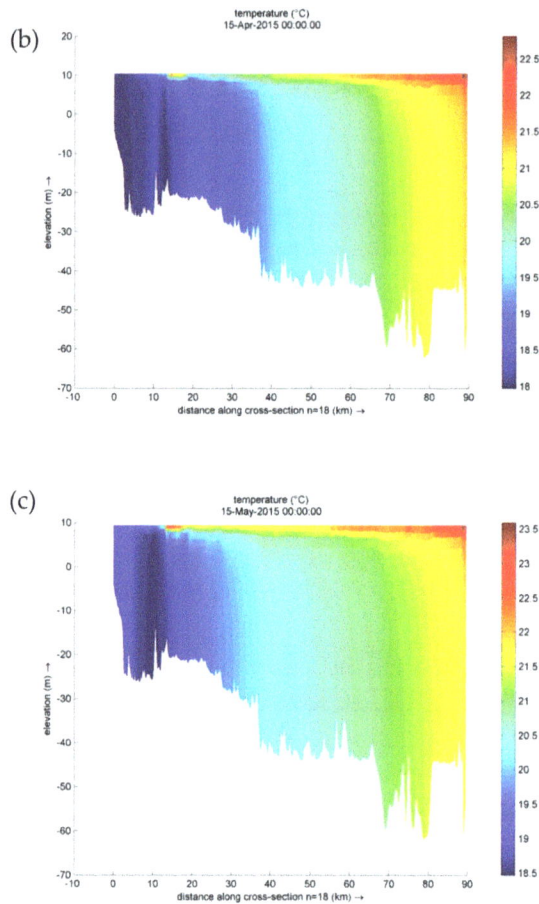

Figure 9. Simulated water temperature profile along the reservoir on (**a**) 15th March, (**b**) 15th April, and (**c**) 15th May of 2015.

4.2. Impact of Solar Radiation and Air Temperature

The absorption of solar radiation and atmosphere-water heat conduction are the two main heat exchange mechanisms governing a water body and the environment. Therefore, the contribution to water temperature alteration provided by solar radiation and atmosphere-water heat conduction can be revealed by comparing the model results of S2 and S3. By removing solar radiation, scenario S2, a notable water temperature drop with a mean value of 3 °C compared with CTL (Figure 10) has been observed. In fact, the water temperature simulated by S2 is rather close to the one measured at the NHS with a mean difference of only 0.4 °C. This indicates that the water parcel does not take in or emit substantial energy while traveling from the NHS to YHS. By further neglecting atmosphere-water heat conduction, S3 produces similar water temperature profiles to the ones of S2. The mean simulated temperature in these two scenarios differs about 0.3 °C, which suggests that the solar radiation, rather than the atmosphere-water heat conduction, plays the dominant role in altering water temperature. Moreover, the difference between S2 and S3 (S2 minus S3) varies with season. The difference reached its maximum (about 0.8 °C) in January and February when air temperature was significantly lower than the water temperature, signifying the water energy loss to the atmosphere. However, in May, when air temperature was substantially higher than the water temperature, the difference is positive (about 0.2 °C), indicating the overlying air heats up the water body.

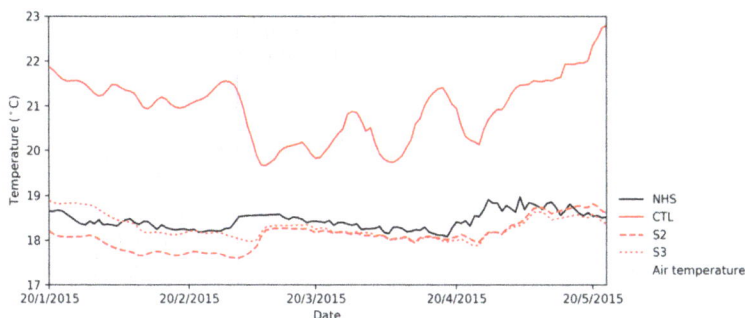

Figure 10. Simulated water temperature by CTL, S2, and S3 at the YHS. The measured water temperature at the NHS and air temperature at the Nuozhadu Dam are shown for reference. The vertical blue line indicates the boundary of the period when air temperature was significantly lower than water temperature at the YHS.

The difference between the simulated water temperature in scenarios CTL and S3 at the YHS was taken as the total impact of solar radiation and air temperature and the weekly contribution (percentage) of both factors were evaluated (Figure 11). Overall, the solar radiation's contribution to the water temperature increment was 106.7%. However, the contribution of air temperature was −6.7%, which suggests that atmosphere-water heat conduction has a decreasing effect on water temperature during the study period. In addition, it is worth observing that the contribution of air temperature is largely dependent on the air-water temperature difference and may vary with simulation periods. When the air temperature was significantly lower than the reservoir water temperature in January and February 2015 (left side of the vertical blue line in Figure 10), its contribution could reach as much as −24%. When the air temperature approached and eventually exceeded the water temperature, the air temperature's contribution to the water temperature increment, became more significant.

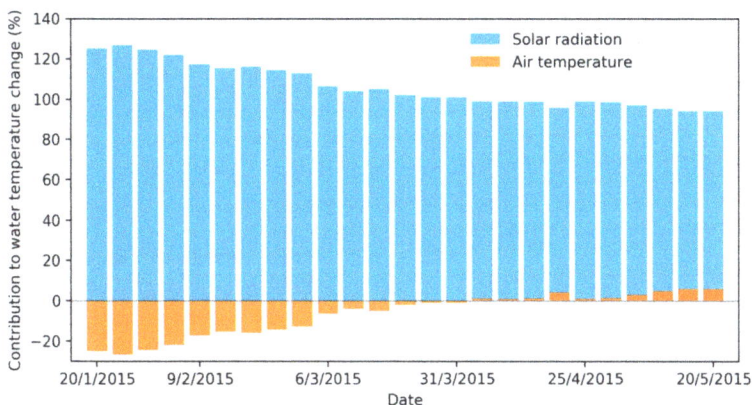

Figure 11. Respective contributions of solar radiation and atmosphere-water heat exchange to the water temperature at the YHS on a weekly scale. Negative contribution values mean that the factor had a decreasing effect on water temperature. Contribution values are computed based on the extent of water temperature changes, taking the water temperature difference between CTL and S3 as 100%.

4.3. Impact of Outflow Rates

A negative correlation between outflow rates and outflow temperature has been observed in Section 2.2.2. For reservoirs with small capacities, a negative relationship between outflow rates and outflow temperature has been reported [92,93]. To further analyze the impact of outflow rates, we carried out eight scenario experiments with different outflow rates ranging from 500 to 4000 $m^3\,s^{-1}$.

One and a half period of the tracer concentration has been designed so as to easily differentiate the respective peak concentration value from outflow, which would increase the readability of the residence time. However, the temporal length of the period is not much restrained, as long as it is longer than the longest residence time studied and shorter than the simulation period, it would be sufficient in explaining the problem. Using seven different concentrations instead of a single concentration would also increase the readability. Additionally, under different scenarios, the diffusion effect is also associated with the residence time. By using seven different concentrations, the tracer effect could be observed from both temporal and amplitude points of view.

Outflow rates regulate the residence time of the reservoir. By definition, the residence time refers to how long a parcel, starting from a specific location within a water body, will remain in the water body before exiting [94]. Generally, the residence time decreases as the outflow increases, which results in a shorter duration available for the water parcel to interact with heat sources/sinks (mainly solar radiation and ambient). Thus, its temperature would result closer to the one of the upstream water body. Residence time has been widely used to describe the variability of the lake thermal structure, isotopic composition, alkalinity, dissolved organic carbon concentration, elemental ratios of heavy metals and nutrients, mineralization rates of organic matter, and primary production [95–100].

The tracer methodology mentioned in Section 3.4 was applied to measure the residence time in different scenarios. The tracer was released at the Nuozhadu Reservoir with cyclical concentration peaks ranging from 1 to 7 $mg\,m^{-3}$ at 1 $mg\,m^{-3}$ intervals (Figure 12a). Tracer concentrations at the YHS were recorded. We computed the residence time by:

$$T_{re} = \frac{1}{2}[(t_{o1} - t_{i1}) + (t_{o7} - t_{i7})],$$

(9)

where T_{re} is the residence time, t_{i1} is the time of inflow tracer peak of 1 $mg\,m^{-3}$, t_{i7} is that of 7 $mg\,m^{-3}$, t_{o1} is the time of outflow tracer peak corresponding to t_{i1} and t_{o7} is that corresponding to t_{i7}. When the outflow rate was extremely high at 4000 $m^3\,s^{-1}$, the residence time was only about 4 days. On the other hand, when the outflow rate was relatively low at 1000 $m^3\,s^{-1}$, it took about 11 days for a water parcel to exit the water body. Generally, the residence time decreased exponentially with the outflow rate and the two variables were highly correlated with a correlation coefficient (R^2) of 0.9999 (Figure 12b).

Afterwards, the simulated outflow water temperature at the YHS was averaged during the study period. We used the temperature difference, the averaged outflow temperature minus the averaged inflow temperature, to indicate the temperature changes under different scenarios (F1–F4). Results based on the eight scenario experiments suggested that water temperature difference and outflow rate were also exponentially correlated with an R^2 of 0.9997 (Figure 13a). Thus, water temperature difference increased linearly with residence time with an R^2 of over 0.9977 (Figure 13b).

To sum up, in the case of the Jinghong Reservoir, outflow rates regulate the residence time. The larger residence time means that the water body is able to absorb more solar energy and, thus, gain a larger temperature increment. Nevertheless, this conclusion is only valid under certain conditions: On one hand, the solar radiation and the atmosphere play a positive role in regulating water temperature; on the other hand, the reservoir capacity is relatively small (for example, the discharge/capacity ratio is larger than 20) so that the outflow rate is able to strongly influence the outflow temperature. The Jinghong Reservoir matches both requirements mentioned above.

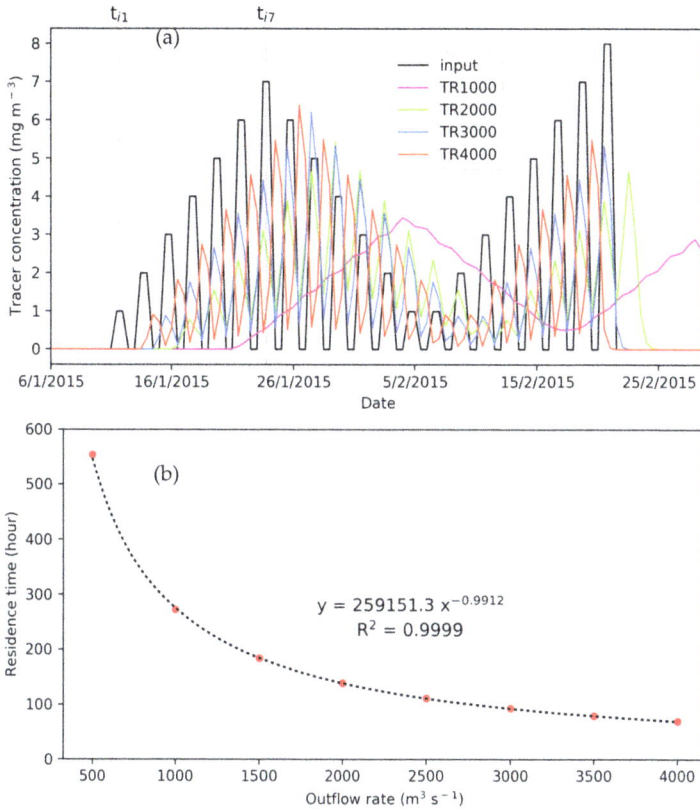

Figure 12. (**a**) Tracer concentration variations at the YHS corresponding to various outflow rates (F2, F4, F6, and F8 are demonstrated in the figure); and (**b**) the exponential relationship between residence time and outflow rate of the Jinghong Reservoir.

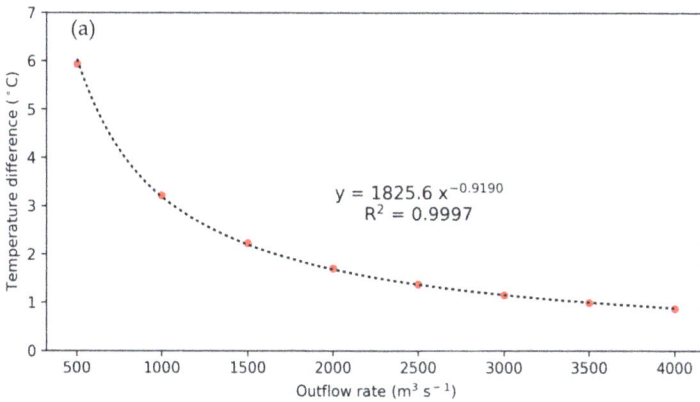

Figure 13. *Cont.*

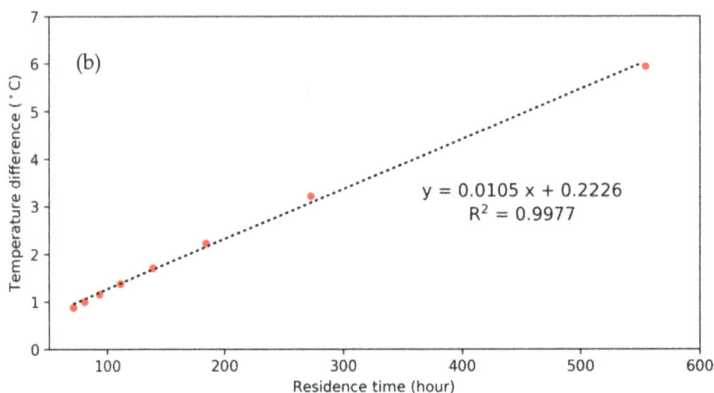

Figure 13. (**a**) Relationship between outflow rate and water temperature difference (outflow minus inflow); and (**b**) relationship between residence time and water temperature difference at the YHS.

4.4. Impact of the Impoundment

In order to examine the overall impact of the impoundment of the Jinghong Reservoir, we established scenario S1, where the Jinghong Dam was removed and the reservoir recovers to the pre-damming state. During the study period, the water temperature for CTL warmed up by about 3 °C on average. However, when the reservoir returned to the natural river state, the temperature increment was not as significant as for the post-impoundment, with an increment of only 1.7 °C, which was about half the case during the reservoir's operation (Figure 14). That is to say, the impoundment of the Jinghong Reservoir alone can explain nearly half (1.3 °C) of the water temperature increment from the NHS to the YHS. Based on the findings of Section 4.3, after the impoundment, the residence time extended from 3 days to 11 days. The period available to absorb solar radiation has been prolonged. Thus, a higher water temperature increment is reached.

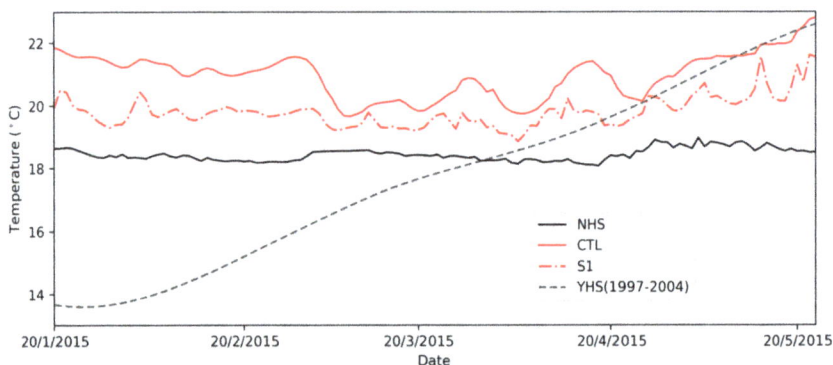

Figure 14. Simulated water temperature during the study period by CTL and S1 at the YHS, and measured water temperature at the NHS and historical water temperature (1997–2004) at the YHS.

5. Research Limitations

Though thorough investigation has been conducted, research limitations due to data availability are not ignorable. Firstly, no continuously measured water temperature data on the Lancang River can be found in the previous studies. Therefore, we conducted a field measurement so as to obtain the water temperature information from November 2014 to February 2016. On the other hand, we only obtained daily solar radiation values from January 2014 to May 2015. Therefore, the study period has been greatly shortened. However, the focus of this study is the explanation of the phenomenon, and a study duration shorter than a full year would not undermine its credibility.

In addition, the influence of the operation rules of the dam have not been directly assessed in this research. In this study, the operation rules have been lumped inside the outflow rate, which is only juxtaposed with outflow water temperature so as to indicate the relation among the two. In which way the operation rules would be most favorable to the downstream environment will be another direction of study in our future research.

6. Conclusions

The temperature characteristics of a river are important due to their strong influence on environmental conditions and aquatic creatures. The impact of reservoirs on downstream water temperature has triggered researchers' concerns worldwide as reservoirs usually release water from deep outlets, in most cases, resulting in a cold water thermal regime downstream. However, the water temperature downstream of the Jinghong Reservoir has increased by 3.0 °C annually compared with historical conditions, according to the observational data. In this study, a Jinghong Reservoir 3D model was established using the Delft3D-FLOW model to explore the unique heating processes of the Jinghong Reservoir on water temperature.

The simulated water temperatures at the outlet showed good agreement with measurements at the Jinghong dam. Comparing the actual operation scenario (CTL) with the pre-damming scenario (S1), it was indicated that the impoundment of the Jinghong Reservoir could explain 1.3 °C (about a half) of the water temperature increase compared to historical values, while the rest might be an accumulated effect by the impoundment of the upstream cascade reservoirs.

Further, this study examined the major factors (solar radiation, atmosphere-water heat conduction) which can potentially influence the water temperature. Numerical experiment results show the contribution of solar radiation to the water temperature increment is up to 106.7%, which consists in the dominating factor. On the other hand, the contribution of the air temperature to the water temperature increment is about −6.7%, which implies that the atmosphere-water heat conduction would result in a decrease of the water temperature during the study period.

Experiment results with different outflow rates (F1–F8) showed that the outflow rate could substantially influence the outflow water temperature because of the variations in the residence time. The residence time is determined by the outflow rate; higher residence time means that the water body absorbs more solar energy and results in a larger temperature increment. After the impoundment of the Jinghong Reservoir, the residence time extended from 3 days to 11 days. The prolonged solar radiation absorption duration is the main reason for its temperature raise.

In conclusion, the Jinghong Reservoir alone could not fully explain the temperature increases compared to historical data since this is also dependent on the accumulated effect by the cascade of dams along the mainstream Lancang River. In the future, a model of the entire cascade of reservoirs should be implemented, in order to improve the assessments of water temperature in pre-dam and post-dam eras.

Author Contributions: Conceptualization: B.J. and G.N.; methodology, validation, and software: B.J.; investigation: F.W. and B.J.; writing—original draft preparation: F.W.; writing—review and editing: B.J., G.N., and F.W.; visualization: G.N.; supervision: G.N.; and funding acquisition: G.N.

funding: This research was funded by the Ministry of Science and Technology of the People's Republic of China grant number [2016YFA0601603].

Acknowledgments: We would like to thank Daming He and Ying Lu of Yunnan University for their contribution in collecting water temperature data. Also, we would like to express sincere thanks to the editor and the three anonymous reviewers whose comments led to great improvement of this paper.

Conflicts of Interest: The authors declare no conflict of interest.

References

1. Carron, J.C.; Rajaram, H. Impact of variable reservoir releases on management of downstream water temperatures. *Water Resour. Res.* **2001**, *37*, 1733–1744. [CrossRef]
2. Olden, J.D.; Naiman, R.J. Incorporating thermal regimes into environmental flows assessments: Modifying dam operations to restore freshwater ecosystem integrity. *Freshw. Biol.* **2010**, *55*, 86–107. [CrossRef]
3. Rheinheimer, D.E.; Null, S.E.; Lund, J.R. Optimizing selective withdrawal from reservoirs to manage downstream temperatures with climate warming. *J. Water Resour. Plan. Manag.* **2015**, *141*, 04014063. [CrossRef]
4. Hynes, H.B.N. Lotic biology. (Book reviews: The ecology of running waters). *Science* **1971**, *172*, 251.
5. Sullivan, K.; Martin, D.J.; Cardwell, R.D.; Inc, P.; Toll, J.E.; Inc, P.; Duke, S. An analysis of the effects of temperature on salmonids of the pacific northwest with implications for selecting temperature criteria. Sustainable ecosystems institute. *Histochemistry* **2000**, *90*, 85–97.
6. Poole, G.C.; Berman, C.H. An ecological perspective on in-stream temperature: Natural heat dynamics and mechanisms of human-caused thermal degradation. *Environ. Manag.* **2001**, *27*, 787–802. [CrossRef]
7. Lessard, J.A.L.; Hayes, D.B. Effects of elevated water temperature on fish and macroinvertebrate communities below small dams. *River Res. Appl.* **2010**, *19*, 721–732. [CrossRef]
8. Murchie, K.J.; Hair, K.P.E.; Pullen, C.E.; Redpath, T.D.; Stephens, H.R.; Cooke, S.J. Fish response to modified flow regimes in regulated rivers: Research methods, effects and opportunities. *River Res. Appl.* **2008**, *24*, 197–217. [CrossRef]
9. Ward, J.V. Thermal characteristics of running waters. *Hydrobiologia* **1985**, *125*, 31–46. [CrossRef]
10. Brooker, M.P. The impact of impoundments on the downstream fisheries and general ecology of rivers. *Adv. Appl. Boil.* **1981**, *6*, 91–152.
11. Horne, B.D.; Rutherford, E.S.; Wehrly, K.E. Simulating effects of hydro-dam alteration on thermal regime and wild steelhead recruitment in a stable-flow lake michigan tributary. *River Res. Appl.* **2004**, *20*, 185–203. [CrossRef]
12. Krause, C.W.; Newcomb, T.J.; Orth, D.J. Thermal habitat assessment of alternative flow scenarios in a tailwater fishery. *River Res. Appl.* **2005**, *21*, 581–593. [CrossRef]
13. Ryan, T.F.; Webb, J.A.; Lennie, R.; Lyon, J. *Satus of Cold Water Releases from Victorian Dam*; Dept. of Natural Resources and Environment: Heidelberg, Australia, 2001.
14. Preece, R. *Cold Water Pollution below Dams in New South Wales: A Desktop Assessment*; Water management Division, Dept. of Infrastructure, Planning and Natural Resources: Sydney, Australia, 2004.
15. Jackson, H.M.; Gibbins, C.N.; Soulsby, C. Role of discharge and temperature variation in determining invertebrate community structure in a regulated river. *River Res. Appl.* **2007**, *23*, 651–669. [CrossRef]
16. Angilletta, M.J.; Steel, E.A.; Bartz, K.K.; Kingsolver, J.G.; Scheuerell, M.D.; Beckman, B.R.; Crozier, L.G. Big dams and salmon evolution: Changes in thermal regimes and their potential evolutionary consequences. *Evol. Appl.* **2008**, *1*, 286–299. [CrossRef] [PubMed]
17. Campbell, I.C. *The Mekong: Biophysical Environment of an International River Basin*; Elsevier Academic Press: Amsterdam, The Netherlands, 2009.
18. Zhao, A.G. Planning and development of lower and middle Lancang River's water power resources. *Pearl River* **2000**, *21*, 5–8.
19. Fan, H.; He, D.; Wang, H. Environmental consequences of damming the mainstream Lancang-Mekong River: A review. *Earth-Sci. Rev.* **2015**, *146*, 77–91. [CrossRef]
20. Lu, X.X.; Siew, R.Y. Water discharge and sediment flux changes over the past decades in the lower Mekong River: Possible impacts of the Chinese dams. *Hydrol. Earth Syst. Sci.* **2006**, *10*, 181–195. [CrossRef]
21. Campbell, I.C. Perceptions, data, and river management: Lessons from the Mekong River. *Water Resour. Res.* **2007**, *43*, 329–335. [CrossRef]
22. Lu, X.X.; Li, S.; Kummu, M.; Padawangi, R.; Wang, J.J. Observed changes in the water flow at Chiang Saen in the lower Mekong: Impacts of Chinese dams? *Quat. Int.* **2014**, *336*, 145–157. [CrossRef]
23. Haxton, T.J.; Findlay, C.S. Meta-analysis of the impacts of water management on aquatic communities. *J. Can. Sci. Halieutiques Aquat.* **2008**, *65*, 437–447. [CrossRef]
24. Gao, X.; Zhang, S.; Zhang, C. 3-D numerical simulation of water temperature released from the multi-level intake of Nuozhadu hydropower station. *J. Hydroelectr. Eng.* **2012**, *31*, 195–201+207.

25. Wurbs. *Computer Models for Water Resources Planning and Management*; S Army Corps of Engineers, Institute for Water Resources: Alexandria, VA, USA, 1994.

26. Imteaz, M.A.; Asaeda, T.; Lockington, D.A. Modelling the effects of inflow parameters on lake water quality. *Environ. Model. Assess.* **2003**, *8*, 63–70. [CrossRef]

27. Peeters, F.; Livingstone, D.M.; Goudsmit, G.; Kipfer, R.; Forster, R. Modeling 50 years of historical temperature profiles in a large central European lake. *Limnol. Ocean.* **2002**, *47*, 186–197. [CrossRef]

28. Kraus, E.B.; Turner, J.S. A one-dimensional model of the seasonal thermocline II. The general theory and its consequences. *Tellus* **1967**, *19*, 98–106. [CrossRef]

29. Imberger, J.; Patterson, J.; Hebbert, B.; Loh, I. Dynamics of reservoir of medium size. *J. Hydraul. Div.* **1978**, *104*, 725–743.

30. Svensson, U. A Mathematical Model of the Seasonal Thermocline. Ph.D. Thesis, University of Lund, Lund, Sweden, 1978.

31. Hostetler, S.W.; Bartlein, P.J. Simulation of lake evaporation with application to modeling lake level variations of Harney-Malheur Lake, oregon. *Water Resour. Res.* **1990**, *26*, 2603–2612.

32. Hostetler, S.W.; Giorgi, F.; Bates, G.T.; Bartlein, P.J. Lake-atmosphere feedbacks associated with paleolakes Bonneville and Lahontan. *Science* **1994**, *263*, 665. [CrossRef] [PubMed]

33. Zamboni, F.; Barbieri, A.; Polli, B.; Salvadè, G.; Simona, M. The dynamic model seemod applied to the Southern Basin of Lake Lugano. *Aquat. Sci.* **1992**, *54*, 367–380. [CrossRef]

34. Karagounis, I.; Trösch, J.; Zamboni, F. A coupled physical-biochemical lake model for forecasting water quality. *Aquat. Sci.* **1993**, *55*, 87–102. [CrossRef]

35. Ulrich, M. Modeling of Chemicals in Lakes-Development and Application of User-Friendly Simulation Software (MASAS & CHEMSEE) on Personal Computers. Ph.D. Thesis, Swiss Federal Institute of Technology, Zürich, Switzerland, 1991.

36. Fang, X.; Stefan, H.G. Long-term lake water temperature and ice cover simulations/measurements. *Cold Reg. Sci. Technol.* **1996**, *24*, 289–304. [CrossRef]

37. Burchard, H.; Bolding, K.; Villarreal, M.R. *GOTM, a General Ocean Turbulence Model: Theory, Implementation and Test Cases*; Space Applications Institute: Ispra, Italy, 1999.

38. Goudsmit, G.H.; Burchard, H.; Peeters, F.; Wüest, A. Application of k-ε turbulence models to enclosed basins: The role of internal seiches. *J. Geophys. Res. Ocean.* **2002**, *107*, C12. [CrossRef]

39. Stepanenko, V.M.; Lykossov, V.N. Numerical modeling of heat and moisture transfer processes in a system lake—Soil. *Russ. J. Meteorol. Hydrol.* **2005**, *3*, 95–104.

40. Stepanenko, V.M.; Machul'Skaya, E.E.; Glagolev, M.V.; Lykossov, V.N. Numerical modeling of methane emissions from lakes in the permafrost zone. *Izv. Atmos. Ocean. Phys.* **2011**, *47*, 252–264. [CrossRef]

41. Subin, Z.M.; Riley, W.J.; Mironov, D. An improved lake model for climate simulations: Model structure, evaluation, and sensitivity analyses in CESM1. *J. Adv. Model. Earth Syst.* **2012**, *4*, 1. [CrossRef]

42. Gu, H.; Jin, J.; Wu, Y.; Ek, M.B.; Subin, Z.M. Calibration and validation of lake surface temperature simulations with the coupled WRF-lake model. *Clim. Chang.* **2015**, *129*, 471–483. [CrossRef]

43. Karpik, S.R.; Raithby, G.D. Laterally averaged hydrodynamics model for reservoir predictions. *J. Hydraul. Eng.* **1990**, *116*, 783–798. [CrossRef]

44. Brown, R.T. *BETTER, A Two-Dimensional Reservoir Water Quality Model: Model User's Guide*; Tennessee Technological University: Cookeville, TN, USA, 1985.

45. Waldrop, W.R.; Ungate, C.D.; Harper, W.L. *Computer Simulation of Hydrodynamics and Temperatures of Tellico Reservoir*; TVA Water Systems Development Branch: Norris, TN, USA, 1980.

46. Edinger, J.E.; Buchak, E.M. *Developments in LARM2: A Longitudinal-Vertical, Time-Varying Hydrodynamic Reservoir Model*; US Army Engineer Waterways Experimental Station: Visburg, MS, USA, 1983.

47. Buchak, E.M.; Edinger, J.E. *Generalized Longitudinal-Vertical Hydrodynamics and Transport: Development, Programming and Application*; US Army Engineer Waterways Experimental Station: Visburg, MS, USA, 1984.

48. Cole, T.M.; Buchak, E.M. *CE-QUAL-W2: A Two-Dimensional, Laterally Averaged, Hydrodynamic and Water Quality Model, Version 2.0. User Manual*; US Army Engineer Waterways Experimental Station: Visburg, MS, USA, 1995.

49. Chen, X.F.; Wang, G.X. Mike 21 software and its application on the offshore reconstruction engineering of Changxing Islands. *J. Dalian Univ.* **2007**, *28*, 93–98. (In Chinese)

50. Wang, Z. Application of Mike 21 in ecological design of artificial lake. *Water Res. Power* **2008**, *26*, 124–127. (In Chinese)

51. Cheng, Y.U.; Ren, X.; Ban, X.A.; Yun, D.U. Application of two-dimensional water quality model in the project of the water diversion in East Lake, Wuhan. *J. Lake Sci.* **2012**, *24*, 43–50. [CrossRef]

52. Liang, Y.; Yin, J.; Zhu, X.; Huang, X. Application of Mike 21 hydrodynamic model in water level simulation of Hongze Lake. *Water Res. Power* **2013**, *31*, 135–137.

53. Leendertse, J.J. *Aspects of a Computational Model for Long-Period Water-Wave Propagation*; Rand Corporation: Santa Monica, LA, USA, 1967.

54. Kuipers, J.; Vreugdeuhil, C.B. *Calculations of Two-Dimensional Horizontal Flow*; Delft Hydraulics Laboratory Report: Delft, The Netherlands, 1973.

55. McGuirk, J.J.; Rodi, W. A depth-averaged mathematical model for the near field of side discharges into open-channel flow. *J. Fluid Mech.* **1978**, *86*, 761–781. [CrossRef]

56. Lauwerier, H.A. Some recent work of the Amsterdam Mathematical Centre on the hydrodynamics of the North Sea. *Sticht. Math. Cent. Toegep. Wiskund.* **1961**, *1*, 1–12.

57. Isozaki, I.; Unoki, S. The numerical computation of the Tsunami in Tokyo Bay caused by the Chilean Earthquake in May 1960. In *Studies on Oceanography—A Collection of Papers Dedicated to Koji Hidaka*; Tokyo University Press: Tokyo, Japan, 1964.

58. Ziegler, C.K.; Nisbet, B. Fine-Grained Sediment Transport in Pawtuxet River, Rhode Island. *J. Hydraul. Eng.* **1994**, *120*, 561–576. [CrossRef]

59. Ziegler, C.K.; Nisbet, B.S. Long-Term Simulation of Fine-Grained Sediment Transport in Large Reservoir. *J. Hydraul. Eng.* **1995**, *121*, 773–781. [CrossRef]

60. Chung, S.; Gu, R. Two-Dimensional Simulations of Contaminant Currents in Stratified Reservoir. *J. Hydraul. Eng.* **1998**, *124*, 704–711. [CrossRef]

61. Tufford, D.L.; Mckellar, H.N. Spatial and temporal hydrodynamic and water quality modeling analysis of a large reservoir on the South Carolina (U.S.) coastal plain. *Ecol. Model.* **1999**, *114*, 137–173. [CrossRef]

62. Ji, Z.G. *Hydrodynamics and Water Quality: Modeling Rivers, Lakes, and Estuaries*; John Wiley & Sons: Hoboken, NJ, USA, 2008.

63. Sheng, Y.P. *A Three-Dimensional Mathematical Model of Coastal, Estuarine and Lake Currents Using Bounary Fitted Grid*; Report No. 585; Aeronautical Research Associates of Princeton: Princeton, NJ, USA, 1986.

64. Blumberg, A.F.; Mellor, G.L. A description of a three-dimensional coastal ocean circulation model. In *Three-Dimensional Coastal Ocean Models Coastal and Estuarine Sciences*; American Geophysical Union: Washington, DC, USA, 1987.

65. Hamrick, J.M. *User's Manual for the Environmental Fluid Dynamics Computer Code*; College of William and Mary: Gloucester Point, VA, USA, 1996.

66. Cerco, C.F.; Cole, T.M. Three-dimensional eutrophication model of Chesapeake Bay. Volume 1: Main report. *Am. Soc. Civ. Eng.* **1994**, *119*, 1006–1025. [CrossRef]

67. Wool, T.A.; Ambrose, R.B.; Martin, J.L.; Cormer, E.A. *Water Quality Analysis Simulation Program (WASP), Version 6.0*; United States Environmental Protection Agency: Washington, DC, USA, 1995.

68. Hydraulics. *Delft3D-FLOW: Simulation of Multi-Dimensional Hydrodynamic Flows and Transport Phenomena, Including Sediments—User Manual*; WL | Delft Hydraulics: Delft, The Netherlands, 2003.

69. HydroQual. *User's Guide for RCA*; Technical Report; HydroQual, Inc.: Mahwah, NJ, USA, 2004.

70. Emma, J.; Lars-Göran, G.; Sten, B.; Tobias, L.; Jan-Olof, S.; Lillemor, C.L.; Destouni, G. Data evaluation and numerical modeling of hydrological interactions between active layer, lake and talik in a permafrost catchment, Western Greenland. *J. Hydrol.* **2015**, *527*, 688–703.

71. Harcourt-Baldwin, J.L.; Diedericks, G.P.J. Numerical modelling and analysis of temperature controlled density currents in Tomales Bay, California. *Estuar. Coast. Shelf Sci.* **2006**, *66*, 417–428. [CrossRef]

72. Maren, D.S.V. Grain size and sediment concentration effects on channel patterns of silt-laden rivers. *Sediment. Geol.* **2007**, *202*, 297–316. [CrossRef]

73. Bouma, T.J.; Duren, L.A.V.; Temmerman, S.; Claverie, T.; Blanco-Garcia, A.; Ysebaert, T.; Herman, P.M.J. Spatial flow and sedimentation patterns within patches of epibenthic structures: Combining field, flume and modelling experiments. *Contin. Shelf Res.* **2007**, *27*, 1020–1045. [CrossRef]

74. Tonnon, P.K.; Rijn, L.C.V.; Walstra, D.J.R. The morphodynamic modelling of tidal sand waves on the shoreface. *Coast. Eng.* **2007**, *54*, 279–296. [CrossRef]

75. Allard, R.; Dykes, J.; Hsu, Y.L.; Kaihatu, J.; Conley, D. A real-time nearshore wave and current prediction system. *J. Mar. Syst.* **2008**, *69*, 37–58. [CrossRef]

76. Lam, N.T. Hydrodynamics and morphodynamics of a seasonally forced tidal inlet system. *J. Water Res. Environ. Eng.* **2008**, *1*, 114–124.

77. Leeuwen, B.V.; Augustijn, D.C.M.; Wesenbeeck, B.K.V.; Hulscher, S.J.M.H.; Vries, M.B.D. Modeling the influence of a young mussel bed on fine sediment dynamics on an intertidal flat in the Wadden Sea. *Ecol. Eng.* **2010**, *36*, 145–153. [CrossRef]

78. Souliotis, D.; Prinos, P. Effect of a vegetation patch on turbulent channel flow. *J. Hydraul. Res.* **2011**, *49*, 157–167. [CrossRef]

79. Department of Water Resource of Yunnan Province, China. Available online: http://www.wcb.yn.gov.cn/ (accessed on 1 July 2015).

80. Khan, F.; Johnson, G.E.; Royer, I.M.; Phillips, N.R.; Hughes, J.S.; Fischer, E.S.; Ham, K.D.; Ploskey, G.R. *Imaging Evaluation of Juvenile Salmonid Behavior in the Immediate Forebay of the Water Temperature Control Tower at Cougar Dam, 2010*; Pacific Northwest National Laboratory: Richland, WA, USA, 2012.

81. Chanudet, V.; Fabre, V.; Kaaij, T.V.D. Application of a three-dimensional hydrodynamic model to the Nam Theun 2 Reservoir (Lao PDR). *J. Great Lakes Res.* **2012**, *38*, 260–269. [CrossRef]

82. Phillips, N.A. A coordinate system having some special advantages for numerical forecasting. *J. Atmos. Sci.* **1957**, *14*, 184–185. [CrossRef]

83. Rodi, W. *Turbulence Models and Their Application in Hydraulics—A State of the Art Review*; International Association for Hydraulic Research: Rotterdam, The Netherlands, 1993.

84. Hu, K.; Ding, P.; Wang, Z.; Yang, S. A 2D/3D hydrodynamic and sediment transport model for the Yangtze Estuary, China. *J. Mar. Syst.* **2009**, *77*, 114–136. [CrossRef]

85. Carrivick, J.L.; Brown, L.E.; Hannah, D.M.; Turner, A.G. Numerical modelling of spatio-temporal thermal heterogeneity in a complex river system. *J. Hydrol.* **2012**, *414–415*, 491–502. [CrossRef]

86. Majerova, M.; Neilson, B.T.; Schmadel, N.M.; Wheaton, J.M.; Snow, C.J. Impacts of beaver dams on hydrologic and temperature regimes in a mountain stream. *Hydrol. Earth Syst. Sci.* **2015**, *19*, 3541–3556. [CrossRef]

87. Chow, V.T. *Open-Channel Hydraulics*; McGraw Hill Book Co.: New York, NY, USA, 1959.

88. Elzawahry, A.E. Advection, Diffusion and Settling in the Coastal Boundary Layer of Lake Erie. Ph.D. Thesis, McMaster University, Hamilton, ON, Canada, 1985.

89. Tsanis, I.K.; Wu, J. Application and verification of a three-dimensional hydrodynamic model to Hamilton Harbour, Canada. *Glob. Nest Int. J.* **2000**, *2*, 77–89.

90. Chen, S. *Accumulation Impact on Water Temperature in a South-North Dammed River: A Case Study in the Middle and Lower Reaches of Lancang River*; Tsinghua University: Beijing, China, 2017. (In Chinese)

91. Huang, Z.; Wu, B. *Three Gorges Dam: Environmental Monitoring Network and Practice*; Science Press: Beijing, China, 2018.

92. Huang, T.; Li, X.; Rijnaarts, H.; Grotenhuis, T.; Ma, W.; Sun, X.; Xu, J. Effects of storm runoff on the thermal regime and water quality of a deep, stratified reservoir in a temperate monsoon zone, in Northwest China. *Sci. Total Environ.* **2014**, *485–486*, 820–827. [CrossRef] [PubMed]

93. Monsen, N.E.; Cloern, J.E.; Lucas, L.V.; Monismith, S.G. A comment on the use of flushing time, residence time, and age as transport time scales. *Limnol. Ocean.* **2002**, *47*, 1545–1553. [CrossRef]

94. Casamitjana, X.; Serra, T.; Colomer, J.; Baserba, C.; Pérez-Losada, J. Effects of the water withdrawal in the stratification patterns of a reservoir. *Hydrobiologia* **2003**, *504*, 21–28. [CrossRef]

95. Hamilton, S.K.; Lewis, W.M. Causes of seasonality in the chemistry of a lake on the Orinoco River floodplain, Venezuela1. *Limnol. Ocean.* **1987**, *32*, 1277–1290. [CrossRef]

96. Herczeg, A.L.; Imboden, D.M. Tritium hydrologic studies in four closed-basin lakes in the Great Basin, U.S.A. *Limnol. Ocean.* **1988**, *33*, 157–173. [CrossRef]

97. Eshleman, K.N.; Hemond, H.F. Alkalinity and major ion budgets for a Massachusetts reservoir and watershed1. *Limnol. Ocean.* **1988**, *33*, 174–185. [CrossRef]

98. Christensen, D.L.; Carpenter, S.R.; Cottingham, K.L.; Knight, S.E.; Lebouton, J.P.; Schindler, D.E.; Pace, M.L. Pelagic responses to changes in dissolved organic carbon following division of a seepage lake. *Limnol. Ocean.* **1996**, *41*, 553–559. [CrossRef]

99. Hecky, R.E.; Campbell, P.; Hendzel, L.L. The stoichiometry of carbon, nitrogen, and phosphorus in particulate matter of lakes and oceans. *Limnol. Ocean.* **1993**, *38*, 709–724. [CrossRef]

100. Jassby, A.D.; Powell, T.M.; Goldman, C.R. Interannual fluctuations in primary production: Direct physical effects and the trohic cascade at Castle Lake, California. *Limnol. Ocean.* **1990**, *35*, 1021–1038. [CrossRef]

water

MDPI

Article

Decline of N and P Uptake in the Inner Protection Zone of a Terminal Reservoir during Inter-Basin Water Transfers

Shengtian Yang [1], Juan Bai [2,3], Changsen Zhao [1], Hezhen Lou [1,*], Zhiwei Wang [2,4], Yabing Guan [2], Yichi Zhang [2], Chunbin Zhang [2] and Xinyi Yu [2]

[1]   College of Water Sciences, Beijing Normal University, Beijing 100875, China;
      yangshengtian@bnu.edu.cn (S.Y.); hzjohnson@gmail.com (C.Z.)
[2]   State Key Laboratory of Remote Sensing Science, Faculty of Geographical Science, Beijing Normal
      University, Beijing Key Laboratory for Remote Sensing of Environment and Digital Cities, Beijing 100875,
      China; baijuanaction@163.com (J.B.); wangzhiweisci@gmail.com (Z.W.); guanyabing2013@163.com (Y.G.);
      yichizhang@mail.bnu.edu.cn (Y.Z.); zhangchunbin1102@163.com (C.Z.); yuxy0106@sina.com (X.Y.)
[3]   College of Geographical Science, Shanxi Normal University, Linfen 041004, China
[4]   Key Laboratory of Radiometric Calibration and Validation for Environmental Satellites, National Satellite
      Meteorological Center, China Meteorological Administration, Beijing 100081, China
*   Correspondence: louhezhenbj@163.com; Tel.: +86-10-5880-5034

Received: 23 December 2017; Accepted: 6 February 2018; Published: 9 February 2018

Abstract: Inter-basin water transfer projects are designed to relieve water scarcity around the world. However, ecological problems relating to reductions in protection zone functions can occur during inter-basin transfers. This paper uses the largest inter-basin water transfer project in the world, namely, the South-to-North Water Transfer Project (SNWTP) in China, as an example to analyze the variation of Miyun Reservoir's inner protection zone functions when water is transferred. Specifically, a riparian model (RIPAM) coupled with remote sensing data were used to calculate the nitrogen (N) and phosphorus (P) losses due to plant uptake, and these results were validated by in situ survey data. Then, correlations between water levels and N and P removal were analyzed. The results show that water table disturbances resulting from elevated water levels strongly influence the growth of plants and have obvious negative impacts on N and P removal in the inner protection zone. With the implementation of the middle route of the SNWTP, the water level of Miyun will rise to 150 m in 2020, and subsequently, the total net primary productivity (NPP) could decline by more than 40.90% from the level in 2015, while the N and P uptake could decline by more than 53.03% and 43.49%, respectively, from the levels in 2015, according to the modeling results. This will lead to declines in the inner protection zone's defense effectiveness for N and P interception and increases in risks to the security of water resources. The results of this study provide useful knowledge for managing the defense function of the terminal reservoir's inner protection zone and for ensuring that water quality is maintained during the diversion process.

Keywords: protection zone; nutrient uptake; NPP; South-to-North Water Transfer Project; Miyun Reservoir

1. Introduction

Inter-basin water transfers can relieve water scarcity and are a popular water-resource topic. To date, many famous projects have been built to alleviate uneven water resource distributions, such as the Central Arizona Project, the Colorado River Projects and California North-to-South Water Transfer Project in the United States [1,2], the Siberian Rivers Diversion in Russia [3], the Snowy Mountains Scheme in Australia [4], the National River Linking Project in India [5], and the South-to-North Water Transfer Project (SNWTP) in China [6]. However, inter-basin water transfers usually are associated with complex environmental changes [7–11], just as Davies et al. [7] proposed: "Any transfer of water within or between basins will have physical, chemical, hydrological and biological implications for both donor and recipient systems, as well as for their estuaries and local marine environments." Therefore, the environmental responses to inter-basin water transfers should be studied.

Numerous ecological changes during inter-basin water transfers have been proposed, including the introduction of nonindigenous species, secondary salinization, disturbances in water chemistry and quality, changes in hydrologic regimes, and alterations of habitats [8,11–18]. The major concerns of these studies were about the protection of water source regions and transfer canals so as to ensure water safety. However, more attention should be devoted to the problems faced in receiving areas of inter-basin water projects because artificial projects can raise the water level of a receiving area and flood its drinking water protection zones, which may result in a decline of surface water quality.

Drinking water protection zones have been established by law to maintain the quality of surface waters and aquatic ecosystems [19,20]. They consist of the following three parts: inner, middle, and outer protection zones [20]. The inner protection zone is located near the drinking water source, and its boundaries are closely related to the riparian buffer boundaries. The middle protection zone runs along the inner protection zone and covers certain buffer areas. The outer protection zone covers the remaining catchment area not covered by the inner protection zone and middle protection zone [19,20]. Among them, the inner protection zone is known for its ability to control non-point source pollution (NPS) by intercepting and retaining excess particulates and dissolved nutrients originating from the surrounding uplands. Plants in the inner protection zone play a significant role in nitrogen (N) and phosphorus (P) cycling, and they can reduce N and P concentrations in soils [21,22].

The absorption of nutrients by plants largely depends upon vegetation production and other environmental factors [23]. Even small changes in the water level may induce detectable changes in vegetation production and associated N and P removal efficiencies [24,25]. When an inter-basin water transfer project is implemented, significant amounts of water will be diverted to the terminal reservoir. The elevated water level will flood the current riparian buffer, and a new one will form with a new vegetation circumstance. Such variations have the potential to render the riparian buffer ineffective at intercepting NPS and may change the inner protection zone from a nutrient sink into a nutrient source [26]. However, though studies on the responses of riparian vegetation to flooding have been done in natural lakes and rivers with some success [27,28], the impacts of inter-basin water transfers on the N and P uptake functions of a terminal reservoir's protection zone have not yet been investigated.

In this study, we use the SNWTP in China, which is the largest inter-basin transfer project in the world, as an example to assess the impacts of water transfers on N and P uptake in a terminal reservoir's inner protection zone. The total N and P losses due to plant uptake were calculated by the combined use of an N and P model and remote sensing data, and then, the relationship between the water level and N and P removal was examined. Finally, the potential risks posed by water level fluctuations and associated impacts on N and P uptake in the inner protection zone were evaluated.

2. Materials and Methods

2.1. Study Area

The SNWTP in China consists of three routes (i.e., the eastern, middle, and western routes) (Figure 1); each route covers a distance of more than 1000 km, and in total they deliver 44.8 billion m^3 of fresh water per year to northern China. It is the largest inter-basin transfer scheme in the world; therefore, it is often taken as an example of inter-basin transfers and has been intensively studied in terms of the ecological and environmental consequences since it was proposed [12,13,29–34].

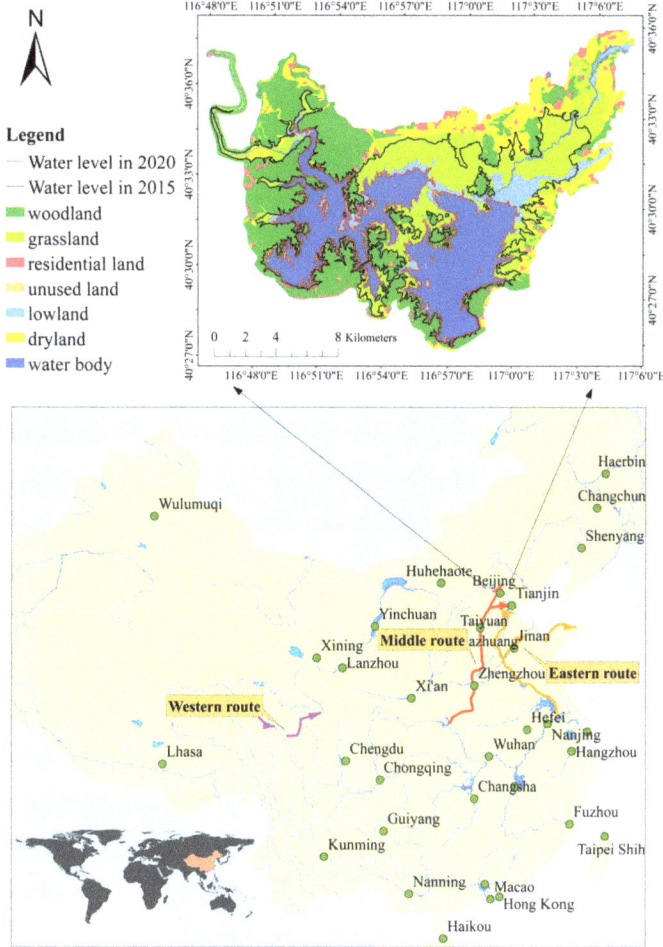

Figure 1. Schema of the South-to-North Water Transfer Project in China and land use information for the terminal Miyun Reservoir of the middle route.

Along the middle route, the Miyun Reservoir serves as the terminal point and stores surplus water transferred from the Dangjiangkou Reservoir. The middle route of the SNWTP was implemented in 2014. Surplus water is lifted 133 m in elevation so that it can enter the Miyun Reservoir through the Jingmi diversion canal with the help of nine pumping stations. The Miyun Reservoir is situated in the mountainous area of Miyun County, and the reservoir extends from 116°47′ E to 117°05′ E and from 40°26′ N to 40°35′ N (Figure 1). It has a large water surface area of 188 km² and a watershed area of 15,788 km²; the total storage capacity is 4.375 billion m³. The study area features a temperate continental monsoon climate, with a mean annual precipitation amount of 628 mm and mean annual air temperature of 11.3 °C. There is an uneven seasonal distribution of precipitation, and more than 70% of the annual precipitation occurs between June and September. The main land use types are dry land, woodland, grassland, water bodies, flood land, and residential land (Figure 1). The inner protection zone along the Miyun Reservoir mainly consists of the area encompassed by the highway surrounding the reservoir and other nearby water zones as designated by the government.

The Miyun Reservoir is the main surface drinking water source for Beijing, and it has played an important role in the social and economic development of the capital [35]. Issues related to water quality in the reservoir have been receiving increasing attention. The upper catchment of the Miyun Reservoir has exhibited eutrophication trends in recent years [36]. Thus, there is an urgent need to strengthen the management of water protection zones. The water level of the Miyun Reservoir was 136 m in 2016. With the implementation of the middle route of the SNWTP, the capacity of the terminal reservoir will be increased from 1.2 billion m^3 to 2 billion m^3 in 2020, which will raise the water level to 150 m. The transferred water will inundate part of the original inner protection zone. Vegetation in the remaining part will likely be affected by the rising water level, and these changes could pose water quality risks for the terminal reservoir.

2.2. Simulation of N and P Uptake

The effects of water level variation on the inner protection zone's defense functions were evaluated through modeling the N and P removal of vegetation during five years (1999, 2000, 2009, 2010, and 2015). The Miyun Reservoir began to receive transferred water in 2015, so we chose 2015 as the current year. In the following years, its water level will increase to 150 m, which is almost equal to the water level in 1999; hence, we chose 1999 as the objective year for reference. Besides, the difference of water levels between 1999 and 2000 was quite significant (149 m in 1999 and 142 m in 2000), as well as the difference between 2009 and 2010 (137 m in 2009 and 134 m in 2010), so these years were chosen as typical years. N and P losses were calculated by coupling remote sensing data with a RIPArian Model (RIPAM, EcoHAT, Beijing, China). RIPAM couples simple process-based modules and remote sensing data to estimate soil denitrification rates, soil nitrogen emissions, phosphorus removed by soil erosion, and nitrogen and phosphorus removal by vegetation uptake [37–40]. The framework of RIPAM that relates to N and P uptake is shown in Figure 2, and its main functions and parameters are listed in Table 1.

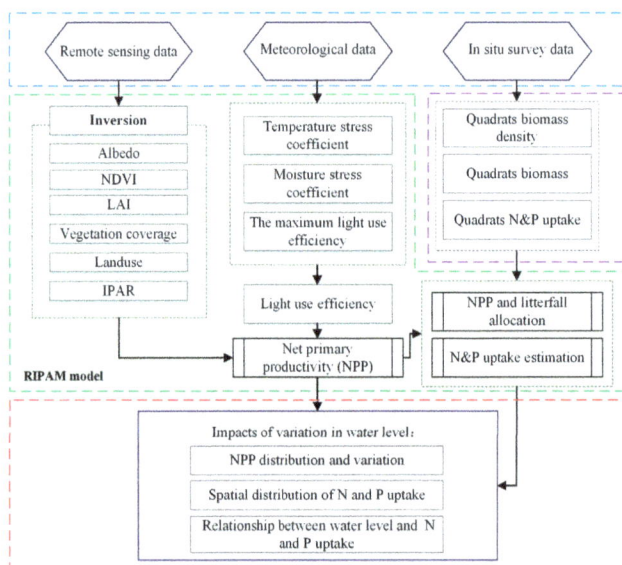

Figure 2. Framework of the N and P uptake simulation. IPAR: the amount of photosynthetically active radiation intercepted by plants; LAI: leaf area index; NPP: net primary productivity; NDVI: normalized difference vegetation index.

Table 1. Functions and parameters related to N and P uptake in the RIPAM model.

No.	Module Name	Equations	Reference
1	NPP simulation	$NPP = GPP - R_a$ $GPP = \varepsilon \times IPAR \times f_1(T) \times f_2(\beta)$ $R_a = GPP \times (7.825 + 1.145T_a)/100$	GLO-PEM model [41]
2	NPP allocation	$F_B = dB_L/d_t + dF_{lit}/d_t$ $B_L = LAI/SLA$ $R_B = K_{ra}(NPP - F_B)$ $W_B = NPP - F_B - R_B$	ForNBM model [42]
3	Nutrient absorption	$X_{uptake} = min(X_{avail}, X_{dem})$ $X_{dem} = (1 - K_{retra}) \times folX \times F_B + X_W \times W_B + X_r \times R_B$ $X_{dem} = NPP \times X_{cont}$	ForNBM model [42,43]
4	Litterfall decomposition	$F_{lit} = \begin{cases} a_{fh}F_B, \ T_{air} > T_{fall}, \ \text{deciduous species} \\ F_B, \ T_{air} < T_{fall}, \ \text{deciduous species} \\ a_{fs}F_B, \ \text{evergreen species} \end{cases}$	ForNBM model [42]

Annotation: NPP: net primary productivity (g C·m^{-2}); GPP: gross primary productivity (g C·m^{-2}); R$_a$: primary production for autotrophic respiration (g C·m^{-2}); IPAR: the amount of photosynthetically active radiation intercepted by plants (MJ·m^{-2}); ε: the real efficiency of radiation utilization (g C·MJ^{-1}); f$_1$(T) is a function of temperature stress; f$_2$(β) is a function of water stress; F$_B$: the NPP foliage obtains (g C·m^{-2}); B$_L$: the foliage biomass (g·m^{-2}); F$_{lit}$: the leaf litter (g C·m^{-2}); t: the time step (month); SLA: specific leaf area; R$_B$: the NPP roots obtain (g C·m^{-2}); K$_{ra}$: return coefficient of leaf nutrients, constant; W$_B$: the NPP wood obtains (g C·m^{-2}); X$_{uptake}$: the nutrient uptake rate (g·m^{-2}); X$_{avail}$: the amount of available nutrients (X = N or P) (g·m^{-2}); X$_{dem}$: the nutrients demanded by biomass growth (g·m^{-2}); K$_{retra}$: the dimensionless foliage nutrient re-translocation proportion (Kretra = 0.35 for this study); folX: the nutrient (X = N or P) concentration of the foliage (g·g^{-1}); X$_W$: the nutrient concentration of the wood (g·g^{-1}); Xr: the nutrient concentration of the roots (g·g^{-1}); a$_{fh}$: the coefficient for deciduous species; afs: the coefficient for evergreen species; T$_{air}$: air temperature (°C); and T$_{fall}$: the threshold temperature for deciduous species litter (°C).

In other published papers [37–40], the RIPAM model was tested in the Guanting Reservoir, and the simulations of net primary productivity (NPP), NPP allocation, and nutrient uptake all had considerable accuracy during the growing season (take N uptake rates for example, its R^2 value is 0.95). Since the Miyun Reservoir and Guanting Reservoir are located within a distance of less than 100 km and their geographical conditions and climate are similar [36,39], we adopted the model parameters of the Guanting Reservoir to apply to the Miyun Reservoir [37,39,44]. The database required included remote sensing data, meteorological data, and in situ survey data. Remote sensing data and meteorological data were used to drive the RIPAM model and simulate NPP and N and P uptake. In situ survey data were used to validate the simulation results. The relationship between water levels and N and P uptake in the inner protection zone were analyzed by implementing zonal statistics. Then the impacts of N and P uptake were evaluated due to rising water levels.

2.3. Multi-Source Data for Driving the RIPAM

Input data for the RIPAM model included meteorological data, remote sensing data, and soil data (Table 2). (1) The daily meteorological data in July of 1999, 2000, 2009, 2010, and 2015 were collected from the national basic weather station at Miyun Reservoir; (2) The remote sensing data included Landsat data and Advanced Spaceborne Thermal Emission and Reflection Radiometer (ASTER)–Global Digital Elevation Model (GDEM) data. Five time series of Landsat images in July or August of 1999, 2000, 2009, 2010, and 2015 were collected to infer land surface parameters. ASTER–GDEM was used to calculate the contour map with intervals of 10 m, and the whole study area was zoned into the following four belts: 130–140 m, 140–150 m, 150–160 m, and 160–170 m; (3) Soil data were derived from the Second National Soil Survey results and field investigations in the study area. All data were reproduced at 30 m × 30 m spatial resolutions and projected to Albers by using the World Geodetic System-84 (WGS 84) datum. Some data were resolved by using ENVI4.8 (Harris Corporation, Melbourne, FL, USA), ArcGIS10.1 (Environmental Systems Research Institute, Redlands, CA, USA), or programs written in IDL (Interactive Data Language).

Table 2. Information for the multi-source data.

Data Type	Data Name	Resolution	Data Source
Meteorological data	Precipitation, wind speed, air pressure, air temperature, daily max/min temperature, relative humidity, sunshine duration, sun radiation, surface temperature	-	National basic weather station (http://data.cma.cn/)
Remote sensing data	Digital elevation model (DEM)	30 m	ASTER-GDEM (http://www.gscloud.cn/)
	Albedo, NDVI, LAI, land use, vegetation coverage, land surface temperature (LST)	30 m	Landsat (http://www.gscloud.cn/)
Soil data	Bulk density, soil texture, soil nutrient content (TN, NO_3-N, NH_4-N), pH	-	Second National Soil Survey field investigations

2.4. In Situ Survey Data for Validation

In practice, biomass is seldom estimated over a large scale. In the present paper, biomass data in situ were collected at the times of Landsat image capture (from 5 to 20 July 2015) within the inner protection zone. The whole zone can be classified into the following three phases: reservoir, terrestrial, and water phase. The water phase is located between the low-water mark and the reservoir bank, the riparian phase contains the zone between the low-water mark and high-water mark, and the remainder is the terrestrial phase. Several 90 m × 90 m quadrants were selected in each phase, and Figure 3 shows some examples of the in situ quadrants. The location of each quadrant was recorded with the Global Positioning System (GPS) to integrate field and Landsat data. The harvesting method [45] was used to measure the average biomass density of grasses and shrubs. In each quadrant, we selected 36 sample areas of 1 m × 1 m for grass or nine sample areas of 5 m × 5 m for shrubs, and the average biomass density was calculated. Forest biomass was estimated using diameter–height models [46] based on the sample trees. In this study, 39 quadrants were selected for the analysis.

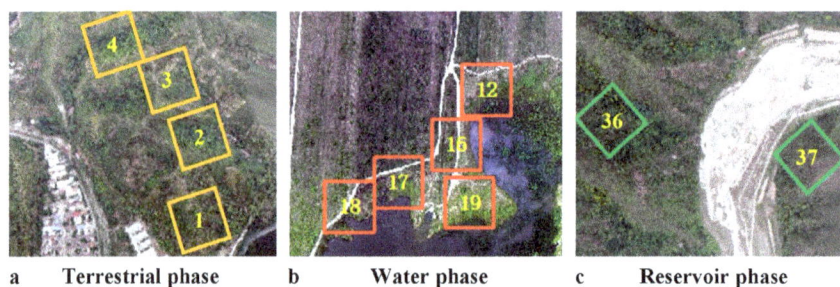

a   **Terrestrial phase**   b   **Water phase**   c   **Reservoir phase**

Figure 3. Examples of the in situ quadrants in different phases. (**a**) Quadrants selected in the terrestrial phase; (**b**) quadrants selected in the water phase; (**c**) quadrants selected in the reservoir phase.

The average biomass density of each quadrant was calculated, as shown in Table 3. Additionally, the data were converted into N and P uptake by using a simple ratio coefficient to validate the accuracy of the RIPAM model (see Section 3.4 for more details).

Table 3. Information for the biomass density of each quadrant. Quadrants 1–11 were selected in the terrestrial phase; quadrants 12–20 were selected in the water phase; quadrants 21–39 were selected in the reservoir phase.

Quadrant ID	Biomass Density ($g \cdot m^{-2}$)	Quadrant ID	Biomass Density ($g \cdot m^{-2}$)	Quadrant ID	Biomass Density ($g \cdot m^{-2}$)
1	1222.13	14	172.35	27	4699.23
2	1114.83	15	188.35	28	3139.40
3	773.61	16	171.57	29	4275.75
4	1321.03	17	192.58	30	13,825.23
5	1375.40	18	184.78	31	10,893.78
6	1185.17	19	177.98	32	7510.89
7	1332.65	20	142.32	33	12,186.66
8	1453.43	21	5291.21	34	10,159.57
9	859.52	22	3275.53	35	11,741.42
10	1001.63	23	1796.98	36	12,123.08
11	1488.09	24	2043.25	37	5930.64
12	172.73	25	1528.09	38	10,361.92
13	152.52	26	2679.09	39	7192.59

3. Results

3.1. NPP Distribution and Variation

The NPP of Miyun's inner protection zone during the growing season of 1999, 2000, 2009, 2010, and 2015 was simulated by the RIPAM model. According to previous research in this study area [47], it is changes in water level but not climate or land use that influence the condition of vegetation in the inner protection zone. The total NPP of the whole zone amounted to 8454.96 t in 1999, 11,922.37 t in 2000, 17,271.43 t in 2009, 13,653.78 t in 2010, and 14,306.82 t in 2015. The spatial distribution and average NPP value during these five years are shown in Figure 4a–f. Higher values were mainly located in the northern part of the study area, where the values were more than 100 g C·m^{-2}, while the NPP of the vegetation surrounding the reservoir was relatively low. The numerical growth of the NPP was significant in the northwestern and southwestern parts of the study area. As the water level rose, the average NPP value showed a decreasing trend, with values ranging from 98.66 g C·m^{-2} in 2015 at a water level of 133 m to 75.10 g C·m^{-2} in 1999 at a water level of 149 m (Figure 4f). The average water level in 2009, 2010, and 2015 was 135 m, and the average total NPP was 102.02 g C·m^{-2}. The average water level in 1999 and 2000 was 145 m, and the corresponding average total NPP decreased to 83.40 g C· m^{-2}, which amounts to a decline rate of 18.26%.

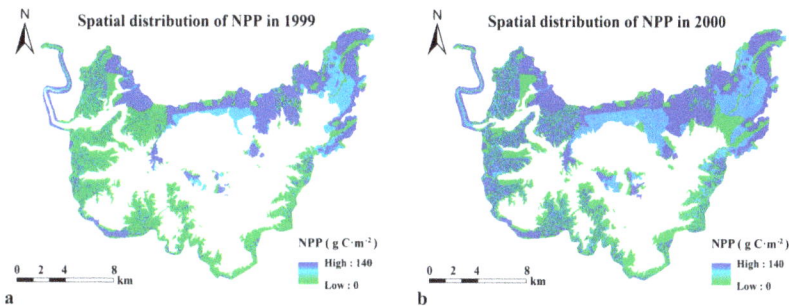

Figure 4. *Cont.*

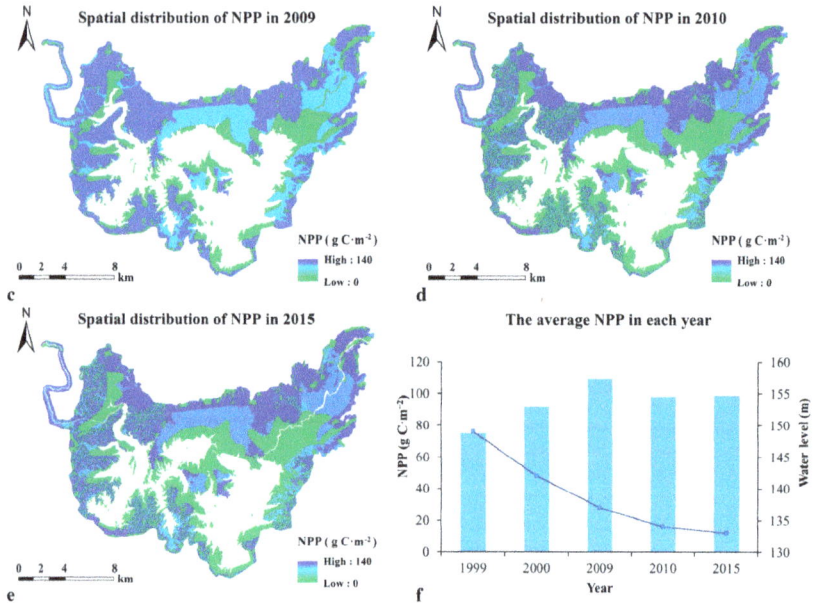

Figure 4. Spatial distribution of NPP in Miyun's inner protection zone during different years.

3.2. Spatiotemporal Variation of N and P Uptake

The RIPAM model was used to simulate the spatial distribution of N and P uptake in the inner protection zone during the growing season of 1999, 2000, 2009, 2010, and 2015. Total N uptake of the whole zone amounted to 0.57 t in 1999, 0.74 t in 2000, 0.86 t in 2009, 0.78 t in 2010, and 1.22 t in 2015, while total P uptake of the whole zone amounted to 3.79 t in 1999, 5.93 t in 2000, 6.96 t in 2009, 6.33 t in 2010, and 6.71 t in 2015. The spatial distribution of N uptake for each year is shown in Figure 5a–e. The modeling results show that plants located in the western and northern parts of the study area had a stronger ability to absorb N; average monthly N uptake rates in the inner protection zone were around 2.62 $g \cdot m^{-2}$. The western part is mountainous, and the major land cover is forest. For the northern part, which has a vast area of grassland, N consumption by plants was relatively weaker, except for dry land, which had higher values. In addition, areas adjacent to the water body displayed a consistently low ability to consume N. The statistics in Figure 5f show that the average N removal due to plant uptake increased from 1999 to 2015. This result is consistent with the NPP data. The value of N consumed by plants increased from 1.80 $g \cdot m^{-2}$ in 1999 to 3.83 $g \cdot m^{-2}$ in 2015.

The spatial distribution of P uptake for each year is shown in Figure 6a–e. The results show that plants located in the northwestern, northeastern, and southwestern parts of the study area had a stronger ability to absorb P; average monthly P uptake rates in the inner protection zone were around 0.041 $g \cdot m^{-2}$. These three areas are all mountainous, and the major land cover is forest. For the northern part, which has a vast area of grassland, the ability of plants to consume P was relatively weaker. Also, areas adjacent to the water body displayed a consistently low ability to consume P. The statistics in Figure 6f show that average P removal due to plant uptake increased from 1999 to 2015. This result is consistent with the NPP and N uptake data. The value of P increased from 0.026 $g \cdot m^{-2}$ in 1999 to 0.046 $g \cdot m^{-2}$ in 2015.

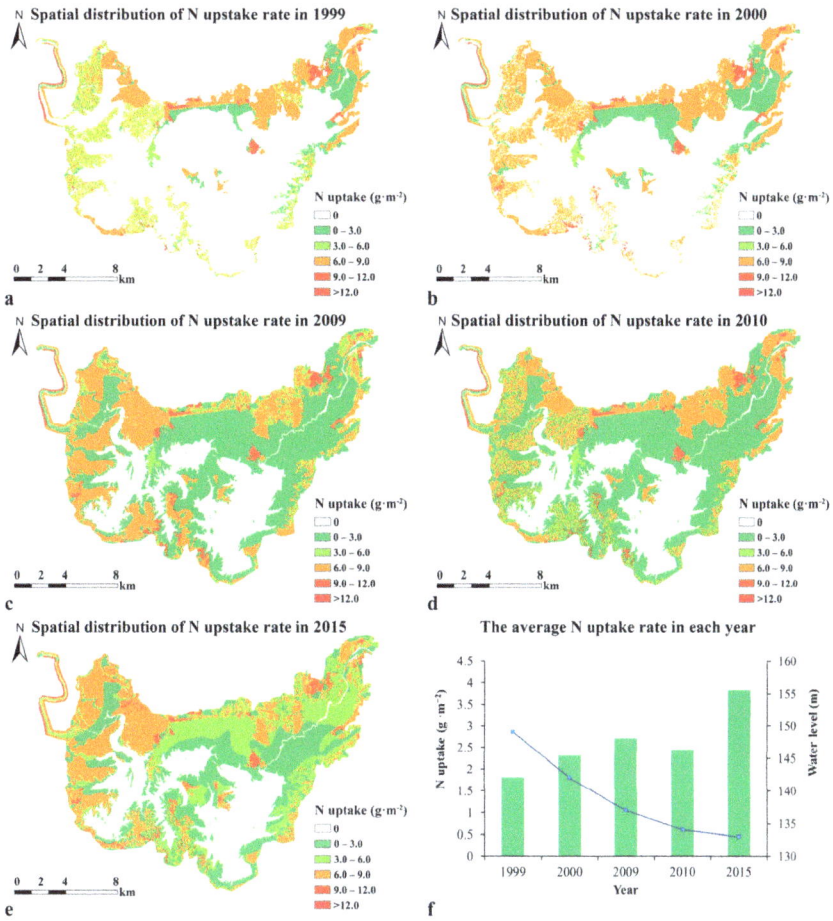

Figure 5. Spatial distribution of the N uptake rate in Miyun's inner protection zone during different years.

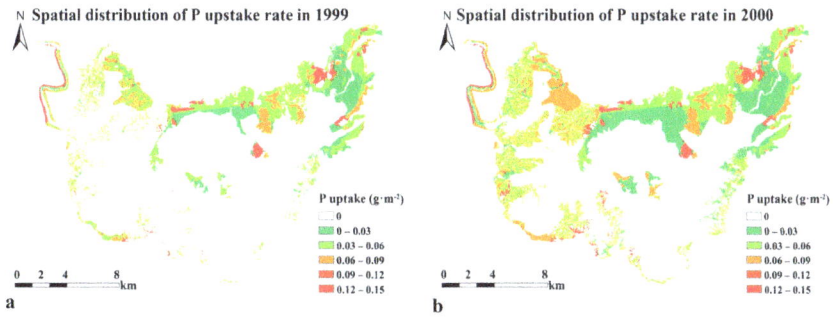

Figure 6. *Cont.*

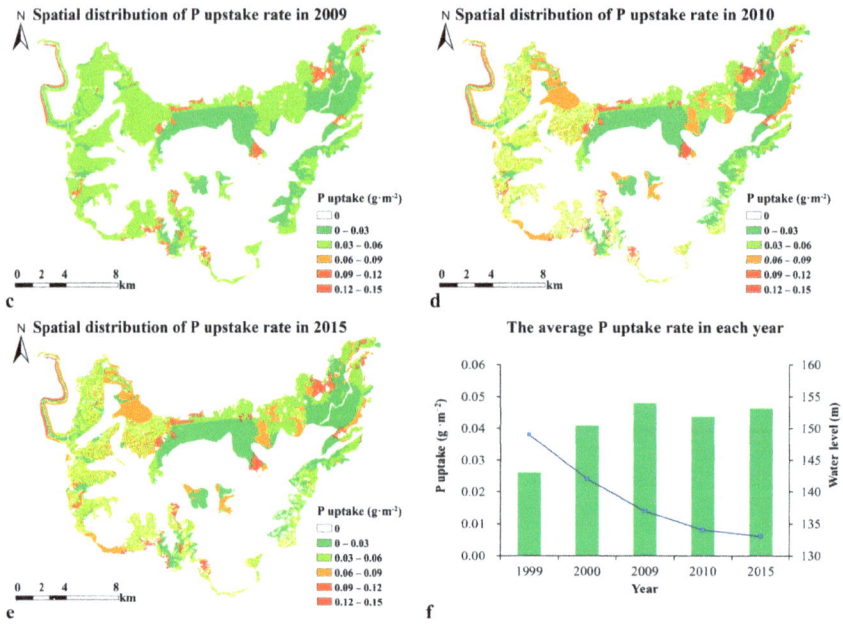

Figure 6. Spatial distribution of the P uptake rate in Miyun's inner protection zone during different years.

3.3. Negative Relationship between the Water Level and N and P Uptake

Water table dynamics strongly influence the N and P removal rates by plants [25] in the inner protection zone and are impacted by both landscape and stream water levels directly [48]. By overlaying the contour map on the spatial distribution of N and P removal by plants and implementing zonal statistics (130–140 m, 140–150 m, 150–160 m, and 160–170 m), we found that the N and P uptake increased synchronously with the elevation gradient (Figure 7a,b). The mean N and P uptake were generally low between elevations of 130–160 m, where values for N varied from 16.44 t to 124.74 t and values for P varied from 16.44 t to 124.74 t. Meanwhile, the values were higher at 160–170 m, where the mean consumed N increased from 124.74 t to 427.20 t and the mean consumed P increased from 124.74 t to 427.20 t. Thus, the uptake of N and P by vegetation presented a positive correlation with elevation.

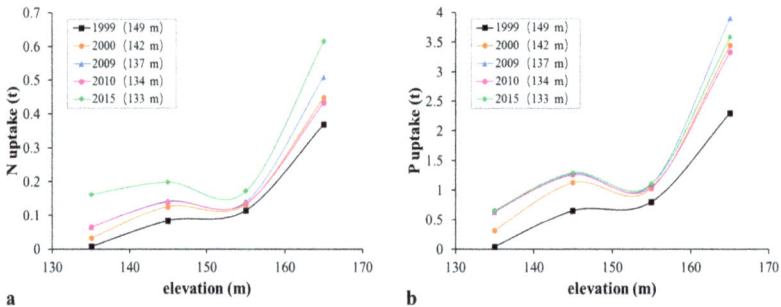

Figure 7. Zonal statistics for N (**a**) and P (**b**) consumed by plants during different years. The water level of each year is shown in parentheses following the year.

The amount of N and P removal by plants was also influenced by the water level. The lowest water level appeared in 2015, and the absorption of N and P by plants was strong in all elevation zones during this time. In 2009 and 2010, the water levels were similar, and the vegetation exhibited a similar N and P uptake capacity. The water level in 1999 was significantly higher than that in 2000, and the N and P uptake capacity of vegetation in 1999 was obviously lower than that of 2000. Thus, the increased water level had an obvious negative impact on N and P removal in the inner protection zone.

The response lines associating elevation with N and P uptake in Figure 7 can be classified into the following two categories based on the water level: high-level periods with an average water level of 145 m (1999, 2000) and low-level periods with an average water level of 134 m (2009, 2010, and 2015). The trend-line slopes of N and P uptake responses to different water levels are shown in Figure 8a,b. The slopes of N and P uptake during high-level periods were 0.145 and 0.0014, respectively, whereas they were 0.102 and 0.0009 during low-level periods, respectively. This shows that the N and P removal capacity of vegetation is more sensitive and weaker in high-level periods than in low-level periods. Therefore, the water levels of the reservoir along with the landscape have a significant impact on nutrient interception in the inner protection zone.

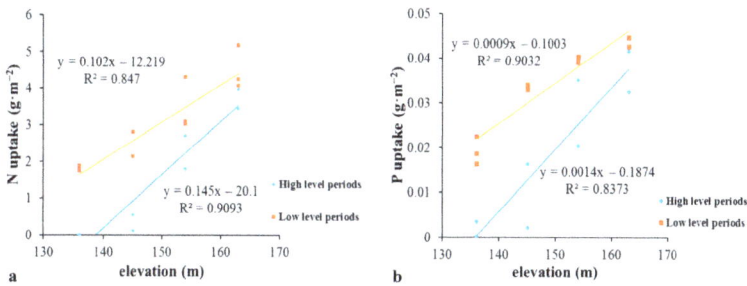

Figure 8. The distribution of N (**a**) and P (**b**) uptake rate responses to different water levels.

3.4. Validation of N and P Uptake Based on In Situ Survey Data

Generally, there are correlations between vegetation biomass density and N and P uptake and biomass can be used as a surrogate for nutrient removal potential [49,50]. To validate the accuracy of the RIPAM model, we converted the biomass density to N and P uptake by using a simple ratio coefficient. Xiao [51] did research on the ratios between biomass and N and P uptake for woodland areas in the Miyun Reservoir watershed, and the average ratio between biomass and N uptake was 0.00487, while that between biomass and P was 0.000579 [51]. For grasslands, Song [52] found that the ratios between biomass and N and P uptake were 0.0152 and 0.0012, respectively [52]. In this study, 39 quadrants were selected (shown in Section 2.4), and the N and P uptake rate of each quadrant was obtained by multiplying the biomass density with the corresponding correlation as follows:

$$X_{uptake} = Biomass \times X_{rate},\qquad(1)$$

where X_{uptake} refers to the N or P uptake rate of each quadrant ($g \cdot m^{-2}$), Biomass is the biomass density of each quadrant ($g \cdot m^{-2}$) (Table 3), and X_{rate} is the ratio between the biomass and N and P uptake according to Xiao [50] and Song [51]. It should be noted that the X_{rate} is varied with elevation and water level. Because of the limitation of the measured data, we used a fixed N or P uptake rate here and this would influence the results.

As each quadrant had a size of 90 m × 90 m, which corresponds to 3 pixels × 3 pixels in the RIPAM model, the average value of each 3 pixel × 3 pixel region in the simulated N and P uptake data was calculated. Then, scatter diagrams of N and P uptake between the RIPAM model and in situ quadrants were obtained, and linear regression was applied to the simulated data and in situ survey data (Figure 9a,b). We found that the R^2 and relation coefficient for N uptake were 0.88 and 0.78, respectively; the R^2 and relation coefficient for P uptake were 0.82 and 0.6, respectively. Both were a good fit. This shows that the two simulations both had considerable accuracy.

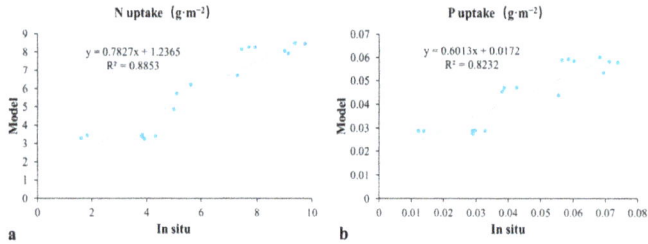

Figure 9. The N (**a**) and P (**b**) uptake rate of quadrant scatter diagrams between the RIPAM model and in situ survey.

4. Discussion

4.1. Reason for the Decline of N and P Uptake

As mentioned in Section 3.3, water level dynamics can have obvious negative impacts on N and P uptake. The growth conditions are a primary influencing factor for nutrient uptake by plants. For vegetation in the inner protection zone, soil moisture is no longer a limiting factor to its growth. In high-water-level periods, prolonged inundation may cause mechanical injury to plants and adversely affect the growth of vegetation. Additionally, high soil moisture will have more inhibitory effects than stimulatory effects on vegetation development. These both will lead to low NPP [24,28].

According to the theory described above, NPP in the inner protection zone was evaluated zonally. It was found that the variation trend of NPP was similar with that of the N and P uptake by plants during the five years that were analyzed (Figure 10a,b). Generally, there was a rising trend in NPP with increasing altitude. The NPP increased slowly from 130 m to 150 m, with mean values varying from 149.43 t to 2780.30 t. At an altitude of 160 m, there was a decline in the NPP. Then, it rose exponentially from 160 m to 170 m, whereby the mean NPP increased from 2301.45 t to 6604.80 t. The NPP was also influenced by the water level. The mean NPP in each elevation zone was higher during the low water level years (corresponding to 2009, 2010, and 2015) than that during the high water level years (corresponding to 1999 and 2000). This can be explained by the differences in the vegetation distribution, which were driven largely by the tolerances of species to specific geomorphic processes. As shown in Figure 1, grasslands and dry lands are the dominant land use types at elevations of 130–160 m; they both are home to herbaceous species and are easily influenced by elevated water tables or flooding. Meanwhile, forests, which are the dominant land use type at 160–170 m, are less affected.

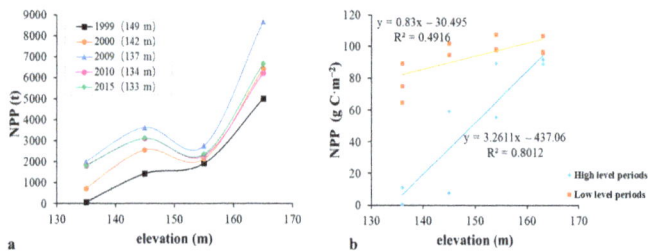

Figure 10. (**a**) Zonal statistics for NPP in different years and (**b**) NPP responses to different water levels. The numbers in parentheses following the years are the water level of each year.

4.2. Prediction of the Inner Protection Zone's Defense for N and P Interception

With the implementation of the middle route of the SNWTP, the storage capacity of the Miyun Reservoir is expected to increase from 1 billion m^3 to 2 billion m^3, and the water level will rise to 150 m in 2020, which is higher than the level in 1999. According to the contour map, the extent of the inner protection zone will decrease to 122.23 km^2, which is less than 56% of the area in 2015, and changes will be especially prominent in the northern floodplain, which represents one of the most important sections that curtails N and P losses to the water source from the upstream watershed.

The rest of the zone will also be influenced by the elevated water tables. According to the formula in Figure 10b, NPP will decrease, especially for vegetation at elevations of 130–150 m, which is sensitive to flooding. The total NPP will likely be less than the amount of 8454.96 t detected in 1999 when the water level was 149 m, and it is predicted to decline by more than 40.90% compared to the level in 2015. According to the formulas in Figure 8a,b, N and P uptake will be less than the amounts of 0.57 t N and 3.79 t P detected in 1999, and each will decline by more than 53.03% and 43.49%, respectively, compared to the levels in 2015. Thus, the ecological functions of the original inner protection zone will be impaired and this will lead to a decline of the inner protection zone's defense effectiveness in terms of N and P interception.

Although some studies indicate that transferred water is safe and has little impact on the trophic state of the receiving water [14], its ecological effectiveness as a nutrient sink providing pollution control and sediment retention will be impacted if the extant inner protection zone is damaged. The inner protection zone may even be turned from an NPS sink to an NPS source [26]. This could pose risks to the security of the water resource. Thus, vegetation cover in the extant inner protection zone of the terminal reservoir will need to be restored if the protection zone is to maintain its usual effectiveness.

5. Conclusions

As the end point of the middle route of the South-to-North Water Transfer Project (SNWTP), the Miyun Reservoir is the most important water source for Beijing. Importantly, water quality in the reservoir can have a direct impact on the wellbeing of people in Beijing and therefore warrants protection. In particular, steps will need to be taken to avoid eutrophication problems. This study used the RIPAM model to simulate N and P uptake by plants in Miyun Reservoir's inner protection zone during five years (1999, 2000, 2009, 2010, and 2015), and the results were validated by in situ survey data (R^2 of N uptake was 0.88, and R^2 of P uptake was 0.82). The correlations between the water level and the N and P uptake capacities showed that water table dynamics strongly influence the N and P removal by plants; thus, variations in the water level can have obvious negative impacts on N and P removal in the inner protection zone.

With the implementation of the middle route of the SNWTP, the water level of the Miyun Reservoir will rise to 150 m in 2020, and the total NPP is predicted to decline by more than 40.90% compared to the level in 2015; additionally, the N and P uptake is predicted to decline by more than 53.03% and 43.49%, respectively, compared to the levels in 2015. This will lead to a decline in the inner protection zone's defense effectiveness in terms of N and P interception and pose risks to the security of the water resource unless steps are taken to avoid such negative consequences. In summary, this study provides useful knowledge for assessing the risks of nutrient removal reductions in a terminal reservoir from a large inter-basin water transfer project before its implementation.

Acknowledgments: We acknowledge all of the editors and reviewers for their valuable advice. The authors are thankful for A-Xing Zhu and Guiming Zhang in the Department of Geography at UW-Madison for their valuable advice. The authors are also thankful for the financial support from the National Natural Science Foundation of China (Grant Nos. U1603241, 51779099), the National Key Project for R&D (Nos. 2016YFC0402403, 2016YFC0402409), the project titled Impact of Rising Water Levels on the Riparian Ecological Environment in Miyun Reservoir (230200019), and the Central non-profit research institutes fundamental research of Yellow River Institute of Hydraulic Research (HKY-JBYW-2017-10).

Author Contributions: Shengtian Yang, Juan Bai and Zhiwei Wang conceived and designed the experiments; Juan Bai, Zhiwei Wang, Hezhen Lou, Chunbin Zhang, Yabing Guan and Xinyi Yu performed the experiments;

Shengtian Yang, Zhiwei Wang, and Hezhen Lou analyzed the data; Chunbin Zhang contributed analysis tools; Yichi Zhang prepared the figures; Shengtian Yang, Juan Bai, and Zhiwei Wang wrote the main manuscript text. All authors discussed the structure and commented on the manuscript at all stages.

Conflicts of Interest: The authors declare no conflict of interest. The founding sponsors had no role in the design of the study; in the collection, analyses, or interpretation of data; in the writing of the manuscript; and in the decision to publish the results.

References

1. Cantor, L.M. The California state water project-a reassessment. *J. Geogr.* **1980**, *79*, 133–140.
2. Gastelum, J.R.; Cullom, C. Application of the Colorado River simulation system model to evaluate water shortage conditions in the central Arizona project. *Water Resour. Manag.* **2013**, *27*, 2369–2389. [CrossRef]
3. Voropaev, G.; Velikanov, A. Partial southward diversion of Northern and Siberian Rivers. *Int. J. Water Resour. Dev.* **1984**, *2*, 67–83. [CrossRef]
4. Hudson, J.W. *The Role of Water in the Biology of the Antelope Ground Squirrel Citellus Leucurus*; University of California Press: Berkeley, CA, USA, 1962.
5. Misra, A.K.; Saxena, A.; Yaduvanshi, M.; Mishra, A.; Bhadauriya, Y.; Thakur, A. Proposed river-linking project of India: A boon or bane to nature. *Environ. Geol.* **2007**, *51*, 1361–1376. [CrossRef]
6. Liu, C.; Zheng, H. South-to-North water transfer schemes for China. *Int. J. Water Resour. Dev.* **2002**, *18*, 453–471. [CrossRef]
7. Davies, B.R.; Thoms, M.; Meador, M. An assessment of the ecological impacts of inter-basin water transfers, and their threats to river basin integrity and conservation. *Aquat. Conserv.-Mar. Freshw. Ecosyst.* **1992**, *2*, 325–349. [CrossRef]
8. Yang, J.; Zhang, S.; Chang, L.; Li, F.; Li, T.; Gao, Y. Gully erosion regionalization of black soil area in Northeastern China. *Chin. Geogr. Sci.* **2017**, *27*, 78–87. [CrossRef]
9. Liu, K.; Ding, H.; Tang, G.; Na, J.; Huang, X.; Xue, Z.; Yang, X.; Li, F. Detection of catchment-scale gully-affected areas using unmanned aerial vehicle (uav) on the Chinese loess plateau. *ISPRS Int. J. Geo-Inf.* **2016**, *5*, 238. [CrossRef]
10. Zhuang, W. Eco-environmental impact of inter-basin water transfer projects: A review. *Environ. Sci. Pollut. Res. Int.* **2016**, *23*, 12867–12879. [CrossRef] [PubMed]
11. Li, Z.; Zhang, Y.; Zhu, Q.; Yang, S.; Li, H.; Ma, H. A gully erosion assessment model for the Chinese loess plateau based on changes in gully length and area. *Catena* **2017**, *148*, 195–203. [CrossRef]
12. Tang, C.; Yi, Y.; Yang, Z.; Cheng, X. Water pollution risk simulation and prediction in the main canal of the South-to-North water transfer project. *J. Hydrol.* **2014**, *519*, 2111–2120. [CrossRef]
13. Yang, W.; Zhang, L.; Shan, L.; Chen, X.; Chen, S. Response of extreme hydrological events to climate change in the water source area for the middle route of south-to-north water diversion project. *Adv. Meteorol.* **2016**, *2016*, 1–15. [CrossRef]
14. Zeng, Q.; Qin, L.; Li, X. The potential impact of an inter-basin water transfer project on nutrients (nitrogen and phosphorous) and chlorophyll a of the receiving water system. *Sci. Total Environ.* **2015**, *536*, 675–686. [CrossRef] [PubMed]
15. Zhang, Q. The South-to-North water transfer project of China: Environmental implications and monitoring strategy1. *J. Am. Water Resour. Assoc.* **2009**, *45*, 1238–1247. [CrossRef]
16. Zhang, Y.; Li, G.-M. Influence of south-to-north water diversion on major cones of depression in north China plain. *Environ. Earth Sci.* **2013**, *71*, 3845–3853. [CrossRef]
17. Li, Z.; Fang, H. Impacts of climate change on water erosion: A review. *Earth-Sci. Rev.* **2016**, *163*, 94–117. [CrossRef]
18. Pandey, A.; Himanshu, S.K.; Mishra, S.K.; Singh, V.P. Physically based soil erosion and sediment yield models revisited. *Catena* **2016**, *147*, 595–620. [CrossRef]
19. Walcher, M.; Bormann, H. On the transferability of the concept of drinking water protection zones from EU to Latin American Countries. *Water Resour. Manag.* **2015**, *29*, 1803–1822. [CrossRef]
20. Brenčič, M.; Prestor, J.; Kompare, B.; Matoz, H.; Kranjc, S. Integrated approach to delineation of drinking water protection zones. *Geologija* **2009**, *52*, 175–182. [CrossRef]

21. Sabater, S.; Butturini, A.; Clement, J.-C.; Burt, T.; Dowrick, D.; Hefting, M.; Matre, V.; Pinay, G.; Postolache, C.; Rzepecki, M. Nitrogen removal by riparian buffers along a European climatic gradient: Patterns and factors of variation. *Ecosystems* **2003**, *6*, 0020–0030. [CrossRef]

22. Correll, D. Buffer zones and water quality protection: General principles, Buffer zones: Their processes and potential in water protection. In Proceedings of the International Conference on Buffer Zones, Harpenden, Hertfordshire, UK, September 1996; pp. 7–20.

23. Klemas, V. Remote sensing of riparian and wetland buffers: An overview. *J. Coast. Res.* **2014**, *297*, 869–880. [CrossRef]

24. Nilsson, C.; Svedmark, M. Basic principles and ecological consequences of changing water regimes: Riparian plant communities. *Environ. Manag.* **2002**, *30*, 468–480. [CrossRef]

25. Liu, X.; Vidon, P.; Jacinthe, P.A.; Fisher, K.; Baker, M. Seasonal and geomorphic controls on n and P removal in riparian zones of the US Midwest. *Biogeochemistry* **2014**, *119*, 245–257. [CrossRef]

26. Schilling, K.E.; Li, Z.; Zhang, Y.-K. Groundwater–surface water interaction in the riparian zone of an incised channel, Walnut Creek, Iowa. *J. Hydrol.* **2006**, *327*, 140–150. [CrossRef]

27. Džubáková, K.; Molnar, P.; Schindler, K.; Trizna, M. Monitoring of riparian vegetation response to flood disturbances using terrestrial photography. *Hydrol. Earth Syst. Sci.* **2015**, *19*, 195–208. [CrossRef]

28. Kozlowski, T.T. Physiological-ecological impacts of flooding on riparian forest ecosystems. *Wetlands* **2002**. [CrossRef]

29. Stone, R.; Jia, H. Going against the flow. *Science* **2006**, *313*, 1034–1037. [CrossRef] [PubMed]

30. Zhang, J.; Meng, F.; Lu, Y.; Jing, Y.; Zhang, H.; Zhang, B.; Zhang, C. Ecological assessment of lakeshore wetland rehabilitation on eastern route of south-to-north water transfer project. *Front. Environ. Sci. Eng. China* **2008**, *2*, 306–310. [CrossRef]

31. Cheng, S.; Song, H. Conservation buffer systems for water quality security in south to north water transfer project in China: An approach review. *Front. For. China* **2009**, *4*, 394–401. [CrossRef]

32. Gu, W.; Shao, D.; Jiang, Y. Risk evaluation of water shortage in source area of middle route project for south-to-north water transfer in China. *Water Resour. Manag.* **2012**, *26*, 3479–3493. [CrossRef]

33. Liang, Y.-S.; Wang, W.; Li, H.-J.; Shen, X.-H.; Xu, Y.-L.; Dai, J.-R. The South-to-North water diversion project: Effect of the water diversion pattern on transmission of oncomelania hupensis, the intermediate host of *Schistosoma japonicum* in China. *Parasites Vectors* **2012**, *5*, 52. [CrossRef] [PubMed]

34. Barnett, J.; Rogers, S.; Webber, M.; Finlayson, B.; Wang, M. Sustainability: Transfer project cannot meet China's water needs. *Nature* **2015**, *527*, 295–297. [CrossRef] [PubMed]

35. Li, X.-S.; Wu, B.-F.; Zhang, L. Dynamic monitoring of soil erosion for upper stream of Miyun reservoir in the last 30 years. *J. Mt. Sci.* **2013**, *10*, 801–811. [CrossRef]

36. Jiao, J.; Du, P.; Lang, C. Nutrient concentrations and fluxes in the upper catchment of the Miyun reservoir, China, and potential nutrient reduction strategies. *Environ. Monit. Assess.* **2015**, *187*, 110. [CrossRef] [PubMed]

37. Dong, G.; Yang, S.; Gao, Y.; Bai, J.; Wang, X.; Zheng, D. Spatial evaluation of phosphorus retention in riparian zones using remote sensing data. *Environ. Earth Sci.* **2014**, *72*, 1643–1657. [CrossRef]

38. Wang, X.; Mannaerts, C.M.; Yang, S.; Gao, Y.; Zheng, D. Evaluation of soil nitrogen emissions from riparian zones coupling simple process-oriented models with remote sensing data. *Sci. Total Environ.* **2010**, *408*, 3310–3318. [CrossRef] [PubMed]

39. Wang, X.; Wang, Q.; Yang, S.; Zheng, D.; Wu, C.; Mannaerts, C.M. Evaluating nitrogen removal by vegetation uptake using satellite image time series in riparian catchments. *Sci. Total Environ.* **2011**, *409*, 2567–2576. [CrossRef] [PubMed]

40. Wang, X.; Yang, S.; Mannaerts, C.M.; Gao, Y.; Guo, J. Spatially explicit estimation of soil denitrification rates and land use effects in the riparian buffer zone of the large Guanting reservoir. *Geoderma* **2009**, *150*, 240–252. [CrossRef]

41. Prince, S.D.; Goward, S.N. Global primary production: A remote sensing approach. *J. Biogeogr.* **1995**, *22*, 815–835. [CrossRef]

42. Zhu, Z.X.; Arp, P.A.; Meng, F.R.; Bourque, C.P.A.; Foster, N.W. A forest nutrient cycling and biomass model (FORNBM) based on year-round, monthly weather conditions, part 1: Assumption, structure and processing. *Ecol. Model.* **2003**, *169*, 347–360. [CrossRef]

43. Arp, P.A.; Oja, T. A forest soil vegetation atmosphere model (ForSVA) 1. Concepts. *Ecol. Model.* **1997**, *95*, 211–224. [CrossRef]

44. Zhang, L.-T.; Li, Z.-B.; Wang, S.-S. Spatial scale effect on sediment dynamics in basin-wide floods within a typical agro-watershed: A case study in the hilly loess region of the Chinese loess plateau. *Sci. Total Environ.* **2016**, *572*, 476–486. [CrossRef] [PubMed]

45. Husson, E.; Lindgren, F.; Ecke, F. Assessing biomass and metal contents in riparian vegetation along a pollution gradient using an unmanned aircraft system. *Water Air Soil Pollut.* **2014**, *225*. [CrossRef]

46. Kachamba, D.J.; Orka, H.O.; Gobakken, T.; Eid, T.; Mwase, W. Biomass estimation using 3D data from unmanned aerial vehicle imagery in a tropical woodland. *Remote Sens.* **2016**, *8*, 968. [CrossRef]

47. Yang, S.; Bai, J.; Zhao, C.; Lou, H.; Zhang, C.; Guan, Y.; Zhang, Y.; Wang, Z.; Yu, X. The assessment of the changes of biomass and riparian buffer width in the terminal reservoir under the impact of the South-to-North water diversion project in China. *Ecol. Indic.* **2018**, *85*, 932–943. [CrossRef]

48. Jung, M.; Burt, T.; Bates, P. Toward a conceptual model of floodplain water table response. *Water Resour. Res.* **2004**, *40*. [CrossRef]

49. Rheinhardt, R.; Brinson, M.; Meyer, G.; Miller, K. Integrating forest biomass and distance from channel to develop an indicator of riparian condition. *Ecol. Indic.* **2012**, *23*, 46–55. [CrossRef]

50. Jiang, F.Y.; Chen, X.; Luo, A.C. A comparative study on the growth and nitrogen and phosphorus uptake characteristics of 15 wetland species. *Chem. Ecol.* **2011**, *27*, 263–272. [CrossRef]

51. Xiao, Y. Study on the Adjustment and Control Mechanism of Forest to Nonpoint Source Pollution in Beijing Mountain Area. Ph.D. Thesis, Beijing Forestry University, Beijing, China, 29 June 2008.

52. Song, X. Study on the Variation of N, P, K Nutrients Pool of Different Type of Alpine Meadow in Eastern Qilian Mountains. Master's Thesis, Gansu Agricultural University, Lanzhou, China, 10 June 2008.

water

MDPI

Article

Estimating Carbon Dioxide (CO$_2$) Emissions from Reservoirs Using Artificial Neural Networks

Zhonghan Chen, Xiaoqian Ye and Ping Huang *

Department of Environmental Science, School of Environmental Science and Engineering,
Sun Yat-sen University, Guangzhou 510275, China; chenzhh43@mail2.sysu.edu.cn (Z.C.);
yexqian@mail2.sysu.edu.cn (X.Y.)
* Correspondence: eeshping@mail.sysu.edu.cn or pinghuang43@foxmail.com; Tel.: +86-132-4974-8826

Received: 3 November 2017; Accepted: 28 December 2017; Published: 1 January 2018

Abstract: Freshwater reservoirs are considered as the source of atmospheric greenhouse gas (GHG), but more than 96% of global reservoirs have never been monitored. Compared to the difficulty and high cost of field measurements, statistical models are a better choice to simulate the carbon emissions from reservoirs. In this study, two types of Artificial Neural Networks (ANNs), Back Propagation Neural Network (BPNN) and Generalized Regression Neural Network (GRNN), were used to predict carbon dioxide (CO$_2$) flux emissions from reservoirs based on the published data. Input variables, which were latitude, age, the potential net primary productivity, and mean depth, were selected by Spearman correlation analysis, and then the rationality of these inputs was proved by sensitivity analysis. Besides this, a Multiple Non-Linear Regression (MNLR) and a Multiple Linear Regression (MLR) were used for comparison with ANNs. The performance of models was assessed by statistical metrics both in training and testing phases. The results indicated that ANNs gave more accurate results than regression models and GRNN provided the best performance. With the help of this GRNN, the total CO$_2$ emitted by global reservoirs was estimated and possible CO$_2$ flux emissions from a planned reservoir was assessed, which illustrated the potential application of GRNN.

Keywords: CO$_2$; reservoirs; general regression neural network; back propagation neural network

1. Introduction

Since the problem of greenhouse gases (GHGs) emissions from hydroelectric reservoirs was first addressed in the publication in 1993 [1], it has been the focus of research around the world [2], especially in Canada [3,4], Brazil [5], and the United States [6,7]. Over the past two decades, a growing amount of work has documented reservoirs' roles as GHG sources [8,9], after extensive research was carried out in various reservoirs. However, the magnitude of global flux of GHG from reservoirs is still highly debatable [10]. According to the current estimations of global carbon dioxide (CO$_2$) emissions, the hydroelectric reservoirs were responsible for emitting 48 Tg C yr^{-1} as CO$_2$ [11], while Demmer et al. [12] estimated that GHG emissions accounted for 36.8 Tg C yr^{-1} as CO$_2$ ignoring the types of reservoirs. These estimates corresponded roughly to 2% of global carbon emissions from inland waters that reported a flux of 2100 Tg C yr^{-1} as CO$_2$ [13]. Although there was a minor difference between the estimated global CO$_2$ flux from hydroelectric systems and all reservoir systems, any significant difference between the areal emissions of CO$_2$ from hydroelectric and non-hydroelectric reservoirs was not detected by statistical analysis [12]. Depending on reservoir type, GHG emissions from reservoirs are related to various factors, which is crucial for understanding the mechanism and the control over GHG emissions. In a single reservoir system, both depth and temperature might be the important predictors of carbon emissions, and also reflect the spatial and seasonal variability [5,14]. GHG emissions tend to decrease with the increase of reservoirs' latitude and age [11]. Considering the internal source of carbon emissions, the initial organic carbon in the flooded area is another key factor, especially in the young reservoirs [15,16]. Besides, the GHG emissions are also related to dam operation regime [5] and water quality, such as pH [17] and Chlorophyll-*a* [18,19].

Despite significance and uncertainty, there is still a lack of measured CO_2 emissions from reservoirs in many regions, which leaves a critical gap in the global CO_2 budgets. To resolve this problem, statistical models are the appropriate methods to extrapolate the flux of reservoirs without measured data and then derive regional or global estimations relying on a limited number of measurements [20]. One of the most common models is the statistical regression model, which can demonstrate the relationship between CO_2 emissions and one [9] or several [11,16] factors by the regression equations. Other models that can identify more complex non-linear responses, such as Random Forests [6] and Monte Carlo simulation [21], were also used to elucidate the relationship between CO_2 emissions and the factors of environment or dams, and also to predict CO_2 emissions from other under-sampled reservoirs in the nearby geographic region. Unlike these models with the inputs of environmental factors, a mathematical model with the theory of Self-Organized Criticality (SOC) was employed to extrapolate values from one reservoir to another directly without any other features [22]. These pioneering models showed a low degree of precision and regional limitations; therefore, developing and optimizing models of reservoir CO_2 emissions is still one of the future research directions [6,12].

Artificial neural networks (ANNs) were frequently being used for the simulation of both water quality [23,24] and GHG emissions caused by energy consumption [25,26], and showed great potential for prediction. However, few research has been conducted to simulate GHG emissions from reservoirs using ANNs. Actually, ANNs have several advantages that are suitable for this study. ANNs have robust learning and generalization ability, after simulating the learning and decision-making process of human beings. Therefore, ANNs can describe the linear or non-linear relationships precisely even with limited input variables [27,28] and identify complex patterns in dataset without adequate understanding of the interaction among variables [24]. Besides this, because the learning mechanism in ANNs is non-parametric, the structure and distribution of data are not limitations [29]. The precision of fitting by ANNs largely depends on the quantity and quality of data [30]. CO_2 fluxes have been measured in at least 229 reservoir systems in the world until 2016 [12], which supplies the sufficient amount of data for ANNs. Besides, many commonly employed techniques for measurement focus on quantifying the diffusive flux of gases across the air–water interface, which is suitable for CO_2 because of its solubility [12]. There are some other key factors that affect model performance, such as the architecture selection and parameter settings [28]. However, it is difficult to reach any conclusion of which model architecture is absolutely suitable to a particular circumstance. Therefore, ad hoc approaches, such as a trial-and-error approach, might be acceptable to determine the parameters, following the principle that the optimal network structure generally keeps a balance between generalization ability and network complexity [31].

In this study, we make the first attempt to simulate the CO_2 flux emission from reservoirs by ANNs, back propagation neural network (BPNN) and general regression neural network (GRNN), based on the published data from various types of reservoirs with a global distribution. The input variables of models is selected through both the correlation analysis and domain knowledge. The rationality of selected sets of inputs is tested by sensitivity analysis. The model parameters selection are described in detail and the model performance is evaluated using statistical indices. Since the models have the ability to predict CO_2 fluxes from a reservoir without measurements, the global fluxes of CO_2 emissions from reservoirs can be estimated. Besides this, the possible magnitude of CO_2 emissions from a planned reservoir can also be assessed by models based on some reservoirs' features, which gives guidance for dam's construction.

2. Materials and Methods

2.1. Data Collection

In this study, most CO_2 emission fluxes from reservoir surface were based on data collection by Barros et al. [11] in 2011 and Deemer et al. [12] in 2016. CO_2 emission monitoring data from reservoirs in some latest literature were also assembled. Some essential parameters about reservoirs and monitoring were collected from the relevant literature and part of missing values were completed from Global Reservoir and Dam (GRanD) database [19]. Another important parameter, the primary productivity of potential vegetation (NPP0) in the reservoir's location, was extracted from the map of the human appropriation of net primary production (HANPP) [32]. In cases where more than one study measured CO_2 fluxes from the system in the same year, the arithmetic mean of these parameters was used for statistic, and CO_2 fluxes from same reservoir measured in different years were kept at the original values. Therefore, 277 data sets were collected based on the studies in 235 reservoirs, and the information of data set can be found in supplementary Table S1.

In total, we assembled 10 parameters of 235 reservoirs with a global distribution, including latitude (Lat), age, chlorophyll-*a* (Chl-*a*), water temperature (WT), mean depth (MD), residence time (RT), dissolved organic carbon (DOC), total phosphorus (TP), the potential net primary productivity of the area (NPP0), and CO_2 flux. CO_2 flux was used as the output of models in this study. The statistical parameters including minimum value, maximum value, mean value, median, standard deviation, and variation coefficient are given in Table 1.

Table 1. The statistical parameters for data sets.

Parameters	Unit	Min	Max	Mean	Median	SD	VC
Lat	°	−42.93	68.00	31.96	38.17	26.36	0.83
Age	yrs	1.00	95.00	39.09	36.00	24.55	0.63
Chl-*a*	$\mu g\,L^{-1}$	0.20	137.50	12.03	4.13	24.78	1.96
WT	°C	6.30	35.00	17.88	17.40	5.52	0.30
MD	m	0.30	400.00	26.58	15.00	40.26	1.52
RT	days	1.25	13,140.00	665.75	180.00	1689.54	2.54
DOC	$mg\,L^{-1}$	1.25	30.00	4.79	3.82	4.01	0.84
TP	$\mu g\,L^{-1}$	1.40	500.00	62.61	29.00	96.89	1.55
NPP0	$mg\,C\,m^{-2}\,d^{-1}$	151.90	3200.68	1529.21	1574.50	604.70	0.40
CO_2 flux	$mg\,C\,m^{-2}\,d^{-1}$	−356.00	3800.00	400.90	254.75	569.89	1.42

Note: SD, the standard deviation; VC, the coefficient of variation; Lat, latitude; Chl-*a*, chlorophyll-*a*; WT, water temperature; MD, mean depth; RT, residence time; DOC, dissolved organic carbon; TP, total phosphorus; NPP0, potential net primary productivity.

2.2. Input Variables and Data Processing

The selection of a set of appropriate input variables is the precondition of developing a satisfactory ANN for prediction [31]. There are two basic principles for selection: one is the confirmation of input–output relationship, and the other is the independence among input variables [26]. This characteristic of independence is very important since correlated data can provide redundant information, which increases the likelihood of overfitting and the difficulty to find optimal weights [31]. There are two general categories of techniques, model-free and model-based approaches, to examine the relationship between alternative inputs and outputs, such as the correlation analysis and the stepwise analysis [31]. In this study, the correlation analysis, sensitivity analysis, the availability of data, and domain knowledge are combined to select an optimal set of input variables for models.

Since the variables have different units, the variables should be scaled to a uniform range before passing them through the ANNs to avoid convergence problems and extremely small weighting factors [33]. There are no fixed rules for the standardization in literature. In this study, normalization was done by the following equation:

$$Y = (y_{max} - y_{min}) \left(\frac{x - x_{min}}{x_{max} - x_{min}} \right) + y_{min} \tag{1}$$

where, Y denotes the data value after normalization, x_{max} and x_{min} are the maximum and the minimum values of each variable, and y_{max} and y_{min} denote the boundary values of the specific range, which are 1 and −1 respectively in this study.

The complete CO_2 emissions data set comprises 235 reservoirs in 21 countries. In this study, 70% of samples with the whole selected input variables were randomly chosen to build and train models, while the remaining data was used to test the models.

2.3. Artificial Neural Networks (ANNs)

Artificial neural networks (ANNs) are the abstract computational models that follow the behavior of the human brain. ANNs have been used widely for prediction and forecasting in the environmental area, because they are believed to approximate any finite non-linear function with high accuracy [34]. Traditionally, ANNs are divided into feed-forward and recurrent networks. Feed-forward architectures, which have many types, such as Multilayer Perceptions (MLP) and Radial Basis Function (RBFNN), are the most popular architectures used in researches [31]. In feed-forward ANNs, the input signal propagates through network in a forward direction, from the input layer to the hidden layer and then to the output neurons [33]. Among many types of feed-forward networks, BPNN and GRNN were chosen for this study.

2.3.1. Back Propagation Neural Networks (BPNNs)

Back Propagation Neural Networks (BPNNs) are typical multi-layer perceptron neural networks (MLPs) with error back-propagation (BP) algorithm for network learning [34]. BPNNs are organized as hierarchical networks with several layers including an input layer, hidden layer(s), and an output layer. Generally, it is believed that BPNNs with one hidden layer are able to describe any finite non-linear relationship with acceptable performance [35]. Each layer has several neurons that transmit input values and process to the next layer by a set of weights (Figure 1). In each neuron, the sum of the products of input variables and their weights is transformed to an output value by an activation function. In this study, the classical *tansig* function was selected as the activation function from the input layer to the hidden layer (Equation (2)). The *purelin* linear function was applied from the hidden layer to the output layer (Equation (3)). There is a wide variety of algorithms available for training a network and adjusting its weights, and Levenberg–Marquardt algorithm (LMA) was used in this study.

$$X_j = tansig \left(\sum_{i=1}^{m} x_i w'_{ji} + b'_j \right) \tag{2}$$

$$Y = \sum_{j=1}^{n} \left(X_j w''_j + b'' \right) \tag{3}$$

where, m and n are the numbers of neurons in the input and hidden layers, respectively. x_i is the values of the input variables; X_j is the result obtained by activation function (*tansig*) from the input neurons; w'_{ji} is the connection weight between the ith input neuron and jth hidden neuron, and w''_j is the connection weight between the jth hidden neuron and the output neuron; b'_j and b'' are the bias for the jth hidden neuron and the output neuron, respectively.

The reason for selecting BPNN is that it is the most commonly used for simulation among ANNs. Besides, the BP algorithm can efficiently minimize network error by dynamically searching for the optimal weights. The optimal number of hidden nodes for BPNN was selected by trying different integers in a reasonable range separately, which created the balance between complexity and accuracy.

Feed-forward of Information

Input layer   **Hidden layer**   **Output layer**

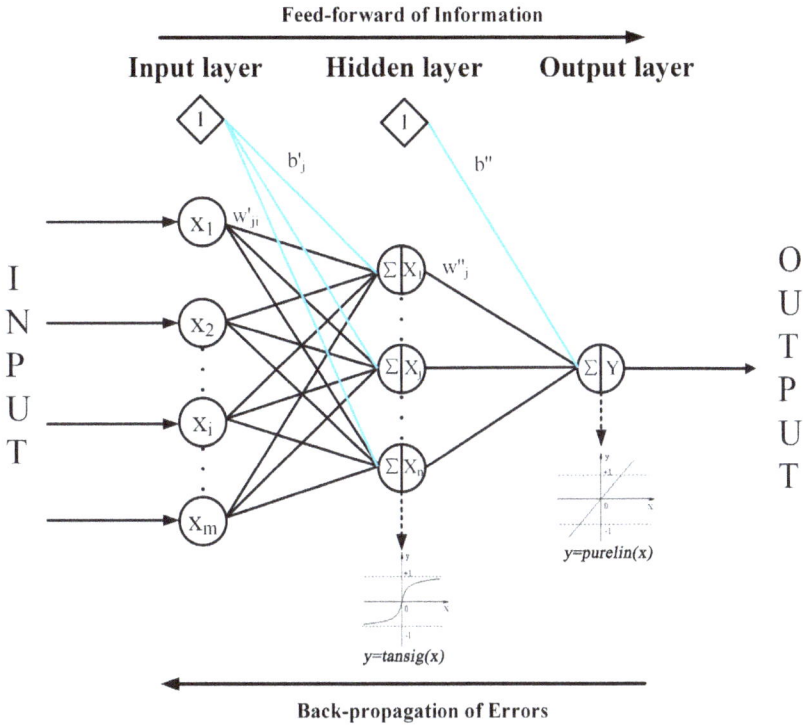

Figure 1. Architecture of Back-Propagation Neural Network (BPNN).

2.3.2. General Regression Neural Networks (GRNNs)

Unlike BPNNs, General Regression Neural Networks (GRNNs) do not rely on iterative procedures for their training but rely on a standard statistical technique called kernel regression [36]. GRNNs consist of four consequent layers, namely input, pattern, summation, and output layers (Figure 2).

Input Layer   **Pattern Layer**   **Summation Layer**   **Output Layer**

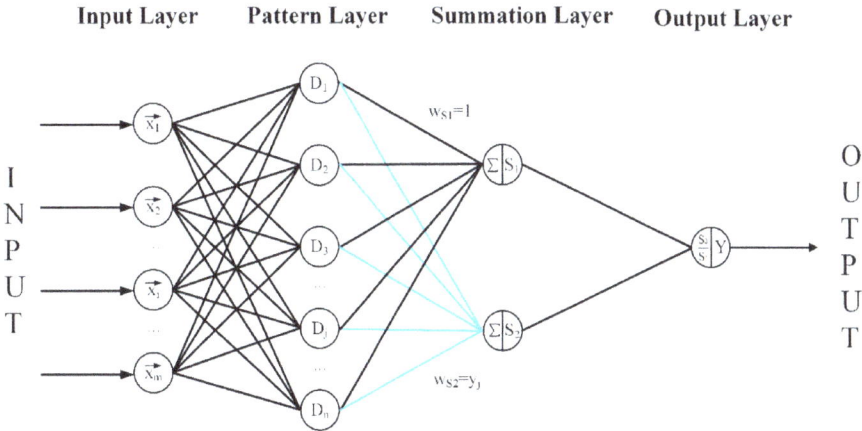

Figure 2. Architecture of General Regression Neural Network (GRNN).

In the first (input) layer, the input units pass input variables to pattern layer through input weights. In the second (pattern) layer, each neuron presents a training pattern, and the similarity between input patterns is calculated using a distance function (Equation (4)). The third (summation) layer consists two different types of summation, including single division and summation units. Each neuron in the pattern layer is connected to both S-summation and D-summation neurons in the summation layer. The S-summation neuron calculates the sum of the weighted responses in pattern layer, whereas the D-summation neuron computes the unweighted output neuron in the pattern layer. The final (output) layer just divides each S-summation neuron on D-summation neuron and represents the network prediction (Equations (5)) [37,38].

$$D_j(x, x_j) = \sum_{i=1}^{m} \left(\frac{x_j - x_{ji}}{\sigma} \right)^2 \tag{4}$$

$$Y = \frac{\sum_{j=1}^{n} y_j exp\left[-D_j(x, x_j)\right]}{\sum_{j=1}^{n} exp\left[-D_j(x, x_j)\right]} \tag{5}$$

where, m and n are the number of elements of an input vector and the number of training patterns, respectively. D is the Gaussian function. The term of $(x_j - x_{ji})$ denotes the difference between the ith training data x_{ji} and the point of estimation x_j. σ is the spread (smoothness) parameter whose optimal value can be experimentally evaluated. y_i represents the weight relationship between ith neuron in the pattern layer and the S-summation neuron in the summation layer.

The GRNN method was also selected because of its fast learning and convergence to the optimal regression surface [39]. Besides, it has another advantage that the network architecture is fixed and only one single parameter named spread needs to be optimized [39]. Different values of spread were also tried separately for each GRNN model in this study.

2.4. Statistical Regression Models

The performance of ANNs was compared with those of the multiple linear regression model (MLR) and multiple non-linear regression model (MNLR). These models had the same inputs as ANNs and were developed and tested using the same data. The MLR is shown in Equation (6). To identify a suitable MNLR, the various numerical transformations were tried, such as reciprocal, logarithm, and square root (Equation (7)).

$$Y = \alpha_0 + \alpha_1 x_1 + \alpha x_2 + \ldots + \alpha_m x_m \tag{6}$$

$$Y = \beta_0 + \beta_1 f_1(x_1) + \beta_1 f_2(x_2) + \ldots + \beta_m f_m(x_m) \tag{7}$$

where, x_m is the input, α_m is the coefficients of first degree inputs, f_m represents the transfer function for the input, and β_m is the coefficients of transferred input.

2.5. Performance Metrics

The results of created models were analyzed using multiple performance metrics. The root mean squared error (RMSE) and the mean absolute error (MAE) measure residual errors, which show the difference between the modeled and observed values. The determination coefficient (R^2) is identical to the square of the correlation coefficient (R) in some cases, which evaluates the degree of variability that can be explained by the model. Nash–Sutcliffe efficiency (NSE) [40] is a popular likelihood function to define the goodness of fit between monitoring data and model outputs. Smaller RMSE and MAE values and larger R^2 and NSE values indicate better model performance.

3. Results

3.1. Alternative Input Variable Selection

The original input dataset was selected through the results of Spearman's correlation analysis between the CO_2 flux and other nine variables of these reservoirs. Spearman correlation coefficient was selected because the data of CO_2 flux was not normally distributed (Kolmogorov-Smirnov test, $p < 0.001$), even after several different attempts at transformation. Some reservoirs were measured many times in different years, which means these reservoirs have more than one set of data with the same parameters except age and CO_2 flux. Therefore, Spearman correlation analysis between CO_2 flux and age was carried out by the original values, while Spearman correlation analysis between CO_2 flux and other parameters (Lat, Chl-a, WT, MD, RT, DOC, TP, and NPP0) was made based on the arithmetic mean.

Based on the results of correlation analysis (Table 2), the variables that have the high absolute value of correlation coefficient and low value of significance coefficient were age, mean depth, and NPP0. Moreover, latitude was reported as a key factor in previous studies [8,11] and the available sample size is sufficient. Therefore, latitude is also considered as an input and tested the rationality by sensitive analysis. Since they are obviously independent, it is unnecessary to analyze the correlation among these parameters. Consequently, the four features were chosen as alternative input variables of the models for CO_2 prediction. Only the reservoir data where the four parameters' records are effective and available were used, otherwise, this set of data would be removed. After deleting the invalid data, a dataset containing 251 sets of data was selected for simulation.

Table 2. Spearman correlation coefficient between CO2 flux and parameters used in present study.

Variables	*n*	Correlation	Sig.	Variables	*n*	Correlation	Sig.
Lat	236	−0.025	0.69	RT	98	0.055	0.59
Age	266	**−0.307**	0.00	DOC	51	0.129	0.36
Chl-a	69	−0.115	0.35	TP	47	0.005	0.98
WT	158	−0.118	0.13	NPP0	234	**0.153**	0.02
MD	217	**−0.151**	0.02				

Note: (1) *n*, the number of samples; Correlation, Spearman Correlation; Sig., Significance coefficient. (2) Latitude was calculated by the absolute value. (3) Bold font: correlation is significant at the 0.05 level (two-tailed).

3.2. Model Parameters Selection

A three-layer BPNN model was used in this study. To determine the topology of the network, diverse numbers of hidden nodes that range from 1 to 15 were tried for the BPNN, and their performance was compared by RMSE. The BPNN with nine hidden neurons showed the best fit performance (RMSE = 399.01 mg C m^{-2} d^{-1}). As a result, the topology of BPNN used in this study has four input neurons, nine hidden neurons, and one output neuron.

For the GRNN model, the spread constant was attempted by comparing the RMSE values obtained for each model with different spread constant varies from 0.1 to 1. The RMSE is lowest at 0.1 spread constant (RMSE = 279.69 mg C m^{-2} d^{-1}).

3.3. Model Performances

Among 251 sets of data, 175 (about 70%) sets are selected randomly for training, and the rest (76 sets) are testing data. The statistical parameters of CO_2 flux for training and testing are given in Table 3.

Table 3. The statistical parameters for CO_2 flux in training and testing phase.

Statistical Parameters	Unit	Training Set	Testing Set
n		76	175
Min	mg C m^{-2} d^{-1}	-325.90	-356.00
Max	mg C m^{-2} d^{-1}	3776.00	3800.00
Mean	mg C m^{-2} d^{-1}	390.82	486.40
Median	mg C m^{-2} d^{-1}	243.29	312.03
SD	mg C m^{-2} d^{-1}	549.75	664.02

The statistical performance measures for MLR, MNLR, BPNN, and GRNN are presented in Table 4 for both training and testing data sets. The GRNN model achieved the best performance in both training and testing phase according to mean performance statistics.

Table 4. The performances of each model during training and testing phases.

Model	Training Data Set				Testing Data Set			
	RMSE	MAE	R^2	NSE	RMSE	MAE	R^2	NSE
MLR	476.42	313.22	0.25	0.25	625.36	429.96	0.12	0.11
MNLR	417.26	282.00	0.43	0.42	529.53	391.46	0.40	0.36
BPNN	396.59	268.53	0.52	0.48	505.43	395.33	0.47	0.42
GRNN	272.50	147.62	0.76	0.75	418.48	295.34	**0.61**	0.60

Note: (1) The unit of RMSE is mg C m^{-1} d^{-1}; the unit of MAE is mg C m^{-1} d^{-1}. RMSE, root mean squared error; MAE, mean absolute error; NSE, Nash–Sutcliffe efficiency; MLR, multiple linear regression; MNLR, multiple non-linear regression; BPNN, Back Propagation Neural Network; GRNN, Generalized Regression Neural Network.

The observed and predicted CO_2 flux values by MLR, MNLR, BPNN, and GRNN models in training and testing stages are plotted in Figure 3. Comparisons among three models above indicate that GRNN generally gives better accuracy than the BPNN, MNLR, and MLR models. This can also be clearly observed from the fit line equations.

The equations of MLR and MNLR are shown in Equations (8) and (9) respectively.

$$CO2flux = 678.70 - 6.34 \cdot |Lat| - 5.71 \cdot Age + 0.42 \cdot NPP0 - 2.21 \cdot MD. \tag{8}$$

$$CO2flux = 471.21 - 3763 \cdot \left(\frac{1}{|Lat|}\right) - 168.97 \cdot \ln(Age) + 61.23 \cdot \ln(NPP0) - 2.04 \cdot MD. \tag{9}$$

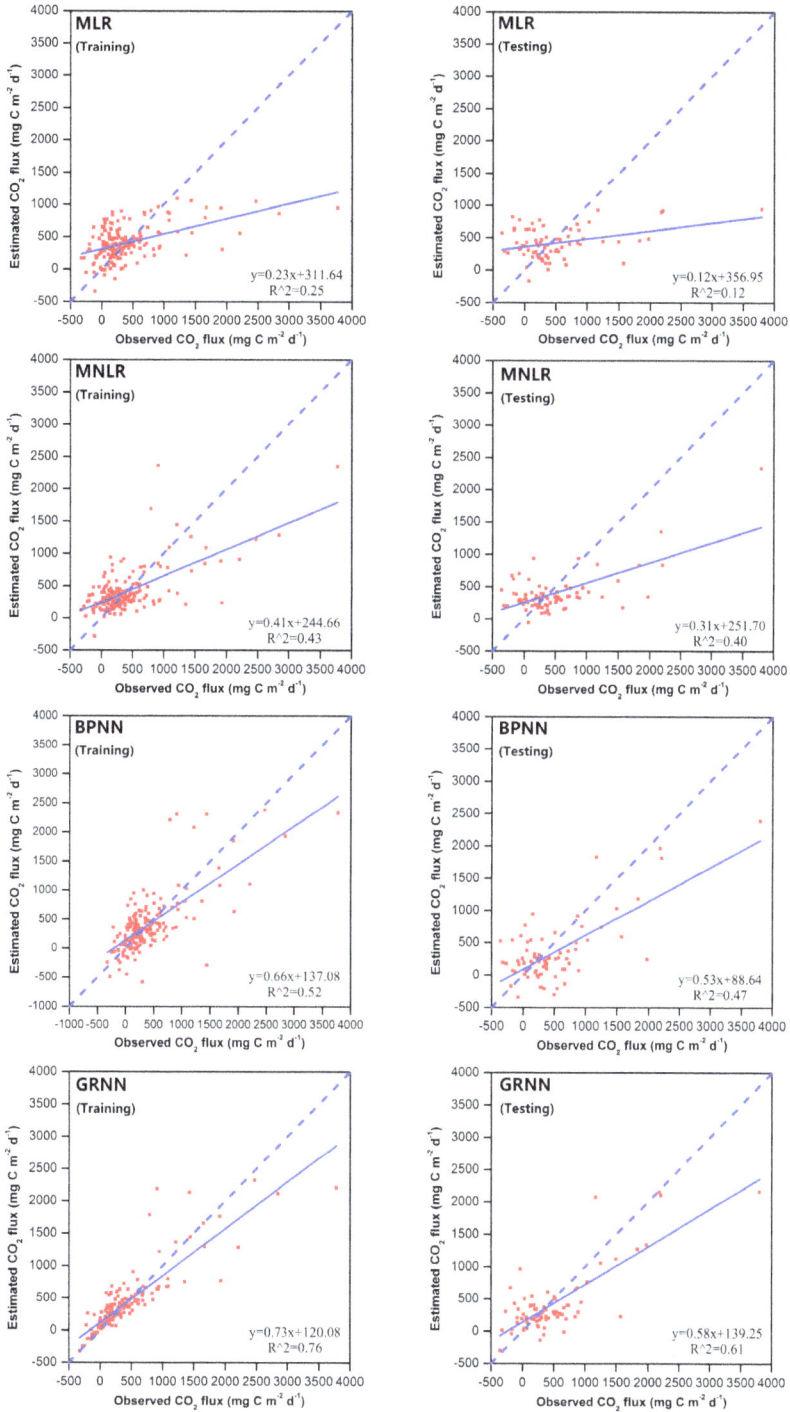

Figure 3. Comparison between CO_2 flux and predicted using MLR, MNLR, BPNN, and GRNN.

3.4. Sensitivity Analysis

Sensitivity analysis was applied to identify whether the selected set of inputs is suitable and which of them is the most important one in simulating CO_2 flux. We built new models based on different combination of input variables and compared their RMSE and R^2 values. These four sets of inputs were made up by omitting a parameter on each run. It was obvious that the omission of the most important parameter could have the highest influence on model performance, which was reflected in higher RMSE values and lower R^2 values [41]. Results of sensitivity analysis are presented in Table 5. The results demonstrated that the performance of the models with four input variables is better than other models', and the latitude parameter is an essential input for GRNN, BPNN, and MNLR.

Table 5. Results of sensitivity analysis for GRNN, BPNN, MNLR, and MLR in the testing phase.

Model	RMSE (mg C m^{-2} d^{-1})				R^2			
	GRNN	BPNN	MNLR	MLR	GRNN	BPNN	MNLR	MLR
All	418.48	505.43	529.53	625.36	0.61	0.47	0.40	0.12
Skip MD	432.14	552.55	530.95	629.10	0.59	0.35	0.39	0.11
Skip NPP0	463.63	567.08	532.26	635.25	0.52	0.38	0.38	0.09
Skip Age	462.65	519.62	535.68	633.44	0.57	0.45	0.39	0.10
Skip Lat	469.06	555.51	628.93	620.69	0.51	0.32	0.11	0.13

4. Discussion

4.1. Comparison of Results Obtained by Models

The results of the training and testing phase in models indicated that ANNs performed superior to the regression models. The GRNN model was found to have preferred accuracies over the BPNN model, while MNLR was superior to MLR in predicting CO_2 emission from reservoirs.

This outcome is consistent with other relevant literature, for example, MLR, MNLR, and GRNN models were used to forecast GHG emissions at national level and the results showed that the GRNN model gave the most preferable results [25]. When Firat et al. [42] predicted the scour depth around circular bridge piers based on the data from various studies, the GRNN model performed superior to BPNN and MLR. Among different types of ANNs, BPNN often shows a fair performance because the best model could be obtained by iterant parameters adjustment after calibration in the models [43]. However, the problem of overtraining often accompanies accuracy, because a large number of weights and biases is generated from many iterations [34]. In contrast, GRNN is a one-pass learning network, which does not require an iterative strategy as in BPNN. Therefore, GRNN can avoid this problem of overfitting to a large extent. Meanwhile, the problem of initial values determination and local minima often occurs in training stage of BPNN, while this problem does not exist in the GRNN procedure [44]. Thus, the GRNN can be preferred over the BPNN.

The regression lines in Figure 3 showed the same tendencies that the slopes were less than 1 and the slops in testing phases were less than ones in corresponding training phases. The first tendency reflects the problems of underestimation for high values of CO_2 flux in training and testing data sets, which is general in statistical prediction models [24,45]. Because the inputs cannot totally explain the outputs especially for extreme values, this tendency can also partly attribute to the non-homogeneous nature of data. Since CO_2 fluxes were measured from various reservoirs in different years, the lower slope in testing phases is due to the differences in training and testing data ranges. As listed in Table 3, the median and standard deviation of the testing data set were higher than those of the training data set.

In previous studies, St. Louis et al. [8] made a unary regression analysis between CO_2 and age based on datasets of 15 reservoirs ($R^2 = 0.35$), and Deemer et al. [12] built the relationship between CO_2 and mean annual precipitation based on datasets of 31 reservoirs ($R^2 = 0.11$). Barros et al. [11] used the multiple regression analysis to describe the relationship between three dependent variables (age, latitude, and DOC) and CO_2 flux based on datasets of 73 reservoirs (R^2 = 0.40). Compared with these regression models, our GRNN showed higher R^2 in both training (R^2 = 0.61) and testing (R^2 = 0.76) phases. The superior performance benefits from not only the advantages of GRNN, but also the larger database. Besides, without a testing process in the previous regression models, it is difficult to evaluate their generalization ability and apply in other reservoirs credibly. This study demonstrates that GRNN models could be an appropriate approach for prediction CO_2 emissions from reservoirs in other study systems.

4.2. Sensitivity Analysis

The results of sensitivity analysis demonstrated that the input variables of NPP0, age, and depth is the best set for MLR, which conformed to the results of Spearman correlation analysis. Moreover, the results also emphasized the importance of the latitude parameter in ANNs and MNLR that aim to fit non-linear functions, which proved the rationality of the selection of input variables. However, the significant relationship between latitude and CO_2 flux was not reflected adequately in the results of Spearman correlation analysis. The possible reason is that the relationship between latitude and CO_2 flux is non-linear. Consequently, the use of non-linear statistical dependence measures is more appropriate for determining inputs to ANN models [31].

4.3. Application of Established Model

4.3.1. Estimation of the Global Magnitude of the CO_2 Fluxes from Reservoirs

With the help of this GRNN, global magnitude of CO_2 emissions from documented reservoirs can be estimated. This estimation was based on the GRanD, which contains 6862 records of reservoirs updated in 2011 [19], and HANPP [32]. We selected latitude, the year of construction and the average depth from GRanD and extracted NPP0 from HANPP following the site of these reservoirs.

To predict CO_2 fluxes from global reservoirs by the tested GRNN, the confidence interval should be given together. However, ANNs are black box models that cannot be described as particular equations. Therefore, the potential predictive confidence interval was given based on the statistical analysis between observed and predicted CO_2 fluxes in testing phase. The methods to calculate confidence interval are as follows [46]: (a) the errors between observed and predicted values in testing phase were calculated; (b) the Bootstrap samples were created by resampling from the errors on 100,000 replicates; (c) the medians of Bootstrap error samples are computed; (d) the $1 - \alpha$ Bootstrap Pivotal Confidence Intervals (CIs) for median of errors are estimated by Equation (10):

$$C_n = \left(2\hat{\theta} - \hat{\theta}^* \left((1 - \alpha/2)B \right), \, 2\hat{\theta} - \hat{\theta}^* \left((\alpha/2)B \right) \right) \tag{10}$$

where, $\hat{\theta}$ is an estimator of parameter θ; $\hat{\theta}^* \left((\alpha/2)B \right)$ is the $\alpha/2$ sample quantile of Bootstrap $\hat{\theta}$ samples. In this study, parameter θ is median of errors between observed and predicted values; α is 0.05, which means 95% confidence level. Bootstrap samples of medians were generated by Matlab R2013a.

After inputting the variables, the absolute values of latitude, the age of reservoirs in 2011, NPP0, and the average depth, we got the CO_2 emission fluxes from 6862 reservoirs in 2011. Because 95% CI of error from GRNN is $(-68.11, 60.86)$, these fluxes were updated into intervals for subsequent estimations. Then we multiplied these fluxes, corresponding area, and the number of days in a year that CO_2 can diffuse on the surface of reservoirs. Considering the influence of seasonality, especially the ice cover, we made an assumption that the temperate reservoirs which located higher than 30° N or 30° S latitudes are ice-free for 200 days on average. This refers to the one of the study in St. Louis et al. [8], since only this study takes ice cover into account among pervious estimations listed in Table 6. As a result, we estimate that global reservoirs emit 40.03 Tg C yr^{-1} as CO_2 (5th and 95th confidence interval: 32.03–47.18 Tg C yr^{-1} as CO_2).

Compared to previous estimations with the same magnitude of area (Table 6), our estimation is moderate and more fairly accurate. However, the CO_2 flux estimated by St. Louis et al. [8] is larger, which might be caused by the overestimation of global reservoirs' area and the young age of sampled reservoirs. Only three tropical reservoirs with the average age of 7.70 years were used to estimate CO_2 fluxes from all tropical reservoirs. The estimation by Hertwich [16] is also a little larger, because CO_2 flux is multiplied by an uncertainty factor of 2. The previous estimations derived from the product of the average of CO_2 flux in database multiplied with the global surface area of reservoirs, while the estimation in this study took into account the annual-scale flux variability of a special reservoir and the difference in geographical position among global reservoirs. However, further refinement is still required for more precise estimates. Since there are no real data of time-series, this model cannot predict potential CO_2 fluxes in long temporal scale. The cause–effect relationships between water quality and CO_2 flux received little attention in this study because of the limited data. The direct and indirect influences from ice cover especially in the boreal region should be fully studied and accurately calculated in future estimations.

Table 6. The global CO_2 flux estimates and parameters of estimations.

Studies		Sample Size	Type of Dataset	Method	Area (10^5 km²)	CO_2 (Tg C yr⁻¹)
This study		251	All reservoirs	Individual [1]	4.47	40.03 [3]
Previous studies	Deemer et al. [12]	229	All reservoirs	Average [2]	3.1	36.8
	Hertwich [16]	142	Hydroelectric	Average [2]	3.3	76
	Barros et al. [11]	85	Hydroelectric	Average [2]	3.4	48
	St. Louis et al. [8]	19	All reservoirs	Average [2]	15.0	272.2

Note: (1) [1] Individual method, calculated by the sum of CO_2 flux from each individual reservoir; [2] Average method, calculated by the multiplication of average CO_2 flux of database and total reservoirs area. (2) [3] The 95% confidence interval is (32.03, 47.18) Tg C yr⁻¹.

4.3.2. Estimations of CO_2 Emissions from a Planned Reservoir

Considering the input variables of this GRNN, the possible CO_2 emission flux during a fixed period of time from a proposed reservoir with planned location and depth can be estimated by this model. Therefore, it is possible to give guidance for dam construction. A hypothetical case, which was not based on any actual events, was used to show this potential utilization of GRNN in this study. We assumed that a reservoir would be constructed, and the geographical coordinates could be selected from 23.5° N to 26.4° N with the same longitude (115° E), and the mean depth could vary from 5 m to 34 m. The possible CO_2 fluxes emission from this reservoirs in the first 20 years with different features were shown in Figure 4. Although the construction of reservoirs should consider many realistic questions, the possible carbon emissions from new reservoirs might also be used as an important index of reservoir construction in the future.

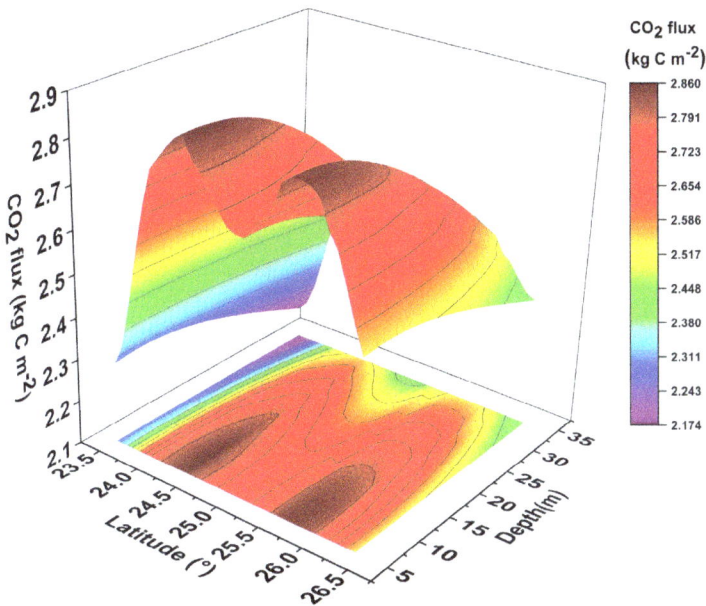

Figure 4. CO_2 emissions from reservoirs with different locations and depths in the first 20 years.

4.4. Future Research Directions

In this study, ANNs were built based on the data of gross CO_2 emissions from existing reservoirs, ignoring the potential CO_2 emissions from that land before impoundment, which might overestimate greenhouse effect from reservoirs. Recent study shows that the net carbon emissions from reservoirs are determined by different types of areas before flooding and provides a simple approach to quantify the net CO_2 emissions [47]. Future studies are therefore necessary to simulate and predict net CO_2 flux emits from reservoirs. Moreover, the combination of classification and regression machine learning can be a promising approach.

As the first attempt to apply ANNs to GHG emissions from reservoirs, CO_2 emission was chosen to simulate because of the quantity and quality of the monitoring data. However, CH_4 is a more powerful GHG than CO_2 [12]. Unlike CO_2 emissions, CH_4 emissions from reservoirs are new and anthropogenic [47]. Therefore, CH_4 footprint should be simulated and the magnitude needs to be estimated based on various released pathways in the future.

5. Conclusions

In this study, ANNs (GRNN and BPNN) and multiple regression models (MNLR and MLR) are applied to predict CO_2 emissions from reservoirs based on data records collected from published various field studies. Input variables used in models were selected by both Spearman correlation analysis and domain knowledge. The performance of models and observation was compared and evaluated by the indexes of RMSE, MAE, R^2, and NSE. It appears that the performance of ANNs is superior to the one of regression models. The GRNN's performance is better than BPNN's, while MNLR is superior to MLR. Sensitivity analysis of these four models confirmed that latitude-value is an important parameter in predicting CO_2 flux. The results demonstrate that GRNNs have great potential to estimate CO_2 emission from reservoirs when it is hard to acquire the monitoring data. The statistical models deserve more attention, because they are effective tools to assess global GHG emissions from reservoirs and provide new insights into the consideration of reservoir's construction during the planning stage. However, since the accuracy and generalization of statistical models largely depend on the measured data, more monitoring will be required in global reservoirs systematically. For example, the global CO_2 flux can be predicted in a longer time scale with the data of continuous monitoring on special reservoirs located in different latitude. Moreover, the mechanism models should be built to understand the relationship between CO_2 emission and other environmental factors clearly in the future.

Supplementary Materials: The following are available online at www.mdpi.com/2073-4441/10/1/26/s1, Table S1: CO_2 emission measurements and other data of reservoirs analyzed in the paper.

Author Contributions: Zhonghan Chen and Ping Huang conceived the idea; Zhonghan Chen and Xiaoqian Ye collected and checked the dataset; Zhonghan Chen built the model and analyzed the data; Zhonghan Chen and Xiaoqian Ye wrote the paper. All authors reviewed the manuscript.

Conflicts of Interest: The authors declare no conflict of interest.

References

1. Rudd, J.W.M.; Hecky, R.E. Are hydroelectric reservoirs significant sources of greenhouse gases? *Ambio* **1993**, *22*, 246–248.
2. Tremblay, A.; Varfalvy, L.; Roehm, C.; Garneau, M. *Greenhouse Gas Emissions-Fluxes and Processes: Hydroelectric Reservoirs and Natural Environments*; Springer Science & Business Media: Berlin/Heidelberg, Germany, 2005; ISBN 3-540-23455-1.
3. Teodoru, C.R.; Bastien, J.; Bonneville, M.C.; del Giorgio, P.A.; Demarty, M.; Garneau, M.; Hélie, J.F.; Pelletier, L.; Prairie, Y.T.; Roulet, N.T. The net carbon footprint of a newly created boreal hydroelectric reservoir. *Glob. Biogeochem. Cycles* **2012**, *26*. [CrossRef]
4. Demarty, M.; Bastien, J.; Tremblay, A.; Hesslein, R.H.; Gill, R. Greenhouse gas emissions from boreal reservoirs in Manitoba and Québec, Canada, measured with automated systems. *Environ. Sci. Technol.* **2009**, *43*, 8908–8915. [CrossRef] [PubMed]
5. Roland, F.; Vidal, L.O.; Pacheco, F.S.; Barros, N.O.; Assireu, A.; Ometto, J.P.H.B.; Cimbleris, A.C.P.; Cole, J.J. Variability of carbon dioxide flux from tropical (Cerrado) hydroelectric reservoirs. *Aquat. Sci.* **2010**, *72*, 283–293. [CrossRef]
6. Mosher, J.J.; Fortner, A.M.; Phillips, J.R.; Bevelhimer, M.S.; Stewart, A.J.; Troia, M.J. Spatial and temporal correlates of greenhouse gas diffusion from a hydropower reservoir in the Southern United States. *Water* **2015**, *7*, 5910–5927. [CrossRef]
7. Bevelhimer, M.S.; Stewart, A.J.; Fortner, A.M.; Phillips, J.R.; Mosher, J.J. CO_2 is dominant greenhouse gas emitted from six hydropower reservoirs in Southeastern United States during peak summer emissions. *Water* **2016**, *8*, 15. [CrossRef]
8. St. Louis, V.L.; Kelly, C.A.; Duchemin, É.; Rudd, J.W.M.; Rosenberg, D.M. Reservoir surfaces as sources of greenhouse gases to the atmosphere: A global estimate. *Bioscience* **2000**, *50*, 766–775. [CrossRef]
9. Saidi, H.; Koschorreck, M. CO_2 emissions from German drinking water reservoirs. *Sci. Total Environ.* **2017**, *581*, 10–18. [CrossRef] [PubMed]
10. Fearnside, P.M.; Pueyo, S. Greenhouse gas emissions from tropical dams. *Nat. Clim. Chang.* **2012**, *2*, 382–384. [CrossRef]
11. Barros, N.; Cole, J.J.; Tranvik, L.J.; Prairie, Y.T.; Bastviken, D.; Huszar, V.L.M.; del Giorgio, P.; Roland, F. Carbon emission from hydroelectric reservoirs linked to reservoir age and latitude. *Nat. Geosci.* **2011**, *4*, 593–596. [CrossRef]
12. Deemer, B.R.; Harrison, J.A.; Li, S.; Beaulieu, J.J.; Delsontro, T.; Barros, N.; Bezerra-Neto, J.F.; Powers, S.M.; dos Santos, M.A.; Vonk, J.A. Greenhouse gas emissions from reservoir water surfaces: A new global synthesis. *Bioscience* **2016**, *66*, 949–964. [CrossRef]
13. Raymond, P.A.; Hartmann, J.; Lauerwald, R.; Sobek, S.; McDonald, C.; Hoover, M.; Butman, D.; Striegl, R.; Mayorga, E.; Humborg, C. Global carbon dioxide emissions from inland waters. *Nature* **2013**, *503*, 355–359. [CrossRef] [PubMed]
14. Zhao, Y.; Wu, B.F.; Zeng, Y. Spatial and temporal patterns of greenhouse gas emissions from three gorges reservoir of china. *Biogeosciences* **2013**, *10*, 1219–1230. [CrossRef]

15. Teodoru, C.R.; Prairie, Y.T.; del Giorgio, P.A. Spatial heterogeneity of surface CO_2 fluxes in a newly created eastmain-1 reservoir in Northern Quebec, Canada. *Ecosystems* **2011**, *14*, 28–46. [CrossRef]

16. Hertwich, E.G. Addressing biogenic greenhouse gas emissions from hydropower in LCA. *Environ. Sci. Technol.* **2013**, *47*, 9604–9611. [CrossRef] [PubMed]

17. Soumis, N.; Duchemin, É.; Canuel, R.; Lucotte, M. Greenhouse gas emissions from reservoirs of the Western United States. *Glob. Biogeochem. Cycles* **2004**, *18*. [CrossRef]

18. Li, S.Y.; Zhang, Q.F.; Bush, R.T.; Sullivan, L.A. Methane and CO_2 emissions from China's hydroelectric reservoirs: A new quantitative synthesis. *Environ. Sci. Pollut. Res.* **2015**, *22*, 5325–5339. [CrossRef] [PubMed]

19. Lehner, B.; Liermann, C.R.; Revenga, C.; Vörösmarty, C.; Fekete, B.; Crouzet, P.; Döll, P.; Endejan, M.; Frenken, K.; Magome, J.; et al. High-resolution mapping of the world's reservoirs and dams for sustainable river-flow management. *Front. Ecol. Environ.* **2011**, *9*, 494–502. [CrossRef]

20. Musenze, R.S.; Grinham, A.; Werner, U.; Gale, D.; Sturm, K.; Udy, J.; Yuan, Z.G. Assessing the spatial and temporal variability of diffusive methane and nitrous oxide emissions from subtropical freshwater reservoirs. *Environ. Sci. Technol.* **2014**, *48*, 14499–14507. [CrossRef] [PubMed]

21. De Faria, F.A.M.; Jaramillo, P.; Sawakuchi, H.O.; Richey, J.E.; Barros, N. Estimating greenhouse gas emissions from future Amazonian hydroelectric reservoirs. *Environ. Res. Lett.* **2015**, *10*, 1–13. [CrossRef]

22. Rosa, L.P.; dos Santos, M.A.; Gesteira, C.; Xavier, A.E. A model for the data extrapolation of greenhouse gas emissions in the Brazilian hydroelectric system. *Environ. Res. Lett.* **2016**, *11*. [CrossRef]

23. Dogan, E.; Sengorur, B.; Koklu, R. Modeling biological oxygen demand of the Melen River in Turkey using an artificial neural network technique. *J. Environ. Manag.* **2009**, *90*, 1229–1235. [CrossRef] [PubMed]

24. Lu, F.; Chen, Z.; Liu, W.Q.; Shao, H.B. Modeling chlorophyll-a concentrations using an artificial neural network for precisely eco-restoring lake basin. *Ecol. Eng.* **2016**, *95*, 422–429. [CrossRef]

25. Antanasijević, D.; Pocajt, V.; Ristić, M.; Perić-Grujić, A. Modeling of energy consumption and related GHG (greenhouse gas) intensity and emissions in Europe using general regression neural networks. *Energy* **2015**, *84*, 816–824. [CrossRef]

26. Antanasijević, D.Z.; Ristić, M.Đ.; Perić-Grujić, A.A.; Pocajt, V.V. Forecasting GHG emissions using an optimized artificial neural network model based on correlation and principal component analysis. *Int. J. Greenh. Gas Control* **2014**, *20*, 244–253. [CrossRef]

27. Antonopoulos, V.Z.; Antonopoulos, A.V. Daily reference evapotranspiration estimates by artificial neural networks technique and empirical equations using limited input climate variables. *Comput. Electron. Agric.* **2017**, *132*, 86–96. [CrossRef]

28. Vakili, M.; Sabbagh-Yazdi, S.R.; Khosrojerdi, S.; Kalhor, K. Evaluating the effect of particulate matter pollution on estimation of daily global solar radiation using artificial neural network modeling based on meteorological data. *J. Clean. Prod.* **2017**, *141*, 1275–1285. [CrossRef]

29. Ding, W.F.; Zhang, J.S.; Leung, Y. Prediction of air pollutant concentration based on sparse response back-propagation training feedforward neural networks. *Environ. Sci. Pollut. Res.* **2016**, *23*, 19481–19494. [CrossRef] [PubMed]

30. Salami, E.S.; Salari, M.; Ehteshami, M.; Bidokhti, N.T.; Ghadimi, H. Application of artificial neural networks and mathematical modeling for the prediction of water quality variables (case study: Southwest of Iran). *Desalin. Water Treat.* **2016**, *57*, 27073–27084. [CrossRef]

31. Maier, H.R.; Jain, A.; Dandy, G.C.; Sudheer, K.P. Methods used for the development of neural networks for the prediction of water resource variables in river systems: Current status and future directions. *Environ. Model. Softw.* **2010**, *25*, 891–909. [CrossRef]

32. Haberl, H.; Erb, K.H.; Krausmann, F.; Gaube, V.; Bondeau, A.; Plutzar, C.; Gingrich, S.; Lucht, W.; Fischer-Kowalski, M. Quantifying and mapping the human appropriation of net primary production in earth's terrestrial ecosystems. *Proc. Natl. Acad. Sci. USA* **2007**, *104*, 12942–12947. [CrossRef] [PubMed]

33. Antonopoulos, V.Z.; Gianniou, S.K.; Antonopoulos, A.V. Artificial neural networks and empirical equations to estimate daily evaporation: Application to Lake Vegoritis, Greece. *Hydrol. Sci. J.* **2016**, *61*, 2590–2599. [CrossRef]

34. Nielsen, M.A. *Neural Networks and Deep Learning*; Determination Press: USA, 2015. Available online: http://neuralnetworksanddeeplearning.com/ (accessed on 29 December 2017).

35. He, B.; Oki, T.; Sun, F.B.; Komori, D.; Kanae, S.; Wang, Y.; Kim, H.; Yamazaki, D. Estimating monthly total nitrogen concentration in streams by using artificial neural network. *J. Environ. Manag.* **2011**, *92*, 172–177. [CrossRef] [PubMed]

36. Specht, D.F. A general regression neural network. *IEEE Trans. Neural Netw.* **1991**, *2*, 568–576. [CrossRef] [PubMed]

37. Kisi, O. Pan evaporation modeling using least square support vector machine, multivariate adaptive regression splines and M5 model tree. *J. Hydrol.* **2015**, *528*, 312–320. [CrossRef]

38. Sammen, S.S.H.; Mohamed, T.A.; Ghazali, A.H.; El-Shafie, A.H.; Sidek, L.M. Generalized regression neural network for prediction of peak outflow from dam breach. *Water Resour. Manag.* **2017**, *31*, 549–562. [CrossRef]

39. Yaseen, Z.M.; El-Shafie, A.; Jaafar, O.; Afan, H.A.; Sayl, K.N. Artificial intelligence based models for stream-flow forecasting: 2000–2015. *J. Hydrol.* **2015**, *530*, 829–844. [CrossRef]

40. Zhao, D.Q.; Chen, J.N.; Wang, H.Z.; Tong, Q.Y.; Cao, S.B.; Sheng, Z. Gis-based urban rainfall-runoff modeling using an automatic catchment-discretization approach: A case study in Macau. *Environ. Earth Sci.* **2009**, *59*, 465–472. [CrossRef]

41. Csábrági, A.; Molnár, S.; Tanos, P.; Kovács, J. Application of artificial neural networks to the forecasting of dissolved oxygen content in the hungarian section of the river danube. *Ecol. Eng.* **2017**, *100*, 63–72. [CrossRef]

42. Firat, M.; Gungor, M. Generalized regression neural networks and feed forward neural networks for prediction of scour depth around bridge piers. *Adv. Eng. Softw.* **2009**, *40*, 731–737. [CrossRef]

43. Safari, M.-J.-S.; Aksoy, H.; Mohammadi, M. Artificial neural network and regression models for flow velocity at sediment incipient deposition. *J. Hydrol.* **2016**, *541*, 1420–1429. [CrossRef]

44. Stamenković, L.J.; Antanasijević, D.Z.; Ristić, M.Đ.; Perić-Grujić, A.A.; Pocajt, V.V. Modeling of methane emissions using artificial neural network approach. *J. Serbian Chem. Soc.* **2015**, *80*, 421–433. [CrossRef]

45. Wang, L.; Kisi, O.; Zounemat-Kermani, M.; Salazar, G.A.; Zhu, Z.; Gong, W. Solar radiation prediction using different techniques: Model evaluation and comparison. *Renew. Sustan. Enery Rev.* **2016**, *61*, 384–397. [CrossRef]

46. Wassermann, L. *All of Nonparametric Statistics*, 3rd ed.; Springer: New York, NY, USA, 2006; pp. 30–35, ISBN 978-0-387-30623-05.

47. Prairie, Y.T.; Alm, J.; Beaulieu, J.; Barros, N.; Battin, T.; Cole, J.; Giorgio, P.D.; DelSontro, T.; Guérin, F.; Harby, A.; et al. Greenhouse gas emissions from freshwater reservoirs: What does the atmosphere see? *Ecosystems* **2017**, 1–14. [CrossRef]

water

MDPI

Article

Multi-Objective Optimization of Resilient Design of the Multipurpose Reservoir in Conditions of Uncertain Climate Change

Stanislav Paseka [1],*, Zoran Kapelan [2] and Daniel Marton [1]

[1] Institute of Landscape Water Management, Faculty of Civil Engineering, Brno University of Technology, 602 00 Brno, Czech Republic; marton.d@fce.vutbr.cz

[2] Centre for Water Systems, College of Engineering, Mathematics and Physical Sciences, University of Exeter, Exeter EX4 4QF, UK; Z.Kapelan@exeter.ac.uk

* Correspondence: paseka.s@fce.vutbr.cz; Tel.: +42-054-114-7774

Received: 18 June 2018; Accepted: 16 August 2018; Published: 21 August 2018

Abstract: This paper presents and assesses a new approach to decision-making methods for the design of new reservoirs due in times of decreasing water resources. The methods used in this case are decision theory, Resilience and Robustness method. The methods have been selected primarily to analyze different design parameters of a new dam, mainly dam heights leading to different reservoir volumes. The study presents a novel approach to the optimal design of a multipurpose reservoir that would provide enough water for downstream environmental flow, residential and industrial water supply, agricultural water supply, and hydropower generation in the current conditions of climate uncertainty. Uncertainties are interpreted as possible future changes in the climate system using outputs from regional climatic models. In the case study, a simulation model was developed which is able to quantify long-term water balance and use this data to quantify resilience and robustness of its water supply. The simulation model was correlated to the GANetXL software in order to perform Genetic Algorithms based optimization of the reservoir's operation. The simulation–optimization model was then applied to a real-life case study in the Czech Republic, in the Morava River Basin where a new dam with the multipurpose reservoir is planned to be built in the future. The results obtained in this way were analyzed in detail to identify the overall best solution consist of dam height and the total reservoir monthly outflow and new operational rules for the analyzed multipurpose reservoir.

Keywords: design and operation of the multipurpose reservoir; water deficit; reservoir simulation model; climate change; multi-objective optimization NSGA II; resilience and robustness; costs and benefits; water energy

1. Introduction

In Czechia, climate change has caused change of the hydrological cycle due to redistribution of precipitation during the hydrological year in the last decade. This phenomenon results in more frequent occurrences of extremes in the form of floods and droughts. But the most urgent problem is that the values of the long-term mean flows are decreasing in rivers as well as the yield of groundwater sources.

Moreover, the years 2015 and 2017 were considered extremely dry in many regions of Czechia. It is becoming a trend of the last couple of years that meteorologists start to call the previous year as the warmest year in the history of meteorological measurement, this confirms WMO Statement on the State of the Global Climate 2017 [1].

The above-mentioned factors create pressure on the effective management of the surface water resources which will become stronger as well as the Czech Republic is sometimes called "roof of

Central Europe", as there is no major inflow to the territory, but only outflows. The limited availability of water resources in this country requires the government to prepare new climate change adaptations and focus on increasing the retention capacity of water in the landscape, long-term water management planning, and environmental politics on global level.

The government of the Czech Republic has issued several documents on adaptation strategies to climate change [2–5], which have been developed by the research institutes: VÚV TGM, v.v.i; EKOTOXA s.r.o.; ČHMÚ in cooperation with Ministry of the Environment of the Czech Republic. These documents indicate many measures to achieve the objectives of protection against the negative impacts of drought namely adaptive measures leading to saving drinking water, increasing of water retention capacity of the soil, stabilizing of water regimes in river basins, restoration of small water reservoirs and increasing their reliability, optimization of the existing reservoir volumes, and importantly, identification of other locations suitable for new water reservoirs and strategic technical option is allocating new water resources like open water reservoirs. Under conditions of climate change uncertainties currently there are several locations surveyed in the process of feasibility studies. These new multipurpose reservoirs will be primarily ensuring drinking water supply, keep ecological flows in the rivers, ensuring water supply for industry and agriculture, and may help to resist the drought problems.

The overall objective of this paper is to introduce a novel approach to the multipurpose reservoir design. The evolutionary multi-objective optimization method was utilized to find the optimal design parameter values of the dam, mainly dam height leading to different reservoir volumes. For the design of dam height and reservoir management, a new level so-called water excess level (I) and the critical threshold level (II) were introduced. In order to complete the main goal, it is necessary that the design must follow the individual objectives:

(a) the goal is to apply a suitable metric which is defined by the resilience of reservoir storage capacity and find optimal economic solutions (to maximize resilience and benefits from the reservoir) using the multi-objective optimization method with required robustness,

(b) the objective of this method is to analyze a range of the dam heights and volumes of the multipurpose dam in conditions of uncertain climate change on the basis of given future climate scenarios using hydrological outputs from 15 regional climate models and potential demand scenarios.

This new approach responding to the actual demands of water under climate change can be a benefit for the design of the multipurpose reservoir under similar conditions. The methods are applied and tested on the case study in the Hanusovice dam profile.

2. Background

In the Czech Republic water reservoirs have always been designed according to the historical hydrological or generated time series. Czech National Standard [6] classifies open water reservoirs based on their strategic importance. Strategic classification is evaluated by temporal reliability R_T which is defined as the ratio of the months without fault and the number of all months for the given time series [7,8]. The class A is the most strategic (A $\geq$ 99.5% R_T) and D is the least strategic (D $\geq$ 95.0% R_T). The water deficit is defined when the active storage capacity is in an unsatisfactory state. Water supply reservoirs have generally been designed and operated by a set of predetermined rules, not taking into consideration climate uncertainty. These rules were formulated based on hydrology, the ability of storage capacity, and yield criteria. Reliability has been the main objective in the sense of achieving the long-term target demand and the prevention against floods. The practice for the design and operation of the reservoir used reliability and the concepts of resilience, which experimental setting is different, were not used. The problem is that this approach does not take into consideration the consequences of possible future hydrological and climate changes when there is an increased social demand, increased needs of agriculture, and long-term decreasing flows. Climate change is creating a potentially large range of uncertainty of possible futures that could threaten the reliability of water supplies.

When modeling water resources systems, we need to determine reliability, vulnerability, resilience, and risk. First Kritskiy and Menkel [7] described the definition of evaluation criteria of water resources; reliability. Hashimoto et al. [8] were among the first to propose the use of the terms of water resource system performance evaluation. These performance criteria refer to (i) how probable is it that a system can get into a unsatisfactory or failure state (reliability), (ii) how severe the consequences of failure might be (vulnerability), and (iii) how quickly it can bounce back, which is the recovery from a failure (resilience) [8]. Using multi-objective linear programming, the trade-offs between reliability, vulnerability, and resilience were researched in water supply reservoir operation in a previous study [9].

Resilience for water resources has generally been quantified as the duration of time (maximum or average) when temporary restrictions are in place due to low supply availability. But its calculation is varied in the literature. Relationships among resilience, reliability, and vulnerability of water supply using many-objective analysis were explored in a past paper [10]. For example, resilience was used as a performance criterion metric [11], in a past work [12] resilience is calculated as the average duration of time when a system is under a temporary restriction, or in another previous work [13] resilience was calculated as a fraction of the total future time a system is under an unsatisfactory state. Roach [14,15] characterized and tested several potential resilience metrics and looked for the resilience of the water system and robustness of the water supply based on adaptation strategies. Watts et al. [16] and Amarasinghe et al. [17,18] have undertaken studies on the resilience of water supply systems. The reservoir resilience as a function of time (static and dynamic) for a multipurpose reservoir operation was compared a past study [19]. The resilience along with robustness to system design wastewater treatment plant control were used previously [20].

The risk and uncertainty were first described by Knight [21], who was the first to distinguish between risk and uncertainty. The risk assessment and risk analysis for water management purposes were described by Kaplan in [22]. Uncertainties used in hydrology have been presented by Beven and Binley [23]. Uncertainties can be categorized by the generation of multiple future scenarios that represent alternative likely conditions under different assumptions [24]. Uncertainties are the factors which can affect reservoir design or operation, as well as the accuracy of results. Uncertainty analysis focusing on reservoir capacity performance with various sizes and types of input uncertainties were tested for the design of the active storage capacity and reliability in the articles [25–28]. Based on the achieved results it is possible that both values active storage capacities and reliability, determined without considering input data uncertainty, may be undervalued. This can lead to failure of storage capacity.

Using Monte Carlo experiments the Markov model reproduced relationships between resilience and reliability for a wide class of water supply systems and provided a general theoretical foundation for understanding the trade-offs among reservoir system storage, yield, reliability, and resilience [29]. In a past paper [30] the behavior of statistical performance indices (reliability, resilience, and vulnerability for a multipurpose storage reservoir) using simulations Monte Carlo on a real reservoir in India was performed, and it was confirmed that traditional reliability for reservoir design and operation cannot completely describe the strengths and weaknesses of a given issue. A past paper [31] showed the optimized operation of the largest multipurpose reservoir in Vietnam using the complex evolution algorithm and the MIKE 11 simulation model. The optimization puts focus on the trade-off between flood control and hydropower generation for the reservoir operation in the flood season and the reservoir level at the beginning of the dry season. In another past paper [32] a novel multi-objective optimization modeling framework for the operation of multipurpose simple reservoirs was presented. This model uses genetic algorithms as optimization techniques. The main objective function minimizes the cost of the annual water shortage for irrigation and the secondary objective maximizes energy production. A previous work [33] presents an evolutionary multi-objective optimization approach

NSGA-II for the study of multiple water usages in multiple interlinked reservoirs. The case studies involve, primarily, the objectives of power generation and navigability on the river.

One of the newest statistical concepts could be used to evaluate results under uncertainty is robustness. Robustness can be defined as the degree to which a water supply system performs at a satisfactory level across a range of future scenarios or conditions [34]. One of the types of satisficing criteria may be based on the proportion of possible future conditions under which system performs correctly. Developing robust long-term water resources plans described by the above definition have generally been assessed previously [35]. The robustness water resources system under deep uncertainty has been used in many other cases and the results from past papers [20,36–42] show that the using of the robustness concept is beneficial for the analysis of results under uncertainty.

For the case study, the optimizing method Non-dominated Sorting Genetic Algorithm II (NSGA II) described previously [43] was used. The NSGA II was successfully used in handling multi-objective optimization problems in water management in several past works [44–50]. The NSGA II was developed from the multi-objective algorithm NSGA [51], which was formulated based on the suggestions made by Goldberg in 1989 [52]. The relation between risk, reliability, water shortages, and hydropower energy in an open reservoir using the NSGA II optimization was tested previously [53]. The NSGA II algorithm and reservoir simulation model were applied in GENetXL [54]. The GENetXL was presented by a simple water supply hypothetical reservoir operation model with two objectives: maximize yield and maximize recreational benefit, and a large combinatorial optimization problem of pump scheduling in water distribution systems.

There is not much research on optimal reservoir design and management under uncertainty. In the world, complex water management planning and optimization of extensive water resources systems in the context of climatic uncertainties are mostly solved [40,55–61]. Given that case studies of water management systems are solved complexly and on a large scale on the other hand, we focus on the individual study in small scale but in more detail. A solution, in the sense of the whole reservoir capacity under climate change, has not been well-studied as an individual target in the design of reservoirs. In addition, the increasing water demand and distribution of water are not sufficiently safeguarded today. Current issues lead to use the uncertainty of climate change, multi-objective optimization, and concepts of resilience and robustness. All of these creating optimal effective design and new operational rules of a multipurpose water reservoir.

3. Case Study

The case study is based on the intended profile in Hanusovice in the Czech Republic, 200 km east of the capital city Prague, see Figure 1. The new dam is planned in the Morava River Basin, maintained by the manager of the Morava River Basin, State Enterprise (PMO), and the Krupa River is the main inflow into the reservoir. The long-term mean river flow Q_a is 2.12 m^3 s^{-1}. The elevation volume and area curves for a given profile were determined by a digital terrain model using GIS software. In this software, Figure 1 has been also generated showing one of the possible new dam locations.

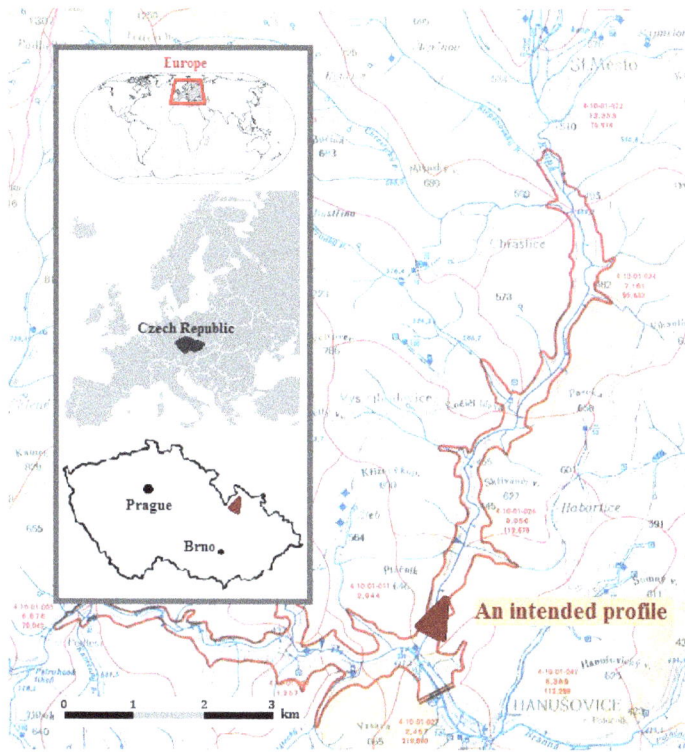

Figure 1. The locality of the plan reservoir to be built in the future.

Hydrologic data in the form of water inflows were derived from 15 regional climate models for the period 2015 to 2100. These are scenarios with a resolution of 25 km × 25 km which are controlled by one of the four global climate models. The results of the climate models are based on the IPCC SRES AR4 [62], only for the conservative emission scenario A1B which represents a very fast future economy growth and development of new technologies with a balance in the use of all fuels as energy sources. In the paper, each regional model represents 1 input of climate scenario. Data were taken from the project RSCN VUV [63]. The data were modified from hydrometric profile Morava/Raskov to hydrometric profile Krupa/Habartice using the analogy method for the river flows were determined.

Total water demand was determined as a sum of the Q_{ECO} (ecological outflow to the river), $Q_{WS:drink}$ (drinking water supply), and $Q_{WS:ind.+agric.}$ (water supply for industry and agriculture). $Q_{ECO} = 0.54$ m^3 s^{-1} and is constant for all months, $Q_{WS:drink.}$ and $Q_{WS:ind.+agric.}$ are based on 25% of total current demand for the part of the Olomouc Region from the plan for the development of water supply [64]. Mean annual demand was divided to monthly demands, see Table 1. Potential monthly redistribution of water demand was set according to the data from the PMO.

Table 1. Potential monthly water demands.

Demands [m^3 s^{-1}]	January	February	March	April	May	June	July	August	September	October	November	December
Q_{ECO}	0.54	0.54	0.54	0.54	0.54	0.54	0.54	0.54	0.54	0.54	0.54	0.54
$Q_{WS:drink.}$	0.54	0.56	0.56	0.58	0.58	0.60	0.56	0.55	0.55	0.56	0.55	0.53
$Q_{WS:ind.+agric.}$	0.35	0.35	0.36	0.37	0.38	0.40	0.40	0.40	0.38	0.36	0.35	0.35
Total Demand	1.43	1.45	1.46	1.49	1.50	1.54	1.50	1.49	1.47	1.46	1.44	1.42

The permanent storage capacity V_P was simplified and evaluated to height 15.0 m and corresponds to 0.44 mil. m^3. The flood storage capacity V_F was calculated for flood hydrograph $Q_{10.000}$. The Klemes method [65] was used to transform the controlled flood and to evaluate V_F. The total flood storage capacity $V_F = 7.72$ mil. m^3. The safety reserve 2.0 m was used to prevent overflow of the dam.

The location of the power plant is situated just below the dam and water supply for industry and agriculture is taken downstream the power plant, while drinking water supply is taken in the reservoir itself.

For the case study, an earth dam was used. Parameters of the dam body are; width of the dam crown is 5.0 m, bottom length is 100.0 m, an upstream slope is 1:3, downstream slope is 1:2, and terrain slopes are 1:1 (left side) and 1:1.8 (right side). The height of dam was calculated ranging from 80.0 to 100.0 m.

The target levels of resilience were set for three targets the duration of the longest water deficit period is up to 3 months (main target), up to 5 months, and 0 months. Two target levels of robustness 80% and more and 90% and more were tested.

All prices in Equations (4)–(6) corresponded to the present trade prices in Euro. The $PRICE_{WATER}$ was 0.245 € m^{-3} charged by PMO last year and the $PRICE_{RED,ELE}$ was 0.10 € kWh^{-1}. The discount rate r according to a past paper [66] for 2015–2044 is presumed 0.035, 2045–2089 is presumed 0.03, and 2090–2100 is presumed 0.025. The total coefficient of efficiency η for the calculation of the energy from the hydropower in Equation (7) is 0.80. The cost of the dam ZCU in Equation (10), respectively (9), is around 26.2 € m^{-3} for the earth dam in this location.

The following configuration of parameters NSGA II optimization in GENetXL was used. Population size was 100 genes. The number of generations was 200. Selection the crowded tournament was set up. The crossover was set up the simple one point and the rate was 0.90. Mutation type was set up as simple with mutation rate 0.05. The number of chromosomes (decision variables) was 12 for $i = 1, \ldots, 12$ months of year with boundary conditions $Q_{EXC,i} \in \langle 0.0 \text{ m}^3 \text{ s}^{-1}; 2.0 \text{ m}^3 \text{ s}^{-1} \rangle$.

4. Methodology

4.1. Problem Formulation

Where designing the new multipurpose dam, the objective functions have to be defined to satisfy all requests at the same time and lead to a new approach for using a multi-objective optimization.

The resilience of storage capacity must be set to a target level to be considered as acceptable under given future climate scenarios and water demand scenario, as well as robustness. That means if the target levels of resilience and robustness are satisfactory then the designed solutions are acceptable.

In the analysis, 12 decision variables were created. Each of them was described as month excess water Q_{EXC} that was added to the total water demand.

In total, two objective functions were used, net present value and resilience. The results obtained from the 12 decision variables for four chosen climate scenarios. Then the robustness was evaluated based on decision variables and resilience.

4.2. Resilience and Robustness

In this case every kind of potential resilience (*RES*) [month] metrics, defined previously [14], were tested, and finally, was traditional metric of resilience was selected with the duration of the longest water deficit period:

$$RES = \max(\text{duration of water deficit}), \tag{1}$$

The robustness (*ROB*) [%] is defined in this case as the percentage ratio of future supply and demand scenarios [36,39]:

$$ROB = \frac{S_A}{S_T} \times 100, \tag{2}$$

where S_A is the number of scenarios in which the reservoir simulation model performs at an acceptable level of resilience and S_T is the total number of scenario combinations of supply and demand.

4.3. Costs and Benefit

One of the objectives is to minimize the cost of dam construction and maximize benefits from the dam. This is expressed by the net present value (NPV) [€], which is calculated according to the following Equation (3):

$$NPV = \sum_{t=1}^{T} \left(\frac{BEN_{WS:drink,t}}{(1+r)^{t-1}} + \frac{BEN_{WS:ind.+agric.,t}}{(1+r)^{t-1}} + \frac{BEN_{HPP,t}}{(1+r)^{t-1}} \right) - C_{DAM}, \tag{3}$$

where BEN [mil €] are benefits from the utilization of water for hydropower and sales of water, C_{DAM} is the cost of the dam construction, r is discount rate and $t = 1, \dots , T$ for $T = 86$ (the total number of years until 2100).

We have three possible benefits: drinking water supply Equation (4), water supply for industry and agriculture Equation (5), and benefits from hydropower plant Equation (6):

$$BEN_{WS:drink,t} = \sum_{i=1}^{n} (Q_{WS:drink,i}) \times PRICE_{WATER}, \tag{4}$$

$$BEN_{WS:ind.+agric.,t} = \sum_{i=1}^{n} \left(Q_{WS:ind.+agric.,i} \right) \times PRICE_{WATER}, \tag{5}$$

$$BEN_{HPP,t} = E_{HPP,t} \times PRICE_{RED,ELE}, \tag{6}$$

where $Q_{WS:drink.,i}$ is drinking water for residential water supply and $Q_{WS:ind.+agric.,i}$ is water for industrial and agricultural water supply; $PRICE_{WATER}$ is the price for water consumption. $PRICE_{RED,ELE}$ is the redemption price of electricity from the hydropower according to the Czech Energy Regulatory Office, $i = 1, \dots , n$ for $n = 12$ (number months of the t-year) and $E_{HPP,t}$ [Wh·month^{-1}] is generated energy from hydropower for a year Equation (7):

$$E_{HPP,t} = \sum_{i=1}^{n} (\rho \times g \times Q_{HPP,i} \times H_i \times \eta) \times \Delta t, \tag{7}$$

where ρ is density of water, g is gravity, H_i is the height of water level in given i-month, η is total coefficient of efficiency, $i = 1, \dots , n$ for $n = 12$, Δt is one year and $Q_{HPP,i}$ [m^3 s^{-1}] is flowed through a hydropower in given i-month and is calculated according to the Equation (8):

$$Q_{HPP,i} = Q_{ECO,i} + Q_{WS:ind.+agric.,i} + (Q_{EXC,i}), \tag{8}$$

where $Q_{ECO,i}$ is ecological outflow for downstream environmental flow, $Q_{WS:ind.+agric.,i}$ is water supply for industry, and $Q_{EXC,i}$ is the water excess for hydropower and to a river in given i-month.

The last parameter of Equation (3) is C_{DAM} [€], which describes the cost of the dam construction and is determined by following Equation (9):

$$C_{DAM} = V_{DAM} \times ZCU, \tag{9}$$

where V_{DAM} is the volume of the dam structure and ZCU [€m^{-3}] is the cost of the dam calculated according to Czech property valuation decree [67]:

$$ZCU = ZC \times K_5 \times K_i, \tag{10}$$

where ZC is a cost according to the type of material of the dam and K are coefficients according to the location of the dam.

4.4. Reservoir Simulation Model

A reservoir simulation model was created based on emptying and filling of reservoir storage capacity. The newly introduced water excess level (I) and critical threshold (II) were used to calculate the resilience. This principle can be described by the inequality (11) which proceeds from the mass balance equation [68]:

$$Water\ lvl.(0) \leq \sum_{i=0}^{n-1}(Q_{OUT,i} - Q_{IN,i})\Delta t + (Q_{OUT,i+1} - Q_{IN,i+1})\Delta t \leq Water\ lvl.(I),(II),(III), \quad (11)$$

where $Q_{IN,i}$ [m^3 s^{-1}] is inflow water to the reservoir for $i = 0, \ldots, n-1$. $Q_{IN,i}$ is generated by future scenarios of regional climate models from [63]. $Q_{OUT,i}$ [m^3 s^{-1}] is the reservoir outflow for $i = 0, \ldots, n - 1$. The $Q_{OUT,i}$ is formulated by the sum of the $Q_{ECO,i}$, $Q_{WS:drink.,i}$, $Q_{WS:ind.+agric.,i}$, and $Q_{EXC,i}$ (water excess for hydropower). Δt is the time step of calculation (one month). For $i = 0$ it is necessary to enter the initial condition (full active storage capacity). Inequality (11) is limited from the left side by water level (0) which characterizes full active storage capacity and from the right as active storage capacity divided by level (I–III) as shown in Figure 2. Each level of active storage capacity is characterized by given outflow $Q_{OUT,i}$.

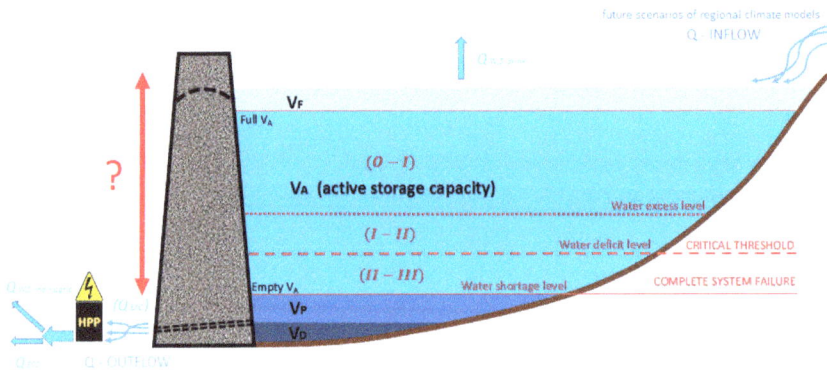

Figure 2. The scheme of the model set up.

The resulting scheme of dam water storage in Figure 2 shows variants of water levels, where V_F is flood storage capacity, V_A is active storage capacity, V_P is permanent storage capacity, and V_D characterizes dead storage capacity.

When the current step i emptying and filling of storage capacity is going up from level (0–III) in (11), then the $Q_{OUT,i}$ [m^3 s^{-1}] is equal to the $Q_{IN,i}$ [m^3 s^{-1}] otherwise new operating rules and restrictions are created according to the following control Equations (12)–(14):

0–I: $\qquad Q_{OUT,i} = Q_{ECO,i} + Q_{WS:drink,i} + Q_{WS:ind.+agric.,i} + Q_{EXC,i}$, $\qquad$ (12)

I–II: $\qquad Q_{OUT,i} = Q_{ECO,i} + Q_{WS:drink,i} + Q_{WS:ind.+agric.,i}$, $\qquad$ (13)

II–III: $\qquad Q_{OUT,i} = Q_{ECO,i} + (Q_{WS:drink,i}) \times 0.7 + \left(Q_{WS:ind.+agric.,i}\right) \times 0.3$, $\qquad$ (14)

As mentioned above, the water level (0) which characterizes full active storage capacity and water level (III) is empty active storage capacity. The volume between levels I–III is set up 11.6 mil. m^3. This volume corresponds to three months of estimated total demand. The volume between levels II–III correspond to the total demand in given month of the year.

When the volume of water is between level I–II, water excess $Q_{EXC,i}$ is null. When the volume of water drops down the critical threshold (II) the deficit begins. At this moment, the restriction of reservoir management by priority roles begins. The main rule is to guarantee Q_{ECO} and next rule is to ensure $Q_{WS:drink.}$ and $Q_{WS:ind.+agric.}$ in the ratio of 70:30, see Equation (14).

4.5. Optimization Method

For the case study, the model in cooperation with the NSGA II algorithm was used in order to create decision variables population of an excess of water $Q_{EXC,i}$.

The key parameters for NSGA II are possible decision variables, dam heights and analyzes of reservoir performance across all climate scenarios in the simulation model. Data output of the optimization method are presented in a form of Pareto sets showing relations between net present value and resilience of the reservoir.

5. Result and Discussion

The obtained relation between resilience and *NPV* in the form of Pareto sets is in Figure 3. For each dam height, 15 climate scenarios from 2015 to 2100 were used from [63] for the planned profile.

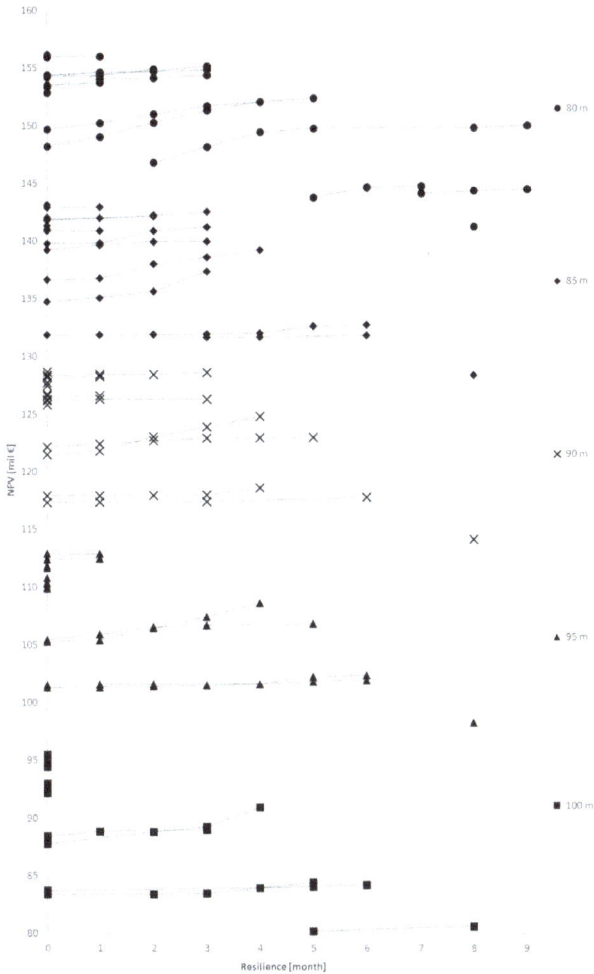

Figure 3. The Pareto optimal sets for resilience vs. *NPV* for varying dam heights and all scenarios.

Then each obtained point of these Pareto optimal sets presents the best solution and same a possible value of excess water Q_{EXC} (decision variables) or the total reservoir outflow Q_{OUT}. All these values can be applied to the design and operation of the dam.

Figure 3 shows increasing resilience (short deficit) in relation to reducing *NPV*. That means less resilient solution (longer deficit duration) representing higher *NPV* (higher benefits of the dam). Pareto sets are relatively flat (only slightly curved), this is mainly due to the setting of present prices for electric power and water supply. It could be expected that future water price increasing would make the Pareto sets more curved because the benefits will increase. Another factor that influences the shape of the Pareto sets could be the dam type.

As it can be seen from Figure 3, the results are better for the lower dam heights, as there are lower costs for construction of the dam. It is obvious that the increasing dam height increases the cost for construction of the dam, but benefits are lower than costs. Lower dam heights bring better results. In addition, building a bigger dam generally induces higher environmental costs, but it was not considered in this case study due to the difficult price quantification

Only Pareto points with 0, 3, and 5 months of resilience, for the lower suitable dam heights of the 80 m, 85 m, and 90 m and only 4 from 15 climate scenarios with different mean annual flows Q_a: as (Sc1) the most water ($Q_a = 2.29$ m^3 s^{-1}), as (Sc2) more water ($Q_a = 2.17$ m^3 s^{-1}), as (Sc3) less water ($Q_a = 2.09$ m^3 s^{-1}), and as (Sc4) the least water ($Q_a = 1.84$ m^3 s^{-1}) were chosen, see Figure 4.

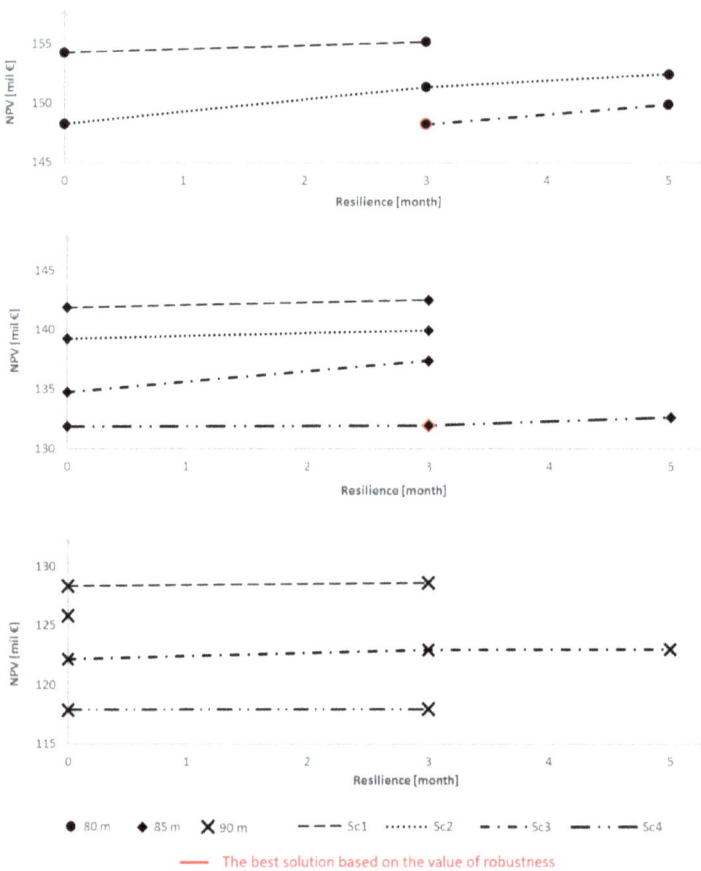

Figure 4. The Pareto optimal sets from four selected climatic scenarios for resilience vs. *NPV*.

Figure 4 shows the Pareto sets for lower dam heights of Figure 3 and only for selected climate scenarios Sc1, Sc2, Sc3, and Sc4. It is clear, that the more water scenarios (Sc1 and Sc2) produce higher the NPV, because the benefits are higher for the same cost and RES.

Figure 5 specifically shows the relation between the present value of benefit $PV(benefit)$ and the present value of cost $PV(cost)$ for the selected four climate scenarios with different mean annual flows Q_a, resilience of the dam from 0 to 5 months and heights of dam 80, 85, and 90 m. As mentioned above, Figure 5 confirms that the more water scenarios (Sc1 and Sc2) cause higher the NPV (higher benefit) than less water scenarios (Sc3 and Sc4).

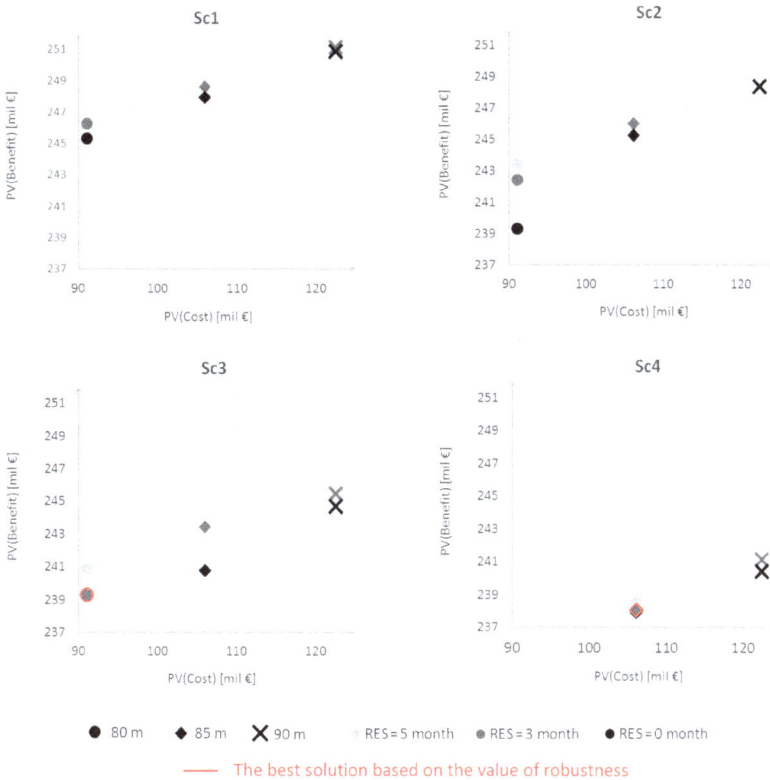

Figure 5. The relations between PV (benefit) and PV (cost) for selected climate scenarios.

Lower points in Figure 5 are the points for the most resilient design of the dam ($RES = 0$ months) and the lowest benefits. The higher points refer to the lowest resilience ($RES = 5$ months) and the highest benefits.

The points in Figures 4 and 5 mark the selected results which were chosen for testing of robustness. The results were in the form of decision variables (Q_{EXC}) achieved by optimization for four climatic scenarios. Then the decision variables were tested in the reservoir simulation model for 15 climatic scenarios and different dam heights variants to determine the resilience. After that robustness evaluation was finally made. The robustness was calculated as the fraction number of scenarios with $RES = 3$ months or less (5 months or less, 0 months) and the total number of climate scenarios, according to Equation (2). In Table 2, we can see the resulting robustness, and NPV for the points from Figures 4 and 5 ($RES = 0$, 3, and 5 months or less).

Table 2. Robustness for selected points.

		Height 80.0 m				Height 85.0 m				Height 90.0 m			
C_{DAM} [mil. €]		91.05				106.03				122.51			
V_A [m³]		34,480,000				44,940,000				57,270,000			
RES [month]	Scenario:	Sc1	Sc2	Sc3	Sc4	Sc1	Sc2	Sc3	Sc4	Sc1	Sc2	Sc3	Sc4
0	ROB [%]	66.7	73.3	-	-	66.7	73.3	**80.0**	**93.3**	66.7	66.7	**80.0**	**93.3**
	⌀ NPV [mil. €]	151.4	147.3	-	-	138.0	137.2	134.3	132.2	123.5	123.4	121.6	118.5
3	ROB [%]	53.3	73.3	**80.0**	-	60.0	66.7	73.3	**93.3**	66.7	-	73.3	**93.3**
	⌀ NPV [mil. €]	151.0	149.7	147.8	-	137.9	138.1	137.1	132.5	123.5	-	122.8	118.6
5	ROB [%]	-	60.0	60.0	-	-	-	-	**80.0**	-	-	66.7	-
	⌀ NPV [mil. €]	-	150.9	149.8	-	-	-	-	134.3	-	-	123.1	-

In Table 2 we can see, if the decision variables for the less water scenarios (Sc3 and Sc4) are selected, than robustness for given solutions will be higher, because decision variables (Q_{EXC}) were lower than decision variables from the more water scenarios. According to Table 2, it is obvious that one of the best results is for the decision variables in Sc3, resilience $RES \leq 3$ months, robustness $ROB \geq 80.0\%$, and then the dam height is 80.0 m. The average *NPV* is 147.8 mil. €, the cost of the construction of dam is 91.05 mil. €, and the benefits from the hydropower and water supply are on average 238.8 mil. €. The active storage capacity V_A is 34.48 mil. m³ and total volume of the dam is 42.63 mil. m³.

Another optimal solution is for the decision variables in Sc4, $RES \leq 3$ months and $ROB \geq 90.0\%$, and then the dam height is 85.0 m. The average *NPV* is 132.5 mil. €, the $C_{DAM} = 106.03$ mil. €, and the *BEN* are on average 238.5 mil. €. The V_A is 44.94 mil. m³ and total volume of the dam is 53.09 mil. m³. If we compare chosen results cost for construction, the dam in more robust solution will be about 15 mil. € (16.5%) more expensive than a less robust solution. Benefits are 0.3 mil. € (0.1%) lower and the *NPV* is lower about 15.3 mil. € (11.5%).

The results of the decision variables in form monthly water excess for hydropower Q_{EXC} and following the total reservoir outflow Q_{OUT} for each month and for $ROB = 80.0\%$ and 90.0% are presented in Table 3.

Table 3. Selected optimal results of the monthly water excess Q_{EXC} and the total reservoir monthly outflow Q_{OUT} for two target levels of *ROB*.

		January	February	March	April	May	June	July	August	September	October	November	December
ROB ≥ 80%	Q_{EXC} [m³ s⁻¹]	0.93	1.12	0.06	0.00	0.13	0.00	0.01	0.00	0.00	0.00	0.00	0.02
(h = 80 m)	Q_{OUT} [m³ s⁻¹]	2.36	2.57	1.52	1.49	1.63	1.54	1.50	1.50	1.47	1.46	1.44	1.44
ROB ≥ 90%	Q_{EXC} [m³ s⁻¹]	0.18	0.01	0.00	0.01	0.00	0.00	0.00	0.00	0.00	0.00	0.09	0.02
(h = 85 m)	Q_{OUT} [m³ s⁻¹]	1.61	1.46	1.46	1.50	1.50	1.54	1.50	1.49	1.47	1.46	1.53	1.44

In Table 3 we can see that for $ROB \geq 80\%$, the dam height is only 80.0 m and water excess Q_{EXC}, and thus total reservoir outflow Q_{OUT} are higher than for the $ROB \geq 90\%$.

The decision variables (water excess Q_{EXC}) could increase in the future by the rise of water and electricity prices. The question is what the future evolution of water and electricity demand will be. The model is using only the long-term scenarios for hydrology changes and current prices for water and electricity. There is a huge amount of uncertainties in the prediction of water and electricity demands as well as market prices of these commodities in the upcoming decades. Thus, in future, it would be useful to add several future demands or socioeconomic scenarios. It should be noted that the climate scenarios in this study are based on only one conservative emission scenario A1B. Further, it could be very interesting to test the robustness effect on several percentage sets of an increase in water and electricity demand and increase in water and electricity prices for a given result.

6. Conclusions

This article describes a new possible attitude to support decision-making in the future for a new reservoir management within a decrease water resources scenario. A simulation model has been coupled with a cost model and an NSGA II multi-objective optimization algorithm to quantify resilience and robustness under a range of uncertain future climate supply scenarios and one possible demand scenario. The results obtained lead to the following key conclusions:

- The analysis of the different dam heights produced different recommendations for the multipurpose reservoir design. This approach has recommended a specific design of the dam height of 80 m for $ROB = 80\%$ and $RES = 3$ months or 85 m for $ROB = 90\%$ and $RES = 3$ months.
- As a result, recommended dam height, and also recommended total reservoir monthly outflows that, in combination with the control equations, determine new operational rules of a multipurpose water reservoir.
- The new operating rules have been created: The water management in the reservoir should be set according to the current month and recommended reservoir monthly outflows Q_{OUT} in Table 3. In the case of dry periods when the water level falls below level (I) or more as it is shown in Figure 1, the Q_{OUT} have to be restricted according to the control Equation (13) or (14).
- All potential resilience metrics are based on those in a past paper [14] which were tested in this analysis and gave similar shapes of Pareto sets, and therefore, the traditional metric of resilience with duration of the longest water deficit period was selected.
- The model clearly shows that the higher dam heights increase the cost for construction of the dam, but the benefits are lower than costs, therefore the results are better for lower dam height.
- Although, in the summer months, due to the higher water demands and lower flow in the river, the water excess is minimal or null, the minimal demand is guaranteed for targets resilience and robustness.
- A more robust solution generally produced lower benefits respectively lower NPV due to a lower water excess. Under the current conditions, it may seem as little profitable, but in the future, with the development of climate change uncertainty, the price of water is expected to increase, and the benefits will be higher.
- The key conclusions, based on the results obtained, serve only as recommendations, but a final decision on the safeness and economy of a new reservoir is on-site of decision-makers.

Based on the case study including climate scenarios, we can recommend using the created universal model considering drought prevention. This new approach to design and operation of the multipurpose reservoir in conditions of uncertain climate change has been tested on the case study and can be applicable to cases in similar conditions with the problem of decreasing water sources. However, several points of the methodological section are specific to the case study and cannot be generalized to all dams.

Author Contributions: S.P. performed the design, collected the data, made the calculations and analysis, and wrote the paper. Z.K. set the objectives of the optimization method, provided useful advice, contributed to the interpretation of results, and made a revision. D.M. helped with setting the reservoir simulation model, contributed to the interpretation of results, and text corrections.

funding: This research was funded by [Design of the Flood Storage Capacity under Uncertainty Conditions] junior research project number [FAST-J-18-5044] and by [Climate Change Uncertainty Propagation in Hydrologic and Water Management Applications] research project number [FAST-S-18-5341].

Conflicts of Interest: The authors declare no conflict of interest.

References

1. World Meteorological Organization. *WMO Statement on the State of the Global Climate in 2017*; Chairperson, Publications Board: Geneva, Switzerland, 2018; ISBN 978-92-63-11212-5.
2. Czech Document: Příprava Realizace Opatření pro Zmírnění Negativních Dopadů Sucha a Nedostatku Vody. *Meziresortní Komise VODA-SUCHO*. 2015. Available online: http://eagri.cz/public/web/file/417667/_3_material_VLADA.pdf (accessed on 1 April 2018).

3.   Czech Government Document: Komplexní Studie Dopadů, Zranitelnosti a Zdrojů Rizik Souvisejících se Změnou Klimatu v ČR. 2015. EKOTOXA s.r.o.. Available online: https://www.mzp.cz/C1257458002F0DC7/cz/studie_dopadu_zmena_klimatu/$FILE/OEOK-Komplexni_studie_dopady_klima-20151201.pdf (accessed on 10 April 2018).

4.   Czech Government Document: Národní Akční Plán Adaptace na Změnu Klimatu. Ministerstvo Životního Prostředí. 2015. Available online: https://www.mzp.cz/C1257458002F0DC7/cz/narodni_akcni_plan_zmena_klimatu/$FILE/OEOK-NAP_cely_20170127.pdf (accessed on 3 April 2018).

5.   Czech Government Document: Strategie Přizpůsobení se Změně Klimatu v Podmínkách ČR. Ministerstvo Životního Prostředí. 2015. Available online: http://www.mzp.cz/C1257458002F0DC7/cz/zmena_klimatu_adaptacni_strategie/$FILE/OEOK-Adaptacni_strategie-20151029.pdf (accessed on 1 April 2018).

6.   Czech Technical Standard ČSN 75 2405 Reservoir Storage Capacity Analysis, ICS 93.160. Available online: http://seznamcsn.unmz.cz/Detailnormy.aspx?k=69792 (accessed on 3 April 2018).

7.   Kritskiy, S.N.; Menkel, M.F. *Water Management Computations*; GIMIZ: Leningrad, Russia, 1952. (In Russian)

8.   Hashimoto, T.; Stedinger, J.R.; Loucks, D.P. Reliability, resiliency, and vulnerability criteria for water resource system performance evaluation. *Water Resour. Res.* **1982**, *18*, 14–20. [CrossRef]

9.   Moy, W.-S.; Cohon, J.L.; ReVelle, C.S. A Programming Model for Analysis of the Reliability, Resilience, and Vulnerability of a Water Supply. *Water Resour. Res.* **1986**, *22*, 489–498. [CrossRef]

10. Zhang, C.; Xu, B.; Li, Y.; Fu, G. Exploring the relationships among reliability, resilience, and vulnerability of water supply using many-objective analysis. *J. Water Resour. Plan. Manag.* **2017**, *143*. [CrossRef]

11. Matrosov, E.S.; Padula, S.; Harou, J.J. Selecting portfolios of water supply and demand management strategies under uncertainty-contrasting economic optimisation and "Robust decision making" approaches. *Water Resour. Manag.* **2012**, *27*, 1123–1148. [CrossRef]

12. Paton, F.L.; Dandy, G.C.; Maier, H.R. Integrated framework for assessing urban water supply security of systems with non-traditional sources under climate change. *Environ. Model. Softw.* **2014**, *60*, 302–319. [CrossRef]

13. Fowler, H.J.; Kilsby, C.G.; O'Connell, P.E. Modeling the impacts of climatic change and variability on the reliability, resilience, and vulnerability of a water resource system. *Water Resour. Res.* **2003**, *39*, 1222. [CrossRef]

14. Roach, T. Decision Making Methods for Water Resources Management Under Deep Uncertainty. Ph.D. Theses, University of Exeter, Exeter, UK, 2016.

15. Roach, T.; Kapelan, Z.; Ledbetter, R. Comparison of Info-gap and Robust Optimisation Methods for Integrated Water Resource Management under Severe Uncertainty. *Procedia Eng.* **2015**, *119*, 874–883. [CrossRef]

16. Watts, G.; Christierson, B.; Hannaford, J.; Lonsdale, K. Testing the resilience of water supply systems to long droughts. *J. Hydraul.* **2012**, *414–415*, 255–267. [CrossRef]

17. Amarasinghe, P.; Liu, A.; Egodawatta, P.; Bernes, P.; McGree, J.; Goonetilleke, A. Quantitative assessment of resilience of a water supply system under rainfall reduction due to climate change. *J. Hydraul.* **2016**, *540*, 1043–1052. [CrossRef]

18. Li, Y.; Lence, B.J. Estimating resilience for water resources systems. *Water Resour. Res.* **2007**, *43*, W07422. [CrossRef]

19. Simonovic, S.P.; Arunkumar, R. Comparison of static and dynamic resilience for a multipurpose reservoir operation. *Water Resour. Res.* **2016**, *52*, 8630–8649. [CrossRef]

20. Sweetapple, C.; Fu, G.; Butler, D. Reliable, Robust, and Resilient System Design Framework with Application to Wastewater-Treatment Plant Control. *J. Environ. Eng.* **2017**, *143*, 04016086. [CrossRef]

21. Knight, F. *Risk, Uncertainty and Profit*; Hart, Schaffner & Marx: Boston, MA, USA; Houghton Mifflin Co.: Boston, MA, USA, 1921.

22. Kaplan, S. *Risk Assessment and Risk Management—Basic Concepts and Terminology. Risk Management: Expanding Horizons in Nuclear Power and other Industries*; Hemisphere Publ. Corp.: Boston, MA, USA, 1991; pp. 11–28.

23. Beven, K.J.; Binley, A.M. The future of distributed models: Model calibration and uncertainty prediction. *Hydrol. Proceaaes* **1992**, *6*, 279–298. [CrossRef]

24. Mahmoud, M.; Liu, Y.; Hartmann, H.; Stewart, S.; Wagener, T.; Semmens, D.; Stewart, R.; Gupta, H.; Dominguez, D.; Dominguez, F.; et al. A formal framework for scenario development in support of environmental decision-making. *Environ. Model. Softw.* **2009**, *24*, 798–808. [CrossRef]

25. Marton, D.; Paseka, S. Uncertainty Impact on Water Management Analysis of Open Water Reservoir. *Environments* **2017**, *4*, 10. [CrossRef]

26. Marton, D.; Starý, M.; Menšík, P. Analysis of the influence of input data uncertainties on determining the reliability of reservoir storage capacity. *J. Hydrol. Hydromech.* **2015**, *63*, 287–294. [CrossRef]

27. Marton, D.; Starý, M.; Menšík, P.; Paseka, S. Hydrological Reliability Assessment of Water Management Solution of Reservoir Storage Capacity in Conditions of Uncertainty. In *Drought: Research and Science-Policy Interfacing*; CRC Press Taylor & Francis Group: Leiden, The Netherlands, 2015; pp. 377–382. ISBN 978-1-138-02779-4.

28. Paseka, S.; Marton, D.; Menšík, P. Uncertainties of reservoir storage capacity during low water period. In Proceedings of the SGEM International Multidisciplinary Geoconference: Hydrology and Water Resources; STEF92 Technology Ltd.: Sofia, Bulgaria, 2016; pp. 789–796, ISBN 978-619-7105-61-2.

29. Vogel, R.M.; Bolognes, A. Storage-reliability-resilience-yield relations for over-year water supply systems. *Water Resour. Res.* **1995**, *31*, 645–654. [CrossRef]

30. Jain, K.S.; Bhunya, K.P. Reliability, resilience and vulnerability of a multipurpose storage reservoir. *Hydrol. Sci. J.* **2008**, *53*, 434–447. [CrossRef]

31. Ngo, L.L.; Madsen, H.; Rosbjerg, D. Simulation and optimisation modelling approach for operation of the Hoa Binh reservoir, Vietnam. *J. Hydrol.* **2007**, *336*, 269–281. [CrossRef]

32. Villa, I.R.; Rodríguez, J.B.M.; Molina, J.L.; Tarragó, J.C.P. Multiobjective Optimization Modeling Approach for Multipurpose Single Reservoir Operation. *Water* **2018**, *10*, 427. [CrossRef]

33. Scola, L.A.; Takahashi, R.H.C.; Cerqueira, S.A.A.G. Multipurpose Water Reservoir Management: An Evolutionary Multiobjective Optimization Approach. *Math. Probl. Eng.* **2014**, *2014*, 638259. [CrossRef]

34. Groves, D.G.; Yates, D.; Tebaldi, C. Developing and applying uncertain global climate change projections for regional water management planning. *Water Resour. Res.* **2008**, *44*, W12413. [CrossRef]

35. Herman, J.D.; Zeff, H.B.; Reed, P.M.; Characklis, G.W. Beyond optimality: Multistakeholder robustness tradeoffs for regional water portfolio planning under deep uncertainty. *Water Resour. Res.* **2014**, *50*, 7692–7713. [CrossRef]

36. Beh, E.H.Y.; Maier, H.R.; Dandy, G.C. Adaptive, multiobjective optimal sequencing approach for urban water supply augmentation under deep uncertainty. *Water Resour. Res.* **2015**, *51*, 1529–1551. [CrossRef]

37. Haasnoot, M.; Kwakkel, J.H.; Walker, W.E.; Maat, T.J. Dynamic adaptive policy pathways: A method for crafting robust decisions for a deeply uncertain world. *Glob. Environ. Chang.* **2013**, *23*, 485–498. [CrossRef]

38. Jeuland, M.; Whittington, D. Water resources planning under climate change: Assessing the robustness of real options for the Blue Nile. *Water Resour. Res.* **2014**, *50*, 2086–2107. [CrossRef]

39. Paton, F.L.; Maier, H.R.; Dandy, G.C. Including adaptation and mitigation responses to climate change in a multiobjective evolutionary algorithm framework for urban water supply systems incorporating GHG emissions. *Water Resour. Res.* **2014**, *50*, 6285–6304. [CrossRef]

40. Roach, T.; Kapelan, Z.; Ledbetter, R.; Ledbetter, M. Comparison of Robust Optimization and Info-Gap Methods for Water Resource Management under Deep Uncertainty. *J. Water Resour. Plan. Manag.* **2017**, *143*. [CrossRef]

41. Whateley, S.; Steinschneider, S.; Brown, C. A climate change range-based method for estimating robustness for water resources supply. *Water Resour. Res.* **2014**, *50*. [CrossRef]

42. Borgomeo, E.; Mortazavi-Naeini, M.; Hall, J.W.; Guillod, B.P. Risk, Robustness and Water Resources Planning Under Uncertainty. *Earth's Future* **2018**, *6*. [CrossRef]

43. Deb, K.; Agrawal, S.; Pratap, A.; Meyarivan, T.; Amrit, P.; Meyarivan, T. *A Fast Elitist Non-Dominated Sorting Genetic Algorithm for Multi-Objective Optimization: NSGA-II*; Schoenauer, M., Deb, K., Rudolph, G., Yao, X., Lutton, E., Merelo, J.J., Schwefel, H.-P., Eds.; Parallel Problem Solving from Nature PPSN VI; Springer: Berlin/Heidelberg, Germany, 2000; pp. 849–858.

44. Kollat, J.B.; Reed, P.M. Comparing state-of-the-art evolutionary multiobjective algorithms for long-term groundwater monitoring design. *Adv. Water Res.* **2006**, *29*, 792–807. [CrossRef]

45. Nicklow, J.; Reed, P.; Savic, D.; Dessalegne, T.; Harrell, L.; Chan-Hilton, A.; Karamouz, M.; Minsker, B.; Ostfeld, A.; Singh, A.; et al. State of the art for genetic algorithms and beyond in Water resources planning and management. *J. Water Resour. Plan. Manag.* **2010**, *136*, 412–432. [CrossRef]

46. Tukimat, N.N.A.; Harun, S. Optimization of water supply reservoir in the framework of climate variation. *Inter. J. Softw. Eng. Appl.* **2014**, *8*, 361–378. [CrossRef]

47. Fu, G.; Kapelan, Z.; Kasprzyk, J.R.; Reed, P. Optimal Design of Water Distribution Systems Using Many-Objective Visual Analytics. *J. Water Resour. Plan. Manag.* **2013**, *139*. [CrossRef]

48. Perelman, L.; Ostfeld, A.; Salomons, E. Cross Entropy multiobjective optimization for water distribution systems design. *Water Resour. Res.* **2008**, *44*. [CrossRef]

49. Zheng, F.; Simpson, A.R.; Zecchin, A.C. An efficient hybrid approach for multiobjective optimization of water distribution systems. *Water Resour. Res.* **2014**, *50*. [CrossRef]

50. Yang, G.; Guo, S.; Liu, P.; Li, L.; Liu, Z. Multiobjective Cascade Reservoir Operation Rules and Uncertainty Analysis Based on PA-DDS Algorithm. *J. Water Resour. Plan. Manag.* **2017**, *143*. [CrossRef]

51. Srinivas, N.; Deb, K. Muiltiobjective Optimization Using Nondominated Sorting in Genetic Algorithms. *Evol. Comput.* **1994**, *2*, 221–248. [CrossRef]

52. Goldberg, D.E. *Genetic Algorithms in Search, Optimization, and Machine Learning*; Addison-Wesley Publishing Company: Reading, MA, USA, 1989; ISBN 0-201-15767-5.

53. Marton, D.; Kapelan, Z. Risk and reliability analysis of open reservoir water shortages using optimization. *Procedia Eng.* **2014**, *89*, 1478–1485. [CrossRef]

54. Savić, D.A.; Bicik, J.; Morley, M.S. A DSS Generator for Multiobjective Optimisation of Spreadsheet-Based Models. *Environ. Model. Softw.* **2011**, *26*, 551–561. [CrossRef]

55. Borgomeo, E.; Hall, J.W.; Fung, F.; Watts, G.; Colquhoun, K.; Lambert, C. Risk-based water resources planning: Incorporating probabilistic nonstationary climate uncertainties. *Water Resour. Res.* **2014**, *50*, 6850–6873. [CrossRef]

56. Deepashree, R.; Mujumdar, P.P. Reservoir performance under uncertainty in hydrologic impacts of climate change. *Adv. Water Res.* **2010**, *33*, 312–326. [CrossRef]

57. Kling, H.; Stanzel, P.; Preishuber, M. Impact modelling of water resources development and climate scenarios on Zambezi River discharge. *J. Hydrol.* **2014**, *1*, 17–43. [CrossRef]

58. Majone, B.; Bovolo, C.I.; Bellin, A.; Blenkinsop, S.; Fowler, H.J. Modeling the impacts of future climate change on water resources for the Gállego river basin (Spain). *Water Resour. Res.* **2012**, *48*, W01512. [CrossRef]

59. Mantel, S.K.; Hughes, D.A.; Slaughter, A.S. Water resources management in the context of future climate and development changes: A South African case study. *J. Water Clim. Chang.* **2015**, *6*, 772–786. [CrossRef]

60. Maiolo, M.; Mendicino, G.; Senatore, A.; Pantusa, D. Optimization of drinking water distribution systems in relation to the effects of climate change. *Water* **2017**, *9*, 803. [CrossRef]

61. Kapelan, Z.; Savić, D.; Mahmoud, H. A Response Methodology for Reducing Impacts of Failure Events in Water Distribution Networks. *Procedia Eng.* **2017**, *186*, 218–227. [CrossRef]

62. IPCC Secretariat. *Climate Change 2007: Impacts, Adaptation and Vulnerability: Summary for Policymakers*; IPCC Secretariat: Geneva, Switzerland, 2007; ISBN 92-9169-121-6.

63. Czech Project. *Development of Information and Data Support for Design of Adaptation Measures and Long-Term Planning of Water Resources Considering the Climate Change Effects*; T. G. Masaryk Water Research Institute: Praha, Czech Republic; Czech Hydro-Meteorological Institute: Praha, Czech Republic; Water Management Development, RSCN: Jubeiha, Jordan, 2012–2014.

64. Czech Document PRVKUK. *The Plan for the Development of Water Supply and Sewerage Systems in the Olomouc Region*; The Regional Office of the Olomouc Region, Ministry of Agriculture of the Czech Republic: Olomouc, Czech Republic, 2006.

65. Klemeš, V. A simlified solution of the flood-routing problem. *Vodohosp. Časopis* **1960**, *8*, 317–326.

66. Government Finance Function. *The Green Book: Appraisal and Evaluation in Central Government*; Government Finance Function: London, UK, 2016.

67. Czech Property Valuation Decree. *Vyhlaska č. 441/2013 Sb., Vyhlaska k Provedeni Zakona o Ocenovani Majetku*; Ministry of Finance of the Czech Republic: Praha, Czech Republic, 2013.

68. Starý, M. *Reservoir and Reservoir System*; Education Tutorial, Faculty of Civil Engineering, Brno University of Technology: Brno, Czech Republic, 2006.

water

MDPI

Article

Piecewise-Linear Hedging Rules for Reservoir Operation with Economic and Ecologic Objectives

Yueyi Liu [1], Jianshi Zhao [2] and Hang Zheng [1,*]

1   School of Environment and Civil Engineering, Dongguan University of Technology,
    Dongguan 523808, China; liuyueyi@dgut.edu.cn
2   State Key Laboratory of Hydro-Science and Engineering, Department of Hydraulic Engineering,
    Tsinghua University, Beijing 100086, China; zhaojianshi@tsinghua.edu.cn
*   Correspondence: zhengh@dgut.edu.cn; Tel.: +86-0769-2286-1232

Received: 10 May 2018; Accepted: 25 June 2018; Published: 29 June 2018

Abstract: The construction of dams and operation of reservoirs have a significant impact on the interruption of aquatic and riparian ecological systems, by altering natural stream flows in river courses. Recently, the ecological requirements are included as an additional objective in reservoir operation in order to restore the natural stream flows and reduce the negative impacts of reservoir operations on ecosystems that rely on the natural flows. The key challenge involving ecological requirements is to balance the ecological and economic objectives by operation rules, on the basis of quantitatively identifying the objective of ecological flows required to maintain the natural flow regime for ecosystem. This study develops a piecewise-linear multi-objective hedging rule (PMHR) for reservoir operations, with ecologic flow objectives represented by 33 hydrologic parameters from the indicators of hydrologic alteration (IHA). Variables of the PMHR are obtained through optimization using a vector evaluated genetic algorithm. The results show that the PMHR improves the ecological water releases without reducing economic water supplies in the case river in the context of hydrological uncertainty. It can offer technological references for improving the utility of water resource management under competitive conditions of water resources.

Keywords: natural flow regime; multi-objective model; uncertainty; genetic algorithm

1. Introduction

For decades, dams have been built and operated mostly for economic purposes that require a reliable water supply for human needs, such as hydropower electricity, irrigation, living and industry water supply, and navigation. Water withdrawn for increasing economic demands has led to conflicts between human water use and ecosystem water needs [1–3]. The operation of dams leads to negative effects on aquatic and riparian ecosystem systems [4–6]. The alteration of the flow regime caused by reservoir operation is recognized as a major driving factor that threatens the integrity of the river ecosystem [7–11]. Bunn and Arithington reviewed the studies focused on the relationship between hydrological flow regime and river ecosystems, and illustrate four critical negative impacts of the alteration: (1) alteration of the magnitude of flow as the determinant of physical habitat and biotic composition; (2) alteration of the timing and frequency of peak flow as key factors of the life history of aquatic species; (3) destruction of the patterns of longitudinal and lateral connectivity for riverine species; and (4) the extinguishing of extreme low flow avoidance, meant to prevent the invasion of exotic species [12].

How to balance the trade-off between humans and ecosystems has been discussed for decades regarding integrated river basin management [13–16]. Many investigations have been conducted to address the conflicts between environmental flow release and the economic water supply over the past decades. To evaluate the ecological objective for aquatic and riparian ecosystems' sustainability, four kinds of approaches are often applied, including (1) estimating the minimum flow requirement for downstream habitats to maintain the survival of specific species [17–19]; (2) determining a flow regime based on fish diversity information [20]; (3) providing a regime-based, prescribed flow duration curve that considers floods and droughts for species and morphological needs [21–23]; and (4) minimizing the degree of flow alterations in order to maintain the stability of river ecosystems and population structures of species [16,24–26]. In these studies, it is can be found that the adopted ecological objectives have been shifted from an emphasis on a minimum flow requirement or single species to the development of a regime-based comprehensive approach. With the first two methods mentioned above, the flow fluctuation, which decides the ecological integrity, is generally neglected. Therefore, ecologists currently prefer to recommend the last two methods, in which the ecological integrity can be sustained by accounting for the functional and structural requirements for aquatic and riparian ecosystems.

Suen and Eheart proposed a multi-objective model based on the ecological flow regime paradigm, which incorporates the intermediate disturbance hypothesis to address the ecological and anthropic need [16]. Their study uses parts of the Taiwan Eco-Hydrology Indicators System (TEIS) as the criteria to represent ecosystem objectives. Lane et al. proposed a multi-step re-operation methodology to address environmental flow requirements and human water use objectives [23]. This model summarizes the environmental flow requirements on the basis of empirical streamflow thresholds for the maintenance of an ecosystem with geomorphic functions. Using a one-dimensional water routing model by the Water Evaluation and Planning System (WEAP), the model develops an alternative reservoir rule curve and suggests the new timing of releases to sustain key ecological and geomorphic functions. Generally, these studies have used a monthly-based interval and have therefore failed to reflect important hydrological factors with daily variations, which may play an important role for sustaining the health of an ecosystem [7,27–29]. Chen et al. proposed a time-nested approach, in order to scale down the decision variables from a monthly to a 10-day basis, and further to a daily basis for reservoir operations [30]. They used a daily flow hydrograph as a constraint in the optimization model. However, the model was extremely complex, with 730 decision variables, leading to a high cost of calculation in the optimization process. Considering daily scale indicators in reservoir operations, such as the timing and frequency of peak flow, the rising and falling rates requires a large number of decision variables in the optimization process, and lead to a computing-heavy task.

It has been noted that involving a set of fixed operation rules in reservoir modelling can reduce the number of decision variables as well as the computational costs [31]. Standard operating policy (SOP) and hedging rules (HR) are the most commonly used rules in reservoir operations. SOP, which is the traditional and simplest operating rule, releases water as close to the delivery targets as possible in order to meet the demands of the current stage [32]. By contrast, HR prefers to keep a certain proportion of available water in the reservoir, in order to minimize the possible losses caused by water shortages in the future. Usually, SOP is recommended only if the objective function is linear, while the hedging rules have been proven to be a more efficient way to optimize the reservoir operations when the supply benefits are nonlinear [33,34]. In real operations, the trade-off analysis of multiple objectives is often complicated with nonlinear systems and uncertain inflow conditions [35–37]. Therefore, HR is often used to cope with uncertain future inflows among different nonlinear objectives in practice [38]. Taghian et al. employed HR to achieve the optimal water allocation, in order to reduce the intensity of severe water shortages [39]. A hybrid model was developed to simultaneously optimize both the conventional rule curve and the hedging rule. Shiau developed a parameterization–simulation–optimization approach for optimal hedging for a water supply reservoir by considering the balance between beneficial release and carryover storage value. The proposed methodology was applied to the Shihmen Reservoir in northern Taiwan to illustrate the effects of derived optimal hedging on reservoir performance in terms of shortage-related indices and hedging uncertainty [40]. Yin et al. used a typical approach that uses three limit curves to divide the reservoir storage and the reservoir inflow into several zones [24]. A series of water supply rules and water releasing rules were derived by the actual reservoir storage and the reservoir inflows to provide the water supply and to satisfy the ecological needs of seasonally variables, such as base flows, dry season flow recessions, high-flow pulses, etc. However, daily interval variations were not exactly considered in the above-mentioned studies.

Accounting for the daily variation of downstream flows, Yang et al. proposed a set of linear rules to generate daily reservoir releases, with the aim of satisfying the needs of ecological assessment [20]. This method can explicitly engage flow regime variation based on real-time inflow information [41], while short-term inflow information becomes crucial for reservoir operation when daily scale ecologic indicators are considered. However, the proposed operation rules were based on traditional operation curves for flood control and ecologic objectives, without further considerations regarding the balance between consumptive water supply and ecological flow needs. Moreover, the ecologic objective in this paper was calculated by seven indicators, including Q_3 and Q_5, the average discharges in March and May, respectively; $Q_{3day\ min}$ and $Q_{7day\ min}$, which represent the annual minimum average of three- and seven-day discharges; $Q_{3day\ max}$, which is the annual maximum average 3-day flow; and D_{min}, representing the Julian date of the annual minimum daily flow. Further information, such as the changing trends of daily flow and the statistics of daily flow reversals, are required to establish a more comprehensive function for the ecological objective.

The purpose of this study is to develop practical hedging rules for reservoir operation with economic and ecologic objectives, which will be able to reflect daily interval variations and engage the real-time daily inflow forecast and storage information simultaneously into decision-making. In this paper, piecewise-linear hedging rules are proposed in order to generate the daily release for ecological and economic objectives, through which the parameters are optimized based on historical and synthetic streamflow series. In additional, the objective of ecological flow requirements is quantified by 33 hydrologic parameters from Indicators of Hydrologic Alteration (IHA). The proposed methodology is applied to a realistic reservoir as a performance test. The rest of this paper is organized as follows. Section 2 briefly introduces the structure of the model. Section 3 presents a real world case study of Baiguishan reservoir, China. Section 4 compares the proposed HRs with SOP and analyzes the value of forecast information. Finally, conclusions are drawn in Section 5.

2. Methodology

The framework of the method for optimizing the reservoir's target water levels and flows released to the economic and ecological objectives is shown in Figure 1. It is composed of the following three steps: (1) generating daily water release based on daily inflow and pre-defined hedging rules, (2) multi-objective optimization for releasing water to economic and ecologic objectives, (3) Monte Carlo simulation and probability analysis to identify the variations of the optimal water release in various hydrological years.

The aim of this framework is to identify a set of practical operating rules for water release. A set of pre-defined, piecewise-linear hedging rules are used to generate water release under the conditions of various inflow and water levels of the reservoirs. A multi-objective genetic algorithm is then applied to optimize the parameters of the rules' curves iteratively, by balancing the economic and ecological objectives of water release. The statistical characteristics of the parameters are obtained through Monte Carlo simulations, in which the historical and synthetic daily inflows are used as the inputs.

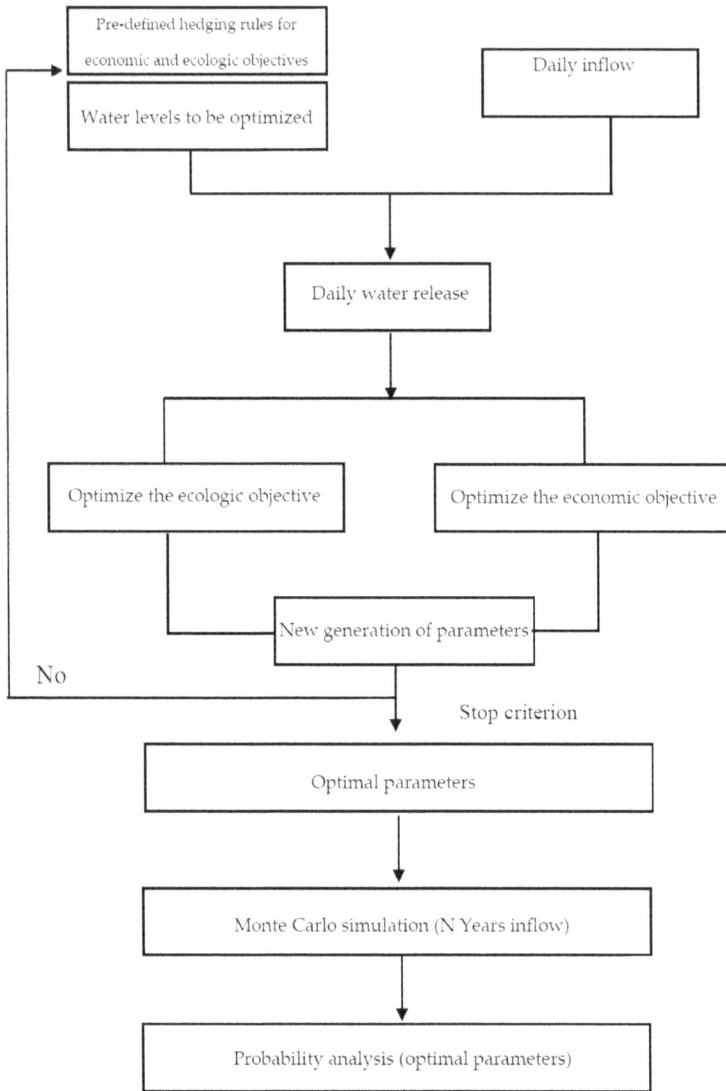

Figure 1. Process diagram for optimizing the operating rules.

2.1. Pre-Defined Piecewise-Linear Hedging Rules

In this paper, piecewise-linear hedging rules considering both economic and ecological objectives are pre-defined for the dry and wet seasons separately. As shown in Figure 2, zones 1–3 represent the rules in the dry season, and zones 4–6 stand for the rules during the wet season. Red lines show the maximum and minimum water level of different periods following the traditional operation rules currently used by the reservoir.

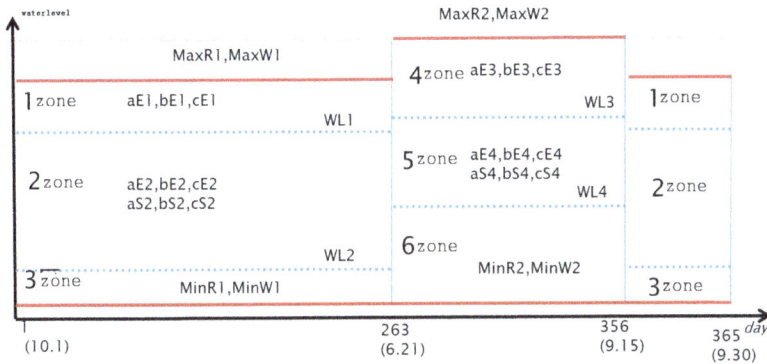

Figure 2. Release functions in different zones characterized by seasons.

For the dry season, daily water release for economic objectives can be defined as a set of piecewise-linear hedging rule functions incorporating actual storage and future inflow while considering the effect of daily inflow forecast [42], as follows:

$$
\begin{cases}
W_t = D_t & L_t \geq WL_1 \\
W_t = Min(D_t, Max(MinW_1, aS_1 \times I_t + bS_1 \times L_{t-1} + cS_1)) & WL_2 \leq L_t < WL_1 \\
W_t = MinW_1 & L_t < WL_2
\end{cases}
\tag{1}
$$

where W_t is the water release for economic objective of day t; D_t is the target release for economic objective of day t; I_t is the inflow of day t; L_t and L_{t-1} are water levels of day t and day $t-1$, respectively; WL_1 and WL_2 stand for upper and lower limits of the water level, respectively; $MinW_1$ represents the minimum release for the economic objective of day t; aS_1, bS_1, and cS_1 are the coefficients of the linear functions of hedging rules.

Specifically, when the actual water level (L_t) is higher than the upper limit (WL_1), it implies that the stored water is enough to defend the area from droughts in the future. In this case, the economic water release W_t can be described as the amount of water that is needed (D_t). When the actual water level (L_t) is between the upper limit (WL_1) and the lower limit (WL_2), it means that there is insufficient storage to defend against future droughts, and thus the water release W_t can be described as a linear function related to the forecasted inflow and current water level. When the current water level is lower than the lower limit (WL_2), it means that storage is rare and there is a huge drought risk in future. In this case, W_t is defined as a fixed minimum value, representing the minimum or basic water release to the economic system.

Similar to Equation (1), the daily water release for an ecologic objective is defined as follows:

$$
\begin{cases}
R_t = Min(MaxR_1, Max(MinR_1, aE_1 \times I_t + bE_1 \times L_{t-1} + cE_1)) & L_t \geq WL_1 \\
R_t = Min(MaxR_1, Max(MinR_1, aE_2 \times I_t + bE_2 \times L_{t-1} + cE_2)) & WL_2 \leq L_t < WL_1 \\
R_t = MinR_1 & L_t < WL_2
\end{cases}
\tag{2}
$$

where R_t represents the ecological water release of day t; aE_1, bE_1, cE_1, aE_2, bE_2, and cE_2 are the parameters of the HR, which will be obtained through the optimizations illustrated in Figure 1; $MaxR_1$ and $MinR_1$ are the maximum and minimum water release that need to be optimized, respectively.

Similarly, economic and ecological water releases in wet seasons can be defined as follows:

$$
\begin{cases}
W_t = D_t & L_t \geq WL_3 \\
W_t = Min(D_t, Max(MinW_2, aS_2 \times I_t + bS_2 \times L_{t-1} + cS_2)) & WL_4 \leq L_t < WL_3 \\
W_t = MinW_2 & L_t < WL_4
\end{cases}
\tag{3}
$$

$$\begin{cases} R_t = Min(MaxR_2, Max(MinR_2, aE_3 \times I_t + bE_3 \times L_{t-1} + cE_3)) & L_t \geq WL_3 \\ R_t = Min(MaxR_2, Max(MinR_2, aE_4 \times I_t + bE_4 \times L_{t-1} + cE_4)) & WL_4 \leq L_t < WL_3 \\ \qquad\qquad R_t = MinR_2 & L_t < WL_4 \end{cases} \quad (4)$$

where WL_3 and WL_4 are the upper and lower limit, respectively; $MaxR_2$ and $MinR_2$ are maximum and minimum water release to be optimized, respectively; and $aS2$, $bS2$, $cS2$, $aE3$, $bE3$, $cE3$, $aE4$, $bE4$, and $cE4$ are the parameters of the HR during wet seasons.

Consequently, the decision variables consist of the following three parts: (1) water level limits ($WL1$ and $WL2$ in dry seasons and $WL3$ and $WL4$ in wet seasons); (2) ecological and economic hedging coefficients ($aE1$, $bE1$, $cE1$, $aE2$, $bE2$, $cE2$, $aE3$, $bE3$, $cE3$, $aE4$, $bE4$, $cE4$ $aS2$, $bS2$, $cS2$, $aS4$, $bS4$, and $cS4$); and (3) maximum and minimum water release ($MinW1$, $MinW2$, $MinR1$, $MinR2$, $MaxW1$, $MaxW2$, $MaxR1$, $MaxR2$).

2.2. Ecological Objective

The ecological objective is to minimize reservoirs' alterations on rivers' natural flows, which have been adapted to by aquatic and riparian species over thousands years of evolution by maintaining the stability of ecosystems and population structures of species. The Indicators of Hydrologic Alteration (IHA) program, proposed by Richter, has been adopted to represent the ecological objective in this paper [4]. It contains 33 hydrologic parameters involving five groups of characteristics, including (1) the magnitude of monthly stream flow; (2) the magnitude of annual extreme flows at different time durations; (3) the timing of annual extreme flows; (4) the frequency and duration of high and low pulses; (5) the rate of change.

For each eco-hydrological indicator, a spectrum of values could be set as the target range, which could reflect the changes of environmental flow regime that a species can adapt to. If the actual value of an indicator falls in that range, it means that the alteration of the natural flow regime is acceptable by the aquatic or riparian species. Those acceptable ranges of all the indicators have been investigated by many researchers [9], among which the findings by Richter et al. are mostly widely applied through the range of variability approach (RVA) [4]. RVA is an efficient and convenient method to evaluate the degree of flow regime alteration. According to Richter and in this paper, the 25th and 75th percentiles of the historical annual value of the indicators are set as the upper and lower limits of the target range of environmental flow regime, respectively.

Here, let $f_i(r)$ represent the effect function of the i-th indicator of IHA. In addition, $iha_i(r)$ represents the value of the i-th indicator. $f_i(r)$ equals 0 if $iha_i(r)$ falls in the target range, which implies that the ecological release and flow alternation compared to the natural flow regime are acceptable. When the value of the indicator falls outside of the target range, $f_i(r)$ is calculated as the distance from the target range, as follows:

$$f_i(r) = \begin{cases} 0 & iha_{ip75} \leq iha_i(r) \leq iha_{ip25} \\ \left(\dfrac{iha_i(r) - iha_{ip25}}{iha_{ip25} - iha_{ip75}}\right)^2 & iha_i(r) \geq iha_{ip25} \\ \left(\dfrac{iha_{ip75} - iha_i(r)}{iha_{ip25} - iha_{ip75}}\right)^2 & iha_i(r) \leq iha_{ip75} \end{cases} \quad (5)$$

where r is the series of ecological water release generated by Equations (2) and (4); iha_{ip25} and iha_{ip75} are the upper and lower limits of the annual values of the indicators, respectively; and $iha_i(r)$ represents the value of the i-th indicator. The ecological objective can be calculated by the sum of the 33 indicators of the IHA, as

$$f_1 = \sum_{i=1}^{33} f_i(r) \quad (6)$$

The value of f_1, the range of which is greater than zero, represents the ecosystem objective to be minimized in the multi-objective optimization model.

2.3. Economic Objective

The economic objective is represented in a target-hitting form, in which the water demands of agriculture, industry, and domestic water use are considered. For equalizing the water supply shortages across time intervals throughout the year, the economic objective is employed through quantifying the water supply deficits, as follows:

$$f_2 = -g(w) = f(w) = \sum_{i=1}^{T} (\frac{d_i - w_i}{d_i})^2 \tag{7}$$

where T is the total number of time periods; d_i is the economic water demand of the i-th day, including the water demands of agriculture, industry and domestic sectors; and w_i is the water supplied in the i-th day. The value of f_2, range from 0 to 1, and represents the economic objective to be minimized in the multi-objective optimization model.

2.4. Constraints

Constraints within the optimization model are set as follows.
(1) Water balance constraint:

$$V_i = V_{i-1} + I_i \Delta t - q_i \Delta t - E_i \tag{8}$$

where V_i and V_{i-1} represent the reservoir storage available at the end of the i-th day and $i-1$ day, respectively; I_i is the inflow of the i-th day; q_i is the total release during the i-th day; and E_i is the evaporation of the reservoir in the i-th day.
(2) Reservoir storage capacity constraint:

$$L_i^{min} \leq L_i \leq L_i^{max} \tag{9}$$

where L_i^{min} and L_i^{max} are the minimum and maximum water level limits at the end of the i-th day, respectively; and L_i is the water level of the i-th day.
(3) Initial and end storage constraint:
Considering the initial water storage of next year, here we set up initial storage and end storage equally as

$$V_0 = V_a \tag{10}$$

where V_0 is the initial storage of the reservoir and V_a is its end storage.
The multiple objectives of the model are set to meet both the ecosystem and human demand, which can be expressed as follows:

$$Obj.\ Min\{f1(r), f2(w)\} \tag{11}$$

2.5. Vector Evaluated Genetic Algorithm

This multi-objective model is solved by the vector evaluated genetic algorithm (VEGA). Different from the standard procedure of a genetic algorithm, VEGA focuses on the optimization of several objectives [43]. Through the VEGA, the initial population is divided into a number of objectives, and the elite of each group is selected for the next generation while the others are put in crossover and mutation pools. A population with a higher fitness value has a greater probability of persevering to the next generation. The procedure of VEGA is presented in Figure 3.

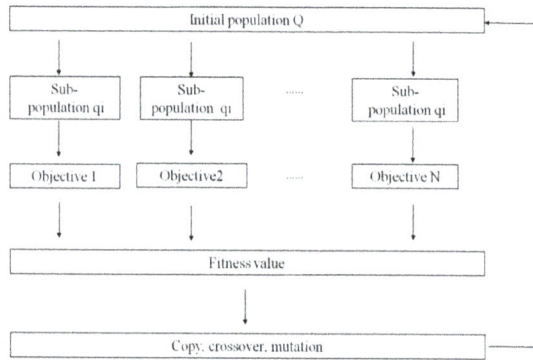

Figure 3. Procedure of the vector evaluated genetic algorithm (VEGA).

3. Results

3.1. Study Site

The Baiguishan Reservoir, which is located on the Shahe River (Figure 4) in the Huaihe River Basin in China, was selected as a case study to test the proposed model. It is a multipurpose reservoir serving as a water supply, flood control, a source of recreation, and ecologic purposes, with a total storage of 0.92 billion m^3. Historical daily inflow series from 1976–2005 were used to design the operation rules. Affected by pacific monsoons, 80% of the rainfall to the reservoir occurs in summer, and this divides the year of operation into two seasons, including the dry season (16 September–20 June) and the wet season (21 June–15 September). The maximum water level used in the current operation is 103 m and 105.9 m in the dry and wet seasons, respectively; the minimum water level is limited to 92 m.

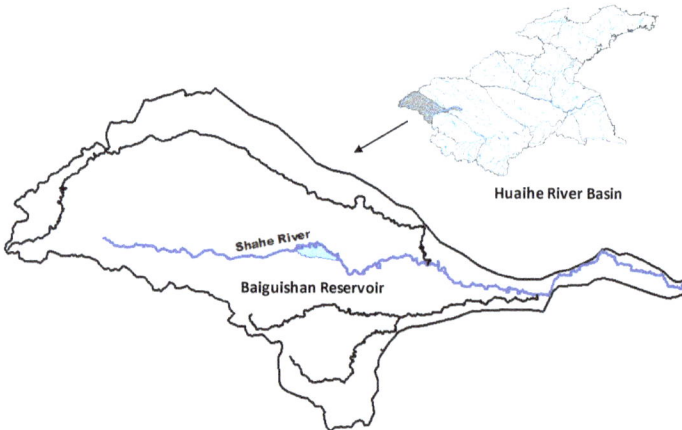

Figure 4. Location of the Baiguishan Reservoir.

3.2. Ecological Management Target Range

Generally, IHA should be calculated based on historically natural flow without intervention. In this paper, the ecological objective is defined based on the historical inflow data of the Baiguishan Reservoir instead, due to the lack of historical daily hydrologic data for the Shahe River before the

1950s—in other words, when the stream flow had not yet been altered by human beings in China. For each eco-hydrological indicator, the target range was calculated by the values of the 25th and 75th percentile of the historical series. Table 1 shows the ecological target ranges calculated by data from 1976–2005. Moreover, the expected value and the standard deviation (SD) are investigated for further discussion.

Table 1. Ecological target ranges and statistical analysis of eco-hydrologic indicators.

No.	Indicator	Unit	Ecological Target Range		Statistical Analysis	
			P25%	P75%	Expected	SD
IHA1	October	m^3/s	19.81	8.21	17.88	20.91
IHA2	November	m^3/s	20.52	4.26	12.68	11.3
IHA3	December	m^3/s	20.71	5.02	11.41	9.03
IHA4	January	m^3/s	14.59	6.6	9.05	6.33
IHA5	February	m^3/s	17.25	5.02	11.44	7.62
IHA6	March	m^3/s	21.48	2.57	13.5	9.11
IHA7	April	m^3/s	17.81	2.16	11.57	10.8
IHA8	May	m^3/s	23.1	8.2	17.36	17.06
IHA9	June	m^3/s	36.65	9.09	24.4	17.86
IHA10	July	m^3/s	38.91	15.97	36.42	42.28
IHA11	August	m^3/s	48.18	19.29	39.97	32.7
IHA12	September	m^3/s	25.75	7.79	21.18	20.34
IHA13	1day-min	m^3/s	2.56	0.3	0.63	1.43
IHA14	3 day-min	m^3/s	3.19	0.24	0.85	1.97
IHA15	7 day-min	m^3/s	3.56	0.29	1.15	2.21
IHA16	30 day-min	m^3/s	7.28	0.84	2.84	3.36
IHA17	90 day-min	m^3/s	10.08	0.91	6.68	5.15
IHA18	1 day-max	m^3/s	435.98	66.1	286.19	260.18
IHA19	3 day-max	m^3/s	329.97	55.33	205.25	192.94
IHA20	7 day-max	m^3/s	219.25	49.42	141.09	133.33
IHA21	30 day-max	m^3/s	81.72	31.73	67.62	50.98
IHA22	90 day-max	m^3/s	51.79	23.8	41.1	24
IHA23	Zero days	days	93	6	44.1	63.79
IHA24	Base flow	/	0.24	0.03	0.05	0.1
IHA25	Date of max	/	276.2	122	274.93	61.4
IHA26	Date of min	/	260	159.6	222.9	117.792
IHA27	Low count	/	12	4	4.55	3.25
IHA28	Low duration	days	16	7.22	22.94	23.62
IHA29	High count	/	13	2	7.69	3.6
IHA30	High duration	days	28.75	6	15.77	18.58
IHA31	Fall rate	/	0.52	0.19	0.63	0.39
IHA32	Rise rate	/	−0.11	−0.18	−0.76	0.62
IHA33	Reversal	/	135	81	121.8	32.62

In Table 1, eco-hydrological indictors are ordered from IHA1 to IHA33. IHA1–IHA12 are the annual mean values of monthly flow from January to December. IHA13–IHA22 are the annual mean values of the maximum or minimum *t*-day (*t* = 1, 3, 7, 9, 30, 90) flows. IHA23 and IHA24 are the number of zero flow days and the base flow, respectively. IHA25 and IHA26 are the Julian date of each annual one-day maximum and minimum, respectively; IHA27 and IHA29 are the number of high or low pulses, respectively, within each year (days); IHA28 and IHA30 are the mean duration of high or low pulses, respectively, within each year; IHA31 and IHA32 are the means of all positive and negative differences between consecutive days, respectively; and IHA33 is the total number of reversals [4].

3.3. Pareto-Optimum Solutions

To deal with the trade-off between economic and ecological objectives, VEGA with a population size of 1000 was adopted to solve the model within one year. With the series of daily inflow from

1976–2005, a Pareto-optimal frontier for every year can be obtained. Using three typical years (frequency = 25%, 50%, 75%) as examples, after running 100 generations, the algorithm is stopped; the result of the Pareto-optimal frontiers of the typical years is shown in Figure 5. Each individual solution represents one possible trade-off between the economic and ecological objectives with different inflow scenarios. Here, a frequency of 75% means the annual runoff for the year will be exceeded in 75 years out of 100. In Figure 5, each frontier solution is calculated by a searching direction in the genetic algorithm, and prioritization between the objectives is decided by the decision-maker. For example, for the scenario of frequency being 75%, the solution at point A represents an emphasis on the ecological objective, while the solution at point D represents an emphasis on the economic objective. Solutions at points B and C are the destinations with two sub-population evolution directions, which represent the trade-offs of the two competitive objectives.

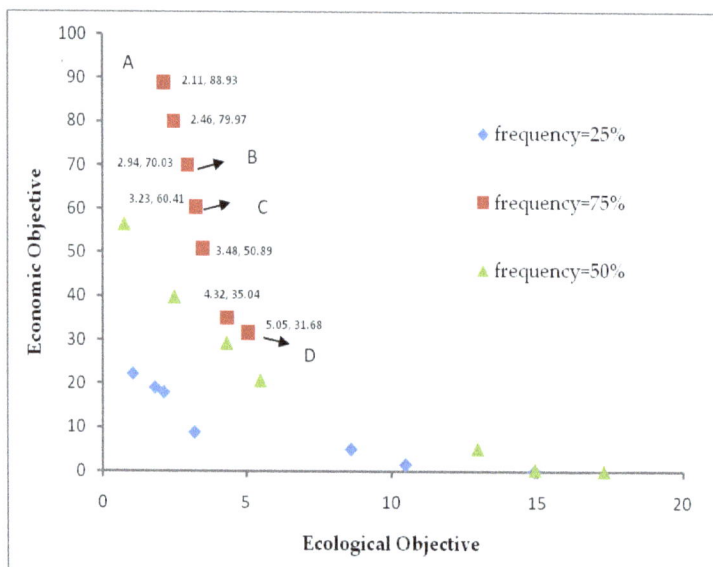

Figure 5. Trade-offs between economic and ecosystem objectives.

The corresponding decision variables of points A, B, C, and D are presented in Table 2.

Table 2. Decision variables of points of A, B, C, and D of the year with 75% frequency.

Pt.	MinW1	MinR1	MaxW1	MaxR1	aE1	bE1	cE1	aS1	bS1	cS1	SL1	MinW2
A	5.14	3.14	8.09	729.09	0.6	0.44	2.89	0.7	−0.95	1.89	102.51	5.13
B	5.9	5.58	8.03	764.83	0.93	−0.35	1.83	0.74	0.99	0.82	102.99	5.33
C	5.48	1.71	8.09	705.23	0.62	−0.8	1.83	0.32	0.05	4.81	102.99	5.66
D	5.35	2.72	8.09	732.35	0.47	0.38	2.32	0.4	−0.07	2.61	102.55	5.4

Pt.	MinR2	MaxW2	MaxR2	aE2	bE2	cE2	aS2	bS2	cS2	SL2	aE3	bE3
A	8.09	8.06	712.42	0.35	0.01	2.5	0.52	−0.95	0.61	97.72	0.2	−0.7
B	6.5	8.09	714.88	0.79	−0.01	2.5	0.63	−0.7	3.09	98.03	0.34	0.83
C	7.6	8.1	776.57	0.82	0	1.08	0.76	0.16	3.69	99.14	0.83	0.1
D	5.53	8.07	737.35	0.63	−0.45	2.55	0.47	0.17	2.45	98.49	0.51	0

Pt.	cE3	aS3	bS3	cS3	SL3	aE4	bE4	cE4	aS4	bS4	cS4	SL4
A	5.06	0.79	0.23	1.03	103.44	0.15	−0.97	3.58	0.05	−0.39	2.19	96.99
B	3.77	0.04	0.38	4.64	102.71	0.71	−0.08	2.77	0.49	−0.05	3.99	98.17
C	1.54	0.66	−0.38	2.65	102.97	0.49	−0.78	2.92	0.77	−0.44	1.26	97.09
D	2.04	0.48	0	2.13	62.4	0.23	−0.18	1.27	0.13	0.22	0.74	97.65

3.4. Monte Carlo Simulation

In the above section, we use the inflow data of a given year to demonstrate the efficiency of the model by generating a set of optimized parameters. However, the obtained optimal parameters for the piecewise-linear multi-objective hedging rule (PMHR) are not applicable for other streamflow conditions. To obtain the optimal PMHR in the context of hydrological uncertainty, a Monte Carlo simulation is applied in this section to deal with different inflow conditions, rather than the inflow of one year.

Based on 29 years of historical hydrological data, a synthetic daily inflow for 100 years was generated, according to the statistical characteristics of the 29-year inflow. To be more specific, the annual mean discharge of these 100 years were obtained by a P-III curve based on samplings from the 29-year historical inflow from 1976–2005. The daily inflow of a synthetic year was decomposed from its annual discharge, according to the historical daily inflow of the year in which the frequency is equal to the synthetic year. The statistical properties of the historical and synthetic flow series is shown in Table 3. Here, the coefficient of variation (C_V) is defined as the ratio of the standard deviation to the mean. It reflects the extent of variability in relation to the mean of inflow. The deviation coefficient (C_S) is measured as the ratio of the difference between the mean and median value of standard deviation. It reflects the skewness of the mean of the inflow.

Table 3. Statistical parameters of the historical inflow series and the synthetic flow series.

	Average Annual Discharge	C_v	C_s
historical inflow series	24.37	0.33	0.83
synthetic flow series	18.70	0.40	0.74

In this paper, the main purposes of synthetic inflow generation are (1) to generate more input scenarios that are close to the historical inflow for the Monte Carlo simulation of the optimal operation model, and (2) to test the effectiveness of the model under the typical frequency of these inflow scenarios. The synthetic flow series can be seen as the input series of a mathematical experiment, while results and discussions are mostly based on the synthetic flow series.

For 100 years of daily inflow data, the optimization model is applied 100 times, and 100 sets of Pareto-optimal frontiers and 100 sets of optimized parameters are calculated correspondingly. In this way, through a Monte Carlo simulation, each set of the optimized parameter forms a possible operation rule for the case reservoir. These 100 groups of parameters represent the optimal operation rules under 100 various inflow scenarios. For each Pareto-optimal frontier, the expected and median values represent the entire probability distribution under different conditions. By using the expected and median values of parameters as the operation rules, optimal economic and ecological water release can be derived under any inflow conditions for decision-makers.

Searching directions used to obtain points B and C in Figure 5 are applied in the model with 100 years' synthetic daily inflow. In this way, 200 groups of optimal decision variables with different searching directions (similar to the method used to obtain points B and C) are obtained. For each searching direction, we calculated the expected and median values of the optimal decision variables (PMHR$_B$ and PMHR$_C$), which represent the optimal operation rules for the case reservoir. Table 4 shows the statistical values of the optimal parameters for 100 years.

Table 4. Statistical Analysis for the Parameters of Counterbalance Solutions.

Parameters		MinW1	MinR1	MaxW1	MaxR1	aE1	bE1	cE1	aS1	bS1	cS1	SL1	MinW2
PMHR$_B$	Expected	5.43	4.84	8.08	736.31	0.55	0.07	2.37	0.58	0.05	3.2	102.69	5.46
	Median	5.44	5.06	8.08	737.19	0.57	0.24	2.52	0.57	0.1	3.41	102.71	5.43
	SD	0.23	2.13	0.02	19.63	0.25	0.47	1.21	0.17	0.49	1.4	0.44	0.23
	Skew	0.26	−0.16	0.09	0.17	−0.18	−0.38	0.3	0.1	0.15	−0.35	−0.64	0.01
PMHR$_C$	Expected	5.44	3.84	8.08	743.23	0.54	0.18	2.66	0.52	0.02	2.85	102.68	5.41
	Median	5.44	4.27	8.08	741.9	0.56	0.28	2.78	0.57	0.1	3	102.67	5.43
	SD	0.25	2.05	0.02	18.96	0.27	0.46	1.16	0.24	0.42	1.26	0.43	0.23
	Skew	0.18	0.05	0.01	0.29	−0.36	−0.53	0.25	−0.46	−0.77	0.04	−0.67	0.05

Parameters		MinR2	MaxW2	MaxR2	aE2	bE2	cE2	aS2	bS2	cS2	SL2	aE3	bE3
PMHR$_B$	Expected	5.21	8.08	739.83	0.55	−0.13	2.69	0.48	0.1	3.22	98.48	0.45	0.02
	Median	5.47	8.08	739.72	0.59	−0.03	2.63	0.44	0.19	3.2	98.55	0.4	−0.01
	SD	2	0.02	19.61	0.24	0.2	1.33	0.24	0.5	0.35	0.35	0.21	0.49
	Skew	−0.32	−0.34	−0.18	−0.25	−1.41	−0.06	−0.01	−0.32	−0.43	−0.43	0.74	0.18
PMHR$_C$	Expected	4.47	8.08	743.41	0.54	−0.27	2.58	0.46	0.33	2.67	98.45	0.45	0.08
	Median	4.33	8.08	742.96	0.53	−0.2	2.49	0.44	0.39	2.72	98.51	0.45	0.07
	SD	2.07	0.03	16.29	0.25	0.3	1.11	0.21	0.41	1.26	0.38	0.21	0.45
	Skew	0.13	0.23	−0.14	−0.17	−0.61	0.11	−0.08	−1.27	0.09	−0.29	0.65	−0.04

Parameters		cE3	aS3	bS3	cS3	SL3	aE4	bE4	cE4	aS4	bS4	cS4	SL4
PMHR$_B$	Expected	2.57	0.5	0.03	2.96	102.71	0.4	−0.36	1.87	0.42	0.04	2	97.68
	Median	2.46	0.52	0.02	3.01	102.06	0.32	−0.24	1.82	0.42	0.11	1.96	97.7
	SD	0.98	0.27	0.48	1.29	0.36	0.28	0.37	1.19	0.27	0.43	1.29	0.48
	Skew	0.63	0.23	−0.02	0.1	−0.83	0.5	−0.36	0.63	0.18	−0.07	0.47	−0.36
PMHR$_C$	Expected	2.76	0.47	0.08	2.82	102.08	0.35	−0.37	2	0.4	0.25	1.93	97.55
	Median	2.65	0.43	0.15	2.99	102.2	0.23	−0.34	1.75	0.37	0.18	1.78	97.65
	SD	0.87	0.24	0.54	1.22	0.36	0.26	0.28	1.34	0.25	0.38	1.31	0.45
	Skew	0.16	0.57	−0.47	0	−0.58	0.72	−0.22	0.55	0.24	−0.4	0.49	−0.53

4. Discussion

To test the effectiveness and the robustness of the proposed model, $PMHR_B$ and $PMHR_C$ in Table 4 were used to generate the economic and ecological water release by which the corresponding objectives were derived. Using the 100 years of synthetic inflows generated in Section 3.4, the economic and ecological water release processes, as well as the corresponding objectives under different inflow scenarios by the two PMHR rules, were calculated. For comparative study, the economic and ecological water release and their objectives were calculated by the traditional and simplest operating rule SOP, which meant releasing water as close to the delivery targets as possible in order to meet the demand.

4.1. Economic Versus Ecological Objectives under Different Inflow Scenarios

To demonstrate the effectiveness of the HR rules comparing with the SOP rule, the values of economic and ecological objectives optimized by the $PMHR_B$, the $PMHR_C$, and the SOP are compared in Table 5.

Table 5. Comparison of economic and ecological objectives.

No.	Frequency	Rules	Economic Objective		Ecological Objective	
			Expected	Median	Expected	Median
1	95%	SOP	0.99		238.65	
		$PMHR_B$	0.57	0.58	17.89	20.38
		$PMHR_C$	0.6	0.64	88.02	229.25
2	90%	SOP	1.00		165.64	
		$PMHR_B$	0.89	0.83	12.86	11.97
		$PMHR_C$	0.92	0.95	43.29	61.25
3	75%	SOP	1.00		148.68	
		$PMHR_B$	0.90	0.82	7.15	10.1
		$PMHR_C$	0.95	0.97	13.43	21.36
4	50%	SOP	1.00		114.05	
		$PMHR_B$	0.99	0.99	3.53	7.23
		$PMHR_C$	1.00	1.00	4.97	18.08
5	25%	SOP	1.00		193.8	
		$PMHR_B$	0.99	0.99	3.77	6.37
		$PMHR_C$	1.00	1.00	2.66	7.23
6	10%	SOP	1.00		137.92	
		$PMHR_B$	1.00	0.99	5.21	3.81
		$PMHR_C$	1.00	1.00	28.01	52.21

According to Equations (5) and (6), the range of ecological objectives is greater than or equal to 0, and a greater value means a less satisfied ecologic objective. According to Equation (7), the range of economic objectives is from 0 to 1, where a greater value means a higher water supply deficit. With the annual inflow data ranging from a frequency of 10% to 95%, two objective values are compared, as shown in Table 3.

The SOP can satisfy almost all of the economic water demands under a majority of scenarios, but with a greater alteration to the natural flow regime. For the water supply objective (economic objective), the SOP guarantees meeting 100% of the water demands, except in extremely dry years (frequency = 95%). At the same time, the corresponding ecologic objective is greater than 100, which implies a highly altered flow regime. Meanwhile, $PMHR_B$ and $PMHR_C$ are able to effectively improve the ecological objective by decreasing the ecologic objectives to the range of 2.66–52.21, and satisfy the economic water demands in wet years (frequency < 50%). In dry years, $PMHR_B$ and $PMHR_C$ have to reduce the water supply for human demand in order to meet the demands of the ecosystem. For a typical dry year (when the inflow frequency is 75%), $PMHR_B$ and $PMHR_C$ each reduce nearly 10% of the water supply for human demands in order to balance the ecological objective. When it comes to the extreme dry year (frequency > 95%), PMHR$_B$ and PMHR$_C$ each reduce about 50% of human water supply to meet the demands of ecosystem.

4.2. Ecological Release

Under the synthetic inflow conditions, the processes of ecological release in a typical dry year (75% frequency), generated by three rules (SOP, PMHR$_B$ and PMHR$_C$), were compared, as shown in Figures 6 and 7. The processes generated by the expected values of PMHR$_B$ and PMHR$_C$ are shown in Figure 6; processes generated by the median values of PMHR$_B$ and PMHR$_C$ are shown in Figure 7. Compared to the SOP, the ecological release processes generated by PMHR$_B$ and PMHR$_C$ are closer to the natural inflow process. The similarity of the ecological water release process to the inflow process can be quantified by the correlation coefficient. The higher the correlation coefficient is, the closer the release is to the inflow of the reservoir. This means that the smaller hydrologic alteration is made through reservoir operations. Using Figure 6 as an example, the correlation coefficient of the inflow and water release under the SOP is 0.54, while the correlation coefficient of inflow and water release under the PMHR$_B$ and PMHR$_C$ are 0.81 and 0.66, respectively.

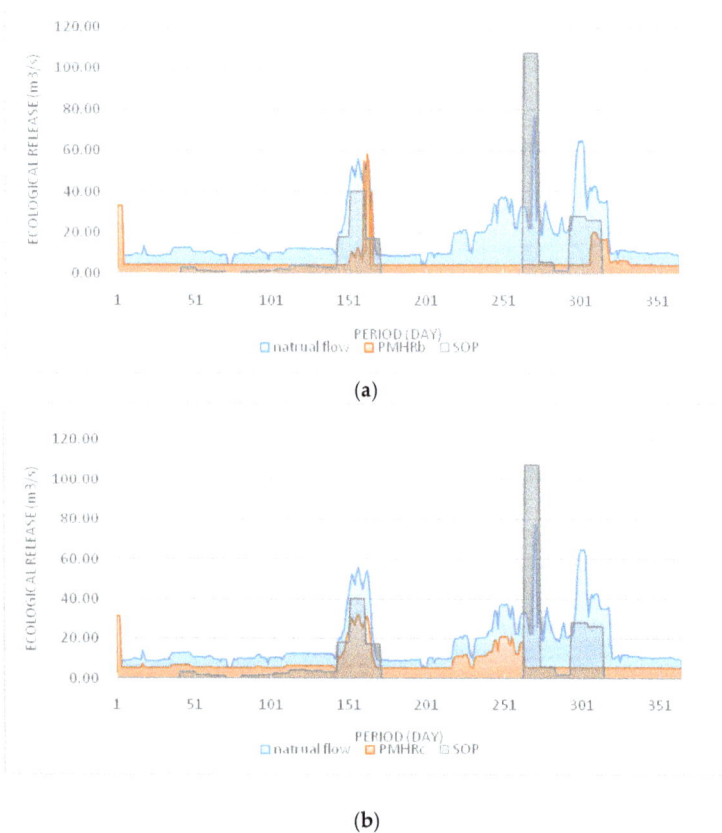

(a)

(b)

Figure 6. Comparison of release processes between different operating rules, calculated by expectation. (a) Comparision of release process of natural flow, PMHR$_B$ and SOP; (b) Comparision of release process of natural flow, PMHR$_C$ and SOP.

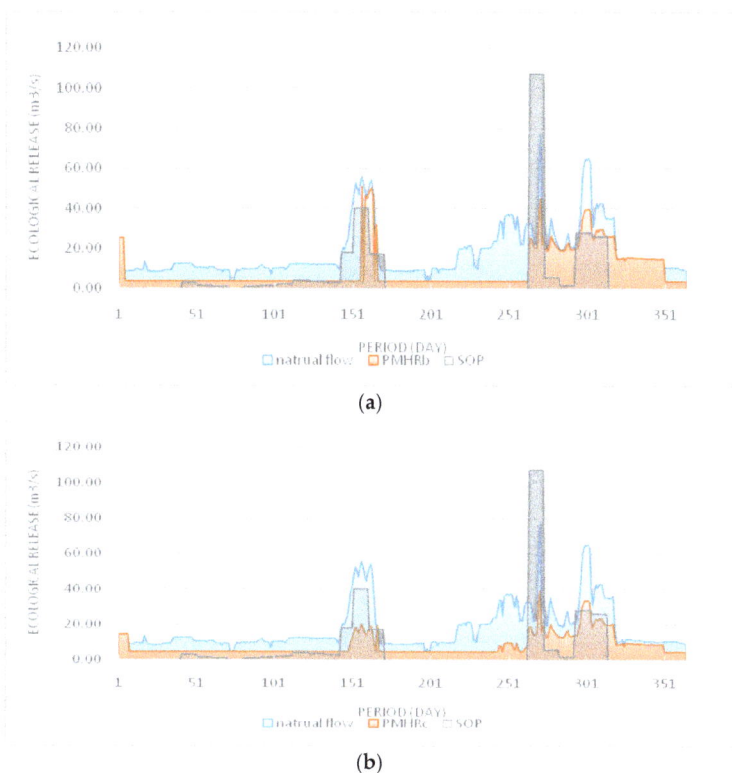

Figure 7. Comparison of release processes between different operating rules, calculated by median values. (**a**) Comparision of release process of natural flow, PMHR$_B$ and SOP; (**b**) Comparision of release process of natural flow, PMHR$_C$ and SOP.

4.3. Monte Carlo Simulation Key Indicators Analysis

Figures 6 and 7 demonstrate that reservoir water release alters the flow regime of the river course. This will threaten fish communities and the integrity of river ecosystems downstream. Compared to the operation results under the SOP, the ecological release under the PMHR$_B$ and the PMHR$_C$ can recover the altered flow regime significantly.

The most altered IHA indicators under the SOP are in Table 6, as well as the indicators under PMHR$_B$ and PMHR$_C$. The upper and lower limits (25% and 75% frequency of natural distribution, respectively) of each indicator are also listed for reference. If the value of indicators falls within the range of the upper and lower limits, it means a more acceptable ecological condition for sustainability (as the bold number in Table 6).

Table 6. Comparison of indicator analysis of standard operating policy (SOP) and the piecewise-linear multi-objective hedging rule (PMHR).

	October	November	December	January	April	May	July	August	September	30 Day-Min	90 Day-Min	Low Duration	Reversal	Rise Rate	Fall Rate
SOP	0	1.69	0.92	2.92	0	0	12.6	8.49	0	0	0	74	5	3.37	−0.57
PMHR$_B$	7.03	4.84	4.84	4.84	4.84	4.84	18.74	14.8	6.94	4.84	4.84	20.31	68	0.2	−0.14
PMHR$_C$	6.65	3.84	3.84	3.84	3.84	3.84	25.03	21.04	9.53	3.84	3.84	21.5	47	0.78	−0.14
Upper limit	19.81	20.52	20.71	14.59	17.81	23.1	38.91	48.18	25.75	7.28	10.08	28.75	135	0.52	−0.11
Lower limit	8.21	4.26	5.02	6.6	2.16	8.2	15.97	19.29	7.79	0.84	0.91	6	81	0.19	−0.18

4.4. Extended Analysis: Impact of Long-Term Forecast Information

In this section, the impacts of using long-term annual forecast inflows in an operation are discussed preliminarily. We suppose that the annual forecast information, which simply indicates the inflow of the next year and is either larger (wet year) or smaller (dry year) than the average level, is known in advance. Thus, the synthetic inflows used by the Monte Carlo simulation can be divided into two parts: the inflow in wet years and the inflow in dry years. By using the inflow of wet or dry years instead of all the synthetic inflows, the parameters of PMHR can be optimized in two groups: parameters adapted to wet years and parameters adapted to dry years. The expected value of these parameters can be used for wet and dry years, respectively.

Figure 8 shows the environmental flow release using rules in both wet and dry years with long-term forecast information, as well as the release using rules of all possible hydrological years without long-term forecast information. Figure 8a presents the results of a typical wet year (frequency = 25%), in which the correlation coefficient between the water release process and the inflow process is 0.95 with forecast information, but 0.90 without forecast. Figure 8b presents the results of a typical dry year, in which the correlation coefficient is 0.66 with forecast information and 0.61 without forecast information. This indicates that the PMHR combined with forecast information is able to improve the restoration of an ecological flow regime compared to one without the forecast.

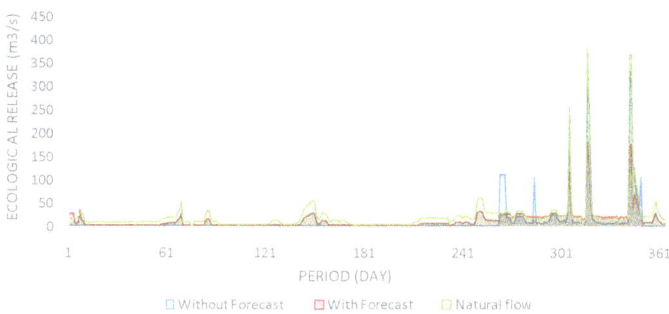

(a)

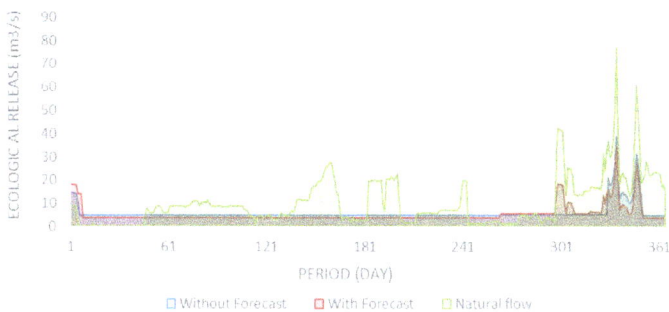

(b)

Figure 8. Comparison of ecological releases, between those with and without a long-term forecast. (a) Comparison of wet years; (b) Comparison of dry years.

5. Conclusions

This study adopts the restoration of natural stream flows as an objective of reservoir operation. An optimization model is proposed by using piecewise-linear multi-objective hedging rules (PMHR) to balance the ecological objective of recovering the natural flow regime and the economic objective of satisfying human consumptive water demand. The proposed model can reflect the daily interval variations of the flow by engaging real-time daily inflow forecasts and storage information into decision-making. The parameters of the PMHR were optimized, and Pareto frontiers were discovered by a vector evaluated genetic algorithm. A Monte Carlo simulation was used to deal with inflow uncertainty during real operation.

The results of Baiguishan Reservoir show the balance between ecological and economic objectives. With the consideration of ecological objectives, the ecological release of a reservoir was restored, reflected by the decrease of IHA to different degrees. To deal with inflow uncertainty, synthetic inflow was generated and used in a Monte Carlo simulation under uncertain hydrological conditions. To test the effectiveness of the model, a set of typical frequency inflows was selected for analysis. The model with PMHR has improved the ecological objective and guaranteed the water supply under most of the hydrological conditions in the case reservoir. The alteration degree of the hydrologic indicators, which has been seriously altered, was recovered into the acceptable range. The impact of involving long-term hydrologic forecast information on improving reservoir operation is also demonstrated by the comparisons of two hydrological conditions (wet and dry years) in the case reservoir. The application of forecast inflow information can obviously improve the scheme of ecological water release.

Author Contributions: The conceptualization and framework of this article are derived from J.Z. Model computation and result analysis are completed by Y.L., supervised by J.Z. and H.Z. The article is modified and reviewed by H.Z.

funding: This research was funded by the National Key Research and Development Program of China (2017YFC0404403 and 2016YFC0401302) and the National Natural Science Foundation of China (91747208 and 51579129).

Acknowledgments: We are grateful to the editors and the three anonymous reviewers for their constructive comments and detailed suggestions, which helped us to substantially improve the paper.

Conflicts of Interest: The authors declare no conflicts of interest.

References

1. Zhong, Y.; Power, G. Environmental impacts of hydroelectric projects on fish resources in China. *Regul. River* **1996**, *12*, 81–98. [CrossRef]
2. Joy, M.K.; Death, R.G. Control of fish and cray fish community structure in Taranaki, New Zealand: Dams, Diadromy or Habitat Structure. *Freshw. Biol.* **2001**, *46*, 417–429. [CrossRef]
3. Homa, E.S.; Vogel, R.M.; Smith, M.P.; Apse, C.D.; Huber-Lee, A.; Seiber, J. An optimization approach for balancing human and ecological flow needs. In *World Water and Environmental Resources Congress*; ASCE: Anchorage, AK, USA, 2005.
4. Richter, B.D.; Baumgartner, J.V.; Wigington, R.; Braun, D.P. How much water does a river need? *Freshw. Biol.* **1997**, *37*, 231–249. [CrossRef]
5. Rehn, A.C. Benthic macroinvertebrates as indicators of biological condition below hydropower dams on West Slope Sierra Nevada Streams, California, USA. *River Res. Appl.* **2009**, *25*, 208–228. [CrossRef]
6. Yin, X.A.; Yang, Z.F.; Petts, G.E. A new method to assess the flow regime alterations in riverine ecosystems. *River Res. Appl.* **2015**, *31*, 497–504. [CrossRef]
7. Poff, N.L.R.; Allan, J.D.; Bain, M.B.; Karr, J.R.; Prestegaard, K.L.; Richter, B.D.; Sparks, R.E.; Stromberg, J.C. The natural flow regime. *BioScience* **1997**, *47*, 769–784. [CrossRef]
8. Tharme, R.E. A global perspective on environmental flow assessment: Emerging trends in the development and application of environmental flow methodologies for rivers. *River Res. Appl.* **2003**, *19*, 397–441. [CrossRef]

9. Olden, J.D.; Poff, N.L. Redundancy and the choice of hydrologic indices for characterizing streamflow regimes. *River Res. Appl.* **2003**, *19*, 101–121. [CrossRef]

10. Petts, G.E. Instream flow science for sustainable river management. *J. Am. Water Resour. Assoc.* **2009**, *45*, 1071–1086. [CrossRef]

11. Poff, N.L. Managing for variability to sustain freshwater ecosystems. *J. Am. Water Resour. Plan. Manag.* **2009**, *135*, 1–4. [CrossRef]

12. Bunn, S.E.; Arthington, A.H. Basic principles and ecological consequences of altered flow regimes for aquatic biodiversity. *Environ. Manag.* **2002**, *30*, 492–507. [CrossRef]

13. Bayley, P.B. The Flood Pulse Advantage and the Restoration of River-floodplain Systems. *River Res. Appl.* **1991**, *6*, 75–86. [CrossRef]

14. Sparks, R.E. Need for ecosystem management of large rivers and floodplains. *BioScience* **1995**, *45*, 168–182. [CrossRef]

15. Jager, H.I.; Smith, B.T. Sustainable reservoir operation: Can we generate hydropower and preserve ecosystem values? *River Res. Appl.* **2008**, *24*, 340–352. [CrossRef]

16. Suen, J.P.; Eheart, J.W. Reservoir management to balance ecosystem and human needs: Incorporating the paradigm of the ecological flow regime. *Water Resour. Res.* **2006**, *42*, W03417. [CrossRef]

17. Cardwell, H.; Jager, H.I.; Sale, M.J. Designing instream flows to satisfy fish and human water needs. *J. Am. Water Resour. Plan. Manag.* **1996**, *122*, 356–363. [CrossRef]

18. Ripo, G.C.; Jacobs, J.M.; Good, J.C. An algorithm to integrate ecological indicators with streamflow withdrawals. In *ASCE/EWRI World Water and Environmental Resources Congress*; American Society of Civil Engineers: Philadelphia, PA, USA, 2003.

19. Guo, W.X.; Wang, H.X.; Xu, J.X.; Xia, Z.Q. Ecological operation for Three Gorges Reservoir. *Water Sci. Eng.* **2011**, *4*, 143–156. [CrossRef]

20. Yang, Y.C.E.; Cai, X.M. Reservoir reoperation for fish ecosystem restoration using daily inflows-case study of Lake Shelbyvile. *J. Am. Water Resour. Plan. Manag.* **2011**, *137*, 470–480. [CrossRef]

21. Shiau, J.T.; Wu, F.C. Optimizing environmental flows for multiple reaches affected by a multipurpose reservoir system in Taiwan: Restoring natural flow regimes at multiple temporal scales. *Water Resour. Res.* **2013**, *49*, 565–584. [CrossRef]

22. Cai, W.J.; Zhang, L.L.; Zhu, X.P.; Zhang, A.J.; Yin, J.X.; Wang, H. Optimized reservoir operation to balance human and environmental requirements: A case study for the Three Gorges and Gezhouba Dams, Yangtze River basin, China. *Econ. Inform.* **2013**, *18*, 40–48. [CrossRef]

23. Lane, B.A.; Sandoval-Solis, S.; Porse, E.C. Environmental flows in a human-dominated system: Integrated water management strategies for the Rio Grande/Bravo Basin. *River Res. Appl.* **2015**, *31*, 1053–1065. [CrossRef]

24. Yin, X.A.; Yang, Z.F.; Petts, G.E. Optimizing environmental flows below dams. *River Res. Appl.* **2012**, *28*, 703–716. [CrossRef]

25. Haghighi, A.T.; Klove, B. Development of monthly optimal flow regimes for allocated environmental flow considering natural flow regimes and several surface water protection targets. *Ecol. Eng.* **2015**, *82*, 390–399. [CrossRef]

26. Hui, W.; Brill, E.D.; Ranjithan, R.S.; Sankarasubramanian, A. A framework for incorporating ecological release in single reservoir operation. *Adv. Water Resour.* **2015**, *78*, 9–21. [CrossRef]

27. Yin, X.A.; Yang, Z.F.; Petts, G.E. Reservoir operating rules to sustain environmental flows in regulated rivers. *Water Resour. Res.* **2011**, *47*, W08509. [CrossRef]

28. Zhu, X.P.; Zhang, C.; Yin, J.X.; Zhou, H.C.; Jiang, Y.Z. Optimization of Water Diversion Based on Reservoir Operating rules: Analysis of the Biliu River Reservoir, China. *J. Hydrol. Eng.* **2014**, *19*, 411–421. [CrossRef]

29. Tilmant, A.; Beevers, L.; Muyunda, B. Restoring a flow regime through the coordinated operation of a multireservoir system: The case of the Zambezi River basin. *Water Resour. Res.* **2010**, *46*, W07533. [CrossRef]

30. Chen, Q.; Chen, D.; Li, R.; Ma, J.; Blanckaert, K. Adapting the operation of two cascaded reservoirs for ecological flow requirement of a de-watered river channel due to diversion-type hydropower stations. *Ecol. Model.* **2013**, *252*, 266–272. [CrossRef]

31. Koutsoyiannis, D.; Economou, A. Evaluation of the parameterization-simulation-optimization approach for the control of reservoir systems. *Water Resour. Res.* **2003**, *39*, WR00214. [CrossRef]

32. Hashimoto, T.; Stedinger, J.R.; Loucks, D.P. Reliability, resilience vulnerability criteria for water resources system performance evaluation. *Water Resour. Res.* **1982**, *18*, 14–20. [CrossRef]

33. Drapper, A.J.; Lund, J.R. Optimal hedging and carryover storage value. *J. Am. Water Resour. Plan. Manag.* **2004**, *130*, 83–87. [CrossRef]

34. Zhao, J.S.; Cai, X.M.; Wang, Z.J. Optimality conditions for a two-stage reservoir operation problem. *Water Resour. Res.* **2011**, *47*, W08503. [CrossRef]

35. Loucks, D.P.; Stedinger, J.R.; Haith, D.A. *Water Resource Systems Planning and Analysis*; Prentice-Hall: Englewood Cliffs, NJ, USA, 1981.

36. Tejada-Guibert, J.A.; Johnson, S.A.; Stedinger, J.R. Comparison of two approaches for implementing multireservoir operating policies derived using stochastic dynamic programming. *Water Resour. Res.* **1993**, *29*, 3969–3980. [CrossRef]

37. Karamouz, M.; Houck, M.H.; Delleur, J.W. Optimization and simulation of multiple reservoir system. *J. Am. Water Resour. Plan. Manag.* **1992**, *118*, 71–81. [CrossRef]

38. Tu, M.Y.; Hsu, N.S.; Tsai, T.C.; Yeh, W.W.G. Optimization of Hedging Rules for Reservoir Operations. *J. Am. Water Resour. Plan. Manag.* **2008**, *134*, 3–13. [CrossRef]

39. Taghian, M.; Dan, R.; Haghighi, A.; Madsen, H. Optimization of conventional rule curves coupled with hedging rules for reservoir operation. *J. Am. Water Resour. Plan. Manag.* **2014**, *140*, 693–698. [CrossRef]

40. Shiau, J.T. Analytical optimal hedging with explicit incorporation of reservoir release and carryover storage targets. *Water Resour. Res.* **2011**, *47*, 238–247. [CrossRef]

41. Gao, Y.X.; Vogel, R.M.; Kroll, C.N.; Poff, N.L.; Olden, J.D. Development of representative indicators of hydrologic alteration. *J. Hydrol.* **2009**, *374*, 136–147. [CrossRef]

42. Zhao, T.; Zhao, J.S.; Yang, D.W.; Wang, H. Generalized martingale model of the uncertainty evolution of streamflow forecast. *Adv. Water Resour.* **2013**, *57*, 41–45. [CrossRef]

43. Schaffer, J. Multiple objective optimization with vector evaluated genetic algorithms. In Proceedings of the 1st International Conference on Genetic Algorithms, Pittsburgh, PA, USA, 24–26 July 1985; Psychology Press: Hillsdale, NJ, USA, 1985; pp. 93–100.

water

MDPI

Article

Long-Term Scheduling of Large-Scale Cascade Hydropower Stations Using Improved Differential Evolution Algorithm

Xiaohao Wen [1,2], Jianzhong Zhou [1,3,*], Zhongzheng He [1,3] and Chao Wang [4]

[1]   School of Hydropower and Information Engineering, Huazhong University of Science and Technology, Wuhan 430074, China; wen_xiaohao@ctg.com.cn (X.W.); hezz_hae@hust.edu.cn (Z.H.)
[2]   China Three Gorges Corporation, Beijing 100038, China
[3]   Hubei Key Laboratory of Digital Valley Science and Technology, Wuhan 430074, China
[4]   China Institute of Water Resources and Hydropower Research, Beijing 100038, China; wangchao@iwhr.com
*   Correspondence: jz.zhou@mail.hust.edu.cn; Tel.: +86-027-8754-3127

Received: 1 March 2018; Accepted: 20 March 2018; Published: 26 March 2018

Abstract: Long-term scheduling of large cascade hydropower stations (LSLCHS) is a complex problem of high dimension, nonlinearity, coupling and complex constraint. In view of the above problem, we present an improved differential evolution (iLSHADE) algorithm based on LSHADE, a state-of-the-art evolutionary algorithm. iLSHADE uses new mutation strategies "current to pbest/2-rand" to obtain wider search range and accelerate convergence with the preventing individual repeated failure evolution (PIRFE) strategy. The handling of complicated constraints strategy of ε-constrained method is presented to handle outflow, water level and output constraints in the cascade reservoir operation. Numerical experiments of 10 benchmark functions have been done, showing that iLSHADE has stable convergence and high efficiency. Furthermore, we demonstrate the performance of the iLSHADE algorithm by comparing it with other improved differential evolution algorithms for LSLCHS in four large hydropower stations of the Jinsha River. With the applications of iLSHADE in reservoir operation, LSLCHS can obtain more power generation benefit than other alternatives in dry, normal, and wet years. The results of numerical experiments and case studies show that the iLSHADE has a distinct optimization effect and good stability, and it is a valid and reliable tool to solve LSLCHS problem.

Keywords: energy; hydropower stations; differential evolution algorithm; optimal scheduling; ε-constrained method

1. Introduction

Hydropower has a significant share on the total energy consumption as it is renewable, clean, and cheap. Therefore, many countries have been working on the development of the utility of hydropower [1], and many hydropower plants have been put into operation in the past few decades [2–6]. Large cascade hydropower stations (LHS) play an increasingly important role in energy production. Many scholars have conducted a lot of research on the water resources management of LHS. Zhou et al. [7] proposed a joint optimal refill rules for cascade reservoirs to solve the conflict between the flood control and refill operation. The energy storage operation chart combined with discriminant coefficient method was put forward by Jiang [8], which was successfully applied to cascade reservoirs of Li Xianjiang River in southwest China. Regarding the input (e.g., inflow) imprecision and uncertainties, Chen et al. [9–12] analyzed the influence of the uncertainty in water resources management and the distribution of flood forecasting error. Djebou et al. [13,14] presented the interactions between these hydrologic factors that interplay at the watershed scale using the entropy-based index. Aiming to determine the optimal strategy that hedges the risk of energy shortfall, Xu et al. [15] develops a multi-objective stochastic programming model for informing hedging decisions for hydropower operations. Due to complex hydrodynamic relation, various complex constraints and diversified interest demand, the long-term scheduling of large cascade hydropower stations (LSLCHS) has developed into a multi-dimensional, non-convex and non-linear optimization problem. Correspondingly, optimization of LSLCHS has become a fairly challenging theoretical and practical problem, which is urgent for optimization techniques and constraints treatment [16–18]. Over many years, to solve this problem, various optimization algorithms have been applied. Usually there are two categories of methods: traditional optimizers and modern heuristic algorithms. Traditional optimizers include linear programming (LP) [19,20], nonlinear programming (NLP) [21,22], dynamic programs (DP) [23,24], progressive optimality algorithm (POA) [25,26], etc. These algorithms have rigorous mathematical foundations but low convergence efficiency. They suffer from curse of dimensionality. Modern heuristic algorithms use intelligent strategies to guide search to better areas, such as particle swarm optimization (PSO) [27,28], genetic algorithm (GA) [29], cultural algorithm (CA) [30], binary artificial sheep algorithm (BASA) [31], ant colony optimization (ACO) [18,32], etc. Compared with traditional optimizers, modern meta-heuristics are significantly more flexible and have high search efficiency as the meta-heuristics are inspired by different nature principles from biology, ethology, or physics. However, the common heuristic algorithms have some disadvantages such as premature convergence because of local fast convergence, and bad local search capability owing to many global searches. Moreover, they lack effective measures to handle complex constraints, making it difficult to be applied to solve complex optimal problems with high dimensions such as LSLCHS.

Differential evolution (DE) is a simple yet practical modern heuristic algorithm for global optimization over continuous spaces introduced by Price and Storn [33]. The DE algorithm has been used in many practical cases [34,35] and gradually become more popular. Similar to all other modern heuristic algorithms, the evolutionary process of DE uses mutations, crossover, and selection operators at each generation to reach the global optimum. The performance of DE basically depends on the mutation strategy, the crossover operator. Besides, the intrinsic control parameters (population size NP, scaling factor F, the crossover rate Cr) play a vital role in balancing the diversity of population and convergence speed of the algorithm. Therefore, Brest et al. [36] proposed a self-adaptive DE (jDE), in which both F and Cr are applied at random with probability $\tau 1$ and $\tau 2$. SaDE is proposed by Qin et al. [37] adaptively adjusts its trial vector generation strategies and control parameters simultaneously by learning from the previous search. JADE [38] is a well-known, effective DE variant which employs a control parameter adaptation mechanism and puts forward mutation strategy "current-to-pbest/1", differential evolution with composite trial vector generation strategies, control parameters (CoDE) [39], differential evolution with ensemble of parameters and mutation strategies (EPSDE) [40]. Success-History-based Adaptive DE (SHADE) [41] is an improved version of JADE which uses a different parameter adaptation mechanism. LSHADE [42] further extends SHADE with Linear Population Size Reduction (LPSR), which continually decreases the population size according to a linear function. In addition, LSHADE is the best ranked DE algorithm on CEC2014 Competition on Real-Parameter Single Objective Optimization [43] (see http://www.ntu.edu.sg/home/EPNSugan/index_files/CEC2014/-CEC2014.htm).

To avoid premature convergence and to accelerate convergence, we present an improved version of the LSHADE algorithm in this paper, called iLSHADE. The main improvement of iLSHADE is that a new mutation strategy "current to pbest/2-rand" is put forward for wider search range to improve search ability and prevent individual repeated failure evolution (PIRFE) strategy applied in the population evolution process. Finally, iLSHADE is applied to LSLCHS in Jinsha River combined with the improved constraints handling technique. Results of the study demonstrates its superiority in dealing with LSLCHS problem.

The remainder of this paper is organized as follows: Section 2 introduces the formulation of LSLCHS problem. In Section 3, a brief view of DE framework and improvement strategies of iLSHADE is presented. Section 4 presents numerical simulation experiment of iLSHADE. In Section 5, implementation of iLSHADE in LSLCHS is shown in detail. In Section 6, iLSHADE is applied to solve LSLCHS, and the results are analyzed. Finally, conclusions are summarized in Section 7.

2. Optimization Model

The primary objective of LSLCHS problem is to maximize the total power generation of LHS over the whole operation periods, while subjecting to kinds of equality and inequality constraints. The objective formula and constraints are described as follows.

2.1. Objective Function

$$\text{obj} = max\sum\nolimits_{i=1}^{M}\sum\nolimits_{t=1}^{T}N_{it}\Delta t, \; N_{it} = A_iH_{it}Q_{it} \tag{1}$$

where obj is the total power generation of LHS over the whole operation periods, M is the number of hydro plants; T is the whole periods; A_i is output coefficient of the i-th hydro plant; Δt shows interval of scheduling term; N_{it}, H_{it} and Q_{it} denote output, pure water head and water discharge through hydro-turbine of the i-th hydro plant in the t-th period, respectively. Moreover, H_{it} is calculated by upstream water level, trail water level and head loss shown in formula (8).

2.2. Constraints

In the process of long-term optimal dispatch, various complex equality and inequality constraints, such as water level, output, and hydraulic connection, should be considered for restricting the total power generation optimization. The constraints of LHS are described as follows:

1. Water balance constraint.

$$\begin{aligned} V_{i,t+1} &= V_{i,t} + (I_{i,t} - Q_{i,t} - S_{i,t})\Delta t, \\ I_{i,t} &= q_{i,t} + Q_{i-1,t} + S_{i-1,t} \end{aligned} \tag{2}$$

$V_{i,t}$ is reservoir storage of the i-th hydropower station at the beginning of period t, $I_{i,t}$ is inflow, $q_{i,t}$ stands for local inflow and $S_{i,t}$ is deserted outflow.

2. Hydraulic connection.

$$Z_{i,t}^{down} = \begin{cases} F(Q_{i,t} + S_{i,t}) & \text{without backwater effect,} \\ F(Q_{i,t} + S_{i,t}, Z_{i+1,t}) & \text{with backwater effect.} \end{cases} \tag{3}$$

where $Z_{i,t}$ stands for upstream water level, $Z_{i,t}^{down}$ is trail water level. Function F represents the hydraulic connection between upstream and downstream hydropower stations. Generally, the trail water level is a function of outflow. However, when the hydropower station is located at the backwater region of its downstream hydropower station, the upstream water level of the downstream hydropower station must be taken into consideration in the function.

3. Water level constraint.

$$Z_{i,t}^{min} \le Z_{i,t} \le Z_{i,t}^{max} \tag{4}$$

$$|Z_{i,t} - Z_{i,t+1}| \le \Delta Z_i \tag{5}$$

$Z_{i,t}^{min}$ and $Z_{i,t}^{max}$ are the upper and lower water level limits and ΔZ_i is the maximum amplitude of water level variation.

4. Power generating constraint.

$$N_{i,t}^{min} \le N_{i,t} \le N_{i,t}^{max}(H_{i,t}) \tag{6}$$

$N_{i,t}^{max}(H_{i,t})$ represents the maximum output. The maximum output is a function of pure water head. $N_{i,t}^{min}$ is the lower output limit, which is generally called Guaranteed output.

5. Outflow constraint.

$$Q_{i,t}^{min} \leq Q_{i,t} + S_{i,t} \leq Q_{i,t}^{max} \tag{7}$$

$Q_{i,t}^{max}$ is the maximum outflow limit and $Q_{i,t}^{min}$ is the minimum outflow limit.

6. Water head equation.

$$H_{i,t} = (Z_{i,t} + Z_{i,t+1})/2 - Z_{i,t}^{down} - H_{i,t}^{loss}(Q_{i,t}) \tag{8}$$

$H_{i,t}$ stands for the pure water head. $Z_{i,t}^{down}$ is trail water level described in formula (3). $H_{i,t}^{loss}(Q_{i,t})$ represents water head loss, which is a function of outflow through hydro-turbines.

7. Boundary condition.

$$Z_{i,0} = Z_i^{begin}, \ Z_{i,T} = Z_i^{end} \tag{9}$$

where Z_i^{begin} and Z_i^{end} are initial water level and terminal water level of the i-th hydro plant, respectively.

3. Overview of iLSHADE

3.1. DE

DE is a group-based evolutionary algorithm. It is used to solve the following continuous domain global optimization problem:

$$\begin{aligned} & minimize \ f(\vec{x}), \ \vec{x} = (x_1, \dots, x_D) \\ & x_i \in [\overline{x}_i, \underline{x}_i] \ \forall i \in \{1, \dots, D\}, \ -\infty < \overline{x}_i < \underline{x}_i < +\infty \end{aligned} \tag{10}$$

where $f(\vec{x})$ is continuous fitness evaluation function, D is the dimension of the problem. DE has three control parameters that need to be set before the calculation: F is scaling factor, CR is crossover control parameter, and NP is population size. The framework of DE is as follows (Figure 1):

```
1:  Set the generation number  G  = 0
2:  Sample  NP  x⃗_{i,G} ,..., x⃗_{NP,G}  form an initial population
3:  while the termination criteria are not meet do
4:     for i  = 1 to  NP  do
5:        Mutation: generate a mutant individual  v⃗_{i,G}  using a DE mutation strategy
6:        Repair: correct  v⃗_{i,G}  to make it feasible if  v⃗_{i,G}  is not feasible,
7:        Crossover: mix  x⃗_{i,G}  and  v⃗_{i,G}  to generate a trial individual u⃗_{i,G}
8:        Selection: If  f(u⃗_{i,G})≤f(x⃗_{i,G}) , set  x⃗_{i,G+1} = u⃗_{i,G} , otherwise, set  x⃗_{i,G+1} = x⃗_{i,G}
9:     end for
10:    G = G + 1
11: end while
Output: the individual with the smallest objective function value in population
```

Figure 1. Pseudocode of DE.

The mutation strategy in original DE is "rand/1", which is expressed in formula (11):

$$\vec{v}_{i,G} = \vec{x}_{r1,G} + F \cdot (\vec{x}_{r2,G} - \vec{x}_{r3,G}) \tag{11}$$

Other common DE mutation strategies are as follows:

- "rand/2":

$$\vec{v}_{i,G} = \vec{x}_{r1,G} + F \cdot (\vec{x}_{r2,G} - \vec{x}_{r3,G}) + F \cdot (\vec{x}_{r4,G} - \vec{x}_{r5,G}) \tag{12}$$

- "best/1":

$$\vec{v}_{i,G} = \vec{x}_{best,G} + F \cdot (\vec{x}_{r1,G} - \vec{x}_{r2,G}) \tag{13}$$

- "best/2":

$$\vec{v}_{i,G} = \vec{x}_{i,G} + F \cdot (\vec{x}_{best,G} - \vec{x}_{i,G}) + F \cdot (\vec{x}_{r1,G} - \vec{x}_{r2,G}) \tag{14}$$

- "current to best/1":

$$\vec{v}_{i,G} = \vec{x}_{i,G} + F \cdot (\vec{x}_{best,G} - \vec{x}_{i,G}) + F \cdot (\vec{x}_{r1,G} - \vec{x}_{r2,G}) \tag{15}$$

where the indexes $r1 - r5$ represent the random and mutually different integers generated within the range $\{1, NP\}$, and also different from index i. $\vec{x}_{best,G}$ is the best individual in a current generation. Each strategy has a different ability to maintain the diversity of the population, which may increase/reduce the rate of convergence in the process of evolution.

3.2. iLSHADE

An improved LSHADE (iLSHADE) with new mutation strategy "current to best/2-rand" and the PIRFE strategy is proposed. The details of these strategies and algorithm procedure are shown below.

3.2.1. Mutation Strategy "Current to pbest/2-rand"

The mutation strategy "current to pbest/1" was proposed by in the framework of JADE (which is expressed in formula (16)).

$$\vec{v}_{i,G} = \vec{x}_{i,G} + F_i \cdot (\vec{x}_{pbest,G} - \vec{x}_{i,G}) + F_i \cdot (\vec{x}_{r1,G} - \vec{x}_{r2,G}) \tag{16}$$

In Equation (16), the individual $\vec{x}_{pbest,G}$ is randomly selected from the top $N \times p(p \in [0,1])$ members in the G-th generation. "current to pbest/1" depends on the control parameter p to balance exploitation and exploration (small p behaves more greedily). $\vec{x}_{i,G}$ and $\vec{x}_{r1,G}$ are selected from P in the same way as in Equation (12), while $\vec{x}_{r2,G}$ is randomly chosen from the union $P \cup A$, of the current population and the archive. We present an improved mutation strategy "current to pbest/2-rand" to improve the search range based on mutation strategy "current to pbest/1", which is expressed as follows:

$$\vec{v}_{i,G} = \vec{x}_{i,G} + F_i \cdot (\vec{x}_{pbest,G} - \vec{x}_{i,G}) + F_i \cdot [(\vec{x}_{r1,G} - \vec{x}_{r2,G}) \cdot rand_i + (\vec{x}_{r3,G} - \vec{x}_{r4,G}) \cdot (1 - rand_i)] \tag{17}$$

where $rand_i$ is a uniformly distributed random number between [0,1]. $\vec{x}_{i,G}$, $\vec{x}_{r1,G}$ and $\vec{x}_{r3,G}$ are selected randomly and different within the range $\{1, NP\}$ from P in the same way as in formula (16), while $\vec{x}_{r2,G}$ and $\vec{x}_{r4,G}$ is randomly chosen from the union, $P \cup A$, of the current population and the archive. The two mutation strategies are illustrated in Figure 2.

As seen in Figure 2a, $\vec{v}_{i,G}$ is the mutation individual generated for individual $\vec{x}_{i,G}$. According to the principle of vector addition, the position of $\vec{v}_{i,G}$ changes with the associated mutation factor F_i, and its position only exists on this "Search Line". Mutation strategy "current to pbest/2-rand" uses $rand_i$ and linear combination of $(\vec{x}_{r1,G} - \vec{x}_{r2,G}) \cdot rand_i + (\vec{x}_{r3,G} - \vec{x}_{r4,G}) \cdot (1 - rand_i)$ to expand the search range. By the varying value $rand_i$, $\vec{v}_{i,G}$ can search anywhere in the shaded triangle area with the change of the F_i and $rand_i$ (see Figure 2b). Obviously, the search range of "current to pbest/2-rand" is much larger than that of "current to pbest/1".

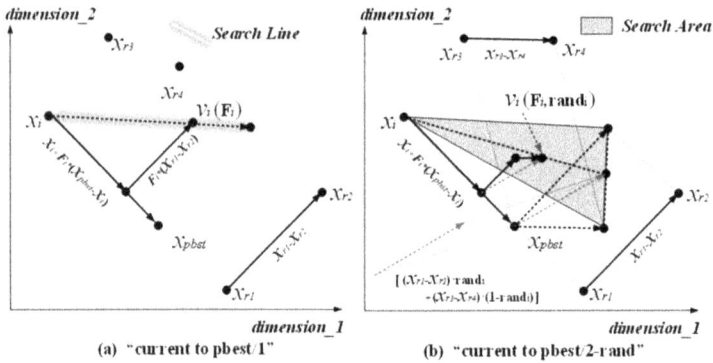

Figure 2. Illustration of the DE mutation strategy in two dimensions.

3.2.2. The PIRFE Strategy

The PIRFE strategy is proposed to avoid individuals in a local optimum lead useless evolution. When an individual falls into a local optimal, it is extremely difficult to produce an effective mutation to jump out of the local optimal. The individual failing evolution number (IFEN) is defined to record the number of individual failing evolution, and it is initialized to zero. $\vec{x}_{i,G}$ generates $\vec{u}_{i,G}$ through mutation and crossover. If $\vec{x}_{i,G+1}$ produces a failing evolution and $\vec{x}_{i,G+1}$ is better than $\vec{u}_{i,G+1}$, $IFEN_{i,G+1}$ is equal to $IFEN_{i,G}$ plus one, otherwise $IFEN_{i,G+1}$ will reset to zero. $IFEN_{i,G+1}$ is illustrated as follows:

$$IFEN_{i,G+1} = \begin{cases} 0 & \text{if } \vec{u}_{i,G+1} \text{ is better than } \vec{x}_{i,G+1} \\ IFEN_{i,G} + 1 & \text{otherwise} \end{cases} \tag{18}$$

If $\vec{x}_i$ produces failing evolutions for $IFEN_i$ generations, and $IFEN_i$ is more than limit evolution generations (see formula (18)), $\vec{x}_i$ falls into local optimal. The *LEG* represents the largest evolution generation allowing failure evolutions. In the next evolutionary process, we should avoid involving $\vec{x}_i$. $\vec{x}_i$ should be replaced by a new individual randomly selected in population. The framework of the PIRFE strategy is as follows (Figure 3):

Initialization- IFEN

1:   $IFEN_{i,G}$ is initialized to zero, ($i \in \{1, \cdots, NP\}$, $G = 1$) and set *LG*

Update-IFEN *$IFEN_{i,G}$*

1:   #DE Selection

2:   **If** $\vec{u}_{i,G}$ is better than $\vec{x}_{i,G}$

3:      $IFEN_{i,G+1} = 0$

4:   **Else**

5:      $IFEN_{i,G+1} = IFEN_{i,G} + 1$

6:   **End**

Mutation-PIRFE

1:   **If** $IFMN_{i,G} \geq LEG$

2:      random select $\vec{x}_{j,G}$ instead of $\vec{x}_{i,G}$ in "current to pbest/2-rand",

and $j \neq i, pbest, r1, r2, r3, r4$, generate $\vec{v}_{i,G}$.

3:   **Else**

4:      generate $\vec{v}_{i,G}$ using mutation strategy "current to pbest/2-rand"

5:   **End**

Figure 3. Pseudocode of PIRFE strategy.

3.2.3. Control Parameters Assignments

SHADE maintains a historical memory with H entries for both DE control parameters CR, F, M_{CR} and M_F. The scaling factor $F \in [0, 1]$ controls the magnitude of the differential mutation operator and $CR \in [0, 1]$ is the crossover rate. In the beginning, the contents of $M_{CR,k}$, $M_{F,k}$ ($k = 1, \cdots, H$) are all initialized to 0.5. In each generation G, the control parameters CR_i and F_i used by each individual x_i are generated by randomly selecting an index r_i from $[1, H]$, and then applying the formulas (19) and (20):

$$CR_i = \begin{cases} 0 & \text{if } M_{CR,r_i} = \perp \\ \text{randn}_i(M_{CR,r_i}, 0.1) & \text{otherwise} \end{cases} \tag{19}$$

$$F_i = \text{randc}_i(M_{F,r_i}, 0.1) \tag{20}$$

In case a value for CR_i outside of $[0, 1]$ is generated, it is replaced by the limit value (0 or 1) closest to the generated value. When $F_i > 1$, is truncated to 1, and when $F_i \leq 0$, formula (20) is repeatedly applied to generate a valid value. These manners are determined according to the procedure for JADE [38]. In formula (19), if M_{CR,r_i} has been assigned the "terminal value" $\perp$, CR_i is set to 0.

In each generation, in formula (17), CR_i and F_i values that succeed in generating a trial individual $\vec{u}_{i,G}$ better than the parent individual $\vec{x}_{i,G}$ are recorded as S_{CR}, S_F. At the end of the generation, the contents of memory are updated as follows:

$$M_{CR,k,G+1} = \begin{cases} \perp & \text{if } M_{CR,k,G} = \perp \text{ or } \max(S_{CR}) = 0 \\ \text{mean}_{WA}(S_{CR}) & \text{if } S_{CR} \neq \varnothing \\ M_{CR,k,G} & \text{otherwise} \end{cases} \tag{21}$$

$$M_{F,k,G+1} = \begin{cases} \text{mean}_{WL}(S_F) & \text{if } S_F \neq \varnothing \\ M_{F,k,G} & \text{otherwise} \end{cases} \tag{22}$$

An index k ($1 < k < H$) determines the position in the memory to update. At the beginning of the search k is initialized to 1. k is incremented whenever a new element is inserted into the history. If $> H$, k is set to 1. In generation G, the k-th element in the memory is updated. In the update formula (21) and (22), when all individuals in generation G fail to generate an individual better than the parent, i.e., $S_{CR} = S_F = \varnothing$, the memory is not updated. Also, the weighted mean $\text{mean}_{WA}(S_{CR})$ is computed according to formula (23) by Peng et al. [44]. The weighted Lehmer mean $\text{mean}_{WL}(S_F)$ is computed using the formula below, and as with $\text{mean}_{WA}(S_{CR})$:

$$\text{mean}_{WL}(S_F) = \sum_{k}^{|S_F|} w_k \cdot S_{F,k}^2 / \sum_{k}^{|S_F|} w_k \cdot S_{F,k} \tag{23}$$

$$\text{mean}_{WA}(S_{CR}) = \sum_{k=1}^{|S_{CR}|} w_k \cdot S_{CR,k} \tag{24}$$

$$w_k = \triangle f_k / \sum_{k=1}^{|S_{CR}|} \triangle f_k \tag{25}$$

where $\triangle f_k = \left| f(\vec{u}_{i,G}) - f(\vec{x}_{i,G}) \right|$. In the same paper, they also proposed a restart strategy for JADE [38].

LSHADE put forward a new method of NP setting—LPSR which reduces the population linearly. The population size at generation 1 is N_{init}, and the population at the end of the run is N_{min}. After each generation G, the population size in the next generation, N_{G+1} is computed according to formula (26):

$$N_{G+1} = \text{round}[((N_{min} - N_{init})/MAXNFE) \cdot NFE + N_{init}] \tag{26}$$

If N_{min} is set to the smallest possible value, the evolutionary operators can be applied in the case of iLSHADE, $N_{min} = 6$ because the mutation strategy "current to pbest/2-rand" showed as formula (17) requires 4 individuals. NFE is the current number of fitness evaluations, and $MAXNFE$ is the maximum number of fitness evaluations. Whenever $N_{G+1} < N_G$, the $(N_G - N_{G+1})$ worst-ranking individuals are deleted from the population. Similarly, the external archive size $|A|$ is set to N_{init} multiplied by a parameter r^{arc}, $|A| = \text{round}(N_{init} \times r_{arc})$. In addition, external archive A is same as population, whenever $|A|_{G+1} < |A|_G$, the $(|A|_G - |A|_{G+1})$ worst-ranking individuals are deleted from the A. The p value for "current to pbest/2-rand" in each generation G is computed as follows:

$$p = (p_{max} - p_{min}) \cdot \text{rand}(0,1) + p_{min} \tag{27}$$

where $\text{rand}(0,1)$ returns a uniformly distributed random number in $[0,1]$, p is a random value in $[p_{max}, p_{min}]$. Finally, the pseudo-code of the iLSHADE algorithm is given in Figure 4.

```
 1:  G ← 1, FES ← 0 ,Archive  A ← ∅ ;
 2:  Initialize population  P_g = (x̄_{1,G}, ⋯, x̄_{NP,G})  randomly
 3:  Initialization-IFEN // described in Figure 3
 4:  Set  M_F[H] = 0.2 , Set other values in  M_F  to 0.5,
 5:  Set  M_CR[H] = 0.8 , Set other values in  M_CR  to 0.8,
 6:  k ← 1  // index counter
 7:  while the termination criteria are not meet do
 8:      S_CR ← ∅ , S_F ← ∅ ,
 9:      for  i = 1 to  NP  do
10:          r_i ← select from  [1, H]  randomly
11:          if  M_{CR,ri} < 0 then  CR_{i,g} ← 0
12:          else  CR_{i,g} ← randn_i(M_{CR,i}, 0.1)
13:          end if
14:          F_{i,g} ← randc_i(M_{F,i}, 0.1)
15:          ū_{i,G} ← Mutation-PIRFE with "current to pbest/2-rand" // described in Figure 3
16:      end for
17:      for i = 1 to  NP  do
18:          if  ū_{i,G} is better than  x̄_{i,G}  then
19:              x̄_{i,G+1} ← ū_{i,G} ;
20:              x̄_{i,G} → A ,  CR_{i,g} → S_CR , F_{i,g} → S_F
21:          else
22:              x̄_{i,G+1} ← x̄_{i,G} ;
23:          end if
24:          Update-IFEN  IFEN_{i,G} // described in Figure 3
25:      end for
26:      Shrink  A , if necessary
27:      Update  M_{CR,i}  and  M_{F,i} except  i = H
28:      Apply LPSR strategy using formula (26)// linear population size reduction
29:      Update  p  using formula (27)
30:      G ← G + 1, FES ← FES + NP
31: end while
```

Figure 4. Pseudocode of iLSHADE.

4. Numerical Experiment

The iLSHADE algorithm was tested in both low and high dimension on a set of 10 benchmark functions demonstrated in Table 1. Table 1 indicates benchmark problems with different structures and characteristics. In the table, "O-V" means the optimum fitness and "O-S" stands for the optimum solution. f1, f2, f3, f5 and f6 are unimodal optimization problems to evaluate the convergence, while f4, f7, f8, f9 and f10 are multimodal optimization problems with a huge number of local optima to test the convergence precision [18].

Table 1. Details of benchmark problems.

Benchmark Function	Name	Domain	O-V	O-S
$f_1 = \sum_{i=1}^{n} x_i^2$	Sphere	$[-100, 100]^n$	0	$\{0, 0, \cdots, 0\}$
$f_2 = \sum_{i=1}^{n} |x_i| + \prod_i^n |x_i|$	Schwefel (2.2)	$[-100, 100]^n$	0	$\{0, 0, \cdots, 0\}$
$f_3 = \sum_{i=1}^{n} \left(\sum_{j=1}^{i} x_j \right)^2$	Schwefel (1.2)	$[-100, 100]^n$	0	$\{0, 0, \cdots, 0\}$
$f_4 = \sum_{i=1}^{n-1} \left[100 (x_{i+1} - x_i^2)^2 + (x_i - 1)^2 \right]$	Rosenbrock	$[-30, 30]^n$	0	$\{0, 0, \cdots, 0\}$
$f_5 = \sum_{i=1}^{n} \left(\lfloor x_j + 0.5 \rfloor \right)^2$	Step	$[-100, 100]^n$	0	$\{0, 0, \cdots, 0\}$
$f_6 = \sum_{i=1}^{n} i x_i^4$	Quartic	$[-1.28, 1.28]^n$	0	$\{0, 0, \cdots, 0\}$
$f_7 = \sum_{i=1}^{n} -x_i \sin\left(\sqrt{|x_i|} \right)$	Schwefel (2.26)	$[-500, 500]^n$	$-418.9n$	*
$f_8 = \sum_{i=1}^{n} \left[x_i^2 - 10 \cos(2\pi x_i) + 10 \right]$	Rastrigin	$[-5.12, 5.12]^n$	0	$\{0, 0, \cdots, 0\}$
$f_9 = \begin{array}{l} -20 \exp\left(-0.2 \sqrt{1/n \sum_{i=1}^{n} x_i^2} \right) \\ - \exp(1/n \cos(2\pi x_i)) + 20 + e \end{array}$	Ackley	$[-32, 32]^n$	0	$\{0, 0, \cdots, 0\}$
$f_{10} = 1 + \sum_{i=1}^{n} \frac{x_i^2}{4000} + \prod_i^n \cos\left(\frac{x_i}{\sqrt{i}} \right)$	Griewank	$[-600, 600]^n$	0	$\{0, 0, \cdots, 0\}$

Note: * (420.9876, 420.9876, ... , 420.9876).

The iLSHADE is compared to DE and other improved DE like LSHADE, JADE, CoDE and jDE. The number of function evaluations is used to appraise the convergence. These experiments are made on a personal computer, Windows10, Intel(R) Core(TM) i7-5500U CPU@ 2.40GHZ, RAM 8.00 GB. The dimension of benchmark functions is D = 10 in low dimension and 30 in high dimension, and 51 runs of an algorithm were needed for each function. The maximum number of objective function evaluations is D × 10,000. The optimal values are known for all benchmark functions.

In the experiments, the parameters in LSHADE, JADE, CoDE, jDE and DE were kept unchanged refer to [33,36,38,39,42], and the parameter setting in the iLSHADE is same as LSHADE except the following parameters:

- Using "current to pbest/2-rand" mutation strategy,
- The p value for mutation strategy is computed as $p_G = \text{rand}[p_{min}, p_{max}]$, where $p_{min} = 2/NP$ is set such that when $\vec{x}_{pbest,G}$ is selected, at least 2 individuals are needed, and $p_{max} = 0.25$.
- Initial population size $N_{init} = 15 log(D) \sqrt{D}$, the control parameter of external archive size $r^{arc} = 2$.
- Historical memory size H = 6; set a final pair of parameters $M_F[H] = 0.2$ and $M_{CR}[H] = 0.8$, other M_F values are initialized to 0.5 and other M_{CR} are initialized to 0.8.
- PIRFE parameter $LEG = 50$.

The aggregate results of statistical testing $(+, -, \approx)$ on 10 functions are shown in Tables 2 and 3. The symbols $+, -, \approx$ indicate that a given algorithm performed significantly better $(+)$, significantly worse $(-)$, or not significantly different better or worse $(\approx)$ compared to iLSHADE using the Wilcoxon rank-sum test [45] (significantly, $p < 0.05$).

Table 2. Experimental results of 10 test functions in low dimension.

f	LSHADE Mean (Std Dev)	JADE Mean (Std Dev)	CoDE Mean (Std Dev)	jDE Mean (Std Dev)	DE Mean (Std Dev)	iLSHADE Mean (Std Dev)
f_1	0.00×10^0 (0.00×10^0) ≈	8.61×10^{-36} (7.18×10^{-36}) −	5.93×10^{-38} (7.34×10^{-38}) −	1.15×10^{-38} (1.18×10^{-38}) −	2.73×10^{-46} (1.42×10^{-45}) −	0.00×10^0 (0.00×10^0)
f_2	6.49×10^{-49} (4.38×10^{-48}) −	8.75×10^{-20} (4.68×10^{-20}) −	7.88×10^{-22} (6.58×10^{-22}) −	1.38×10^{-22} (1.01×10^{-22}) −	8.03×10^{-25} (1.65×10^{-24}) −	3.36×10^{-64} (2.05×10^{-63})
f_3	1.10×10^{-91} (5.90×10^{-91}) −	2.81×10^{-35} (2.75×10^{-35}) −	2.43×10^{-39} (4.02×10^{-39}) −	1.24×10^{-40} (1.79×10^{-40}) −	4.58×10^{-45} (2.03×10^{-44}) −	0.00×10^0 (0.00×10^0)
f_4	0.00×10^0 (0.00×10^0) ≈	0.00×10^0 (0.00×10^0) ≈	1.47×10^{-05} (1.96×10^{-05}) −	7.39×10^{-03} (9.11×10^{-03}) −	7.82×10^{-02} (5.53×10^{-01}) −	0.00×10^0 (0.00×10^0)
f_5	0.00×10^0 (0.00×10^0) ≈	0.00×10^0 (0.00×10^0) ≈	0.00×10^0 (0.00×10^0) ≈	0.00×10^0 (0.00×10^0) ≈	0.00×10^0 (0.00×10^0)≈	0.00×10^0 (0.00×10^0)
f_6	0.00×10^0 (0.00×10^0) ≈	1.40×10^{-71} (3.27×10^{-71}) −	6.47×10^{-71} (1.52×10^{-70}) −	1.40×10^{-72} (3.66×10^{-72}) −	8.39×10^{-88} (4.17×10^{-87}) −	0.00×10^0 (0.00×10^0)
f_7	-4189.83 (2.73×10^{-12}) ≈	-4189.83 (2.73×10^{-12}) ≈	-4189.83 (2.73×10^{-12}) ≈	-4189.83 (2.73×10^{-12}) ≈	-4189.83 (2.73×10^{-12}) ≈	-4189.83 (2.73×10^{-12})
f_8	0.00×10^0 (0.00×10^0) ≈	0.00×10^0 (0.00×10^0) ≈	0.00×10^0 (0.00×10^0) ≈	0.00×10^0 (0.00×10^0) ≈	3.06×10^0 (2.47×10^0) −	0.00×10^0 (0.00×10^0)
f_9	3.72×10^{-15} (9.55×10^{-16}) −	3.86×10^{-15} (6.90×10^{-16}) −	4.00×10^{-15} (2.37×10^{-30}) −	3.93×10^{-15} (4.93×10^{-16}) −	3.93×10^{-15} (4.93×10^{-16}) −	3.72×10^{-15} (9.55×10^{-16})
f_{10}	0.00×10^0 (0.00×10^0) ≈	2.42×10^{-12} (6.51×10^{-12}) −	0.00×10^0 (0.00×10^0) ≈	3.22×10^{-04} (2.28×10^{-03}) −	8.58×10^{-02} (6.17×10^{-02}) −	0.00×10^0 (0.00×10^0)
−	2	6	6	7	8	
+	0	0	0	0	0	
≈	8	4	4	3	2	

Table 3. Experimental results of 10 test functions in high dimension.

f	LSHADE Mean (Std Dev)	JADE Mean (Std Dev)	CoDE Mean (Std Dev)	jDE Mean (Std Dev)	DE Mean (Std Dev)	iLSHADE Mean (Std Dev)
f_1	1.12×10^{-90} (6.44×10^{-90}) −	0.00×10^0 (0.00×10^0) ≈	9.85×10^{-19} (7.04×10^{-19}) −	4.31×10^{-41} (4.35×10^{-41}) −	9.34×10^{-44} (2.74×10^{-43}) −	0.00×10^0 (0.00×10^0)
f_2	2.09×10^{-42} (1.03×10^{-41}) −	4.11×10^{-27} (4.89×10^{-27}) −	4.01×10^{-12} (1.34×10^{-12}) −	3.48×10^{-24} (1.96×10^{-24}) −	1.16×10^{-05} (8.17×10^{-05}) −	4.88×10^{-58} (1.55×10^{-57})
f_3	3.85×10^{-81} (1.74×10^{-80}) −	3.58×10^{-49} (7.53×10^{-49}) −	4.22×10^{-19} (3.41×10^{-19}) −	7.27×10^{-43} (1.05×10^{-42}) −	1.16×10^{-46} (6.85×10^{-46}) −	0.00×10^0 (0.00×10^0)
f_4	1.40×10^{-25} (9.70×10^{-25}) −	$1.85 \times 10^{+01}$ ($1.01 \times 10^{+01}$) −	$1.82 \times 10^{+01}$ (3.29×10^0) −	$1.15 \times 10^{+01}$ (8.23×10^0) −	2.62×10^0 (2.60×10^0) −	0.00×10^0 (0.00×10^0)
f_5	0.00×10^0 (0.00×10^0) ≈	1.96×10^{-02} (1.39×10^{-01}) −	0.00×10^0 (0.00×10^0) ≈	0.00×10^0 (0.00×10^0) ≈	4.22×10^0 (7.61×10^0) −	0.00×10^0 (0.00×10^0)
f_6	0.00×10^0 (0.00×10^0) ≈	0.00×10^0 (0.00×10^0) ≈	2.35×10^{-33} (2.93×10^{-33}) −	1.79×10^{-69} (3.86×10^{-69}) −	1.74×10^{-59} (1.10×10^{-58}) −	0.00×10^0 (0.00×10^0)
f_7	$-12,569.49$ (1.82×10^{-12}) ≈	$-12,567.16$ ($1.64 \times 10^{+01}$) −	$-12,569.49$ (1.82×10^{-12}) ≈	$-12,569.49$ (1.82×10^{-12}) ≈	-11552.14 ($3.68 \times 10^{+02}$) −	$-12,569.49$ (1.88×10^{-05})
f_8	1.74×10^{-16} (6.35×10^{-16}) +	0.00×10^0 (0.00×10^0) ≈	8.38×10^{-12} (9.18×10^{-12}) +	4.83×10^0 (3.86×10^0) −	$3.62 \times 10^{+01}$ ($1.44 \times 10^{+01}$) −	3.16×10^{-11} (1.66×10^{-10})
f_9	4.00×10^{-15} (2.37×10^{-30}) ≈	4.76×10^{-15} (1.46×10^{-15}) −	2.74×10^{-10} (1.08×10^{-10}) −	5.60×10^{-15} (1.77×10^{-15}) −	2.64×10^{-01} (5.21×10^{-01}) −	4.00×10^{-15} (2.37×10^{-30})
f_{10}	0.00×10^0 (0.00×10^0) ≈	1.55×10^{-03} (3.81×10^{-03}) −	3.05×10^{-17} (1.08×10^{-16}) −	0.00×10^0 (0.00×10^0) ≈	7.99×10^{-03} (1.46×10^{-02}) −	0.00×10^0 (0.00×10^0)
−	4	6	7	7	10	
+	1	1	1	0	0	
≈	5	3	2	3	0	

1   Simulation in low dimension.

Table 2 summarizes the experimental results in low dimension. It shows that iLSHADE eventually converges to optimum for 8 test functions except f_2 and f_9. Moreover, iLSHADE significantly outperforms JADE, CoDE, jDE, DE. Compared to LSHADE, iLSHADE beats LSHADE on two test functions f_2 and f_3. Especially in f_3, only iLSHADE converges to the global optimal value. Overall, iLSHADE performs better than other algorithms on low dimension optimization problems.

2   Simulation in high dimension.

The experimental results in high dimension are summarized in Table 3. iLSHADE eventually converges to optimum for 7 test functions except f_2, f_8 and f_9. iLSHADE performs the best in f_2 and f_9, although it does not converge to the global optimal. While on f_8, JADE performs the best. The proposed algorithm is not good as expected. Above all, iLSHADE has an obvious advantage over other algorithms on high dimension optimization problems.

To sum up, iLSHADE is suitable for both low dimension and high dimension, meaning that the improvement proposed in this paper is effective.

5. Implementation of iLSHADE for LSLCHS

5.1. Solution Structure and Initial Population

To handle constraints and calculate objective function, solution structure for the LSLCHS comprises a group of monthly water levels as the decision variables shown as follows.

$$X = \begin{bmatrix} X^1 \\ X^2 \\ \vdots \\ X^M \end{bmatrix} = \begin{bmatrix} x_1^1, x_2^1, \cdots x_T^1 \\ x_1^2, x_2^2, \cdots x_T^2 \\ \vdots \\ x_1^M, x_2^M, \cdots x_T^M \end{bmatrix} \tag{28}$$

where M is the number of hydropower stations, T (12 month in a year) is the number of intervals. In algorithms relying on heuristic search, initial population is an important issue to convergence speed and population diversity. The iLSHADE has a large initial population size based on LPSR and ensures the diversity of the population by random initialization.

5.2. Constraint Handling

It is multiple and complicate for flow constraint, power generating constraint, amplitude of water level variation and hydraulic connection of cascade. The handling measure currently used for water balance constraint, water level constraint and boundary condition is often corrected to the boundary [18,46]. It has defects such as: (1) The direction of the entering feasible domain is relatively simple and centered on the boundary because of the excessive attention to the rapid into the feasible area; (2) When there are multiple feasible domains, it is easy to ignore small feasible areas;

The ε-constrained method is first proposed by Takahama et al. [47], which relaxes the greed of the feasibility criterion to the constraint conditions. The ε value is set as threshold value in ε-constrained method. In general, constrained optimization maximum problems can be mathematically formulated as follows:

$$\max f\left(\vec{x}\right), \ \vec{x} = (x_1, \ldots, x_D) \tag{29}$$

$$\text{subject to} \begin{cases} g_i\left(\vec{x}\right) \leq 0 \ i = 1, 2, \cdots, m \\ h_j\left(\vec{x}\right) = 0 \ j = 1, 2, \cdots, n \end{cases} \tag{30}$$

where $g_i\left(\overrightarrow{x}\right)$ is inequality constraints, m is the total number of inequality constraint. $h_j\left(\overrightarrow{x}\right)$ is equality constraints and n is the total number of equality constraints. The value of constraint violation $\varphi\left(\overrightarrow{x}\right)$ can be calculated in the following formula (30), and $\varphi\left(\overrightarrow{x}\right)$ of infeasible solutions is bigger than 0.

$$\varphi\left(\overrightarrow{x}\right) = \sum_{i=0}^{m} \max\left(0, g_i\left(\overrightarrow{x}\right)\right) + \sum_{j=1}^{n}\left(\left|h_j\left(\overrightarrow{x}\right)\right|, 0\right) \tag{31}$$

When the constraint violation values of both solutions are smaller than ε value, the one with better objective function value is selected. Otherwise, the one with smaller constraint violation value is selected. Overall, when any of the following conditions are met, $\overrightarrow{x}_i$ is superior to $\overrightarrow{x}_j$:

$$\begin{cases} f\left(\overrightarrow{x}_i\right) < f\left(\overrightarrow{x}_j\right), & \text{if } g\left(\overrightarrow{x}_i\right) \le \varepsilon \cap g\left(\overrightarrow{x}_j\right) \le \varepsilon \\ f\left(\overrightarrow{x}_i\right) < f\left(\overrightarrow{x}_j\right), & \text{if } g\left(\overrightarrow{x}_i\right) = g\left(\overrightarrow{x}_j\right) \\ g\left(\overrightarrow{x}_i\right) < g\left(\overrightarrow{x}_j\right), & \text{otherwise} \end{cases} \tag{32}$$

and the final effectiveness of e-constrained method strongly depends on the control method of ε value. Takahama et al. [47] proposed the following method,

$$\varepsilon(0) = \varphi\left(P_0^\theta\right) \tag{33}$$

$$\varepsilon(t) = \begin{cases} \varepsilon(0)(1 - t/T_c)^{cp} & 0 < t < T_c \\ 0 & t > T_c \end{cases} \tag{34}$$

where P_0^θ is the top θ-th individual in the initial population, cp is a control parameter. If the number of iterations t is less than a given threshold value T_c, the ε value declines in an exponential way. Otherwise, ε is set to 0 (see formula (33) and (34)).

The ε-constrained method can expand search space, avoid the constraint correction for unfeasible solutions and enable to search infeasible region that is around feasible region. However, the constraints consist of flow, power generating and amplitude of water level. Operating water level in LSLCHS problem is multiple and complex. Their units are not integrated, and the physical quantities corresponding to the same level of different reservoirs are different. To solve the above problems, we proposed ε-constrained in cascade reservoir operation method (ε-CRO) with unify different physical quantities constraint violation. The ε-CRO chooses water to unify different physical quantities constraint violation because all constraints can be converted to outflow constraint and flow accumulated over time as water. The feasible range of outflow $Q_{i,t}$ is expressed in formula (35) and (36).

$$\overline{Q}_{i,t+1} = \min \begin{cases} Q_{i,t+1}^{\max} \\ \left(V(Z_{i,t}) - V(Z_{i,t+1}^{\min})\right)/\Delta t \\ (V(Z_{i,t}) - V(Z_{i,t} - \Delta Z_i))/\Delta t \end{cases} \tag{35}$$

$$\underline{Q}_{i,t+1} = \max \begin{cases} Q_{i,t+1}^{\min} \\ I_{i,t} + \left(V(Z_{i,t}) - V(Z_{i,t+1}^{\max})\right)/\Delta t \\ I_{i,t} + (V(Z_{i,t}) - V(Z_{i,t} + \Delta Z_i))/\Delta t \\ Q_{i,t+1}^{\min N} \end{cases} \tag{36}$$

where $Z_{i,t}$ represents the i-th reservoir water level at the -th period, $Z(V)$ is the relationship between water level and storage. $\overline{Q}_{i,t+1}$ and $\underline{Q}_{i,t+1}$ stand for the minimum and maximum outflow under all constraints. $Q_{i,t+1}^{\min,N}$ is the outflow for guaranteed output and Δt stands for time horizon. The value of constraint violation in LSLCHS problem $\varphi_{CRO}\left(\vec{x}\right)$ can be calculated as follows:

$$\varphi_{CRO}\left(\vec{x}\right) = \sum_{i=1}^{M} \sum_{t=1}^{T} \left[\max\left(0, Q_{i,t} - \overline{Q}_{i,t+1}\right) + \max\left(0, \underline{Q}_{i,t+1} - Q_{i,t}\right)\right] \tag{37}$$

$\varepsilon_{CRO}(0)$ and $\varepsilon_{CRO}(t)$ are calculated like in [47] except $\theta = 0.5$, $cp = 1$ and $T_c = 0.5 * MAXNFE$.

6. Case Study

6.1. Description of Case Study

The Jinsha River, the upper stretch of the Yangtze River, is 2290 km long with a 485,000 km^2 basin area flowing through the provinces of Qinghai, Sichuan, and Yunnan in western China (See in Figure 5). Along the river, there are four large hydropower stations with large installed capacity, huge regulating storage and high water head. The total installed capacity of the four large hydropower stations is twice more than the Three Gorges Project (the largest hydropower station in the world). The main parameters of these hydropower stations are listed in Table 4.

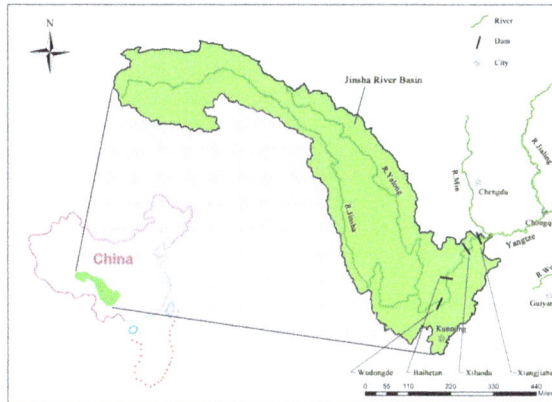

Figure 5. The location of the Jinsha River Basin in China.

Table 4. The main parameters of four large hydropower stations in Jinsha River.

Parameter	Wudongde	Baihetan	Xiluodu	Xiangjiaba
Adjustment ability	Season	Annual	Annual	Season
Regulating storage (billion m^3)	2.60	10.40	6.46	0.90
Hydro plant discharge range (m^3/s)	[49,400, 906]	[49,700, 905]	[43,700, 1500]	[49,800, 1500]
Upriver water level range (m)	[975, 945]	[825, 765]	[600, 540]	[380, 370]
Installed capacity (MW)	12000	16000	13860	6400
Normal water level (m)	975	825	600	380

6.2. Results and Analysis

In the case, the four large hydropower stations are all taken into consideration. According to historical runoff from 1959 to 2014 in the basin, three typical years are chosen to be the inflow conditions: wet year (historical runoff of 1999), normal year (2008) and dry year (1969). Simulation results of iLSHADE are compared to LSHADE, JADE and CoDE in three typical years. The initial water level and terminal water level of all the hydropower stations are set to the normal water level. The schedule period consists of 12 intervals with one month for each interval. In addition, the parameters in all algorithms are the same as those mentioned in Section 4. The maximum evaluation time is set to 40,000.

Table 5 represents that iLSHADE gains the best benefit of power production in three typical years. The convergence process of different algorithms in dry year is shown in Figure 6. Compared to LSHADE, JADE and CoDE on average optimal benefit of 51 independent simulations are illustrated in Table 5, iLSHADE increases the power production by 2.02, 4.04, 2.39 (10^8 KWh) in wet year, 3.03, 7.37, 3.76 (10^8 KWh) in normal year, 3.48, 5.68, 1.96 (10^8 KWh) in dry year. Obviously, the proposed iLSHADE is superior when solving LSLCHS problem by obtaining the maximal benefit of power production efficiently. In particular, the standard deviation of 51 independent simulations in iLSHADE is 0.02 in wet year, 0.01 in normal year, 0.01 in dry year, which shows that the convergence stability of iLSHADE is better than other algorithms. Meanwhile, it can be seen easily from Figure 6a that iLSHADE can avoid premature convergence effectively, at the same evaluation times keep a fast convergence speed compared to LSHADE, CoDE and JADE. Figure 6b depicts that the $\varphi_{CRO}\left(\vec{x}\right)$ of iLSHADE and LSHADE frequent changes and always lower than $\varepsilon_{CRO}(t)$, until evaluation times is greater than T_c, $\varphi_{CRO}\left(\vec{x}\right)$ is limited to 0.

Table 5. Results of 51 independent simulations on generated energy optimization (10^8 KWh).

Method	Wet Year (1999)			Normal Year (2008)			Dry Year (1969)		
	Max	Mean	Std	Max	Mean	Std	Max	Mean	Std
iLSHADE	2425.03	2425.01	0.02	2268.13	2268.11	0.01	1814.36	1814.35	0.01
LSHADE	2423.86	2422.99	0.48	2266.88	2265.08	0.94	1812.64	1810.87	0.99
Diff	1.17	2.02		1.25	3.03		1.72	3.48	
JADE	2424.43	2420.97	1.32	2267.51	2260.74	2.43	1814.00	1808.67	2.234
Diff	0.6	4.04		0.62	7.37		0.36	5.68	
CoDE	2423.39	2422.62	0.30	2265.62	2264.35	0.64	1813.08	1812.39	0.39
Diff	1.64	2.39		2.51	3.76		1.28	1.96	

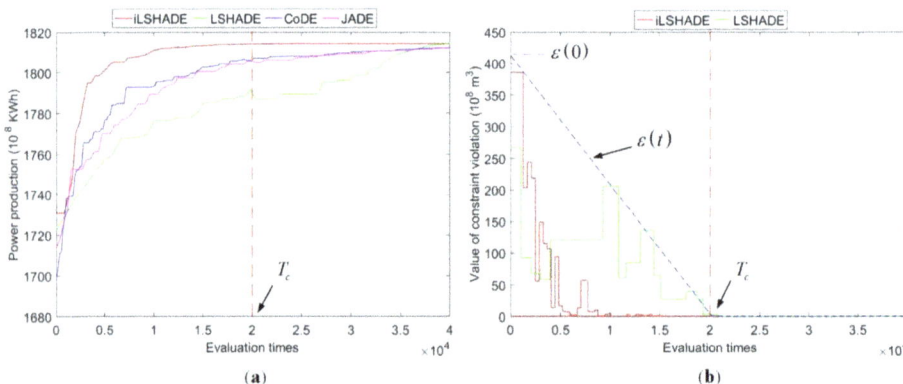

Figure 6. Convergence process of different algorithms.

The monthly reservoir water levels and outflow, as well as the optimal schedule result of Wudongde, Baihetan, Xiluodu and Xiangjiaba result in normal year obtained by iLSHADE are shown in Figure 7. Due to sufficient inflow, the cascade reservoir does not need to release storage capacity to meet the constraints of minimum outflow limit during the dry season from January to March. The inflow of Wudongde is very low in April, so the cascade reservoir needs to release storage capacity to meet the constraint requirements. As one of the upstream reservoirs, Wudongde first lowers the water level. To reduce the water spillage before flood season, Baihetan will lower its water level below the flood control level in advance and impound some water at the last period. Xiluodu and Xiangjiaba lower the water level in the last period. During the impoundment period, Wudongde and Baihetan store water to normal water level before Xiluodu and Xiangjiaba in October. In this way, the water in upstream reservoirs can utilize the downstream high hydraulic head to generate more electric power.

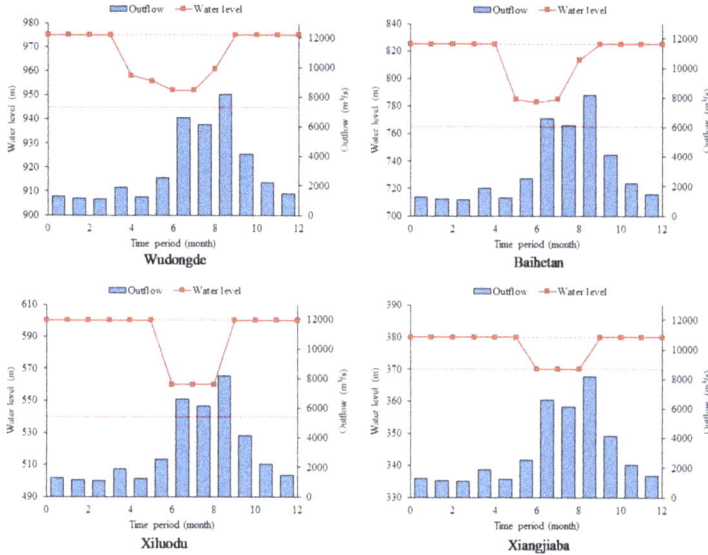

Figure 7. Optimal results in normal year by month.

Furthermore, the historical runoff data from 1959 to 2014 are selected for long sequence calculation. The parameters in all algorithms are the same as those mentioned in Section 6.2 and the maximum evaluation time is set to 40,000. The average adding annual power production that iLSHADE compares to LSHADE, JADE and CoDE is presented in Figure 8. It can be seen clearly from Figure 8 that iLSHADE is superior compared to other algorithm in solving the LSLCHS problem with different types of historical runoff from 1959 to 2014.

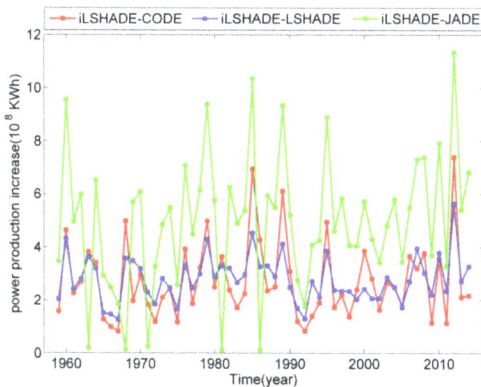

Figure 8. Historical runoff data from 1959 to 2014 for the annual power production increase that iLSHADE compares to LSHADE, JADE and CoDE.

According to the above analysis, proposed iLSHADE is superior when solving LSLCHS problem in different types of runoff by obtaining the maximize benefit of electric power production. All these experiment results fully demonstrate that iLSHADE is a competitive method to solve the LSLCHS problem.

7. Conclusions

An iLSHADE algorithm with new mutation strategy "current to pbest/2-rand" and PIRFE strategy has been developed in this paper to solve the LSLCHS problem. The significant modifications are mainly focused on preventing premature convergence and accelerating convergence. To verify the performance of iLSHADE, numerical simulation in both low and high dimension on a set of 10 benchmark functions has been done. Compared with other improved differential evolution algorithms, iLSHADE obtains better performance with all ten functions in the low dimension and nine functions in the high dimension. This indicates that the proposed new mutation strategy "current to pbest/2-rand" and PIRFE strategy in iLSHADE enhance the performance of original algorithm LSHADE effectively. Then iLSHADE is applied to solve LSLCHS problem for four large hydropower stations in Jinsha River. Compared to LSHADE, JADE and CoDE on average optimal benefit, iLSHADE increases the power production by 3.03, 7.37, 3.76 (10^8 KWh) in a normal year. In particular, the standard deviation of 51 independent simulations in iLSHADE is far lower than other algorithms. Moreover, according to its successful simulation performance with the historical runoff data from 1959 to 2014, iLSHADE can obtain better schedule results with lager generation benefits and better convergence property compared to LSHADE, JADE and CoDE. Above all, iLSHADE is a valid and reliable tool in solving the LSLCHS problem. Future research should consider the iLSHADE algorithm combined with other methods when solving multi objective scheduling problems in LSLCHS problem.

Acknowledgments: This work is supported by the National Natural Science Foundation Key Project of China (No. 91547208), the National Natural Science Foundation of China (No. 51579107), the National Key R&D Program of China (No. 2016YFC0402205) and the Foundation of Huazhong University of Science and Technology (No. 2017KFYXJJ209). Special thanks are given to the anonymous reviewers and editors for their constructive comments.

Author Contributions: Xiaohao Wen and Zhongzheng He designed and conducted the experiments. Xiaohao Wen wrote the draft of the paper. Zhongzheng He prepared the figures for this paper. Jianzhong Zhou proposed the main structure of this study. Chao Wang provided useful advice and made some corrections. All authors read and approved the final manuscript.

Conflicts of Interest: The authors declare no conflict of interest.

References

1. Lior, N. Sustainable energy development: The present (2011) situation and possible paths to the future. *Energy* **2012**, *43*, 174–191. [CrossRef]
2. Wu, X.-Y.; Cheng, C.-T.; Shen, J.-J.; Luo, B.; Liao, S.-L.; Li, G. A multi-objective short term hydropower scheduling model for peak shaving. *Int. J. Electr. Power Energy Syst.* **2015**, *68*, 278–293. [CrossRef]
3. Madani, K. Hydropower licensing and climate change: Insights from cooperative game theory. *Adv. Water Resour.* **2011**, *34*, 174–183. [CrossRef]
4. Xu, J.; Ni, T.; Zheng, B. Hydropower development trends from a technological paradigm perspective. *Energy Convers. Manag.* **2015**, *90*, 195–206. [CrossRef]
5. Chang, X.; Liu, X.; Zhou, W. Hydropower in china at present and its further development. *Energy* **2010**, *35*, 4400–4406. [CrossRef]
6. Tasdemiroglu, E. Development of small hydropower in Türkiye. *Energy* **1993**, *18*, 699–702. [CrossRef]
7. Zhou, Y.; Guo, S.; Xu, C.-Y.; Liu, P.; Qin, H. Deriving joint optimal refill rules for cascade reservoirs with multi-objective evaluation. *J. Hydrol.* **2015**, *524*, 166–181. [CrossRef]
8. Jiang, Z.; Li, A.; Ji, C.; Qin, H.; Yu, S.; Li, Y. Research and application of key technologies in drawing energy storage operation chart by discriminant coefficient method. *Energy* **2016**, *114*, 774–786. [CrossRef]

9. Chen, L.; Singh, V.P. Entropy-based derivation of generalized distributions for hydrometeorological frequency analysis. *J. Hydrol.* **2017**. [CrossRef]
10. Chen, L.; Singh, V.P.; Huang, K. Bayesian technique for the selection of probability distributions for frequency analyses of hydrometeorological extremes. *Entropy* **2018**, *20*, 117. [CrossRef]
11. Chen, L.; Singh, V.P.; Guo, S.; Zhou, J.; Zhang, J.; Liu, P. An objective method for partitioning the entire flood season into multiple sub-seasons. *J. Hydrol.* **2015**, *528*, 621–630. [CrossRef]
12. Chen, L.; Singh, V.P.; Lu, W.; Zhang, J.; Zhou, J.; Guo, S. Streamflow forecast uncertainty evolution and its effect on real-time reservoir operation. *J. Hydrol.* **2016**, *540*, 712–726. [CrossRef]
13. Djebou, D.C.S. Spectrum of climate change and streamflow alteration at a watershed scale. *Environ. Earth Sci.* **2017**, *76*, 653. [CrossRef]
14. Sohoulande Djebou, D.C.; Singh, V.P. Entropy-based index for spatiotemporal analysis of streamflow, precipitation, and land-cover. *J. Hydrol. Eng.* **2016**, *21*, 05016024. [CrossRef]
15. Xu, B.; Zhong, P.-A.; Wu, Y.; Fu, F.; Chen, Y.; Zhao, Y. A multiobjective stochastic programming model for hydropower hedging operations under inexact information. *Water Resour. Manag.* **2017**, *31*, 4649–4667. [CrossRef]
16. Mahmoud, M.; Dutton, K.; Denman, M. Dynamical modelling and simulation of a cascaded reserevoirs hydropower plant. *Electr. Power Syst. Res.* **2004**, *70*, 129–139. [CrossRef]
17. Cârdu, M.; Bara, T. Romanian achievement in hydro-power plants. *Energy Convers. Manag.* **1998**, *39*, 1193–1201. [CrossRef]
18. Wang, C.; Zhou, J.; Lu, P.; Yuan, L. Long-term scheduling of large cascade hydropower stations in Jinsha river, China. *Energy Convers. Manag.* **2015**, *90*, 476–487. [CrossRef]
19. Windsor, J.S. Optimization model for the operation of flood control systems. *Water Resour. Res.* **1973**, *9*, 1219–1226. [CrossRef]
20. Yoo, J.-H. Maximization of hydropower generation through the application of a linear programming model. *J. Hydrol.* **2009**, *376*, 182–187. [CrossRef]
21. Brandao, J.L.B. Performance of the equivalent reservoir modelling technique for multi-reservoir hydropower systems. *Water Resour. Manag.* **2010**, *24*, 3101–3114. [CrossRef]
22. Catalão, J.P.D.S.; Pousinho, H.M.I.; Mendes, V.M.F. Scheduling of head-dependent cascaded hydro systems: Mixed-integer quadratic programming approach. *Energy Convers. Manag.* **2010**, *51*, 524–530. [CrossRef]
23. Jiang, Z.; Qin, H.; Ji, C.; Feng, Z.; Zhou, J. Two dimension reduction methods for multi-dimensional dynamic programming and its application in cascade reservoirs operation optimization. *Water* **2017**, *9*, 634. [CrossRef]
24. Lee, J.H.; Labadie, J.W. Stochastic optimization of multireservoir systems via reinforcement learning. *Water Resour. Res.* **2007**, *43*. [CrossRef]
25. Nanda, J.; Bijwe, P.R. Optimal hydrothermal scheduling with cascaded plants using progressive optimality algorithm. *IEEE Trans. Power Appar. Syst.* **1981**, *PAS-100*, 2093–2099. [CrossRef]
26. Jiang, Z.; Ji, C.; Qin, H.; Feng, Z. Multi-stage progressive optimality algorithm and its application in energy storage operation chart optimization of cascade reservoirs. *Energy* **2018**, *148*, 309–323. [CrossRef]
27. Jiang, C.; Bompard, E. A self-adaptive chaotic particle swarm algorithm for short term hydroelectric system scheduling in deregulated environment. *Energy Convers. Manag.* **2005**, *46*, 2689–2696.
28. Cheng, C.T.; Liao, S.L.; Tang, Z.T.; Zhao, M.Y. Comparison of particle swarm optimization and dynamic programming for large scale hydro unit load dispatch. *Energy Convers. Manag.* **2009**, *50*, 3007–3014. [CrossRef]
29. Hınçal, O.; Altan-Sakarya, A.B.; Ger, A.M. Optimization of multireservoir systems by genetic algorithm. *Water Resour. Manag.* **2011**, *25*, 1465–1487. [CrossRef]
30. Yuan, X.; Yuan, Y. Application of cultural algorithm to generation scheduling of hydrothermal systems. *Energy Convers. Manag.* **2006**, *47*, 2192–2201. [CrossRef]
31. Wang, W.; Li, C.; Liao, X.; Qin, H. Study on unit commitment problem considering pumped storage and renewable energy via a novel binary artificial sheep algorithm. *Appl. Energy* **2017**, *187*, 612–626. [CrossRef]
32. Jalali, M.R.; Afshar, A.; Mariño, M.A. Multi-colony ant algorithm for continuous multi-reservoir operation optimization problem. *Water Resour. Manag.* **2007**, *21*, 1429–1447. [CrossRef]
33. Storn, R.; Price, K. Differential evolution—A simple and efficient heuristic for global optimization over continuous spaces. *J. Glob. Optim.* **1997**, *11*, 341–359. [CrossRef]
34. Lu, Y.; Zhou, J.; Qin, H.; Li, Y.; Zhang, Y. An adaptive hybrid differential evolution algorithm for dynamic economic dispatch with valve-point effects. *Expert Syst. Appl. Int. J.* **2010**, *37*, 4842–4849. [CrossRef]

35. Yuan, X.; Cao, B.; Yang, B.; Yuan, Y. Hydrothermal scheduling using chaotic hybrid differential evolution. *Energy Convers. Manag.* **2008**, *49*, 3627–3633. [CrossRef]

36. Brest, J.; Greiner, S.; Boskovic, B.; Mernik, M.; Zumer, V. Self-adapting control parameters in differential evolution: A comparative study on numerical benchmark problems. *IEEE Trans. Evolut. Comput.* **2006**, *10*, 646–657. [CrossRef]

37. Qin, A.K.; Huang, V.L.; Suganthan, P.N. Differential evolution algorithm with strategy adaptation for global numerical optimization. *IEEE Trans. Evolut. Comput.* **2009**, *13*, 398–417. [CrossRef]

38. Zhang, J.; Sanderson, A.C. Jade: Adaptive differential evolution with optional external archive. *IEEE Trans. Evolut. Comput.* **2009**, *13*, 945–958. [CrossRef]

39. Wang, Y.; Cai, Z.; Zhang, Q. Differential evolution with composite trial vector generation strategies and control parameters. *IEEE Trans. Evolut. Comput.* **2011**, *15*, 55–66. [CrossRef]

40. Mallipeddi, R.; Suganthan, P.N.; Pan, Q.K.; Tasgetiren, M.F. Differential evolution algorithm with ensemble of parameters and mutation strategies. *Appl. Soft Comput.* **2011**, *11*, 1679–1696. [CrossRef]

41. Tanabe, R.; Fukunaga, A. Success-History Based Parameter Adaptation for Differential Evolution. In Proceedings of the 2013 IEEE Congress on Evolutionary Computation, Cancun, Mexico, 20–23 June 2013; pp. 71–78.

42. Tanabe, R.; Fukunaga, A.S. Improving the search performance of shade using linear population size reduction. In Proceedings of the 2014 IEEE Congress on Evolutionary Computation, Beijing, China, 6–11 July 2014; pp. 1658–1665.

43. Liang, J.J.; Qu, B.Y.; Suganthan, P.N. *Problem Definitions and Evaluation Criteria for the CEC 2014 Special Session and Competition on Single Objective Real-Parameter Numerical Optimization*; Computational Intelligence Laboratory, Zhengzhou University, Zhengzhou China and Technical Report; Nanyang Technological University: Singapore, 2013.

44. Peng, F.; Tang, K.; Chen, G.; Yao, X. Multi-start jade with knowledge transfer for numerical optimization. In Proceedings of the Eleventh Conference on Congress on Evolutionary Computation, Trondheim, Norway, 18–21 May 2009; pp. 1889–1895.

45. Cortina-Borja, M. *Handbook of Parametric and Nonparametric Statistical Procedures*, 5th ed.; Sheskin, J., Ed.; CRC Press: Boca Raton, FL, USA, 2012; p. 382.

46. Liao, X.; Zhou, J.; Zhang, R.; Zhang, Y. An adaptive artificial bee colony algorithm for long-term economic dispatch in cascaded hydropower systems. *Int. J. Electr. Power Energy Syst.* **2012**, *43*, 1340–1345. [CrossRef]

47. Takahama, T.; Sakai, S.; Iwane, N. Constrained optimization by the ε constrained hybrid algorithm of particle swarm optimization and genetic algorithm. In Proceedings of the Australasian Joint Conference on Artificial Intelligence, Sydney, Australia, 5–9 December 2005; pp. 389–400.

water

MDPI

Article

Optimal Operation of Cascade Reservoirs for Flood Control of Multiple Areas Downstream: A Case Study in the Upper Yangtze River Basin

Chao Zhou [1], Na Sun [2], Lu Chen [2,*], Yi Ding [1], Jianzhong Zhou [2], Gang Zha [2], Guanglei Luo [2], Ling Dai [2] and Xin Yang [2]

[1] Changjiang Institute of Survey, Planning, Design and Research, Wuhan 430010, China; zhouchao2@cjwsjy.com.cn (C.Z.); dingyi@cjwsjy.com.cn (Y.D.)

[2] School of Hydropower and Information Engineering, Huazhong University of Science and Technology, Wuhan 430074, China; sunna@hust.edu.cn (N.S.); jz.zhou@hust.edu.cn (J.Z.); zhagang@hust.edu.cn (G.Z.); m201773774@hust.edu.cn (G.L.); dailing2016@hust.edu.cn (L.D.); d201880948@hust.edu.cn (X.Y.)

* Correspondence: chen_lu@hust.edu.cn; Tel.: +86-139-8605-1604

Received: 20 June 2018; Accepted: 11 September 2018; Published: 14 September 2018

Abstract: The purpose of a flood control reservoir operation is to prevent flood damage downstream of the reservoir and the safety of the reservoir itself. When a single reservoir cannot provide enough storage capacity for certain flood control points downstream, cascade reservoirs should be operated together to protect these areas from flooding. In this study, for efficient use of the reservoir storage, an optimal flood control operation model of cascade reservoirs for certain flood control points downstream was proposed. In the proposed model, the upstream reservoirs with the optimal operation strategy were considered to reduce the inflow of the reservoir downstream. For a large river basin, the flood routing and time-lag cannot be neglected. So, dynamic programming (DP) combined with the progressive optimality algorithm (POA) method, DP-POA, was proposed. Thus, the innovation of this study is to propose a two-stage optimal reservoir operation model with a DP-POA algorithm to solve the problem of optimal co-operation of cascade reservoirs for multiple flood control points downstream during the flood season. The upper Yangtze River was selected as a case study. Three reservoirs from upstream to downstream, Xiluodu, Xiangjiaba and the Three Gorges reservoirs (TGR) in the upper Yangtze River, were taken into account. Results demonstrate that the two-stage optimization algorithm has a good performance in solving the cascade reservoirs optimization problem, because the inflow of reservoir downstream and the division volumes were largely reduced. After the optimal operation of Xiluodu and Xiangjiaba reservoirs, the average reduction of flood peak for all these 13 typical flood hydrographs (TFHs) is 13.6%. Meanwhile, the cascade reservoirs can also store much more storm water during a flood event, and the maximum volumes stored in those two reservoirs upstream in this study can reach 25.2 billion m^3 during a flood event. Comprising the proposed method with the current operation method, results demonstrate that the flood diversion volumes at the flood control points along the river decrease significantly.

Keywords: optimal flood control operation; cascade reservoirs; dynamic programming with progressive optimality algorithm (DP-POA); the upper Yangtze River Basin

1. Introduction

For safety of lives and property from risks related to floods, proper planning of relevant applications is important for development [1–13]. To date, dams have become numerous and have had a profound effect on social and economic development. As of 2007, China was the world's leader in the construction of large dams [14]. Reservoirs are often operated with a number of purposes related to environmental, economic and public services. Generally, these purposes include flood control, hydropower generation, navigation, sediment control, water supply, recreation, and fisheries, among which flood control is the most significant function for many of the reservoirs [15].

The main purpose of flood control is to prevent flood damage downstream of the reservoir and the safety of the reservoir itself. The key variables governing the operation of flood control reservoirs are the available storage capacity and the expected inflow magnitude from an incoming flood [15]. Some of the research on multi-reservoir flood control operations is for design of the reservoirs [16], in which the objective functions are the minimum of the initial investment cost and potential flood damage costs. Until now, most of the research on multi-reservoir flood control operation concerns minimizing the maximal water level for the dam itself or the releases at the downstream flood control points, such as the research carried out by Lee et al. [17], Qi et al. [18] and Chen et al. [11]. The other aspect for multi-reservoir flood control operation is to address the conflicts between flood control and conservation. A great number of research works on dynamic control flood limiting water level have been carried out in recent years to balance the conflict between flood control and conservation, such as the work done by Li et al. [19], Chen et al. [20], and Zhou et al. [9,21].

Accordingly, releases are restricted by the maximum allowable non-damaging channel capacity at the downstream flood control points [15]. Sometimes the available storage capacity of one reservoir is not enough to protect the flood control points downstream. In this situation, more reservoirs are needed to work together to protect these flood control points from flooding, especially the downstream control points with great importance. Thus, a new optimal reservoir operation model, which can make efficient use of reservoir storage to protect downstream flood control points from flooding during the flood season is needed. He et al. [22] studied the flood control operation of Jinsha River cascade reservoirs combined with the Three Gorges for flood control of the Chenglingji areas downstream based on the equal water storage method, in which the optimal operation technique was not considered. The storage capacities of these reservoirs are underutilized. Zhou et al. [9] proposed a virtual reservoir approach that aggregates multi-reservoir systems into a virtual reservoir for flood control decision making. However, in their method, two reservoirs were taken as a virtual reservoir, which cannot reflect the hydraulic connection between reservoirs. Therefore, in this paper, a two-stage optimal operation of cascade reservoirs for flood control of multiple areas downstream, which considers the optimization and hydraulic connections among cascade reservoirs, was proposed.

Take three reservoirs, as shown in Figure 1, for example. A, B, and C are three reservoirs from upstream to downstream, and points 1, 2, and 3 are three flood control points downstream of the reservoirs, whose locations can also be seen in Figure 1. Points 1 and 2 are protected by Reservoirs A and B. Point 3 is protected by Reservoirs A, B and C, since Reservoir C cannot have enough capacity for flood control of Point 3. Reservoirs A and B can help Reservoir C realize the flood control by minimizing the inflow of Reservoir C. Therefore, two objective functions are needed. For Reservoirs A and B, the objective function is to minimize the inflow of Reservoir C; for Reservoir C, the objective function is to minimize the excess flood volumes diverted to the floodplain at point C.

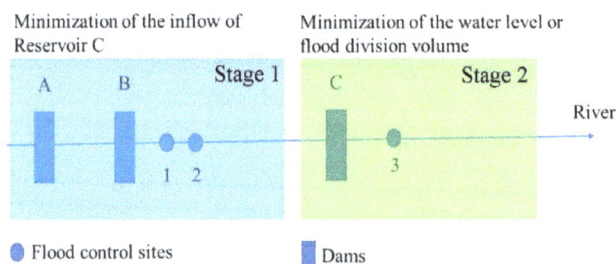

Figure 1. Diagram of dams and flood control points.

In addition, usually the optimal flood control reservoir operation model for a large river basin does not consider the flood routing during the processes for obtaining the optimal solution. When the river basin is large, these factors cannot be ignored. Since the Muskingum method was used for flood routing [23–25], the delayed time is usually considered, which causes the time-lags issue during the optimal processes [10]. In this case, the number of decision variables and constraints is increased exponentially, and it is difficult for a single traditional optimization method or modern intelligence algorithm to tackle the time-lags problem [10]. In the broad area of reservoir operation, development and use of prescriptive models based on optimization techniques have been extensively used [26,27]. An adaptive optimal algorithm also should be considered for the large river basins with flood routing and time-delayed issues.

There are various studies on optimal operations of single reservoir and multi-reservoir systems for flood control in the literature [9,28–30]. The dynamic programming (DP) method is the most widely used optimization technique for reservoir operation. It also has been used for reservoir operation by [31–34]. DP can be used for solving problems containing discrete variables, non-convex, non-continuous, and non-differentiable objective functions. However, the "curse of dimensionality", the phenomenon that when the number of reservoirs rises, the computation scale of DP increases exponentially, limits its application of DP. To deal with the dimensionality problems, modified DP algorithms, such as discrete differential dynamic programming (DDDP) [35–37], dynamic programming successive approximation (DPSA) [38,39] and the progressive optimality algorithm (POA) have been widely used for finding the optimal or near-optimal reservoir release hydrographs for flood control [40].

The POA, proposed by Howson and Sancho [41], was used to resolve the *n*th single reservoir operation problems, which divides the multi-stage problem into multiple two-stage sub-problems. Turgeon [42] used the principle of progressive optimality to minimize the total production cost of a multi-dimensional and multistage optimization problem. Guo et al. [43] applied the POA algorithm to solve the joint operation of the multi-reservoir systems of the Three Gorges and the Qingjiang cascade reservoirs. However, the disadvantages of the POA method are that the initial solutions have great influence on the final optimal results. In this study, a combined algorithm of DP and POA was proposed to solve the problem of flood routing and the influence of initial solutions.

The objective of this study was, therefore, to establish an optimal flood control operation model of cascade reservoirs for multiple flood control points downstream. The objective of the whole model is to minimize the flood diversion volumes at the flood control points. An adaptive combined algorithm is also proposed to solve the problem of flood routing. The upper Yangtze River, China, was selected as a case study. Three reservoirs were considered in this study.

2. Optimal Flood Control Reservoir Operation Model

A two-stage optimal reservoir operation model was proposed. Take the three reservoirs for an example, as shown in Figure 1. In the first stage, the optimal operation models of Reservoirs A and B were established, in which the objective function is to reduce the inflow of the downstream reservoir. In the second stage, the optimal operation model of Reservoir C is established, in which the objective function is to minimize the highest water level and the flood diversion volumes at flood control point 3.

2.1. Optimal Operation in the First Stage

2.1.1. Objective Function

Take the three cascade reservoirs as an example. Reservoirs A and B are expected to store much more water during the flood season and reduce the inflow of Reservoir C as much as possible. In other words, the best situation is that the inflow of Reservoir C is as small as possible, after the utilization of the two reservoirs upstream.

Thus, the objective function of Reservoirs A and B can be defined as:

$$\min F = \min\left\{\sum_{t=1}^{T}[Q'(t) + q(t)]^2\right\} \tag{1}$$

where $Q'(t)$ is the discharge of Reservoir B considering the flood routing along the river; and $q(t)$ is the discharge of the interval basin, which also includes the flow from other tributaries of the river.

2.1.2. Constraints for Model in the First Stage

Four constraints were considered in this study. The first one is water balance equation, which must obey during the reservoir operation process. The second one is the constraint concerning water level, which gives the allowable maximum and minimum water levels during the flood season. The third one is the discharge limit, which gives the allowable maximum and minimum discharges during the flood season. The fourth is the limit of the discharge for downstream flood control point, which protects downstream areas from flooding.

(a) Water balance equation:

$$V_{i,t} = V_{i,t-1} + (I_{i,t} - Q_{i,t})\Delta t \tag{2}$$

where $V_{i,t}$ is the reservoir storage at time t for the ith reservoir; and $I_{i,t}$ and $Q_{i,t}$ are the inflow and outflow at time t for the ith reservoir, respectively.

(b) Constraint of water level:

$$Z_{i,t}^{\min} \leq Z_{i,t} \leq Z_{i,t}^{\max} \tag{3}$$

where $Z_{i,t}$ represents the water level of the ith reservoir at time t; $Z_{i,t}^{\min}$ and $Z_{i,t}^{\max}$ represent the minimum and maximum water level of the ith reservoir at time t, respectively.

(c) Constraint of reservoir discharge:

$$Q_i^{\min} \leq Q_{i,t} \leq Q_i^{\max}(Z_t) \tag{4}$$

where $Q_{i,t}$ is the discharge of the ith reservoir at time t; $Q_i^{\max}(Z_t)$ means the maximum discharge ability corresponding to the water lever Z_t at time t for the ith reservoir; and $Q_i^{\min}$ means the minimum discharge for the ith reservoir. The maximum discharge ability is related to the water level, and the reservoir discharge $Q_{i,t}$ cannot exceed the maximum discharge ability.

Besides Equation (4), the discharge variation between adjacent time periods should be also considered by:

$$|Q_{i,t} - Q_{i,t-1}| \leq \Delta Q_i \tag{5}$$

where ΔQ_i is the allowed discharge variation between adjacent time periods for the ith reservoir.

(d) Constraint of discharge for downstream flood control points

$$q_{j,t} = Q'_{i,t} + \Delta q_t \leq q_j^{\max}, j = 1, 2, 3 \tag{6}$$

where $q_{j,t}$ is the discharge at time t for the jth flood control point downstream; $Q'_{i,t}$ is the discharge of the ith reservoir considering the flood routing at the jth flood control point; Δq_t is the inflow in the interval zone between the ith reservoir and the flood control point downstream; and $q_j^{\max}$ is the allowed maximum discharge for the jth flood control point downstream.

2.2. Optimal Operation in the Second Stage

2.2.1. Objective Function

Flood control reservoir operation is generally a complex problem, which needs to consider the safety of a reservoir and its upstream and downstream areas. This paper established a flood control operation model aiming to control the highest water level of Reservoir C and reduce the flood diversion magnitudes of the downstream areas. Two objective functions were considered and given as follows. When the release does not exceed the safety water level downstream, the highest water level of Reservoir C is taken as an objective function. Although those reservoirs operate together, it is possible that the water level at flood control points still exceeds the safety water level for a high return period flood (e.g., 1000-year). In this situation, the objective function is to minimize the flood diversion magnitude in the downstream areas. These two objective functions are given as follows.

(1) Minimize the highest water level to ensure the safety of the reservoir itself and its upstream areas

$$\min F'_1 = \min\{\max Z_t\}, t = 1, 2, \cdots, T_d \tag{7}$$

where Z_t is the water level at time t; and T_d is the time duration for the flood control operation.

(2) Minimize the flood diversion magnitudes for downstream areas:

$$\min F'_2 = \min\left\{\sum_{t=1}^{T_s} \Delta q(t) \cdot \Delta t\right\} \tag{8}$$

where $\Delta q(t)$ is the excess water volume into the flood diversion areas around the downstream control site at time t; Δt is the interval time; and T_s is the time duration for flood diversion, when the water level exceeds the safety water level downstream.

2.2.2. Constraints for the Model in the Second Stage

The constraints include the water balance, water level limits, reservoir discharge limits and allowable discharges for downstream flood control points. Equations for the constraints are the same as those in stage one (Section 2.1).

2.3. Optimization Algorithm

Considering the "curse of dimensionality", the time-lag issue for multi-reservoir operations, and the influence of the initial solutions, the dynamic programming combined with the progressive optimality algorithm (DP-POA) method was proposed to find the optimal operation polices for multi-reservoir systems, in which the DP method was used first, and the output of the DP method was the input of the POA method. The skeleton of this algorithm is shown in Figure 2. The equations for DP and POA methods are given below.

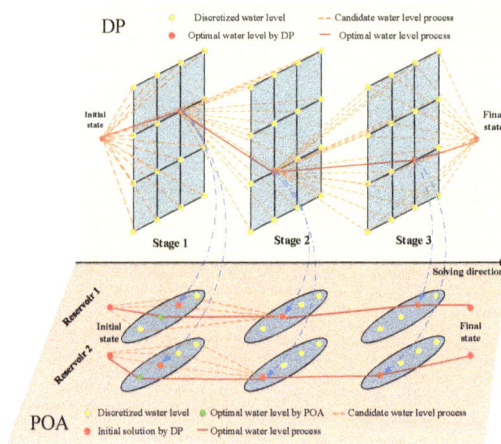

Figure 2. The skeleton of the dynamic programming combined with progressive optimality (DP-POA) algorithm.

2.3.1. Dynamic Programming (DP)

A multiple-period reservoir operation is formulated as a recursive function in DP:

$$\max \ G_t(s_t) = g_t(s_t, r_t) + G_{t+1}(s_{t+1})$$
$$s.t. \quad s_t + q_t = r_t + s_{t+1} \tag{9}$$

where s_t, r_t, and q_t are storage, release, and inflow at period t, respectively; $g_t(s_t, r_t)$ is single-period utility function; and $G_t(s_t)$ is the maximum cumulative utility from period t (current period) to period T (end of the operation period). Equation (9) is a generalized formula that characterizes the reservoir operations of hydropower, water supply, flood control, and so on.

2.3.2. Progressive Optimality Algorithm (POA)

POA often starts with an initial solution obtained by engineering judgment, heuristic search algorithms and other methods. In this method, the initial solution was given based on the results of the DP method. Then, POA solves all the two-stage optimization sub-problems in a serial way. For the jth sub-problem, use Equation (10) to find an improved decision vector between stages $j-1$ and $j+1$.

$$F^*(S_{j-1}, S_j, S_{j+1}) = \min_{S_j \in C_j} \{F(S_{j-1}, S_j) + F(S_j, S_{j+1})\} \tag{10}$$

where S_j is the possible water level at the jth period, C_j is the set of water level at the jth period, $F(S_{j-1}, S_j)$ is the objective water level between stages $j-1$ and j, $F^*(S_{j-1}, S_j, S_{j+1})$ denotes the optimal objective water level between stages $j-1$ and $j+1$.

3. Study Area

3.1. Introduction of Study Area

The upper Yangtze River is selected as a case study. The largest reservoir in China is created by the Three Gorges Dam (TGD), which stores 39.3 billion m^3 of water and has a surface area of 1045 km^2. The TGD is located on the Yangtze River, which is about 6300 km long and occupies 18.8% of China's land area. The function of the TGD includes flood control, power generation and navigation and so on. Since the Yangtze River includes many devastating floods over the centuries killing thousands of people and causing millions of dollars in damage, among these functions, flood control is the most important one.

Floods in the middle and lower reaches of the Yangtze River mainly stem from the upper region of the Yichang gauging station, which is also the control site of TGD [44]. Usually the flood volume of the upper Yichang site is about 50% of the total flow volume of the Yangtze River, about 90% of the Jingjiang River reach that is part of the Yangtze River from the Zhicheng to Chenglinji gauging station and regarded as the most key area for flood prevention [44–46]. The Three Gorges Reservoir (TGR), which uses the available empty space of the reservoir to absorb the flood, can reduce the downstream flooding around the Jingjiang River and Dongting Lake areas and guarantee the safety of the Jingjiang River downstream when a 100-year flood occurs in TGD.

However, the flood control capacity of TGR with a capacity of 22.2 billion m^3 is still not enough. If a flood event throughout the Yangtze River Basin is to occur, the TGD cannot hold back floods downstream of the dam (such as the 1954 flood that killed 30,000 people outright and more through starvation and disease). Thus, more reservoirs located upstream are needed to help TGD reduce the flood. The Xiluodu and Xiajiaba are two dams built in the Jinsha River, which is part of the Yangtze River upstream as shown in Figure 3.

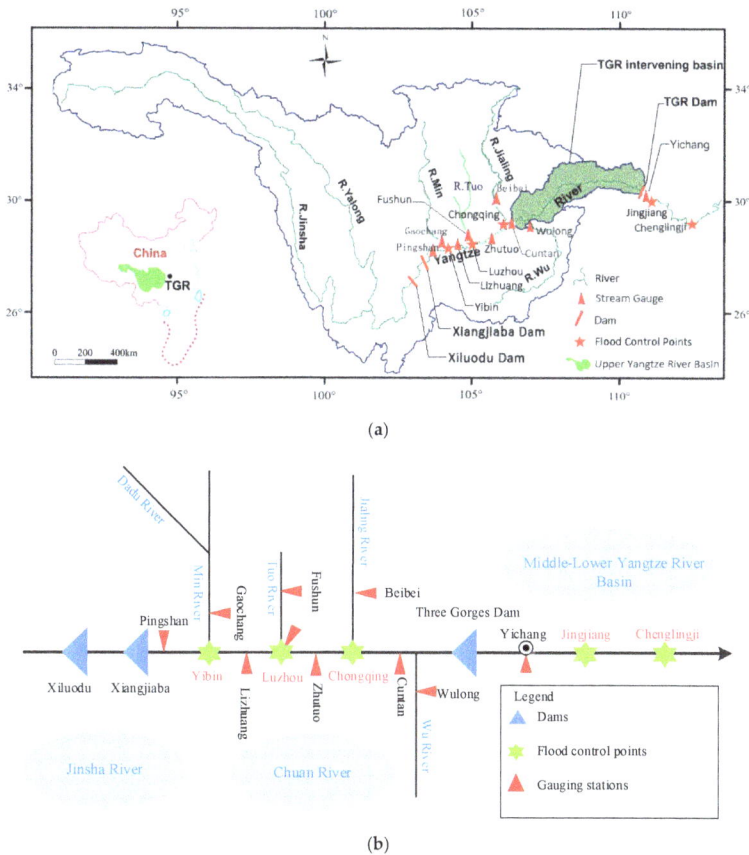

(a)

(b)

Figure 3. Locations of the dams, flood control points and gauging stations in the Yangtze River Basin. (**a**) The upper Yangtze River Basin; (**b**) skeleton of the upper Yangtze River Basin.

The Xiluodu dam, located on the upper reach of the Yangtze River between Yunnan and Sichuan Provinces, is the third tallest arch dam in the world and a key component of the Jinsha River Project. It is also the second largest dam in China, next only to the TGD in terms of size, construction cost and generating capacity. The overall elevation of the dam crest is 600 m, with the concrete double-curvature arch being 285.5 m high and 700 m long. The reservoir capacity is about 12.7 billion m^3 with 4.7 billion m^3 for flood control. The Xiangjiaba dam is located downstream of the Xiluodu dam, which is only 157 km far from it. It is a large gravity dam on the Jinsha River. The flood control capacity is 0.9 billion m^3. The total flood control capacity of both of the two dams are nearly 5.6 billion m^3. The characteristics of the three dams are summarized in Table 1.

Five flood control areas considered in this study from upstream to downstream are Yinbin, Luzhou, Chongqing, Jingjiang and Chenglingji. The Xiluodu and Xiangjiaba reservoirs need to protect the cities Yibin and Luzhou from flooding, when the flood with a 20-year return period occurs, and protect the city Chongqing from flooding, when the 50-year return period flood occurs. The Xiluodu, Xiangjiaba and TGR operate together to protect the Jingjiang area from flooding, when a 100-year flood event occurs. The design standard for the TGR is 1000 years corresponding to the water level of 175 m. Therefore, the flood events with 50-year, 100-year and 1000-year return periods were considered in this study. Three major cities from upstream to downstream (Yibin, Luzhou, and Chongqing) are located along the Yangtze River. Among these cities, Chongqing is the most populous Chinese municipality, and also the largest directly controlled municipality in China, and comprises 26 districts, eight counties, and four autonomous counties. Yibin and Luzhou are two prefecture-level cities in the south-eastern part of Sichuan province, China. Yibin is located at the junctions of the Min River and Yangtze Rivers. Luzhou, situated at the confluence of the Tuo River and the Yangtze River, and is an important port on the Yangtze River. According to the 2015 census, these three major cities have a total population of about 41 million.

In recent years, the economy of the region around the Jingjiang River reach has continued to develop. Concretely, the current agricultural acreage area and aquatic area are 37,000 and 6500 square hectometers, respectively, while fixed assets in the district are valued at 9 billion yuan, and gross domestic product is 8.7 billion yuan. This area has become an important grain, cotton, and aquaculture base in China. Therefore, once a catastrophic flood occurred, the affected population is very large, and the economic losses are extremely high. Another flood control points is Chenglingji station which is the outlet of the Dongting Lake. There are many retention basins around the Chenglingji station, which is home to a lot of people. According to the Chinese flood control regulations, as the backbone of flood management measures for the mid-downstream Yangtze River Basin, the Three Gorges Reservoir provides enormous benefits to flood control for the Yangtze River. The operation of TGR has improved the flood control capacity of the Jingjiang River Reach downstream from 20-year to 100-year, and prevents devastating damage through combining the use of retention basins for the case of floods of 100–1000 years' frequency. Additionally, it also reduces the possibility of using retention basins near the Chenglingji area to ensure the safety of life and property.

Table 1. Characteristics of the Xiluodu, Xiangjiaba and Three Gorges Dam (TGD).

Dam	Drainage Area (10^4 km^2)	Normal Water Level (m)	Flood Control Water Level (m)	Dead Water Level (m)	Regulating Storage Capacity (Billion m^3)	Flood Control Reservoir Capacity (Billion m^3)
Xiluodu	45.4	600	560	540	6.5	4.7
Xiangjiaba	45.9	380	370	370	0.9	0.9
TGD	100	175	145	145	16.5	2.2

3.2. Data of Study Area

The Xiluodu and Xiajiaba reservoirs can provide certain flood capacity to store the flood occurring in the upper Yangtze River, especially the flood in the Jinsha River. However, under the condition of preventing the downstream cities—Yibin, Luzhou and Chongqing—from flooding as shown in

Figure 3, how to use the two reservoirs rationally and effectively to reduce the inflow of TGR is a challenging task. In addition, besides those two reservoirs, how to use the flood capacity of TGR to realize the flood control of the Jingjiang River downstream effectively is another important task. Thus, since the flood control problems are much more complicated for cascade reservoirs with multiple flood control points, judicious operation methods during a flood event are needed.

The gauging station used in this study is shown in Figure 3 as well. From Xiangjiaba to Yichang, there are four main tributaries, namely Min, Tuo, Jialing, and Wu Rivers. The control sites for Min, Tuo, Jiangling, and Wu Rivers are Gaochang, Fushun, Beibei, and Wulong, respectively. The Yangtze River upper than Yibin is also named the Jinsha River, and the Yangtze River from Yibin to Yichang is also named the Chuan River because these areas belonged to Sichuang Province before. There are three flood control points located on the Chuang River reach, namely Yibin City, Luzhou City and Chongqing City; and there are two flood control points located downstream of the TGD, namely Jingjiang and Chenglingji areas. The Xiluodu and Xiangjiaba dams were employed to protect the cities of Yibin, Luzhou, and Chongqing from flooding. The TGD combined with the two dams operated together to guarantee the safety of the Jingjiang and Chenglingji areas.

Since the Xiluodu and Xiangjiaba dams are on the Jinsha River, the characteristics of the flood in the Jinsha River were discussed. The runoff in the Jinsha River Basin mainly concentrate in the period of June to October, which can account for 80% of the total annual runoff in the upper Jinsha River and 75% in the lower Jinisha River. The mean runoff and its percentage for each month at the Pingshan gauging station in the Jinsha River are given in Table 2. It is indicated from Table 2 that the river flow in July, August and September accounting for 54.0% of the total amount of the annual flow, and in June to November accounting for 81.2%.

Table 2. Runoff and its percentage for each month at the Pingshan gauging station.

Months	Jan.	Feb.	Mar.	Apr.	May	Jun.	Jul.	Aug.	Sept.	Oct.	Nov.	Dec.
Runoff (m^3/s)	1660	1420	1350	1540	2320	5080	9580	10,200	10,000	6540	3450	2180
Volume ($10^8 \ m^3$)	44.5	34.7	36.2	39.8	62.1	132	257	274	260	175	89.5	58.3
Percentage (%)	3.04	2.38	2.47	2.72	4.25	9.00	17.5	18.7	17.8	12.0	6.12	3.99

4. Application

4.1. Design Flood Hydrographs

During the flood control reservoir operation, usually the extreme or large flood events were considered. Therefore, the input of the model is the T-year design flood hydrograph. One of the methods to derive the design flood hydrographs (DFH) is the typical flood hydrograph (TFH) method which has been widely used by practitioners [47,48]. The DFH can be obtained according to the following steps.

Step 1: The annual maximum (AM) sampling method was used in this study, and the AM peak or n-day volumes data were derived.

Step 2: The Pearson type three distribution was used to fit the AM data sets. The design flood peak or volumes for certain return periods, such as 50 years, 100 years and 1000 years, were obtained.

Step 3: The typical flood hydrographs (TFH) were determined. Usually, the observed flood hydrograph with highest peak or largest volume was selected as a TFH [49].

Step 4: The DFH was constructed by multiplying each discharge ordinate of the TFH by an amplifier [12,13]. Certain amplitude method was used to make the flood peak or volumes of TFH equaling to the design values. In this study, only the volume amplitude method was considered, in accordance with the practical situation. Assume that the T-year design and observed n-day flood volumes are W_d^T and W_O respectively, and the amplifier with return period T is defined as k_T, $k_T = W_d^T/W_O$. The TFH was transformed to DFH by multiplying the parameter k_T. However, some other attributes, such as duration, were also considered for derivation of DFHs in some research works (e.g., Serinaldi and Grimaldi [50]; Vandenberghe et al. [51]; Gräler et al. [52]; Brunner et al. [53]), which were not considered in this study. The design flood hydrographs (DHFs) for the whole season can be obtained finally.

The floods in the years 1954, 1968, 1969, 1980, 1981, 1982, 1983, 1988, 1996, 1998, 1999, 2002 and 2010 were selected as TFHs in this study. Since the summer rainfall in the 1980s and 1990s is much higher than in the previous decades, eight flood hydrographs were selected during these periods [54]. The 1954 and 1998 floods are two serious flood events that happened in the past 100 years, which should be selected for flood control analysis. In the years of 1968 and 2002, the Wu River, which is a main tributary of the upper Yangtze river, experienced a severe flood [55]. Thus, the flood hydrographs of these two years were selected. In the year of 1969, extreme flood events happen simultaneously on the three main tributaries: the Jinsha, Min, and Tuo Rivers. In the year of 2010, the entire Yangtze River basin suffered from tremendous flooding, which led to the number of affected people as about 1.4 billion, and the direct economic loss of 197.6 billion Chinese yuan. According to the shape of these flood hydrographs, they can be divided into 3 types, namely the long duration type, single flood peak type and multiple flood peak type, which contain all the potential characteristics of flood events occurring in the upper Yangtze River Basin. The floods of 1954, 1996 and 1998 show long duration, which lasted for more than one month. For example, the 1954 and 1998 China floods lasted from middle of June to the beginning of September. There is only one main flood peak in the years of 1980, 1981, 1982, 1999 and 2002. In addition, there are multiple flood peaks in the years of 1968, 1969, 1983, 1988 and 2010. All the selected THFs are shown in Figure 4.

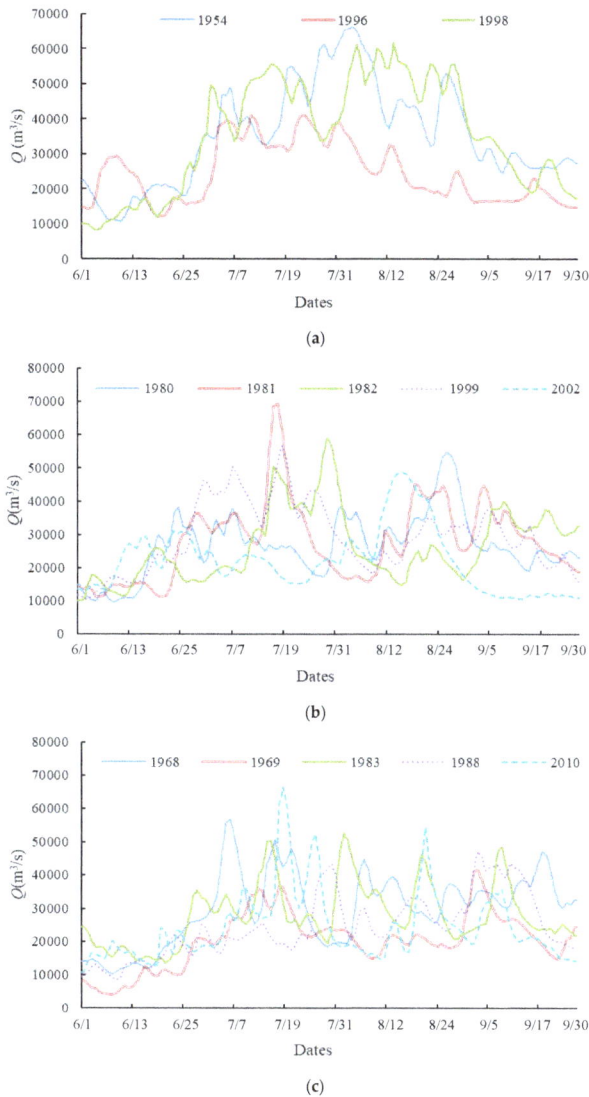

Figure 4. All the typical flood hydrographs (THFs) used in this study. (**a**) Observed flood hydrographs of the years 1954, 1996, and 1998; (**b**) observed flood hydrographs of the years 1980, 1981, 1982, 1999 and 2002; (**c**) observed flood hydrographs of the years 1968, 1969, 1983, 1988 and 2010.

The duration of the flood hydrograph plays a key role in determination of the DFH. In this study, the hydrograph lasts for the whole flood season (1 June to 30 September) [56–58]. All typical flood hydrographs mentioned above were used to derive the DFHs. The volume amplitude method, namely the 30-day volume amplifier, was selected in this study to obtain the DFHs, in accordance with practical use. In other words, the 30-day volume amplifier was employed for amplifying the hydrograph of the whole flood season. Considering the length of the paper, only the design flood hydrographs based on the TFH of 1954 were shown in Figure 5.

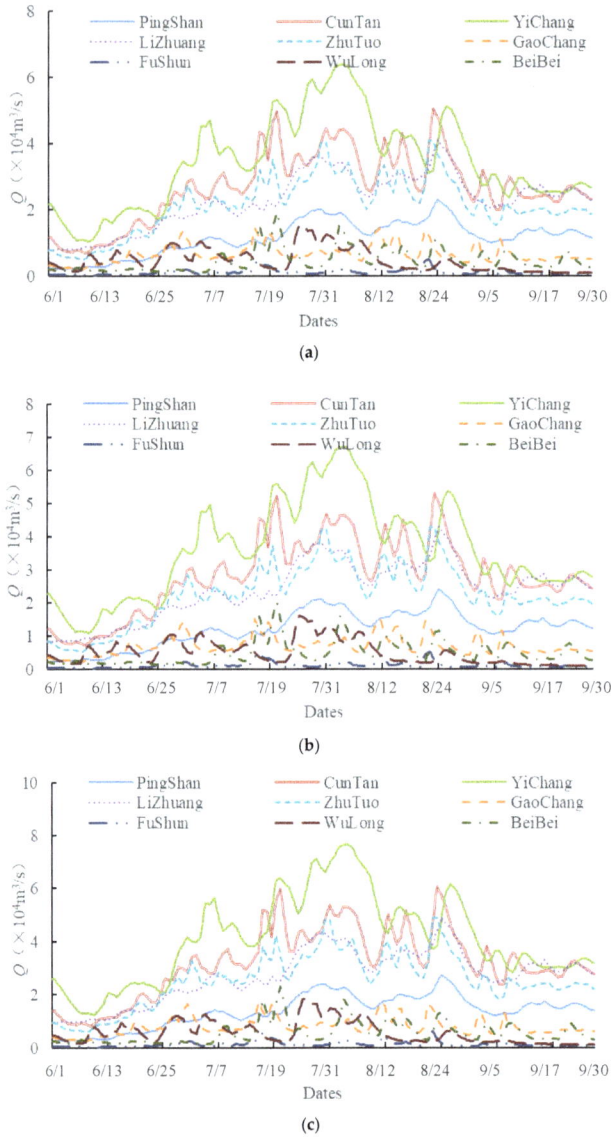

Figure 5. The 50-year, 100-year and 1000-year design flood hydrographs based on the typical flood hydrograph of 1954. (**a**) 50-year; (**b**) 100-year; (**c**) 1000-year.

4.2. Optimal Reservoir Operation Results

4.2.1. Operation Results of the Xiluodu and Xiangjiaba Reservoirs

The Xiluodu and Xiangjiaba Dams were used to reduce the inflow of TGR, which is also an effective method to alleviate the flood control burden of the Jingjiang and Chenglingji areas. A two-stage optimization algorithm was employed to solve the optimization model of the Xiluodu and Xiangjiaba cascade reservoirs, because of the "curse of dimensionality" and time-lag issue for multi-reservoir systems. First, the DP method was used to obtain the optimal solution set. Second, the output of the DP method was the input of the POA method, which avoids the influence of initial solutions on the optimal solutions. The years of 1954, 1968, 1969, 1980, 1981, 1982, 1983, 1988, 1996, 1998, 1999, 2002, and 2010 were selected as TFHs. The real-time water level and discharge at each time period were calculated by the flood hydrograph routing method.

Operation results of each DFH are given in Table 3, in which the inflow values of TGR and the flood stored in the Xiluodu and Xiangjiaba Dams are shown. The flood reductions are also given in Table 3.

As shown in Table 3, the Xiluodu and Xiangjiaba cascade reservoirs can obviously reduce the the inflow of TGR. For the 1000-year design flood, amplified based on the TFH of 1998, the maximum inflow of the Three Gorges decreased from 109,705.2 m^3/s to 92,570.6 m^3/s. The flood peak was reduced by 17,134.6 m^3/s and its corresponding reducing rate was 15.6%. The average reduction of flood peak for all these 13 TFHs can reach 13.6%. Therefore, the decrease of the peak inflow can relieve the flood control burdens of TGD. Additionally, the cascade reservoirs can also store much more water during a flood event, and the maximum volumes stored in those two reservoirs upstream in this study can reach 252.1 × 10^8 m^3. It also can be seen from Table 3 that the two-stage optimization algorithm has a good performance in solving the cascade reservoirs optimization model. Furthermore, when the POA method was used in the second stage, the model shows higher efficiency. More water was stored in the two reservoirs, and the inflow of TGR was further reduced. For instance, the maximum inflow of the 100-year design flood based on 1998 TFH reduced to 59,905.7 m^3/s in the first stage, and to 57,890.4 m^3/s in the second stage.

Figure 6 presents the 1000-year design flood hydrographs of 1954, and the inflow of TGR after the reservoir operation. The inflows regulated by the cascade reservoirs are much flatter with smaller peak values, which means that the inflow is easier to control.

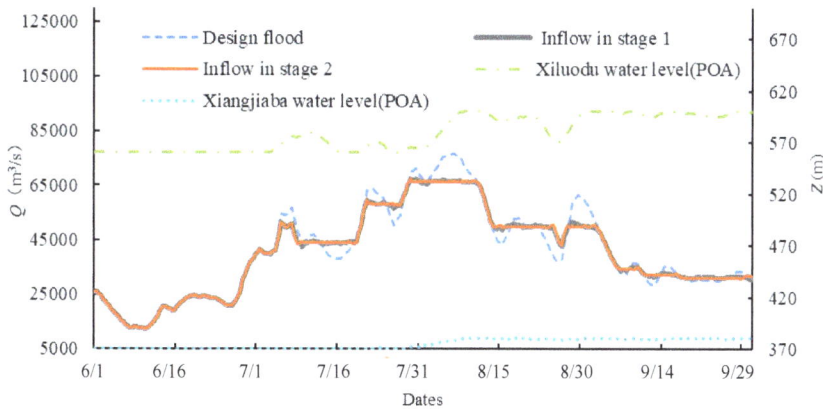

Figure 6. Design flood hydrograph of the year 1954, inflow of TGR and water levels of the Xiluodu and Xiangjiaba Dams.

Table 3. Operation results of the Xiluodu–Xiangjiaba cascade reservoirs.

TFH [1]	Return Periods (Year)	Inflow of the Three Gorges (m³/s)					Flood Stored in the Xiluodu and Xiangjiaba Reservoirs (10⁸ m³)		
		Natural Runoff	Stage 1	Flood Reduction	Stage 2	Flood Reduction	Stage 1	Stage 2	Comparisons
1954	50	76,676	67,500.34	9175.66	67,284.74	215.60	144.28	159.37	15.08
	100	67,025.4	58,388.07	8637.32	58,147.99	240.08	138.16	152.22	14.06
	1000	63,720.4	55,267.93	8452.47	54,993.17	274.75	133.9	148.0	14.1
1968	50	96,390	84,091.62	12,298.37	84,091.0	0.57	240.2	252.1	11.9
	100	83,916	74,102.2	9813.77	74,101.3	0.92	234.2	243.5	9.3
	1000	79,833.6	70,840.2	8993.4	70,839.8	0.35	231.6	239.9	8.3
1969	50	92,599	77,086.3	15,512.6	76,456.3	630.1	160.0	174.3	14.3
	100	80,908.9	64,984.7	15,924.1	64,588.8	395.9	150.9	162.7	11.8
	1000	76,928.4	61,000.5	15,927.9	60,846.7	153.7	147.6	158.6	11.0
1980	50	91,182	79,087.6	12,094.38	78,731.4	356.17	163.8	175.7	11.9
	100	79,661.4	66,305.01	13,356.39	66,116.25	188.75	167.32	179.13	11.8
	1000	75,730.2	63,189.1	12,541.04	63,160.1	29.03	165.79	176.63	10.8
1981	50	111,200	96,768.65	14,431.35	96,689.93	78.71	177.64	188.79	11.1
	100	97,091.5	88,907.63	8183.8	88,408.22	499.4	151.5	161.02	9.4
	1000	92,921.5	81,542.78	11,378.72	81,504.39	38.38	156.96	165.87	8.9
1982	50	92,040	79,071.08	12,968.92	79,030.08	41	162.40	174.88	12.48
	100	80,417	69,906.41	10,510.5	69,889.25	17.16	149.2	163.11	13.9
	1000	76,523	66,847.61	9675.39	66,832.96	14.64	143.3	157.4	14.06
1983	50	95,732	87,083.95	8648.05	87,083.9	0.05	182.1	195.1	13.0
	100	82,582	74,997.29	7584.7	74,997.29	0	167.09	186.42	19.3
	1000	78,479.2	70,371.11	8108.08	70,371.1	0	188.91	206.8	17.89
1988	50	80,106	69,897.8	10,208.17	68,171.5	1726.338	155.74	169.56	13.8
	100	69,772.8	59,948.75	9824.05	59,915.72	33.03	144.4	156.49	12.04
	1000	66,360	57,210.63	9149.37	57,210.63	0	142.0	153.2	11.2
1996	50	78,501	69,472.82	9028.18	69,058.74	414.1	153.67	181.93	28.25
	100	68,554.8	60,624.35	7930.4	60,622.61	1.74	140.67	174.7	34.02
	1000	65,184.6	57,982.89	7201.7	57,982.76	0.13	136.29	170.24	33.9
1998	50	109,705.2	92,571.2	17,133.9	92,570.6	0.6	192.3	201.2	8.8
	100	67,438.1	59,905.7	7532.3	57,890.4	2015.3	159.4	176.6	17.2
	1000	64,168	56,802.3	7365.6	54,815.2	1987.0	157.7	174.3	16.6
1999	50	90,153	73,305.4	16,847.5	73,306.6	-1.1	180.9	197.6	16.7
	100	78,642.9	64,300.6	14,342.6	64,143.8	156.4	168.5	185.0	16.4
	1000	74,787.3	61,446.0	13,341.2	61,318.9	127.1	164.1	181.5	17.4
2002	50	81,648	70,413.2	11,234.7	69,874.5	538.7	127.5	136.5	9.0
	100	71,344.8	60,551.9	10,792.8	60,001.7	550.2	117.2	124.5	7.3
	1000	67,845.6	57,273.2	10,572.3	56,824.3	448.8	113.5	120.7	7.2
2010	50	109,705.2	99,823.4	9881.8	99,823.0	0.39	190.3	201.8	11.4
	100	96,407.6	88,551.2	7856.3	88,550.7	0.52	176.4	189.3	12.9
	1000	91,088.56	84,048.3	7040.2	84,047.9	0.5	167.5	179.7	12.2

[1] TFH: typical flood hydrograph.

4.2.2. Operation Results of the Three Gorges Reservoir

Considering the flood control in the middle and lower reaches of the Yangtze River, this paper compared two operation methods, namely single reservoir operation model (TGR) and the cascade operation model (Xiluodu, Xiangjiaba and TGR).

Tables 4 and 5 present the flood diversion volumes of each DFH at the Jingjiang and Chenglingji flood control points, respectively, when different operation models were used. As shown in Table 4, the flood in the Jingjiang area does not need to be diverged for all 50-year DFHs. That means both the single and cascade models can ensure the water level of the Jingjiang area not exceeding the safety water level, when a 50-year flood event occurs. For 100-year DHFs, the single model can protect the Jingjiang areas from flooding for most of DFHs, except for the year of 1968, which means that when a single reservoir is used, the 100-year flood control standard for the Jingjiang area cannot always be guaranteed. When the cascade model used, there is no need to diverse the flood volumes and the safety of this areas can be ensured for the flood events with the 100-year return period. Generally, the differences between single reservoir operation model and the cascade models are not so obvious for the 50-year and 100-year DFHs. But for the 1000-year DFHs, it is clear from Table 4 that the flood diversion volumes obtained by the cascade operation model are less than those by the single reservoir model. For example, for the 1000-year design flood amplified based on the TFH of 1988, the flood diversion at the Jingjiang flood control point is 20.1 billion m^3 calculated by the single operation model, but reduced to zero when the Xiluodu and Xiangjiaba reservoirs were taken into accounts. It can be seen from Table 5 that the volumes decrease significantly when the cascade reservoir models are used. Tables 4 and 5 show that the cascade optimal operation model works better than single operation model in reducing the flood diversion volumes in the Jingjiang and Chenglingji areas.

Table 4. Flood diversion volumes of each design flood hydrograph (DFH) at the Jingjiang flood control point (billion m^3).

TFH	Return Periods (Year)	Without Operations	For the Jingjiang Area		For the Chenglingji Area	
			Single	Cascade	Single	Cascade
1954	50	4.8	0	0	0	0
	100	8.4	0	0	0	0
	1000	23.7	0	0	0	0
1968	50	17.0	0	0	0	0
	100	24.0	4.2	0	4.2	0
	1000	56.5	33.9	37.9	56.4	37.9
1969	50	12.6	0	0	0	0
	100	17.7	0	0	0	0
	1000	37.5	17.3	12.2	17.2	12.2
1980	50	8.5	0	0	0	0
	100	11.1	0	0	0	0
	1000	24.8	0	0	0	0
1981	50	14.0	0	0	0	0
	100	18.5	0	0	0	0
	1000	38.1	0	0	0	0
1982	50	8.3	0	0	0	0
	100	11.3	0	0	0	0
	1000	26.0	17.8	0	16.9	0
1983	50	19.0	0	0	0	0
	100	25.5	0	0	0	0
	1000	55.2	55.2	39.6	55.1	39.7
1988	50	5.9	0	0	0	0
	100	10.5	0	0	0	0
	1000	26.7	20.1	0	20.1	0

Table 4. *Cont.*

TFH	Return Periods (Year)	Without Operations	For the Jingjiang Area		For the Chenglingji Area	
			Single	Cascade	Single	Cascade
1996	50	9.7	0	0	0	0
	100	15.9	0	0	0	0
	1000	42.1	18.9	13.4	18.9	13.4
1998	50	6.0	0	0	0	0
	100	12.9	0	0	0	0
	1000	29.4	0	0	0	0
1999	50	9.4	0	0	0	0
	100	15.2	0	0	0	0
	1000	37.5	14.3	8.9	14.3	8.9
2002	50	6.8	0	0	0	0
	100	9.8	0	0	0	0
	1000	20.0	0	0	0	0
2010	50	13.4	0	0	0	0
	100	17.2	0	0	0	0
	1000	29.4	0	0	0	0

Table 5. Flood diversion volumes of each design flood hydrograph (DFH) at the Chenglingji flood control point (billion m^3).

TFH	Return Periods (Year)	Without Operations	For the Jingjiang Area		For the Chenglingji Area	
			Single	Cascade	Single	Cascade
1954	50	31.2	35.4	26.4	9.6	4.0
	100	37.1	45.3	36.6	19.3	13.7
	1000	54.3	74.2	70.1	59.8	54.2
1968	50	59.9	73.5	63.8	64.5	59.0
	100	69.8	88.4	80.8	79.7	78.4
	1000	93.9	111.1	101.7	93.9	102.1
1969	50	37.7	50.0	38.8	33.2	28.3
	100	45.2	61.7	50.5	50.5	45.6
	1000	70.8	88.8	87.3	88.9	87.4
1980	50	28.7	35.7	26.0	12.1	6.6
	100	36.2	46.1	35.9	23.5	18.0
	1000	61.6	83.9	74.2	77.2	71.6
1981	50	39.8	49.4	38.8	20.5	14.4
	100	46.3	59.1	48.2	35.8	30.3
	1000	68.3	98.0	91.2	89.9	84.4
1982	50	28.9	35.1	25.6	14.8	3.7
	100	35.5	43.2	34.3	24.6	13.6
	1000	51.7	67.9	61.9	56.4	52.5
1983	50	47.6	64.4	52.5	52.4	46.9
	100	55.5	78.9	69.9	72.5	66.9
	1000	83.0	83.0	92.2	82.8	91.7
1988	50	30.5	34.4	25.4	17.8	12.4
	100	34.4	41.1	33.6	30.3	24.8
	1000	48.2	54.1	62.8	52.4	62.6
1996	50	47.6	57.1	51.5	32.5	26.9
	100	54.0	68.0	63.9	43.7	38.1
	1000	68.7	89.4	88.8	63.1	63.0
1998	50	55.5	61.2	54.1	37.5	32.0
	100	62.0	73.7	68.1	51.9	46.4
	1000	41.5	61.5	53.0	39.3	33.7

Table 5. *Cont.*

TFH	Return Periods (Year)	Without Operations	For the Jingjiang Area		For the Chenglingji Area	
			Single	Cascade	Single	Cascade
1999	50	43.5	52.2	41.1	24.7	19.1
	100	50.6	62.3	52.9	42.6	37.1
	1000	71.7	89.6	88.9	82.6	82.5
2002	50	18.2	22.9	19.4	1.6	0.0
	100	20.8	27.1	24.0	5.6	0.1
	1000	32.0	43.1	40.5	19.8	14.6
2010	50	23.2	32.1	23.1	3.0	0.0
	100	27.9	39.3	29.3	9.0	3.3
	1000	41.5	61.5	52.7	39.3	33.7

Table 6 gives the highest water level of TGR of each TFH with different return periods, for the flood control of the Jingjiang area, and Table 7 for the flood control of both the Jiangjiang and Chenglingji areas. Tables 6 and 7 illustrate that the highest water level of TGR generally is lower, when the cascade optimal operation model was considered instead of the single operation model. For the cascade optimal operation model considering the Jingjiang flood control point only, the highest water level nearly reached 170 m for the 1000-year DFH amplified by three TFHs. For a 100-year design flood, the maximum highest water level was 164.1 m, and the minimum was only 150.8 m. These results indicate that when the cascade optimal operation model is in use, the safety of TGR can be ensured, but the reservoir's flood control capacity is underutilized only considering the Jingjiang flood control area. Furthermore, as shown in Table 7, when both the Jingjiang and Chenglingji flood control points were taken into account, the highest water level of TGR rises obviously, which proves that the flood control capacity of TGR is further utilized. Thus, both the Jingjiang and Chenglingji flood control areas can be protected in a practical flood control operation.

Table 6. The highest water level of Three Gorges Reservoir considering the flood control for the Jingjiang area (m).

TFH		1954			1981			1998	
Return Periods (Year)	1000	100	50	1000	100	50	1000	100	50
Single mode	175.0	159.3	153.9	174.4	166.1	163.9	174.7	160.7	153.4
Cascade mode	169.3	150.8	145.0	169.4	164.1	156.4	169.8	152.4	145.3
Comparison	5.7	8.5	8.9	5.0	2.0	7.5	4.9	8.3	8.1

Table 7. The highest water levelof Three Gorges Reservoir considering the flood control for the Jingjiang and Chenglingji areas (m).

TFH		1954			1981			1998	
Return Periods (Year)	1000	100	50	1000	100	50	1000	100	50
Single mode	175	175	175	175	175	175	175	175	175
Cascade mode	171.9	175	175	170.8	172.0	173.0	172.3	175	175
Comparison	3.1	0	0	4.2	3.0	2.0	2.7	0	0

Figure 7 presents the discharge and water level hydrographs of TGR calculated by single and cascade optimal operation models, considering the flood control in the Chenglingji area, for the 1000-year design flood of the year 1954. This figure demonstrates that results obtained by the cascade optimal operation model is generally better than that obtained by single operation, since the peak value is much smaller and the discharge curve is much flatter. Especially, the highest water level of TGR was also lower.

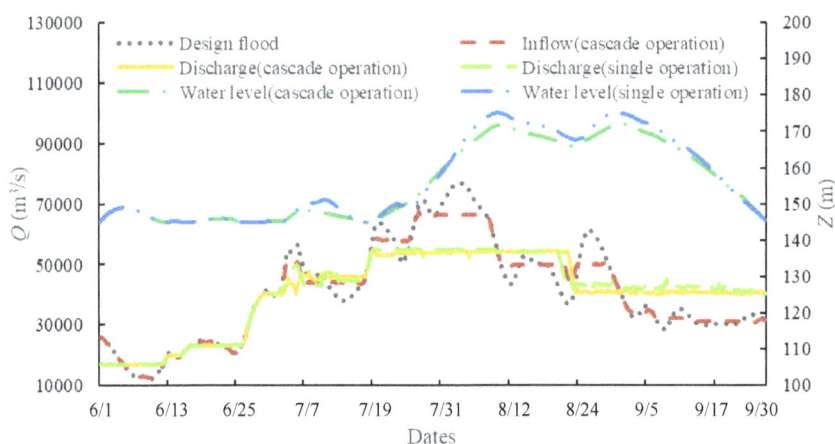

Figure 7. Discharge and water level hydrographs of TGR obtained by single and cascade optimal operation models for the 1954 DFH.

4.3. Comparisons with the Current Operation Rules of Xiluodu, Xiangjiaba and TGD in the Flood Season

To testify the feasibility of the cascade optimal operation model, this paper also compares the optimal operation model with current operation method of these three dams, where the TFHs in use are the same as those mentioned in Section 4.1 in this paper.

The flood diversion volumes at the Jingjiang and Chenglingji flood control points were calculated by the optimal and current operation models, since the objective of the reservoir operation is to minimize the flood diversion volumes at the control points. As shown in Tables 8 and 9, compared with the current operation model, the flood diversion volumes calculated by the optimal operation rules decrease significantly. From this point of view, the optimal model, which has the smaller flood diversion volumes, performs better than the current operation model. For the 50-year and 100-year design flood, when both of the two methods are used, the safety of the Jingjiang area can be ensured. But for the 1000-year DFHs the cascade optimal operation method has a better performance than the current used model, obviously. For instance, for the 1000-year design flood based on the TFH of 1999, the flood diversion volumes in the Jingjiang area obtained by the current operation method are 13.8 billion m^3 (considering the Jingjiang flood control point only) and 18.2 billion m^3 (considering the Jingjiang and Chenglingji flood control points); while they are the optimal operation models used, these diversion values can be reduced to 8.9 billion m^3 for both of the flood control points. This improvement is much more significant for the Chenglingji area. Generally, the cascade optimal operation model can be with less flood diversion volumes than the current operation model for all DFHs with different return periods. Hence, it can be concluded that the cascade optimal operation model can improve the flood control capacity in the middle and lower reaches of the Yangtze River, compared with the current operation model. All the results obtained by the two models do not consider the uncertainties in flood control reservoir operation, which can be discussed in future research.

Table 8. Flood diversion volumes of some design flood hydrographs at the Jingjiang flood control point (billion m^3).

TFH	Return Periods (Year)	Without Operations	For the Jingjiang Area		For both the Jingjiang and Chenglingji Areas	
			Current	Optimal	Current	Optimal
1954	50	4.8	0	0	0	0
	100	8.4	0	0	0	0
	1000	23.7	0	0	5.7	0
1980	50	8.5	0	0	0	0
	100	11.1	0	0	0	0
	1000	24.8	0	0	4.1	0
1981	50	14.0	0	0	0	0
	100	18.5	0	0	0	0
	1000	38.1	0	0	5.2	0
1996	50	9.7	0	0	0	0
	100	15.9	0	0	0	0
	1000	42.1	18.7	13.4	22.8	133.8
1999	50	9.4	0	0	0	0
	100	15.2	0	0	0	0
	1000	37.5	13.8	8.9	18.2	8.9
2002	50	6.8	0	0	0	0
	100	9.8	0	0	0	0
	1000	20.0	0	0	2.8	0

Table 9. Flood diversion volumes of some design flood hydrographs at the Chenglingji flood control point (billion m^3).

TFH	Return Periods (Year)	Without Operations	For the Jingjiang Area		For both the Jingjiang and Chenglingji Areas	
			Current	Optimal	Current	Optimal
1954	50	31.2	35.5	26.4	26.8	4.0
	100	37.1	45.2	36.6	36.2	13.7
	1000	54.3	75.1	70.1	65.3	54.2
1980	50	28.7	35.7	26.0	21.1	6.6
	100	36.2	46.2	35.9	32.7	18.0
	1000	61.6	84.2	74.2	69.5	71.6
1981	50	39.8	50.7	38.8	38.2	14.4
	100	46.3	60.4	48.2	48.2	30.3
	1000	68.3	99.2	91.2	86.4	84.4
1996	50	47.6	57.1	51.5	49.8	26.9
	100	54.0	68.6	63.9	59.9	38.1
	1000	68.7	89.8	88.8	74.0	63.0
1999	50	43.5	51.9	41.1	41.5	19.1
	100	50.6	62.6	52.9	52.4	37.1
	1000	71.7	90.4	88.9	79.4	82.5
2002	50	18.2	23.7	19.4	17.8	0
	100	20.8	28.3	24.0	20.7	0.1
	1000	32.0	44.8	40.5	32.3	14.6

Figure 8 displays the 1000-year design inflow flood hydrograph of 1954, in which one hydrograph is operated by the current operation model and the other by the Xiluodu and Xiangjiaba cascade optimal operation model. Figure 9 shows the discharge and water level curves of TGR obtained respectively by current and cascade optimal operation models considering the flood control of the Chenglingji area. It is obvious that when using the cascade optimal operation model, both the inflow and discharge hydrographs of TGR are flatter and have smaller peak values than those obtained by the current model. For example, the maximum discharge value of 1980 DHF with the return period 1000 years reduces from 76,000 m³/s to 54,995 m³/s, which can ensure the safety of the Jingjiang River reach. These results also indicate that the capacities of the Xiluodu and Xiangjiaba reservoirs were further utilized when the cascade optimal operation model was used, which alleviates the flood control burdens of TGR.

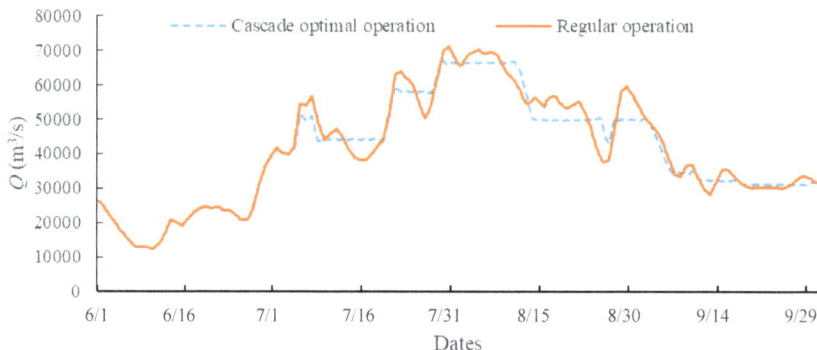

Figure 8. Inflow of TGR obtained by the two operation models for the 1000-year DHF amplified by the 1954 TFH.

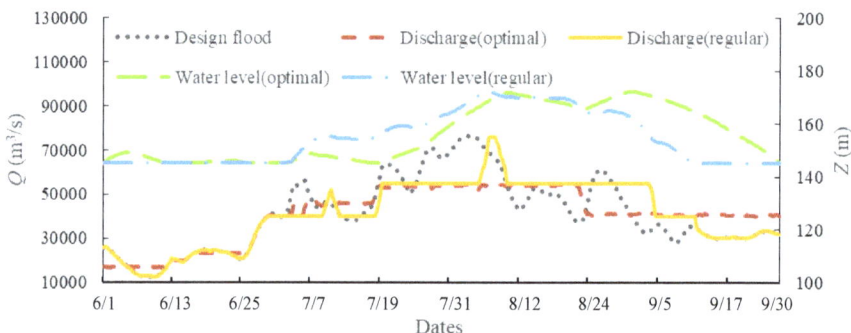

Figure 9. Discharge and water level hydrographs of Three Gorges Reservoir (TGR) obtained by regular and optimal operation models.

In order to further testify the reliability of the model, the historical observations, including the flood hydrographs of the years 1954, 1968, and 1998, were considered in this section and taken as the inputs of the current and optimal models. The highest water level of TGR routed by the proposed and current models are shown in Table 10, which indicates that the highest water level of TGR generally is lower based on the optimal model.

Table 10. The highest water level of TGR by two models (m).

TFH	1954	1968	1998
Cascade regular operation model	157.3	145.3	147.9
Cascade optimal operation model	148.8	145	145.2
Comparisons	8.5	0.3	2.7

According to the comparisons given in Table 10, it can be seen that the cascade optimal operation model can produce superior results to the current operation model. Therefore, the flood control capacity of the middle and lower reaches of the Yangtze river can be enhanced. The cascade optimal operation model built in this paper worked well in a multi-reservoir flood control operation system, and can be considered in practical application.

5. Conclusions and Discussions

This study established optimal flood control operation models of cascade reservoirs for flood control of multiple points downstream. The current multi-reservoir flood control co-operation models mainly concern minimizing the maximal water level for the dam itself or the releases at the downstream flood control points, or balancing the conflict between the flood control and conservation during the flood season. In this study, a two-stage optimal operation model of cascade reservoirs was proposed to ensure the safety of multiple downstream areas, in which the objective functions are to minimize the inflow of the reservoir C downstream in the first stage and the water level or flood diversion volume in the second stage. For a large river basin, the river reach is very long, the time-lag issue and Muskingum flood routing along the river cannot be neglected. A DP-POA optimal algorithm was proposed in this study for the establishment of the optimal model to avoid the curse of dimensionality and influence of initial solutions. The upper Yangtze River was selected as a case study. The flood capacity of TGR cannot hold back floods downstream, when extreme flooding occurred throughout the whole upper Yangtze River Basin. Therefore, more reservoirs are needed to help TGR realize the purpose of flood control. The two reservoirs, Xiluodu and Xiangjiaba located in the Jinsha River Basin, were considered in this study to reduce the inflow of TGR. The main conclusions of this paper are summarized as follows.

(1) The first innovation of this paper is the proposed two-stage optimal operation model with different objective functions. In the first stage, the optimal operation model of Reservoirs A and B was established to reduce the inflow of the downstream reservoir. In the second stage, the optimal operation model of the reservoir downstream was established to minimize the highest water level and the flood diversion volumes at flood control points. The second innovation of this paper is the proposed DP-POA algorithm, which can solve the curse of dimensionality and influence initial solutions effectively. The results of Figures 6 and 7 indicate that optimal reservoir operation processes can be obtained. The Muskingum flood routing method combined with the time delay makes the operation's results much more objective. The two-stage optimization algorithm has a good performance in solving the cascade reservoirs optimization model. When the POA method was used in the second stage, the model shows higher efficiency as more water is stored in the two reservoirs and the inflow of the reservoir downstream is further reduced.

(2) Based on the proposed model, the Xiluodu and Xiangjiaba Reservoirs can effectively reduce the peak inflow and volumes of TGR. After the optimal reservoir operation of Xiluodu and Xiangjiaba, the average reduction of flood peaks for all these 13 TFHs are 13.6%. Meanwhile, the cascade reservoirs can also store much more water during a flood event, and the maximum volumes stored in those two reservoirs upstream in this study can reach 25.2 billion m^3 during a flood event.

(3) The cascade optimal operation model works better than the single operation model. The highest water level of TGR is generally lower when the cascade optimal operation model is considered instead of the single operation model. For the 1000-year DFHs, it is obvious that the flood diversion volumes obtained by the cascade operation model are less than those by the single reservoir model. After the optimal operation of cascade reservoirs, the peak value is much smaller and the discharge curve is much flatter, which is beneficial for the flood control of the middle and lower reaches of the Yangtze River.

(4) Compared with the current operation rules, the cascade optimal operation model can produce a superior result to the current operation model. For the 50-year and 100-year design floods, when both of the two methods were used, the safety of the Jingjiang area can be ensured. But for the 1000-year

DFHs, the cascade optimal operation method has a better performance than the current one obviously. For the 1000-year design flood amplified based on the TFH of 1999, the diversion values can decline to 8.9 billion m^3 for both of the Jingjiang and Chengliangji flood control points, when the cascade optimal operation model was used.

This paper only discussed the downstream flood control points that are protected by cascade reservoirs upstream. The proposed work can give some guidance for real-time flood management. Flood control reservoir operation rules can be developed based on the optimal results. In future, more reservoirs, including cascade reservoirs and parallel reservoirs, will be considered, and a more efficient algorithm will be proposed. Furthermore, the main purpose of this paper is to discuss the protection of the flood control areas downstream based on the multi-reservoir systems. The overtopping risk of the dam is not discussed in the paper.

In addition, many uncertainties also present in this complex problem have an influence on the flood control reservoir operation. Shi et al. [59], Zhang et al. [60] and Chen et al. [61] indicated that the most important uncertainty factor that affects the flood control reservoir operation was the flood forecast error. The predicted flow is the input of the flood control reservoir operation model, which has a great influence on the operation results. Besides flood forecast errors, Li et al. [19] and Huang et al. [62] demonstrated that the uncertainties in flood hydrograph shape also had a significant influence on the highest water level and maximum discharge. Besides these, Chen et al. [63] indicated that the uncertainties in the discharge capacity curve and water level-storage curve also influenced reservoir operation greatly. Onyutha and Willems [4] assessed the uncertainties in estimation of flood quantiles and/or hydrographs, especially of very high return periods which is useful for various types of risk-based water management applications related to floods. These uncertainties were not considered in this deterministic modelling, because of the burden of the optimal calculation. Stochastic frameworks, which can work better for these uncertainty problems, will be studied in the future research.

Author Contributions: C.Z. proposed the ideas of this paper and guided the writing and calculation work; L.C. designed the experiments and wrote most parts of the paper; N.S. prepared the manuscript and made some corrections; Y.D. and J.Z. made some corrections; G.Z., L.D., G.L. and X.Y. together did the calculation work and draw the figures.

funding: This research was funded by the National Key R&D Program of China, grant number [2017YFC0405900, 2016YFC0402210, 2016YFC0402205]; the National Natural Science Foundation of China, grant number [51679094]; and the project with the title of "impact of flood control multi-reservoir co-operation in the upper Yangtze River Basin on the Three Gorges Reservoir and its corresponding risk analysis". The APC was funded by the National Key R&D Program of China, grant number [2017YFC0405900].

Acknowledgments: This research was founded by the National Key R&D Program of China, grant number [2017YFC0405900, 2016YFC0402210; 2016YFC0402205]; the National Natural Science Foundation of China, grant number [51679094]; and the project with the title of "impact of flood control multi-reservoir co-operation in the upper Yangtze River Basin on the Three Gorges Reservoir and its corresponding risk analysis".

Conflicts of Interest: The authors declare no conflict of interest.

References

1. Webster, T.; Mcguigan, K.; Collins, K.; Macdonald, C. Integrated river and coastal hydrodynamic flood risk mapping of the lahave river estuary and town of Bridgewater, Nova Scotia, Canada. *Water* **2014**, *6*, 517–546. [CrossRef]

2. Schnebele, E.; Cervone, G.; Kumar, S.; Waters, N. Real time estimation of the calgary floods using limited remote sensing data. *Water* **2014**, *6*, 381–398. [CrossRef]

3. Morita, M. Flood risk impact factor for comparatively evaluating the main causes that contribute to flood risk in urban drainage areas. *Water* **2015**, *6*, 253–270. [CrossRef]

4. Onyutha, C.; Willems, P. Uncertainties in flow-duration-frequency relationships of high and low flow extremes in lake victoria basin. *Water* **2013**, *5*, 1561–1579. [CrossRef]

5. Assani, A.; Landry, R.; Labrèche, M.; Frenette, J.J.; Gratton, D. Temporal variability of monthly daily extreme water levels in the st. Lawrence river at the sorel station from 1912 to 2010. *Water* **2014**, *6*, 196–212. [CrossRef]

6. Joo, J.; Lee, J.; Kim, J.H.; Jun, H.; Jo, D. Inter-event time definition setting procedure for urban drainage systems. *Water* **2013**, *6*, 45–58. [CrossRef]

7. Keast, D.; Ellison, J. Magnitude frequency analysis of small floods using the annual and partial series. *Water* **2013**, *5*, 1816–1829. [CrossRef]

8. Box, P.; Thomalla, F.; van den Honert, R. Flood risk in australia: Whose responsibility is it, anyway? *Water* **2013**, *5*, 1580–1597. [CrossRef]

9. Zhou, Y.L.; Guo, S.L.; Chang, F.J.; Liu, P.; Chen, A.B. Methodology that improves water utilization and hydropower generation without increasing flood risk in mega cascade reservoirs. *Energy* **2018**, *143*, 785–796. [CrossRef]

10. Zhang, W.; Liu, P.; Chen, X.; Wang, L.; Ai, X.; Feng, M.; Liu, D.; Liu, Y. Optimal operation of multi-reservoir systems considering time-lags of flood routing. *Water Resour. Manag.* **2016**, *30*, 523–540. [CrossRef]

11. Chen, C.; Yuan, Y.; Yuan, X. An improved nsga-iii algorithm for reservoir flood control operation. *Water Resour. Manag.* **2017**, *31*, 4469–4483. [CrossRef]

12. Chen, L.; Singh, V.; Huang, K. Bayesian technique for the selection of probability distributions for frequency analyses of hydrometeorological extremes. *Entropy* **2018**, *20*, 117. [CrossRef]

13. Chen, L.; Singh, V.P. Entropy-based derivation of generalized distributions for hydrometeorological frequency analysis. *J. Hydrol.* **2018**, *557*, 699–712. [CrossRef]

14. Narasaiah, M.L. *Microcredit and Agricultural Development*; Discovery Publish House: New Delhi, India, 2007.

15. Liu, X.; Chen, L.; Zhu, Y.; Singh, V.P.; Qu, G.; Guo, X. Multi-objective reservoir operation during flood season considering spillway optimization. *J. Hydrol.* **2017**, *552*, 554–563. [CrossRef]

16. Yazdi, J.; Neyshabouri, S.A.A.S. Optimal design of flood-control multi-reservoir system on a watershed scale. *Nat. Hazards* **2012**, *63*, 629–646. [CrossRef]

17. Lee, S.Y.; Hamlet, A.F.; Fitzgerald, C.J.; Burges, S.J. Optimized flood control in the columbia river basin for a global warming scenario. *J. Water Resour. Plan. Manag. ASCE* **2009**, *135*, 440–450. [CrossRef]

18. Qi, Y.; Yu, J.; Li, X.; Wei, Y.; Miao, Q. Reservoir flood control operation using multi-objective evolutionary algorithm with decomposition and preferences. *Appl. Soft Comput.* **2017**, *50*, 21–33. [CrossRef]

19. Li, X.; Guo, S.; Liu, P.; Chen, G. Dynamic control of flood limited water level for reservoir operation by considering inflow uncertainty. *J. Hydrol.* **2010**, *391*, 126–134. [CrossRef]

20. Chen, J.; Guo, S.; Li, Y.; Liu, P.; Zhou, Y. Joint operation and dynamic control of flood limiting water levels for cascade reservoirs. *Water Resour. Manag.* **2013**, *27*, 749–763. [CrossRef]

21. Zhou, Y.; Guo, S.; Liu, P.; Xu, C. Joint operation and dynamic control of flood limiting water levels for mixed cascade reservoir systems. *J. Hydrol.* **2014**, *519*, 248–257. [CrossRef]

22. He, X.; Zhang, L.; Liu, L. Study on flood control operation of jinsha river cascade reservoirs combined with the three gorges along the yangtze river. *Mater. Sci. Eng.* **2017**, *199*, 012032.

23. Sahoo, B. Field application of the multilinear muskingum discharge routing method. *Water Resour. Manag.* **2013**, *27*, 1193–1205. [CrossRef]

24. Khan, M.H. Muskingum flood routing model for multiple tributaries. *Water Resour. Res.* **1993**, *29*, 1057–1062. [CrossRef]

25. Gill, M.A. Flood routing by the muskingum method. *J. Hydrol.* **1978**, *36*, 353–363. [CrossRef]

26. Yeh, W.W.G. Reservoir management and operations models: A state-of-the-art review. *Water Resour. Res.* **1985**, *21*, 1797–1818. [CrossRef]

27. Choudhury, P. Reservoir flood control operation model incorporating multiple uncontrolled water flows. *Lakes Reserv. Res. Manag.* **2010**, *15*, 153–163. [CrossRef]

28. Hall, W.A.; Butcher, W.S.; Esogbue, A. Optimization of the operation of a multiple-purpose reservoir by dynamic programming. *Water Resour. Res.* **1968**, *4*, 471–477. [CrossRef]

29. Loucks, D.P.; Sigvaldason, O.T. *Multiple-Reservoir Operation in North America*; American Society of Civil Engineers (ASCE): Reston, VA, USA, 1982; pp. 711–728.

30. Labadie, J.W. Optimal operation of multireservoir systems: State-of-the-art review. *J. Water Resour. Plan. Manag.* **2004**, *130*, 93–111. [CrossRef]

31. Butcher, W.S. Stochastic dynamic programming for optimum reservoir operation. *Water Resour. Bull.* **1971**, *7*, 115–123. [CrossRef]

32. Yeh, W.W.G.; Becker, L.; Chu, W.-S. Real-time hourly reservoir operation. *J. Water Resour. Plan. Manag. Div. Am. Soc. Civ. Eng.* **1979**, *105*, 187–203.

33. Yakowitz, S. Dynamic programming applications in water resources. *Water Resour. Res.* **1982**, *18*, 673–696. [CrossRef]

34. Kelman, J.; Stedinger, J.R.; Cooper, L.A.; Hsu, E.; Sun-Quan, Y. Sampling stochastic dynamic programming applied to reservoir operation. *Water Resour. Res.* **1990**, *26*, 447–454. [CrossRef]

35. Heidari, M.; Chow, V.T.; Kokotović, P.V.; Meredith, D.D. Discrete differential dynamic programing approach to water resources systems optimization. *Water Resour. Res.* **1971**, *7*, 273–282. [CrossRef]

36. Chow, V.T.; Cortes-Rivera, G. *Application of DDDP in Water Resources Planning*; University of Illinois at Urbana-Champaign. Water Resources Center: Saint Paul, MN, USA, 1974.

37. Cheng, C.T.; Wang, S.; Chau, K.W.; Wu, X.Y. Parallel discrete differential dynamic programming for multireservoir operation. *Environ. Model. Softw.* **2014**, *57*, 152–164. [CrossRef]

38. Larson, R.E.; Korsak, A.J. *A Dynamic Programming Successive Approximations Technique with Convergence Proofs*; Pergamon Press, Inc.: Oxford, UK, 1970; pp. 245–252.

39. Erkmen, I.; Karatas, B. Short-term hydrothermal coordination by using multi-pass dynamic programming with successive approximation. In Proceedings of the Mediterranean Electrotechnical Conference (MELECON '94), Antalya, Turkey, 12–14 April 1994; Volume 923, pp. 925–928.

40. Liu, P.; Cai, X.M.; Guo, S.L. Deriving multiple near-optimal solutions to deterministic reservoir operation problems. *Water Resour. Res.* **2011**, *47*, 20. [CrossRef]

41. Howson, H.R.; Sancho, N.G.F. A new algorithm for the solution of multi-state dynamic programming problems. *Math. Program.* **1975**, *8*, 114–116. [CrossRef]

42. Turgeon, A. Optimal short-term hydro scheduling from the principle of progressive optimality. *Water Resour. Res.* **1981**, *17*, 481–486. [CrossRef]

43. Guo, S.L.; Chen, J.H.; Li, Y.; Liu, P.; Li, T.Y. Joint operation of the multi-reservoir system of the three gorges and the qingjiang cascade reservoirs. *Energies* **2011**, *4*, 1036–1050. [CrossRef]

44. Chen, L.; Singh, V.P.; Guo, S.L.; Hao, Z.C.; Li, T.Y. Flood coincidence risk analysis using multivariate copula functions. *J. Hydrol. Eng.* **2012**, *17*, 742–755. [CrossRef]

45. Chen, L.; Ye, L.; Singh, V.; Asce, F.; Zhou, J.; Guo, S. Determination of input for artificial neural networks for flood forecasting using the copula entropy method. *J. Hydrol. Eng.* **2014**, *19*, 217–226. [CrossRef]

46. Chen, L.; Singh, V.P.; Guo, S.; Zhou, J.; Zhang, J. Copula-based method for multisite monthly and daily streamflow simulation. *J. Hydrol.* **2015**, *528*, 369–384. [CrossRef]

47. Nezhikhovsky, R.A. *Channel Network of the Basin and Runoff Formation*; Hydrometeorological Publishing: Leningrad, Russia, 1971.

48. Yue, S.; Ouarda, T.B.M.J.; Bobée, B.; Legendre, P.; Bruneau, P. Approach for describing statistical properties of flood hydrograph. *J. Hydrol. Eng.* **2002**, *7*, 147–153. [CrossRef]

49. Chen, L.; Guo, S.L.; Yan, B.W.; Liu, P.; Fang, B. A new seasonal design flood method based on bivariate joint distribution of flood magnitude and date of occurrence. *Hydrol. Sci. J.* **2010**, *55*, 1264–1280. [CrossRef]

50. Serinaldi, F.; Grimaldi, S. Synthetic design hydrographs based on distribution functions with finite support. *J. Hydrol. Eng.* **2011**, *16*, 434–446. [CrossRef]

51. Vandenberghe, S.; Berg, M.J.; Gräler, B.; Petroselli, A.; Grimaldi, S.; De Baets, B.; Verhoest, N.E.C. Joint return periods in hydrology: A critical and practical review focusing on synthetic design hydrograph estimation. *Hydrol. Earth Syst. Sci. Discuss.* **2012**, *9*, 6781–6828. [CrossRef]

52. Graeler, B.; van den Berg, M.J.; Vandenberghe, S.; Petroselli, A.; Grimaldi, S.; De Baets, B.; Verhoest, N.E.C. Multivariate return periods in hydrology: A critical and practical review focusing on synthetic design hydrograph estimation. *Hydrol. Earth Syst. Sci.* **2013**, *17*, 1281–1296. [CrossRef]

53. Brunner, M.I.; Viviroli, D.; Sikorska, A.E.; Vannier, O.; Favre, A.C.; Seibert, J. Flood type specific construction of synthetic design hydrographs. *Water Resour. Res.* **2017**, *53*, 1390–1406. [CrossRef]

54. GONG. Flooding 1990s along the yangtze river, has it concern of global warming? *J. Geogr. Sci.* **2001**, *11*, 43–52. [CrossRef]

55. Wang, L.N.; Chen, X.H.; Shao, Q.X.; Li, Y. Flood indicators and their clustering features in wujiang river, south china. *Ecol. Eng.* **2015**, *76*, 66–74. [CrossRef]

56. Huang, K.; Lu, C.; Zhou, J.; Zhang, J.; Singh, V.P. Flood hydrograph coincidence analysis for mainstream and its tributaries. *J. Hydrol.* **2018**, *565*, 341–353. [CrossRef]

57. Chen, L.; Singh, V.P.; Guo, S.; Zhou, J.; Zhang, J.; Liu, P. An objective method for partitioning the entire flood season into multiple sub-seasons. *J. Hydrol.* **2015**, *528*, 621–630. [CrossRef]

58. Chen, L.; Singh, V.P.; Guo, S.L.; Fang, B.; Liu, P. A new method for identification of flood seasons using directional statistics. *Hydrol. Sci. J.* **2013**, *58*, 28–40. [CrossRef]

59. Shi, H.; Li, T.; Liu, R.; Chen, J.; Li, J.; Zhang, A.; Wang, G. A service-oriented architecture for ensemble flood forecast from numerical weather prediction. *J. Hydrol.* **2015**, *527*, 933–942. [CrossRef]

60. Zhang, J.; Chen, L.; Singh, V.P.; Cao, W.; Wang, D. Erratum to: Determination of the distribution of flood forecasting error. *Nat. Hazards* **2015**, *75*, 2065. [CrossRef]

61. Chen, L.; Singh, V.P.; Lu, W.; Zhang, J.; Zhou, J.; Guo, S. Streamflow forecast uncertainty evolution and its effect on real-time reservoir operation. *J. Hydrol.* **2016**, *540*, 712–726. [CrossRef]

62. Huang, K.; Chen, L.; Wang, Q.; Dai, L.; Zhou, J.; Singh, V.P.; Huang, M.; Zhang, J. Risk analysis of flood control reservoir operation considering multiple uncertainties. *J. Hydrol.* **2018**. [CrossRef]

63. Chen, J.; Zhong, P.A.; Zhao, Y.F. Research on a layered coupling optimal operation model of the three gorges and gezhouba cascade hydropower stations. *Energy Convers. Manag.* **2014**, *86*, 756–763. [CrossRef]

Article

Stochastic Linear Programming for Reservoir Operation with Constraints on Reliability and Vulnerability

Cheng Chen, Chuanxiong Kang and Jinwen Wang *

School of Hydropower and Information Engineering, Huazhong University of Science and Technology, Wuhan 430074, China; chencheng410@gmail.com (C.C.); kachxi@gmail.com (C.K.)
* Correspondence: jinwen.wang@hust.edu.cn; Tel.: +86-027-8754-3437

Received: 16 December 2017; Accepted: 6 February 2018; Published: 9 February 2018

Abstract: Reliability and vulnerability (RV) are two very important performance measures but, due to their stage-inseparable nature, they cannot be explicitly incorporated in stochastic dynamic programming (SDP), which is extensively used in reservoir operation. With inflows described as a Markov chain, a stochastic linear programming (SLP) model is formulated in this paper to explicitly incorporate the RV constraints in the reservoir operation, aimed at maximizing the expected power generation by determining the optimal scheduling decisions and their probabilities. Simulation results of the SLP and SDP models indicate the equivalence of the proposed SLP and SDP models without considering the RV constraints, as well as the strength of the SLP in explicitly incorporating the RV constraints. A simulated scheduling solution also reveals a reduction of power generation fluctuation, with the reservoir capacity emptied in advance to meet given reliability and vulnerability.

Keywords: stochastic linear programming; Markov chain; reliability; vulnerability; reservoir operation; stochastic dynamic programming

---

1. Introduction

Generally being "high-dimensional, dynamic, nonlinear, and stochastic" [1], a water resource system is difficult to optimize. To solve such a complex problem, many researchers have attempted a variety of methods, mainly focused on models and algorithms, which are complementary in different situations. As the most mature mathematical programming, linear programming (LP) and dynamic programming (DP) provide two different ways of solving water resource planning problems. LP is one of the most widely used mathematical programming methods owing to its strict mathematic theory and general solution in hydropower management. Generally, a standard form is required for an LP model to be solved by an efficient generic solver. Based upon Bellman's optimality principle, DP is only applicable to stage-separable processes and encounters difficulties of so-called "curse of dimensionality" for exponential growth of variables in the growth of dimensionality.

Since randomness always accompanies the application of LP and DP, the stochastic characteristics generally cannot be ignored. Moreover, the handling of random characteristics has a profound influence on optimization or simulation performance. According to different treatments of stochastic characteristics, the deterministic and stochastic models represent two patterns. The deterministic model regards the stochastic hydrologic parameters as known quantities. This treatment reduces the complexity of the model to some extent. Nevertheless, the simplification cannot retain all the essential characteristics of the original data, and may lead to unsatisfactory results. Integrating the stochastic information represented in different ways into an LP or DP model, a variety of stochastic models has been developed.

With the randomness involved, stochastic dynamic programming (SDP) has been so extensively used in reservoir operation, and its selection of state and decision variables has brought different models. Among them, Bras et al. [2] incorporated current hydrologic forecast information in an SDP model. Stedinger et al. [3] incorporated available hydrologic information into an SDP decision model by using the streamflow forecast as the hydrologic state variable. Kelman et al. [4] introduced sampling stochastic dynamic programming (SSDP) to optimize water system operations on the Feather River in California (USA) using multiple historical streamflow time series as scenarios to capture the variability of streamflow processes by example. Another method to incorporate stochastic information into SDP is Bayesian SDP. Karamouz and Vasiliadis [5] described streamflow with a discrete lag 1 Markov process and updated the probabilities with new information; it showed that the forecast of the next period's flow as state variables performed better than that used for the forecast of the current period's flow. Xu et al. [6] proposed a new two-stage Bayesian stochastic dynamic programming model which partitions the forecast horizon into two periods.

The SDP requires an objective and constraints to be stage-separable, which prevents it from explicitly incorporating many performance criteria into its formulation. Chen [7] developed a fuzzy dynamic programming approach by applying the fuzzy iteration model to evaluate the decisions at each stage. In many cases, however, these criteria of a system are the most critical aspects that should be considered. The fundamental performance indices include the mean, variance, and deficit of statistical values. For example, the mean, variance of statistical power generations is used to evaluate the performance of a hydropower station. Other evaluation indices have also been put forward, including the robustness by Hashimoto et al. [8] and sustainability in a water resource system by Loucks [9] and Sandoval-Solis [10]. Hashimoto et al., in 1982 [11], introduced systematically three criteria: the reliability (how likely the system is to fail), vulnerability (how severe the consequences of failure may be), and resilience (how quickly a system recovers from failure), abbreviated as RRV criteria.

The RRV criteria were the most widely used in different water resource systems. Moy et al. [12] optimized these three measures and revealed relationships among them regarding a water supply reservoir using multi-objective mixed-integer linear programming. Kundzewicz et al. [13] found that conflicts between particular criteria might arise. In the literature of Zhang et al. [14], both conflicting and synergetic relationships were found between reliability, vulnerability and resilience using many-objective analysis. Based on the assumption of stationary hydrology, a joint analysis of multi-criteria was carried out by Jain and Bhunya [15], in which the behavior of statistical performance indices, namely the RRV, of a multipurpose storage reservoir was examined. They used a probabilistic approach to interpret the behavior of these indices and computed several performance indices for each sequence simulated by the Monte Carlo method and presented a framework for a multi-criteria approach that computed all the relevant quantities for each time period in reservoir design and management. The statistical behaviors of three indices were examined using input data by the Monte Carlo method. Some researchers combined these three performances to evaluate the performance of water resources systems. Loucks [9] combined RRV into an index to quantify sustainability. After that, the sustainability index (SI) was utilized in Kay's research [16] to help identify decisions. New combinations of RRV, such as, a combined reliability–vulnerability index and a robustness index were utilized via fuzzy set theory by Ibrahim El-Baroudy et al. [17]. The RRV or sustainability index tend to be applied in the field of climate change on water resources systems [18,19].

The possibility of explicitly incorporating the RRV constraints into a stochastic linear programming (SLP) has not been investigated in previous works, especially hydropower system operation. Applications of the SLP in reservoir operation are not new, though only a small amount of literature is available on the pure application of the SLP. Among the scarce previous literature, Loucks [20] presented an SLP model, where the random and serially correlated inflows were described with a first-order Markov chain. Lee et al. [21] developed a two-stage SLP model based on the form of a "fan of individual scenarios" to coordinate the multi-reservoir operation. Using inflow scenarios generated by a multivariate periodic AR(1) model considering serial and spatial connections, a stochastic model indicates its advantages over a deterministic model and its effectiveness in real-time operation. Baliarsingh and Kumar [22] developed a form of stochastic linear programming in which the decision variables were the joint probabilities of the reservoir release given an initial reservoir storage volume, an inflow, and the final storage volume for a given time period. System performance was set as the sum of squared deviations from target storage and target release.

In view of the lack of an SLP model which takes the reliability and vulnerability criteria into consideration, this paper will explore the possibility to incorporate the constraints on reliability and vulnerability (RV) into an SLP model to maximize the expected power generation in reservoir operation. The model will introduce a new variable, namely, the probability of reservoir capacity at the end of the time period when reservoir capacity at the beginning of the time period and inflow during the time period have been given. The inflows will be described as a Markov chain because of their uncertainty. The IBM CPLEX software (IBM, Armonk, NK, USA), a convenient and excellent LP solver, will be employed to solve the SLP model, which is expected to have numerous combinations of discrete representative values. Since both the SLP and SDP can handle the reservoir operation problem when the RV constraints are not involved, their results will be compared to illustrate their equivalence in this situation. This work will also investigate the capability of the SLP to explicitly incorporate the RV constraints, showing its advantage over the SDP.

2. Problem Formulation

The problem is formulated to determine a closed-loop feedback operational policy that maximizes the expected energy production while meeting the power yield with required minimum reliability and maximum vulnerability. The inflows are described as a Markov chain and represented with representative values and transition probability matrices. The conditional probabilities of final reservoir volumes, given the initial volumes and representative inflows for each time period, as a variable to be solved, are introduced. Simultaneously, a binary variable whose value can only be taken as 0 or 1, representing whether the decision is made or not, is set as another variable to be solved.

2.1. Objective

Mathematically, the objective is to maximize the expected generation, expressed as

$$
\max \sum_{t=1}^{T} \sum_{(i,k,l) \in \Omega(t)} (P_{iklt} \cdot G_{iklt} \cdot \Delta t) \tag{1}
$$

where T is the number of the time period; t is the time period index; Δt is the time length in hours of time period t; i is the index representing the discretization values of storage at the beginning of time period t; k is the index representing the characteristic values of inflow during time period t; l is the index representing the discretization values of storage at the end of time period t, and also the index representing the value of storage at the beginning of time period $t+1$; P_{iklt} is the probability of being at state (i,k) and making the decision (l) at the beginning of t, expressed as $l = \ell(i,k,t)$, which stands for the target storage at the end of time period t and is determined by the inflow k in t and storage i at the beginning of t; $\Omega(t)$ is the feasible combinations of state and decision variables; and G_{iklt} is the power generation in GW produced in time period t,

$$
G_{iklt} = G\left[S_t^{(i)}, Q_t^{(k)}, S_{t+1}^{(l)} \right] \tag{2}
$$

which is calculated with the approximate formula proposed in Wang's paper [23],

$$
G_{iklt} = G\left[S_t^{(i)}, Q_t^{(k)}, S_{t+1}^{(l)} \right] = \eta \cdot \left[h^{\min} + \alpha \sqrt{0.5(S_t^{(i)} + S_{t+1}^{(l)})} \right] \cdot \min[R_t(i,k,l), U^{\max}] \tag{3}
$$

Here, η is the energy conversion factor used to convert m^4/s into GW; $h^{\min}$ is the minimal water head in m; α is a coefficient associated with the reservoir shape, which in this paper is determined by solving

$$h_i^{\max} = h_i^{\min} + \alpha \cdot \sqrt[3]{S_i^{\max}} \tag{4}$$

$S_t^{(i)}$ is the ith characteristic storage value in m^3 at the beginning of t; $S_{t+1}^{(l)}$ is the jth characteristic storage value in m^3 at the end of t, also the characteristic storage value in m^3 at the beginning of $t+1$; $Q_t^{(k)}$ is the kth characteristic inflow value in m^3/s in t; $U^{\max}$ is the outflow capacity in m^3/s for generation; and $R_t(i,k,l)$ is the release in m^3/s in time period t, determined by

$$R_t(i,k,l) = \frac{1}{\Delta t}\left[S_t^{(i)} - S_{t+1}^{(l)}\right] + Q_t^{(k)} \tag{5}$$

2.2. General Constraints

The constraints include the following:
(1) The joint probabilities for any time period must sum to 1.0,

$$\sum_{(i,k,l)\in\Omega(t)} P_{iklt} = 1 (\forall t) \tag{6}$$

which means that the decision in time period t must be one of the feasible decisions.
(2) The probability transition satisfies

$$\sum_{(i,k)\in\Omega(t)} P_{iklt} \cdot P_{kj}^{[t]} = \sum_{m\in\Omega(t+1)} P_{l,j,m,t+1} \; for \; (l,j) \in \Omega(t+1) \tag{7}$$

where $P_{kj}^{[t]}$ is the transition probability, defined as the probability that the jth characteristic inflow in time period $t+1$ occurs given that the kth inflow has been observed in time period t. This constraint implies that any state in the next time period $t+1$ is ruled by the decision and by the transition probabilities in the current time period t.
(3) Decision uniqueness should be enforced on

$$\sum_{l\in\Omega(t)} u_{iklt} = 1 \; for \; (i,k) \in \Omega(t) \tag{8}$$

and

$$P_{iklt} \leq u_{iklt} \; for \; \forall(i,k,l) \in \Omega(t) \tag{9}$$

where u_{iklt} is a binary variable that decides whether or not decision l is made at state (i,k) in time period t, which implies that only one of the optional target storage values at the end of a time period must be selected given the state at the beginning of the time period.

2.3. Reliability and Vulnerability Constraints

Owing to the introduction of conditional probability variables, as well as to the expression of reliability and vulnerability corresponding to the probability of meeting requirements and expected violation, the two performances can be explicitly expressed in the SLP model.

2.3.1. Reliability

Reliability has been a primary performance metric in evaluating operational system stability. In reservoir operation, reliability is measured by the probability of a system being in a satisfactory state. In this paper, since joint probability variables have been introduced, the average value of the sum of all probabilities satisfying generation requirements is used to assess reliability performance logically. The reliability performance should be no less than a preset percentage β, expressed as

$$\frac{1}{T}\sum_{t=1}^{T}\sum_{(i,k,l)\in\Theta(t)} P_{iklt} \geq \beta \tag{10}$$

where $\Theta(t)$ is a subset of $\Omega(t)$ and consists of the feasible combinations (i,k,l) in time period t that meet the generation yield, i.e.,

$$\Theta(t) = \{(i,k,l)|G_{iklt} \geq Y; R_t(i,k,l) \geq 0\} \tag{11}$$

in which Y is the generation yield in GW. Here, the reliability is defined as the probability that the generation yield is met.

2.3.2. Vulnerability

Reliability is without question a criterion that should not be ignored during water resources system operation. Generally speaking, higher reliability of an operational system is always desired. However, from the reliability criterion, we can only evaluate how much the system can operate to fulfill the requirements. A failure cannot be avoided. When the failure occurs, to what extent the damage can cause an unsatisfactory situation, or what the likely magnitude of such a situation is, is always of concern. Similarly, the expected quantification of violations in all time periods is used to quantify the vulnerability performance.

The vulnerability should be less than a preset value v,

$$\frac{1}{T}\sum_{t=1}^{T}\sum_{(i,k,l)\in\overline{\Theta}(t)} P_{iklt}(Y - G_{iklt}) \leq v \tag{12}$$

where $\overline{\Theta}(t)$ is the relative complementary set of $\Theta(t)$ with respect to $\Omega(t)$, v is given before the calculation which indicates how much the average power generation yield can violate. Here, the vulnerability is defined as the expected violation of the generation yield when the violations occur.

3. Solution Procedure

Evidently, in view of the fact that the probability transition matrices for each time period can be derived, objectives and constraints in the model are linear. It can therefore be solved using an LP solver. Without considering the RV constraints, the SDP model is applied to obtain the optimal strategy compared with the SLP model.

3.1. Inflow Description and Probability Transition Matrices

The reservoir inflow is described as a periodic first-order Markov chain. State and transition probability are two critical concepts in a Markov chain for describing inflow in each particular time period. The inflow is divided into m intervals for each time period, and each state is represented by an average value Q_t^k for each interval. Thus, the probability transition matrices $P_{kj}^{[t]}$ that are used to describe the probabilities of a representative value Q_{t+1}^j in time period $t+1$ given a representative value Q_t^k in time period t are derived by frequency analysis. The reservoir capacity discrete value takes on arithmetic sequences between dead reservoir volume and maximum volume.

3.2. Solving the SLP Model

With the sets $\Omega(t)$, $\Theta(t)$, and G_{iklt} predetermined, as well as inflow transition probability $P_{kj}^{[t]}$ derived, the model becomes a mixed integer linear programming with objective (1) subject to (6)–(10) and (12). Owing to the existence of 0–1 binary variables u_{iklt}, the model is complicated to solve. However, benefiting from the standard mathematical form of linear programming and the improvement of computer performance, many efficient and convenient solvers are available, including CPLEX (IBM, Armonk, NY, USA), YALMIP (Johan Löfberg, Jekyll & Minimal Mistakes), and GUROBI (GUROBI, Houston, TX, USA), which make solving large-scale optimization problems with thousands of variables possible. In this work, the goal is to determine the following decision variables: the binary u_{iklt} and the probabilities P_{iklt}.

3.3. Solving the SDP Model

Though it is very difficult, if not impossible, for the SDP to incorporate the RV constraints, the SDP is actually very powerful in dealing with stochastic sequential decision-making process problems. Here, a universal reservoir operation problem is formulated as an SDP model, where the state variables are as follows: the storage S_t at the beginning of time period t and inflow Q_t into a reservoir during time period t; and the decision variables are as follows: storage values S_{t+1} at the end of time period t.

The recursive equation is expressed as

$$f_t(S_t, Q_t) = \max_{S_{t+1}}[G(S_t, Q_t, S_{t+1}) + \underset{\langle Q_{t+1}|Q_t\rangle}{E} f_{t+1}(S_{t+1}, Q_{t+1})] \tag{13}$$

where $f_t(S_t, Q_t)$ is the maximum expected power generation corresponding to a set of S_t and Q_t over the remaining time periods of the time horizon; and $G(\cdot)$ is the immediate power generation in time period t; $\underset{\langle Q_{t+1}|Q_t\rangle}{E}$ is the conditional expectation operator over Q_{t+1} given Q_t.

In order to obtain a stationary operation policy, a backward recursion should be implemented in several cycles, starting with $f_T(S_T, Q_T) = 0$. The maximum expected benefit $f_t(S_t, Q_t)$ over the remaining time period from t to T is calculated for each given S_t and Q_t. The aforementioned recursive calculation continues until a convergent and stable policy is obtained.

4. Case Studies

The stochastic dynamic programming cannot explicitly incorporate the RV constraints. The Xiaowan and Nuozhadu hydropower reservoirs in Yunnan Province, China, are used as case studies to verify the effectiveness of the stochastic linear model in obtaining an operating policy and the equivalence between the proposed SLP without RV constraints and the SDP model.

Using 56 years of one-third-monthly (at an interval about 10 days) historical inflows of Xiaowan and Nuozhadu Reservoirs during 1953–2008 and also 1490 years of one-third-monthly inflows of Xiaowan Reservoir simulated by AR(2), the representative inflows are divided into seven states, and the 7 × 7-dimensional matrices representing the transition probability are calculated based on the previously mentioned first-order Markov chain for each time period. Karamouz and Houck [24] showed that 20 storage values could be adequate for a reservoir with storage capacity of up to 170% of the mean annual flow. For convenience of calculation, a number of 21 representative discrete storage values are set evenly.

In view of its rapid solving ability and language-support features, CPLEX12.6 (IBM, Armonk, New York, USA) was employed to solve the SLP model in this work.

4.1. Comparison between the SLP and SDP Models without the RV Constraints

Without considering the RV constraints, the unknown decision variables u_{iklt} and corresponding decision probabilities P_{iklt} can be derived by solving the SLP model. There are two ways to calculate statistical values, one based on parameters and variables in the SLP model, and another one by operation simulation with sample inflows following an optimal decision rule derived either from the SLP or SDP results. Figure 1 shows the expected one-third-monthly power generations derived by the SLP model, and the statistical annual power generation trajectory obtained by simulation based on SLP and SDP decisions. The results indicate that the expected generation trajectory calculated in the SLP and the averaged generation trajectory when simulated based on the SLP and SDP solutions are almost the same. The SDP has long been proven capable of producing the optimal decision rule, which, then, can also be achieved by the SLP model.

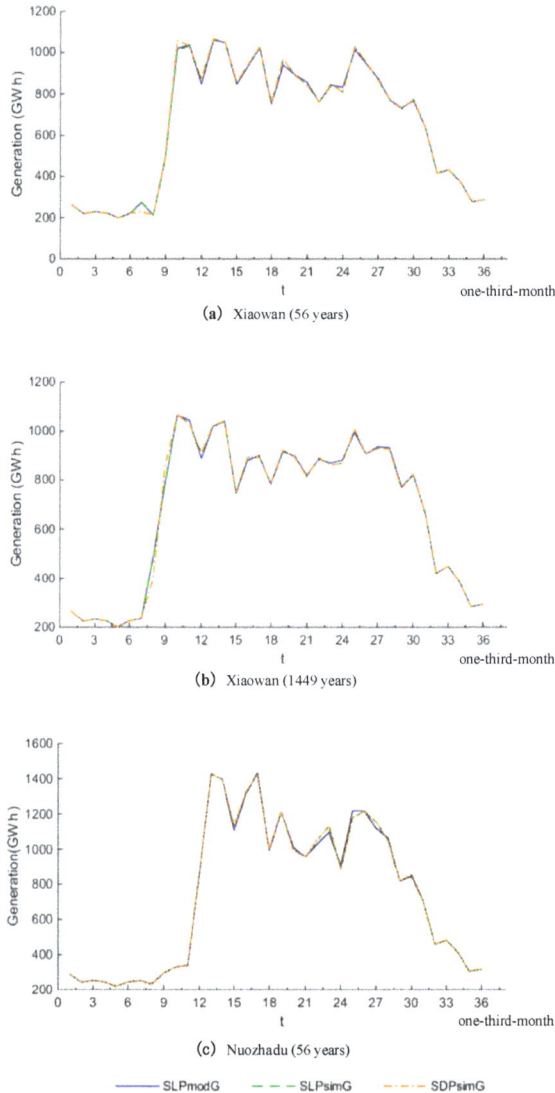

(a) Xiaowan (56 years)

(b) Xiaowan (1449 years)

(c) Nuozhadu (56 years)

——— SLPmodG   – – – SLPsimG   –·–·– SDPsimG

Figure 1. Comparison of the expected results from stochastic linear programming (SLP) with the simulated statistical results from the SLP and stochastic dynamic programming (SDP) models without reliability and vulnerability (RV) constraints for annual power generation trajectory using: (**a**) 56 years of one-third-monthly historical inflows of Xiaowan Reservoir; (**b**) 1449 years of one-third-monthly simulated inflows of Xiaowan Reservoir; (**c**) 56 years of one-third-monthly historical inflows of Nuozhadu Reservoir.

Table 1 gives the statistical annual energy production when adopting different sample inflows, namely 56 years of one-third-monthly inflows historically observed for Xiaowan and Nuozhadu reservoirs and 1449 years of one-third-monthly inflows simulated for the Xiaowan Reservoir. Statistical annual energy productions derived by simulating the SLP and SDP decision rules are almost identical, either for the 56 years of historical inflows or the 1449 years of simulated inflows. Simulating the decision rules, when employing the 56 years of historical inflows, gives averaged annual energy productions approximately 0.2–0.3% greater than those theoretically determined in the SLP model. The gap, however, can be reduced to within 0.007% by using the 1449 years of simulated inflows to diminish the sample errors. It is thus concluded that the SLP and SDP models without considering the reliability and vulnerability constraints can produce an identical solution to optimally operate a reservoir.

Table 1. Annual energy production in GWh by the SLP model (SLPmod), and by simulation for the SLP (SLPsim) and SDP (SDPsim) without RV constraints.

Calculation Ways	Xiaowan His.	Xiaowan Sim.	Nuozhadu His.
SLP_{mod}	23,504.5	24,389.7	26,619.8
SLP_{sim}	23,568.3	24,391.4	26,684.2
SDP_{sim}	23,568.6	24,390.5	26,684.1

4.2. Result of the SLP Model with the RV Constraints

With 1449 years of artificially generated one-third-monthly inflows of Xiaowan reservoir, the SLP model is formulated and solved to evaluate the influence of the RV constraints on the behavior of reservoir operation.

Table 2 presents the results detailed in each time period when parameters Y, β and v are set to 1.778, 0.98 and 0.01, respectively. Again, the expected energy production, the reliability and the vulnerability at each time period are all calculated in two ways, one by the SLP model itself, and another one through operation simulation following the optimal decision rule derived by the SLP model.

Table 2 shows that the power generation trajectory, reliability, and vulnerability in each time period, by both the model and simulation, are very close, though the gap of vulnerability is slightly larger relative to the other two indices. Similarity is also found in the cumulative power generation, averaged reliability and vulnerability over a representative year. The closeness suggests that the discretization resolution of both the storage capacity and the possible inflow is good enough for the SLP model to obtain an optimal decision rule that can achieve what the SLP model expects for many years of reservoir operation.

Changing the power generation yield Y, reliability β, and vulnerability v will yield different solutions by solving the SLP model, with some feasible and some infeasible. Table 3 gives five feasible combinations of Y, β and v, as well as the annual energy productions derived by the model itself and by the operation simulation following the optimal decision rule. There is a very small gap between the model and simulated results for cumulative power generation. As shown in Table 3, the model and simulated results derived with 1449 years of simulated inflows are very close to each other.

Table 2. Comparison of model and simulated (Sim) values of expected power generation (Exp. G), reliability, and vulnerability for each time period with $(Y, \beta, v) = (1.778, 0.98, 0.01)$.

Time Period	Exp. G (GWh)		Reliability		Vulnerability	
	Model	Sim	Model	Sim	Model	Sim
1	565.52	565.91	1.000	0.999	0.0008	0.0005
2	521.67	521.77	1.000	0.997	0.0000	0.0007
3	528.33	528.44	1.000	0.999	0.0000	0.0006
4	518.05	518.17	1.000	0.999	0.0000	0.0006
5	492.88	493.01	1.000	0.983	0.0000	0.0012
6	513.34	513.50	1.000	0.998	0.0000	0.0006
7	518.37	518.54	1.000	0.990	0.0000	0.0009
8	531.56	525.57	0.963	0.946	0.0126	0.0242
9	549.91	550.26	1.000	0.994	0.0000	0.0006
10	575.74	576.35	0.999	0.995	0.0011	0.0006
11	578.61	579.10	0.998	0.967	0.0007	0.0040
12	607.22	609.45	0.988	0.977	0.0067	0.0098
13	547.20	543.84	0.921	0.912	0.0705	0.0702
14	570.52	572.71	0.980	0.948	0.0059	0.0085
15	616.19	623.16	0.931	0.951	0.0051	0.0082
16	643.97	644.45	1.000	0.968	0.0000	0.0059
17	673.59	667.30	0.965	0.966	0.0152	0.0100
18	767.28	774.60	1.000	0.996	0.0000	0.0006
19	746.00	745.96	1.000	0.954	0.0000	0.0170
20	825.48	814.86	0.948	0.948	0.0055	0.0220
21	809.98	806.48	1.000	0.983	0.0000	0.0046
22	851.21	855.25	1.000	0.984	0.0000	0.0066
23	864.13	858.83	0.961	0.946	0.0433	0.0343
24	857.16	847.16	1.000	0.978	0.0000	0.0087
25	888.24	890.25	1.000	0.979	0.0000	0.0083
26	902.43	901.37	0.948	0.944	0.0525	0.0306
27	890.18	889.46	0.989	0.939	0.0035	0.0230
28	906.55	915.71	0.943	0.944	0.0395	0.0349
29	865.72	856.29	0.999	0.987	0.0002	0.0013
30	802.89	808.26	0.982	0.963	0.0032	0.0071
31	674.16	670.77	0.910	0.940	0.0373	0.0497
32	583.68	586.13	0.966	0.963	0.0098	0.0056
33	555.31	553.79	0.936	0.936	0.0299	0.0325
34	591.11	600.74	0.955	0.992	0.0163	0.0037
35	588.57	588.92	1.000	1.000	0.0002	0.0000
36	596.25	596.50	1.000	0.999	0.0003	0.0007
Sum	24,118.998	24,112.841	-	-	-	-
Avg.	669.97216	669.80115	0.97999	0.97118	0.01	0.01216

Table 3. Annual power generation (GWh) by the SLP model itself and by operation simulation with different parameters.

No.	Combinations (Y, β, v)	Model	Simulation
1	(1.778, 0.98, 0.01)	24,119	24,112
2	(1.778, 0.98, 0)	24,169	24,167
3	(1.778, 0, 0.01)	24,196	24,183
4	(2.5, 0.7, 0.3)	24,076	24,061
5	(2.5, 0.6, 0.3)	24,258	24,247

Figure 2 shows the statistical results of the power generation trajectory by 1449 years of operation simulations when the power yield Y, reliability β, and vulnerability v are set to $(0, 0, 0)$, $(1.778, 0.98, 0.01)$, and $(2.5, 0.7, 0.3)$, respectively. The results indicate that enforcing a power yield with certain RV constraints will achieve a more even power generation over time despite less energy production occurring. By trial and error, the power yield can be improved from 1.778 GW to 2.5 GW only to the detriment of the reliability by 28% = 98% − 70%. With the power yield and reliability set to 2.5 GW and 70% respectively, the minimum vulnerability can be achieved at 0.3. It seems that in this case study, the higher reliability also benefits a lower vulnerability.

Figure 3 shows the maximum, average and minimum of one-third-monthly storages in a representative year, calculated by operation simulation with 1449 years of artificially generated inflows for combinations (Y, β, v) = (0, 0, 0), (1.778, 0.98, 0.01), and (2.5, 0.7, 0.3), respectively. Without enforcement on the power yield and its reliability and vulnerability for (Y, β, v) = (0, 0, 0), the SLP model will derive an optimal decision rule that maximizes the energy production by keeping the storage level as high as possible unless it is beneficial to empty the reservoir to alleviate spillages. With a power yield Y = 1.778 GW, the reservoir is utilizing its storage during the dry seasons to achieve a high reliability β = 98% and a low vulnerability v = 0.01 to the detriment of energy production as a whole. Comparing (b) and (c) in Figure 3, it is interesting to find that increasing the power yield from 1.778 GW to 2.5 GW narrows the operational space of the reservoir, which operates in a smaller corridor confined by the maximum and minimum storages.

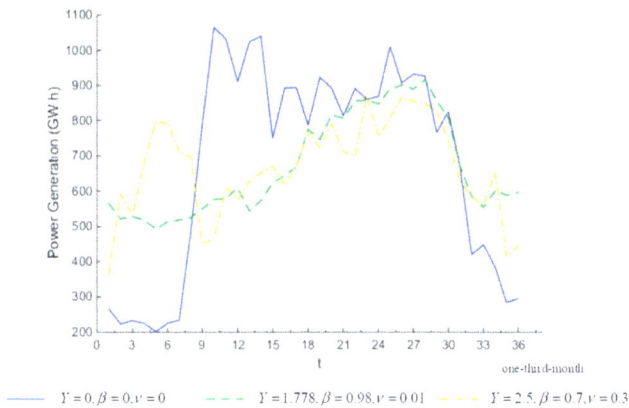

Figure 2. Averaged power generation trajectory by simulation with (Y, β, v) = (0, 0, 0), (1.778, 0.98, 0.01), and (2.5, 0.7, 0.3), respectively.

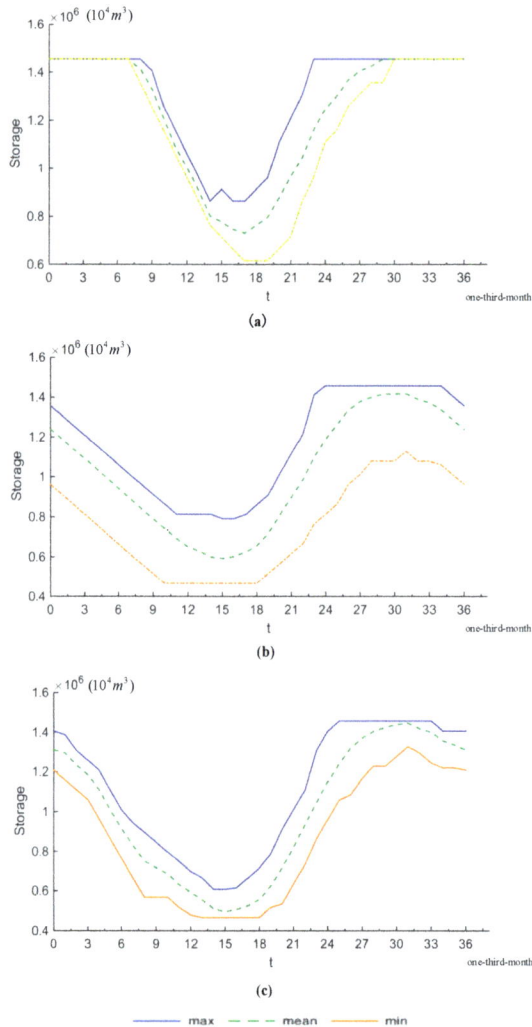

Figure 3. Max, mean, and min storage trajectory for 1449 years of simulations.

5. Conclusions

This paper presents a stochastic linear programming (SLP) model with strength over SDP in that it explicitly incorporates reliability and vulnerability (RV) constraints. The stochastic characteristics of inflow are captured by Markov transition matrices. The IBM's CPLEX LP solver (IBM, Armonk, NK, USA) is employed to solve the SLP model, which, without enforcing the RV constraints, produces solutions equivalent to those by the SDP model, which has long been proved effectively in reservoir operation. With 1449 years of one-third-monthly inflows artificially generated by an AR(2) model, the SLP model evinces its capability to take into account the RV constraints. The case studies also show that a higher power yield can evidently result in difficulty to meet stricter reliability and vulnerability requirements. Meanwhile, adding the RV constraints decreases the power generation, but can make the power generation more reliable and less vulnerable.

Water **2018**, *10*, 175

The proposed SLP model can also be used to conduct a tradeoff analysis among the power yield, reliability and vulnerability. The performance resilience has a more complicated expression than the reliability and vulnerability, and its possibility to be explicitly incorporated into the SLP model needs to be investigated in future works.

Acknowledgments: This paper is supported by the National Key Research and Development Program of China under grant No. 2016YFC0401910, and the Fundamental Research Funds for the Central Universities under grant No. 2017KFYXJJ207.

Author Contributions: C.C. and J.W. conceived and designed the experiments; C.C. performed the experiments; C.C. and J.W. analyzed the data; C.K. contributed reagents/materials/analysis tools; C.C. wrote the paper.

Conflicts of Interest: The authors declare no conflict of interest.

References

1. Labadie, J.W. Optimal operation of multireservoir systems: State-of-the-art review. *J. Water Resour. Plan. Manag.* **2004**, *130*, 93–111. [CrossRef]
2. Bras, R.; Buchanan, R.; Curry, K. Real time adaptive closed loop control of reservoirs with the High Aswan Dam as a case study. *Water Resour. Res.* **1983**, *19*, 33–52. [CrossRef]
3. Stedinger, J.R.; Sule, B.F.; Loucks, D.P. Stochastic dynamic programming models for reservoir operation optimization. *Water Resour. Res.* **1984**, *20*, 1499–1505. [CrossRef]
4. Kelman, J.; Stedinger, J.R.; Cooper, L.A.; Hsu, E.; Yuan, S.Q. Sampling stochastic dynamic programming applied to reservoir operation. *Water Resour. Res.* **1990**, *26*, 447–454. [CrossRef]
5. Karamouz, M.; Vasiliadis, H.V. Bayesian stochastic optimization of reservoir operation using uncertain forecasts. *Water Resour. Res.* **1992**, *28*, 1221–1232. [CrossRef]
6. Xu, W.; Zhang, C.; Peng, Y.; Zhou, H. A two stage Bayesian stochastic optimization model for cascaded hydropower systems considering varying uncertainty of flow forecasts. *Water Resour. Res.* **2014**, *50*, 9267–9286. [CrossRef]
7. Chen, S.; Fu, G. Combining fuzzy iteration model with dynamic programming to solve multi-objective multistage decision making problems. *Fuzzy Sets Syst.* **2005**, *152*, 499–512. [CrossRef]
8. Hashimoto, T.; Loucks, D.P.; Stedinger, J.R. Robustness of water resources systems. *Water Resour. Res.* **1982**, *18*, 21–26. [CrossRef]
9. Loucks, D.P. Quantifying trends in system sustainability. *Hydrol. Sci. J.* **1997**, *42*, 513–530. [CrossRef]
10. Sandoval-Solis, S.; McKinney, D.; Loucks, D. Sustainability index for water resources planning and management. *J. Water Resour. Plan. Manag.* **2010**, *137*, 381–390. [CrossRef]
11. Hashimoto, T.; Stedinger, J.R.; Loucks, D.P. Reliability, resiliency, and vulnerability criteria for water resource system performance evaluation. *Water Resour. Res.* **1982**, *18*, 14–20. [CrossRef]
12. Moy, W.S.; Cohon, J.L.; ReVelle, C.S. A programming model for analysis of the reliability, resilience, and vulnerability of a water supply reservoir. *Water Resour. Res.* **1986**, *22*, 489–498. [CrossRef]
13. Kundzewicz, Z.W.; Kindler, J. Multiple criteria for evaluation of reliability aspects of water resource systems. *IAHS Publ. Ser. Proc. Rep. Intern. Assoc. Hydrol. Sci.* **1995**, *231*, 217–224.
14. Zhang, C.; Xu, B.; Li, Y.; Fu, G. Exploring the relationships among reliability, resilience, and vulnerability of water supply using many-objective analysis. *J. Water Resour. Plan. Manag.* **2017**, *143*, 04017044. [CrossRef]
15. Jain, S.; Bhunya, P. Reliability, resilience and vulnerability of a multipurpose storage reservoir/Confiance, résilience et vulnérabilité d'un barrage multi-objectifs. *Hydrol. Sci. J.* **2008**, *53*, 434–447. [CrossRef]
16. Kay, P.A. Measuring sustainability in Israel's water system. *Water Int.* **2000**, *25*, 617–623. [CrossRef]
17. El-Baroudy, I.; Simonovic, S.P. Fuzzy criteria for the evaluation of water resource systems performance. *Water Resour. Res.* **2004**, *40*. [CrossRef]
18. Hernández-Bedolla, J.; Solera, A.; Paredes-Arquiola, J.; Pedro-Monzonís, M.; Andreu, J.; Sánchez-Quispe, S.T. The Assessment of Sustainability Indexes and Climate Change Impacts on Integrated Water Resource Management. *Water* **2017**, *9*, 213. [CrossRef]
19. Mateus, M.C.; Tullos, D. Reliability, Sensitivity, and Vulnerability of Reservoir Operations under Climate Change. *J. Water Resour. Plan. Manag.* **2016**, *143*. [CrossRef]
20. Loucks, D.P. Computer models for reservoir regulation. *J. Sanit. Eng. Div.* **1968**, *94*, 657–670.

21. Lee, Y.; Kim, S.-K.; Ko, I.H. Two-Stage Stochastic Linear Programming Model for Coordinated Multi-Reservoir Operation. *J. Hydroinform.* **2008**, *10*, 400–410. [CrossRef]
22. Baliarsingh, F.; Kumar, D.N. Stochastic linear programming for optimal reservoir Operation: A case study. In Proceedings of the International Conference on Large Scale Water Resources Development in Developing Countries: New Dimensions of Prospects and Problems, Kathmandu, Nepal, 20–23 October 1997; pp. 124–130.
23. Wang, J. A new stochastic control approach to multireservoir operation problems with uncertain forecasts. *Water Resour. Res.* **2010**, *46*. [CrossRef]
24. Karamouz, M.; Houck, M.H. Annual and monthly reservoir operating rules generated by deterministic optimization. *Water Resour. Res.* **1982**, *18*. [CrossRef]

Article

Real-Time Flood Control by Tree-Based Model Predictive Control Including Forecast Uncertainty: A Case Study Reservoir in Turkey

Gökçen Uysal [1,2,*], Rodolfo Alvarado-Montero [1,3], Dirk Schwanenberg [1,4] and Aynur Şensoy [2]

[1] Institute of Hydraulic Engineering and Water Resources Management, University of Duisburg-Essen, 45141 Essen, Germany; Rodolfo.AlvaradoMontero@deltares.nl (R.A.-M.); Dirk.Schwanenberg@kisters.de (D.S.)

[2] Department of Civil Engineering, Anadolu University, 26555 Eskişehir, Turkey; asensoy@anadolu.edu.tr

[3] Deltares, Operational Water Management, Deltares, Rotterdamseweg 185, 26 MH Delft, The Netherlands

[4] KISTERS AG, Business Unit, Pascalstraße, 52076 Aachen, Germany

* Correspondence: gokcenuysal@anadolu.edu.tr; Tel.: +90-222-321-35-50

Received: 18 December 2017; Accepted: 9 March 2018; Published: 19 March 2018

Abstract: Optimal control of reservoirs is a challenging task due to conflicting objectives, complex system structure, and uncertainties in the system. Real time control decisions suffer from streamflow forecast uncertainty. This study aims to use Probabilistic Streamflow Forecasts (PSFs) having a lead-time up to 48 h as input for the recurrent reservoir operation problem. A related technique for decision making is multi-stage stochastic optimization using scenario trees, referred to as Tree-based Model Predictive Control (TB-MPC). Deterministic Streamflow Forecasts (DSFs) are provided by applying random perturbations on perfect data. PSFs are synthetically generated from DSFs by a new approach which explicitly presents dynamic uncertainty evolution. We assessed different variables in the generation of stochasticity and compared the results using different scenarios. The developed real-time hourly flood control was applied to a test case which had limited reservoir storage and restricted downstream condition. According to hindcasting closed-loop experiment results, TB-MPC outperforms the deterministic counterpart in terms of decreased downstream flood risk according to different independent forecast scenarios. TB-MPC was also tested considering different number of tree branches, forecast horizons, and different inflow conditions. We conclude that using synthetic PSFs in TB-MPC can provide more robust solutions against forecast uncertainty by resolution of uncertainty in trees.

Keywords: reservoir operation; multi-stage stochastic optimization; TB-MPC; flood control; real-time control

---

1. Introduction

Reservoirs are one of the main components of integrated water resources systems. Their control poses a challenging problem, since it must cope with different conflicting objectives as water supply, hydropower, and flood control [1–5]. Operating with guide curves is a common practice in typical reservoir operation [6–8]. These operating strategies however, rely on long term records that can filter and underestimate extreme events such as drought and flood conditions [9]. Flood events strain reservoir operators to refill and keep maximum water levels. In response to short-term fluctuations, the operators need to anticipate actions and release the excess amount of water in order to have sufficient flood storage volume in the reservoir [10]. However, the storage after the event must be recovered in order to satisfy water supply during the following dry season [11]. Therefore, flood control and water conservation require carefully planned strategies for short-term operation. On the other hand, system optimization on this time scale becomes more complex due to the high-dimensionality, non-linearity of the system, and dynamic structure of the control process.

In general, a reservoir operation is assumed to be a dynamic system in which the future states are a function of the current states. Throughout optimization of sequential decision processes, Dynamic Programming (DP) developed by Bellman [12] is an important milestone. The main problem can be divided into sub-problems, separately solved successively over each stage to get an overall optimum solution. Later, researchers contributed on this technique in different aspects such as using sampling stochastic DP, incremental DP, stochastic dual DP [13–15]. Though the solution is well suited for highly nonlinear, nonconvex problems, the main difficulty arises with the so-called "curse of dimensionality". This means more (exponentially growing) computational time with increasing dimensions of states, decisions, and disturbances [12,16].

Operation of the reservoir system is an optimal control problem, thus an alternative solution is proposed by adapting Model Predictive Control (MPC, also known as Receding Horizon Control) [17]. The concept has been tested in different application areas such as rivers, reservoirs, and irrigation canals [18–20] in water resources management. The system is optimized for a forecast horizon by solving an open-loop control problem simultaneously in every time step, then only the first time step control value of the computed sequence is applied and the rest are discarded. At the next step, system states are updated and the process is repeated again. This feedback mechanism is called closed-loop optimization. In the real-world, the reservoir system is operated according to forecast based control decisions, but updated to the realization of the disturbance (inflows). Thus, in the study closed-loop experiments are preferred due to the feedback mechanism to mimic the real-world reservoir operation decisions with respect to open-loop experiments.

MPC requires prediction of disturbance in the real-time control of a water structure, but streamflow forecasts always bring along forecast uncertainty. In real-time forecasting applications, forecasts can be biased and tend to over- or underestimate the actual streamflow [21] and it is hard to fully avoid this [10]. In practice, there are several sources of uncertainty that reduce the accuracy of peak flow estimation. Essentially, these are the hydrological forecast model structure [22], model external forcing e.g., precipitation [23,24], model parameters [25], and initial conditions [26].

Although it is not possible to eliminate all types of uncertainties, one solution might be the consideration of forecast uncertainty using ensemble forecasts in short-term operation. Labadie [16] sorted the hindrances of reservoir system operations, among them he emphasized the need to incorporate uncertainty and risk issues in optimization models. Recently, different researchers showed how probabilistic flood forecasts are more robust and effective than the traditional deterministic ones; correspondingly systems are improved by models that quantify forecast uncertainty [27–29].

In short-term management, Numerical Weather Prediction (NWP) based Ensemble Prediction Systems (EPS) can provide probabilistic streamflow forecasts. The idea behind it is to represent future states of the atmosphere by perturbing the initial conditions (uncertainty). Uncertainty becomes much larger when managing small basins and small rivers [30]. In the case of non-availability or inadequacy of the numerical weather forecast data, synthetic streamflow generation is a common practice for reservoir simulation and optimization studies especially in hypothetical studies [28]. There are various approaches to produce synthetic data e.g., the Thomas-Fiering model [28], Fractional Gaussian noise model [31], artificial neural network [32] or hybrid models [33]. Among them, only a limited number of studies [28,31] demonstrate the forecast uncertainty evolution process such that future periods have larger uncertainty and that there is a correlation between uncertainties.

While MPC seeks optimum trajectories based on a single disturbance, i.e., the inflow into a reservoir, its stochastic counterparts try to find an optimum solution for the entire ensemble and incorporate forecast uncertainty. However, this might yield an infeasible solution due to the burden of multiple disturbance in combination with hard constraints. An innovative technique to deliberate ensemble forecasts with an adaptive control is proposed as Tree-Based MPC (TB-MPC) [34]. This approach reduces the number of ensemble members by its tree generation algorithms using all trajectories and then proper problem formulation is set by Multi-Stage Stochastic Programming. The method is relatively new in reservoir operation [34–37], especially closed-loop hindcasting experiments and its assessment is quite rare [24] in the literature.

This paper demonstrates a real-time flood control in consideration of streamflow forecast uncertainty especially for limited storage multi-purpose reservoirs using synthetic ensemble inflows and the mass-conservative TB-MPC method. While we focus on the trade-off between flood risks and water conservation benefits, its implementations into the test case follow: (1) development of a novel Probabilistic Streamflow Forecasts (PSFs) method to produce ensemble inflows, (2) the configuration of an hourly (deterministic and stochastic) predictive reservoir optimization model, (3) testing of different PSF scenarios in closed-loop hindcasting mode in comparison with deterministic counterparts (perfect data and Deterministic Streamflow Forecasts, DSF) and (4) the assessment of TB-MPC model features considering different number of tree branches, different inflow conditions, and different forecast horizons in the area of interest.

In this study, the methodology is applied to the Yuvacik dam reservoir, fed by a catchment area of 258 km^2 and located in Turkey, owing to its challenging operation due to downstream flow constraints. The reservoir serves as a main water supply for Kocaeli province. First, a hydro-meteorological rule based decision support system is developed for daily and hourly operation [38]. Later, it is shown that underestimated daily inflow forecasts during a flood event operation in critical reservoir level may result in underestimated spillage releases when using deterministic MPC [39]. Therefore, the case establishes a precedent for similar relatively small reservoirs with multi-purpose operational characteristics. At this point, this paper complements deterministic methods by PSF integrated TB-MPC including forecast uncertainty.

2. Materials and Methods

Hindcasting experiments are the representation of a real-time system by an iterative process: (i) MPC anticipates required spillages with the optimization model for a given time length (so-called forecast horizon). (ii) The reservoir simulation model updates with actual inflows and anticipated spillages from step (i) to find updated forebay elevations. (iii) This process is repeated every hour (so-called receding horizon) until the whole hindcasting period (96 h in this study) has been completed. The general methodology is outlined in Figure 1. We apply closed-loop hindcasting experiments by the following three modes:

1. Perfect Hindcast Experiments: The flood hydrograph corresponding to a 100 years return period (Q_{100}) of the basin is utilized as perfect forecast data in deterministic MPC. This represents the best solution since it exhibits the optimized releases and forebay elevations under perfect knowledge of the future inflows, and it is evaluated as reference case for comparative analyses with forecast based models.

2. Deterministic Hindcast Experiments: This represents the skill of single value DSF evaluation by deterministic MPC. Randomly perturbed inflows in each receding horizon are employed in deterministic MPC mode. The random perturbation having a forecast horizon up to 48 h is updated in each hour.

3. Probabilistic Hindcast Experiments: This represents the skill of ensemble PSF evaluation by multi-stage stochastic TB-MPC. Synthetically generated ensemble inflows in each receding horizon are employed in stochastic MPC mode (Exp-A). This provides forecast uncertainty consideration in the application. Moreover, features of TB-MPC are investigated with selection of different forecast horizons, tree branching numbers, and different inflow conditions (Exp-B).

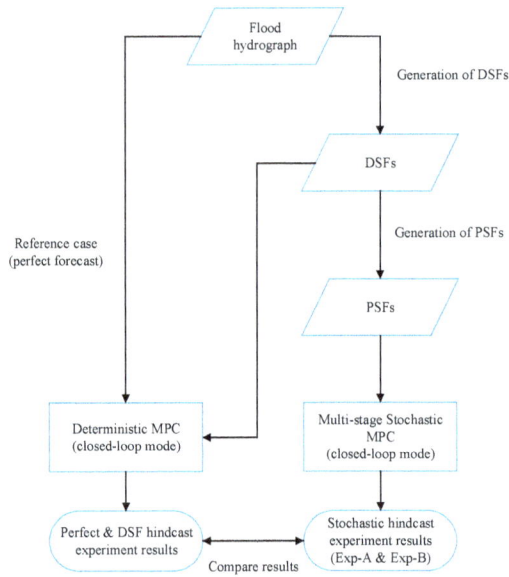

Figure 1. The general framework of the experiments. DSFs stands for Deterministic Streamflow Forecasts. PSFs stands for Probabilistic Streamflow Forecasts. MPC stands for Model Predictive Control.

2.1. Deterministic and Probabilistic Synthetic Streamflow Generation

Since forecast inflows include uncertainties, they can be represented by the relative inflow forecasting error [10]. In this paper, we refer to them as DSF scenarios which are perturbed from original flood hydrographs. While the DSF sequence is a single value, the PSFs are generated by applying a dispersion around the DSF with a specified probability distribution (Figure 2). They are eventually repeated in each time step (depending on selected receding horizon) to mimic a dynamic real-time system.

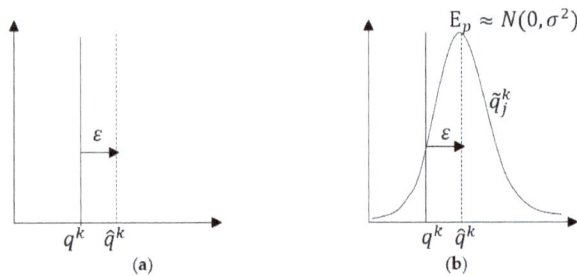

Figure 2. Schematic of single time-step streamflow forecast uncertainty: (a) DSF schematization; (b) PSF schematization. q^k stands for observed inflow. ε stands for relative inflow forecasting error. $\hat{q}^k$ stands for DSF member. $\tilde{q}_j^k$ stands for j^{th} PSF member. k stands for time index.

DSF is represented by random perturbation using relative inflow forecasting error (ε).

$$\varepsilon^k = \frac{\hat{q}^k - q^k}{q^k} \tag{1}$$

$$DSF: \hat{q}^k = q^k \left(1 + \varepsilon^k\right) \tag{2}$$

where ε, $\hat{q}$, q stands for relative inflow forecasting error, forecasted inflows, observed inflows, respectively over a forecast horizon represented by $k = 1, \dots, N$ time instants.

Even though a range of relative inflow forecasting errors can be detected in the literature [40], acceptable ε during a flood forecasting is within a 0.15–0.30 interval in real-world hydrological forecasting studies [4,10,41]. These studies cover different project scales and climatology. In this study ε values are randomly selected from uniform distribution and range between -0.3 and 0.3 to generate individual DSF values from observed inflows. Also, ε values change for each k time instant.

We produce a set of synthetized members (ensembles, PSF) that are spread around single valued DSF. Conventionally, uncertainty is defined by relative standard error which is expected to be independent (also referred as white noise), usually Gaussian with zero-mean. Therefore, the error pdf is assumed to be $E_p \approx N\left(0, \sigma^2\right)$ for all time steps [21,28,42,43].

We introduced a forecast evolution process which is similar to that of Zhao et al. [28] with increasing forecast uncertainty through the forecast horizon. Our implementation of correlation between forecast errors however, is defined by the algorithm below. Synthetic PSFs are generated as an empirical conditional distribution of the perturbed DSF. In other words, they have certain mean (μ) from DSF and relative standard error ($\hat{\sigma} = \sigma * \mu$), considered as an uncertainty level.

We consider the following notation of probabilistic scenario in an ensemble forecast (PSF) matrix as:

$$PSF = \begin{bmatrix} \tilde{q}_1^1 & \tilde{q}_2^1 & \cdots & \tilde{q}_j^1 & \cdots & \tilde{q}_M^1 \\ \tilde{q}_1^2 & \tilde{q}_2^2 & \cdots & \tilde{q}_j^2 & \cdots & \tilde{q}_M^2 \\ \cdots & \cdots & \cdots & \cdots & \cdots & \cdots \\ \tilde{q}_1^k & \tilde{q}_2^k & \cdots & \tilde{q}_j^k & \cdots & \tilde{q}_M^k \\ \tilde{q}_1^N & \tilde{q}_2^N & \cdots & \tilde{q}_j^N & \cdots & \tilde{q}_M^N \end{bmatrix} \tag{3}$$

where N denotes the length of forecast data (so-called forecast lead-time), M denotes the number of ensemble members, $\tilde{q}_j^k$ denotes an arbitrary ensemble member in matrix, and k, j correspond to lead-time and ensemble member indexes, respectively

The procedure to generate synthetic PSFs is accomplished mainly in two successive steps:

1. In the initial time step, PSF ($\tilde{q}_j^1$) members are generated as:

$$\tilde{q}_j^k = N \sim \left(\hat{q}^k, \hat{q}^k * \hat{\sigma}^k\right), \quad k = 1 \ \& \ j = 1, 2, \dots, M \tag{4}$$

2. The PSF should be correlated with previous members, therefore, the differences between successive DSF values (referred to k time instants for the same DSF sequence) are calculated, then normally distributed errors having mean and standard error from the previous time step are generated and added to the differences. Maximum function is added in order to eliminate negative values, and the remaining PSF members ($\tilde{q}$) are formulated as:

$$\tilde{q}_j^k = \left(\hat{q}^k - \hat{q}^{k-1}\right) + \max\left(N \sim \left(\hat{q}^{k-1}, \hat{q}^{k-1} * \hat{\sigma}^k\right), 0\right)$$
$$k = 2, \dots, N \ \& \ j = 1, 2, \dots, M \tag{5}$$

According to intuition, a longer forecast lead-time results in a less reliable forecast. The level of forecast uncertainty (in terms of relative standard error) is assumed to be the same for all members but increasing towards the forecast horizon:

$$\hat{\sigma}^{k-1} \le \hat{\sigma}^k \le \hat{\sigma}^{k+1} \tag{6}$$

Uncertainty in the precipitation is typically much greater (even the largest source of uncertainty from input data) compared to the other meteorological variables [44]. Since higher precipitation leads to higher discharge data, total uncertainty of the forecast chain is expected to be higher during rainfall events, and less during no rain condition. Therefore, this condition is considered in the application by updated relative standard errors depending on DSF values as:

$$
\hat{\sigma}^k = \left\{ \begin{array}{ll} \hat{\sigma}^k = \hat{\sigma}^{k-1} & if \quad \hat{q}^k \leq \hat{q}^{k-1} \\ \hat{\sigma}^k > \hat{\sigma}^{k-1} & if \quad \hat{q}^k > \hat{q}^{k-1} \end{array} \right\}
\tag{7}
$$

In the application, a cumulative forecast uncertainty (relative standard error of the final time step, $\hat{\sigma}^N$) is selected beforehand and incremental standard error values are calculated for each time instant. This is a posteriori information based on trial-and-error by the assessment of forecast uncertainty.

2.2. Deterministic Model Predictive Control (MPC)

Contrary to feedback controls where control actions are determined with the current state of the system, MPC takes the future state of the system into account. MPC is commonly implemented by having several components in a receding horizon strategy. These components are a system description (model), a set of disturbances (water inflow into the system, water extractions etc.), objective and physical or operational constraints to represent physical operational management objectives and constraints of the system. Deterministic MPC considers a discrete time-dynamic system according to

$$
x^k = f\left(x^{k-1}, x^k, u^k, d^k\right)
\tag{8}
$$

$$
y^k = g\left(x^k, u^k, d^k\right)
\tag{9}
$$

where x, y, u, d are, respectively, the state, dependent variable (output), control, and disturbance vectors, $f(.)$ and $g(.)$ are linear or non-linear, time-variant vector functions representing an arbitrary water resources model [45]. If being applied in MPC, Equations (8) and (9) are used for predicting future trajectories of the state x and dependent variable y over a finite time horizon represented by $k = 1, \ldots, N$ time instants, in order to determine the optimal set of control variables u by an optimization algorithm. Under the assumption of knowing the realization of the disturbance d^k over the time horizon N, for example the inflows into the reservoir system $\left\{ d^k \right\}_1^N$, the simultaneous (or collocated) MPC [45] becomes:

$$
\min_{u, x \in \{0, \ldots, T\}} \sum_{k=1}^{N-1} J\left(x^k, u^k, d^k\right) + E\left(x^N, u^N, d^k\right)
\tag{10}
$$

$$
s.t. : h\left(x^k, y^k, u^k, d^k\right) \leq 0, k = 1, \ldots, N
\tag{11}
$$

$$
x^k - f\left(x^{k-1}, x^k, u^k, d^k\right) = 0
\tag{12}
$$

where $J()$ is a cost function associated with each state transition, $E()$ is an additional cost function related to the final state condition, and $h()$ are hard constraints on control variables and states, respectively. It receives a single disturbance vector, so that it might be either an observed or deterministic forecast. In a simultaneous or collocated mode, the related process model (Equation (8)) becomes an equality constraint of the optimization problem at dedicated time steps in Equation (12) [46]. The optimization problem is solved using an efficient gradient-based optimizer such as the nonlinear optimizers IPOPT [47] with MA27 solver of the HSL library in combination with adjoint modeling. Adjoint modeling is essential for the efficient computation of the derivative of the objective function with respect to the controlled variables ($dJ = d\left(x^k, u^k, d^k\right)/du^k$). The model itself is implemented in RTC-Tools [45].

2.3. Multi-Stage Stochastic Tree-Based MPC (TB-MPC)

In this part, the deterministic model is extended to a multi-stage stochastic optimization model. The stochastic nature of inflows is reflected by a probabilistic forecasts ensemble, d_j^k, where j denotes the ensemble index ($j = 1, \ldots, M$) and k denotes the time instant ($k = 1, \ldots, N$). Hence, Equation (10) is rearranged for J and E, then it becomes the probability-weighted sum of the objective function terms of the individual ensemble. The formula is shown below as:

$$\min_{u, x \in \{0, \ldots, T\}} \sum_{j=1}^{M} p_j \sum_{k=1}^{N-1} J(x_j^k, u_j^k, d_j^k) + E\left(x_j^N, u_j^N, d_j^N\right) \tag{13}$$

where p_j stands for the probability of the ensemble member (equal for each one), M stands for the number of the ensembles.

The key factor of the approach depends on the control variable (u_j^k) definition for setting up of a stochastic optimization problem due to the limitation of the usable storage. A simple way would be to find the control trajectory which is optimal for the whole ensemble (on average, worst case, chance constrained, etc.). In case of an average definition, it might give feasible solutions on consideration of the ensemble having small variations. However, if the operation of the reservoir system is highly constrained due to a limitation of storage, the solution of the optimization can be dominated by these constraints or it becomes infeasible.

To avoid this problem, multi-stage stochastic optimization is proposed and successfully applied using scenario trees for disturbance, states, and control trajectories [34,48]. Thus, the approach becomes adaptive, the control trajectory does take into account the resolution of uncertainty. This means uncertainty is gradually resolved by showing tree branches at specific time instants along the forecast horizon. While scenario reduction is employed in various ways and represented by a tree nodal partition matrix $P(j,k) \in \mathbb{Z}^{MxN}$, a recently proposed novel mass conservative approach which keeps the probability-weighted sum of a quantity of the original ensemble is used in this study. Details are available in [24,37]. We present the first implementation of the TB-MPC in application to a synthetically generated ensemble streamflow forecast for a limited storage multi-purpose reservoir under flood case.

3. A Real World Test Case and Model Set-Up

3.1. Study Area

The proposed methodology was tested in the Yuvacik Dam reservoir, with a 258 km^2 catchment area, located in the east part of Marmara Region, Turkey (Figure 3). The earth-filled dam was constructed in 1999 for the water supply of Kocaeli city. An annual 142 hm^3 of drinking, domestic and industrial water for 1.5 million inhabitants should be supplied. The relatively limited reservoir has an active storage capacity of approximately 51.2 hm^3 at maximum operating level of 169.3 m whereas 169.8 m is the maximum water level. The spillway is controlled by four radial gates. It should be noted that while a volume of 36.60 hm^3 is stored between minimum operation level of 112.50 m and spillway crest elevation of 159.95 m, a volume of 14.60 hm^3 is kept above spillway crest elevation and behind the radial gates. On the other hand, a Dam Management System (DMS) which has been developed as a part of a Supervisory Control and Data Acquisition (SCADA) system by the reservoir operators provides data collection and transmission from automatic gauges.

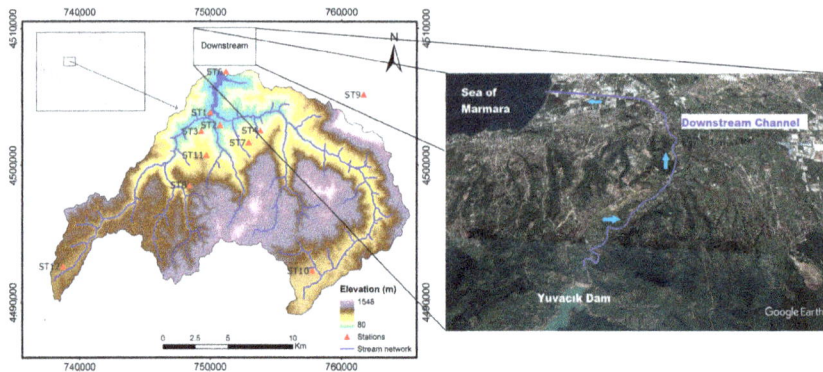

Figure 3. Yuvacık Basin DEM, station network and downstream region.

3.2. MPC Inputs

The maximum amount of water to be released during daily operation is set as 100 m^3/s by the regional water authority taking the drainage discharge conditions of the downstream canal and lateral flows into consideration. The main reason for that is a 12 km long downstream reach that passes from a narrow valley near a rural district and flows into the Marmara Sea after a sharp curvature by a manmade channel next to industrial and urban areas. The dam is built to protect the downstream region against extreme flood events, and the maximum spillway release is set to 200 m^3/s during hourly flood control. This value is the maximum allowable flood limit without severe damage in the downstream. The operation of the dam is multi-purpose subject to two main objectives: (i) water supply and (ii) flood control.

Since available data are insufficient to precisely assess extreme events [49], design flood hydrographs are generally derived based on maximum instantaneous records at the dam site, selected probability distributions using several parameters [50,51], and considered to be representative of extreme conditions expected in the future. In the Yuvacık catchment, the annual peak flow values over 19 years were statistically analyzed and it was detected that the series is in line with the Gumbel dispersion function according to the DSI (State Hydraulic Works, the main governmental water organization in Turkey) report [52]. The peak flow flood values are derived for different return periods (Table 1).

Table 1. Flood hydrograph peak values.

Return Periods (Years)	Project Value (m^3/s)
5	208
10	297
25	410
50	506
100	597

The study adopted a 24 hourly flood hydrograph having an expected occurrence of 100 years that is utilized as a perfect streamflow forecast (Q$_{100}$ flood hydrograph) in the main test application. The flood peak-occurrence time from the beginning is 6 h (i.e., the response time of the catchment to the rainfall event) and the peak flow corresponds to 597 m^3/s. The total flood volume equals to 17.1 hm^3. Being an extreme case, the peak flow of the selected flood hydrograph is three times greater than the downstream channel capacity (200 m^3/s). The hindcasting experiments were conducted in an arbitrary year during a critical operation period (May) when the initial forebay elevation was high. The whole closed-loop hindcasting period covers 96 h, from [1-May-2012 00:00:00] to [4-May-2012 23:00:00]. It is assumed that a 24-h long flood hydrograph occurs between [3-May-2012 00:00:00] and [3-May-2012 23:00:00]. Hourly forecast data are produced for 48 h lead-time. This means that in each 1 h, 48-h long DSF data (with 1 member) and 48-h long PSF data (with M = 50 ensemble members) are generated throughout the whole hindcasting period. Given the lack of probabilistic hydrological forecasts for this case study, we decided to recreate the stochasticity by assuming a normally distributed noise around the deterministic forecast. This in fact becomes an innovation on its own in the paper, since an objective approach has not been reported in the literature before. Although it was desired to increase the number of PSF members in order to cover much more possibilities, this ended up with the reverse situation in the optimization model due to the high-dimensional data space. Different ensemble sizes were also tried in the control model, and it was observed that increasing member number estimates higher uncertainty range, 50 members are considered to provide enough spread to capture the major uncertainties in the forecasts.

In this study, DSFs with error range between −0.3 and 0.3 and PSFs with final relative standard error ($\hat{\sigma}^{48} = 0.2$) are depicted with three different randomized scenarios (Sce-Q100a, Sce-Q100b and Sce-Q100c). Here, a scenario corresponds to randomly generated and independent forecasting data sets (both DSFs and PSFs) to be employed in the same hindcasting period. In each scenario, randomly perturbed DSFs from perfect data are produced in each hour for a given lead-time and an error interval. Then, similarly (in each scenario and in each hour), ensemble PSFs are produced from the mother DSFs with a given uncertainty. Since the generated streamflows are produced in each hour, the receding horizon is always set to one hour in all deterministic and stochastic MPC hindcasting experiments. In terms of operational point of view, the control trajectory changes at each time step and is not a fixed one for the complete event (it changes depending on the forecast and its uncertainty, as well as the forecast horizon).

As explanatory example from two scenarios (Sce-Q100a and Sce-Q100b) is shown in Figure 4 for a lead-time of 48 h. Fifty members are generated for PSFs, but only two members (min and max values) are given in the figure to show the uncertainty ranges (referred to uncertainty band). In the figure, [T0] corresponds to "time zero" namely the starting time of the optimization. Two time instants [T0: 01-May-2012 12:00:00] (Figure 4a) and [T0: 01-May-2012 13:00:00] (Figure 4b) are exampled here, but the process is dynamically repeated in each one hour throughout the hindcasting period in closed-loop experiments for each scenario. Finally, each scenario is composed of all generated 96 forecast sequences (both in DSFs and their PSFs) having a 48 h lead-time in the hindcasting period.

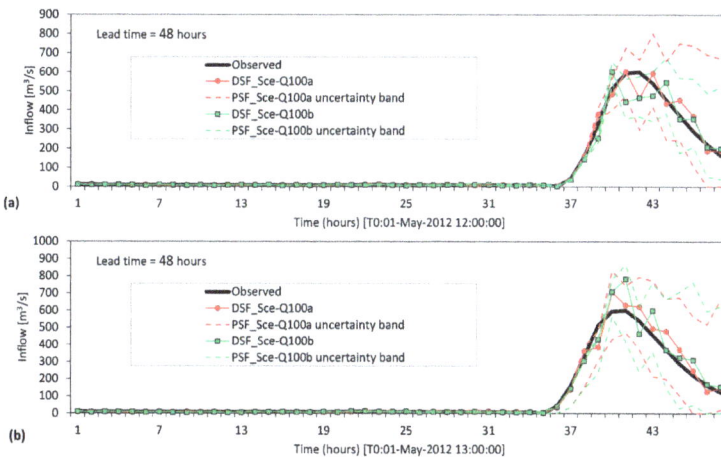

Figure 4. Graphical representation of the scenarios (Sce-Q100a and Sce-Q100b) for: (**a**) T0: 01-May-2012 12:00:00; (**b**) T0: 01-May-2012 13:00:00.

Different inflow conditions such as Q_{25} and Q_{50} (from Table 1) are also tested under Section 4.3.3. Characteristics of the PSF scenarios are presented with the performance assessment by using a mean Continuous Ranked Probability Score (CRPS) which generalizes the Mean Absolute Error (MAE) in the case of probabilistic forecasts (Supplementary Figure S1). Mean CRPS summarizes the quality of a probability forecast into a number by comparing the integrated square difference between the cumulative distribution function of forecasts and observations [24]. According to that, mean CRPS increases with forecast lead-time, while each scenario shows a different performance. The scenario number is not critical in the study because the focal idea is to develop an objective approach for stochasticity of the flows, to use them in stochastic optimization set-up and to compare the results with a deterministic equivalent. Thus, scenarios can be deliberated as different source based forecast data sets.

An example of generated forecast data sets (Sce-Q100a) is given for different time steps shown by T0 in Figure 5. Similarly, [T0] corresponds to "time zero" namely the starting time of the optimization in the figure. The red dashed line shows DSFs which are derived from a perfect forecast by 30% random noise. Synthesized PSF data are shown in gray ink. While generating PSF, updated forecast data have "different starting time" and "different duration to the peak flow" which affect the uncertainty band in the end. For example, the starting time of the 01-May-2012 20:00:00 (fourth panel) and 02-May-2012 00:00:00 (fifth panel) have less duration to peak flow, thus they have higher dispersion at peak flow compared to the last panel and subsequently uncertainty is kept high in the falling limb in the fourth and fifth panels. However moving forward in time, the duration to peak flow at T0 decreases, so that the relative standard error cannot increase too much due to the flow condition. That is the reason why the last panel has less spread in the end. It is directly related to the approach used to generate the ensembles.

Contrary imperfect data (DSFs and PSFs), perfect forecast (shown in Figure 5 as a black line) is a single chain and utilized in the initial MPC hindcasting experiment. This test allows the performance of the developed real-time reservoir system control model to be assessed without forecast uncertainty; as a result a sufficiently long forecast horizon is decided before starting imperfect forecast tests. During TB-MPC hindcasting experiments, PSF data are converted to trees depending on the selected branch numbers. Hereby, previously generated forecast data (from Sce-Q100a) PSFs with 50 members (Figure 5) are transformed into optimization tree inflows with 16 branches (members) in Figure 6. Thereby, the number of generated PSF data are limited by this reduction, but still capture the major uncertainty, and the tree is used in the optimization algorithm.

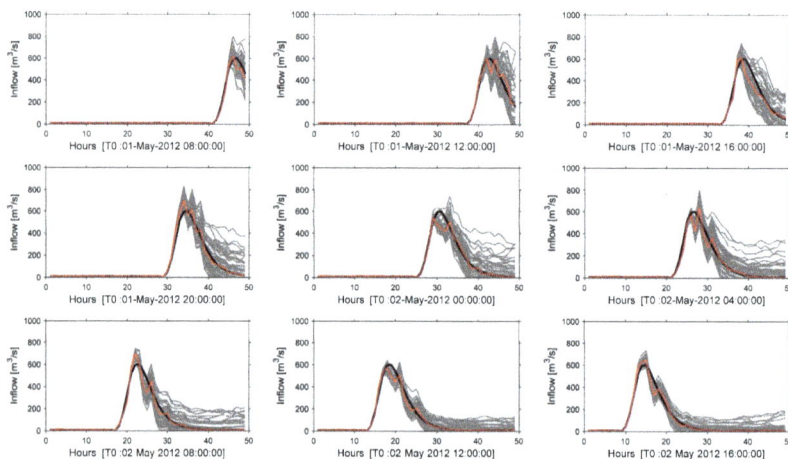

Figure 5. Perfect forecast, DSF vs. PSF for different receding horizons (Sce-Q100a).

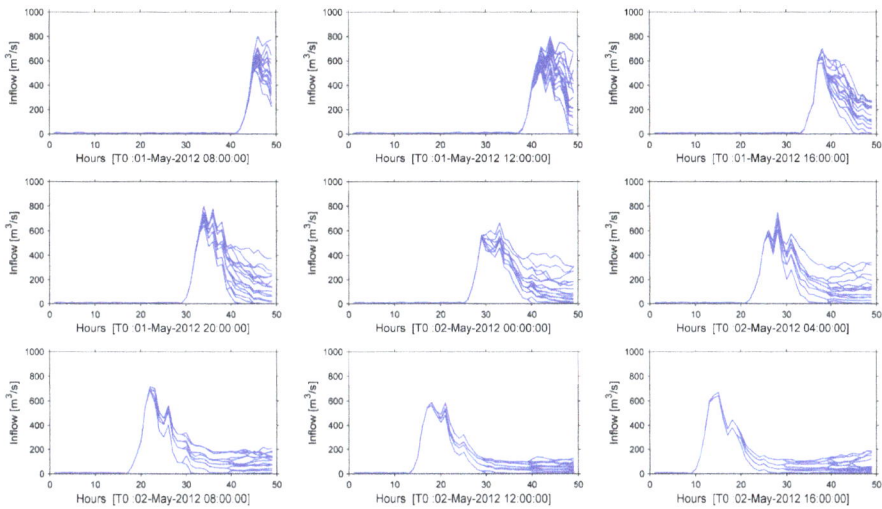

Figure 6. PSF ensemble members (50 members) transformed into optimization trees (16 branches).

Assessment of an effective reservoir operation in the short term is done by quantifying the minimization of downstream damage, the dam safety and maximization of the forebay elevation at the end of the flood event [53]. Similarly, four objective criteria are used in this study to test the experiments. These are:

- Forebay elevation at the end of a flood event: Forebay elevation should be same at the end of a flood event in order to provide long term water supply targets,
- Flooding threshold value: Spillway discharges should be less than channel capacity, thus this is considered as the flooding threshold value and the maximum discharge at the dam outlet is checked,
- Total flood volume at the downstream area: The cumulative volume of the released flood water (only above the maximum flood limit of 200 m^3/s) should be zero for the best flood management,
- Flood storage index (FSI): It is essential to have enough flood pool in the reservoir to attenuate the hourly extremes. To measure this, FSI is defined by the ratio of the total effective flood storage over the total volume of storage corresponding to Flood Control Levels (FCLs) [4] as shown in Equations (14) and (15) . The reservoir level should be kept at FCL as suggested to reserve space for flood attenuation. FSI ranges between zero and one. While zero indicates the reservoir level is always kept above FCL, one indicates operation is totally based on FCL. Higher FSI ensures more reliable flood operation (under forecast uncertainty) by having a high empty reservoir volume against flood risk.

$$FSI_f = \frac{\sum_{k=1}^{N} v_f^k}{\sum_{k=1}^{N} v_{FCL}^k} \tag{14}$$

$$v_f^k = \begin{cases} v_{act}^k \ if \ v^k \le v_{FCL}^k \\ v_{FCL}^k \ if \ v^k > v_{FCL}^k \end{cases} \quad for \ k = 1, 2, \dots N \tag{15}$$

where v_f^k is the effective flood storage, v_{FCL}^k is the storage corresponding to FCL, v_{act}^k is the actual volume of the flood control pool; v^k is the current storage by $k = 1, \dots, N$ time instants. If v^k is below FCL, it is equal to the volume of the flood control pool; otherwise, it is the actual available storage.

While the "flooding threshold value" gives the maximum instantaneous damage in the downstream, "total flood volume at downstream area" reflects the cumulative total penalty from the target set point of operational constraint within a hindcast period. It is also important to note that the perfect data based model results neglect uncertainty completely and serve as reference. Any deviation from these results objectively presents the inefficiency of the DSF and PSF based model results.

3.3. Model Set-Up

The reservoir model is based on the continuity equation for the reservoir. The water balance equation for a single reservoir can be arranged as:

$$s^k = s^{k-1} + \Delta t\left(Q_I^k - Q_S^k - Q_{WS}^k\right) \tag{16}$$

where, s is the storage volume (m^3), Δt is the time difference between k^{th} and $(k-1)^{th}$ time steps, Q_I, Q_S, and Q_{WS} are the reservoir inflows, spillway flow and water supply (m^3/s), respectively. Also, forebay elevation, fb could be computed by:

$$fb^k = f_{ls}\left(s^k\right) \tag{17}$$

where f_{sl} is a piecewise-linear level-storage relation.

In order to satisfy physical conditions capacity curves, discharge limits, and boundary elevations are introduced in constraints.

3.4. Physical and Operational Constraints

The constraints that form the equalities and inequalities are the continuity equation and the physical boundaries. The continuity equation ensures a consistent mass balance definition, accordingly residuum (r^k) is introduced as an equality constraint and it must be zero as formulated below:

$$r^k = s^k - s^{k-1} - \Delta t\left(Q_I^k - Q_S^k - Q_{WS}^k\right) = 0 \tag{18}$$

The system's physical limits should be met and they are defined as hard constraints. First, forebay level physical constraints are defined as:

$$fb_{min} \leq fb^k \leq fb_{max} \tag{19}$$

where fb_{min} is the minimum reservoir elevation and fb_{max} is the maximum reservoir elevation. Beyond the operational targets, the spillway discharge flow should be within physical limits depending on its capacity curve and is defined as the constraint below:

$$Qs_{min} \leq Qs^k \leq Qs_{max}^k \tag{20}$$

$$Qs_{max}^k = f_{sdc}\left(fb^k\right) \tag{21}$$

where Qs_{min} is the minimum spillway flow and is set to zero and Qs_{max} is the maximum allowable spillway flow and it is determined by the spillway discharge curve (f_{sdc}) i.e., zero at spillway crest elevation and variable above elevations depending on the current forebay elevation (as a function of fb).

3.5. Objective Function

An objective function determines the target of the operation by its terms. These terms are sometimes called soft constraints and an overall optimum of all terms is targeted. Optimization of a multi-purpose reservoir is aimed to mitigate flooding while maximizing water benefit or hydropower assets as its own objective function. Controlled variables are the forebay elevation and spillway discharges. The term sensitivity is set by the weights and the power of the equation. These weights are case-specific and depend on the significance of the soft constraints. For example, in this study higher penalizations are assigned for w_3 and w_4 in order to prevent downstream flooding and provide the long term water supply target, respectively. The weights are determined by a trial-and-error approach.

A reservoir having a limited capacity should include the terms below for hourly management:

- Differences between optimized forebay elevation and maximum operating elevation are minimized in Equation (22),
- Spillway discharges are minimized in Equation (23),
- Spillway releases above a specified discharge (200 m³/s) are constrained in Equation (24). This has a high weight in order to prevent damage in the downstream,
- The case of a set point for forebay elevation is given i.e., a variable guide curve (it is the same for each hour in a day) for long term targets [39], this term stands to minimize deviations from it in Equation (25),
- Spillway discharges between two consecutive time steps are constrained by consideration of the mechanical gate operation efficiency (against wear and tear) in Equation (26).

$$J1(fb) = w_1 \sum_{k=1}^{N} \left(f_{max} - fb^k \right) \tag{22}$$

$$J2(Qs) = w_2 \sum_{k=1}^{N} \left(Qs^k \right) \tag{23}$$

$$J3(Qs) = w_3 \sum_{k=1}^{N} \max \left(Qs^k - Q_{set}, 0 \right)^2 \tag{24}$$

$$J4(fb) = w_4 \sum_{k=1}^{N} \max \left(fb^k - fb_{set}, 0 \right)^2 \tag{25}$$

$$J5(Qs) = w_5 \sum_{k=1}^{N} \left(Qs^{k+1} - Qs^k \right)^2 \tag{26}$$

where, $w_j (j = 1, \dots, 5)$ stands for weights associated with each term ($w_j > 0$), Q_{set} stands for the maximum channel capacity without downstream region flooding, fb_{set} stands for the time-dependent variable guide curve, and N stands for forecast horizon.

Finally, closed-loop optimization minimizes the main objective function (J), which equals the summation of different objectives, in Equation (27) for optimum forebay elevations and spillway discharges agreement with pre-defined constraints in Equations (18)–(21). Optimizations are conducted in each hour (due to the selection of a one hour receding horizon in this study) for N forecast horizon throughout the entire hindcasting period. Only the first value of the computed control sequence (spillway discharge) is applied to the system and the rest is discarded. In the next time step, the state (forebay elevation) is updated and the optimization is repeated with updated forecast data.

$$\begin{aligned} \min J(fb, Qs) \\ k \in \{0, \dots, T\} \end{aligned} = J1 + J2 + J3 + J4 + J5 \tag{27}$$

4. Numerical Experiments and Results

4.1. Deterministic MPC Hindcasts Using Perfect Forecasts

It is assumed that selecting a sufficiently long forecast horizon in the closed-loop mode provides an approximately actual infinite horizon solution [54]. In the first trial, hindcasting experiments were employed using perfect forecasts with respect to different forecast horizons. The tests were conducted for 6, 12, 18, 24, 36, and 48 h to estimate a sufficiently long forecast horizon, and named as PER6, PER12, PER18, PER24, PER36, and PER48, respectively. The results of the experiments are presented in terms of spillway discharges and forebay elevation in Figure 7. Short forecast horizons (such as 6 h and 12 h) produce delayed releases which in return increase peakflow at the dam outlet. This is an important indicator of flood risk assessment since higher discharges create a larger extension of damage in the downstream area. Therefore, MPC needs a longer forecast horizon to handle the problem even with perfect inflow forecasts. Forecast horizons of 18 h or longer provide downstream safety by reaching reasonable maximum discharges under 200 m^3/s limit at the outlet. Longer forecast horizons than 18 h e.g., 24, 36, and 48 h, result in a similar response. Therefore, the experiment results of PER24, PER36, and PER48 overlap in the figure. According to the results, one can note that the mitigation of a major flood even with maximum operating levels and 200 m^3/s downstream channel constraint can be achieved under perfect future knowledge of flood inflows at least 18 h beforehand.

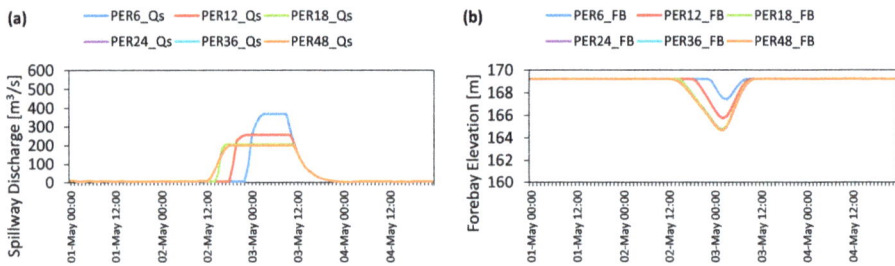

Figure 7. Comparison of closed-loop MPC forecast horizon performance using perfect streamflow forecasts: (**a**) Spillway discharge (m^3/s); (**b**) forebay elevation (m).

4.2. Deterministic MPC Hindcasts Using DSFs

DSFs are produced as single forecast trajectories by adding random perturbations on observations, thus they under/over-estimate the actual flood inflows. This situation becomes more significant for the refilling season in spring when the initial reservoir forebay elevation is at a critical level. DSF data based MPC hindcasting experiment results using Sce-Q100a data are presented in Figure 8. Likewise, longer forecast horizons than 18 h e.g., 24, 36, and 48 h result in a similar response, thus the experiment results of DSF24, DSF36, and DSF overlap in the figure. According to the results, spillway discharges are greater than the upper limit of 200 m^3/s and create flooding in the downstream. This is considered as lower reliability compared to perfect data based experiments, and mainly attributed to the forecast disturbance which introduces 30% bias to the control strategy. Longer forecast horizons (such as 18, 24, 36, 48 h) perform better and releases are shifted to earlier time steps. However, it is not possible to mitigate the flood event even for forecast horizons longer than 18 h due to the given bias in the inflows and the lack of uncertainty in the system optimization. Compared to perfect forecasts based MPC, the variations in spillages are higher due to updated information for each receding horizon.

4.3. Multi-Stage Stochastic TB-MPC Hindcasts Using PSFs

In this part, hindcasting experiments are conducted by means of PSF data and multi-stage stochastic optimization. TB-MPC uses scenario trees for disturbances (inflows), states (forebay elevation) and control trajectories (spillway discharges) to form a related stochastic model. Definition of the multiple stages at branching points using binary trees makes the process a multi-stage stochastic optimization in TB-MPC. Hence, the results become more adaptive to the resolution of uncertainty and have better expected performances.

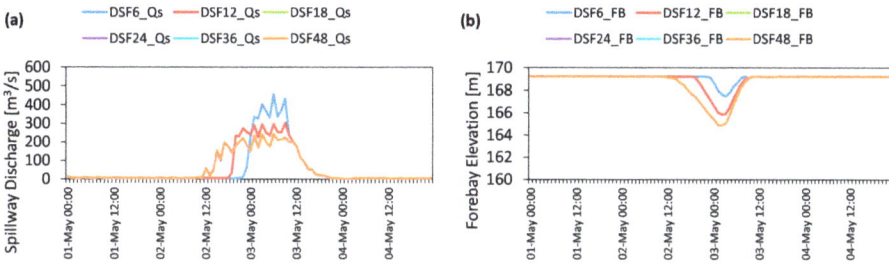

Figure 8. Comparison of closed-loop MPC forecast horizon performances with DSF (Sce-Q100a):
(**a**) Spillway discharge (m³/s); (**b**) forebay elevation (m).

MPC solves an open-loop optimization at every time instant along a forecast horizon and applies only to the first control value to the system. Therefore, an example from a selected single case [T0 = 2-May-2012 14:00:00] of open-loop optimization results is presented with spillway discharge and forebay elevation in smooth trees (Figure 9). One can note that the use of binary trees splits the stochastic variables into two trajectories at each specific branching point. The timing of specific branching points is automatically defined due to the tree-reduction algorithm (check Section 2.3) when the branch number and optimization forecast horizon have been fixed. In this study, 16 tree branches were chosen based on the performance comparison (check Section 4.3.1) for the forecast horizon of 48 h (check Section 4.3.2), thus branching points are fixed every 10 h by the tree reduction algorithm. The root value is an optimal discharge for the entire control sequence considering uncertainty resolution through the forecast horizon. Only single discharge data from a rooted tree of the open-loop optimization is used in the simulation, and the process is shifted to the next time step. At the following time instant, the optimization is reformulated with the updated initial reservoir level and updated forecast data. After applying this procedure through the whole hindcast period (96 h), closed-loop TB-MPC hindcasting experiment results (one single output) are obtained. These results are discussed under the subsections of Section 4.3.

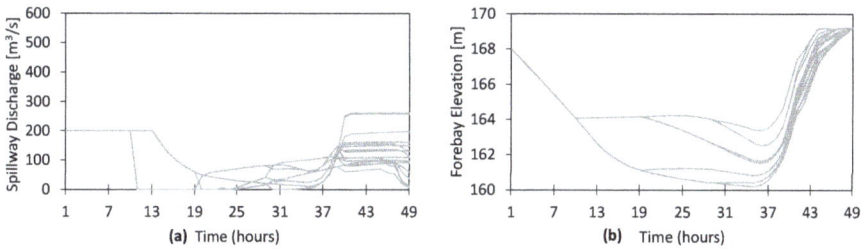

Figure 9. Open-loop optimization results of multi-stage stochastic optimization (from Sce-Q100a):
(**a**) Spillway discharge trees (m³/s); (**b**) Forebay elevation trees (m).

4.3.1. TB-MPC Hindcasts Considering a Different Number of Tree Branches

The number of tree branches used in TB-MPC is user-defined and there is no direct method to select the most appropriate one. In this part, performances of TB-MPC hindcasting experiments using Sce-Q100a data are conducted by selecting a different number of tree branches in each trial and the results are compared with each other. According to TB-MPC methodology, total PSF members can be reduced to 2^x branches, e.g., 1, 2, 4, 8, ... etc. [24]. Therefore, in this study 50 PSF ensembles were reduced to six different branches (1, 2, 4, 8, 16 and 32) and tested in a hindcast test. The experiments were done using Sce-Q100a PSF as input forecast data and comparison of the optimization results from different tree branches numbers is given in Figure 10 in terms of spillway flows and forebay elevation. This experiment shows the effects of the resolution of tree and correspondingly capturing forecast uncertainty in stochastic optimization. If the forecast horizon is set to 48 h, we can get optimum results after 16 branches as shown (Figure 10a) in terms of the spillway discharge. Since the results for 16–32 trees are exactly the same, they overlap with each other in the same figure. Although higher resolution overestimates the inflows which increases the pre-releases, it is still able to restore the forebay elevation target at the end of the flood event (Figure 10b).

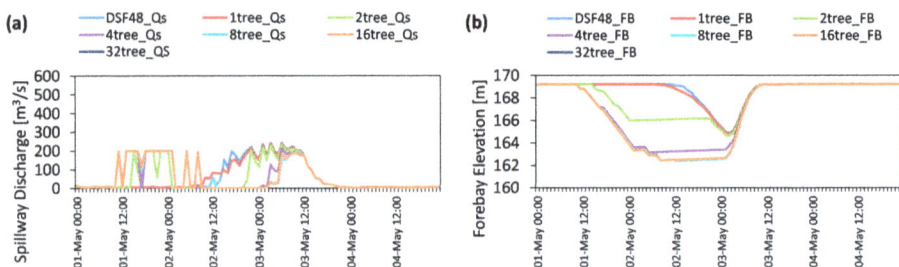

Figure 10. Comparison of closed-loop MPC with different tree reduction branches for 48 h forecast horizon (Sce-Q100a): (**a**) Spillway discharge (m^3/s); (**b**) forebay elevation (m).

The computation time is given for different tree reduction branches in the hindcasting experiments in comparison with deterministic MPC with DSF (Table 2). Note that this is the total time for the entire hindcast period (96 h). The time spent in deterministic MPC is three times less than TB-MPC with one tree branch which can be considered as the ensemble mean. This is mainly attributed to the tree reduction process before the optimization of the model. However, the computation time increases as long as higher tree branches are used in the model. Therefore, the tree number is fixed to 16 branches for the following experiments of the study. These results are also valid for the remaining scenarios.

Table 2. Computation times in MPC hindcasting experiments. MPC stands for Model Predictive Control. DSF stands for Deterministic Streamflow Forecast. TB-MPC stands for Tree-based MPC.

Hindcasting Experiment	Total CPU Time (s)
MPC with DSF	151
TB-MPC with 1 tree branch	491
TB-MPC with 2 tree branches	551
TB-MPC with 4 tree branches	633
TB-MPC with 8 tree branches	677
TB-MPC with 16 tree branches	867
TB-MPC with 32 tree branches	1354

4.3.2. TB-MPC Hindcasts Considering Different Forecast Horizons

Figure 11 presents a comparative visualization of different hindcasting experiments with perfect data and DSF under deterministic configuration and PSF under multi-stage stochastic TB-MPC for 18, 24, 36, and 48 h forecast horizons. Notice that deterministic methods have similar results for a longer forecast horizon whereas TB-MPC start earlier pre-releases. This is mainly attributed to consideration of uncertainty resolution through the forecast horizon in the decision mechanism. When the selected forecast horizon increases in TB-MPC experiments, the uncertainty band in the PSF spread increases and thereby much more water is evacuated from the spillway with the more conservative policy for flood control. It is remarkable to note that even when a longer forecast horizon is selected, forebay elevations at the end of the flood event are always equal to the initial reservoir elevation. These results are also valid for the remaining scenarios.

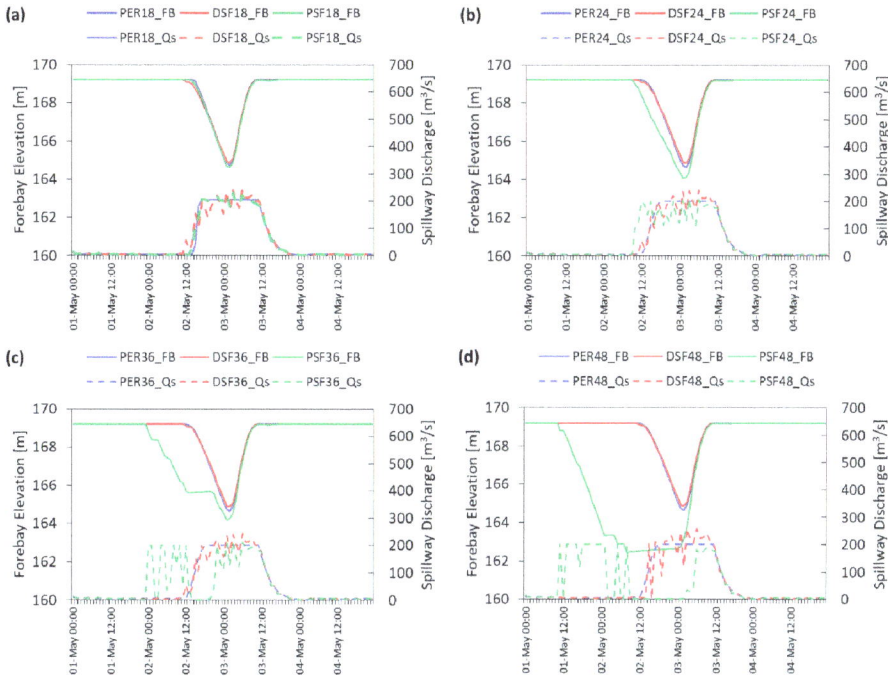

Figure 11. Comparison of deterministic (perfect and DSF) and stochastic (PSF) closed-loop MPC results with different forecast horizons (Sce-Q100a): (**a**) 18 h; (**b**) 24 h; (**c**) 36 h; (**d**) 48 h.

4.3.3. Assessment of the Approach for Different Inflow Conditions and Scenarios

The hindcasting experiments having forecast horizons of 48 h were enriched with different inflow conditions to check the robustness of the developed methodology. To that end, hydrographs characterized by return periods lower than 100 years such as 25 and 50 years, which have peak flows greater than the downstream channel capacity, were also used in the closed-loop hindcasting experiments. All experiments were evaluated by forebay elevation at the end of the flood event, the penalization of the flooding threshold value by the peak flow observed at the Yuvacık outlet, the total flood volume at the downstream area, and FSI. Similar to the previous set-up, flood hydrographs were independently utilized in each hindcasting test as perfect data, DSF and PSF data sets were generated. The tests were also elaborately investigated with three different independent scenarios (a,b,c) for each flow condition. The complete results belonging to the hindcasting experiments are given in Supplementary Figures S2 and S3 for Q_{25} and Q_{50} scenarios, respectively. On the other hand, deterministic (perfect and DSF) and stochastic (PSF) closed-loop MPC results from Sce-Q100a, Sce-Q100b and Sce-Q100c are shown in Figure 12. It is notable that the stochastic set-up always provides pre-releases and takes precautions against flood event. The deterministic model only takes actions over several hours which is similar to the perfect data based reference model, but generates much more spillage above the flood threshold compared to stochastic TM-MPC.

A brief summary of all inflow conditions for the three scenarios are given (Table 3) in terms of peak flow values. Since perfect data based models can give the desired maximum spillway of

200 m^3/s for each condition, they are not shown in the table. According to results, there is always an improvement in spillway discharges for different flows conditions of Q_{25}, Q_{50} and Q_{100}.

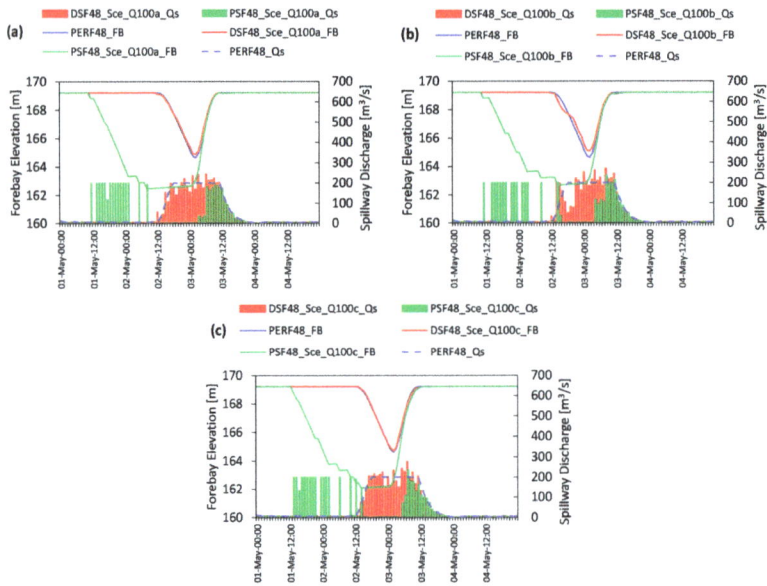

Figure 12. Comparison of deterministic (perfect and DSF) and stochastic (PSF) closed-loop MPC results with different forecast scenarios for Q_{100}: (**a**) Sce-Q100a; (**b**) Q100-Sceb; (**c**) Q100-Scec.

Table 3. Peakflow assessment of deterministic and stochastic closed-loop MPC results for different inflow conditions with forecast horizons of 48 h.

Flood Hydrograph	Scenarios	Peakflow at Yuvacik Outlet (m^3/s)	
		Deterministic MPC	Stochastic MPC
	Sce-Q25a	243	231
Q_{25}	Sce-Q25b	255	243
	Sce-Q25c	248	243
	Sce-Q50a	241	211
Q_{50}	Sce-Q50b	245	200
	Sce-Q50c	246	200
	Sce-Q100a	242	200
Q_{100}	Sce-Q100b	269	235
	Sce-Q100c	278	233

Table 4 presents the total flood volumes at the downstream area calculated in the hindcasting period. Compared to flooding threshold assessment, this indicator provides more insight into spillway operation. The best management (zero volume) is possible by perfect data using a forecast horizon of 48 h. Almost in all inflow cases and scenarios, the total flood volume decreases in the stochastic mode. This change highlights the added value of the stochastic optimization in comparison to the deterministic one especially for different forecast data sets (scenarios) and different flow conditions. FSI values were also calculated for both DSF and PSF scenarios and results are presented in Table 5, respectively. It is important to note that each FSI is calculated according to its own FCL i.e., FCL Q_{100} is used for Q_{100} operation assessment whereas FCL Q_{50} is used for the Q_{50} based flood case. According to this, TB-MPC stands as the more confident by higher FSI but also still can provide a high reservoir level at the end of the event without compromising water supply targets. For an uncertain future, a higher FSI is more reliable and preferable with lower risk for water supply as well.

Table 4. Flood volume assessment of deterministic and stochastic closed-loop MPC for different inflow conditions with forecast horizon of 48 h.

Flood Condition	Scenarios	Total Flood Volume (1×10^6 m^3)	
		Deterministic MPC	Stochastic MPC
Q_{25}	Sce-Q25a	0.507	0.302
	Sce-Q25b	0.549	0.254
	Sce-Q25c	0.438	0.271
Q_{50}	Sce-Q50a	0.666	0.062
	Sce-Q50b	0.471	0.004
	Sce-Q50c	0.331	0.004
Q_{100}	Sce-Q100a	0.690	0.004
	Sce-Q100b	1.256	0.184
	Sce-Q100c	1.018	0.127

Table 5. FSI value assessment of deterministic and stochastic closed-loop MPC according to Flood Control Levels (FCLs) for different inflow conditions with forecast horizon of 48 h.

Flood Condition	Scenarios	Flood Storage Index (FSI)	
		Deterministic MPC	Stochastic MPC
Q_{25}	Sce-Q25a	0.652	0.800
	Sce-Q25b	0.659	0.990
	Sce-Q25c	0.659	0.796
Q_{50}	Sce-Q50a	0.566	0.723
	Sce-Q50b	0.598	0.770
	Sce-Q50c	0.606	0.758
Q_{100}	Sce-Q100a	0.457	0.650
	Sce-Q100b	0.463	0.645
	Sce-Q100c	0.456	0.645

According to results, PSF based multi-stage stochastic MPC optimization has the advantage of including forecast uncertainty in the optimization set-up. Also, the uncertainty in the flows can be represented in the synthesized PSFs by developing the proposed error generation method. There is always performance improvement in the TB-MPC model for flood flows of Q_{25}, Q_{50}, Q_{100} with a forecast horizon up to 48 h with respect to (i) maximum discharge at the dam outlet (compared to flooding threshold value), (ii) total flood volume at the downstream area and (iii) FSI, while keeping the forebay elevation at the desired level for water supply at the end of the flood event. This shows the added value of the approach and provides reasonable outputs compared to the deterministic counterpart. The developed framework also indicates robust solutions against forecast uncertainty along with a different independent hindcasting experiment assessment.

5. Conclusions and Outlook

This study shows the added value of stochastic optimization using a synthetic probabilistic forecast generation and mass-conservative scenario tree reduction technique. TB-MPC provides multi-stage stochastic optimization in comparison to its deterministic counterpart. According to hindcasting experiments, our main conclusions are:

- Forecast uncertainty is indispensable especially for flood management. It is critical for those cases in which wrong or poor decisions may result with loss of life and property. At this point, considering uncertainty provides better management in terms of flood metrics without discarding water supply purposes.

- Independent closed-loop hindcasting experiment scenarios demonstrate the robustness of the system developed against biased information (disturbances).
- Probabilistic data represent forecast uncertainty in comparison to deterministic equivalents. In this study, a new synthetic streamflow generation method is proposed to represent forecast uncertainty for reservoir optimization.
- The synthetic PSF generation model that considers the dynamic evolution of uncertainties is valuable if hydrological model outputs driven by a rainfall and temperature forecast ensemble are not available. This method is very advantageous from the operational standpoint, since it does not require complex computations and is easy to implement while considering conditional (flow dependent) increasing uncertainty through time. It is simple to formulate, comprehend, and easy to repeat.
- Besides the ensemble generation, tree reduction parameters should be carefully investigated in the problem definition phase. In the case of selecting a lower branch and forecast horizon than required, TB-MPC results may converge to deterministic MPC results.
- The system was also tested against different inflow conditions which have greater flood value than the downstream channel capacity. According to the results, the method provides reliable results against different high flood conditions in the hindcasting experiments.

In future work, a hydrological model will be used, thus synthesized PSF might be derived from perturbed hydrological model forcings such as precipitation and temperature. A comparison of these results can contribute to improve synthetic ensemble generation and its consideration in TB-MPC.

Supplementary Materials: The following are available online at http://www.mdpi.com/2073-4441/10/3/340/s1, Figure S1: mean CRPS of ensemble scenarios: (a) Sce-Q25; (b) Sce-Q50; (c) Sce-Q100, Figure S2: Comparison of deterministic (perfect and DSF) and stochastic (PSF) closed-loop MPC results with different forecast scenarios for Q_{25}: (a) Sce-Q25a; (b) Sce-Q25b; (c) Sce-Q25c, Figure S3: Comparison of deterministic (perfect and DSF) and stochastic (PSF) closed-loop MPC results with different forecast scenarios for Q_{50}: (a) Sce-Q50a; (b) Sce-Q50b; (c) Sce-Q50c.

Acknowledgments: The first author would like to thank The Scientific and Technological Research Council of Turkey (TUBITAK) for the scholarship (2214A program). This study is supported by Anadolu University Scientific Research Projects Commission (under the grant No: 1506F502 and No: 1705F189). Graphs were prepared by Daniel's XL Toolbox (www.xltoolbox.net) and MATLAB 2012a (License number: 991708).

Author Contributions: Aynur Şensoy set the problem and background. Gökcen Uysal organized the main structure of the paper. Gökcen Uysal, Rodolfo Alvarado-Montero, and Dirk Schwanenberg designed the methodology. Gökcen Uysal conducted the experiments, and prepared the draft paper. Finally, Dirk Schwanenberg, Rodolfo Alvarado-Montero, and Aynur Şensoy provided useful advice and made some corrections. All authors read and approved the final manuscript.

Conflicts of Interest: The authors declare no conflict of interest.

References

1. Ahmad, S.; Simonovic, S.P. System dynamics modeling of reservoir operations for flood management. *J. Comput. Civ. Eng.* **2000**, *14*, 190–198. [CrossRef]
2. Plate, E.J. Flood risk and flood management. *J. Hydrol.* **2002**, *267*, 2–11. [CrossRef]
3. Wei, C.C.; Hsu, N.S. Optimal tree-based release rules for real-time flood control operations on a multipurpose multireservoir system. *J. Hydrol.* **2009**, *365*, 213–224. [CrossRef]
4. Şensoy, A.; Uysal, G.; Şorman, A.A. Developing a decision support framework for real-time flood management using integrated models. *J. Flood Risk Manag.* **2016**. [CrossRef]
5. Cheng, W.M.; Huang, C.L.; Hsu, N.S.; Wei, C.C. Risk analysis of reservoir operations considering short-term flood control and long-term water supply: A case study for the Da-Han Creek Basin in Taiwan. *Water* **2017**, *9*, 424. [CrossRef]
6. Oliveira, R.; Loucks, D.P. Operating rules for multireservoir systems. *Water Resour. Res.* **1997**, *33*, 839–852. [CrossRef]
7. Chang, F.J.; Chen, L.; Chang, L.C. Optimizing the reservoir operating rule curves by genetic algorithms. *Hydrol. Process.* **2005**, *19*, 2277–2289. [CrossRef]

8. Liu, P.; Guo, S.; Xu, X.; Chen, J. Derivation of aggregation-based joint operating rule curves for cascade hydropower reservoirs. *Water Resour. Manag.* **2011**, *25*, 3177–3200. [CrossRef]

9. Rani, D.; Moreira, M.M. Simulation–optimization modeling: A survey and potential application in reservoir systems operation. *Water Resour. Manag.* **2010**, *24*, 1107–1138. [CrossRef]

10. Li, X.; Guo, S.; Liu, P.; Chen, G. Dynamic control of flood limited water level for reservoir operation by considering inflow uncertainty. *J. Hydrol.* **2010**, *391*, 124–132. [CrossRef]

11. Wan, W.; Zhao, J.; Lund, J.R.; Zhao, T.; Lei, X.; Wang, H. Optimal hedging rule for reservoir refill. *J. Water Resour. Plan. Manag. ASCE* **2016**, *142*, 04016051. [CrossRef]

12. Bellman, R. The theory of dynamic programming. *Bull. Am. Math. Soc.* **1954**, *60*, 503–516. [CrossRef]

13. Faber, B.A.; Stedinger, J.R. Reservoir optimization using sampling SDP with ensemble streamflow prediction (ESP) forecasts. *J. Hydrol.* **2001**, *249*, 113–133. [CrossRef]

14. Yurtal, R.; Seckin, G.; Ardiclioglu, G. Hydropower optimization for the lower Seyhan system in Turkey using dynamic programming. *Water Int.* **2005**, *30*, 522–529. [CrossRef]

15. Raso, L.; Malaterre, P.O.; Bader, J.C. Effective streamflow process modeling for optimal reservoir operation using stochastic dual dynamic programming. *J. Water Resour. Plann. Manage. ASCE* **2017**, *143*, 04017003. [CrossRef]

16. Labadie, J.W. Optimal operation of multireservoir systems: State-of-the-art review. *J. Water Resour. Plan. Manag. ASCE* **2004**, *130*, 93–111. [CrossRef]

17. Morari, M.; Lee, J.H. Model predictive control: Past, present and future. *Comput. Chem. Eng.* **1999**, *23*, 667–682. [CrossRef]

18. Blanco, T.B.; Willems, P.; Chiang, P.K.; Haverbeke, N.; Berlamont, J.; De Moor, B. Flood regulation using nonlinear model predictive control. *Control Eng. Pract.* **2010**, *18*, 1147–1157. [CrossRef]

19. Van Overloop, P.J.; Horváth, K.; Aydin, B.E. Model predictive control based on an integrator resonance model applied to an open water channel. *Control Eng. Pract.* **2014**, *27*, 54–60. [CrossRef]

20. Horváth, K.; Galvis, E.; Valentín, M.G.; Rodellar, J. New offset-free method for model predictive control of open channels. *Control Eng. Pract.* **2015**, *41*, 13–25. [CrossRef]

21. Zhao, T.; Zhao, J.; Yang, D.; Wang, H. Generalized martingale model of the uncertainty evolution of streamflow forecasts. *Adv. Water Resour.* **2013**, *57*, 41–51. [CrossRef]

22. Chen, J.; Brissette, F.P.; Leconte, R. Uncertainty of downscaling method in quantifying the impact of climate change on hydrology. *J. Hydrol.* **2011**, *401*, 190–202. [CrossRef]

23. Roulin, E.; Vannitsem, S. Post-processing of medium-range probabilistic hydrological forecasting: Impact of forcing, initial conditions and model errors. *Hydrol. Process.* **2015**, *29*, 1434–1449. [CrossRef]

24. Fan, F.M.; Schwanenberg, D.; Alvarado, R.; dos Reis, A.A.; Collischonn, W.; Naumman, S. Performance of deterministic and probabilistic hydrological forecasts for the short-term optimization of a tropical hydropower reservoir. *Water Resour. Manag.* **2016**, *30*, 3609–3625. [CrossRef]

25. Kuczera, G.; Parent, E. Monte Carlo assessment of parameter uncertainty in conceptual catchment models: The Metropolis algorithm. *J. Hydrol.* **1998**, *211*, 69–85. [CrossRef]

26. Montero, R.A.; Schwanenberg, D.; Krahe, P.; Lisniak, D.; Sensoy, A.; Sorman, A.A.; Akkol, B. Moving horizon estimation for assimilating H-SAF remote sensing data into the HBV hydrological model. *Adv. Water Resour.* **2016**, *92*, 248–257. [CrossRef]

27. Yao, H.; Georgakakos, A. Assessment of Folsom Lake response to historical and potential future climate scenarios: 2. Reservoir management. *J. Hydrol.* **2001**, *249*, 176–196. [CrossRef]

28. Zhao, T.; Cai, X.; Yang, D. Effect of streamflow forecast uncertainty on real-time reservoir operation. *Adv. Water Resour.* **2011**, *34*, 495–504. [CrossRef]

29. Liu, Z.; Guo, Y.; Wang, L.; Wang, Q. Streamflow forecast errors and their impacts on forecast-based reservoir flood control. *Water Resour. Manag.* **2015**, *29*, 4557–4572. [CrossRef]

30. Todini, E. Role and treatment of uncertainty in real-time flood forecasting. *Hydrol. Process.* **2004**, *18*, 2743–2746. [CrossRef]

31. Koutsoyiannis, D. The Hurst phenomenon and fractional Gaussian noise made easy. *Hydrol. Sci. J.* **2002**, *47*, 573–595. [CrossRef]

32. Ahmed, J.A.; Sarma, A.K. Artificial neural network model for synthetic streamflow generation. *Water Resour. Manag.* **2007**, *21*, 1015. [CrossRef]

33. Ochoa-Rivera, J.C.; García-Bartual, R.; Andreu, J. Multivariate synthetic streamflow generation using a hybrid model based on artificial neural networks. *Hydrol. Earth Syst. Sci.* **2002**, *6*, 641–654. [CrossRef]

34. Raso, L.; Schwanenberg, D.; van de Giesen, N.C.; van Overloop, P.J. Short-term optimal operation of water systems using ensemble forecasts. *Adv. Water Resour.* **2014**, *71*, 200–208. [CrossRef]

35. Stive, P.M. Performance Assessment of Tree-Based Model Predictive Control. Master's Thesis, Delft University of Technology, Delft, The Netherlands, 2011.

36. Ficchi, A.; Raso, L.; Dorchies, D.; Pianosi, F.; Malaterre, P.O.; van Overloop, P.J.; Jay-Allemand, M. Optimal operation of the multireservoir system in the Seine River Basin using deterministic and ensemble forecasts. *J. Water Resour. Plan. Manag.* **2015**, *142*. [CrossRef]

37. Schwanenberg, D.; Fan, F.M.; Naumann, S.; Kuwajima, J.I.; Montero, R.A.; dos Reis, A.A. Short-term reservoir optimization for flood mitigation under meteorological and hydrological forecast uncertainty. *Water Resour. Manag.* **2015**, *29*, 1635–1651. [CrossRef]

38. Uysal, G.; Şensoy, A.; Şorman, A.A.; Akgün, T.; Gezgin, T. Basin/reservoir system integration for real time reservoir operation. *Water Resour. Manag.* **2016**, *30*, 1653–1668. [CrossRef]

39. Uysal, G.; Schwanenberg, D.; Alvarado-Montero, R.; Şensoy, A. Short term optimal operation of water supply reservoir under flood control stress using model predictive control. *Water Resour. Manag.* **2018**, *32*, 583–597. [CrossRef]

40. Hui, P. *Yellow River Group Project, A Subproject of the China-DC WRE Project*; Final Research Report of Cluster 2; Delft University of Technology: Delft, The Netherlands, 2002.

41. Yan, B.; Guo, S.; Chen, L. Estimation of reservoir flood control operation risks with considering inflow forecasting errors. *Stoch. Environ. Res. Risk Assess.* **2014**, *28*, 359–368. [CrossRef]

42. Datta, B.; Burges, S.J. Short-term, single, multiple-purpose reservoir operation: Importance of loss functions and forecast errors. *Water Resour. Res.* **1984**, *20*, 1167–1176. [CrossRef]

43. Pianosi, F.; Raso, L. Dynamic modeling of predictive uncertainty by regression on absolute errors. *Water Resour. Res.* **2012**, *48*. [CrossRef]

44. Biemans, H.; Hutjes, R.W.A.; Kabat, P.; Strengers, B.J.; Gerten, D.; Rost, S. Effects of precipitation uncertainty on discharge calculations for main river basins. *J. Hydrometeorol.* **2009**, *10*, 1011–1025. [CrossRef]

45. Schwanenberg, D.; Becker, B.P.J.; Xu, M. The open real-time control (RTC)-Tools software framework for modeling RTC in water resources sytems. *J. Hydroinform.* **2015**, *17*, 130–148. [CrossRef]

46. Xu, M.; Schwanenberg, D. Comparison of sequential and simultaneous model predictive control of reservoir systems. In Proceedings of the 10th International Conference on Hydroinformatics, Hamburg, Germany, 14–18 July 2012.

47. Wächter, A.; Biegler, L.T. On the implementation of an interior-point filter line-search algorithm for large-scale nonlinear programming. *Math. Program.* **2006**, *106*, 25–57. [CrossRef]

48. Heitsch, H.; Römisch, W. Scenario reduction algorithms in stochastic programming. *Comput. Optim. Appl.* **2003**, *24*, 187–206. [CrossRef]

49. Nadarajah, S.; Shiau, J.T. Analysis of extreme flood events for the Pachang River, Taiwan. *Water Resour. Manag.* **2005**, *19*, 363–374. [CrossRef]

50. Stedinger, J.R.; Vogel, R.M.; Foufoula-Georgiou, E. Frequency analysis of extreme events. In *Handbook of Hydrology*; Maidment, D.R., Ed.; McGraw-Hill: New York, NY, USA, 1993; Chapter 18.

51. Xiao, Y.; Guo, S.; Liu, P.; Yan, B.; Chen, L. Design flood hydrograph based on multicharacteristic synthesis index method. *J. Hydrol. Eng.* **2009**, *14*, 1359–1364. [CrossRef]

52. General Directorate of State Hydraulic Works (DSI). *Kirazdere Dam Engineering Hydrology and Planning Report*; Prepared by 1st Regional Directorate of State Hydraulics Works; Planning and Science Board: Bursa, Turkey, 1983. (In Turkish)

53. Mediero, L.; Garrote, L.; Martin-Carrasco, F. A probabilistic model to support reservoir operation decisions during flash floods. *Hydrol. Sci. J. ASCE* **2007**, *52*, 523–537. [CrossRef]

54. Jørgensen, J.B. Moving Horizon Estimation and Control. Ph.D. Thesis, Technical University of Denmark, Lyngby, Denmark, 2005.

Article

Research on Cascade Reservoirs' Short-Term Optimal Operation under the Effect of Reverse Regulation

Changming Ji [1], Hongjie Yu [1,*], Jiajie Wu [1], Xiaoran Yan [1] and Rui Li [2]

[1] School of Renewable Energy, North China Electric Power University, Beijing 102206, China;
cmji@ncepu.edu.cn (C.J.); 1121550126@ncepu.edu.cn (J.W.); yanxiaoran1014@163.com (X.Y.)

[2] Overseas Exam Service Center, Shijiazhuang Information Engineering Vocational College,
Shijiazhuang 050035, China; liruijoyce@163.com

* Correspondence: yhjyeah@163.com

Received: 7 May 2018; Accepted: 4 June 2018; Published: 19 June 2018

Abstract: Currently research on joint operation of a large reservoir and its re-regulating reservoir focuses on either water quantity regulation or water head regulation. The accuracy of relevant models is in need of improvement if the influence of factors such as water flow hysteresis and the aftereffect of tail water level variation are taken into consideration. In this paper, given the actual production of Pankou-Xiaoxuan cascade hydropower stations that combines two operation modes ('electricity to water' and 'water to electricity'), a coupling model of their short-term optimal operation is developed, which considers Xiaoxuan reservoir's regulating effect on Pankou reservoir's outflow volume and water head. Factors such as water flow hysteresis and the aftereffect of tail water level variation are also considered to enhance the model's accuracy. The Backward Propagation (BP) neural network is employed for precise calculation of the downstream reservoir's inflow and the upstream reservoir's tail water level. Besides, we put forth Accompanying Progressive Optimality Algorithm (APOA) to solve the coupling model with aftereffect. An example is given to verify the scientificity of the proposed model and the advantages of APOA. Through analysis of the model calculation results, the optimal operation rules of the cascade reservoirs are obtained in terms of water quantity regulation and water head regulation, which can provide scientific reference for cascade reservoirs' optimal operation.

Keywords: reverse regulation; coupling model; aftereffect; accompanying progressive optimality algorithm

1. Introduction

In the downstream of a large reservoir, the construction of a small re-regulating reservoir can not only ensure the upstream hydropower station's peak-regulating capability in the power grid, but also exert a positive effect on downstream water supply, ecology, shipping, etc., spawning significant comprehensive utilization benefits. Therefore, this kind of development mode has been widely adopted in water conservancy construction across the world, e.g., the Three Gorges-Gezhouba cascade reservoirs in China [1], the Xiaolangdi-Xixiayuan cascade reservoirs in China [2], the Bigge-Ahausen cascade reservoirs in Germany [3], the Srinagarindra-Tha Thung Na cascade reservoirs in Thailand, and so forth [4].

The downstream reservoir's reverse regulation effect on the upstream reservoir is mainly embodied in two facets: (1) water quantity regulation, i.e., using the downstream reservoir's storage capacity to optimize the upstream reservoir's outflow; (2) water head regulation, i.e., controlling the downstream reservoir's water level to optimize the overall output given its backwater effect on the upstream reservoir's tail water level. To date, many scholars have studied the joint operation of a large reservoir and its re-regulating reservoir. For example, Richter, B. D. proposed that the re-regulating reservoir can readjust the upstream reservoir's outflow to eliminate the unnatural fluctuations caused by hydropower operation, so that water discharge to the downstream channel can move in a way much closer to natural flow, thus protecting the downstream ecological environment [5]. Bai, T. et al. established a model of Xiaolangdi-Xixiayuan cascade hydropower stations' joint peak-regulating operation, which used Xixiayuan reservoir to regulate Xiaolangdi reservoir's outflow volume, thus alleviating the conflict between water regulation and power dispatching during Xiaolangdi hydropower station's operation process [6]. Taking the Three Gorges-Gezhouba cascade reservoirs as an example, Cai, Z. et al. analyzed the law of the influence of Gezhouba reservoir's water level on the cascade hydropower generation during different operation states of the Three Gorges reservoir, and obtained the control table of optimal water levels [7]. However, the above studies have only considered either water quantity regulation or water head regulation. For instance, Bai, T. did not consider Xixiayuan reservoir's influence on Xiaolangdi reservoir's tail water level in calculation. For simplification, Cai, Z. used the 'runoff operation' hypothesis on Gezhouba reservoir, whose water level stayed unchanged during calculation. So far, there has been no research that considers the combination of water quantity regulation and water head regulation; hence the above models somewhat deviate from the actual operation of cascade reservoirs, with limitations in their conclusions.

Additionally, in previous studies, to solve the cascade reservoirs' short-term optimal operation model, the calculation of such variables as downstream reservoir's inflow and upstream reservoir's tail water level was often simplified (e.g., the upstream reservoir's outflow was directly used as the downstream reservoir's inflow [8], the upstream reservoir's tail water level was calculated via the stage-discharge relation method [9], etc.). These simplifications neglect the impact of factors such as water flow hysteresis and the aftereffect of tail water level variation, which makes model variables deviate from reality and reduces model accuracy. Yet, in the study of reverse regulation, downstream reservoir's inflow and upstream reservoir's tail water level are the key variables during calculation, as they can reflect the connection of water quantity and water head between the upstream and the downstream reservoirs. Therefore, it is important to accurately calculate these two variables. Research about water flow hysteresis is focused mainly on flow travel time, which is generally treated as a fixed value in primary studies [10]. In recent years, some scholars have proposed that flow travel time varies dynamically with the change of upstream reservoir outflow [11]. Although this improvement takes into consideration the influence of reservoir outflow's rate on its travel speed, it is still regarded as a steady flow without considering flow attenuation during travel. Consequently, there is still a certain deviation between the study results and the actual situation. Apart from the stage-discharge relation method, research on the variation rule of tail water level mostly adopts hydraulic methods that determine reservoir tail water level by constructing and solving a hydrodynamic model of open-channel unsteady flow [12,13]. However, hydraulic models are unsuitable for practical production due to their stringent requirements for boundary condition data and slow calculation speed. Accordingly, to seek a universal and efficient method for further exploration and accurate simulation of cascade reservoirs' operation process with various complex factors is of great significance in improving model accuracy.

Short-term optimal operation is mostly applied to reservoirs' practical production. Therefore, besides model accuracy, there is also high demand for solving methods of the model in terms of computation time, calculation results, and so on. Among the many methods to work out reservoir optimal operation models, Dynamic Programming (DP) is widely used thanks to its advantages such as global convergence and high stability [14,15]. However, DP is restricted by the 'curse of dimensionality' [16], which means that the amount of calculation increases exponentially with the growing number of reservoirs. Moreover, the no-aftereffect condition is no longer satisfied in cascade reservoirs' short-term optimal operation when factors such as water flow hysteresis and the aftereffect of tail water level variation are taken into account, which makes DP inapplicable. In this regard, Ji, C. et al. tried to solve the model of cascade reservoirs' short-term optimal operation with aftereffect by using Multi-Stage Dynamic Programming (MSDP) [17]. This method has global convergence, but its problem of 'dimensionality curse' is even more serious, which renders it inapplicable to practical production.

Mei, Y. proposed Multi-Dimensional Dynamic Programming Approximation Algorithm [18], which has a small amount of calculation, but may not produce the actual global optimal solution due to approximation in the calculation process. Although modern intelligent algorithms, such as Genetic Algorithm [19,20], Particle Swarm Optimization Algorithm [21,22], and Ant Colony Optimization Algorithm [23,24], can solve problems with aftereffect in a short period of time, they tend to get stuck in the local optimal solution. Until now, there has been no mature method that can fully meet the requirements of actual production, so presently it is necessary to develop a method that can accurately figure out the model's global optimal solution in a relatively short time.

In view of the above problems in the study of short-term optimal operation of cascade reservoirs under the effect of reverse regulation, this paper combines the two operation modes ('electricity to water' and 'water to electricity') of Pankou-Xiaoxuan cascade hydropower stations to build a coupling model of their short-term optimal operation, which considers both Xiaoxuan reservoir's regulating effect on Pankou reservoir's outflow and the former's backwater effect on the latter's tail water level. Through the BP neural network, factors such as water flow hysteresis and the aftereffect of tail water level variation are studied. Furthermore, Accompanying Progressive Optimality Algorithm (APOA) is put forward to solve the proposed coupling model, so as to provide scientific guidance for the short-term optimal operation of cascade reservoirs under the effect of reverse regulation.

2. Coupling Model

2.1. Objective Function

The subject of this paper is Pankou-Xiaoxuan cascade hydropower stations located in Hubei province in China. With an installed capacity of 500 MW, the Pankou hydropower station is an important peak-regulating power station in the province. Its regular power operation is controlled by Hubei Provincial Power Dispatching Center (a typical 'electricity to water' mode). With an installed capacity of 50 MW, Xiaoxuan hydropower station, as the re-regulating hydropower station for Pankou, mainly regulates Pankou's outflow for power generation (a typical 'water to electricity' mode), which is controlled by Han Jiang Hydropower Development Co., Ltd. (hereafter referred to as the Company, Shiyan, China). Although the two hydropower stations have different dispatching superordinates, they both belong to the Company as their decision maker. The Company hopes that through Xianxuan reservoir's reverse regulation over Pankou reservoir, the overall power generation efficiency of the whole cascade hydropower stations can be improved. Long-term production practice has proved that Xiaoxuan reservoir's water level has a certain backwater effect on Pankou reservoir's tail water level. On one hand, while maintaining a high water level of Xiaoxuan reservoir is apparently beneficial to its own power generation, it will raise the tail water level of Pankou reservoir, thus lowering its productive water head and affecting its power generation efficiency, which means consuming more water to complete the same power generation quota. On the other hand, Xiaoxuan reservoir can operate at a low water level to enhance the power generation efficiency of Pankou hydropower station, but at the cost of losing some of its own power generation benefits. Plus, when Pankou hydropower station is in operation to regulate peak load, its outflow is so uneven that Xiaoxuan reservoir has to re-regulate it to meet the demand for downstream ecological flow. Hence, for the economic operation of the cascade hydropower stations, it is of crucial importance to determine the operation mode of Xiaoxuan hydropower station that can reversely regulate Pankou reservoir's outflow and water head in a scientific manner.

Against the backdrop of coupling the two operation modes ('electricity to water' and 'water to electricity'), the aim of the cascade hydropower stations is to balance Pankou's power generation efficiency with Xiaoxuan's generated energy. Accordingly, from the angle of the whole cascade hydropower stations' total energy, two optimization criteria—minimum energy consumption and maximum generated energy—are integrated to develop the objective function in the coupling model of Pankou-Xiaoxuan cascade reservoirs' short-term optimal operation [25,26], as shown by Equation (1).

Unlike the traditional objective function that involves water quantity, this one calculates water energy, with Pankou's power generation efficiency reflected by its hydroenergy consumption for power generation (meaning that the less hydroenergy consumption, the higher its power generation efficiency). In this way, the operation goals of both Pankou and Xiaoxuan are unified in the form of energy, which avoids the emergence of a multi-objective problem. In addition, as the two optimization criteria conflict with each other, the objective function is the subtraction of the former from the latter, finally seeking the maximum power generation benefits of the whole cascade hydropower stations:

$$E = \max \sum_{t=1}^{T} [N_{2,t}(q_{2,t}, H_{2,t})\Delta t - 3600\lambda_{1,t}q_{1,t}(N_{1,t}, H_{1,t})\Delta t] \tag{1}$$

where E is the remainder of Xiaoxuan's generated energy after deducting Pankou's hydroenergy consumption in the operation period, unit: kWh; T is the number of calculation periods over the entire operation period; $N_{2,t}(q_{2,t}, H_{2,t})$ is Xiaoxuan's power output when its power generation flow and water head are $q_{2,t}$ and $H_{2,t}$ respectively in period t, unit: kW, and is obtained according to the power characteristic curve of its generator set; Δt is the length of the calculation period, unit: h, and $\Delta t = 0.25$ h; $\lambda_{1,t}$ is the energy efficiency coefficient of Pankou hydropower station in period t, unit: kWh/m^3, and its physical meaning is the amount of energy contained in every cubic meter of Pankou reservoir's water in period t; $q_{1,t}(N_{1,t}, H_{1,t})$ is Pankou's power generation discharge when its output and water head are $N_{1,t}$ and $H_{1,t}$ respectively in period t, unit: m^3/s, and is also obtained according to the power characteristic curve of its generator set.

2.2. Constraint Conditions

Now that the coupling model combines the two operation modes, the constraint conditions must contain particular constraints specific to these two modes, as shown by Equations (2) and (3), while other regular constraints are expressed by Equations (4)–(10).

(1)   Output command constraint

Because Pankou hydropower station operates in the 'electricity to water' mode, its output in each period is already determined:

$$N_{1,t} = N_t \tag{2}$$

where $N_{i,t}$ is the output of hydropower station i in period t, unit: kW (i = 1 for Pankou, i = 2 for Xiaoxuan and the same hereafter); N_t is the output command from the superior dispatching center for Pankou hydropower station in period t, unit: kW.

(2)   Water level constraint at the end of the operation period

Because Xiaoxuan hydropower station operates in the 'water to electricity' mode, its water level at the end of the operation period is already determined:

$$Z_{2,T+1} = Z_{2,\text{end}} \tag{3}$$

where $Z_{i,t}$ is the water level of reservoir i at moment t, unit: m; $Z_{2,\text{end}}$ is Xiaoxuan reservoir's controlled water level at the end of the operation period, unit: m.

(3)   The upper and lower bounds of water level:

$$Z_i^{\min} \leq Z_{i,t} \leq Z_i^{\max} \tag{4}$$

where $Z_i^{\max}$, $Z_i^{\min}$ are the upper and lower bounds of water level of reservoir i, respectively, unit: m.

(4)   The upper and lower bounds of output:

$$N_i^{\min} \leq N_{i,t} \leq N_i^{\max} \tag{5}$$

where $N_i^{\max}$, $N_i^{\min}$ are the upper and lower bounds of output of hydropower station i, respectively, unit: kW.

(5)   The upper and lower bounds of flow rate:

$$q_i^{\min} \leq q_{i,t} \leq q_i^{\max} \tag{6}$$

where $q_{i,t}$ is the power generation discharge of hydropower station i in period t, unit: m^3/s, and in this paper, it equals to reservoir outflow because water abandonment is not taken into consideration; $q_i^{\max}$, $q_i^{\min}$ are the upper and lower bounds of power generation discharge of hydropower station i, respectively, unit: m^3/s.

(6)   Water balance constraint:

$$V_{i,t+1} = V_{i,t} + 3600(Q_{i,t} - q_{i,t})\Delta t \tag{7}$$

where $V_{i,t}$, $V_{i,t+1}$ are the storage capacity of reservoir i at respectively the beginning and the end of period t (i.e., moment t and moment $t + 1$), unit: m^3; $Q_{i,t}$ is the inflow of reservoir i in period t, unit: m^3/s.

(7)   Vibration zone constraint:

$$N_{i,j} \notin [N_{i,j}^{\min}, N_{i,j}^{\max}] \tag{8}$$

where $N_{i,j}$ is the output of generator unit j in hydropower station i, unit: kW; $N_{i,j}^{\min}$, $N_{i,j}^{\max}$ are the upper and lower bounds of the vibration zone of generator unit j in hydropower station i, unit: kW.

(8)   Ecological flow constraint

To meet the demand for downstream ecological flow [27], Xiaoxuan reservoir's outflow should not be less than 16.7 m^3/s:

$$q_{2,t} \geq q_e \tag{9}$$

where q_e is the demand for downstream ecological flow, unit: m^3/s, and q_e = 16.7 m^3/s.

(9)   Auxiliary power constraint

To prevent the occurrence of inverse power transmission, when Pankou hydropower station halts production, it is Xiaoxuan hydropower station that undertakes the auxiliary power supply mission for the running of both stations:

$$N_{2,t} \geq N_s \text{ when } N_{1,t} = 0 \tag{10}$$

where N_s is the auxiliary power need of the cascade hydropower stations, unit: kW, and N_s = 2000 kW.

The ecological flow constraint and the auxiliary power constraint can well reflect Xiaoxuan reservoir's regulation effect on Pankou reservoir's discharge volume. Since the completion of Xiaoxuan reservoir, Pankou reservoir no longer bears the task of downstream ecological water supply, and the task of auxiliary power supply has been transferred to Xiaoxuan hydropower station. Therefore, when Pankou hydropower station does not undertake peak-regulating tasks, all its generator units can stop running instead of operating in the low efficiency zone just to meet the constraints of ecological flow and auxiliary power supply.

3. Calculation of the Model's Key Variables

Between the upstream and the downstream reservoirs, there is a certain hydraulic connection in terms of water quantity and water head [28], as shown in Figure 1. Water quantity connection refers to the upstream reservoir's outflow turning into the downstream reservoir's inflow after travelling through the interval river channel. Water head connection refers to that the downstream reservoir's backwater raises the water level of the interval channel, causing a jacking effect on the upstream reservoir's tail water level. Due to these two kinds of connections, the re-regulating reservoir can adjust the upstream reservoir's discharge volume and water head. Thus, the downstream reservoir's inflow and the upstream reservoir's tail water level, which can respectively reflect these two types of connections, become the key variables in the coupling model of Pankou-Xiaoxuan cascade reservoirs' short-term optimal operation under the effect of reverse regulation. Computation of the key variables has a critical influence on the model's accuracy in that an error in their calculations will cause the model's results to deviate from reality, losing its instructive significance to practical production. Thus, it is very important to calculate them accurately.

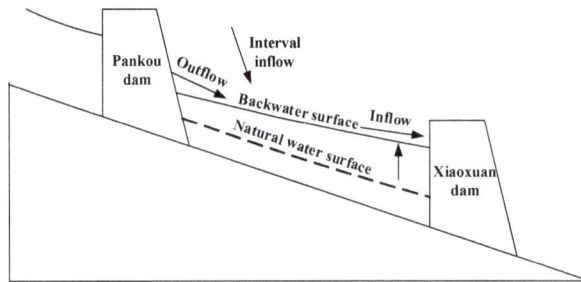

Figure 1. Schematic diagram of the cascade reservoirs' hydraulic connection.

3.1. The Downstream Reservoir's Inflow Considering Water Flow Hysteresis and Interval Inflow

In the cascade reservoirs, the downstream reservoir's inflow is composed of the upstream reservoir's outflow and the interval inflow. The upstream reservoir's outflow arrives at the downstream reservoir with water flow hysteresis that mainly involves two parts. (1) It takes some time for the water flow to travel from the upstream reservoir to the downstream reservoir. (2) During this travel process, the upstream reservoir's outflow, which is an unsteady flow, goes through not only displacement but also attenuation to some degree [29]. Figure 2 illustrates the comparison between Pankou reservoir's outflow and Xiaoxuan reservoir's inflow during a certain power generation process, from which it can be seen that in contrast to Pankou reservoir's outflow, Xiaoxuan reservoir's inflow somewhat lags behind and gets smoother. Water flow hysteresis makes it a complicated hydraulics problem as to when and how Pankou reservoir's outflow will reach Xiaoxuan reservoir.

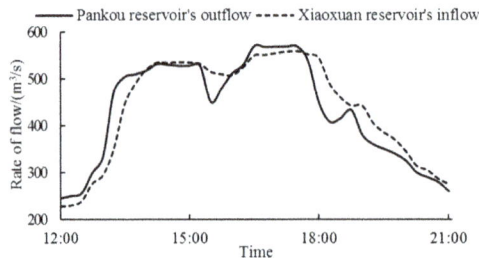

Figure 2. Pankou's outflow and Xiaoxuan's inflow.

The dam sites of Pankou and Xiaoxuan are only 10.4 km apart, without any tributary in between, and the interval catchment area is fairly small, so the interval inflow is often neglected in practical production. However, long-term monitoring data show that on average Xiaoxuan reservoir's inflow is slightly greater than Pankou reservoir's outflow, so the existence of the interval inflow cannot be ignored in order to improve the model's accuracy. The interval inflow between Pankou and Xiaoxuan is mainly formed by runoff generation and confluence in the interval basin, which largely depends on the basin's rainfall [30]. When the interval inflow is taken into consideration, the calculation of Xiaoxuan reservoir's inflow will become more complicated in that not only the evolution of water current in the river but also the hydrologic cycle from precipitation to runoff is involved. Since the physical mechanisms of these natural processes are not fully understood, it is difficult to accurately simulate them. Hence in this paper, with masses of data, the BP neural network is used to figure out the relationship between Xiaoxuan reservoir's inflow and Pankou reservoir's outflow as well as the interval basin's rainfall, so that the accurate value of Xiaoxuan reservoir's inflow can be worked out [31].

The BP neural network is a multilayer feedforward network trained by the error backward propagation algorithm. As one of the most widely-used neural networks, it can approximate any nonlinear function with any precision. Structurally, the BP neural network is divided into one input layer, one output layer, and several hidden layers [32]. In the calculation of Xiaoxuan reservoir's inflow, the output layer is represented by Xiaoxuan reservoir's inflow in the current period ($Q_{2,t}$). As for the input layer, Pankou reservoir's outflow ($q_{1,t} \sim q_{1,t-4}$) as well as the interval basin's precipitation ($P_t \sim P_{t-4}$) in the current and the previous four periods are chosen as the primary data and screened by the correlation coefficient method. It is demonstrated in Table 1 that the correlation coefficient between Pankou reservoir's outflow and Xiaoxuan reservoir's inflow is smaller in the earlier period. For $q_{1,t}$, $q_{1,t-1}$ and $q_{1,t-2}$, their correlation coefficients with $Q_{2,t}$ are similar, whereas the correlation coefficients of $q_{1,t-3}$ and $q_{1,t-4}$ with $Q_{2,t}$ are noticeably smaller. Therefore, $q_{1,t}$, $q_{1,t-1}$ and $q_{1,t-2}$ are selected as the input layer data. Compared with Pankou reservoir's outflow, the interval basin's precipitation has a much smaller correlation coefficient, which changes little over time. Considering that the interval inflow accounts for only a small portion of Xiaoxuan reservoir's inflow, the precipitation P_{t-1}, with the largest correlation coefficient with $Q_{2,t}$ in this regard, is selected as the input layer data. In 1989, Cybenko, G. and Hornik, K. proved that the three-layer network (with one input layer, one output layer, and one hidden layer) can simulate any complex nonlinear problems [33,34]. Therefore, one hidden layer is set up whose node numbers are determined according to Equation (11) [35]:

$$h = \sqrt{u + v} + a \tag{11}$$

where h is the number of nodes in the hidden layer; u is the number of nodes in the input layer; v is the number of nodes in the output layer; a is an adjustment coefficient between 0 and 10, and according to tentative calculation when $a = 4$ the neural network yields the best training results in this study.

After the structure of the neural network is established, the historical data of 2012–2016 are input into the neural network for training. The neuron transfer function is a hyperbolic tangent function, and after 10,000 times of training, the network is saved for later calculation of Xiaoxuan reservoir's inflow. Figure 3 shows the calculated Xiaoxuan reservoir's inflow during the power generation process mentioned before. Compared with the actual inflow, the inflow produced by the BP neural network has a relative error of 1.11%, which is more accurate than that obtained by the translation method (relative error: 3.84%), which only considers the travel time (i.e., Pankou reservoir's outflow merely translates into Xiaoxuan reservoir's inflow).

Table 1. Correlation coefficients between input layer data and the output layer.

Input Layer Data	$q_{1,t}$	$q_{1,t-1}$	$q_{1,t-2}$	$q_{1,t-3}$	$q_{1,t-4}$	P_t	P_{t-1}	P_{t-2}	P_{t-3}	P_{t-4}
Correlation Coefficient	0.91	0.89	0.88	0.83	0.79	0.19	0.20	0.19	0.19	0.18

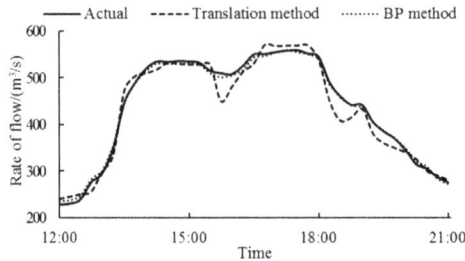

Figure 3. Calculation results of Xiaoxuan's inflow.

3.2. The Upstream Reservoir's Tail Water Level Considering the Influence of Dual Aftereffect Factors

During the operation of cascade reservoirs, under the backwater effect of the downstream reservoir, the upstream reservoir's tail water level is closely related to the downstream reservoir's water level. Different water levels of the downstream reservoir have different degrees of backwater effect on the upstream reservoir's tail water level. Figure 4 is the stage-discharge curve of Pankou reservoir corresponding to different water levels of Xiaoxuan reservoir. Long-term practical production shows that Pankou reservoir's tail water level obtained by the stage-discharge relation method still deviates from the actual value. This kind of deviation can be explained by two reasons. For one thing, the tail water level-discharge curve is based on the assumption that the reservoir outflow is a steady flow. However, in fact, it is an unsteady flow whose characteristics result in a certain aftereffect of tail water level variation [36], that is, the tail water level in the current period is related to that in the previous period. For another, it takes some time for Xiaoxuan reservoir's backwater to reach Pankou dam, meaning that Pankou reservoir's tail water level in the current period is not necessarily influenced by Xiaoxuan reservoir's water level in the current period but the previous several periods [37], which is another aftereffect of Pankou's tail water level variation. Under the influence of such dual aftereffect factors, Pankou's tail water level in the current period is affected by three kinds of variables: Pankou's outflow in the current period, Pankou's tail water level in the previous period, and Xiaoxuan's water level in the previous n (n is determined by the travel time of Xiaoxuan's backwater) periods. As the physical mechanism is very complicated, the BP neural network is again used for calculation.

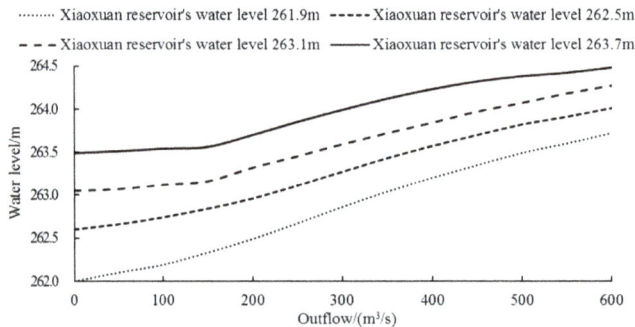

Figure 4. Pankou's stage-discharge curve.

Here, Pankou reservoir's tail water level in the current period ($DZ_{1,t}$) represents the output layer. As for the input layer data, aside from Pankou reservoir's outflow in the current period ($q_{1,t}$) and its tail water level in the previous period ($DZ_{1,t-1}$), it is also needed to decide which periods of Xiaoxuan reservoir's water level are concerned. Similarly, Xiaoxuan reservoir's water levels in the current and the previous four periods ($Z_{2,t}$~$Z_{2,t-4}$) are chosen as the primary data to be filtered by the correlation coefficient method. From Table 2 it can be seen that the correlation coefficient between Xiaoxuan's water level and Pankou's tail water level gradually decreases with time. For $Z_{2,t}$, $Z_{2,t-1}$ and $Z_{2,t-2}$, their correlation coefficients have little difference, while for $Z_{2,t-3}$ and $Z_{2,t-4}$ their correlation coefficients are apparently smaller, which indicates that the time it takes for Xiaoxuan reservoir's backwater to reach Pankou is between 0 and 2 periods. Thus, $Z_{2,t}$, $Z_{2,t-1}$ and $Z_{2,t-2}$ are selected as the input layer data. The configuration of the hidden layer and the training parameters is similar to that described in Section 3.1. After the training is completed, the network is saved for later calculation of Pankou reservoir's tail water level. Figure 5 shows the resulted tail water level variation of Pankou reservoir in the same power generation process mentioned before. Compared with the actual variation, the results of the BP neural network only produce a calculation error of 0.001 m, which is notably smaller than the calculation error (0.027 m) produced by the stage-discharge relation method, thus proving the advantage of the BP neural network.

Table 2. Correlation coefficients between input layer data and the output layer.

Input layer Data	$Z_{2,t}$	$Z_{2,t-1}$	$Z_{2,t-2}$	$Z_{2,t-3}$	$Z_{2,t-4}$
Correlation Coefficient	0.74	0.73	0.73	0.70	0.67

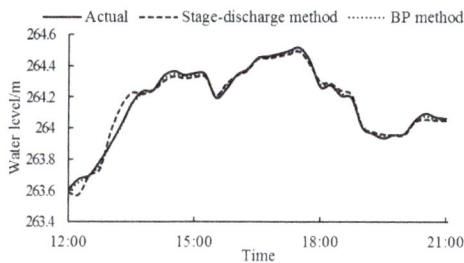

Figure 5. Calculation results of Pankou's tail water level.

4. Model Solution

When factors such as water flow hysteresis and the aftereffects of tail water level variation are taken into account, the decision variable (power generation flow) of the current period for the cascade reservoirs is not only related to the state (reservoir water level) of this period but also the state of the previous several periods. Since the 'no aftereffect' condition is not satisfied, the proposed model of short-term optimal operation cannot be solved by DP. Although MSDP can deal with the problem with aftereffect, it is restricted by a serious 'curse of dimensionality' [17]. The subject of this study is taken as an example for further explanation. The short-term optimal operation of Pankou-Xiaoxuan cascade reservoirs is an issue about the joint operation of two reservoirs. Nonetheless, as Pankou hydropower station operates in the 'electricity to water' model, once the operation state of Xiaoxuan hydropower station is determined, Pankou's operation state will also be settled (by in-plant economic operation calculation), which reduces the issue to an essentially one-dimensional problem. In this one-dimensional scenario, suppose that the number of discrete points of the reservoir water level is M and the number of interrelated periods is N (i.e., the state of the former $N-1$ periods influences the decision of the current period), then in each multi-stage calculation of MSDP there are M^{N+1} combinations of Xiaoxuan's water level and the total amount of calculation is $T \cdot M^{N+1}$ with T times of multi-stage calculation. Since the value of $T \cdot M^{N+1}$ grows exponentially with the increase in N, even in a one-dimensional circumstance the dimensionality curse still emerges due to the interconnection of N periods. As the number of dimensions increases to L ($L > 1$), the total amount of calculation will become $T \cdot (M^{N+1})^L$ with $(M^{N+1})^L$ combinations of water level at each multistage, which is a more serious problem of dimensionality curse. In view of this, we come up with an improved POA (i.e., APOA) to solve the model of the cascade reservoirs' short-term optimal operation with aftereffect.

4.1. Basic Principle of APOA

POA is an improved DP that simplifies a complicated multi-stage decision problem into a series of two-stage decision problems, considerably reducing the complexity and effectively mitigating the dimensionality curse [38]. In contrast to POA, APOA has two facets of improvements.

(1) When Xiaoxuan hydropower station is being optimized, Pankou reservoir's water level is taken as the variable that spatially accompanies Xiaoxuan reservoir's water level for calculation. In each discrete computation where Xiaoxuan's water levels in the current and the previous n periods are known, the magnitude of Xiaoxuan's backwater effect on Pankou's tail water level is definite. In addition, because Pankou's tail water level in the previous one period is already known, in the current period, Pankou's tail water level is only related to Pankou's outflow. Under the condition of a definite output of Pankou hydropower station, its reservoir outflow can be obtained through trial calculation. Its water level can then be figured out according to its inflow and outflow. Hence, for each discrete water level of Xiaoxuan, there is only one corresponding water level of Pankou and accordingly, one hydrograph of Pankou corresponding to each hydrograph of Xiaoxuan in the operation period.

(2) In each two-stage calculation, the variables of the preceding related $N - 1$ periods are taken as the temporally-accompanying variables of the current period for calculation. Considering the interrelation of N periods, the decision of the current two-stage calculation is affected by those decisions made in the previous $N - 1$ periods. APOA divides the problem with T periods into $T - 1$ sub problems, each of which includes two main calculation periods accompanied by $N - 1$ related periods. As shown in Figure 6, in the calculation of period t and period $t + 1$, the state and the decision of other periods (including the accompanying $N - 1$ related periods) are known, thereby effectively tackling the aftereffect problem. In each two-stage calculation, there are M combinations of Xiaoxuan's water level and $T - 1$ times of two-stage calculation for one iteration, so the amount of calculation is $(T - 1) \cdot M$. If the precision requirement is satisfied after K times of iteration, the total amount of calculation will be $K \cdot (T - 1) \cdot M$, evidently smaller than $T \cdot M^{N+1}$ produced by MSDP. As a result, APOA can effectively alleviate the problem of dimensionality curse.

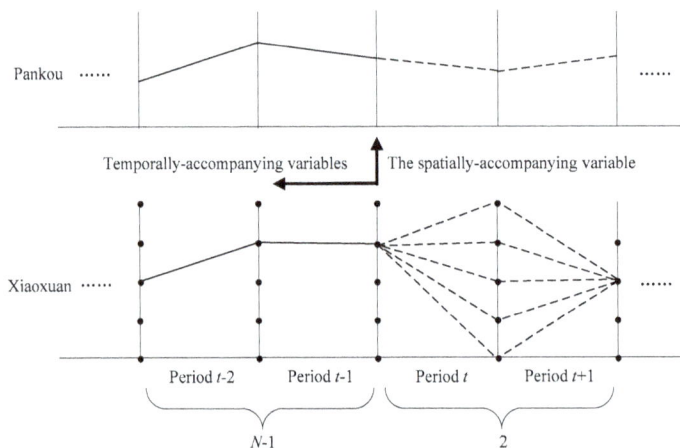

Figure 6. Schematic diagram of APOA.

4.2. Calculation Procedure of APOA

The calculation procedure of APOA is detailed as follows:

Step 1: Determine the number of associated periods. Use the correlation coefficient method to calculate the number of periods connected with the current period in terms of water quantity and water head. If the first m periods are related to the current period in terms of Xiaoxuan reservoir's inflow and the first n periods are related to the current period in terms of Pankou reservoir's tail water level, then the number of related periods can be calculated by Equation (12), where $m = 2$, $n = 2$ and $N = 3$ in this article:

$$N = \max\{m + 1, n + 1\} \tag{12}$$

Step 2: Train the BP neural networks. The BP neural networks are trained by inputting training data, including the type of training data, starting and ending time, as well as the structure and parameters of the neural networks, which are detailed in Sections 3.1 and 3.2. After the training is completed, the neural network for calculating Xiaoxuan reservoir's inflow (net_1) and that for Pankou reservoir's tail water level (net_2) are obtained.

Step 3: Set the initial solution. The selection of initial solutions has a major influence on the convergence speed and computation time of APOA [39]. Long-term experience of the dispatchers is often embodied in the operation rule of the hydropower station, according to which a relatively optimal solution can be obtained. Thus, in this paper, such a solution is used as the initial solution. It should be noted that in the initial solution, the initial water level of Xiaoxuan reservoir ($Z_2^0 = \{Z_{2,1}^0, Z_{2,2}^0, \ldots, Z_{2,T+1}^0\}$) is the most critical variable while that of Pankou reservoir ($Z_1^0 = \{Z_{1,1}^0, Z_{1,2}^0, \ldots, Z_{1,T+1}^0\}$) is only the variable accompanying it.

Step 4: Optimize progressively. Carry out the optimization step by step from period 1 to period T. When the calculation progresses to period t and period $t + 1$ (accompanied by the two related periods of $t - 2$ and $t - 1$), fix Xiaoxuan reservoir's water levels at all moments except that at moment $t + 1$ and calculate all the discrete points of its water level at that moment. Xiaoxuan reservoir's inflow and Pankou reservoir's tail water level are calculated by the BP neural networks, as shown in Equation (13), with Pankou reservoir's water level acting as the variable accompanying Xiaoxuan's water level. After running through all the discrete points, select the optimal water level of Xiaoxuan reservoir ($Z_{2,t+1}^*$) and correspondingly the optimal water level of Pankou ($Z_{1,t+1}^*$) at moment $t + 1$. Make $Z_{2,t+1}^1 = Z_{2,t+1}^*$ and $Z_{1,t+1}^1 = Z_{1,t+1}^*$.

$$Q_{2,t} = net_1(q_{1,t}, q_{1,t-1}, q_{1,t-2}, P_{t-1}), Q_{2,t+1} = net_1(q_{1,t+1}, q_{1,t}, q_{1,t-1}, P_t)$$
$$DZ_{1,t} = net_2(q_{1,t}, DZ_{1,t-1}, Z_{2,t}, Z_{2,t-1}, Z_{2,t-2}), DZ_{1,t+1} = net_2(q_{1,t+1}, DZ_{1,t}, Z_{2,t+1}, Z_{2,t}, Z_{2,t-1}) \tag{13}$$

Step 5: Iterative calculation. After Step 4 is executed, the first iteration completes and produces a new solution $Z_2^1 = \{Z_{2,1}^1, Z_{2,2}^1, \ldots, Z_{2,T+1}^1\}$ accompanied by $Z_1^1 = \{Z_{1,1}^1, Z_{1,2}^1, \ldots, Z_{1,T+1}^1\}$. Compare Z_2^1 with Z_2^0 in terms of the value of the objective function. If the difference between their values is less than the calculation precision (ε), then the convergence condition is satisfied and Z_2^1 (accompanied by Z_1^1) is the final solution. Otherwise replace the initial solution with Z_2^1 and repeat Step 4 until the convergence condition is satisfied.

5. Case Study

The Du River, the largest tributary of the Hanjiang River, is located across Shaanxi province and Hubei province in China. Pankou-Xiaoxuan cascade hydropower stations, whose basic information is listed in Table 3, lie in the upper reaches of the Du River, as shown in Figure 7. Under the 'electricity to water mode, Pankou hydropower station generates electricity in accordance with the daily instructions (in the form of a 96-point load curve) from the superior dispatching center. In this paper, the operation period is one day with 96 calculation periods lasting 15 min each. Considering the different conditions among flood season, dry season, and transition season, 10 representative days are selected from each season of 2016–2017 for calculation. The results are analyzed in terms of model solving methods and the patterns of reverse regulation.

Table 3. Parameters of Pankou-Xiaoxuan cascade hydropower stations.

Items	Unit	Pankou	Xiaoxuan
Normal water level	M	355	264
Dead water level	M	330	261.3
Regulation volume	10^8 m^3	11.2	0.0678
Regulation performance	-	annual regulation	daily regulation
Installed capacity	MW	500	50
Operation mode	-	'electricity to water'	'water to electricity'

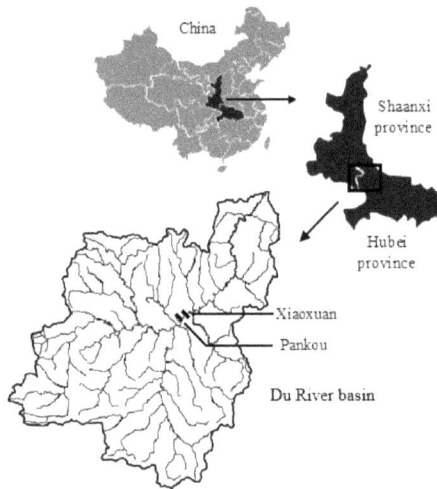

Figure 7. The location of Pankou-Xiaoxuan cascade reservoirs.

5.1. Comparative Analysis of Model Solving Methods

Take 10 December 2016 as an example to develop the coupling model of Pankou-Xiaoxuan cascade reservoirs' short-term optimal operation and use DP, POA, MSDP and APOA, respectively, to work it out. While water flow hysteresis and the aftereffect of tail water level variation are taken into account in MSDP and APOA by using BP neural network to calculate Xiaoxuan reservoir's inflow and Pankou reservoir's tail water level, the two factors are not considered in DP and POA where Pankou's outflow is directly taken as Xiaoxuan's inflow and Pankou's tail water level is obtained by the stage-discharge relation method. The calculated results of each method are statistically summarized in Table 4, in which the optimization margin refers to the ratio of the energy increase after optimization (the difference between the two objective functions) to the actual generated energy.

Table 4. The statistics of calculated results.

Items	Unit	Actual	DP	POA	MSDP	APOA
Pankou's hydroenergy consumption	10^3 kWh	1518.5	1519.4	1519.4	1518.9	1518.9
Xiaoxuan's generated energy	10^3 kWh	267.7	297.5	297.5	307.4	307.4
Objective function	10^3 kWh	−1250.8	−1221.9	−1221.9	−1211.5	−1211.5
Optimization margin	%	-	1.61	1.61	2.19	2.19
Computation time	s	-	673.10	86.56	9.43×10^5	156.37

5.1.1. Comparative Analysis of the Objective Function Value of Each Method

MSDP has been theoretically proved to have global convergence, so on the premise of the same calculation conditions (taking into account the water flow hysteresis, the aftereffect of tail water level variation and the with same amount of discrete points), the objective function calculated by APOA is compared with that calculated by MSDP in order to assess the calculated results of APOA. It can be seen from Table 4 that for 10 December 2016, the objective function of APOA equals that of MSDP. Among the other 29 representative days, there are 20 days where their objective functions are equal and 9 days where their objective functions are very close. According to the results of the 30 representative days, the average optimization margin of APOA is 0.93%, only 0.03% less than that of MSDP, which demonstrates the advantage of APOA.

Similarly, the comparative analysis of POA and DP can also illustrate the advantage of POA. It is shown from Table 4 that the objective functions of DP and POA are different from those of MSDP and APOA. This is due to their different calculation conditions. Thus, there is no point in comparing the objective functions between the former two and the latter two. Instead, we make comparisons in terms of the length of computation time and the calculation accuracy.

5.1.2. Analysis of Computation Time

In this case, the amount of calculation is $T \cdot M^2$ for DP and $T \cdot M^{N+1}$ for MSDP with the aftereffect factors taken into consideration. With the number of discrete points $M = 560$ and the number of related periods $N = 3$, the amount of calculation of MSDP is 313,600 times that of DP. However, Table 4 shows that the computation time of MSDP is about 1400 times that of DP, far fewer than 313,600 times. This is because MSDP imposes constraint on the feasible domain and leaves out those discrete points that do not satisfy the constraints, saving a great deal of calculation time. Nevertheless, its computation time of 262 h still cannot meet the efficiency requirement in practical production.

As an improved algorithm for alleviating the dimensionality curse of DP, POA is characterized by an obviously smaller amount of calculation compared with DP, as demonstrated by the statistical results from Table 4. Unlike the contrast between DP and MSDP, APOA has the same form of calculation amount as POA ($K \cdot (T - 1) \cdot M$) despite considering aftereffect factors. No matter how the number of iterations (K) may change, the change in the amount of calculation is linear rather than exponential, so the amount of calculation for APOA and POA is in the same order of magnitude. Table 4 shows that the calculation time of APOA is 156.37 s, which is 1.8 times that of POA but only 1/6000 that of MSDP. This means that with the same calculation results, APOA substantially shortens the computation time in comparison with MSDP, which can meet the efficiency requirement in actual production.

5.1.3. Analysis of Calculation Accuracy

The accuracy of calculation depends on how close the calculation result is to the actual process. As is shown in Figure 8a, the outflow graphs of Pankou reservoir obtained by APOA and POA basically coincide, with merely a slight difference that can be seen from the partially enlarged detail (the reason will be analyzed in detail in Section 5.2). While the difference of Pankou's outflow processes obtained by the two methods is small, the calculated tail water level graphs differ distinctly, as is displayed in Figure 8b. For POA, the shape of Pankou's tail water level graph resembles that of the reservoir's outflow graph, with steep rises and falls. However, Pankou's tail water level variation is a continuous and slow process. The stage-discharge relation method used in POA only takes into account the impact of Pankou's outflow and Xiaoxuan's water level in the current period and neglects the aftereffect of tail water level variation, making the calculation results inconsistent with the reality. On the contrary, the BP neural network used in APOA considers various factors that influence Pankou's tail water level and produces a relatively smooth graph, which tallies more with the actual situation. Therefore, APOA has higher calculation accuracy than POA with respect to Pankou's tail water level.

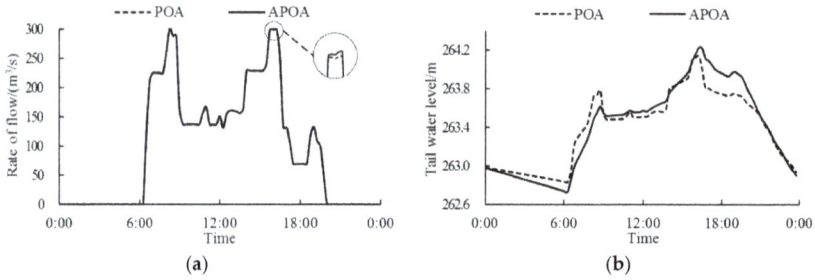

Figure 8. Comparison of Pankou's calculated results between POA and APOA. (**a**) Pankou's outflow process; (**b**) Pankou's tail water level process.

Next, we analyze from the angle of Xiaoxuan reservoir's inflow. As the BP neural network adopted in APOA takes water flow hysteresis into consideration, the corresponding inflow of Xiaoxuan reservoir, as shown in Figure 9a, slightly lags behind and is smoother compared to POA. Due to water flow hysteresis, Xiaoxuan's operation strategy obtained by APOA, as signified by Xiaoxuan's water level process in Figure 9b and its output process in Figure 9c, also lags behind that obtained by POA, especially around the periods when Pankou starts power generation. On that day, it is Pankou hydropower station's downtime until 6:30 when power generation begins, spurring an increase in its outflow. In the calculation of POA, Pankou's outflow is directly used as Xiaoxuan's inflow, causing Xiaoxuan to receive inflow at the exact moment and gradually increase its output as well as water level. Yet actually, when Pankou starts operation, its generator units are at the climbing stage with a relatively small outflow that travels a long time (about two periods) to reach Xiaoxuan. Accordingly, during the two periods of 6:30–6:45 and 6:45–7:00 Xiaoxuan's inflow is too little to keep the reservoir water level rising while discharging water for power generation. Because of the low accuracy in calculating Xiaoxuan reservoir's inflow, the operation scheme produced by POA is against the natural law and impractical in actual production. The calculation results of APOA, however, show that Xiaoxuan reservoir's water level does not rise until 7:00 when Pankou's outflow from the previous two periods has already got to Xiaoxuan. Therefore, Xiaoxuan does follow the rule of water balance, discharging water for power generation and meanwhile raising reservoir water level. To sum up, owing to the excellent performance of the BP neural network in the calculation of Xiaoxuan reservoir's inflow, the accuracy of APOA is higher than that of POA.

5.2. Analysis of the Reverse Regulation Rule

The optimal operation scheme obtained by APOA is compared to the actual operation scheme in order to find out the reverse regulation rule and get the optimal operation mode of the cascade reservoirs. Figure 10 is a comparison chart of the optimal and the actual operation schemes on 10 December 2016. During the actual operation, Pankou generates electricity as instructed, with its water level and output process shown in Figure 10a, while Xiaoxuan's water level and output process are shown in Figure 10b. Between 0:00 and 6:30, the downtime of Pankou hydropower station, Xiaoxuan hydropower station keeps operating with about 2 MW of output to meet the requirements for auxiliary power and ecological flow, and its water level drops steadily and slowly. After 6:30, as Pankou begins power generation, Xiaoxuan also increases its output until Pankou halts operation at 20:00. Then Xiaoxuan's operation returns to about 2 MW of output until the end of the operation period.

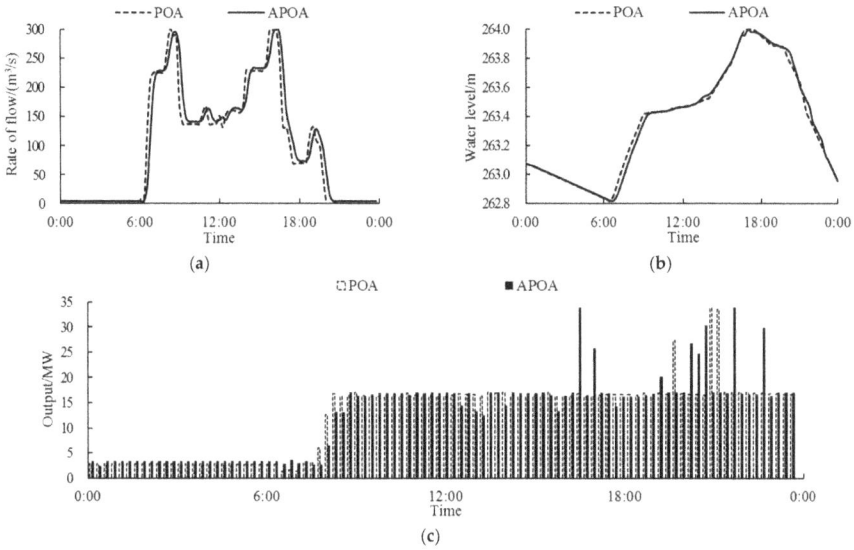

Figure 9. Comparison of Xiaoxuan's calculated results between POA and APOA. (**a**) Xiaoxuan's inflow process; (**b**) Xiaoxuan's water level process; (**c**) Xiaoxuan's output process.

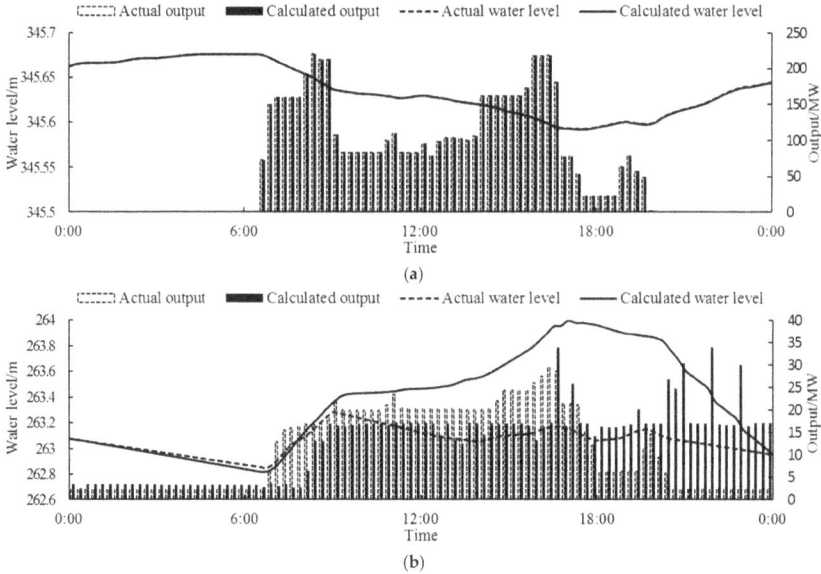

Figure 10. Comparison between the optimal and the actual operation schemes. (**a**) Pankou; (**b**) Xiaoxuan.

In contrast to the actual operation scheme, the optimal operation scheme produces a markedly different operation strategy for Xiaoxuan. For one thing, after Pankou commences power generation, Xiaoxuan's water level steadily rises and nearly reaches the normal water level at 17:00, and keeping a high water level yields greater output with the same flow rate. For another, after Pankou halts operation at 20:00, Xiaoxuan's power generation remains efficient with about 16 MW of output as its water level is still high. As a result, Xiaoxuan's generated energy in the optimal operation scheme is 307.4 thousand kWh, an increase of 39.7 thousand kWh from the actual situation (267.7 thousand kWh).

As for Pankou, the elevated water level of Xiaoxuan increases the backwater effect on its tail water level and affects its power generation efficiency. However, it can be seen from Figure 10a that the calculated hydrograph of Pankou virtually coincides with the actual one. In addition, it can be obtained from Table 4 that the hydroenergy consumption of Pankou in the optimal operation scheme is 1518.9 thousand kWh, only 0.4 thousand kWh more than the actual situation (1518.5 thousand kWh), meaning that the backwater effect of Xiaoxuan on Pankou's power generation efficiency is insignificant.

As far as the whole cascade hydropower stations are concerned, the elevated water level of Xiaoxuan creates a growth in generated energy that is remarkably greater than the hydroenergy consumption increase of Pankou caused by its fall in power generation efficiency. Moreover, the optimization of Xiaoxuan's storage and discharge leads to more efficient utilization of Pankou's outflow. Consequently, the value of objective function in the optimal operation scheme is 39.3 thousand kWh greater than the actual value, with an optimization margin of 2.19%.

A primary conclusion can be drawn from these calculated results—to raise Xiaoxuan's water level (water head regulation) and keep its generator units working within the high-efficiency zone (water quantity regulation) can boost the power generation benefits of the cascade hydropower stations.

Thus, to make the above conclusion more convincing, the other 29 representative days are also analyzed. As the above analyses have involved the representative day in the dry season, Figure 11 only displays the results of 1 July 2017 (flood season) and 12 June 2017 (transition season). Besides, it can be seen from Figure 10a that in the optimal scheme and the actual scheme, Pankou's operation processes are basically the same, and the focus of this article is Xiaoxuan's re-regulating effect on Pankou. Therefore, only Xiaoxuan's operation schemes are displayed in Figure 11. It can be seen from Figures 10b and 11 that whatever the season is, the calculated hydrographs of Xiaoxuan are above the actual ones and that in the optimal operation scheme, Xiaoxuan hydropower station's output is around 16 MW, 32 MW or 48 MW. When such output is assigned to its generator units, the generator units operate in the high-efficiency zone (the rated output of Xiaoxuan's generator unit is 16.7 MW, near which the power generation efficiency is relatively high), which also verifies the above conclusion.

It should be noted that during Pankou's downtime, if Xiaoxuan discharges water by efficiently operating one generator unit, its water level will decline continually because of its small reservoir capacity. If the downtime lasts long, to prevent the situation of a too low water level with no water for use, Xiaoxuan has to generate electricity inefficiently with a flow that just satisfies the needs of ecological flow and auxiliary power.

It is easy to understand that keeping Xiaoxuan's generators operating in the high-efficiency zone benefits the power generation of the cascade hydropower stations. As for the reason why raising Xiaoxuan's water level can boost the power generation benefits of the cascade hydropower stations, it is discussed from two aspects: the efficiency of the generator units and the total water head of the cascade hydropower stations. From the angle of generators' efficiency, while Pankou hydropower station with a high water head is affected mainly by power generation flow and insignificantly by a minor change in water head, Xiaoxuan hydropower station with a low water head is sensitive to changes in water head. Suppose the impact of Xiaoxuan's water level variation on the cascade hydropower stations' total water head is neglected (i.e., when Xiaoxuan's water level is raised by 1 m, Pankou's tail water level will also rise by 1 m correspondingly and the cascade hydropower stations' total water head stays unchanged), then we design two operation schemes as shown in Table 5. On the basis of Scheme 1, Scheme 2 raises Xiaoxuan's water level by 2 m, thus increasing the water head of Xiaoxuan from 12 to 14 m while decreasing that of Pankou from 80 to 78 m. The calculation results show that with the same power generation flow of 100 m^3/s, Xiaoxuan sees an increase of 0.0198 in power generation efficiency, much greater than the reduction of 0.0063 that Pankou suffers. Therefore, to raise Xiaoxuan's water level by 2 m means an increase of 0.0021 in the overall power generation efficiency.

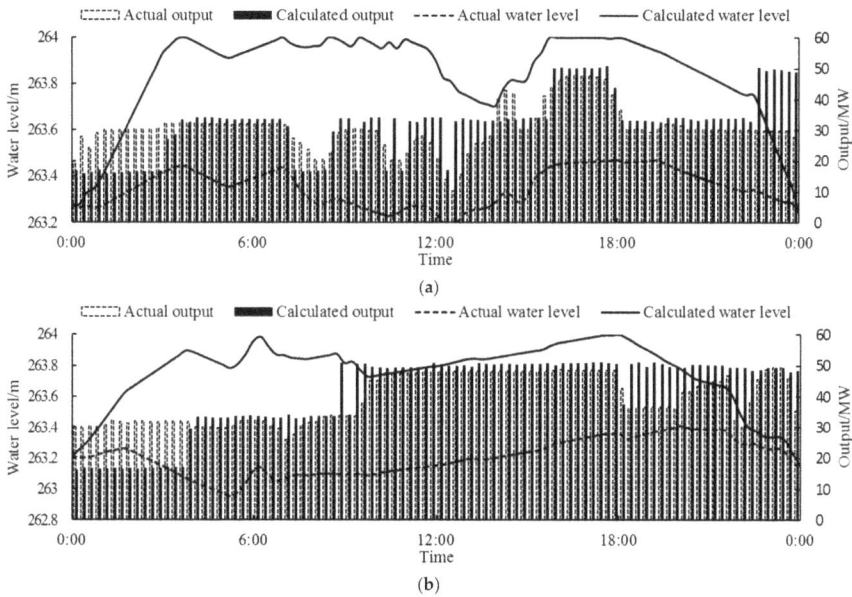

Figure 11. Comparison between optimal and actual operation schemes of Xiaoxuan, (**a**) 1 July; (**b**) 12 June.

In fact, the water level variation of Xiaoxuan reservoir has some influence on the total water head of the cascade hydropower stations. As is shown in Figure 1, the farther the backwater is from Xiaoxuan dam, the closer the backwater surface gets to the natural water surface. It is found from statistical analysis of historical data that an increase of 1 m in Xiaoxuan reservoir's water level will raise Pankou reservoir's tail water level by about 0.6 m. Therefore, Scheme 1 should actually become Scheme 3 after Xiaoxuan's water level is raised by 2 m, as shown in Table 5. From the angle of the cascade hydropower stations' total water head, Scheme 3 uses 0.8 m more of water head than Scheme 1 and further elevates the overall efficiency compared to Scheme 2.

Table 5. Comparison of different operation schemes $\eta = N/(9.81qH)$.

Items	Unit	Scheme 1	Scheme 2	Scheme 3
H_1	m	80	78	78.8
H_2	m	12	14	14
q_1	m³/s	100	100	100
q_2	m³/s	100	100	100
η_1	-	0.6467	0.6404	0.6429
η_2	-	0.8503	0.8701	0.8701
η	-	0.6732	0.6753	0.6771

In summary, when Xiaoxuan hydropower station raises its operation water level, even though Pankou's water head will fall due to the backwater effect, the impact on the efficiency of Pankou's generator units is unnoticeable. Besides, the increment in Xiaoxuan's water head can greatly enhance the efficiency of its generator units and thus considerably increase its generated energy. In contrast, the improvement of Xiaoxuan's power generation benefits is more significant. Therefore, it is beneficial for the cascade hydropower stations to raise Xiaoxuan's operation water level.

6. Conclusions

In this paper, Pankou-Xiaoxuan cascade reservoirs are taken as an example and integrated with their practical production, a coupling model of their short-term optimal operation under the effect of reverse regulation is developed and solved. Through analysis of the model's calculation results, Xiaoxuan reservoir's reverse regulation effect on Pankou reservoir is studied from two facets (water quantity and water head) and the optimal operation rules of the cascade reservoirs are obtained, with the following conclusions:

1. On the basis of considering Xiaoxuan reservoir's regulation on both water quantity and water head of Pankou reservoir, the model takes into account both Pankou's power generation efficiency and Xiaoxuan's generated energy to seek the maximum of overall power generation benefits from the angle of the cascade hydropower stations' total energy, which fits the requirements of actual production. The calculation results show that the model can effectively enhance power generation benefits of the cascade hydropower stations, which also verifies the model's validity.
2. The BP neural network has excellent performance in exploring water flow hysteresis and the aftereffect of tail water level variation, so that the accurate values of downstream reservoir's inflow and upstream reservoir's tail water level can be obtained, which significantly improves the coupling model's accuracy. The proposed APOA can efficiently work out the short-term optimal operation model of cascade reservoirs with aftereffect. With the merits and accuracy of its calculation results demonstrated, APOA is proved to meet the demand of actual production.
3. As for the rule of reverse regulation, from the aspect of water quantity regulation, Xiaoxuan reservoir should strategically store and discharge the inflow from Pankou reservoir and try to discharge flow in the mode where its generator units are in the high-efficiency zone, so that this portion of water can be utilized more efficiently; from the aspect of water head regulation, the increase in Xiaoxuan's generated energy brought by raising its operation water level is greater than Pankou's hydroenergy loss caused by the fall in its power generation efficiency. Therefore, to raise Xiaoxuan's operation water level is beneficial to power generation of the whole cascade hydropower stations.

It should be noted that we need to balance water quantity regulation with water head regulation. Excessive elevation of Xiaoxuan's operation water level may result in lack of space for water quantity regulation and even water abandonment of Xiaoxuan. Thus, it will be the focus of future research to study how high Xiaoxuan's operation water level should be raised, considering the uncertainty of Pankou's output.

Author Contributions: C.J. conceived and designed the experiments; H.Y. and X.Y. performed the experiments; H.Y. and J.W. analyzed the data; H.Y., J.W. and R.L. wrote the paper.

Acknowledgments: This study was financially supported by the National Key Research and Development Program of China (2016YFC0402208) and the National Natural Science Foundation of China (51709105). The authors are grateful to the reviewers for their comments and valuable suggestions.

Conflicts of Interest: The authors declare no conflict of interest.

References

1. Mo, L.; Lu, P.; Wang, C.; Zhou, J. Short-term hydro generation scheduling of Three Gorges-Gezhouba cascaded hydropower plants using hybrid MACS-ADE approach. *Energy Convers. Manag.* **2013**, *76*, 260–273. [CrossRef]
2. Liao, X.; Li, Y.; Zhu, X.; Zhao, K.; Bai, T.; Gao, K. Research on peak regulation capacity of Xiaolangdi-Xixiayuan cascade hydropower station. *Int. J. Hydroelectr. Energy* **2011**, *29*, 29–31.
3. Dickel, K.; Klein, P.; Nielinger, A. Rehabilitation of the weir at Ahausen reservoir. *Hydropower Dam* **2001**, *4*, 94–97.

4.   Tospornsampan, J.; Kita, I.; Ishii, M.; Kitamura, Y. Optimization of a multiple reservoir system operation using a combination of genetic algorithm and discrete differential dynamic programming: A case study in Mae Klong system, Thailand. *Paddy Water Environ.* **2005**, *3*, 29–38. [CrossRef]

5.   Richter, B.D.; Thomas, G.A. Restoring environmental flows by modifying dam operations. *Ecol. Soc.* **2007**, *12*, 181–194. [CrossRef]

6.   Bai, T.; Chang, J.; Fang, J.; Huang, Q.; Zhu, X.; Liao, X. Study on short-term peak load of Xiaolangdi and Xixiayuan cascade hydropower stations. *J. Hydroelectr. Eng.* **2012**, *31*, 83–88.

7.   Cai, Z.; Zhang, A.; Zhang, J. Optimized control on water level between Three Gorges dam and Gezhouba dam and its benefit analysis. *J. Hydroelectr. Eng.* **2010**, *29*, 10–17.

8.   Shawwash, Z.K.; Siu, T.K.; Russell, S.O.D. The B.C. hydro short term hydro scheduling optimization model. *IEEE Trans. Power Syst.* **2000**, *15*, 1125–1131. [CrossRef]

9.   Habert, J.; Ricci, S.; Papa, E.L.; Thual, O.; Piacentini, A.; Goutal, N.; Jonville, G.; Rochoux, M. Reduction of the uncertainties in the water level-discharge relation of a 1D hydraulic model in the context of operational flood forecasting. *J. Hydrol.* **2016**, *532*, 52–64. [CrossRef]

10.   Li, A.; Wang, L.; Lin, W.; Ji, C. Application of immune particle swarm optimization algorithm to short-term optimal dispatch of cascade hydropower stations. *J. Hydraul. Eng.* **2008**, *39*, 426–432.

11.   Ge, X.; Zhang, L.; Shu, J.; Xu, N. Short-term hydropower optimal scheduling considering the optimization of water time delay. *Electr. Power Syst. Res.* **2014**, *110*, 188–197. [CrossRef]

12.   Shang, Y.; Liu, R.; Li, T.; Zhang, C.; Wang, G. Transient flow control for an artificial open channel based on finite difference method. *Sci. China Technol. Sci.* **2011**, *54*, 781–792. [CrossRef]

13.   Wang, N.; Wang, C.; Du, S.; Le, H.; Li, J. Application of dynamic relationship between water level and discharge in hydrodynamic flood forecasting model. *Int. J. Hydroelectr. Energy* **2014**, *32*, 68–72.

14.   Bellman, R. Dynamic programming. *Science* **1966**, *153*, 34–37. [CrossRef] [PubMed]

15.   Saadat, M.; Asghari, K. Reliability improved stochastic dynamic programming for reservoir operation optimization. *Water Resour. Manag.* **2017**, *31*, 1795–1807. [CrossRef]

16.   Chu, W.; Yang, T.; Gao, X. Comment on "High-dimensional posterior exploration of hydrologic models using multiple-try DREAM (ZS) and high-performance computing" by Eric Laloy and Jasper A. Vrugt. *Water Resour. Res.* **2014**, *50*, 2775–2780. [CrossRef]

17.   Ji, C.; Li, C.; Wang, B.; Liu, M.; Wang, L. Multi-stage dynamic programming method for short-term cascade reservoirs optimal operation with flow attenuation. *Water Resour. Manag.* **2017**, *31*, 4571–4586. [CrossRef]

18.   Mei, Y. Dynamic programming model without Markovin property of cascade reservoirs operation and its application. *Adv. Water Sci.* **2000**, *11*, 194–198.

19.   Prasanchum, H.; Kangrang, A. Optimal reservoir rule curves under climatic and land use changes for Lampao Dam using Genetic Algorithm. *KSCE J. Civ. Eng.* **2017**, *22*, 351–364. [CrossRef]

20.   Chang, L.C. Guiding rational reservoir flood operation using penalty-type genetic algorithm. *J. Hydrol.* **2008**, *354*, 65–74. [CrossRef]

21.   Kumar, D.N.; Reddy, M.J. Multipurpose reservoir operation using particle swarm optimization. *J. Water Resour. Plan. Manag.* **2007**, *2006*, 192–201. [CrossRef]

22.   Xie, W.; Ji, C.; Wu, Y.; Li, X. Particle swarm optimization based on cultural algorithm for flood optimal scheduling of hydropower reservoir. *J. Hydraul. Eng.* **2010**, *41*, 452–457.

23.   Afshar, M.H.; Ketabchi, H.; Rasa, E. Elitist continuous ant colony optimization algorithm: Application to reservoir operation problems. *Int. J. Civ. Eng.* **2006**, *4*, 274–285.

24.   Jalali, M.R.; Afshar, A.; Marino, M.A. Multi-reservoir operation by adaptive pheromone re-initiated ant colony optimization algorithm. *Int. J. Civ. Eng.* **2007**, *5*, 284–301.

25.   Chen, J.; Wu, Y.; Hu, B.; Cheng, T. Improved mathematical model of minimum energy consumption for load distribution of cascade hydropower plants. *Automat. Electr. Power Syst.* **2017**, *41*, 155–160.

26.   Jiang, Z.; Qin, H.; Ji, C.; Feng, Z.; Zhou, J. Two dimension reduction methods for multi-dimensional dynamic programming and its application in cascade reservoirs operation optimization. *Water* **2017**, *9*, 634. [CrossRef]

27.   Suen, J.P.; Eheart, J.W. Reservoir management to balance ecosystem and human needs: Incorporating the paradigm of the ecological flow regime. *Water Resour. Res.* **2016**, *42*, 178–196. [CrossRef]

28.   Yuan, B. Short-Term Optimization Dispatching of Cascaded Hydropower Daily-Optimized Dispatching for YILIHE Cascaded Hydropower. Master's Thesis, Kunming University of Science and Technology, Kunming, China, 2012.

29. Barbetta, S.; Moramarco, T.; Perumal, M. A Muskingum-based methodology for river discharge estimation and rating curve development under significant lateral inflow conditions. *J. Hydrol.* **2017**, *554*, 216–232. [CrossRef]

30. Papazafiriou, Z.G. Linear and nonlinear approaches for short-term runoff estimations in time-invariant open hydrologic systems. *J. Hydrol.* **1976**, *30*, 63–80. [CrossRef]

31. Othman, F.; Naseri, M. Reservoir inflow forecasting using artificial neural network. *Int. J. Phys. Sci.* **2011**, *6*, 433–440.

32. Coulibaly, P.; Anctil, F.; Bobée, B. Daily reservoir inflow forecasting using artificial neural network with stopped training approach. *J. Hydrol.* **2000**, *230*, 244–257. [CrossRef]

33. Cybenko, G. Approximation by superposition of sigmoidal function. *Math. Control Signal.* **1989**, *2*, 304–314. [CrossRef]

34. Hornik, K.; Stinchcombe, M.; White, H. Multilayer feedforward networks are universal approximators. *Neural Netw.* **1989**, *2*, 359–366. [CrossRef]

35. Shen, H.; Wang, Z.; Gao, C.; Qin, J.; Yao, F.; Xu, W. Determining the number of BP neural network hidden layer units. *J. Tianjin Univ. Technol.* **2008**, *24*, 13–15.

36. Liu, J.; Wang, J. Exploration of empirical calculation method about tailwater level in Gezhouba hydropower station. *Hydropower Autom. Dam Monit.* **2011**, *35*, 77–80.

37. Tang, H.; Chen, S.; Zhao, Y. Model and algorithm of short-term optimal scheduling of Three Gorges cascaded hydropower stations. *Int. J. Hydroelectr. Energy* **2008**, *26*, 133–136.

38. Lucas, N.J.D.; Perera, P.J. Short-term hydroelectric scheduling using the progressive optimality algorithm. *Water Resour. Res.* **1985**, *21*, 1456–1458. [CrossRef]

39. Zhong, P.; Zhang, J.; Xu, B.; Zhang, M. Daily optimal operation model of cascade reservoirs considering delay of flow propagation. *J. Hydroelectr. Eng.* **2012**, *31*, 34–38.

water MDPI

Article

Season-Dependent Hedging Policies for Reservoir Operation—A Comparison Study

Nikhil Bhatia [1,†], Roshan Srivastav [2,*] and Kasthrirengan Srinivasan [3]

[1] Former Post-Graduate Student, Department of Biological and Agricultural Engineering, Texas A&M University, College Station, TX 77843, USA
[2] Department of Civil Engineering, IIT Tirupati, Renigunta, Tirupati, Andhra Pradesh 517506, India
[3] EWRE Division, Department of Civil Engineering, IIT Madras, Chennai, Tamil Nadu 600036, India; ksrini@iitm.ac.in
* Correspondence: roshan@iittp.ac.in
† Deceased: 29 August 2017

Received: 20 July 2018; Accepted: 16 September 2018; Published: 22 September 2018

Abstract: During periods of significant water shortage or when drought is impending, it is customary to implement some kind of water supply reduction measures with a view to prevent the occurrence of severe shortages (vulnerability) in the near future. In the case of operation of a water supply reservoir, this reduction of water supply is affected by hedging schemes or hedging policies. This research work aims to compare the popular hedging policies: (i) linear two-point hedging; (ii) modified two-point hedging; and, (iii) discrete hedging based on time-varying and constant hedging parameters. A parameterization-simulation-optimization (PSO) framework is employed for the selection of the parameters of the compromising hedging policies. The multi-objective evolutionary search-based technique (Non-dominated Sorting based Genetic Algorithm-II) was used to identify the Pareto-optimal front of hedging policies that seek to obtain the trade-off between shortage ratio and vulnerability. The case example used for illustration is the Hemavathy reservoir in Karnataka, India. It is observed that the Pareto-optimal front that was obtained from time-varying hedging policies show significant improvement in reservoir performance when compared to constant hedging policies. The variation in the monthly parameters of the time-variant hedging policies shows a strong correlation with monthly inflows and available water.

Keywords: parameterization; simulation; optimization; direct policy search; hedging policy; shortage ratio: Vulnerability; NSGA-II

1. Introduction

The rule or policy of any reservoir operation involves deciding the amount of releases to be made from the reservoir to meet the specified demands for different purposes based on the "current storage in the reservoir and the expected (likely) inflows to the reservoir" (available water). The standard operation policy is a simple operating rule for a reservoir, which aims to meet the demand in each period based on the available water in the current period. If the available water is higher than the demand, then the demand is completely satisfied. If the available water is less than the demand, then the available water is released towards meeting the demand. This policy is likely to result in high volumes of deficits in the future periods of operation. In order to avoid severe water deficits during drought periods or when drought is impending, hedging is done, which reduces water supplies proactively and conserves more water for future use [1].

The trigger for the initiation and the termination of hedging, along with the amount of rationing to be done in each time step, typically characterize a hedging rule. The parameters of a hedging rule can be expressed as a function of water available in the reservoir, which is the sum of the current storage and the expected inflows into the reservoir. Bayazit and Unal [2] defined the two-point hedging rule in terms of starting water availability (SWA), i.e., the volume of water availability above which the reservoir release is hedged and ending water availability (EWA), i.e., the hedging is stopped and the normal situation is restored. The effectiveness of hedging rules can be enhanced by having control over the amount of water to be released during hedging. Srinivasan and Philipose [3,4] included the hedging factor as a third parameter in addition to SWA and EWA to define the modified two-point hedging rule. The hedging factor specifies the amount of hedging that is to be done in each time step. They evaluate the trade-off among the reservoir performance indicators based on a large number of pre-defined hedging policies, using Monte-Carlo simulation technique. In addition, these simulation models do not yield optimal hedging rules.

Optimization models that make use of systems techniques have been employed in a number of research works to identify the hedging rules either with regard to the economic outcomes, such as benefit/loss functions [1,5,6] or performance outcomes, such as water supply reliability and vulnerability [7–12]. The optimal appropriation of water can be done by analyzing the benefits of current release against the benefits of storing water for future use as carryover storage [1]. Draper and Lund [1] provided an analytical view of hedging rules and operations by deriving optimal hedging policies, given a pair of benefit functions for current delivery and carry-over storage. You and Cai [5] expanded the theoretical analysis of Draper and Lund [1] to develop a conceptual two-period model for reservoir operation. Since it is difficult to derive the actual benefit/utility functions for current delivery as well as carry-over storage, the water supply characteristics of the reservoirs are used as surrogates to evaluate their performance.

Shih and ReVelle [8] used mixed-integer non-linear programming technique and polytope search procedure to find the optimal linear hedging rule with starting water availability as the only decision vector that is based on minimizing the maximum shortfall (vulnerability). Following this, they also proposed an explicit two-phase discrete hedging rule and implemented the same while using mixed-integer programming model [9]. This formulation was solved for a single critical drought. Oliveira and Loucks [13] proposed a piecewise linear hedging rule to derive the optimal hedging based operating policy for multi-reservoir systems using a genetic algorithm (GA). However, the performance of the hedging rule was evaluated based only on the single objective of minimizing the total deficit. Srinivasan and Kranthi [14] adopted a multi-objective simulation-optimization (S-O) framework for piecewise linear hedging. The pareto-optimal solutions and the computational efficiency of the multi-objective stochastic search-based optimization algorithm were improved by obtaining initial feasible solution from a constant hedging parameter based S-O framework. Liu et al. [15] derived the optimal reservoir operation rules using piecewise linear hedging based environmental flows and economic objectives. Neelakantan and Pundarikanthan [16] developed an ANN-based parameterization-simulation-optimization (PSO) framework while using discrete hedging policies to obtain releases for multi-reservoir system. Sangiorgio and Guariso [17], developed a neural network based implicit stochastic optimization (ISO) framework for multi-reservoir system. They have shown that using ISO, a closed-loop control policy, is possible for multi-reservoir system. Ji et al. [18], proposed a hedging polices for optimal reservoir operation based on a two-period reservoir simulation model under simulation-optimization framework. It is observed that the two-period optimal hedging model is able to improve the overall efficacy of the reservoir operation.

Tu et al. [19] developed a multi-objective mixed-integer quadratic programming model that can simultaneously obtain the water allocation and new hedging rules. They have shown that new hedging rules obtained while using the above method improve the performance of the reservoir. Celeste and Billib [12] compared seven stochastic models to obtain optimal reservoir polices. Further, they have discussed the benefits of parameterization-simulation-optimization (PSO) framework over implicit stochastic optimization (ISO) and explicit stochastic optimization (ESO). Shiau [20], shown the merits of the multi-period ahead hedging method when used in combination with the two-point hedging rule of Srinivasan and Philipose [3,4] associating time-varying hedging parameters into the rule. It is shown that the multi-period ahead hedging improves the results over single period hedging rule. Later, Shiau [21] derived analytical solutions for optimal hedging policies for a water supply reservoir by explicitly incorporating the reservoir release and carryover storage targets. The optimal hedging policy that was obtained from the analytical procedure was carried out for two-point hedging and one-point hedging. Wang and Liu [22] developed a framework to include both the inflow forecast and naïve hedging strategy to evaluate the performance of a water supply reservoir. They have used gridded precipitation forecast from a climate model to obtain reservoir inflow forecasting. Spiliotis et al. [23] adopted particle-swarm-optimization algorithm to derived optimal drought hedging rules that are based on appropriate identification of activation thresholds and rationing factors. The use of predefined activation functions reduces the number of parameters to be adopted in the optimization. Recently, Xu et al. [24] used two criterion namely conditional value-at-risk (CVaR) and forecast uncertainty to improve the efficacy of the reservoir operation under dry and extremely dry hydrological conditions. They found that CVaR based hedging performs better in comparison to the expected value-based hedging policy.

The main objective of this paper is to investigate the improvement in the performance of the reservoir operation when subjected to time-varying hedging parameters in comparison to constant hedging parameters. Most of the studies in the literature have adopted constant hedging parameters to evaluate the performance of the reservoir operation. In this study, we compare three popular hedging rules that are based on time-varying and constant hedging parameters. To the best of our knowledge, a detailed comparison of time-varying (TV) and constant hedging policies have not been reported. A parameterization-simulation-optimization (PSO) [25] or Direct Search Policy (DPS) [26] framework was adopted for obtaining the Pareto-optimal hedging policies for the operation of a single-purpose water supply reservoir. The optimal hedging policies are derived based on the reservoir performance indices proposed by Hashimoto et al. [7]. In this study, two performance indices, namely the shortage ratio and the period vulnerability (or maximum water shortage), are used for the two objective functions. The vulnerability index defines the severity of the system when the system is in a failure state (release is less than demand) [7]. On the other hand, the shortage ratio defines the expected water shortage over the total operation period. These indices are conflicting with another i.e., when the shortage ratio decreases, the maximum water shortage increases and vice-a-versa. The three hedging rules used for reservoir operation form the core of the model (simulation part) and a multi-objective reservoir performance optimization model is the driver of the framework. The decision variables of the optimization model are the time-varying (monthly) parameters of the three hedging rules. Similarly, the parameters of the constant hedging policies are used as the decision variable in the optimization model.

Performance evaluation of the selected hedging policies from the pareto-optimal front is carried out by reservoir simulation while using reservoir performance indicators, such as occurrence reliability, volume reliability, resilience, mean period deficit, and mean event deficit. The case example used for illustration is the Hemavathy reservoir in Karnataka, Southern India. Derivation of the pareto-optimal hedging policies and the detailed evaluation of the same have been done using the observed monthly stream flows into the Hemavathy reservoir for various percentages of demand levels. The multi-objective evolutionary search-based technique (Non-dominated Sorting based Genetic Algorithm (NSGA)-II) was employed to obtain the trade-off solutions. A performance comparison between the three hedging rules is presented for the selected operating policies from the respective Pareto-optimal fronts.

The remainder of the paper is organized as follows. Section 2 describes the case study and the parameterization-simulation-optimization (PSO) framework, including the model formulation, a detailed description of three popular hedging policies adopted in this study, and the basic steps that are involved in multi-objective NSGA-II. Following this, the results and discussion are presented in Section 3 aiming to bring out the efficacy of the time-varying hedging parameters. Section 4, outlines the summary and conclusions of the present study.

2. Methodology and Case Study

2.1. Parametrization-Simulation-Optimization (PSO) Framework

The parameterization-simulation-optimization (PSO) framework is developed in this study to obtain the optimal trade-off between the two surrogate objective functions that are mentioned below, for three of the common hedging rules, namely, (i) Two-point linear hedging; (ii) Modified Two-point hedging; and, (iii) Discrete hedging. In all cases, the PSO framework is performed for three different demand levels (namely 75%, 80%, and 85% mean annual flow). The model formulation corresponding to the PSO framework is described in the following paragraphs.

2.1.1. Objective Functions

The choice of objective functions plays a significant role in improving the efficacy of the reservoir management. The performance measures adopted in reservoir operation are based on [7]: (i) reliability: reducing the number of failure periods and total deficit; (ii) resilience: time to recover the system from a failure state; and, (iii) vulnerability: minimize the large magnitude of deficit either for a period or event. It is to be noted that the maximizing the reliability or minimizing the shortages of the system may lead to a larger magnitude of failure event [7], i.e., these performance measures are found to be conflicting objectives for reservoir operation. In this study, the following conflicting objective functions are adopted to derive optimal hedging parameters.

(i)   Minimize the Period Vulnerability

$$Z_1 = \text{Minimize } \{V_P\} \tag{1}$$

(ii)   Minimize Shortage Ratio

$$Z_2 = \text{Minimize } \{SR\} \tag{2}$$

In Equation (1) Period Vulnerability (V_p) refers to the maximum single period deficit encountered over the operation horizon, i.e.,

$$V_P = \max[D_t - R_t] \tag{3}$$

where D_t denotes the demand during period 't', R_t denotes the release made during period 't'. In Equation (2), the Shortage ratio is computed as the ratio of the sum of total deficits to the sum of total demands.

$$SR = \frac{\sum_{t=1}^{T} [D_t - R_t]}{D_t} \tag{4}$$

where T = total number of periods of operation in the horizon considered.

2.1.2. Two-Point Linear Hedging Rule

Bayzit and Unal [2] developed two-point linear hedging rule (Figure 1), in which, when the water availability falls below the starting water availability (SWA), the available water is released to satisfy the demand, which leads the reservoir storage to zero. If the water availability is greater than the ending water availability (EWA), hedging is stopped and normal operation is resumed. In case of water availability is between SWA and EWA, the hedging is applied and partial demand is satisfied in order to increase the storage (anticipating low flows in the future). Once the available water is more than ending water availability, hedging is stopped and normal operation is resumed.

$$AW_t = S_t + I_t \tag{5}$$

$$R_t = AW_t \quad \text{if} \quad AW_t \leq SWA_t \tag{6}$$

$$R_t = WA_t + (AW_t - SWA_t) \times \left(\frac{D_t - SWA_t}{EWA_t - SWA_t} \right) \quad \text{if} \quad SWA_t \leq AW_t \leq EWA_t \tag{7}$$

$$R_t = D_t \quad \text{if} \quad EWA_t \leq AW_t \leq K + D_t \tag{8}$$

$$R_t = D_t \quad \text{if} \quad K + D_t \leq AW_t \tag{9}$$

$$Spill_t = \begin{cases} AW_t - K - D_t & \text{if} \quad K + D_t \leq AW_t \\ 0 & \text{else} \end{cases} \tag{10}$$

$$S_{t+1} = S_t + I_t - R_t - Spill_t \tag{11}$$

$$SWA_t = \alpha \times D_t \tag{12}$$

$$EWA_t = D_t + (K \times \beta) \tag{13}$$

In Equations (5)–(13), K is the reservoir capacity, AW_t denotes the available water during time period 't', S_t denotes the initial storage, S_{t+1} the final storage, R_t the release, and Q_t the inflows during time period 't'. In the optimization formulation of the two-point hedging rule has two parameters α and β, which represents the starting water availability (SWA_t) and ending water availability (EWA_t), respectively.

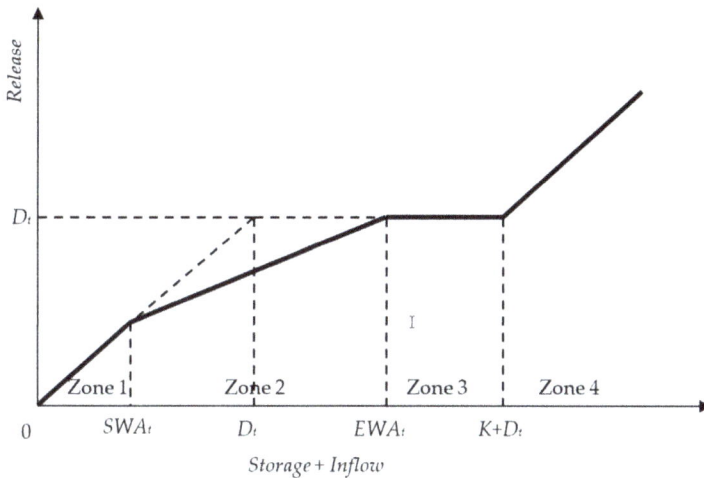

Figure 1. Two-point hedging policy (SWA_t—Starting Water availability; EWA_t—Ending Water availability; D_t—Demand; and, t denotes the time period).

Srinivasan and Philipose [3,4], proposed a modified two-point hedging rule (Figure 2), in which the hedging factor (HF) specifies the amount of rationing to be done in addition to SWA and EWA. This answer the question "how much to hedge?" in addition to the starting and the ending periods of hedging.

$$AW_t = S_t + I_t \tag{14}$$

$$R_t = AW_t \quad \text{if} \quad AW_t \leq SWA_t \tag{15}$$

$$R_t = AW_t (1 - HF) \quad \text{if} \quad SWA_t \leq AW_t \leq D_t \tag{16}$$

$$R_t = D_t (1 - HF) \quad \text{if} \quad D_t \leq AW_t \leq EWA_t \tag{17}$$

$$R_t = D_t \quad \text{if} \quad K + D_t \leq AW_t \tag{18}$$

$$Spill_t = \begin{cases} AW_t - K - D_t & \text{if} \quad K + D_t \leq AW_t \\ 0 & \text{else} \end{cases} \tag{19}$$

$$S_{t+1} = S_t + I_t - R_t - Spill_t \tag{20}$$

$$SWA_t = \alpha \times D_t \tag{21}$$

$$EWA_t = D_t + (K \times \beta) \tag{22}$$

$$0 \leq HF \leq 1 \tag{23}$$

In Equations (14)–(23), K is the reservoir capacity, AW_t denotes the available water during time period 't', S_t denotes the initial storage, S_{t+1} the final storage, R_t the release, and Q_t the inflows during time period 't'. In the optimization formulation of the modified two-point hedging rule, has three decision variables namely α, β, and HF.

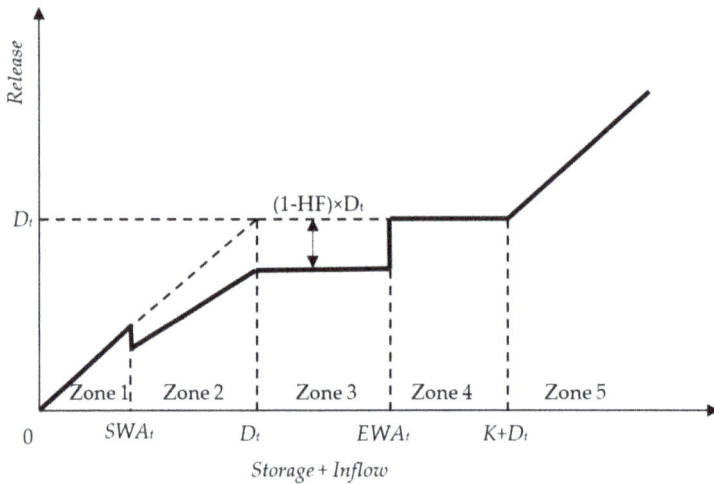

Figure 2. Modified two-point hedging policy (SWA_t—Starting Water availability, EWA_t—Ending Water availability, D_t—Demand, HF–Hedging Factor; and, t denotes the time period). Modified two-point hedging rule.

2.1.3. Discrete Hedging Rule

Shih and Revelle [9] proposed the discrete hedging scheme in which rationing is done on demand two phases based on the available water, as presented in Figure 3. In this rule, the trigger volumes of available water are introduced, where k_1, k_2, k_3 are the coefficients used to calculate V_{1p}, V_{2p}, V_{3p}, respectively.

$$AW_t = S_t + I_t \tag{24}$$

$$R_t = 0 \quad \text{if} \quad AW_t \leq V_{1p} \tag{25}$$

$$R_t = \alpha_1 \times D_t \quad \text{if} \quad V_{1p} \leq AW_t \leq V_{2p} \tag{26}$$

$$R_t = \alpha_2 \times D_t \quad \text{if} \quad V_{2p} \leq AW_t \leq V_{3p} \tag{27}$$

$$R_t = D_t \quad \text{if} \quad V_{3p} \leq AW_t \leq K \tag{28}$$

$$R_t = D_t \quad \text{if} \quad AW_t \geq K \tag{29}$$

$$Spill_t = \begin{cases} AW_t - K - D_t & \text{if } K + D_t \le AW_t \\ 0 & \text{else} \end{cases} \tag{30}$$

$$S_{t+1} = S_t + I_t - R_t - Spill_t \tag{31}$$

$$V_{1p} = k_1 \times D_t \tag{32}$$

$$V_{2p} = k_2 \times D_t \tag{33}$$

$$V_{3p} = D_t + (k_3 \times (K - D_t)) \tag{34}$$

$$0 \le \alpha_1 \le 1 \tag{35}$$

$$0 \le \alpha_2 \le 1 \tag{36}$$

$$k_1 \ge \alpha_1 \tag{37}$$

$$k_2 \ge \alpha_2 \tag{38}$$

$$\alpha_2 \ge \alpha_1 \tag{39}$$

$$0 \le k_1, k_2, k_3 \le 1 \tag{40}$$

$$k_2 \ge k_1 \tag{41}$$

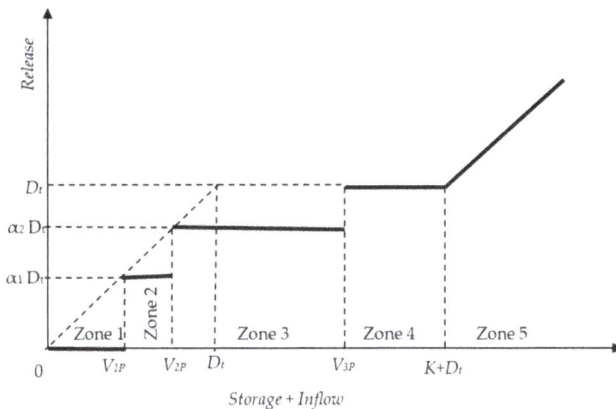

Figure 3. Discrete hedging policy (D_t—demand, α_1, α_2-rationing factor; and, t denotes time period).

2.2. Performance Evaluation

The PSO framework will provide a number of pareto-optimal solutions corresponding to each of the three hedging rules that are invoked. These pareto-optimal solutions that were obtained from the framework need to be evaluated in detail for their operational performance over the time horizon considered while using the reservoir simulation module. For the reservoir performance evaluation over the operation horizon, the following performance indicators are computed.

(i) Occurrence-based reliability, the ratio of the number of times the demand is satisfied to the number of times the reservoir is operated [7].
(ii) Resilience, the ratio of the number of times the system moved from failure to success to the total number of periods the system was in a failure state [7].
(iii) Mean event deficit, the ratio of the total deficit volume encountered during the operation horizon to the total number of failure events. Herein, 'event' denotes a sequence of failure periods.

The high magnitude of event deficit encountered during an irrigation season is detrimental to crop yield.

(iv) Event vulnerability is the maximum event deficit that is encountered during the operation horizon of the reservoir.

2.3. Solution Technique

The technique adopted in this research work to solve the multi-objective optimization problem is the Non-dominated Sorting Genetic Algorithm-II (NSGA-II), as proposed by Deb et al. [27]. This technique is known to be better than the traditional multi-objective optimization methods such as ε -constraint method, weighted sum method in generating near-global pareto-optimal fronts. This technique is suited for handling complex objective functions involving discontinuities, disjoint feasible spaces, and noisy function evaluations [28]. The multi-objective optimization model is the driver and the simulation model that is based on the hedging rules forms the engine of the framework. The decision variables of the optimization model are the hedging rule parameters. The strings that are generated from NSGA-II are evaluated for the two fitness functions while using the simulation model. The near-(global) optimal search is based on the "survival of the fittest" principle of the evolution theory. The improvements in the quality of the solutions are achieved through the genetic operators, selection, crossover, and mutation. Elitist-based Non-dominated sorting, tournament selection, and crowded comparison operator are a few of the special features that were implemented into NSGA-II to enhance its speed, quality and diversity of the non-dominated solutions. The Multi-Objective Genetic Algorithm (MOGA) input requirements are population size, number of generations, crossover probability, mutation probability, and random seed. The other inputs that are required for running the simulation module are inflows into the reservoir irrigation demands and the physical characteristics of the reservoir and the choice of the hedging rule for the operation of the reservoir. The illustration of the framework, including the steps involved in NSGA-II, is presented in Appendix A.

2.4. Case Study—Hemavathy Reservoir

The reservoir performance for the three selected hedging policies based on the PSO framework is evaluated using Hemavathy Reservoir, located in the Upper Cauvery River Basin, in Southern India (Figure 4). The salient features of the reservoir are: (i) total catchment area of 5910 km^2; (ii) gross storage capacity is 1048 Mm^3; and, (iii) live storage capacity of 962.77 Mm^3. In this study, we used monthly inflows and irrigation demands for the reservoir operation model (Table 1). It can be observed from Table 1 that most of the inflows are received between the months of June and November (~93%). While the remaining months receives less than 10% of the total annual flow. Further, it is to be noted that the reservoir storage exhibits a with-in year behavior, i.e., both filling and emptying occurs within the operating year. A data set for a period of 58 years is used for the present study. For more details about the reservoir salient features and inflow and demand characteristics, the readers are referred to [3].

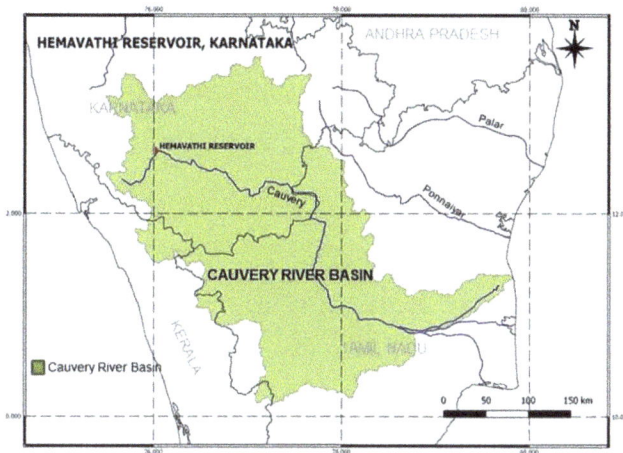

Figure 4. Location of Hemavathi Reservoir – Upper Cauvery Basin. Blue line indicates the river network.

Table 1. Mean monthly inflows and monthly target yields of River Hemavathy.

Month	June	July	August	September	October	November	December	January	February	March	April	May
Mean Monthly Inflow (Mm³)	150	856	665	296	285	127	55	30	18	14	14	36
Target Yield (Mm³)	165	260	275	75	50	120	280	350	225	80	20	10

3. Results and Discussion

In this study, the efficacy of the time-varying (TV) hedging (TVH) is compared with that of the constant hedging (CH) for three selected hedging policies, namely, two-point hedging (TPH), modified two-point hedging (MTPH), and discrete hedging (DH). This results in a total of six cases are used for the comparison studies (Table 2). The performance of each of the hedging policies has been evaluated using various indices, such as period vulnerability, shortage ratio, occurrence reliability, volume reliability, and resilience. In addition, the results are presented for three critical demand levels, namely 75%, 80%, and 85% of the mean annual flow.

Table 2. The list of hedging models (with acronyms) used for comparison.

	Two-Point Hedging (TPH)	Modified Two-Point Hedging (MTPH)	Discrete Hedging (DH)
Time-Varying	TV-TPH	TV-MTPH	TV-DH
Constant	C-TPH	C-MTPH	C-DH

3.1. Selection of GA Parameters

For the three hedging policies that are considered in this study, sensitivity analysis is carried out on NSGA-II parameters, namely, number of generations, population size, mutation probability, cross-over probability, and random seed. Table 2 provides the details of the range of parameters considered and the selected parameters that are based on the inter-comparison of the Pareto-optimal fronts. It is observed from Table 2 that the population size (100), number of generations (300), and mutation probability (0.01) remained constant for all of the hedging policies and across all the demand levels. However, the cross-over probability and random seed are found to be sensitive in obtaining the near optimal pareto-fronts and vary with hedging policies (Table 3).

Table 3. Selected genetic algorithm (GA) parameters based on the sensitivity analysis for time-varying hedging policies (two-point hedging (TPH), modified two-point hedging (MTPH), and discrete hedging (DH)).

GA Parameter	Range	Selected Parameter								
		Two-Point Hedging (TV-TPH)			Modified Two-Point Hedging (TV-MTPH)			Discrete Hedging (TV-DH)		
Demand %		75	80	85	75	80	85	75	80	85
Population	50,100,200	100	100	100	100	100	100	100	100	100
Generation	100,300,500	300	300	300	300	300	300	300	300	300
Cross Over	0.6,0.7,0.8,0.9	0.7	0.8	0.9	0.8	0.7	0.7	0.6	0.7	0.7
Mutation	0.001,0.005,0.01	0.001	0.001	0.001	0.001	0.001	0.001	0.001	0.001	0.001
Random Seed	0.25,0.35,0.45 0.55,0.65,0.75	0.65	0.45	0.45	0.75	0.25	0.25	0.65	0.25	0.45

3.2. Comparison of Time-Varying and Constant Hedging Policies

The pareto-optimal fronts comparing the variation of selected best hedging policies for both the constant and time-varying hedging parameters are presented in Figure 5. It is evident from the Figure 5 that the time-varying hedging policies are found to perform better in comparison to the constant hedging policies at all demand levels that are considered in this study. However, the constant hedging policies produce a wider range of pareto-optimal solutions when compared to the time-varying hedging policies. In the case of time-varying hedging policies, TV-TPH has more range of pareto-optimal solutions when compared to TV-MTPH and TV-DH. Further, the relative performance of the TV-DH decreases as the demand level increases.

The detailed comparison of the performance indicators that were adopted in this study for each of the hedging policies is presented in Tables 4–6. For brevity, the results for the 75% demand level is presented here, as the similar performance was observed for the other demand levels. The results for the 80% and 85% demand levels are provided for the readers as supplementary material (Figure S1–S6 and Table S1–S6). The pareto-front for each of the hedging models contains 100 possible trade-off solutions. For brevity few solutions from the pareto-front are selected for comparison of the hedging policies. The selection includes: (i) two extreme solutions related to the maximum shortage ratio and maximum vulnerability and (ii) three intermediate solutions which includes the solution closest to the origin (i.e., the best trade-off solution). It is to be noted that, in few cases, the number of feasible solutions show limited range due to restricted parameter search space. In such instances, the number of intermediate solutions is restricted to one or two intermediate solutions depending on the available range of pareto-front. In this study, for comparison, three intermediate solutions (TV-A75, TV-B75, TV-C75) and two extreme solutions that include minimum vulnerability (or maximum shortage ratio) (TV-Max S/R) and minimum shortage ratio (or maximum vulnerability) (TV-Max Vul) for the 75% demand level are selected from the pareto-optimal fronts. Further, these results are compared with the performance of the standard operating policy (SOP).

Table 4. Reservoir performance indices at 75% demand level for two point linear—a comparison of time-varying and constant hedging policies for selected compromising solutions (see figure 5).

	Period Vulnerability	Shortage Ratio	Volume Reliability	Occurrence Reliability	Resilience	Mean Event Deficit	Number of Period Deficits
SOP	216.58	0.031	0.969	0.93	0.51	135.2	49
Time-Varying Hedging							
TV-Max S/R	64.44	0.084	0.916	0.461	0.275	89.89	376
TV-Max Vul	136.97	0.031	0.969	0.841	0.387	79.2	111
TV-A75	80.01	0.048	0.952	0.595	0.355	53.24	282
TV-B75	102.08	0.036	0.964	0.728	0.344	61.12	189
TV-C75	119.23	0.032	0.968	0.829	0.479	62.47	119
Constant Hedging							
C-Max S/R	82.81	0.111	0.889	0.389	0.134	215.61	426
C-Max Vul	216.58	0.031	0.969	0.917	0.431	135.23	58
C-A75	82.81	0.111	0.889	0.389	0.134	215.61	426
C-B75	98.78	0.103	0.897	0.428	0.143	200.48	399
C-C75	120.56	0.084	0.916	0.501	0.164	163.29	347

Table 5. Reservoir performance indices at 75% demand level for modified two point linear—a comparison of time-varying and constant hedging policies for selected compromising solutions (see figure 5).

	Period Vulnerability	Shortage Ratio	Volume Reliability	Occurrence Reliability	Resilience	Mean Event Deficit	Number of Period Deficits
SOP	216.58	0.031	0.969	0.93	0.51	135.2	49
Time-Varying Hedging							
TV-Max S/R	84.45	0.041	0.959	0.865	0.606	79.75	94
TV-Max Vul	126.95	0.033	0.967	0.911	0.709	81.96	62
TV-A75	84.45	0.041	0.958	0.865	0.606	79.75	94
TV-B75	100.34	0.039	0.961	0.904	0.716	89.08	67
TV-C75	119.99	0.033	0.967	0.899	0.714	73.36	70
Constant Hedging							
C-Max S/R	67.41	0.115	0.885	0.395	0.133	227.33	422
C-Max Vul	214.72	0.031	0.969	0.917	0.431	135.75	58
C-A75	82.14	0.113	0.887	0.402	0.135	223.16	417
C-B75	100.21	0.108	0.892	0.391	0.13	217.32	425
C-C75	119.05	0.095	0.905	0.579	0.198	180.76	293

Table 6. Reservoir performance indices at 75% demand level for discrete hedging—a comparison of time-varying and constant hedging policies for selected compromising solutions (see figure 5).

	Period Vulnerability	Shortage Ratio	Volume Reliability	Occurrence Reliability	Resilience	Mean Event Deficit	Number of Period Deficits
SOP	216.58	0.031	0.969	0.93	0.51	135.2	49
Time-Varying Hedging							
TV-Max S/R	69.57	0.05	0.95	0.79	0.74	51.52	146
TV-Max Vul	123.61	0.033	0.967	0.856	0.43	84.19	100
TV-A75	78.53	0.044	0.956	0.866	0.7	74.2	93
TV-B75	98.35	0.037	0.963	0.888	0.628	83.76	78
TV-C75	123.61	0.033	0.967	0.856	0.43	84.19	100
Constant Hedging							
C-Max S/R	65.59	0.112	0.888	0.394	0.133	221.6	423
C-Max Vul	216.58	0.03	0.969	0.922	0.5	125.28	54
C-A75	79.18	0.097	0.903	0.395	0.128	198.02	422
C-B75	106.22	0.096	0.904	0.402	0.132	193.44	417
C-C75	123.5	0.084	0.916	0.46	0.152	162.81	377

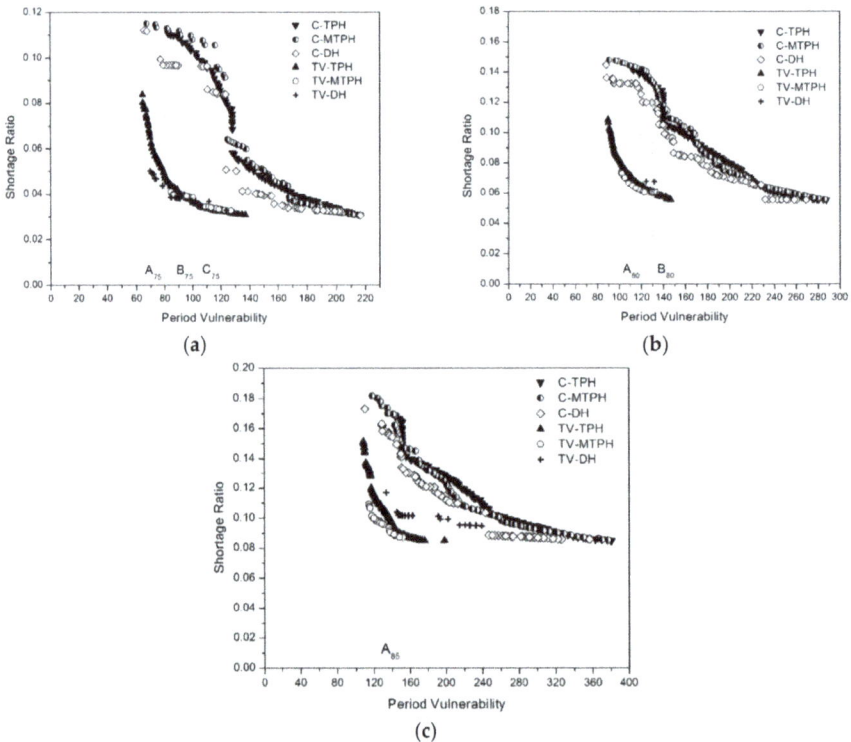

Figure 5. Pareto fronts for two point linear hedging, modified two point hedging and discrete hedging at 75%, 80% and 85% demand levels—Comparison of time-varying and constant hedging.

It is observed from Tables 4–6, that the hedging policies perform better when compared to SOP in terms of reducing the period vulnerability of reservoir operation. It is to be noted that the SOP does not account for the low reservoir inflows, and hence resulting in larger vulnerabilities. In addition, the hedging policies reduce the overall shortages by increasing the number of deficit periods when compared to SOP (which has higher shortages and lower number of deficits). From the Tables 4–6, it is evident that the time-varying hedging policies show considerable improvement in reservoir performance indicators when compared to the constant hedging policies. Further, the time-varying hedging rules produces relatively lower vulnerability, shortage ratio, and mean event deficit when compared to constant hedging rules. For the selected solutions (A, B, C), the time-varying hedging (TVH) policies show (relatively) decrease in shortage ratio by 60% and mean event deficit by 68%. In addition, the TVH policies show an average increase in volume reliability and occurrence reliability by 6% and 65% respectively. However, the resilience of the TVH is more than the CH, which indicates that the length of the events is longer when compared to the CH. It is to be noted that TVH is able to improve the performance of the long-term reservoir operation by having longer low volume (event) deficits when compared to shorter high volume (event) deficits by CH or SOP. The better performance of the TVH polices is due to the significant decrease in number of deficit periods occurred and mean event deficit during the entire reservoir simulation period when compared to CH. It is to be noted that the significant decrease in deficit periods could be explained by the time-varying hedging/rationing parameters of the policies. The variation of hedging parameters for the selected cases is presented in the following paragraphs.

The variation of the three hedging parameters for 75% demand levels are presented in Figures 6–8. The following points are observed

(i)   The CH parameters are higher in many months when compared to TVH parameters, i.e., the hedging factors (rationing as well as storage levels-based factors) are higher. For example, in the case of two-point hedging policy (Figure 6): higher vulnerability solution (C-A_{75}) the rationing is carried out even though the reservoir storage levels are high.

(ii) For TV-TPH (Figure 6) it is observed that for the months April to August, the release is marginally different from SOP, i.e., the deficits are minimized by utilizing the maximum available water from the reservoir. It is evident from Figure 6 that the TVH parameters are adaptable to hedge the available water from high inflow months and carry-over the same during the low-flow months when compared to CH. In CH, although the hedging is carried out during the high-flow months, due to constant parameters, it is forced to continue hedging in low-flow periods, resulting in higher volume of deficits.

(iii) Similarly it is observed from Figure 7, that for MTPH most of the dry months TVH parameters have low hedging factors, indicating that those months are simulated as a SOP. The rationing is carried out during high inflow months and low storage levels as contradictory to constant hedging policies.

(iv) In case of MTPH, the additional rationing factor HF plays a significant role in the variation of parameters alpha and beta. It is observed from Figure 7 that the rationing factor is higher in case of CH when compared to TVH, except for few months. In case of TVH, during October-January and April-May is simulated as two-point hedging rule. It is noted that, due to time-varying parameters in MTPH, it is able to efficiently hedge in demand (HF) and/or storage (alpha and beta), unlike the CH. This could be one of the plausible reasons for MTPH to perform better when compared to TPH. Further it is the variation of beta in both TPH and MTPH are similar, however MTPH alpha is significantly different from TPH. This shows that starting water availability is significantly affected by the rationing factor.

(v)   It is observed from Figure 8, that, for discrete hedging policy, the time-varying parameters are significantly different for all of the months in comparison to constant hedging. The K_3 parameter is has similar trend to beta parameter of TVH and MTPH.

(vi) It is evident that, most of the rationing for CH is carried out in zones 1, 2, and 3. However, the TVH the rationing factors are dominant in high flow months when compared to low flow months. Therefore, the TVH is able reduce the number of failure events when compared to CH.

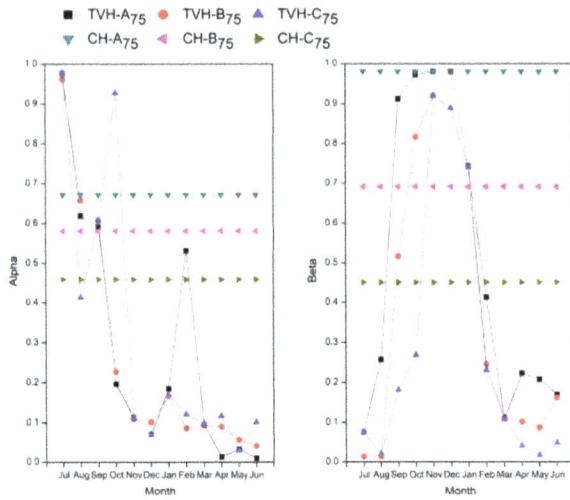

Figure 6. Time-varying hedging (TVH) and constant hedging (CH) parameters (alpha and beta) for two-point hedging policy—A comparison between the selected pareto-optimal solutions for 75% demand level.

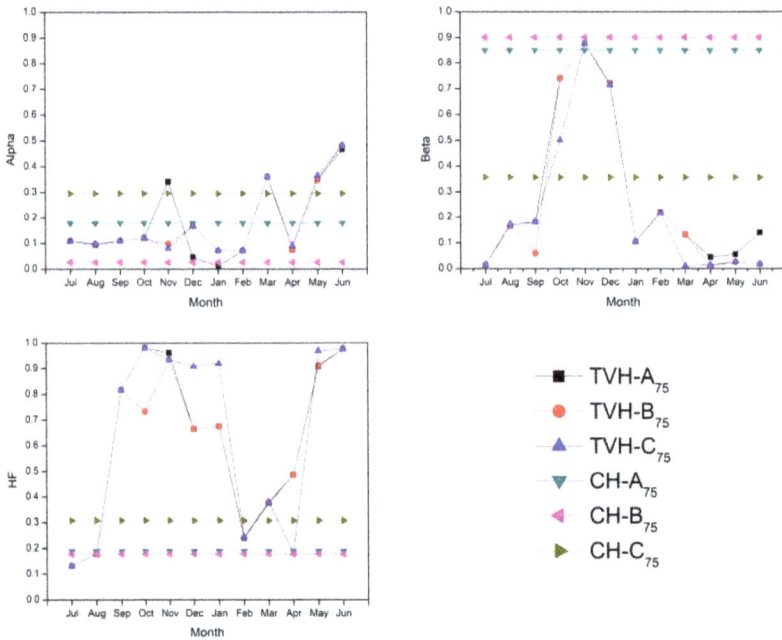

Figure 7. Time-varying hedging (TVH) and constant hedging (CH) parameters (alpha, beta and hedging factor) for modified two-point hedging policy—A comparison between the selected pareto-optimal solutions for 75% demand level.

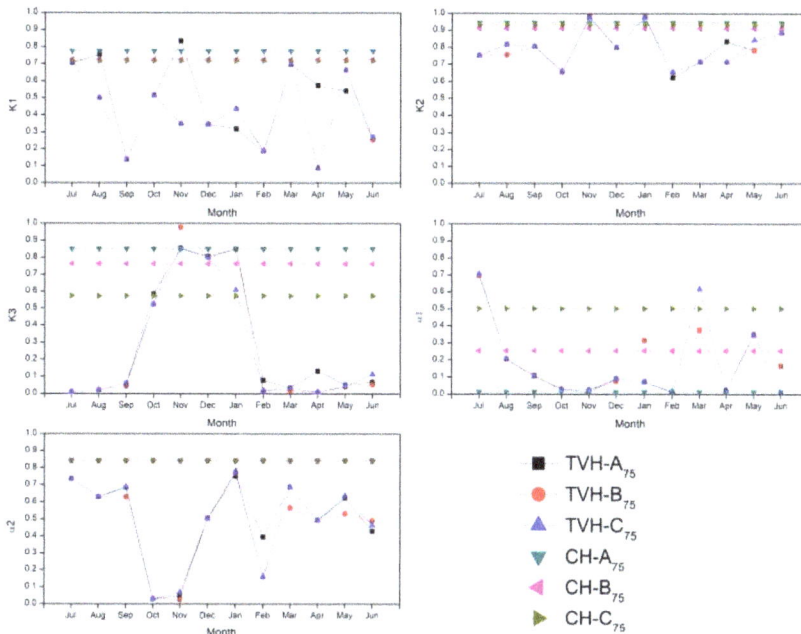

Figure 8. Time-varying hedging (TVH) and constant hedging (CH) parameters (K_1, K_2, K_3, α_1, α_2) for discrete hedging policy—A comparison between the selected pareto-optimal solutions for 75% demand level.

In addition, it is observed that the variation of the parameters for the selected pareto solutions for TVH in most of the dry months is insignificant. However, in high flow months, there is considerable variation in the parameters. Further, the parameters that are related to storage levels have more impact on the performance of the reservoir in comparison to the rationing parameters.

4. Summary and Conclusions

The study examined the performance of the reservoir simulation model while using time-varying hedging (TVH) policies and compared with the constant hedging (CH) policies. A parameterization-simulation-optimization (PSO) framework is used for obtaining compromising hedging policies for the operation of a reservoir. These hedging policies seek to obtain the trade-off between minimizing shortage ratio and minimizing vulnerability, which are the two primary objectives of a water manager for the operation of a reservoir during droughts. The NSGA-II algorithm is adopted as an optimization tool. The performance comparison is carried out for three commonly used hedging rules for reservoir operation. The case example that is used for illustration is the operation of the Hemavathy reservoir, Southern India. The following conclusions are drawn from this research:

(i)   The sensitivity analysis on NSGA-II parameters indicated that the cross-over probability and random seed are found to be sensitive when compared to population size, number of generations, and mutation probability.

(ii) Both the TVH and CH yield better alternative solutions in comparison to SOP, in terms of lower period vulnerabilities and shortage ratios.

(iii) The reservoir performance has significantly increased with TVH when compared to CH.

(iv)   The decrease in number of deficits and mean period vulnerability are the key factors for better performance of the TVH

(v)   The hedging parameters for TVH indicate less rationing in low reservoir inflows and lower storage levels when compared to CH rationing, which is constant irrespective of inflows and storage levels.

Supplementary Materials: The following are available online at http://www.mdpi.com/2073-4441/10/10/1311/s1, Figure S1: Time varying and constant hedging parameters (alpha and beta) for two-point hedging policy–A comparison between the selected pareto-optimal solutions for 80% demand level, Figure S2: Time varying and constant hedging parameters (alpha, beta and hedging factor) for modified two-point hedging policy–A comparison between the selected pareto-optimal solutions for 80% demand level, Figure S3: Time varying and constant hedging parameters (K_1, K_2, K_3, HF_1, HF_2) for discrete hedging policy–A comparison between the selected pareto-optimal solutions for 80% demand level, Figure S4: Time varying and constant hedging parameters (alpha and beta) for two-point hedging policy–A comparison between the selected pareto-optimal solutions for 85% demand level, Figure S5: Time varying and constant hedging parameters (alpha, beta and hedging factor) for modified two-point hedging policy–A comparison between the selected pareto-optimal solutions for 85% demand level, Figure S6: Time varying and constant hedging parameters (K_1, K_2, K_3, HF_1, HF_2) for discrete hedging policy–A comparison between the selected pareto-optimal solutions for 85% demand level, Table S1: Comparison of Time Varying and Constant Hedging at 80% Demand Level for Two Point Linear, Table S2: Comparison of Time Varying and Constant Hedging at 80% Demand Level for Modified Two Point Linear, Table S3: Comparison of Time Varying and Constant Hedging at 80% Demand Level for Discrete, Table S4: Comparison of Time Varying and Constant Hedging at 85% Demand Level for Two Point Linear, Table S5: Comparison of Time Varying and Constant Hedging at 85% Demand Level for Modified Two Point Linear, Table S6: Comparison of Time Varying and Constant Hedging at 85% Demand Level for Discrete.

Author Contributions: The conceptualization and framework of this article are derived from K.S. and R.S. Model computation and result analysis are completed by N.B., supervised by R.S. The article is modified and reviewed by R.S.

funding: This research was funded by Suman Bhatia under NBR funds.

Acknowledgments: The second and third author would like to thank Mr. Varun Raj (former undergraduate student) for the preliminary investigation of the hedging rules and Chirag Kothari, Prashanth Kumar for their technical support. We wish to acknowledge the effort put in by the two anonymous reviewer and the Editor for their words of encouragement, good suggestions, and constructive comments.

Conflicts of Interest: "The authors declare no conflict of interest."

Appendix

The following Figure A1 presents the steps involved in parametrization-simulation-optimization framework. The framework consists of two major components, namely, the optimization algorithm (NSGA-II) and reservoir simulation model.

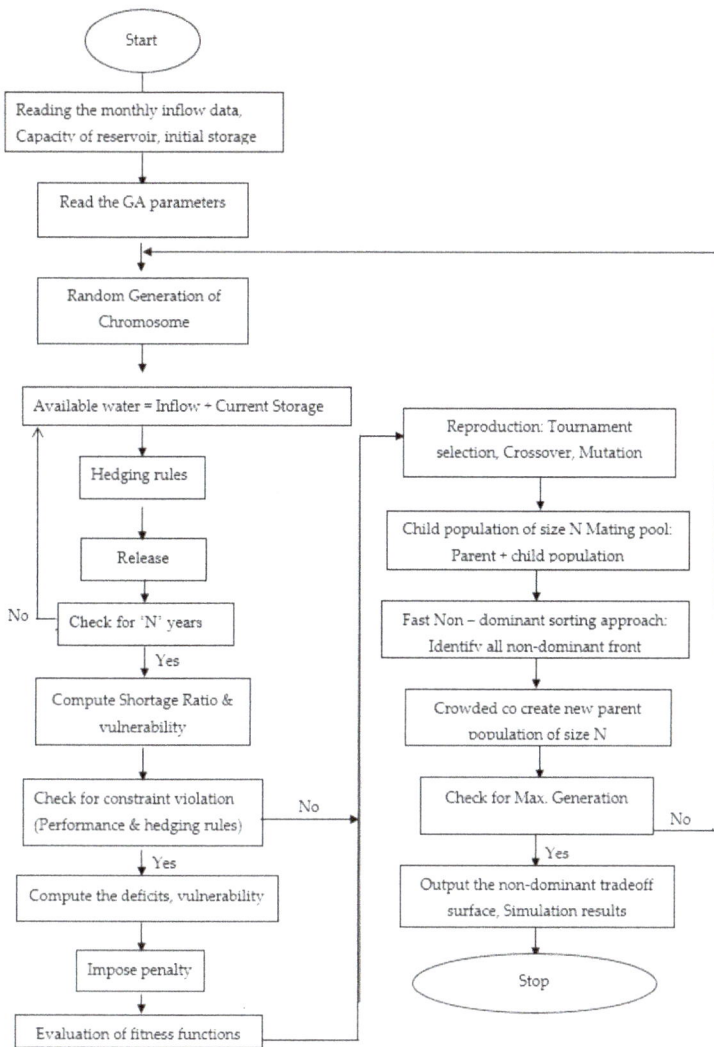

Figure A1. Block diagram of PSO frame work for single purpose reservoir operation.

References

1. Draper, A.J.; Lund, J.R. Optimal hedging and carryover storage value. *J. Water Resour. Plann. Manag.* **2004**, *130*, 83–87. [CrossRef]
2. Bayazit, M.; Unal, N.E. Effects of hedging on reservoir performance. *Water Resour. Res.* **1990**, *26*, 713–719. [CrossRef]
3. Srinivasan, K.; Philipose, M.C. Evaluation and Selection of Hedging policies using stochastic Reservoir Simulation. *Water Resour. Manag.* **1996**, *10*, 163–188. [CrossRef]
4. Srinivasan, K.; Philipose, M.C. Effect of hedging on over-year reservoir performance. *Water Resour. Manag.* **1998**, *12*, 95–120. [CrossRef]
5. You, J.Y.; Cai, X. Hedging Rules for reservoir operations 1. A theoretical analysis. *Water Resour. Res.* **2008**, *44*, W01415. [CrossRef]

6. You, J.Y.; Cai, X. Hedging rule for reservoir operations: 2. A numerical model. *Water Resour. Res.* **2008**, *44*, W01416. [CrossRef]

7. Hashimoto, T.; Stedinger, J.R.; Loucks, D.P. Reliability, resiliency, and vulnerability criteria for water resources system evaluation. *Water Resour. Res.* **1982**, *18*, 14–20. [CrossRef]

8. Shih, J.; ReVelle, C. Water-Supply Operations During Drought: Continuous Hedging Rule. *J. Water Resour. Plan. Manag.* **1994**, *120*, 613–629. [CrossRef]

9. Shih, J.-S.; ReVelle, C. Water Supply Operations During Drought: A Discrete Hedging Rule. *Eur. J. Oper. Res.* **1995**, *82*, 163–175. [CrossRef]

10. Neelakantan, T.R.; Pundarikanthan, N.V. Hedging rule optimization for water supply reservoirs system. *J. Water Resour. Plann. Manag.* **1999**, *13*, 409–426. [CrossRef]

11. Shiau, J.T.; Lee, H.C. Derivation of optimal hedging rules for a water supply reservoir through compromise programming. *Water Resour. Manag.* **2005**, *19*, 111–132. [CrossRef]

12. Celeste, A.B.; Billib, M. Evaluation of stochastic reservoir operation optimization models. *Adv. Water Resour.* **2009**, *32*, 1429–1443. [CrossRef]

13. Oliveira, R.; Loucks, D. Operating rules for multi-reservoir systems. *Water Resour. Res.* **1997**, *33*, 839–852. [CrossRef]

14. Srinivasan, K.; Kranthi, K. Multi-Objective Simulation-Optimization model for long-term reservoir operation using piecewise linear hedging rule. *Water Resour. Manag.* **2018**, *32*, 1901–1911.

15. Liu, Y.; Zhao, J.; Hang, Z. Piecewise-linear hedging rules for reservoir operation with economic and ecologic objectives. *Water* **2018**, *10*, 865. [CrossRef]

16. Neelakantan, T.R.; Pundarikanthan, N.V. Neural network-based simulation-optimization model for reservoir operation. *J. Water Resour. Plan. Manag.* **2000**, *126*, 57–64. [CrossRef]

17. Sangiorgio, M.; Guariso, G. NN-Based Implicit Stochastic Optimization of Multi-Reservoir Systems Management. *Water* **2018**, *10*, 303. [CrossRef]

18. Yi, J.; Lei, X.; Cai, S.; Wang, X. Hedging rules for water supply reservoir based on the model of simulation and optimization. *Water* **2016**, *8*, 249.

19. Tu, M.N.; Hsu, N.S.; William, Y.G. Optimization of Hedging Rules for Reservoir Operations. *J. Water Resour. Plan. Manag.* **2008**, *134*, 3–13. [CrossRef]

20. Shiau, T.J. Optimization of Reservoir Hedging Rules Using Multiobjective Genetic Algorithm. *J. Water Resour. Plan. Manag.* **2009**, *135*, 355. [CrossRef]

21. Shiau, J.T. Analytical optimal hedging with explicit incorporation of reservoir release and carryover storage targets. *Water Resour. Res.* **2011**, *47*. [CrossRef]

22. Wang, H.; Liu, J. Reservoir Operation Incorporating Hedging Rules and Operational Inflow Forecasts. *Water Resour. Manag.* **2013**, *27*, 1427–1438. [CrossRef]

23. Spiliotis, M.; Luis, M.; Luis, G. Optimization of Hedging Rules for Reservoir Operation During Droughts Based on Particle Swarm Optimization. *Water Resour. Manag.* **2016**, *30*, 5759–5778. [CrossRef]

24. Xu, B.; Zhong, P.-A.; Huang, Q.; Wang, J.; Yu, Z.; Zhang, J. Optimal Hedging Rules for Water Supply Reservoir Operations under Forecast Uncertainty and Conditional Value-at-Risk Criterion. *Water* **2017**, *9*, 568. [CrossRef]

25. Koutsoyiannis Demetris; Athanasia Economou. Evaluation of the parameterization-simulation-optimization approach for the control of reservoir systems. *Water Resour. Res.* **2003**, *39*, 1170.

26. Giuliani, M.; Emanuele, M.; Andrea, C.; Francesca, P.; Rodolfo, S. Universal approximators for direct policy search in multi-purpose water reservoir management: A comparative analysis. In Proceedings of the 19th World Congress, The International Federation of Automatic Control, Cape Town, South Africa, 24–29 August 2014; pp. 6234–6239.

27. Deb, K.; Pratap, A.; Agarwal, S.; Meyarivan, T. A Fast and Elitist Multiobjective Genetic Algorithm: NSGA-II. *IEEE Trans. Evolut. Comput.* **2002**, *6*, 182–197. [CrossRef]

28. Fonseca, C.M.; Fleming, P.J. An overview of evolutionary algorithms in multi-objective optimization. *Evolut. Comput.* **1995**, *3*, 1–16. [CrossRef]

water MDPI

Article

Inherent Relationship between Flow Duration Curves at Different Time Scales: A Perspective on Monthly Flow Data Utilization in Daily Flow Duration Curve Estimation

Lei Ye [1], Wei Ding [1], Xiaofan Zeng [2], Zhuohang Xin [1,*], Jian Wu [1] and Chi Zhang [1]

[1] School of Hydraulic Engineering, Dalian University of Technology, Dalian 116023, China; yelei@dlut.edu.cn (L.Y.); weiding@dlut.edu.cn (W.D.); 1455238909@mail.dlut.edu.cn (J.W.); czhang@dlut.edu.cn (C.Z.)

[2] School of Hydropower and Information Engineering, Huazhong University of Science and Technology, Wuhan 430074, China; zengxiaofan@hust.edu.cn

* Correspondence: xinzh@dlut.edu.cn

Received: 12 June 2018; Accepted: 24 July 2018; Published: 30 July 2018

Abstract: Modelling flow duration curves (FDCs) has long been a topic of interest since it is widely used in various hydrological applications. Most studies related to the estimation of FDCs in ungauged or partial gauged basins focus primarily on using climate and catchment characteristics to regionalize FDC at some single time scale. However, the relationship of FDCs at various time scales are rarely analyzed or studied. Here, we propose two methods, which are Modelled FDC Parameter comparison (M-FDC-P) and Empirical FDC Ratio comparison (E-FDC-R), to study the quantitative relationship between daily and monthly FDCs. One method M-FDC-P, selects a Kappa (KAP) distribution to represent the characteristics of the FDCs and then analyzes the relationship between KAP parameters of modelled FDCs at different time scales. Results indicate that three out of four parameters have strong correlations between FDCs at daily and monthly time scales. The other method, E-FDC-R, compares the quantitative relationship between daily and monthly empirical FDCs with given exceedance probabilities. The Power function is used for fitting the ratio-exceedance probability curves. In addition, the simulated daily FDC derived from monthly FDC can be very consistent with the observed daily flow records when the two parameters of power function are quantified precisely. These results clearly indicate that there are strong connections between daily and monthly FDCs, and monthly FDC can provide valuable information for daily FDC estimation. Since flow records at a large time scale are easier to obtain, daily FDC could be derived from monthly FDC by considering the inherent relationships between FDCs at different time scales, which is not sufficiently realized in previous studies.

Keywords: Kappa distribution; parameter relation; partial gauged basin; power function; ratio curve; ungauged basin

1. Introduction

The flow duration curve (FDC), which illustrates the percentage of time with which daily, monthly or some other time interval flows are equaled or exceeded over a historical period for a particular river basin, is recognized as one of the most important and widely used signatures of catchment runoff response [1]. The FDC is often interpreted as the complement of the cumulative distribution function of the flow, which provides a graphical representation of the relationship between the frequency and magnitude of flows, making it a compact signature of a catchment's functioning [2,3].

However, the modeling of FDCs has proven elusive, because of the complicated dynamics existing in flows, such as fractal and multifractal behavior [4], chaos-like dynamics [5–7], and periodicity in the mean, standard deviation, and skewness [8]. For example, at the daily time scale, daily flow is a complex time series with flow values ranging over many orders of magnitude [9]. It is very difficult to characterize the probability distribution that can approximate daily flow precisely.

At present, major studies related to the estimation of FDCs in ungauged or partial gauged basins focus primarily on statistical methods, which can be divided into three types [10,11]. The first one is the regression method, which involves establishing the relationship between flow with exceedance probabilities (or flow durations) and a set of climate and catchment characteristics [12–15]. The second one is the index method, which includes regionalization of parameters that represent the distribution function of FDC [16–18]. When no streamflow records are available, regional FDCs are used to synthesize FDCs [19]. The third one is the geostatistical method, which involves relying on kriging methods to produce predictions of hydrologic phenomena at ungauged sites with the combination of the information collected at neighboring gauging stations [20]. Over the past decade, geostatistical approaches predicting streamflow indices in ungauged basins have become increasingly popular [21–24].

However, as it can be seen, the above researches related with FDC mainly focus on a certain time scale, such as daily or annual. Except very few studies, the existing literature rarely concerns or analyzes the relationships among the FDCs at different time scales. Since FDCs at daily or monthly or annual time scale have different usages in the hydrology [9,25], and flow records at large time scale are easier to obtain, we are curious about the relationship between FDCs at different time scales. Are there any connections among the FDCs at different time scales? Is the information in FDC at large time scale implicit in small time scale? Can large scale FDCs provide some reference for small scale FDC?

To answer these questions, this paper comprehensively studies the correlations between FDCs at different time scales, especially for daily and monthly FDCs. The expected research results will be very important and useful for ungauged or partial gauged basins, which often just have flow records at larger time scale. To achieve this objective, the paper is structured as follows: Section 2 proposes the methodology to calculate the correlations between FDCs at different time scales. Section 3 describes the study area and data used. In Section 4, the variation characteristics of FDCs at different time scales and the correlations between different FDCs are revealed. Section 5 provides a detailed discussion of the results. The last section summarizes the conclusions of this study.

2. Methodology

In the present study, the correlations between FDCs at different time scales are studied by applying two types of methods. One method, called the Modelled FDC Parameter comparison (abbreviated as M-FDC-P), is to calculate the correlations between the parameters of modelled FDCs at different time scales. The other one, called the Empirical FDC Ratio comparison (abbreviated as E-FDC-R), is to compare the ratio relationship of different empirical FDCs with given exceedance probabilities.

2.1. M-FDC-P Method

M-FDC-P method first selects a specific probability distribution for modelling FDCs at different time scales and then compares the correlations between the probability distribution parameters of FDCs. By doing so, it is vivid to see to what degree the parameters are correlated, and which parameter is obviously strongly or weakly correlated.

Previous studies have noted that a complex distribution with at least four parameters is needed to approximate the probability distribution of the daily streamflow [2,16]. Blum et al. [9] demonstrated that four-parameter Kappa (KAP) distribution outperformed other three-parameter distributions and could provide a very good fit to the distribution of daily streamflow across most of the US. KAP distribution [18,25,26] is, therefore, selected as a probability distribution for modelling FDCs at different time scales, and its function is defined as follows:

$$F(x) = \left\{ 1 - h \left[1 - \frac{k}{\alpha}(x - \zeta)^{1/k} \right] \right\}^{1/h} \tag{1}$$

where ζ is a location parameter, α is a scale parameter, and k and h are shape parameters. The corresponding quantile function is described as:

$$x(F) = \zeta + \frac{\alpha}{k} \left\{ 1 - \left(\frac{1 - F^h}{h} \right) \right\}^{1/h} \tag{2}$$

KAP distribution is a base function of many commonly used three-parameter probability distributions. When $k \neq 0$, the general Pareto, generalized extreme value and generalized logistic distributions are all special cases of KAP distribution for $h = 1$, $h = 0$ and $h = -1$, respectively.

KAP distribution parameters are usually estimated by the method of L-moments, in which the first p (number of the unknown parameters in the distribution, 4 for KAP distribution) sample L-moments or L-ratios is equaled to the corresponding population quantities. Estimates of L-moment ratios exhibit substantially less bias than ordinary moment ratio estimators and are resistant to the influence of data outliers [26]. Vogel and Fennessey [1], Hosking and Wallis [26] among others have described and summarized the advantages of L-moments, and therefore they are not reproduced herein. The detail description of the theory of L-moments and parameter estimation of KAP distribution can be found in [27,28].

When the probability distribution function is chosen, parameter is the only factor that affects the variation of the probability distribution. Finally, the correlation among KAP distribution parameters at different time scales are evaluated by the most widely used linear correlation coefficient to analyze the relationship between FDCs at different time scales.

2.2. E-FDC-R Method

The above M-FDC-P method uses theoretical probability distribution to model FDCs and then compares the relationship between parameters at different time scales, which may not be universally applicable due to the subjective choice of the probability distribution function. Therefore, the study also selects another way, called E-FDC-R, which does not consider specifying any probability distribution. The steps of the E-FDC-R method are presented as follows:

1. Estimation of the empirical FDCs. An empirical FDC is constructed by ranking flows at specific time scales from all recorded years and plotting them against an estimate of their exceedance probability, known as a plotting position [1]. The first step in empirical FDC construction is to sort flow data from highest to lowest. For the probability with which each flow is exceeded, the Weibull plotting position is then used, as it provides an unbiased estimate of exceedance probability, regardless of the underlying probability distribution of the ranked observations [1]. The Weibull plotting position is described as follows:

$$p = \frac{m}{N+1} \times 100\% \tag{3}$$

    where p is the exceedance probability for the mth flow data, and N is the number of the total flow records. Plot p on x axis and the corresponding flow (the mth flow value) on y axis. The plotted dots and the x and y axis form the empirical FDC.

2. Calculation of the ratios between empirical FDCs. Firstly, sample a series of flow with pre-selected exceedance probabilities of empirical FDCs at different time scales, and then calculate the ratios of flow values at different time scales with given exceedance probabilities. Thus, the quantitative

relationship of FDCs at different time scales is achieved. It should be noted that the number of pre-selected exceedance probabilities needs to be large enough to sufficiently represent the ratio relation of FDCs. Secondly, the quantitative relationship is analyzed in order to find a certain function to represent the quantitative relationship between FDCs.

3. Evaluation of Modelled FDC. Once the specific function is obtained, the FDC at smaller time scale can be derived via the empirical FDC at larger time scale. To evaluate the performance of FDC at smaller time scale to reproduce observations, a measure of the standardized mean square error commonly referred to as Nash–Sutcliffe efficiency (NSE) is used. The description of NSE is introduced in [9], and hence not reproduced herein.

3. Study Area and Data

To keep the accuracy of the research data and the repeatability of the research, the international Model Parameter Estimation Experiment (MOPEX) data set is used in this study. The data set is described by Duan et al. [29] and can be downloaded from ftp://hydrology.nws.noaa.gov/. This data set includes mean areal precipitation, potential evaporation, daily streamflow, and daily maximum and minimum air temperature with an adequate number of precipitation gauges. As this study focuses on FDCs, only the streamflow for 219 catchments in the MOPEX data set is selected for analysis. Most of the streamflow data are included in the Hydroclimatic Data Network (HCDN) [30], which includes only gauges believed to be unaffected by upstream regulations [31] and therefore considered to represent near-natural hydrological conditions, without the need to worry that modifications to streamflow could have substantial impacts on FDCs [10,32]. The MOPEX data set consists of 56 years of daily streamflow from 1948 to 2003, which is long enough to construct FDCs. The catchment locations and boundaries are shown in Figure 1.

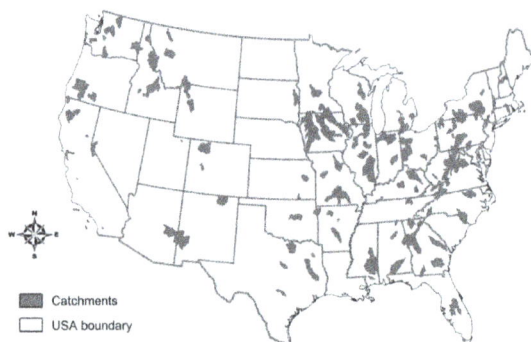

Figure 1. Locations and boundaries of 219 MOPEX catchments.

4. Results and Analysis

4.1. Empirical FDCs Variation with Different Time Scales

To see how the FDC variations with time scales, Figure 2 displays the empirical FDCs for a randomly selected catchment (No. 1064500) at time scales from 1 day to 365 days. It is evident from Figure 2a that the empirical FDCs change slowly with time scales, which means two FDCs could be very alike when their time scales are not much different, indicating that information of FDC at large time scale might be somewhat implicit in small time scales. FDC at a time scale of 1 day is the steepest (having largest slope) with obvious high flow values for low exceedance probabilities and low flow values for high exceedance probabilities, which then becomes flatter as the time scale gets larger. This is due to the fact that extreme flow events usually occur at the smaller time scale, and larger time scale flow will aggregate the difference, therefore resulting in flat FDC.

Figure 2b shows the empirical FDCs at time scales 1, 7, 15, 30, 90, and 365 days, corresponding to daily, weekly, half-monthly, monthly, seasonally, and yearly FDCs. The empirical FDCs at 1 day to 30 day time scales are very similar. Obvious differences occur when the time scale is larger than 90 days. Therefore, it is logical to investigate the possibilities of estimating daily FDC by using monthly flow data, since monthly flow data is far easier to obtain in the real world.

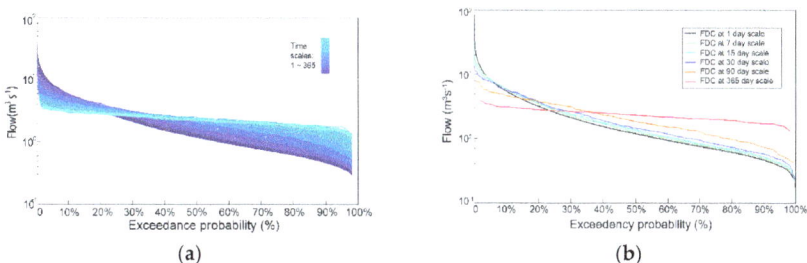

(a)   (b)

Figure 2. Empirical flow duration curve (FDC) variations with time scales for a randomly selected catchment. (**a**) Time scales from 1 day to 365 days. (**b**) Time scales at 1, 7, 15, 30, 90, and 365 days.

4.2. Relationships of FDCs Derived via M-FDC-P

The daily flow data is averaged at different time scales and then fitted to a KAP distribution, and the relationships of the parameters of the KAP distribution is studied. As demonstrated earlier, the empirical FDCs at 1 to 30 day scales are very similar and might have strong correlations, therefore, this paper concentrates research interest on time scales from 1 to 30 days.

To assess how the KAP distribution parameters vary across time scales, Figure 3 presents the average of four KAP parameters of modelled FDCs for 219 catchments at time scales from 1 to 30 days. It can be seen that all the parameters change gradually with these time scales, some parameters even show a distinct changing trend. The maximum ξ ($\sim$$-0.4$) occurs at a time scale of 1 day and then ξ decreases step by step, and finally stays steady around -0.8 after 19 days. For parameter α, it increases with time scales from 1 day to 30 days, with a minimum 1.1 at 1 day scale and a maximum around 2 at the 29 day scale. k is very similar to α; it increases along with all the time scales from negative to positive, and is very closed to zero at a time scale of 16 days. Parameter h is a little complicated compared to other parameters, for it first increases from 1 day to 11 day time scales and then decreases slightly until 30 day time scales.

The correlations of the four KAP distribution parameters for 219 catchments at different time scales are also studied. The linear correlation coefficients between four parameters at 1, 7, 15, and 30 day time scales are calculated in Table 1. For parameter ξ, all the correlation coefficients exceed 0.85, indicating very high correlations exist in parameter ξ at 1 to 30 days. The linear correlation coefficients between 1 day and other days become smaller as the time scale gets larger, so does the other time scales. It can also be seen in Table 1, except that the correlation coefficient for α between the largest time scale difference of 1 day and 30 day is 0.835; all the other correlation coefficients are higher or much closed to 0.9, which also indicates that high correlations exist in parameter α at 1 to 30 day. However, when compared to ξ, the correlations of α at different scales are a little weaker. When looking at parameter k, the linear correlation coefficients are dramatically lower than parameters ξ and α, with lowest correlation coefficients between 1 day and 30 day less than 0.5, indicating very weak correlation. The correlation coefficients for parameter h between 1 day and other days become smaller as the time scale gets larger, just like other parameters. The linear correlation coefficients for h are the highest of all the parameters, with the lowest correlation coefficient between 1 day and 30 day still passing 0.94, indicating an extremely strong correlation.

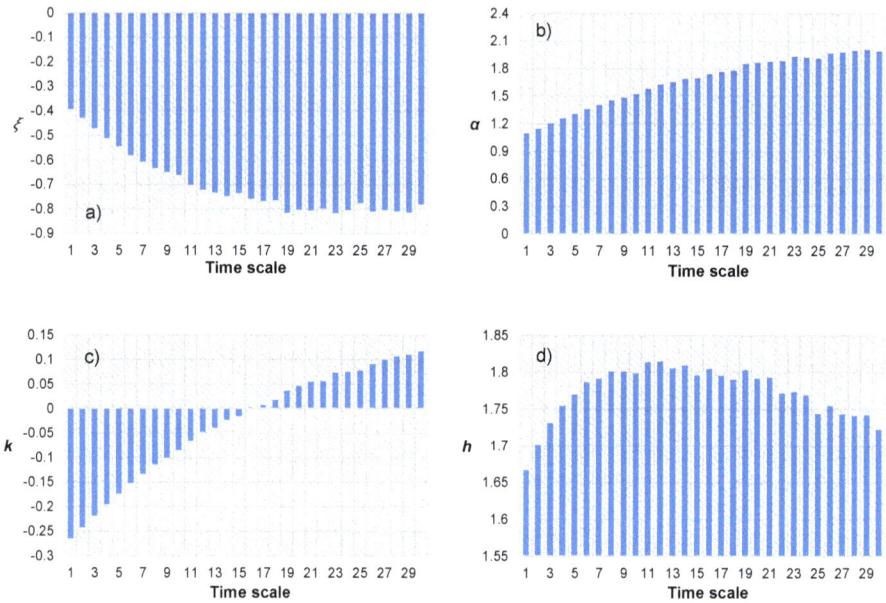

Figure 3. Variations of average four-parameter Kappa (KAP) parameters at time scales from 1 day to 30 day; (**a**) parameter ξ; (**b**); parameter α; (**c**) parameter k; (**d**) parameter h.

In addition, the relationships of the four parameters at different times are also studied by plotting corresponding parameter values for all the catchments. For illustration, the four parameters at the largest time scale difference of 1 day and 30 day are plotted in Figure 4. For parameter ξ, most of the points lie in or very near to the regression line. By comparing the linear correlation coefficients and the plot, it is easily to see the degree to which $\xi1$ can be linear regressed by $\xi30$. It can also be seen in Figure 4, that although most of the points lie in or very near to the regression line, there are more points far away from the regression line compared to parameter ξ. Meanwhile, these points are very centrally distributed, with a large proportion of points falling in the region bounded by $\alpha1$ less than 2 and $\alpha30$ less than 3. When looking at parameter k, the points are much dispersed distributed, and only a few proportion of the points lie in or near to the regression line. Further, these points do not follow a clear relationship, which indicates $k30$ is far less informative to estimate $k1$ and more information should be incorporated. For parameter h, it can be seen that except few points away from the regression line, most of the points lie in or very near to the regression line, which indicates that simple linear regression can have good performance for estimating h1 when only h30 is available.

Table 1. Linear correlation coefficients between four KAP parameters at different time scales.

Parameter	ζ				α				k				h			
Time Scale	1 Day	7 Day	15 Day	30 Day	1 Day	7 Day	15 Day	30 Day	1 Day	7 Day	15 Day	30 Day	1 Day	7 Day	15 Day	30 Day
1 day	1	0.959	0.901	0.858	1	0.953	0.893	0.835	1	0.867	0.654	0.475	1	0.987	0.967	0.940
7 day	0.959	1	0.982	0.954	0.953	1	0.981	0.941	0.867	1	0.931	0.817	0.987	1	0.992	0.973
15 day	0.901	0.982	1	0.990	0.893	0.981	1	0.985	0.654	0.931	1	0.962	0.967	0.992	1	0.992
30 day	0.858	0.954	0.990	1	0.835	0.941	0.985	1	0.475	0.817	0.962	1	0.940	0.973	0.992	1

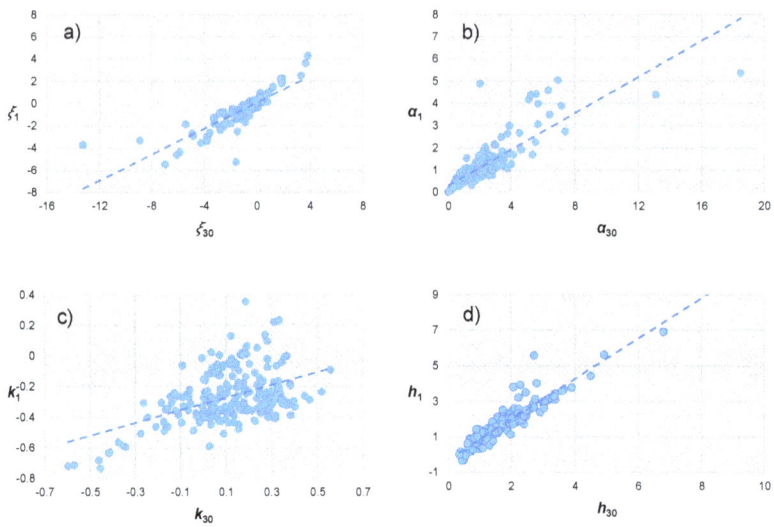

Figure 4. Scatter plots of four parameters at time scales 1 day and 30 day; (**a**) parameter ξ; (**b**); parameter α; (**c**) paramter k; (**d**) parameter h.

4.3. Relationships of FDCs Derived via E-FDC-R

Here we also focus research interest on the correlation between the FDCs at 1 day and 30 day. The flow values corresponding to 1 day and 30 day are sampled from the exceedance probabilities 0.5% to 99.5% with 0.5% as the sample interval. The total number of the sampled value is 199, which is large enough to analyze the quantitative relationship of FDCs at two time scales.

The ratios of FDC flow values at time scale 1 day to 30 day corresponding to sampled exceedance probabilities are shown in Figure 5. It is evident from Figure 5a that the ratio lines have distinct decreasing trend, these lines firstly decrease dramatically within 20% exceedance probability and slightly decrease thereafter. This is easy to understand that in flood season, maximum daily flow is certainly much higher than the monthly average, so the ratios at low exceedance probabilities are very high. While in non-flood season, minimum daily flow is also far lower than the monthly average, so the ratios at low exceedance probabilities are very low. Consequently, the ratio lines clearly show that daily FDC has higher flow values for low exceedance probability and lower flow values for high exceedance probability compared to monthly FDC, which is also consistent with Figure 2. It should also be noted that most of the curves reach 1 at less than 10% exceedance probability, consistent with [33], in which daily and monthly FDC curves normally cross (representing the ratio equals to 1) between the 1% and 10% exceedance probability in South Africa.

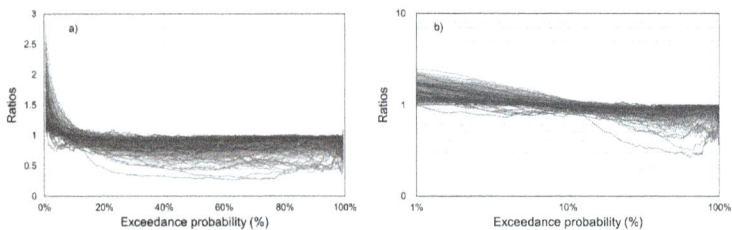

Figure 5. Ratios of daily to monthly FDCs at given exceedance probabilities; (**a**) conventional coordinates; (**b**) double logarithmic coordinates.

Seen from Figure 5a, the ratio relationship appears to be a power function, we apply the log-log transformation to the ratio relationship. It is evident from Figure 5b that the relationship between ratio value and exceedance probability is overall near-linear in the log-log plot, except a very small proportion in the very high exceedance probability. Here, the power function is used to fit the ratio curves as follows:

$$y = ax^{-b} \tag{4}$$

where x is the exceedance probability; y is the ratio value.

For each catchment, the parameter a and b are estimated and shown in Figure 6. The parameter a for all the catchments is positive and less than 1. This is due to the fact that the ratio is less than 1 for medium and has high exceedance probability. Despite for only two catchments, the parameter b for the rest of the catchments is negative; this is simply because the ratio curve is decreasing overall.

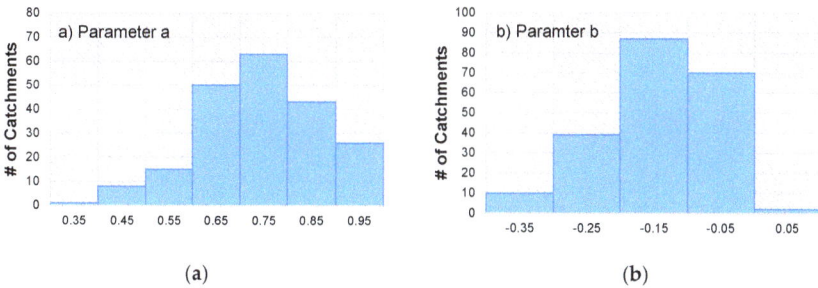

Figure 6. Histograms of parameters (**a**,**b**) in the power function.

By applying the power function to fit the ratio of daily to monthly FDCs, the daily flow at given exceedance probability can be estimated based on the monthly FDC. The simulated daily FDC for a randomly selected catchment is presented in Figure 7. It is evident that an obvious improvement has been made for simulated daily FDC as it is much closer to daily FDC. Especially for the exceedance probability ranging from 10~70%, the simulated daily FDC and the observed daily FDC almost overlap.

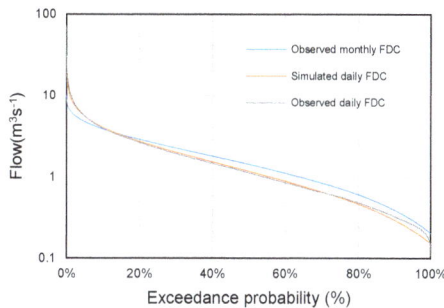

Figure 7. Simulated daily FDC based on monthly FDC.

After graphical comparison with observed daily FDC, the simulated daily FDC is modelled by KAP distribution and then evaluated with observed daily flow data. Figure 8 presents the NSE of modelled daily FDC. It is obvious that most of the catchments have very high NSE with median higher than 0.96. The results confirm again that power function is an appropriate relationship for the ratio of daily to monthly FDCs. After power transformation, the modelled daily FDC with good accuracy can be obtained based on monthly FDC.

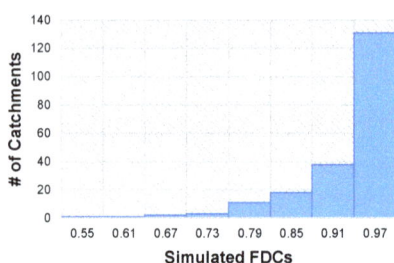

Figure 8. NSE of simulated daily FDCs based on monthly FDC.

5. Discussion

The present study aims to ascertain the relationships between FDCs at different time scales, based on M-FDC-P and R-FDC-P. Note that the two methods use different ways to examine the relationships, and some research results are discussed here.

In previous studies, the value of every single parameter of KAP distribution is rarely presented or discussed, though Parida [34] have done some research on the restrictions of the four parameters. In the present research, the averages of four KAP distribution parameters at time scales from 1 to 30 days are analyzed. The relationship between parameter h at 1 and 30 day is so strong that h_1 can be well estimated by merely h_{30}. The high correlation or insensitive of the h parameter is to be expected, as the generalized Pareto, a special case of KAP when $h = 1$, has been used to describe daily FDCs in many studies [16,18,25]. Even Blum et al. [9] illustrated that there are some regions in the United States such as New England, the Appalachian, and Valley and Ridge regions, for which the generalized Pareto distribution can provide a very good fit to daily FDCs. However, the relationship of the other three parameters are not following a very obvious linear trend or even a single trend (for parameter k). As stated before, even modeling daily FDC in daily gauged sites is very difficult, we are not expecting a single relationship can be found to estimate daily FDC parameters by using only monthly FDC parameters. Therefore, this paper proves that FDC parameters at different time scales are correlated and the relationships found in the paper can be used as an important supplement to the regional regression models to estimate daily FDC at ungauged or partial gauged sites.

As far as the E-FDC-R is concerned, Smakhtin [35] described a method to estimate daily FDCs from monthly flow data, which first assumed that both daily and monthly FDCs are linear in a log-normal space and then established the regression relationship for only two daily FDC indices based on monthly flow data. This method is very ambitious because it established the indices for 200 gauged hydrological stations without any specific parameters representing at-site characteristics. Sugiyama et al. [36] also stated that a distribution of FDC at given exceedance probability was usually represented with a straight line on log-normal probability paper. However, as for the ratio curves of flow values at given exceedance probability, they are very like exponential curves in our work, thus we use the power function to simulate the ratio curves and apply the log-log transformation to the ratio relationship. The results show that the relationship between ratio value and exceedance probability almost tends to be near-linear in the log-log axis, except a very small proportion in the very high exceedance probability. Future studies will consider further improving power function by using a piecewise function to describe the ratio curves for some kind of region (e.g., the stations with positive parameter b after figuring out a way to regionalize parameters in the power function).

6. Conclusions

This study provides a comprehensive assessment of relationships between FDCs at different time scales based on two methods including M-FDC-P and E-FDC-R. Selecting the MOPEX as the research data, the correlations of daily FDC and monthly FDC are deeply analyzed.

As for the M-FDC-P method, the Kappa (KAP) distribution is selected in this work to represent the streamflow distribution characteristics, due to the already existed existing research. The averages of all four KAP parameters tend to change gradually with time scales from 1 to 30 day. Meanwhile, three out of four parameters have linear correlations except the shape parameter k. The relationships found in the paper can be used as an important supplement to the regional regression models to estimate daily FDC at ungauged sites, when observed records with larger time scales are available.

As for the E-FDC-R method, it tries to quantify the relationship between daily and monthly FDCs, based on the ratio of FDCs rather than parameters of a specific distribution. Power function seems to be an effective functions for depicting the relationship between daily and monthly FDCs. The simulated daily FDC derived from monthly FDC can be very consistent with the observed daily flow records, when the two parameters in power function are quantified precisely. There are only two parameters in the power function that need to be regionalized by the site climate and catchment characteristics, which can reduce the uncertainties compared to other regional FDC methods that involve more parameters. Further, the parameter a in power function mainly controls the ratio for high exceedance probability or low flow data; and the parameter b mainly controls the shape of the ratio curve for low exceedance probability or high flow data.

Generally, results based on the above two methods clearly prove that there are strong connections between daily and monthly FDCs, which is not sufficiently realized in previous studies. If a specific probability distribution is applied for modelling FDCs at larger time scales, the gradual change characteristics of the distribution parameters for different FDCs can be considered to provide valuable information for daily FDC estimation. If there is no enough information to ascertain the probability distribution, the simple ratio relationships between different FDCs can help to assess the smaller time scale FDCs. Therefore, the inherent relationship between different FDCs will be very helpful for flow regulation and water resources management, especially in ungauged or partially gauged basins. Nonetheless, our future research will consider the regionalization of the distribution parameters to provide a complete daily FDC estimation using monthly flow data and other information collectively.

Author Contributions: L.Y. designed the study and wrote the manuscript; W.D. assisted with the results and discussion section; X.Z. assisted with the discussions and conclusions section; Z.X. and C.Z. provided feedback on the structure of the manuscript and reviewed the manuscript; J.W. performed the experiments and analyzed the data.

funding: This study was jointly supported by the National Key Research and Development Program (2016YFC0401005, 2016YFC0401004), the National Natural Science Foundation of China (51709033), and the Funds for the Central Universities, HUST (2016YXZD046).

Acknowledgments: Special thanks are given to the two anonymous reviewers for their constructive comments that helped us to greatly improve the manuscript.

Conflicts of Interest: The authors declare no conflict of interest.

References

1. Vogel, R.M.; Fennessey, N.M. L moment diagrams should replace product moment diagrams. *Water Resour. Res.* **1993**, *29*, 1745–1752. [CrossRef]
2. LeBoutillier, D.W.; Waylen, P.R. A stochastic model of flow duration curves. *Water Resour. Res.* **1993**, *29*, 3535–3541. [CrossRef]
3. Cheng, L.; Yaeger, M.; Viglione, A.; Coopersmith, E. Exploring the physical controls of regional patterns of flow duration curves—Part 1: Insights from statistical analyses. *Hydrol. Earth Syst. Sci.* **2012**, *9*, 4435–4446. [CrossRef]
4. Zhang, Q.; Xu, C.Y.; Chen, Y.D.; Yu, Z. Multifractal detrended fluctuation analysis of streamflow series of the Yangtze River basin, China. *Hydrol. Process.* **2008**, *22*, 4997–5003. [CrossRef]
5. Sivakumar, B. Chaos theory in geophysics: Past, present and future. *Chaos Soliton Fract.* **2004**, *19*, 441–462. [CrossRef]
6. Wang, W.; Vrijling, J.K.; van Gelder, P.H.A.J.M.; Ma, J. Testing for nonlinearity of streamflow processes at different timescales. *J. Hydrol.* **2006**, *322*, 247–268. [CrossRef]

7.   Tung, W.W.; Gao, J.; Hu, J.; Yang, L. Detecting chaos in heavy-noise environments. *Phys. Rev. E* **2011**, *83*, 046210. [CrossRef] [PubMed]

8.   Bowers, M.C.; Tung, W.W.; Gao, J.B. On the distributions of seasonal river flows: Lognormal or power law? *Water Resour. Res.* **2012**, *48*, 5536. [CrossRef]

9.   Blum, A.G.; Archfield, S.A.; Vogel, R.M. On the probability distribution of daily streamflow in the United States. *Hydrol. Earth Syst. Sci.* **2017**, *21*, 3093–3103. [CrossRef]

10.   Castellarin, A.; Botter, G.; Hughes, D.A.; Liu, S.; Ouarda, T.B.M.J.; Parajka, J.; Post, D.A.; Sivapalan, M.; Spence, C.; Viglione, A.; et al. Prediction of flow duration curves ungauged basins. In *Runoff Prediction in Ungauged Basins: Synthesis Across Processes, Places and Scales*; Blöschl, G., Sivapalan, M., Wagener, T., Viglione, A., Savenije, H., Eds.; Cambridge University Press: Cambridge, UK, 2013; p. 465.

11.   Hsu, N.S.; Huang, C.J. Estimation of Flow Duration Curve at Ungauged Locations in Taiwan. *J. Hydrol. Eng.* **2017**, *22*, 05017009. [CrossRef]

12.   Chiang, S.M. Hydrologic Regionalization of Watersheds. II: Applications. *J. Water Res. Plan. Manag.* **2002**, *128*, 12–20. [CrossRef]

13.   Mohamoud, Y. Prediction of daily flow duration curves and streamflow for ungauged catchments using regional flow duration curves. *Hydrol. Sci. J.* **2008**, *53*, 706–724. [CrossRef]

14.   Hope, A.; Bart, R. Synthetic monthly flow duration curves for the Cape Floristic Region, South Africa. *Water SA* **2012**, *38*, 191–200. [CrossRef]

15.   Shu, C.; Ouarda, T.B.M.J. Improved methods for daily streamflow estimates at ungauged sites. *Water Resour. Res.* **2012**, *48*, 2523. [CrossRef]

16.   Castellarin, A.; Galeati, G.; Brandimarte, L.; Montanari, A.; Brath, A. Regional flow-duration curves: Reliability for ungauged basins. *Adv. Water Resour.* **2004**, *27*, 953–965. [CrossRef]

17.   Li, M.; Shao, Q.; Zhang, L.; Chiew, F.H.S. A new regionalization approach and its application to predict flow duration curve in ungauged basins. *J. Hydrol.* **2010**, *389*, 137–145. [CrossRef]

18.   Mendicino, G.; Senatore, A. Evaluation of parametric and statistical approaches for the regionalization of flow duration curves in intermittent regimes. *J. Hydrol.* **2013**, *480*, 19–32. [CrossRef]

19.   Yu, P.S.; Yang, T.C.; Liu, C.W. A regional model of low flow for southern Taiwan. *Hydrol. Process.* **2010**, *16*, 2017–2034. [CrossRef]

20.   Pugliese, A.; Farmer, W.H.; Castellarin, A.; Archfield, S.A.; Vogel, R.M. Regional flow duration curves: Geostatistical techniques versus multivariate regression. *Adv. Water Resour.* **2016**, *96*, 11–22. [CrossRef]

21.   Chokmani, K.; Ouarda, T.B.M.J. Physiographical space-based kriging for regional flood frequency estimation at ungauged sites. *Water Resour. Res.* **2004**, *40*, 1–13. [CrossRef]

22.   Skoien, J.O.; Merz, R.; Bloschl, G. Top-kriging-geostatistics on stream networks. *Hydrol. Earth Syst. Sci.* **2006**, *2*, 2253–2286. [CrossRef]

23.   Castiglioni, S.; Castellarin, A.; Montanari, A. Prediction of low-flow indices in ungauged basins through physiographical space-based interpolation. *J. Hydrol.* **2009**, *378*, 272–280. [CrossRef]

24.   Archfield, S.A.; Pugliese, A.; Castellarin, A.; Skøien, J.O. Topological and canonical kriging for design-flood prediction in ungauged catchments: An improvement over traditional regional regression approach? *Hydrol. Earth Syst. Sci.* **2013**, *9*, 1575–1588. [CrossRef]

25.   Castellarin, A.; Camorani, G.; Brath, A. Predicting annual and long-term flow-duration curves in ungauged basins. *Adv. Water Resour.* **2007**, *30*, 937–953. [CrossRef]

26.   Hosking, J.R.M.; Wallis, J.R. *Regional Frequency Analysis: An Approach Based on L-Moments*; Cambridge University Press: Cambridge, UK, 1997; p. 244.

27.   Hosking, J.R.M. L-moments: Analysis and estimation of distributions using linear combinations of order statistics. *J. R. Stat. Soc. B Methodol.* **1990**, *52*, 105–124.

28.   Kjeldsen, T.R.; Ahn, H.; Prosdocimi, I. On the use of a four-parameter kappa distribution in regional frequency analysis. *Hydrol. Sci. J.* **2017**, *62*, 1–10. [CrossRef]

29.   Duan, Q.; Schaakeb, J.; Andréassian, V.; Franks, S.; Gotetie, G.; Gupta, H.V.; Gusev, Y.M.; Habet, F.; Hall, A.; Hay, L.; et al. Model Parameter Estimation Experiment (MOPEX): An overview of science strategy and major results from the second and third workshops. *J. Hydrol.* **2006**, *320*, 3–17.

30.   Slack, J.R.; Lumb, A.M.; Landwehr, J.M. Hydro-climatic data network (HCDN): A U.S. Geological Survey streamflow data set for the United States for the study of climate variation, 1874–1988. *J. Phys. Chem. C* **1992**, *113*, 2538–2544.

31. Wang, D.; Hejazi, M. Quantifying the relative contribution of the climate and direct human impacts on mean annual streamflow in the contiguous United States. *Water Resour. Res.* **2011**, *47*, 411. [CrossRef]

32. Kroll, C.N.; Croteau, K.E.; Vogel, R.M. Hypothesis tests for hydrologic alteration. *J. Hydrol.* **2015**, *530*, 117–126. [CrossRef]

33. Smakhtin, V.Y.; Watkins, D.A. *Low Flow Estimation in South Africa*; Report No 494/1/97; Water Research Commission: Pretoria, South Africa, 1997.

34. Parida, B.P. Modelling of Indian Summer Monsoon Rainfall Using a Four-parameter Kappa Distribution. *Int. J. Climatol.* **1999**, *19*, 1389–1398. [CrossRef]

35. Smakhtin, V.U. Estimating daily flow duration curves from monthly streamflow data. *Water SA* **2000**, *26*, 13–18.

36. Sugiyama, H.; Vudhivanich, V.; Whitaker, A.; Lorsirirat, K. Stochastic Flow Duration Curves for Evaluation of Flow Regimes in Rivers. *J. Am. Water Resour. Assoc.* **2007**, *39*, 47–58. [CrossRef]

water

MDPI

Article

Quantification of Seasonal Precipitation over the upper Chao Phraya River Basin in the Past Fifty Years Based on Monsoon and El Niño/Southern Oscillation Related Climate Indices

Tsuyoshi Kinouchi [1],*, Gakuji Yamamoto [1], Atchara Komsai [1] and Winai Liengcharernsit [2]

[1] Department of Transdisciplinary Science and Engineering, School of Environment and Society, Tokyo Institute of Technology, Yokohama 226-8503, Japan; yamamoto.g.ab@m.titech.ac.jp (G.Y.); komsai.a.aa@m.titech.ac.jp (A.K.)

[2] Department of Environmental Engineering, Faculty of Engineering, Kasetsart University, Bangkok 10900, Thailand; fengwnl@ku.ac.th

*   Correspondence: kinouchi.t.ab@m.titech.ac.jp; Tel.: +81-45-924-5524

Received: 29 April 2018; Accepted: 13 June 2018; Published: 17 June 2018

Abstract: For better water resources management, we proposed a method to estimate basin-scale seasonal rainfall over selected areas of the Chao Phraya River Basin, Thailand, from existing climate indices that represent variations in the Asian summer monsoon, the El Niño/Southern Oscillation, and sea surface temperatures (SST) in the Pacific Ocean. The basin-scale seasonal rainfall between 1965 and 2015 was calculated for the upper Ping River Basin (PRB) and the upper Nan River Basin (NRB) from a gridded rainfall dataset and rainfall data collected at several gauging stations. The corresponding climate indices, i.e., the Equatorial-Southern Oscillation Index (EQ-SOI), Indian Monsoon Index (IMI), and SST-related indices, were examined to quantify seasonal rainfall. Based on variations in the rainfall anomaly and each climate index, we found that IMI is the primary variable that can explain variations in seasonal rainfall when EQ-SOI is negative. Through a multiple regression analysis, we found that EQ-SOI and two SST-related indices, i.e., Pacific Decadal Oscillation Index (PDO) and SST anomalies in the tropical western Pacific (SST_{NW}), can quantify the seasonal rainfall for years with positive EQ-SOI. The seasonal rainfall calculated for 1975 to 2015 based on the proposed method was highly correlated with the observed rainfall, with correlation coefficients of 0.8 and 0.86 for PRB and NRB, respectively. These results suggest that the existing indices are useful for quantifying basin-scale seasonal rainfall, provided a proper classification and combination of the climate indices are introduced. The developed method could forecast seasonal rainfall over the target basins if well-forecasted climate indices are provided with sufficient leading time.

Keywords: seasonal rainfall; upper Chao Phraya River Basin; El Niño/Southern Oscillation; Indian Monsoon; sea surface temperatures

1. Introduction

Recently, many regions have endured significant impacts from extreme floods and droughts, including Southeastern Asian countries [1–4]; disastrous floods and droughts are expected to occur more frequently due to the changing climate [2,5–7]. To mitigate the impacts of extreme hydro-meteorological events, well-prepared water resource management practices are required, for which the quantification of rainfall and resulting runoff is key information.

In Thailand, a devastating flood occurred in the Chao Phraya River Basin in 2011 [8,9], resulting in a change in policy to focus more on flood mitigation. However, the basin suffered a serious drought during 2015 and 2016 due to the reduced water storage of the major reservoirs, i.e., the Bhumibol and Sirikit Dam reservoirs during the preceding years and the limited amount of rainfall in the wet season of 2015. To sustain socio-economic activities in the downstream areas of the Chao Phraya River Basin, sufficient water storage in these reservoirs is required before the dry season while enough empty volume should be maintained for flood-control during the rainy season. Therefore, water release from the reservoirs needs to be well-managed by forecasting the areal rainfall and its resulting inflows to the reservoirs with sufficient leading time.

Different approaches have been adopted to quantify or forecast rainfall in Thailand using climate indices. Most approaches are based on a simple linear correlation method [10,11], and a limited number of works have considered multiple indices using a linear multiple-regression model and a non-parametric method [12]. The linear regression approach is conventional, but not suitable if the rainfall is non-linearly related to climate indices. The non-parametric method is advantageous in this situation, although its implementation is complicated.

Furthermore, previous studies focused on rainfall at a limited number of sites with shorter duration (for example, [12]); however, for water resources management purposes, rainfall must be estimated to represent the entire rainy season at regional or basin-scales, rather than the smaller area represented by individual gauging stations.

In this study, we attempt to develop a method that can relate seasonal rainfall over the two major upstream sub-basins of the Chao Phraya River Basin draining to the Bhumibol and Sirikit Dam reservoirs, i.e., the upper Ping River Basin (PRB) and the upper Nan River Basin (NRB), to multiple climate indices that are predictable prior to the rainy season. The Bhumibol and Sirikit Dam reservoirs are the first and second-largest reservoirs in Thailand and provide water for various economic and environmental purposes [13]. The areal-averaged rainfall data of the rainy seasons between 1965 and 2015 were used to determine a relationship between the total rainfall and selected indices, i.e., the Southern Oscillation index (Equatorial SOI, hereafter EQ-SOI), Indian Monsoon index (IMI), El Niño/Southern Oscillation index (SSTs), and the Pacific Decadal Oscillation index (PDO).

2. Study Area and Data Analysis

2.1. Study Area

Two major sub-basins of the Chao Phraya River Basin (CPRB) draining to the Bhumibol and Sirikit Dam reservoirs, i.e., the PRB and NRB, were selected as study sites, respectively (Figure 1). The PRB and NRB cover 26,300 km^2 and 11,950 km^2, respectively. The Bhumibol and Sirikit Dam reservoirs discharge a significant amount of water for domestic and industrial use, as well as for irrigation and environmental purposes [13,14]. Although domestic usage receives priority in the water allocation of the Chao Phraya River Basin, a large area of farmland requires a large amount of water for irrigation, especially during the dry season, and depends on the two major dams and other medium-to small-scale reservoirs. Therefore, it is highly important to know the volume of water available in these reservoirs before the dry season arrives.

The climate of the study area is characterized by distinct rainy and dry seasons (Figure 2). During the rainy season, the amount of precipitation and its areal extent in the CPRB are caused by the southwest monsoons, together with typhoons, monsoon troughs, and depressions from the South China Sea. The mean monthly rainfall over the study area is large during May and October (PRB) or September (NRB), which corresponds to the southwest monsoon periods [15] (Figure 2).

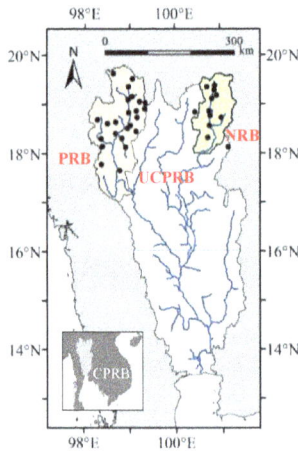

Figure 1. Location of the upper Ping River Basin (PRB) and the upper Nan River Basin (NRB). The inset shows the area of the Chao Phraya River Basin (CPRB). The Upper Chao Phraya River Basin (UCPRB) is defined to cover the area north of 16° N. The black dots indicate the locations of rain gauges used to estimate areal rainfall since 2001.

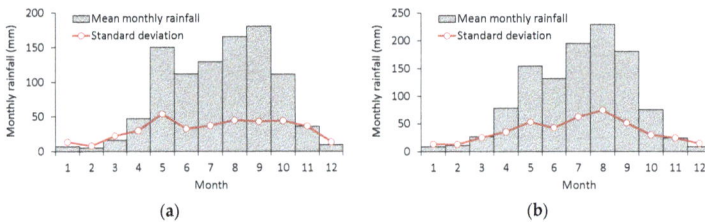

Figure 2. Mean monthly areal-averaged rainfall for the period from 1975 to 2015 over (**a**) PRB and (**b**) NRB.

2.2. Data Source

To determine the relationship between seasonal rainfall and regional climate conditions, we used the areal rainfall obtained from gridded and gauged rainfalls, and indices representing the regional climate conditions, such as the monsoon and El Niño/Southern Oscillation (ENSO). The area-averaged rainfall during the rainy season was calculated for the period between April and October utilizing a gridded rainfall dataset (APHRODITE, [16]) and rain-gauge data collected by the Royal Irrigation Department of Thailand (RID) at multiple sites located in the PRB and NRB (Figure 1). Rainfall in April is included in the rainy season as the onset of the summer rainy season begins earlier in the inland Indochina in late April to May [17], and the monthly rainfall in April is comparable to that of October for NRB (Figure 2). As the APHRODITE dataset only contains data until 2007, the areal rainfall during 2008–2015 was estimated from the original relationship between the areal-mean monthly rainfalls obtained from the APHRODITE dataset and those calculated by the Thiessen Polygon applied to the rain-gauge data of RID for the period from 1980 to 2000 (see the supplement for detailed method information to calculate the areal rainfall).

The climate indices used in the analysis include the EQ-SOI, IMI, SST anomaly over selected domains in the Pacific Ocean, PDO, and the Indian Ocean Dipole Mode index (DMI). The definition of each index is listed in Table 1. EQ-SOI is superior to SOI in representing the strength of trade winds over the tropical Pacific as one of the domains for calculating EQ-SOI (SLP.2 in Figure 3) is closer to our study area while SOI is defined as the difference between the sea level pressure at two specific locations in the Southern Hemisphere (Tahiti and Darwin). IMI is the difference between the 850-hPa zonal wind in domain U850.1 and that in domain U850.2 (Figure 3) [18]. A larger IMI indicates a stronger summer monsoon over the Bay of Bengal. Anomalies in sea surface temperature (SST) are defined for the selected domains over the tropical Pacific (Table 1). If EQ-SOI is positive, which likely corresponds to stronger trade winds than normal, the sea surface temperature in these domains may affect the supply of atmospheric vapor in the leeward region. PDO is an anomaly pattern of SST in the Pacific Ocean (north of 20° N) that shifts phases on an inter-decadal time scale that usually covers over 10 years [19]. A positive (negative) value of PDO indicates that SST is lower (higher) than normal in the northwestern Pacific. The intensity of the Indian Ocean dipole mode (IOD) is represented by the Dipole Mode Index (DMI) as an anomalous SST gradient between two selected domains (Table 1) [20]. The domain for calculating the finally selected indices is illustrated in Figure 3.

Table 1. Description of climate indices used in this study.

Index	Definition	Data Sources
EQ-SOI	The difference between the area-averaged sea level pressure in an area of the eastern equatorial Pacific (80° W–130° W, 5° N–5° S; indicated as SLP.1 in Figure 3) and an area over Indonesia (90° E–140° E, 5° N–5° S; SLP.2 in Figure 3)	[21]
IMI	The difference between the 850-hPa zonal wind in the domain of 40° E–80° E, 5° N–15° N (indicated as U850.1 in Figure 3) and that in the domain of 70° E–90° E, 20° N–30° N (U850.2 in Figure 3)	[22]
SST NINO.WEST NINO.3 NINO.4 NINO.34	Anomalies in sea surface temperature for the domain over the western tropical Pacific (130° E–150° E, 15° N–EQ) the eastern equatorial Pacific (150° W–90° W, 5° N–5° S) the central equatorial Pacific (160° E–150° W, 5° N–5° S) the central/eastern equatorial Pacific (170° W–120° W, 5° N–5° S)	[23]
PDO	The leading principal component of the mean monthly SST in the Pacific Ocean north of 20° N [19]	[24]
DMI	An anomalous SST gradient between the western (50° E–70° E, 10° S–10° N) and southeastern equatorial Indian Ocean (90° E–110° E, 10° S–0° N)	[25]

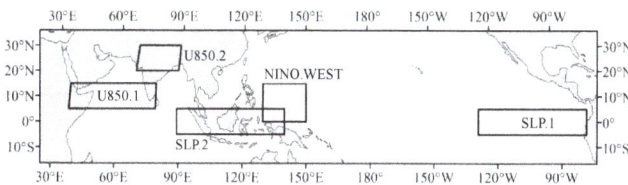

Figure 3. Domains related with each climate index finally employed in the proposed method (the domain related with PDO is not shown).

The EQ-SOI and IMI data were obtained from the NOAA National Weather Service [21] and the International Pacific Research Center (IPRC), University of Hawaii [22], respectively. The SST indices over NINO.WEST, NINO.3, NINO.4, and NINO.34 were obtained from the Japan Meteorological Agency [23]. The PDO and DMI data were obtained from the NOAA Earth System Research Laboratory (ESRL) [24] and the Japan Agency for Marine-Earth Science and Technology (JAMSTEC) [25], respectively.

2.3. Overview of Data Analysis

To quantify the seasonal rainfall from existing climate indices, we assume that a functional relationship between the total rainfall from April to October and the selected climate indices averaged over the same period can be applied, which may take the form of Equation (1):

$$TR = f(EQ - SOI, IMI, PDO, DMI, SST) \tag{1}$$

where TR (mm) is the areal-averaged seasonal rainfall during April and October, and SST represents the SST anomaly over a certain domain in the Pacific Ocean.

Based on our preliminary check for coherent variations in TR and each index, we found that the influential climate factors differed depending on EQ-SOI; thus, we classified the data into two categories with respect to EQ-SOI. When EQ-SOI is negative, we simplified Equation (1) to utilize a single independent variable (IMI), however, when EQ-SOI is positive, multiple regression analysis was used to determine the exact form of Equation (1). We obtained the functional relationship by parameter optimization (when EQ-SOI is negative) and the multiple regression analysis (when EQ-SOI is positive) based on the data for 1975 to 2015, and verified it for data between 1965 and 1974.

3. Results

3.1. Characteristics of Rainfall and Corresponding Climate Indices

Areal rainfall data indicates that over 90% of the annual rainfall occurs during April and October almost every year, with the largest mean monthly rainfall occurring in September (PRB) or August (NRB) (Figure 2). In the NRB, the mean monthly rainfall in April is larger than that in October. The anomaly of seasonal rainfall from April to October ranges from −291 to 367 mm for PRB and from −315 to 444 mm for NRB, with mean rainfall of 890 mm and 1073 mm, and standard deviations of 124 mm and 165 mm for PRB and NRB, respectively (Figure 4).

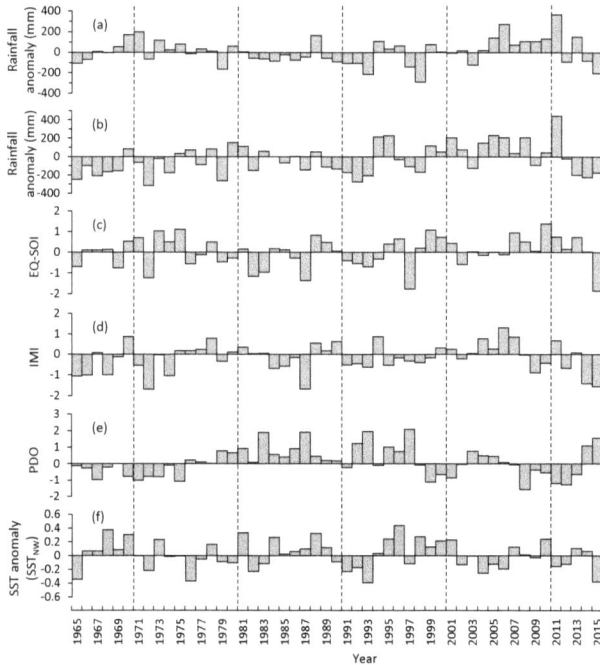

Figure 4. Anomalies of the seasonal rainfall and finally selected climate indices for the period between 1965 and 2015. (**a**) Rainfall anomaly in PRB; (**b**) Rainfall anomaly in NRB; (**c**) EQ-SOI; (**d**) IMI; (**e**) PDO; and (**f**) SST_{NW} (SST anomaly over NINO.WEST). Each decade is separated by dashed lines. The rainfall anomaly is calculated as the deviation from the mean rainfall for the period from 1975 to 2015.

The areal rainfall over PRB tends to be smaller and less variable than that over NRB, which could be due to the effect of the Dawna Range located on the western side of PRB and the limited influence of the easterly wind. The rainfall anomalies for PRB and NRB tended to be positively large between 2004 and 2011 (excluding some years), while negative anomalies were predominant during 1981–1993 and since 2012 (Figure 4). The average of each climate index between April and October is plotted in Figure 4. In years with negative EQ-SOI and IMI (1979, 1991, 1992, 1993, 1997, and 2015), it is likely that the rainfall anomalies are also negative. A positive rainfall anomaly was found in years with positively large IMI (1994 and 2006) and negative EQ-SOI. PDO was large and negative between 2008 and 2013, when the rainfall anomaly is mostly positive. EQ-SOI was negatively correlated with PDO on inter-annual and decadal time scales. Coherent variation is observed between EQ-SOI and the SST anomaly over NINO.WEST (SST_{NW}) in both the negative and positive phases. However, the influence of SST_{NW} on rainfall anomalies is not clear from Figure 4.

3.2. Relationship between Seasonal Rainfall and Climate Indices

Based on the coherent variation of the rainfall anomaly and each climate index described in the previous section, we classified the rainfall dataset and climate indices into two groups depending on EQ-SOI (the threshold value was slightly larger than 0). Seasonal rainfall is primarily related to IMI when EQ-SOI is negative and approaches the lower (upper) bound when IMI is negatively (positively) large (Figure 5). This suggests that, when EQ-SOI is negative, the effect of the monsoon on rainfall over the Indochina Peninsula is predominant due to the lower influence of the weaker surface trade winds. A nonlinear equation based on a sigmoid function (Equation (2)) was applied to the relationship between IMI and seasonal rainfall TR (mm) by calibrating the parameters in Equation (2) to minimize the root mean square error (RMSE) between the actual and calculated rainfalls (Figure 5):

$$TR = \frac{a}{1 + \exp(-\lambda \times IMI)} + b \tag{2}$$

where a, b, and λ are the parameters calibrated for each basin (Table 2).

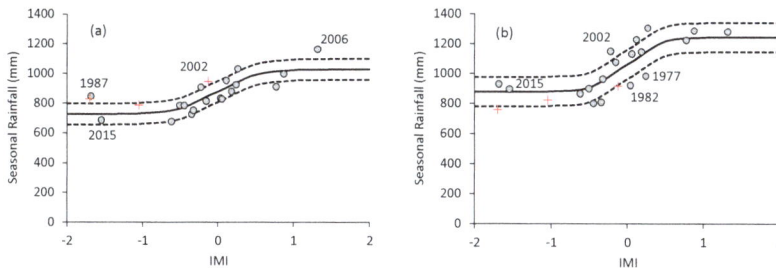

Figure 5. The relation between the seasonal rainfall and IMI for (**a**) PRB and (**b**) NRB. Two dashed lines are separated from the solid line (Equation (1)) by the standard deviation of the error between the observed and estimated seasonal rainfalls. The gray circles are the data used for calibrating the parameters, and the cross marks indicate data for 1965, 1969, and 1972, which were not used for calibration.

Table 2. Calibrated parameters used in Equation (1).

Basin	a (mm)	b (mm)	λ
PRB	302.4	723.0	4.107
NRB	395.1	870.6	4.734

The most recent drought year, 2015, is plotted close to the curve of Equation (2) (Figure 5). The difference between the estimated (solid line) and observed rainfall in some years is large (Figure 6). Although we checked the influence of other factors (EQ-SOI, PDO, and other SST-related indices), none of them were likely to be responsible for this difference. Data for years before 1975 are also plotted in Figure 5, indicating that the seasonal rainfall for these years can also be represented by Equation (2) despite the limited number of gauging stations used to obtain the areal rainfall. Furthermore, it was found that Equation (2) is applicable to the upper Chao Phraya River Basin (UCPRB, Figure 1, the basin area is 109,000 km^2) with the correlation coefficient $r = 0.77$ and RMSE = 83.1 mm for calibration period from 1976 to 2006 (sample number = 17).

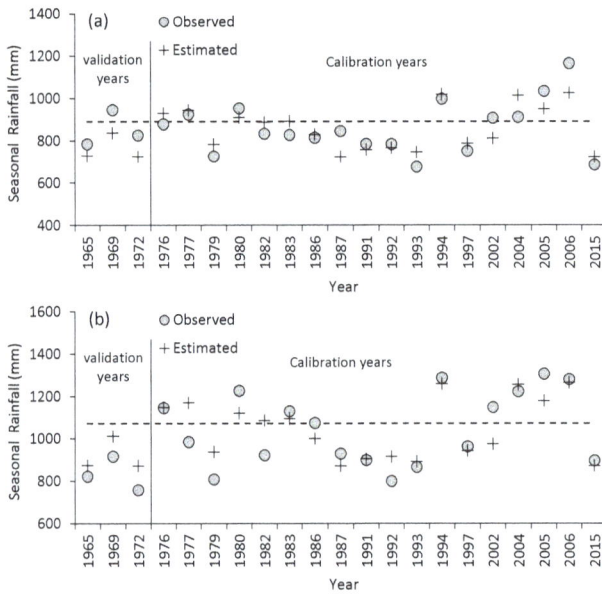

Figure 6. Comparison of observed and estimated seasonal rainfall for (**a**) PRB and (**b**) NRB for calibration and validation years with EQ-SOI < 0.02. Equation (2) with calibrated parameters (Table 2) was applied. The dashed line indicates the mean rainfall for the period from 1975 to 2015.

When EQ-SOI is positive, i.e., when the surface trade winds are larger than normal, seasonal rainfall was not significantly related to any one of the tested indices; however, it was positively correlated to EQ-SOI (Figure 9a). Since the arrival and passage of typhoons and tropical cyclones are major causes of increased rainfall, it is likely that the trade winds and SSTs in the Pacific Ocean contribute to these phenomena, with additional influence on seasonal rainfall from the Indian Ocean and monsoon [26]. In this study, we applied an equation that considers interactive and non-linear relationships between the seasonal rainfall and ENSO and SST-related indices for the Pacific Ocean using quadratic functions (Equation (3)). In addition, we tested the inclusion of IMI and DMI as additional components in Equation (6) to consider the influence of the Indian Ocean and monsoon, respectively; however, there was no significant improvement for rainfall estimation:

$$TR = (g_1 \times PDO^2 + g_2 \times PDO + g_3) \times EQ - SOI + d_0 \tag{3}$$

$$g_1 = d_1 \times SST^2 + d_2 \times SST + d_3 \tag{4}$$

$$g_2 = d_4 \times SST^2 + d_5 \times SST + d_6 \tag{5}$$

$$g_3 = d_7 \times \text{SST}^2 + d_8 \times \text{SST} + d_9 \tag{6}$$

where SST is any one of the four SST anomalies over the domain of NINO.WEST, NINO.3, NINO.4, and NINO.34, and d_0–d_9 are the coefficients of the multiple regression analysis. Equation (3) was evaluated with each SST index by the RMSE, multiple correlation coefficient, the similarity of coefficients d_1–d_9 between PRB and NRB, and the consistency with the perceived physical influences, which will be discussed in the following sections.

On average, comparatively high correlation coefficients and small RMSE were obtained for PRB and NRB by applying Equation (3) with the SST anomaly over NINO.WEST (SST_{NW}). Figure 7 compares the observed and estimated seasonal rainfall during the parameter setting period by the multiple regression analysis. A favorable agreement is found throughout the period, including flood years (1995 and 2011) and drought years (2003), although there are gaps in some years. Thus, we determined that Equation (3) with SST_{NW} is suitable within the range of the given climate indices examined, despite the limited flexibility caused by the form (quadratic function).

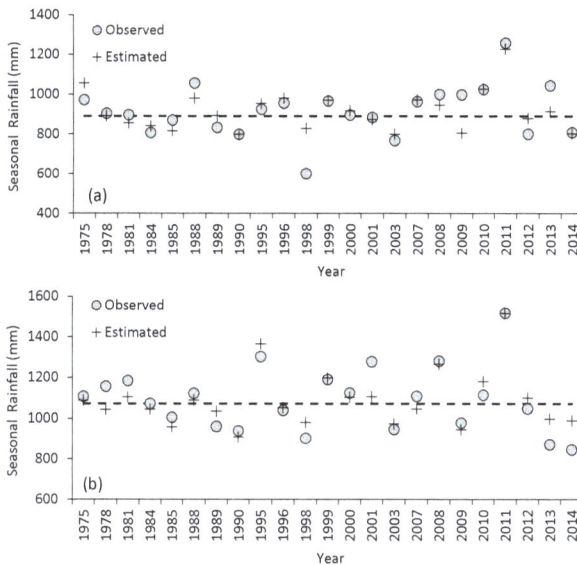

Figure 7. Comparison of observed and estimated seasonal rainfall for (**a**) PRB and (**b**) NRB for years with EQ-SOI > 0.02. Equation (3) with NINO.WEST was applied. The dashed line indicates the mean rainfall for the period from 1975 to 2015.

We validated this method by comparing the observed seasonal rainfall during 1966–1974 (when EQ-SOI is positive) with those estimated using Equation (3), resulting in RMSE of 91.0 and 101.4 mm for PRB and NRB, respectively (Figure 8). These RMSEs are larger than those for the calibration period (during 1975–2014) probably due to the limited number of gauging stations used for APHRODITE products during the period. In addition, we applied multiple regression analysis using Equation (3) to UCPRB (Figure 1), which resulted in satisfactory agreement for both the period of regression analysis between 1975 and 2007 (sample number = 16, r = 0.86, RMSE = 44.6 mm) and the period of validation between 1966 and 1974 (sample number = 6, r = 0.92, RMSE = 40.2 mm). Table 3 summarizes the results of the proposed Equations (2) and (3) applied to all years between 1975 and 2015. The correlation coefficients were greatly improved from those in previous studies (such as [11]), although the domain and duration differed.

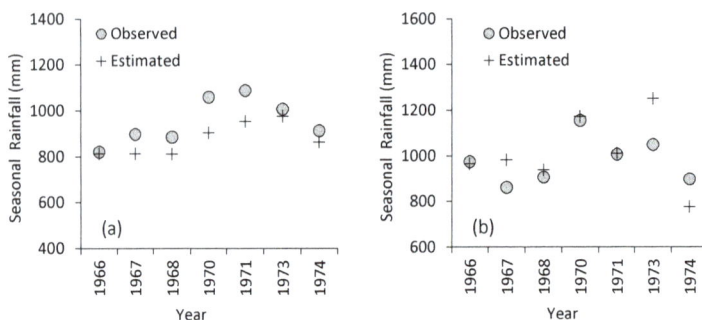

Figure 8. Same as Figure 7 but for validation years between 1966 and 1974 for (**a**) PRB and (**b**) NRB.

Table 3. Correlation coefficient *r* and RMSE for the period from 1975 to 2015 for years of negative EQ-SOI, positive EQ-SOI and all years.

Basin		EQ-SOI < 0.0	EQ-SOI > 0.0	All Years
PRB	*r*	0.83	0.79	0.81
	RMSE (mm)	69.0	78.4	74.4
NRB	*r*	0.82	0.89	0.86
	RMSE (mm)	95.4	72.1	83.1

4. Discussion

The seasonal rainfall data and three climate indices (EQ-SOI, PDO, and SST_{NW}) used in Equation (3) are plotted in Figure 9 to visualize the influence of these indices during years with positive EQ-SOI. Figure 9a shows that EQ-SOI is positively correlated with seasonal rainfall for both PRB and NRB at a significance level less than 5%, while there is no clear difference depending on the PDO phase (Figure 9a). Therefore, it is reasonable to include EQ-SOI as a multiplier of the quadratic function in Equation (3).

In PRB and NRB, the seasonal rainfall during the cool PDO phase (PDO < 0) is 220 mm and 150 mm larger than that in the warm phase (PDO > 0), respectively, under negative SST_{NW} conditions (Figure 9b). However, for years with a positive SST_{NW}, the differences in seasonal rainfall between the negative and positive PDO phases are much smaller for both basins (below 35 mm). There was a negative correlation between PDO and seasonal rainfall in PRB for all plots ($r = -0.42$, $p = 0.02$), while there was no clear correlation for NRB ($r = -0.26$, $p = 0.16$), and seasonal rainfall increases with positive-phase PDO when EQ-SOI exceeds 0.03 (Figure 9b, right). Similar characteristics were observed by Sen Roy and Sen Roy [27], who concluded that monsoon precipitation in eastern Myanmar is positively related to PDO during its warm phase, and negatively correlated during the cold phase. This is considered in the quadratic function.

Seasonal rainfall in both the PRB and NRB is positively correlated with SST_{NW}, with a *p*-value below 5%, when PDO is positive (Figure 9c), while seasonal rainfall negatively related to SST_{NW} when PDO is negative. A previous study found that the positive (negative) phase of PDO is associated with deficit (excess) rainfall over India [28]. The same phase characteristics were found between annual rainfall in Thailand and an ENSO-related index (MEI) and PDO [29]. Our results indicate that seasonal rainfall cannot be solely related to a single index, and multiple indices (including EQ-SOI) need to be considered when quantifying rainfall.

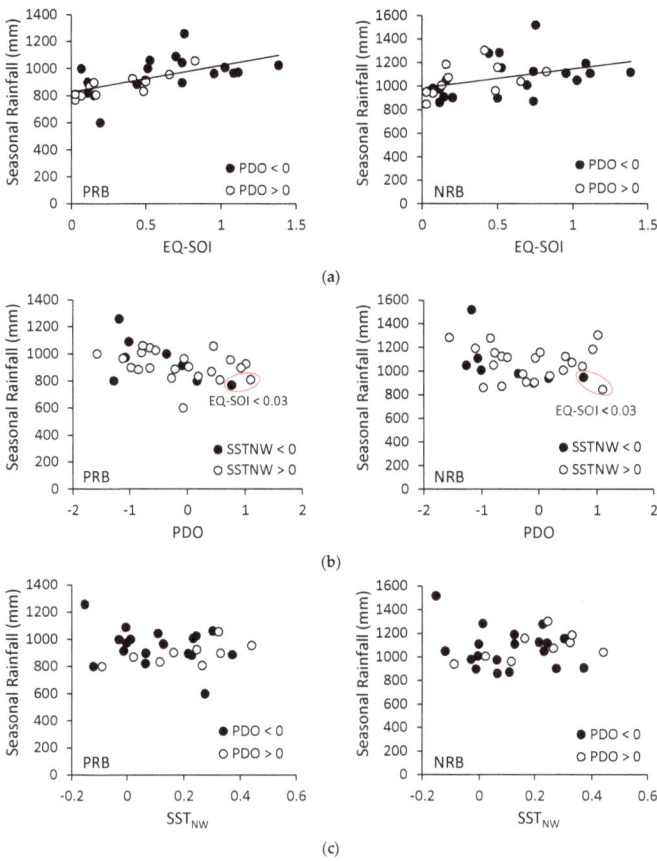

Figure 9. Relationship between seasonal rainfall and (**a**) EQ-SOI; (**b**) PDO and (**c**) SST$_{NW}$ for (left) PRB and (right) NRB between 1965 and 2015 (when EQ-SOI is positive).

To highlight the influences of PDO and SST$_{NW}$, we define the normalized rainfall anomaly (Equation (7)) by rearranging Equation (3):

$$\frac{TR - d_0}{EQ - SOI} = g_1 \times PDO^2 + g_2 \times PDO + g_3 \tag{7}$$

where g_1, g_2, and g_3 are given by Equations (4)–(6) with calibrated parameters. Figure 10 shows the relationships between the normalized rainfall anomaly and PDO and SST$_{NW}$. Each curve in Figure 10 was drawn using Equation (3), and the data plots were limited for years between 1975 and 2015 with EQ-SOI larger than 0.2 for clear indication. Most of the data plots in each class of SST$_{NW}$ (Figure 10a) or PDO (Figure 10b) are consistent with the corresponding curves with similar SST$_{NW}$ or PDO value, respectively, although there are some outliers.

The lines in Figure 10a indicate that PDO increases the normalized rainfall anomaly when it is largely negative or positive for the same or similar SST$_{NW}$. Therefore, we confirmed that the sensitivity of Equation (3) is consistent with the recognized influence of PDO, i.e., abundant precipitation tends to occur during La Niña years and the PDO cool phase [29]. Furthermore, the results in Figure 10a suggest that the seasonal rainfall in PRB and NRB can increase with both negative and positive PDO as there was no clear relationship between EQ-SOI and PDO when EQ-SOI is positive, which is consistent with the data plots in Figure 9b (especially for NRB) as well as previous analysis [27].

Unlike PDO, the influence of SST_{NW} on the normalized rainfall anomaly appears to be more complicated (Figure 10b). For NRB (Figure 10b right), SST_{NW} positively affects the normalized rainfall anomaly when PDO is positive and SST_{NW} is within a certain positive range, while the normalized rainfall anomaly is likely to decrease as SST_{NW} becomes larger (Figure 10b). These characteristics may be applied to seasonal rainfall itself as there is no certain correlation between EQ-SOI and SST_{NW} when EQ-SOI is positive. When PDO is negative, local minima are found for the normalized rainfall anomaly, as indicated by the circle plots and corresponding curve (PDO = −1.0) for NRB in Figure 10b. Similar characteristics were found for PRB, although the variation in the normalized rainfall anomaly is smaller than that for NRB as there was less seasonal rainfall with lower variation over NRB. Although the mechanistic reason for these characteristics is unclear, the normalized rainfall anomaly could be amplified when PDO and SST_{NW} are in the same phase. Hoell and Funk [30] reported that, for La Niña events (when EQ-SOI tends to be positive), precipitation is enhanced over the western Pacific and extended to the Indian Ocean during a strong western Pacific SST gradient (WPG), which is defined as the standardized difference between area-averaged SST over the central Pacific Ocean and the domain similar to NINO.WEST. This suggests that SST over NINO.WEST, combined with PDO, could affect precipitation over the Indochina Peninsula through the SST gradient in the western Pacific.

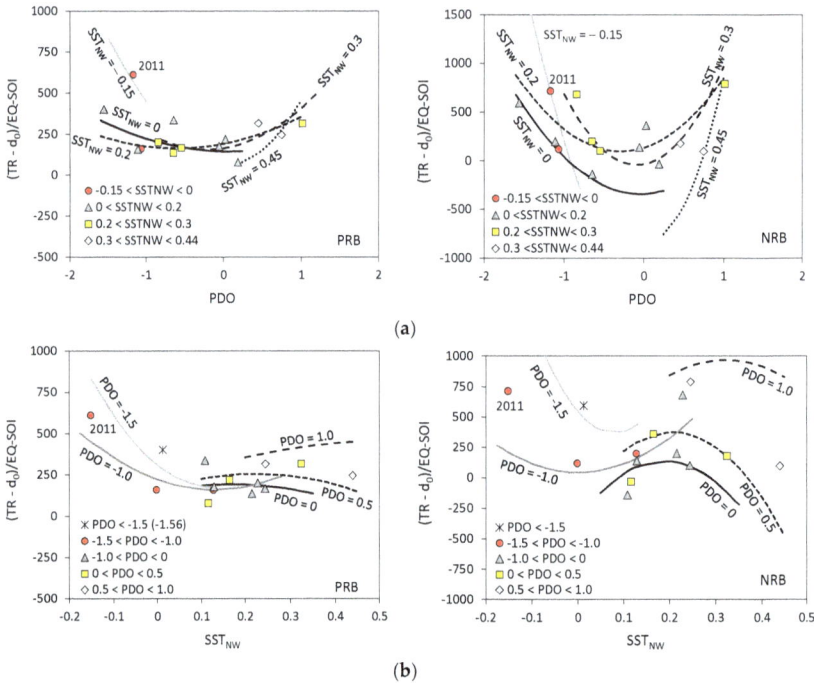

Figure 10. Relationship between seasonal rainfall and (**a**) PDO and (**b**) SST_{NW} for (left) PRB and (right) NRB. The vertical axis indicates the seasonal rainfall anomaly divided by EQ-SOI. The solid and dashed curves in each panel are given by Equation (3). Each symbol represents the data when EQ-SOI is positive.

5. Conclusions

The seasonal rainfall over the 50 years during 1965–2015 over the upper Ping River Basin (PRB) and upper Nan River Basin (NRB) in Thailand were analyzed to elucidate a quantitative relationship with existing climate indices. We showed that, when the Equatorial Southern Oscillation index (EQ-SOI) is negative, seasonal rainfall is predominantly related to the Indian Monsoon index (IMI) using a sigmoid function; however, when EQ-SOI is positive, seasonal rainfall is estimated by the nonlinear function using the El Niño/Southern Oscillation index (SST_{NW}), Pacific Decadal Oscillation index (PDO), and EQ-SOI. The areal rainfall calculated for the period between 1975 and 2015 from the proposed functions exhibits a high correlation with the observed seasonal rainfall, with correlation coefficients of 0.8 and 0.86 for PRB and NRB, respectively.

Although our proposed method exhibited a relatively high performance for estimating seasonal rainfall, further verification and improvement are necessary by updating the rainfall data and climate indices. Furthermore, to apply the method in practical water resource management, such as dam reservoir operation and the assessment of climate change impacts [31], the use of the forecasted or projected results of selected climate indices should be examined to evaluate the accuracy and uncertainty of seasonal rainfall resulting from the proposed method.

Supplementary Materials: The following are available online at http://www.mdpi.com/2073-4441/10/6/800/s1, Figure S1: Relationship between the areal-mean monthly rainfalls obtained from the APHRODITE dataset (Aphrodite-based) and those calculated by the Thiessen Polygon applied to the rain-gauge data of RID (RID-based) for the period from 1980 to 2000 for (left) PRB and (right) NRB; Figure S2. The number of gauging stations used to calculate the RID-based areal-mean monthly rainfall (red line) and that used to generate the APHRODITE dataset (black line) for (top) PRB and (bottom) NRB.

Author Contributions: Conceptualization, T.K.; Methodology, T.K.; Data collection and processing, A.K., G.Y. and W.L.; Data Analysis, T.K. and G.Y.; Writing-Original Draft Preparation, T.K.; Writing-Review & Editing, T.K. and W.L.; Visualization, T.K. and G.Y.; Project Administration, T.K. and W.L.

Funding: This work was supported by JSPS Core-to-Core Program, B. Asia-Africa Science Platforms.

Acknowledgments: The authors are grateful to the Royal Irrigation Department for providing useful data for this research.

Conflicts of Interest: The authors declare no conflict of interest.

References

1. United Nations. *Global Assessment Report on Disaster Risk Reduction*; United Nations: Geneva, Switzerland, 2009; p. 207. ISBN 9789211320282.
2. Intergovernmental Panel on Climate Change (IPCC). *Climate Change 2014: Synthesis Report. Contribution of Working Groups I, II and III to the Fifth Assessment Report of the Intergovernmental Panel on Climate Change*; Core Writing Team, Pachauri, R.K., Meyer, L.A., Eds.; IPCC: Geneva, Switzerland, 2014; p. 151. Available online: https://www.ipcc.ch/pdf/assessment-report/ar5/syr/SYR_AR5_FINAL_full_wcover.pdf (accessed on 29 April 2018).
3. Miyan, M.A. Droughts in Asian Least Developed Countries: Vulnerability and sustainability. *Weather Clim. Extr.* **2015**, *7*, 8–23. [CrossRef]
4. Tanoue, M.; Hirabayashi, Y.; Ikeuchi, H. Global-scale river flood vulnerability in the last 50 years. *Sci. Rep.* **2016**, *6*, 36021. [CrossRef] [PubMed]
5. Dai, A. Increasing drought under global warming in observations and models. *Nat. Clim. Chang.* **2013**, *3*, 52–58. [CrossRef]
6. Cai, W.; Borlace, S.; Lengaigne, M.; van Rensch, P.; Collins, M.; Vecchi, G.; Timmermann, A.; Santoso, A.; McPhaden, M.J.; Wu, L.; et al. Increasing frequency of extreme El Niño events due to greenhouse warming. *Nat. Clim. Chang.* **2014**, *4*, 111–116. [CrossRef]
7. Cai, W.; Wang, G.; Santoso, A.; McPhaden, M.J.; Wu, L.; Jin, F.-F.; Timmermann, A.; Collins, M.; Vecchi, G.; Lengaigne, M.; et al. Increased frequency of extreme La Niña events under greenhouse warming. *Nat. Clim. Chang.* **2015**, *5*, 132–137. [CrossRef]

8. Komori, D.; Nakamura, S.; Kiguchi, M.; Nishijima, A.; Yamazaki, D.; Suzuki, S.; Kawasaki, A.; Oki, K.; Oki, T. Characteristics of the 2011 Chao Phraya River flood in central Thailand. *Hydrol. Res. Lett.* **2012**, *6*, 41–46. [CrossRef]

9. Promchote, P.; Wang, S.S.-Y.; Johnson, P.G. The 2011 Great Flood in Thailand: Climate Diagnostics and Implications from Climate Change. *J. Clim.* **2016**, *29*, 367–379. [CrossRef]

10. Ueangsawat, K.; Jintrawet, A. The Impacts of ENSO Phases on the Variation of Rainfall and Stream Flow in the Upper Ping River Basin, Northern Thailand. *Environ. Nat. Resour. J.* **2013**, *11*, 97–119.

11. Tsai, C.; Behera, S.K.; Waseda, T. Indo-China Monsoon Indices. *Sci. Rep.* **2015**, *5*, 8107. [CrossRef] [PubMed]

12. Singhrattna, N.; Rajagopalan, B.; Clark, M.; Krishna Kumar, K. Seasonal forecasting of Thailand summer monsoon rainfall. *Int. J. Climatol.* **2005**, *25*, 649–664. [CrossRef]

13. Divakar, L.; Babel, M.S.; Perret, S.R.; Das Gupta, A. Optimal allocation of bulk water supplies to competing use sectors based on economic criterion—An application to the Chao Phraya River Basin, Thailand. *J. Hydrol.* **2011**, *401*, 22–35. [CrossRef]

14. Takeda, M.; Laphimsing, A.; Putthividhya, A. Dry season water allocation in the Chao Phraya River basin, Thailand. *Int. J. Water Resour. Dev.* **2015**, *32*, 321–338. [CrossRef]

15. Kripalani, R.H.; Singh, S.V.; Panchawagh, N.; Brikshavana, M. Variability of the summer monsoon rainfall over Thailand—Comparison with features over India. *Int. J. Climatol.* **1995**, *15*, 657–672. [CrossRef]

16. Yatagai, A.; Kamiguchi, K.; Arakawa, O.; Hamada, A.; Yasutomi, N.; Kitoh, A. APHRODITE: Constructing a Long-term Daily Gridded Precipitation Dataset for Asia based on a Dense Network of Rain Gauges. *Bull. Am. Meteorol. Soc.* **2012**, *93*, 1401–1415. [CrossRef]

17. Matsumoto, J. Seasonal transition of summer rainy season over Indochina and adjacent monsoon region. *Adv. Atmos. Sci.* **1997**, *14*, 231–245. [CrossRef]

18. Wang, B.; Fan, Z. Choice of South Asian summer monsoon indices. *Bull. Am. Meteorol. Soc.* **1999**, *80*, 629–638. [CrossRef]

19. Mantua, N.J.; Hare, S.R.; Zhang, Y.; Wallace, J.M.; Francis, R.C. A Pacific decadal climate oscillation with impacts on salmon. *Bull. Am. Meteorol. Soc.* **1997**, *78*, 1069–1079. [CrossRef]

20. Saji, N.H.; Goswami, B.N.; Vinayachandran, P.N.; Yamagata, T. A dipole mode in the tropical Indian Ocean. *Nature* **1999**, *401*, 360–362. [CrossRef] [PubMed]

21. NOAA National Weather Service Climate Prediction Center. Available online: http://www.cpc.ncep.noaa. gov/data/indices/ (accessed on 29 April 2018).

22. University of Hawaii, International Pacific Research Center (IPRC). Available online: http://apdrc.soest. hawaii.edu/projects/monsoon/index.html (accessed on 29 April 2018).

23. Japan Meteorological Agency. Available online: http://ds.data.jma.go.jp/tcc/tcc/products/elnino/index/ (accessed on 29 April 2018).

24. NOAA ESRL Physical Sciences Division. Available online: http://www.esrl.noaa.gov/psd/gcos_wgsp/ Timeseries/PDO/ (accessed on 29 April 2018).

25. JAMSTEC. Available online: http://www.jamstec.go.jp/frsgc/research/d1/iod/iod/dipole_mode_index. html (accessed on 29 April 2018).

26. Meyers, G.; McIntosh, P.; Pigot, L.; Pook, M. The years of El Niño, La Niña, and interactions with the tropical Indian Ocean. *J. Clim.* **2007**, *20*, 2872–2880. [CrossRef]

27. Sen Roy, S.; Sen Roy, N. Influence of Pacific decadal oscillation and El Niño Southern oscillation on the summer monsoon precipitation in Myanmar. *Int. J. Climatol.* **2011**, *31*, 14–21. [CrossRef]

28. Krishnamurthy, L.; Krishnamurthy, V. Influence of PDO on South Asian summer monsoon and monsoon–ENSO relation. *Clim. Dyn.* **2014**, *42*, 2397–2410. [CrossRef]

29. Limsakul, A.; Singhruck, P. Long-term trends and variability of total and extreme precipitation in Thailand. *Atmos. Res.* **2016**, *169*, 301–317. [CrossRef]

30. Hoell, A.; Funk, C. The ENSO-Related West Pacific Sea Surface Temperature Gradient. *J. Clim.* **2013**, *26*, 9545–9562. [CrossRef]

31. Singhrattna, N.; Babel, M.S. Changes in summer monsoon rainfall in the Upper Chao Phraya River Basin, Thailand. *Clim. Res.* **2011**, *49*, 155–168. [CrossRef]

water

MDPI

Article

Multiple Climate Change Scenarios and Runoff Response in Biliu River

Xueping Zhu [1,*], Chi Zhang [2], Wei Qi [3,4,*], Wenjun Cai [1], Xuehua Zhao [1] and Xueni Wang [1]

[1]   College of Water Resources Science and Engineering, Taiyuan University of Technology,
      Taiyuan 030024, China; caiwenjun62620@163.com (W.C.); zhaoxuehua@tyut.edu.cn (X.Z.);
      xnwang@mail.dlut.edu.cn (X.W.)
[2]   School of Hydraulic Engineering, Dalian University of Technology, Dalian 116024, China;
      czhang@dlut.edu.cn
[3]   School of Environmental Science and Engineering, South University of Science and Technology of China,
      Shenzhen 518055, China
[4]   State Key Laboratory of Water Resource & Hydropower Engineering Science, Wuhan University,
      Wuhan 430072, China
*   Correspondence: xpzhu01@163.com (X.Z.); QiWei_waterresources@hotmail.com (W.Q.);
      Tel.: +86-152-9661-8839 (X.Z.); +86-152-4111-0143 (W.Q.)

Received: 19 October 2017; Accepted: 26 January 2018; Published: 30 January 2018

Abstract: The impacts of temperature and precipitation changes on regional evaporation and runoff characteristics have been investigated for the Biliu River basin, which is located in Liaoning Province, northeast China. Multiple climate change scenarios from phase 3 and phase 5 of the Coupled Model Intercomparison Project (CMIP3 and CMIP5) (21 scenarios in total) were utilized. A calibrated hydrologic model—SWAT model—was used to simulate future discharges based on downscaled climate data through a validated morphing method. Results show that both annual temperature and precipitation increase under most of the CMIP3 and CMIP5 scenarios, and increase more in the far future (2041–2065) than in the near future (2016–2040). These changes in precipitation and temperature lead to an increase in evaporation under 19 scenarios and a decrease in runoff under two-thirds of the selected scenarios. Compared to CMIP3, CMIP5 scenarios show higher temperature and wider ranges of changes in precipitation and runoff. The results provide important information on the impacts of global climate change on water resources availability in the Biliu River basin, which is beneficial for the planning and management of water resources in this region.

Keywords: climate change; CMIP3; CMIP5; downscaling; runoff response; SWAT model

1. Introduction

It has been recognized that climate change could have profound impacts on the global water cycle [1–3]. Therefore, it is important to consider potential impacts of climate change in the planning and management of regional water resources.

To quantitate climate change impacts on water resources, the output of general circulation models (GCMs) is commonly used [4–6], coupled with hydrological models or forcing offline hydrological models [7–10]. The results help to understand the impacts of climate change and develop strategies to adapt to or possibly mitigate these impacts [3]. The Coupled Model Intercomparison Project phase 3 (CMIP3) and phase 5 (CMIP5) have provided abundant climate data (e.g., the IPCC's AR5) [11], and both CMIP3 and CMIP5 climate datasets have been widely utilized globally [12–16]. Several studies have revealed that, compared to CMIP3, CMIP5 ensemble simulations have substantially improved the statistical representation of daily mean precipitation and temperature [17,18]. However, few studies have compared the impacts of CMIP3 and CMIP5 on the design of the planning and management infrastructure of water resources [19–21], especially in China [22].

Northeast China is an important food production area in China and it has suffered from droughts in recent years. It has been reported that the inflow of Biliu River Reservoir (an important river in northeast China) declined significantly over the period of 1990–2005 [23]. Because the reservoir is the most important water source for nearby big cities and also because it plays an important role in cropland irrigation, a water transfer project is currently being constructed to transfer water into the reservoir. Zhang et al. optimized water diversion and supply rules of Biliu River Reservoir and attempt to identify resilient water diversion strategies to mitigate the potential impacts of climate change [10]. However, the design of the water transfer project was based on CMIP3 data, which may inaccurately estimate the severity of a water resource shortage [19,20]. Therefore, it is necessary to compare the availability of simulated water resources using CMIP3 and CMIP5, which can provide important information on the differences in the availability of projected water resources and therefore facilitate the design and evaluation of the water transfer project [19–21].

The overall objective of this study is to compare the projected availability of runoff using datasets from CMIP3 and CMIP5 in the Biliu River basin. Three GCM outputs of CMIP3 for three emission scenarios and six GCM outputs of CMIP5 for two representative concentration pathways are utilized. A 'morphing' downscaling method is utilized to derive local climatic parameters on a daily time scale. The downscaled datasets are utilized as the input of the SWAT hydrological model to simulate future runoff.

2. Methodology

2.1. CMIP3 and CMIP5 Datasets

2.1.1. Climate Models and Emission Scenarios

Several studies have evaluated the precipitation and temperature simulation performance of the CMIP3 and CMIP5 historical experiments and found that CMIP5 improved the simulation of basic atmospheric variables compared to CMIP3 [17,18]. Three CMIP3 models and six CMIP5 models are used in this study to consider the uncertainties of global climate prediction. They cover a representative range of projections of the two experiments and perform well in the statistical representation of daily precipitation and temperature [17,18].

Emission scenarios A1B, A2, and B1 from CMIP3, which can capture the emission uncertainties, have been analyzed by other studies [24,25] and are considered in this study. Three GCMs for each emission scenario are utilized (as shown in Table 1). The baseline scenario (20C3M) for the historical period is also considered. The regional monthly temperature and precipitation of the study area are derived by linear interpolation of the nearest GCM data. Although it is simple, its precision is acceptable [26–28].

The formulation of the long-term (century time-scale) simulations is a new part of CMIP5 when compared with CMIP3. The long-term experiment was initiated from the end of freely evolving simulations of the historical period; the experiment was conducted with atmosphere-ocean global climate models (AOGCMs) which in some cases may be coupled with carbon cycle models [11,29]. The long-term experiment includes the pre-industrial control run, the historical run (1850 to at least 2005), and the future scenario run (~2006–2100, or extended to 2300), etc.

For future climate projections, four emission and concentration scenarios called 'representative concentration pathways' (RCPs) were designed that lead to radiative forcing levels of 8.5, 6, 4.5, and 2.6 W/m^2 around the end of the century. Each of the RCPs covers the period 2006–2100, and extensions have been formulated for periods up to 2300 [30]. RCP4.5 and RCP8.5 are core experiments [29]. RCP4.5 is a medium forcing integration and RCP8.5 is a high radiative forcing case. They correspond to a medium mitigation scenario and a high emissions scenario, respectively. RCP6 and RCP2.6 are two carbon cycle feedback experiments, part of tier 1 experiments. The tier 1 and tier 2 experiments explore various aspects of the core experiments in further detail. The core experiments (i.e., RCP4.5 and

RCP8.5) are utilized in this study. Six climate models (as shown in Table 1) from CMIP5 are considered for comparison.

Table 1. Details of CMIP3 and CMIP5 climate models and scenarios used in this study.

Model	Country	Resolution	Scenarios
BCCR_BCM2.0	Norway	$2.81° \times 2.81°$	
CSIRO_MK3.0	Australia	$1.88° \times 1.88°$	A1B, A2, B1 for each model respectively
MIROC3.2m	Japan	$2.81° \times 2.81°$	
ACCESS1.0	Australia	$1.88° \times 2.48°$	
BCC-CSM1.1(m)	China	$1.13° \times 1.13°$	
CESM1(BGC)	USA	$1.3° \times 0.9°$	
CESM1(CAM5)	USA	$1.3° \times 0.9°$	RCP8.5 and RCP4.5 for each model respectively
CMCC-CM	Italy	$0.75° \times 0.75°$	
MPI-ESM-MR	Germany	$1.88° \times 1.88°$	

2.1.2. Downscaling Method

The coarse resolution climate model predictions need to be downscaled to fine spatial and temporal resolutions to facilitate hydrological simulations. There are many downscaling methods, and they can be divided into two categories: dynamical downscaling and statistical downscaling methods. The statistical downscaling methods are relatively simpler than dynamical downscaling and have been widely used [7–9]. One of the statistical downscaling methods called "morphing" [7] is adopted here. The "morphing" approach has two characteristics. First, the 'baseline climate' is reliable, because it is based on observed climate data. Second, the resulting weather sequence is likely to be meteorologically consistent.

Morphing involves three generic operations: a shift, a linear stretch (scaling factor), and a combination of shift and a stretch. A shift by Δx_m is applied to the present-day climate variable x_0 by $x = x_0 + \Delta x_m$. Here, Δx_m is the absolute changes in monthly mean climate for month m. The monthly variance of climate variables is unchanged. A stretch of α_m is applied by $x = \alpha_m x_0$, where α_m is the fractional change in the monthly-mean value for month m. A combination of shift and stretch is obtained by $x = x_0 + \Delta x_m + \alpha_m \times (x_0 - \overline{x}_m^0)$, in which $\overline{x}_m^0$ is the baseline climatological value for month m, and is calculated as $\overline{x}_m^0 = \frac{1}{24 \times d_m \times Y} \sum_{Y \, year} \sum_{month \, m} x_0$, where Y is the number of years and d_m is the number of days in month m.

Precipitation and temperature are two main climate variables considered in previous research in runoff simulation. Therefore, we analyze runoff responses to changes in precipitation and temperature. The precipitation data is downscaled by a combination of shift and stretch morphing, and the temperature data is downscaled by shift morphing by

$$P_1 = (1 + \alpha_m) \times P_0 = \left(1 + \frac{\overline{P}_m - \overline{P}_m^b}{\overline{P}_m^b}\right) \times P_0 \tag{1}$$

$$T_1 = T_0 + \Delta T = T_0 + (\overline{T}_m - \overline{T}_m^b) \tag{2}$$

$$T_1^{max} = T_0^{max} + \Delta T_{max} = T_0^{max} + (\overline{T}_m^{max} - \overline{T}_m^{b,max}) \tag{3}$$

$$T_1^{min} = T_0^{min} + \Delta T_{min} = T_0^{min} + (\overline{T}_m^{min} - \overline{T}_m^{b,min}) \tag{4}$$

where P_1, T_1, T_1^{max} and T_1^{min} are downscaled precipitation, daily mean, and maximum and minimum temperature; P_0, T_0, T_0^{max} and T_0^{min} are historical observation precipitation, daily mean, and maximum and minimum temperature; $\overline{P}_m$, $\overline{T}_m$, $\overline{T}_m^{max}$ and $\overline{T}_m^{min}$ are average values of future precipitation, daily mean, and maximum and minimum temperature over month m; $\overline{P}_m^b$, $\overline{T}_m^b$, $\overline{T}_m^{b,max}$ and $\overline{T}_m^{b,min}$ are average values of precipitation, daily mean, and maximum and minimum temperature over month m for the historical period from GCMs.

2.2. Hydrological Model

The continuous and physically distributed hydrological model—the Soil and Water Assessment Tool (SWAT) [31]—is utilized in this study. The main components of SWAT include hydrology, climate, nutrient cycling, soil temperature, sediment movement, crop growth, agricultural management, and pesticide dynamics [32]. The model has been widely applied to simulate hydrological processes globally [23,33–35] and it can conveniently consider weather adjustment.

Model calibration and validation are necessary to identify the SWAT model parameters before it can be used for prediction. Sensitive parameters were first identified by the LH-OAT (latin hypercube sampling based on one-factor-at-a-time) method [36] incorporated in SWAT. Manual calibration was then carried out for several of the most sensitive parameters. The performance of SWAT was estimated by three performance metrics: the Nash–Sutcliffe model efficiency (NSE), the average relative error (Re) and the coefficient of determination (R^2). These metrics are defined as

$$\text{NSE} = 1 - \sum_{t=1}^{T} (Q_{mt} - Q_{st})^2 / \sum_{t=1}^{T} (Q_{mt} - Q_{mavg})^2 \tag{5}$$

$$\text{Re} = (Q_{mavg} - Q_{savg}) / Q_{savg} \times 100 \tag{6}$$

$$R^2 = \sum_{t=1}^{T} (Q_{st} - Q_{savg})(Q_{mt} - Q_{mavg}) / \left\{ [\sum_{t=1}^{T} (Q_{st} - Q_{savg})^2][\sum_{t=1}^{T} (Q_{mt} - Q_{mavg})^2] \right\}^{\frac{1}{2}} \tag{7}$$

where Q_{mt} and Q_{st} are measured and simulated flow at time t; Q_{mavg} and Q_{savg} are average values of observed flow and simulated flow; T is the total number of time steps. The model performance is considered to be acceptable when NSE > 0.50, R^2 > 0.60, and Re < ±20% according to the study by Hao et al. [37].

3. Study Region and Datasets

3.1. Biliu River Basin

The Biliu River basin is located in the Liaoning Province, northeast China (Figure 1). The study area is about 2085 km^2. The Biliu River Reservoir was built in 1975 and it has a storage capacity of 934×10^6 m^3. The reservoir is the most important water source of nearby big cities and also plays an important role in cropland irrigation. The study area is located in the north temperate zone and is characterized by a moist climate. Forest land and farmland are the main land use types. The main soil types are brown soil and meadow soil. Another reservoir, named Yushi Reservoir, with a drainage area of 313 km^2 and a storage capacity of 89×10^6 m^3, was built upstream in 2001. The reservoir supplies water to the outside of the basin. Therefore, the impact of Yushi Reservoir needs to be considered in the hydrological model. The Biliu River Reservoir has experienced severe water shortage problems recently and therefore future runoff conditions under climate change need to be analyzed for adaptation measures.

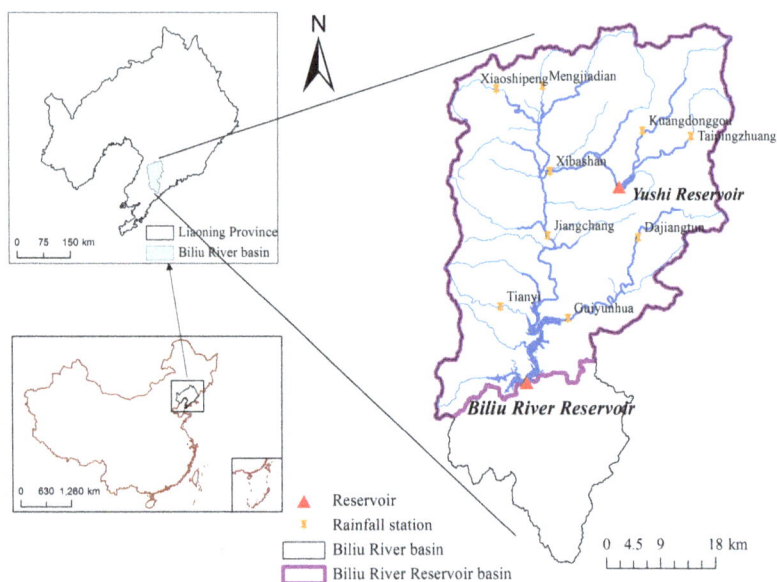

Figure 1. The Biliu River basin.

3.2. Dataset

The climate data in 1901–2099 for A1B, A2, and B1 were downloaded from the National Climate Center (http://ncc.cma.gov.cn). The long-term experiment data of 1850–2100 for the chosen six climate models in CMIP5 were downloaded from the Program for Climate Model Diagnosis and Intercomparison (PCMDI, http://pcmdi3.llnl.gov/esgcet/). The climate data were extracted for 1980–2004 period and two future periods (2016–2040 and 2041–2065). Future precipitation and temperature data output from the climate models are used as the input of the SWAT model to simulate future runoff.

Yearly and monthly precipitation and runoff data in 1958–2011 and daily runoff data in 1978–2004, were obtained from the Biliu River Reservoir administration. Daily precipitation data in 1978–2004 at nine precipitation stations were obtained from the Hydrology Bureau of Liaoning Province. Daily meteorological data—including mean, maximum, and minimum temperature, humidity, wind speed and direction, and solar radiation—were obtained from the China Meteorological Data Sharing Service System (http://cdc.cma.gov.cn/index.jsp). The Digital Elevation Model (DEM) data (90 × 90 m) were obtained from the CGIAR Consortium for Spatial Information (CGIAR-CSI) (http://srtm.csi.cgiar.org). Soil type and land use maps were obtained from the Data Center for Resources and Environmental Sciences, Chinese Academy of Sciences (http://www.resdc.cn/first.asp).

4. Results

This section includes three main parts. First, the SWAT model used is calibrated and validated based on observed data; second, the variations of precipitation and temperature in the study region are studied based on GCMs data from CMIP3 and CMIP5; third, the precipitation and temperature data are used as the input of SWAT model to predict the runoff in this region. In the resulting figures, runoff data are for the location of the Biliu River Reservoir, while precipitation, temperature, and evaporation data are the mean values of the Biliu River Reservoir basin in Figure 1.

4.1. SWAT Model Calibration and Validation Results

The SWAT model is calibrated and validated beforehand. The periods 1980–1994 and 1995–2004 were used as the calibration period and validation period, respectively. Yushi Reservoir runs after 2001 in the model. The simulated and measured monthly runoff are shown in Figure 2. The values of NSE and R^2 exceed 0.90 and 0.95, respectively. Re values are 4.04% and 12.64% in the calibration and validation. Therefore, the SWAT model is applicable for runoff prediction in this area.

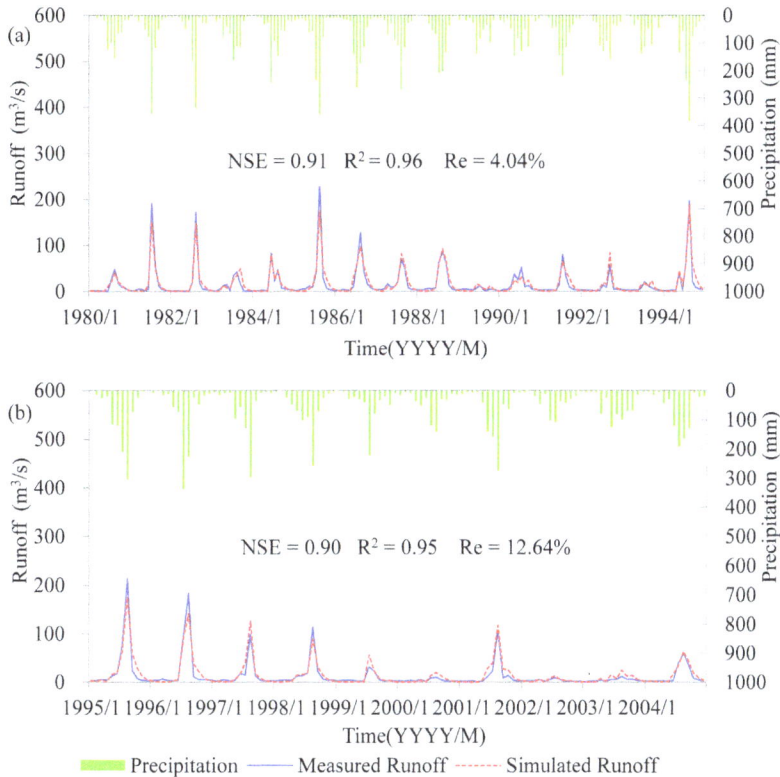

Figure 2. Simulated monthly runoff at Biliu River Reservoir station: (**a**) calibration period of 1980–1989; (**b**) validation period of 1990–1999.

4.2. Precipitation and Temperature Variations

After the 1980s, the Biliu River basin is affected by both climate change and human activities. The influence will exist in the future, and it is difficult to return to the natural state before 1980s. Therefore, the period after 1980s is used for precipitation and temperature downscaling. In addition, daily data are required for hydrological simulation. Thus, 1980–2004 is considered for downscaling since daily scale is not available after 2004. The climate model outputs in 1980–2004 and two future periods, 2016–2040 (near future) and 2041–2065 (far future), are utilized. Precipitation and temperature of the two future periods are downscaled respectively on the basis of 1980–2004 data. The output of the GCMs is bias corrected according to the observation in the historically period, and the same bias correction approach is applied to the output of GCMs in the future.

4.2.1. Temperature Variations

Two time periods, 1985–1994 and 1995–2004, were utilized to validate the morphing method for temperature downscaling. The downscaled temperature of 1995–2004 can be calculated using Equations (2)–(4). The downscaled temperature was compared with the historical observation in Figures 3 and 4. The nine selected models are verified detail by detail because there is no scenario difference in the historical period.

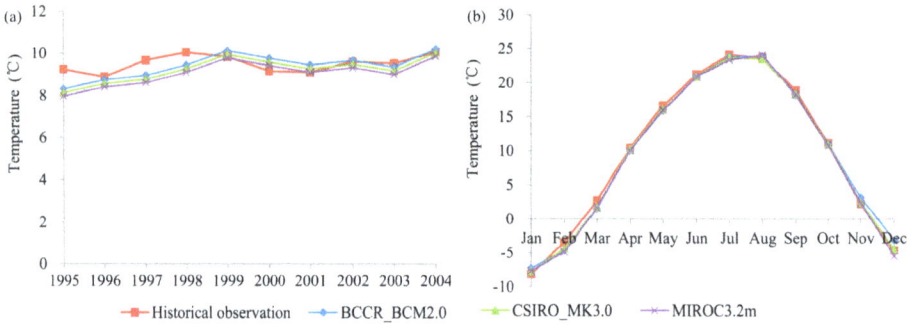

Figure 3. Comparison of the average downscaled temperature from CMIP3 using morphing and historical observation temperature series: (**a**) annual series; (**b**) monthly distribution.

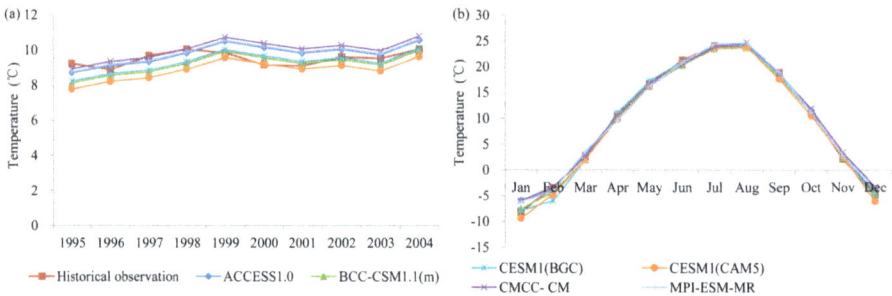

Figure 4. Comparison of the average downscaled temperature from CMIP5 using morphing and historical observation temperature series: (**a**) annual series; (**b**) monthly distribution.

Figure 3a compares the average annual mean temperatures of CMIP3 models. For BCCR_BCM2.0, CSIRO_MK3.0, and MIROC3.2m, the relative errors (Re) are −1.2%, −3.1%, and −4.8% respectively, and correlation coefficients (R^2) all are larger than 0.60. Figure 3b compares monthly temperature variations. It can be seen that they agree well. Figure 4a shows the average annual mean temperature of the six models from CMIP5. Three of them—ACCESS1.0, CMCC-CM and MPI-ESM-MR—slightly overestimate the temperature with Re being 2.7%, 5.2%, and 3.4% respectively. The other three models slightly underestimate the temperature with Re of BCC-CSM1.1(m), CESM1(BGC), and CESM1(CAM5) being −3.4%, −2.3%, and −7.0% respectively. The R^2 of the six models all are larger than 0.60. Figure 4b shows the comparison of monthly temperature variations, and it can be seen that the downscaled temperatures agree well with observation. Overall, the morphing approach shows acceptable performance in the temperature downscaling.

The annual mean temperature changes relative to 1980–2004 period are shown in Figures 5 and 6. The horizontal dash lines refer to the maximum and minimum multi-year mean temperature changes, which show the ranges of change uncertainty, and can be calculated by Equations (8) and (9).

$$z_{MAX} = \max \left\{ \frac{1}{Y} \sum_{j=1}^{Y} \Delta z_i^j, \ i = 1, 2, \cdots, k \right\} \tag{8}$$

$$z_{MIN} = \min \left\{ \frac{1}{Y} \sum_{j=1}^{Y} \Delta z_i^j, \ i = 1, 2, \cdots, k \right\} \tag{9}$$

where z_{MAX} and z_{MIN} represent the maximum and minimum multi-year mean changes, respectively; Δz_i^j is the absolute or fractional changes in annual mean climate for year j based on the ith scenario; Y is the number of years and k is the number of scenarios.

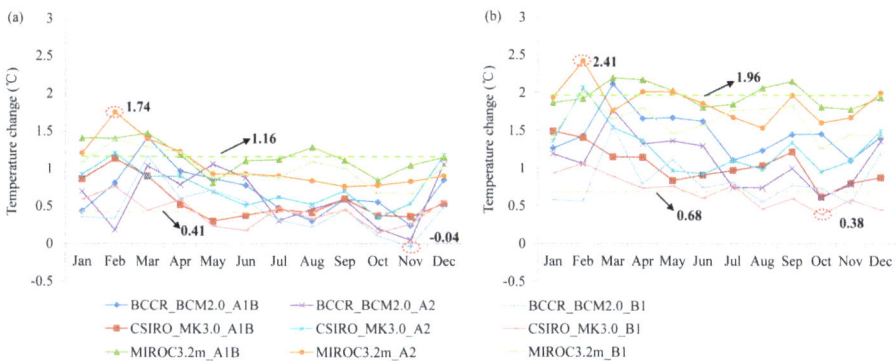

Figure 5. The forecasted temperature differences relative to 1980–2004 period under CMIP3: (**a**) the period 2016–2040; (**b**) the period 2041–2065.

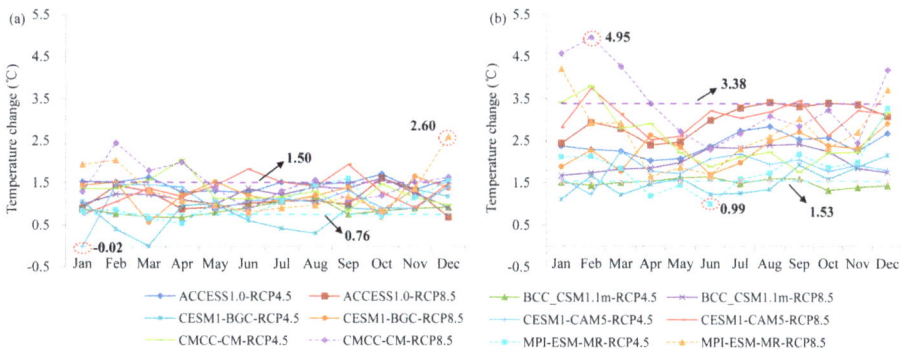

Figure 6. The forecasted temperature differences relative to 1980–2004 period under CMIP5: (**a**) the period 2016–2040; (**b**) the period 2041–2065.

As shown in Figures 5 and 6, the annual mean temperature increases under all of the CMIP3 and CMIP5 scenarios. The annual mean temperature changes range from 0.41 to 1.16 °C under CMIP3 and range from 0.76 to 1.50 °C under CMIP5 in 2016–2040. The temperatures under all of the CMIP3 and CMIP5 scenarios increase even more in 2041–2065 than in 2016–2040, and the increases range from 0.68 to 1.96 °C under CMIP3 and range from 1.53 to 3.38 °C under CMIP5. Overall, future temperature increases range from 0.41 to 1.50 °C in 2016–2040 and 0.68 to 3.38 °C in 2041–2065 relative to 1980–2004 period on the basis of the selected scenarios. The results imply that temperatures are likely to increase in the future and increase more in the far future than in the near future. Moreover, temperature increases more under CMIP5 than under CMIP3.

The temperature differences between emission scenarios are also compared. Among A1B, A2, and B1, the increase in temperature from largest to smallest is A1B > A2 > B1 on the basis of BCCR_BCM2.0 and MIROC3.2m in both future periods, while it is A2 > A1B > B1 based on CSIRO_MK3.0. In addition, it is found that the temperature increases more in RCP8.5 than in RCP4.5 under all the models in both future periods except ACCESS1.0 in 2016–2040. This implies that climate emission scenarios are closely associated with temperature.

4.2.2. Precipitation Variations

Similar to the temperature downscaling validation, two time periods, 1985–1994 and 1995–2004, were used to validate the applicability of the morphing method for precipitation downscaling. Downscaled precipitation of 1995–2004 can be calculated by Equation (1). The downscaled precipitation is compared with the historical observation precipitation to verify its accuracy in Figures 7 and 8.

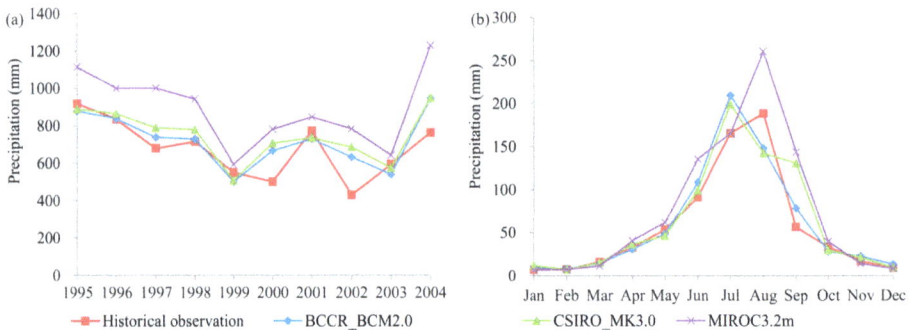

Figure 7. Comparison of the downscaled precipitation from CMIP3 using morphing and historical observation precipitation series: (**a**) annual series; (**b**) monthly distribution.

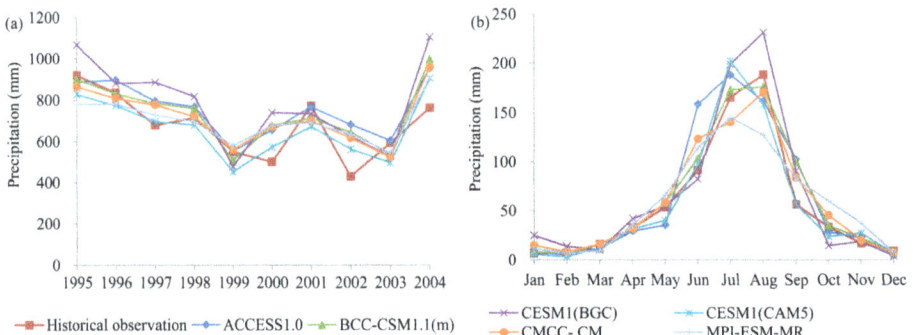

Figure 8. Comparison of the downscaled precipitation from CMIP5 using morphing and historical observation precipitation series: (**a**) annual series; (**b**) monthly distribution.

Figure 7a compares the annual precipitation of CMIP3 models. For BCCR_BCM2.0, CSIRO_MK3.0 and MIROC3.2m, Re are 6.4%, 10.7% and 32.2% respectively, and R^2 are larger than 0.72. Figure 7b compares monthly precipitation variations. It can be seen that they agree well. Figure 8a shows the average annual precipitation of the six models from CMIP5. Only one of them—CESM1(CAM5)—slightly underestimated the precipitation (Re is −1.8%). The other five models all overestimate the precipitation. The Re of ACCESS1.0, BCC-CSM1.1(m), CMCC-CM, CESM1(BGC), and MPI-ESM-MR are 11.9, 8.8, 6.5, 16.3, and 4.0% respectively. The R^2 of the six models all are larger than 0.70. Figure 8b shows a comparison of monthly precipitation variations. The fitting result is accurate: R^2 values are larger than 0.93. Overall, the morphing results basically meet the downscaling requirements, which demonstrates that the morphing method is valid for precipitation downscaling.

The annual mean change percentages relative to 1980–2004 period are shown in Figures 9 and 10. The horizontal dash lines refer to the maximum and minimum multi-year mean precipitation change percentages, which show the ranges of change uncertainty, as calculated by Equations (8) and (9).

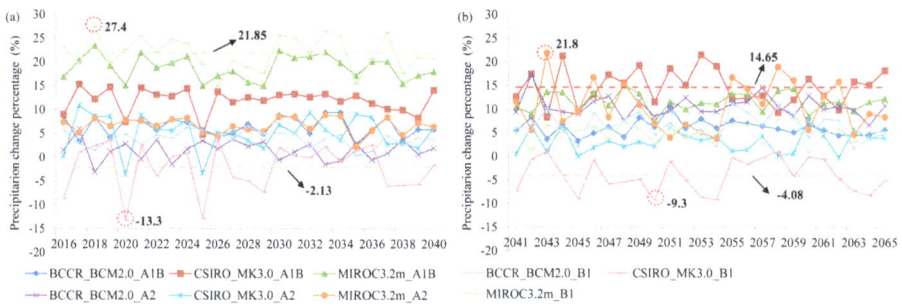

Figure 9. The forecasted precipitation change percentage relative to 1980–2004 period under CMIP3 (**a**) 2016–2040; (**b**) 2041–2065.

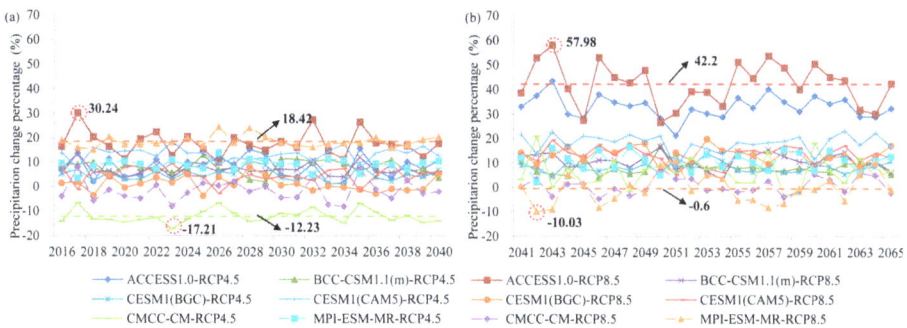

Figure 10. The forecasted precipitation change percentage relative to 1980–2004 period under CMIP5 (**a**) 2016–2040; (**b**) 2041–2065.

As shown in Figure 9, annual precipitation increases under most of the CMIP3 scenarios except CSIRO_MK3.0 (B1). The mean annual change percentage ranges from −2.13% to 21.85% in 2016–2040. The range is slightly smaller in 2041–2065, which ranges from −4.08% to 14.65%. The difference of MIROC3.2m (B1) in the two future periods is the greatest, followed by BCCR_BCM2.0 (A2) and MIROC3.2m (A1B). The other six scenarios differ by less than ±5%. This implies that future precipitation is likely to increase, with similar changes in the near future and far future.

As shown in Figure 10, annual precipitation increases under most of the CMIP5 scenarios except CMCC-CM (RCP4.5 and RCP8.5) in 2016–2040 and MPI-ESM-MR (RCP8.5) in 2041–2065. The annual mean change percentage ranges from −12.23% to 18.42% in 2016–2040. The range is significantly wider in 2041–2065, which ranges from −0.6% to 42.2%. Compared to the results in 2016–2040, the precipitation change percentage in 2041–2065 decreases only under MPI-ESM-MR (RCP8.5), while the other 11 scenarios all increase. This implies that future precipitation is likely to increase. Moreover, precipitation in the far future is likely to increase more significantly than in the near future. On the other hand, future precipitation decreases under a few scenarios, which reveals the uncertainties of future precipitation. Overall, future precipitation changes range from −12.23% to 21.85% in 2016–2040 and −4.08% to 42.2% in 2041–2065 relative to 1980–2004 period on the basis of the selected scenarios.

The precipitation differences between emission scenarios are also compared. Among A1B, A2, and B1, the increase in precipitation from largest to smallest is A1B > A2 > B1 on the basis of CSIRO_MK3.0 in both future periods. However, the pattern is not obvious under the other two models. In addition, it is found that half of the models have larger change percentages under RCP8.5 while the other half have larger change percentages under RCP4.5. This implies that climate emission scenarios are not so closely associated with precipitation.

The historical annual series of 1958–2011 data are considered as the baseline to further analyze the precipitation variations. Compared to the historical mean annual precipitation of the baseline period (739.4 mm), the change percentages of the forecasted long series (2016–2065) are no more than ±10% under CMIP3 scenarios. Specifically, the forecasted mean annual precipitation decreases by 0.56, 1.77, and 8.3% under BCCR_BCM2.0 (B1), CSIRO_MK3.0 (A2), and CSIRO_MK3.0 (B1) respectively, while it increases under the other six CMIP3 scenarios. Among the 12 CMIP5 scenarios, the decline in precipitation only appears in the CMCC-CM model, with declines of 8.67% and 6.56% for RCP4.5 and RCP8.5, respectively. The maximum increase appears in the ACCESS1.0 model, while the other scenarios show smaller increases. The increased proportions of precipitation are mostly below 10%, with two exceptions: the ACCESS1.0 models for RCP4.5 and RCP8.5. A comparison of the results is shown in Table 2.

Table 2. Mean annual precipitation and runoff in the historical period (1958–2011) and future period (2016–2065).

Climate Scenario	Precipitation (mm)	Runoff (mm)	Runoff Coefficient	Change Percentage (%)		
				Precipitation	Runoff	Runoff Coefficient
Historical observation	739.43	275.43	0.37	-	-	-
BCCR_BCM2.0(A1B)	741.76	258.55	0.35	0.31	−6.13	−6.43
BCCR_BCM2.0(A2)	739.93	263.35	0.36	0.07	−4.39	−4.45
BCCR_BCM2.0(B1)	735.28	260.24	0.35	−0.56	−5.52	−4.98
CSIRO_MK3.0(A1B)	790.78	300.37	0.38	6.94	9.06	1.97
CSIRO_MK3.0(A2)	726.37	249.02	0.34	−1.77	−9.59	−7.97
CSIRO_MK3.0(B1)	678.03	217.10	0.32	−8.30	−21.18	−14.04
MIROC3.2m(A1B)	805.40	306.29	0.38	8.92	11.20	2.10
MIROC3.2m(A2)	760.50	276.14	0.36	2.85	0.26	−2.52
MIROC3.2m(B1)	793.18	300.71	0.38	7.27	9.18	1.78
ACCESS1.0 (RCP4.5)	841.15	331.45	0.39	13.76	20.34	5.79
ACCESS1.0 (RCP8.5)	909.90	398.39	0.44	23.05	44.64	17.54
BCC-CSM1.1(m)(RCP4.5)	752.82	261.36	0.35	1.81	−5.11	−6.80
BCC-CSM1.1(m)(RCP8.5)	759.89	269.18	0.35	2.77	−2.27	−4.90
CESM1(BGC) (RCP4.5)	741.93	254.00	0.34	0.34	−7.78	−8.09
CESM1(BGC) (RCP8.5)	748.55	264.06	0.35	1.23	−4.13	−5.30
CESM1(CAM5) (RCP4.5)	807.84	307.95	0.38	9.25	11.81	2.34
CESM1(CAM5) (RCP8.5)	761.03	268.87	0.35	2.92	−2.38	−5.15
CMCC-CM (RCP4.5)	675.31	204.83	0.30	−8.67	−25.63	−18.57
CMCC-CM (RCP8.5)	690.96	204.70	0.30	−6.56	−25.68	−20.47
MPI-ESM-MR (RCP4.5)	766.74	266.63	0.35	3.69	−3.20	−6.64
MPI-ESM-MR (RCP8.5)	756.81	248.18	0.33	2.35	−9.90	−11.97

4.3. Future Evaporation and Runoff Conditions under Climate Changes

The annual evaporation and runoff of the 21 scenarios in the two future periods are predicted by the SWAT model using the downscaled daily precipitation and temperature data. The predicted evaporation and runoff percentage differences relative to the 1980–2004 period are shown in Figures 11–14, respectively. The horizontal dashed lines in the figures refer to the maximum and minimum multi-year mean evaporation and runoff change percentages, which can be calculated by Equations (8) and (9), show the ranges of change uncertainty.

Figure 11 shows the annual evaporation changes under CMIP3 scenarios. The annual mean change percentage ranges from −1.09% to 5.50% in 2016–2040. The range is slightly wider in 2041–2065, which ranges from −0.85% to 7.21%. Figure 12 shows the annual evaporation changes under CMIP5 scenarios. The annual mean change percentage ranges from −2.05% to 8.90% in 2016–2040. The range is slightly smaller in 2041–2065, which ranges from 5.10% to 12.55%. Compared to the results in 2016–2040, the annual evaporation change percentages are similar under CMIP3 but much larger under CMIP5. Future annual evaporation all increase except CSIRO_MK3.0 (B1) for two future periods and CMCC-CM (RCP4.5) for 2016–2040. Overall, future evaporation changes range from −2.05% to 8.90% in 2016–2040 and −2.70% to 12.55% in 2041–2065 relative to 1980–2004 period on the basis of the selected scenarios. The increase in future evaporation is mainly caused by the rising temperature.

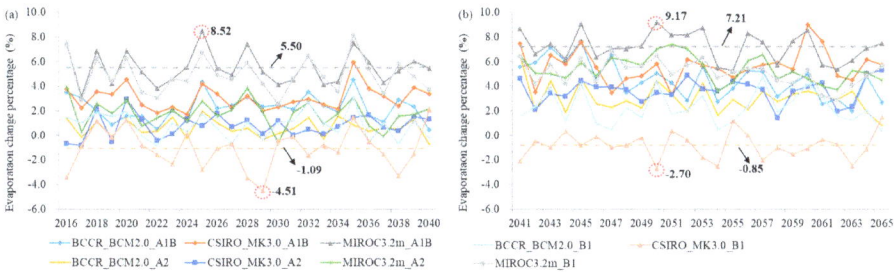

Figure 11. The forecasted evaporation change percentage relative to 1980–2004 period under CMIP3: (**a**) 2016–2040; (**b**) 2041–2065.

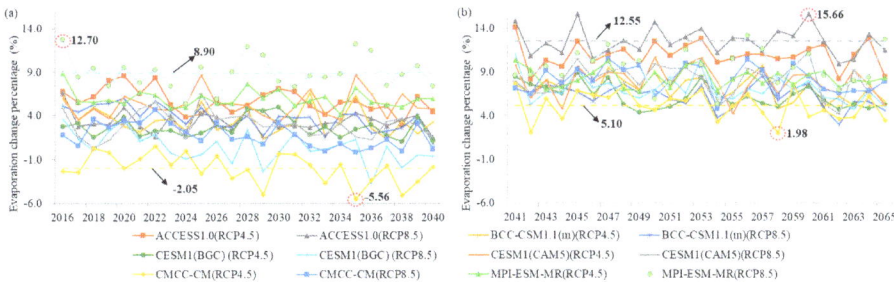

Figure 12. The forecasted evaporation change percentage relative to 1980–2004 period under CMIP5: (**a**) 2016–2040; (**b**) 2041–2065.

Figure 13 shows the annual runoff changes under CMIP3 scenarios. The annual mean change percentage ranges from −13.35% to 46.59% in 2016–2040. The range is slightly smaller in 2041–2065, which ranges from −19.39% to 25.47%. The difference of MIROC3.2m (B1) in the two future periods is the greatest, followed by MIROC3.2m (A1B) and BCCR_BCM2.0 (A2). The other six scenarios differ by less than ±10%.

Figure 14 shows the annual runoff changes under CMIP5 scenarios. The annual mean change percentage ranges from −37.97% to 37.60% in 2016–2040. The range is significantly wider in 2041–2065, which ranges from −24.05% to 90.16%. Compared to the results in 2016–2040, the runoff change percentage in 2041–2065 decreases most under MPI-ESM-MR (RCP8.5), which is also the only scenario under which precipitation decreases. Meanwhile, runoff slightly decreases under BCC-CSM1.1(m) (RCP4.5), CMCC-CM (RCP8.5), and MPI-ESM-MR (RCP4.5), while under the other eight scenarios runoff increases. Overall, future runoff changes range from −37.97% to 46.59% in 2016–2040 and −24.05% to 90.16% in 2041–2065 relative to 1980–2004 period on the basis of the selected scenarios.

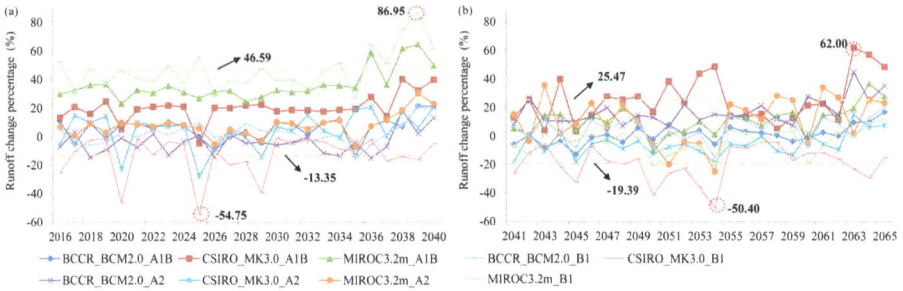

Figure 13. The forecasted runoff change percentage relative to 1980–2004 period under CMIP3: (**a**) 2016–2040; (**b**) 2041–2065.

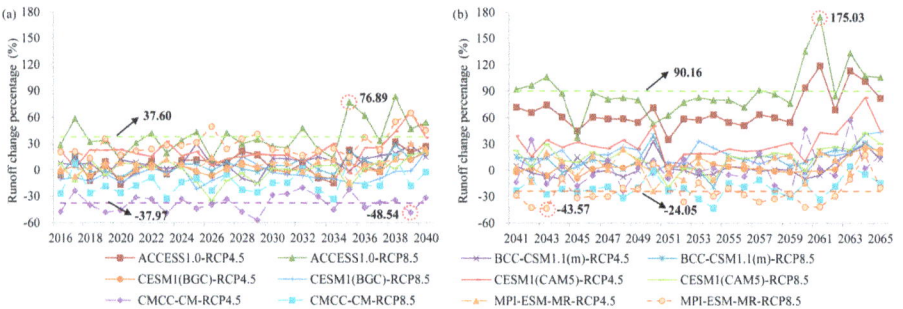

Figure 14. The forecasted runoff change percentage relative to 1980–2004 period under CMIP5: (**a**) 2016–2040; (**b**) 2041–2065.

This implies that changes in future runoff are greater than changes in precipitation under both CMIP3 and CMIP5. However, future runoff changes are relatively similar in the near future and far future under CMIP3, but are much more severe in the far future than in the near future under CMIP5. In addition, the ranges of change in runoff under CMIP5 are much wider than under CMIP3.

Among A1B, A2, and B1, the change of runoff from largest to smallest is A1B > A2 > B1 under CSIRO_MK3.0 in both future periods. However, the pattern is not obvious under the other two models. In addition, it is found that half of the models have larger change percentages under RCP8.5 while the other half have larger change percentages under RCP4.5. The situation is just like that of the precipitation changes, but with different change percentages. This implies that climate emission scenarios are also not so closely associated with runoff.

The forecasted long series (2016–2065) of runoff is compared with the baseline series (1958–2011) in order to thoroughly analyze the future runoff changes which are more practical for decision makers. Compared to the historical mean annual runoff (275.43 mm), the mean annual runoff decreases under five CMIP3 scenarios while it increases under the other four scenarios. Specifically, the scenarios under which mean annual runoff increases include MIROC3.2m for A1B, A2, and B1, and CSIRO_MK3.0 (A1B). The forecasted mean annual runoff increases by up to 11.20% under MIROC3.2m (A1B), while it decreases by up to 21.18% under CSIRO_MK3.0 (B1); this is because the precipitation under these scenarios increases the most and decreases the most respectively. Among the 12 CMIP5 scenarios, runoffs increase in the ACCESS1.0 model for RCP4.5 and RCP8.5, as well as the CESM1 (CAM5) model for RCP4.5, while the other scenarios all have a decline in runoff. Runoff increases by up to 20.34% and 44.64% under the ACCESS1.0 model for RCP4.5 and RCP8.5 respectively, mainly because their precipitation increases the most. On the other hand, runoffs of the CMCC-CM model decrease by up to 25.63% and 25.68% for RCP4.5 and RCP8.5 respectively, which is because the precipitation decreases the most. A comparison of the results is shown in Table 2. Overall, compared to the baseline series, the runoff decreases under two-thirds of the selected scenarios including five CMIP3 scenarios and nine CMIP5 scenarios.

5. Discussion

The morphing results are affected by the characteristics of the 'baseline climate'. However, the morphing results meet the downscaling requirements in this study; in addition, the method is simple and requires only a small computation volume. That is the reason why morphing is utilized in this study.

Compared to 1980–2004 period, both temperature and precipitation increase under most of the scenarios, and they both increase more in the far future than in the near future under both CMIP3 and CMIP5. Moreover, the temperature increases more under CMIP5 than under CMIP3 scenarios. Several studies have shown that CMIP5 presents warmer and wetter predictions [19,20]. Meanwhile, the ranges of change in both precipitation and runoff under CMIP5 are much wider than under CMIP3, which indicates that the selected CMIP5 scenarios are more effective in capturing future runoff uncertainties than the selected CMIP3 scenarios. In addition, the results imply that climate emission scenarios are closely associated with temperature but not so closely associated with precipitation and runoff.

As shown in Table 2, the runoff changes differ from precipitation changes. In most scenarios (15 of the 21 chosen scenarios), the decline in precipitation results in a more significant decline in runoff, while the increase in precipitation results in less increase in runoff. This implies that whether the precipitation increases or decreases, part of the water is failed to form runoff. There are six exceptions when the annual precipitation is more than 790 mm and the increase in runoff is larger than that of precipitation. The six exceptions include three CMIP3 scenarios (CSIRO_MK3.0 (A1B), MIROC3.2m (A1B), and MIROC3.2m (B1)) and three CMIP5 scenarios (the ACCESS1.0 model for RCP4.5, RCP8.5, and CESM1(CAM5) (RCP4.5)), which are the same as the six runoff increase scenarios. Precipitation supplements the soil moisture content when precipitation reaches or exceeds a certain amount. After the soil moisture content is saturated, most of the remaining precipitation forms runoff [31]. The runoff coefficient of precipitation thus increases (as show in Table 2), which may result in the larger increase in runoff than precipitation.

The relatively low flow conditions of the 1980–2004 period may partly result in the decline in future runoff when compared to the baseline series. Three impact factors are further analyzed. First, Yushi Reservoir supplies water to the outside of the basin, about 46 million m^3 each year (about 22.1 mm). This is one cause of the decline in runoff. Second, compared to 1980–2004 period, the simulated future evaporation increased under all the selected scenarios in both future periods except CSIRO_MK3.0 (B1) and CMCC-CM (RCP4.5) with slightly decreases. Evaporation increases greater in the far future than in the near future, which is the same to the temperature increase pattern. The increase in evaporation caused by the rising temperature is another cause of the decline in future runoff. Third, on the basis of water balance in the basin, the upstream water consumption, such as farmers getting water from the stream for irrigation, may be another factor which affects future runoff. Unlike the former two factors, the amount of upstream water consumptions is relatively small, and the impact on runoff is not obvious in these scenarios.

Acknowledging the uncertainty of the simulated runoff using the GCMs in the future [13,20,21,38], it is hard to give an accurate estimate about how much water the reservoir will have. However, the results in this study do show that the runoff in the future could decrease. This decrease could provide important information for the long-term planning of the development of this region. At present, the study region has a few factories that consume much water every year, and the water demand may increase in the future. In addition, the water used for irrigation could also increase in the future. The decreasing runoff (as shown in this study) suggests that this region may suffer from water shortages if industrial and irrigation demands keep increasing in the future. This information is beneficial for regional sustainable development, and indicates that possible adaption measures (e.g., building water transfer projects) should be taken in advance to adapt to climate change.

6. Conclusions

This study investigated the variations of future climate and water resources availability in an important river basin (the Biliu River basin) in northeast China. We found that both temperature and precipitation increase under most of the CMIP3 and CMIP5 scenarios, and CMIP5 shows higher temperature and wider ranges of changes in precipitation and runoff than CMIP3. We also found that the evaporation could increase and the runoff could decrease in the future. The decline in runoff may aggravate water shortages in the Biliu River basin, which may influence the water security in nearby big cities. The results provide very important information for water resources planning and management in this region and could have important implications to the regional sustainable development.

The uncertainties caused by the downscaling method and hydrological model, which are not considered in this article, require further research. In addition, the impact of human activities on future runoff, as well as the comprehensive impact of change conditions, also requires further research. Furthermore, it is necessary to combine these results with the analysis of reservoir regulation rules in order to use this information to make an informed decision about the future water availability for this reservoir.

Acknowledgments: This study was supported by the National Natural Science Foundation of China (grant No. 51509176), Natural Science Foundation of Shanxi Province, China (grant No. 201601D021086 and grant No. 2016011054), and Project of Hydrology Bureau, Shanxi Province (grant No. ZNGZ2015-036). The authors would like to thank the editors and reviewers for their valuable comments and suggestions.

Author Contributions: Xueping Zhu and Chi Zhang conceived and defined the research themes. Xueping Zhu and Wei Qi designed the methods and modeling and conducted the modeling investigations. Xueping Zhu and Wenjun Cai analyzed the data; Wei Qi, Xuehua Zhao, and Xueni Wang contributed to the discussion and analyses. All authors have contributed to the revision and approved the manuscript.

Conflicts of Interest: The authors declare no conflict of interest.

References

1. Menzel, L.; Bürger, G. Climate change scenarios and runoff response in the Mulde catchment (Southern Elbe, Germany). *J. Hydrol.* **2002**, *267*, 53–64. [CrossRef]
2. Hawkins, E.; Sutton, R. The potential to narrow uncertainty in regional climate predictions. *Bull. Am. Meteorol. Soc.* **2009**, *90*, 1095–1107. [CrossRef]
3. Woldemeskel, F.M.; Sharma, A.; Sivakumar, B.; Mehrotra, R. A framework to quantify GCM uncertainties for use in impact assessment studies. *J. Hydrol.* **2014**, *519*, 1453–1465. [CrossRef]
4. Intergovernmental Panel on Climate Change (IPCC). *Climate Change 2007: The Physical Science Basis. Contribution of Working Group I to the Fourth Assessment Report of the Intergovernmental Panel on Climate Change*; Cambridge Universty Press: New York, NY, USA, 2007.
5. Sivakumar, B. Global climate change and its impacts on water resources planning and management: Assessment and challenges. *Stoch. Environ. Res. Risk Assess.* **2011**, *25*, 583–600. [CrossRef]
6. Towler, E.; Rajagopalan, B.; Gilleland, E.; Summers, R.S.; Yates, D.; Katz, R.W. Modeling hydrologic and water quality extremes in a changing climate: A statistical approach based on extreme value theory. *Water Resour. Res.* **2010**, *46*, W11504. [CrossRef]
7. Belcher, S.E.; Hacker, J.N.; Powell, D.S. Constructing design weather data for future climates. *Build. Serv. Eng. Res. Technol.* **2005**, *26*, 49–61. [CrossRef]

8.   Milzow, C.; Burg, V.; Kinzelbach, W. Estimating future ecoregion distributions within the Okavango Delta Wetlands based on hydrological simulations and future climate and development scenarios. *J. Hydrol.* **2010**, *381*, 89–100. [CrossRef]

9.   Chen, S.T.; Yu, P.S.; Tang, Y.H. Statistical downscaling of daily precipitation using support vector machines and multivariate analysis. *J. Hydrol.* **2010**, *385*, 13–22. [CrossRef]

10. Zhang, C.; Zhu, X.P.; Fu, G.T.; Zhou, H.C.; Wang, H. The impacts of climate change on water diversion strategies into a water deficit reservoir. *J. Hydroinform.* **2014**, *16*, 872–889. [CrossRef]

11. Taylor, K.E.; Stouffer, R.J.; Meehl, G.A. A summary of the CMIP5 experiment design. *Bull. Am. Meteorol. Soc.* **2012**, *93*, 485–498. [CrossRef]

12. Brekke, L.D.; Dettinger, M.D.; Maurer, E.P.; Anderson, M. Significance of model credibility in estimating climate projection distributions for regional hydroclimatological risk assessments. *Clim. Chang.* **2008**, *89*, 371–394. [CrossRef]

13. Hoang, L.P.; Lauri, H.; Kummu, M.; Koponen, J.; van Vliet, M.T.H.; Supit, I.; Leemans, R.; Kabat, P.; Ludwig, F. Mekong River flow and hydrological extremes under climate change. *Hydrol. Earth Syst. Sci.* **2016**, *20*, 3027–3041. [CrossRef]

14. Donat, M.G.; Lowry, A.L.; Alexander, L.V.; O'Gorman, P.A.; Maher, N. More extreme precipitation in the world's dry and wet regions. *Nat. Clim. Chang.* **2016**, *6*, 508–513. [CrossRef]

15. Wang, G.L.; Wang, D.G.; Trenberth, K.E.; Erfanian, A.; Yu, M.; Bosilovich, M.G.; Parr, D.T. The peak structure and future changes of the relationships between extreme precipitation and temperature. *Nat. Clim. Chang.* **2017**, *7*, 268–274. [CrossRef]

16. Wang, X.Y.; Yang, T.; Li, X.L.; Shi, P.F.; Zhou, X.D. Spatio-temporal changes of precipitation and temperature over the Pearl River basin based on CMIP5 multi-model ensemble. *Stoch. Environ. Res. Risk Assess.* **2017**, *31*, 1077–1089. [CrossRef]

17. Koutroulis, A.G.; Grillakis, M.G.; Tsanis, I.K.; Papadimitriou, L. Evaluation of precipitation and temperature simulation performance of the CMIP3 and CMIP5 historical experiments. *Clim. Dyn.* **2016**, *47*, 1881–1898. [CrossRef]

18. Sun, Q.; Miao, C.; Duan, Q. Comparative analysis of CMIP3 and CMIP5 global climate models for simulating the daily mean, maximum, and minimum temperatures and daily precipitation over China. *J. Geophys. Res. Atmos.* **2015**, *120*, 4806–4824. [CrossRef]

19. Ayers, J.; Ficklin, D.L.; Stewart, I.T.; Strunk, M. Comparison of CMIP3 and CMIP5 projected hydrologic conditions over the Upper Colorado River Basin. *Int. J. Climatol.* **2016**, *36*, 3807–3818. [CrossRef]

20. Ficklin, D.L.; Letsinger, S.L.; Stewart, I.T.; Maurer, E.P. Assessing differences in snowmelt-dependent hydrologic projections using CMIP3 and CMIP5 climate forcing data for the western United States. *Hydrol. Res.* **2016**, *47*, 483–500. [CrossRef]

21. Shamir, E.; Megdal, S.B.; Carrillo, C.; Castro, C.L.; Chang, H.I.; Chief, K.; Corkhill, F.E.; Eden, S.; Georgakakos, K.P.; Nelson, K.M.; et al. Climate change and water resources management in the Upper Santa Cruz River, Arizona. *J. Hydrol.* **2015**, *521*, 18–33. [CrossRef]

22. Su, F.; Zhang, L.; Ou, T.; Chen, D.; Yao, T.; Tong, K.; Qi, Y. Hydrological response to future climate changes for the major upstream river basins in the Tibetan Plateau. *Glob. Planet. Chang.* **2016**, *136*, 82–95. [CrossRef]

23. Zhang, C.; Shoemaker, C.A.; Woodbury, J.D.; Cao, M.L.; Zhu, X.P. Impact of human activities on stream flow in the Biliu River basin, China. *Hydrol. Process.* **2013**, *27*, 2509–2523. [CrossRef]

24. Raje, D.; Mujumdar, P.P. Reservoir performance under uncertainty in hydrologic impacts of climate change. *Adv. Water Resour.* **2010**, *33*, 312–326. [CrossRef]

25. Vicuna, S.; Dracup, J.A.; Lund, J.R.; Dale, L.L.; Maurer, E.P. Basin-scale water system operations with uncertain future climate conditions: Methodology and case studies. *Water Resour. Res.* **2010**, *46*, W04505. [CrossRef]

26. Paiva, R.C.D.; Durand, M.T.; Hossain, F. Spatiotemporal interpolation of discharge across a river network by using synthetic SWOT satellite data. *Water Resour. Res.* **2015**, *51*, 430–449. [CrossRef]

27. Matin, M.A.; Bourque, C.P.A. Intra- and inter-annual variations in snow-water storage in data sparse desert-mountain regions assessed from remote sensing. *Remote Sens. Environ.* **2013**, *139*, 18–34. [CrossRef]

28. Glenn, J.; Tonina, D.; Morehead, M.D.; Fiedler, F.; Benjankar, R. Effect of transect location, transect spacing and interpolation methods on river bathymetry accuracy. *Earth Surf. Process. Landf.* **2016**, *41*, 1185–1198. [CrossRef]

29. CLIVAR Exchanges, World Climate Research Programme (WCRP). *WCRP Coupled Model Intercomparison Project—Phase 5 (CMIP5)*; CLIVAR Exchanges No. 56, Vol. 16, No. 2; Indigo Press: Southampton, UK, 2011.

30. Van Vuuren, D.P.; Edmonds, J.; Kainuma, M.; Riahi, K.; Thomson, A.; Hibbard, K.A.; Hurtt, G.C.; Kram, T.; Krey, V.; Lamarque, J.F.; et al. The representative concentration pathways: An overview. *Clim. Chang.* **2011**, *109*, 5–31. [CrossRef]

31. Neitsch, S.L.; Arnold, J.G.; Kiniry, J.R.; Williams, J.R. *Soil and Water Assessment Tool Theoretical Documentation*; Version 2005; US Department of Agriculture (USDA) Agricultural Research Service (ARS): Temple, TX, USA, 2005.

32. Arnold, J.G.; Srinivasan, R.; Muttiah, R.S.; Williams, J.R. Large area hydrologic modeling and assessment. Part 1-Model development. *J. Am. Water Resour. Assoc.* **1998**, *34*, 1–17. [CrossRef]

33. U.S. Environmental Protection Agency (EPA). *Protocols for Developing Nutrient TMDLs*; Office of Water EPA 841-B-99-007; U.S. Environmental Protection Agency (EPA): Washington, DC, USA, 1999.

34. Marshall, E.; Randhir, T.O. Spatial modeling of land cover change and watershed response using Markovian cellular automata and simulation. *Water Resour. Res.* **2008**, *44*, W04423. [CrossRef]

35. Li, Z.; Liu, W.Z.; Zhang, X.C.; Zheng, F.L. Impacts of land use change and climate variability on hydrology in an agricultural catchment on the Loess Plateau of China. *J. Hydrol.* **2009**, *377*, 35–42. [CrossRef]

36. Van Griensven, A.; Meixner, T.; Grunwald, S.; Bishop, T.; Diluzio, M.; Srinivasan, R. A global sensitivity analysis tool for the parameters of multi-variable catchment models. *J. Hydrol.* **2006**, *324*, 10–23. [CrossRef]

37. Hao, F.H.; Cheng, T.G.; Yang, S.T. *Non-Point Source Pollution Model*; China Environmental Science Press: Beijing, China, 2006.

38. Mcsweeney, C.F.; Jones, R.G.; Booth, B.B.B. Selecting ensemble members to provide regional climate change information. *J. Clim.* **2012**, *25*, 7100–7121. [CrossRef]

Article

Simulating the Evolution of the Land and Water Resource System under Different Climates in Heilongjiang Province, China

Qiuxiang Jiang, Youzhu Zhao, Zilong Wang *, Qiang Fu, Tian Wang, Zhimei Zhou and Yujie Dong

School of Water Conservancy and Civil Engineering, Northeast Agricultural University, Harbin 150030, China;
jiangqiuxiang2017@163.com (Q.J.); zhaoyouzhu1993@163.com (Y.Z.); fuqiang@neau.edu.cn (Q.F.);
15604610738@163.com (T.W.); 18345062143@139.com (Z.Z.); DYJ1994727@126.com (Y.D.)
* Correspondence: wangzilong2017@126.com; Tel.: +86-0451-5519-1534

Received: 11 May 2018; Accepted: 26 June 2018; Published: 29 June 2018

Abstract: Heilongjiang Province is under the pressure of a water shortage due to climate change, population growth and economic development. To effectively manage regional land and water resources, this paper describes a system dynamics model that was built to simulate the interaction between land and water resources and socioeconomic factors, as well as the evolution of regional land and water resources in different climates in Heilongjiang Province. The results show that the declining trend of unused land area and the water supply–demand ratio will not stop, even under the most optimistic (e.g., humid climate) climate conditions, if the current land use patterns continue. Therefore, measures should be taken to manage the unreasonable usage patterns of land and water resources in this region. This study simulated the evolution of regional land and water resources for five scenarios under an arid climate by changing the net irrigation quota for paddy fields, the water quota for industrial use, forestland area, annual change rate of farmland area, and the growth rate of the gross industrial output value. Further, a combined scenario that can maximally reduce the regional land and water resource sustainable risk was identified. The simulation of the combined scenario showed that it can effectively increase the degree of regional land and water resource use in the region, as well as reduce the risks that threaten these resources. This study provides theoretical support for the efficient use of land and water resources in the future.

Keywords: land and water resources; system dynamics; modeling; scenario analysis; Heilongjiang

1. Introduction

Land and water resources are important for human survival; they not only show a region's resource endowment, but also determine the region's agricultural and industrial development patterns [1–3]. In recent years, the evolution of regional land and water resources has become more dynamic and complex than ever due to climate change, relevant policies, and economic development. More and more scholars have been working on sustainable management of land and water resource systems.

System dynamics (SD) is a method developed by Forrester et al. [4] to study the structure of a system. This method establishes a systematic model with a series of interrelated and feedback variables to study a system's past, present, and future. This model can be used to reproduce and analyze the dynamic behaviors of a system [5–7]. With the increasing complexity, dynamics, and variability of modern society, approaches based on a comprehensive system theory (e.g., SD, cybernetics and information theory) help people to understand nonlinear and time-varying phenomena [8]. The present study uses the SD model because the land and water resource system is the result of the interplay of social, ecological, and economic factors [9], and the model can help us to analyze the evolution of the land and water resource system in the long term [10,11].

At present, land and water resource systems are often used as an important factor of a subsystem in regional environmental [12] or agricultural systems [13]. In a regional environmental system, water and land resources can be used as key components of the resource subsystem; by analyzing the production and consumption of land and water resources, the overall impact of the resource subsystem on the regional environmental system can be analyzed. In agricultural systems, which are based on land and water resources, SD can be used to simulate the dynamic changes in land and water resources to promote the dynamic development of agricultural systems, evaluate the agricultural system as a whole, and analyze the rationality of agricultural models to guarantee the development of agricultural systems.

This paper describes an SD model of the land and water resource system in Heilongjiang Province, which was built to simulate the evolution of the system under different conditions (e.g., economic policies, climate change) and can systematically analyze the land and water use patterns, analyze the interaction between the conditions and results, and identify the potential risks of using regional land and water resources in the region. This study can solve the conflicts between the demand (for future economic development and grain production) and the limited land and water resources, provide guidance for the sustainable use of regional land and water resources, and realize sustainable socioeconomic development.

2. The Study Area

Heilongjiang Province is located in northeastern China; the latitude ranges between 43°26′ and 53°33′ and longitude between 121°11′ and 135°05′ (Figure 1). The province has a temperate continental monsoon climate with high summer temperatures, abundant rainfall, a long winter, and cold and dry weather. The annual sunshine duration is approximately 2300–2800 h, the annual average temperature is approximately −7.9 to 7 °C, and the annual rainfall is approximately 360–830 mm; the solar energy resource is abundant. With its vast territory, the province had farmland area accounting for 11.75% of the country's total in 2015 and is an important grain production base.

Figure 1. Heilongjiang Province location map.

In recent years, the Chinese government has adopted a series of economic policies which have helped the economic development of the province, and the province's grain production had increased over a consecutive 12-year period, from 2003 to 2015. However, with the economic development and grain production increase, the province presented a sharp increase in water demand and a trend of land resource overdevelopment, resulting in an imbalance between the supply and demand of regional land and water resources. Moreover, the imbalance curbed the region's economic development and grain production. The "12 consecutive years of increase in grain production" stopped in 2016.

3. The Model and Data Sources

3.1. System Dynamics

System dynamics has enormous advantages for the analysis of complex dynamic feedback systems. This advantage makes it widely used in nature, resources, society, engineering, and other fields, and its application requires software support. With the development and popularization of computer technology, system dynamics software has been greatly improved. From many versions of the software, this study selected Vensim-DSS, which was developed by Ventana Corporation in the United States. This software can set the flowchart, the writing equation, the feedback loop analysis, and the output graph form. This study used Vensim to model the simulation flowchart and used the Equations function to assign calculation equations and parameters to various variables, and then used the Run a Simulation function to express the changes in each variable in the form of a chart. Additionally, structural analysis and data set analysis of the model were performed, and the changes in other characteristic variables were observed by changing one of the variables; the operating results were compared to provide a comprehensive analysis of the plan.

3.2. Constructing the Model

To analyze and predict the evolution of the land and water resources in Heilongjiang Province, this study built an SD model (Figure 2) including four subsystems: a population subsystem, water subsystem, land subsystem, and economic subsystem. The study time ranges from 2000 to 2030 with 1 year as the time step. The historical years range from 2000 to 2015, and the planning years range from 2016 to 2030. Four subsystems were also introduced in the model: the population subsystem, land subsystem, water subsystem, and economic subsystem. The subsystem interconnection overview is shown in Figure 2, and the workflow with more details of the subsystems is given in Figures 3–6.

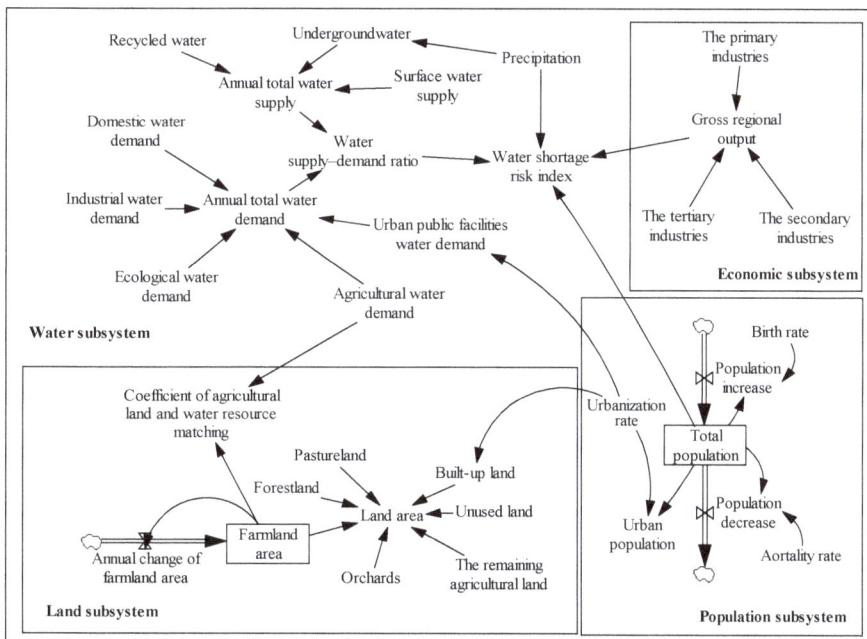

Figure 2. Subsystem interconnection overview.

The population size and structure of the population subsystem are the main driving forces for regional water consumption and land use type conversion. For example, facing the expansion of urban population, the government needs to convert farmland and unused land to built-up land or to reduce the water supply for agricultural land to meet the water demand in urban areas (Figure 3). The parameter that determines the population size and structure is total population, which is calculated using Equation (1)

$$\text{Total population}_{t+1} = \text{Total population}_t + \text{Population increase}_t - \text{Population decrease}_t \quad (1)$$

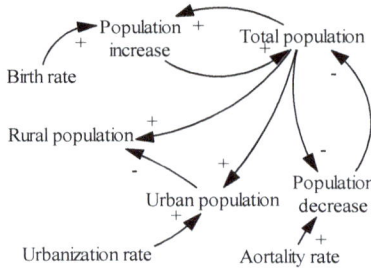

Figure 3. Workflow for the population subsystem.

In the water subsystem (Figure 4), the key variable is the water supply–demand ratio, i.e., the ratio of annual total water supply to annual total water demand; the annual total water supply and annual total water demand are presented as follows:

$$\text{Annual total water supply} = \text{Surface water supply} + \text{Groundwater} + \text{Recycled water} \quad (2)$$

$$\text{Annual total water demand} = \text{Agricultural water demand} + \text{Industrial water demand} + \text{Ecological water demand} + \text{Domestic water demand} + \text{Urban public facilities water demand} \quad (3)$$

To quantitatively evaluate the status quo of the region's water resources, this paper introduced characteristic variables (such as the water shortage risk index (WSRI), water use efficiency index (WUI), and coefficient of agricultural land and water resource matching (CA)) into the water subsystem.

The water shortage risk index depends on a range of factors, as in Equation (4) [14].

$$\text{WSRI} = \frac{\text{Precipitation coefficient} \times \text{Total population} \times \sqrt{\text{Gross region output}}}{\text{Annual total water supply} + \text{Diverter water volume of transit water}} \quad (4)$$

The water use efficiency index reflects the region's water use efficiency and its water-saving potential, which is calculated using Equation (5) [15].

$$\text{WUI} = \frac{\frac{\text{Gross agriculture output}}{\text{Gross region output}} \times \text{Agriculture water demand ratio} + \frac{\text{Gross industry output}}{\text{Gross region output}} \times \text{Industry water demand ratio}}{\text{Agriculture water demand ratio} + \text{Industry water demand ratio}} \quad (5)$$

The coefficient of agricultural land and water resource matching represents the matching of the region's agricultural land and water resources, as in Equation (6) [16].

$$\text{CA} = \frac{\text{Annual total water supply} \times \text{Agricultural water demand ratio}}{\text{Farmland area}} \quad (6)$$

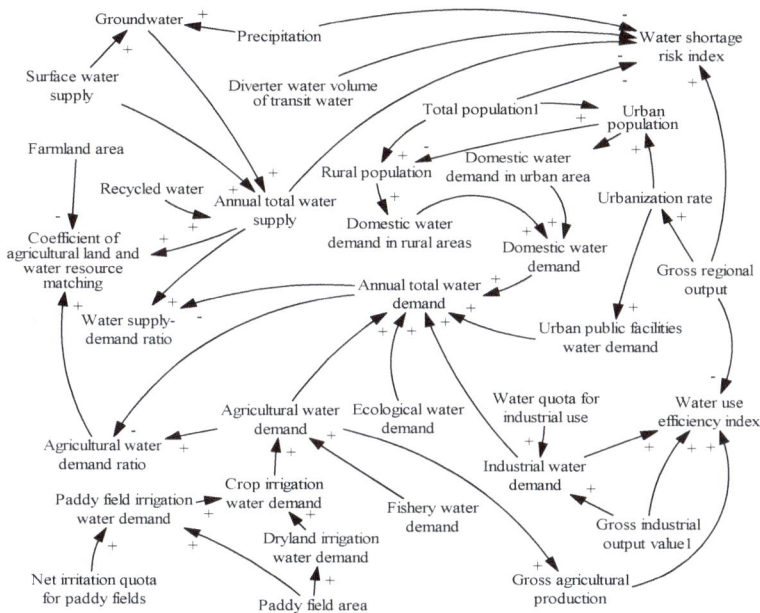

Figure 4. Workflow for the water subsystem.

The land subsystem (Figure 5) divides the region's land area into seven land use types: farmland, orchards, built-up land, forestland, the remaining agricultural land, pastureland, and unused land. Unused land is determined by the other land use types, as shown in Equation (7)

$$\text{Unused land} = \text{Land area} - \text{Orchards} - \text{Built-up land} - \text{Forestland} - \text{Pastureland} - \text{Farmland area} - \text{The remaining agricultural land} \tag{7}$$

The structure of land use is determined by many factors. Therefore, this paper introduces characteristic variables, such as the comprehensive index of the land use degree (CILU), land use diversity index (LUDI), and land use ecological risk index (LUERI).

The comprehensive index of the land use degree measures the land use degree level and the effects of interactions between humans and nature, which is calculated using Equation (8) [17].

$$\text{CILU} = \frac{\text{Unused land} + 2 \times (\text{Forestland} + \text{Pastureland} + \text{Waters}) + 3 \times (\text{Orchards} + \text{Farmland area}) + 4 \times \text{Built-up land}}{\text{Land area}} \tag{8}$$

The land use diversity index represents the diversity of land use types, as shown in Equation (9) [18].

$$\text{LUDI} = -\sum_{i=1}^{7} (\text{Land use type area} \times \log_2(\text{Land use type area})) \tag{9}$$

There are seven land use types: farmland, orchards, built-up land, forestland, remaining agricultural land, pastureland, and unused land.

The land use ecological risk index represents the size and degree of damage caused to land when ecological disasters occur, which is calculated using Equation (10) [19].

$$\text{LUERI} = \sum_{i=1}^{7} \frac{\text{Land use type area} \times W_i}{\text{Land area}} \tag{10}$$

W_i is the ecological risk intensity parameter for the ith land use type, which can be found in Reference [19].

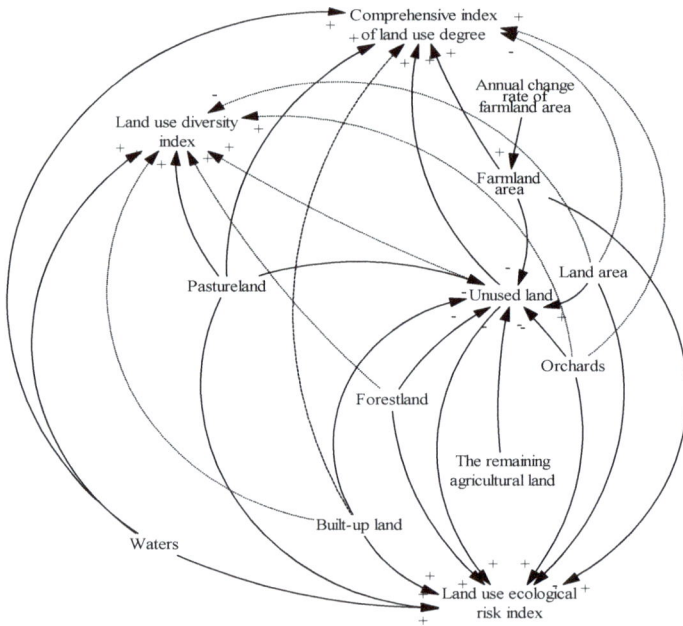

Figure 5. Workflow for the land subsystem.

An economy subsystem is crucial to analyze the regional development. This study used the gross regional output to measure the regional development level (Figure 6). The gross regional output is shown in Equation (11).

Gross regional output = The primary industries + The secondary industries + The tertiary industries   (11)

Without water-saving measures, economic growth always leads to an increase in water demand. Considering the province's agricultural and industrial development status quo, government may manage its water demand by reducing the water use per gross regional output unit.

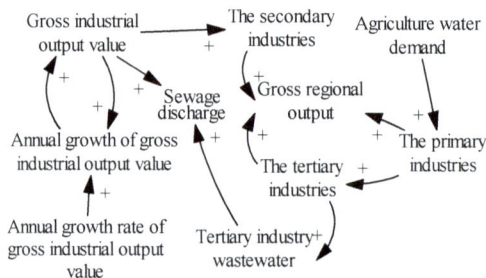

Figure 6. Workflow for the economic subsystem.

3.3. Data Sources

The historical data of the variables in the model were obtained from the China Statistical Yearbook [20] and Heilongjiang Statistical Yearbook [21]. The planning year data were obtained from the country and region's government documents, such as the Thirteenth Five-Year Plan of Heilongjiang Province [22] and the Comprehensive Land Use Planning of Heilongjiang Province [23]. The base year data on the development and use of land and water resources are the same as the status quo.

4. The Simulation Model and Results Analysis

Considering the impact of future climate scenarios on land and water resources [24–26], this study introduced three climate types (i.e., arid, humid, and normal) into the SD model (Vensim DSS software, Ventana Systems Inc, Wiltshire, UK). These climate types were based on precipitation data series from 1965 to 2015 to represent different climate types. An arid climate has a minimum precipitation of 382.9 mm (in year 2001); a normal climate has an average precipitation of 518.6 mm; and a humid climate has a maximum precipitation of 691.7 mm (in year 2013).

4.1. Model Testing

This work used historical data to test the precision and accuracy of the simulation (see Table 1). Comparing the predicted with the actual values, it can be seen that the difference is within ±10%; the relative errors are small. The simulation result can represent the actual situation. Therefore, the model can be used to predict the evolution of the land and water resource system in Heilongjiang Province.

4.2. Simulating the Evolution of the Land and Water Resource System under Different Climates

This study used precipitation to separate different climates (i.e., arid, humid, and normal) (Figure 7). Some variables changed dramatically over time (i.e., water supply–demand ratio, water shortage risk index, coefficient of agricultural land and water resource matching, unused land area, degree index, diversity index, and ecological risk index), while the water use efficiency index did not change much.

From Figure 3, it can be seen that the unused land area and water supply–demand ratio both declined under different climates. Under the humid climate, the ecological risk index and water shortage risk index tend to decrease, while the coefficient of agricultural land and water resource matching, degree index, and diversity index increase slightly. However, under other climates, the ecological risk index and water shortage risk index tend to increase, while the coefficient of agricultural land and water resource matching, degree index, and diversity index tend to decrease.

Therefore, even under the most optimistic condition (i.e., humid climate), the declining trend of the unused land area and water supply–demand ratio will not stop if the current land and water use patterns continue. Moreover, the land and water resources may face various challenges (e.g., ecological risk, shortage risk, mismatching, low efficiency and diversity) unless the humid climate continues in the future. Therefore, the region may face the inevitable crisis of land and water resource shortages unless it improves its current land and water use patterns.

4.3. Simulation of the Land and Water Resource System under Different Scenarios

The combination of different climates and resource use patterns can make the land and water resource system evolve in different directions. Although the future is uncertain, the overall direction will still achieve the desired goals.

The Thirteenth Five-Year Plan of Heilongjiang Province pointed out the following: Heilongjiang Province should have an advantage, focus on industry, optimize and upgrade the industrial structure and ownership structure; accelerate artificial afforestation, build a national reserve forest base;

and improve the effective coefficient of irrigative water utilization for farmland. The Comprehensive Land Use Planning of Heilongjiang Province pointed out that Heilongjiang Province should adhere to the most stringent farmland protection system, strictly protect basic farmland; increase the supply of farmland; strictly protect forestland, prohibit deforestation; actively develop water conservancy projects; and promote water-saving and drought-resistance technologies. Considering the possible temperature rise due to global warming, this study used the arid climate as an example to adjust the variables within each subsystem according to the regional development plan to simulate the setting of land and water resource-use strategies under different scenarios.

Table 1. The model verification results using historical data.

Index	Variable	Unit	2000	2002	2004	2006	2008	2010	2012	2014	2015
Annual total water supply	Predicted value	10^8 m^3	287	249	251	282	300	324	350	348	348
	Actual value	10^8 m^3	297	252	259	286	297	325	359	364	355
	Error	%	−3.48	−1.20	−3.19	−1.42	1.00	−0.31	−2.57	−4.60	−2.01
Groundwater resources	Predicted value	10^8 m^3	275	247	262	268	236	285	285	282	282
	Actual value	10^8 m^3	268	270	274	279	248	278	290	295	283
	Error	%	2.55	−9.31	−4.58	−4.10	−5.08	2.46	−1.75	−4.61	−0.35
Gross regional output	Predicted value	10^8 Yuan	2963	3375	4551	6233	8161	9976	14,398	15,247	15,440
	Actual value	10^8 Yuan	3151	3637	4751	6212	8314	10,369	13,692	15,039	15,490
	Error	%	−6.34	−7.76	−4.39	0.34	−1.87	−3.94	4.90	1.36	−0.32
Urbanizat-ion rate	Predicted value	%	52	52	52	53	55	56	57	58	58
	Actual value	%	52	53	53	53	55	56	57	58	59
	Error	%	0.00	−1.92	−1.92	0.00	0.00	0.00	0.00	0.00	−1.72

4.3.1. The Subsystems' Variable Selection and the Scenarios

In the water subsystem, reducing the water quota for industries can reduce the overall water demand, and the factors influencing the annual total water demand can be examined (Figure 8). As the figure shows, the trend of the annual total water demand matches that of the water demand for paddy fields. Therefore, the net irrigation quota can be used for paddy fields to manage the water quota for agricultural use. Furthermore, the region should also introduce advanced water-saving technologies to properly reduce the water quota for industrial use.

In the land subsystem, inappropriate land development may result in a decrease in the unused land area, the diversity index, and degree index but an increase in the ecological risk. Forestland accounts for the largest area of Heilongjiang Province, followed by farmland. Considering the importance of forestland and farmland, the region may effectively control the region's land development by controlling the conversion between the two land use types. Through the annual change rate of farmland area in the land subsystem, the government can control the conversion between forestland and farmland.

In the economic subsystem, this paper uses the gross regional output to represent the region's economic situation. Heilongjiang is largely an agricultural province and part of the industrial base of the country. Since agriculture is the region's base industry, changing agricultural land use types may lead to social chaos. Therefore, this study adjusted the region's economic factors without changing anything in the agricultural industry. This model adjusted the annual growth rate of the gross regional output to evaluate its impact on the gross regional output and the land and water resources.

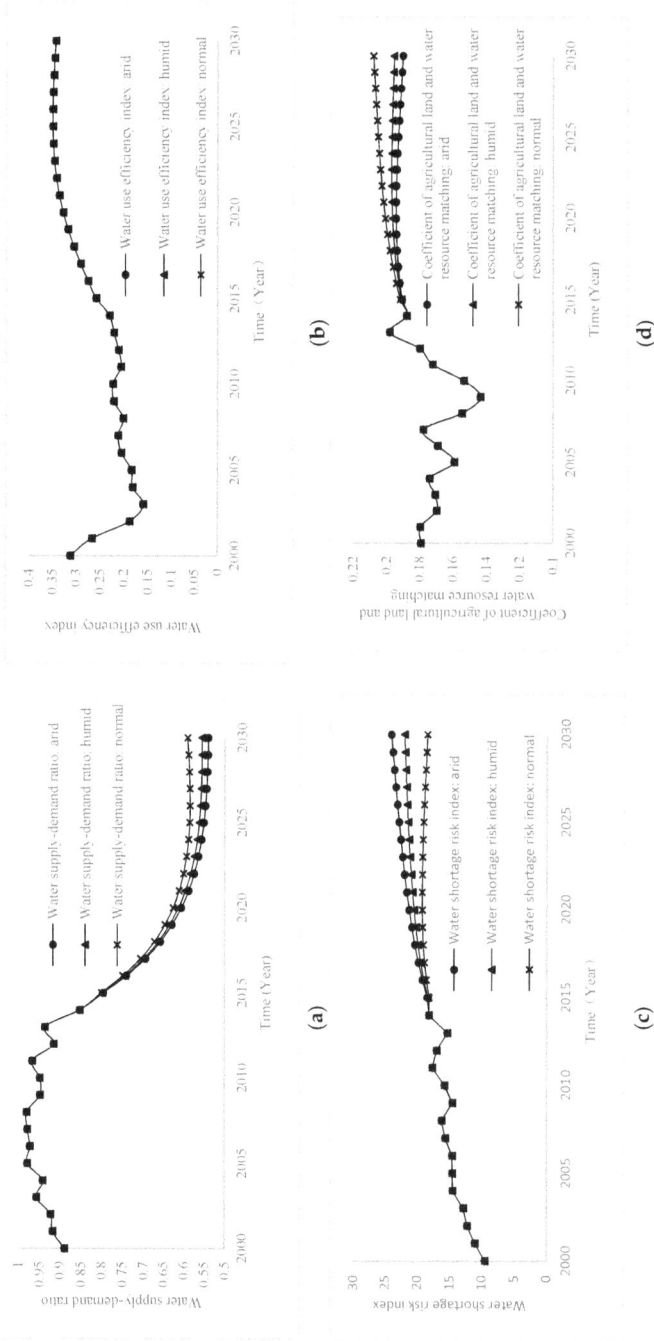

Figure 7. *Cont.*

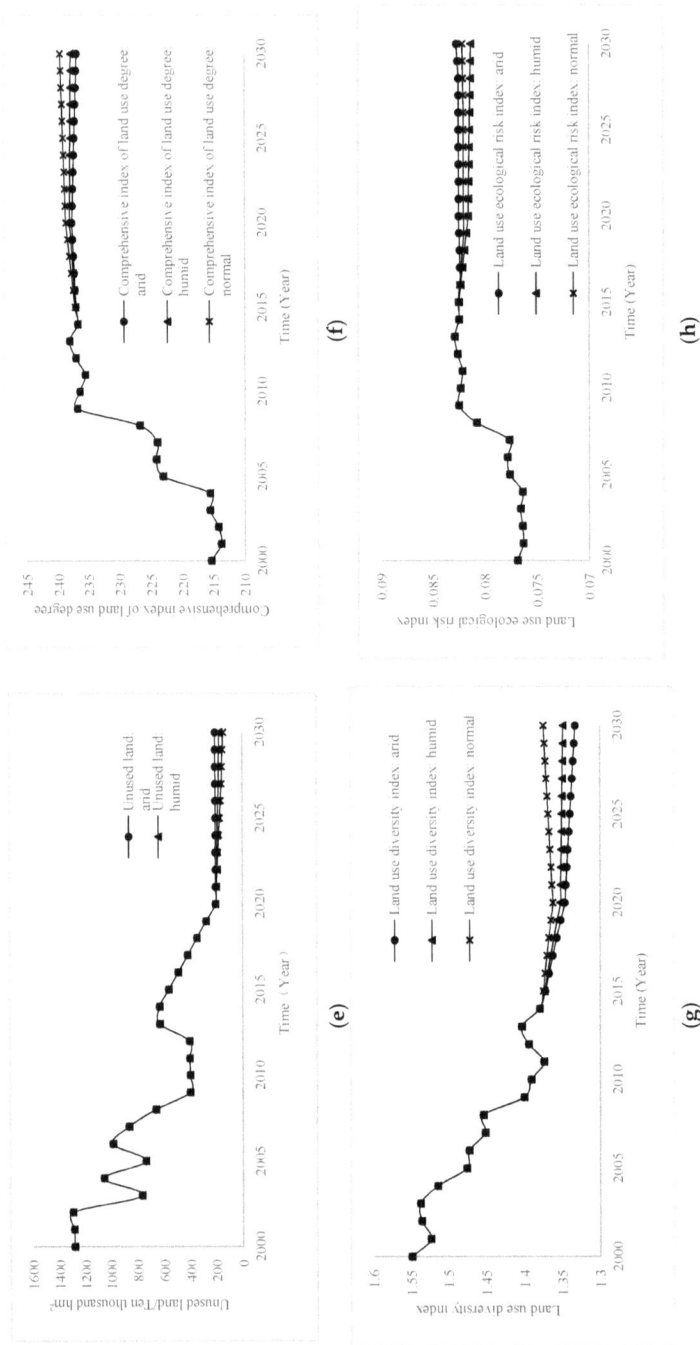

Figure 7. Simulation of land and water resource system dynamics model of Heilongjiang Province under different climate conditions. (**a**) Water supply-demand ratio; (**b**) Water use efficiency index; (**c**) Water shortage risk index; (**d**) Coefficient of agricultural land and water resource matching; (**e**) Unused land/Ten thousand hm²; (**f**) Comprehensive index of land use degree; (**g**) Land use diversity index; (**h**) Land use ecological risk index.

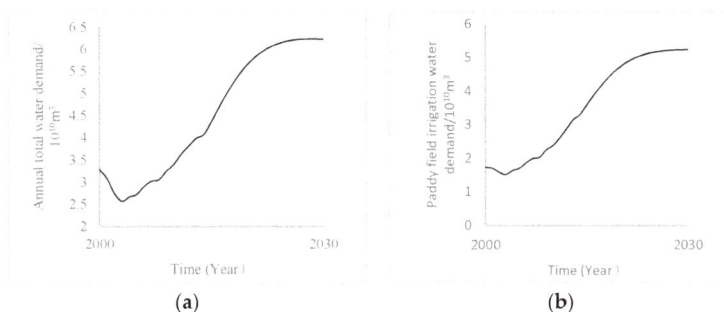

Figure 8. Influential factors of the annual total water demand. (**a**) Annual total water demand; (**b**) Paddy field irrigation water.

Based on the above analysis, this work selected the net irrigation quota for paddy fields, the water quota for industrial use, forestland area, annual change rate of farmland area, and annual growth rate of gross industrial output value to set different scenarios within a reasonable range, according to government programs at the planning horizon, to simulate the evolution of the land and water resource system in the planning years. This model also considered the province's existing problems and socioeconomic development status quo with respect to land and water resources to set the planning years' conditions. In Scenario 1, the model reduced the net irrigation quota for paddy fields and the water quota for industrial use, given the implementation of water-saving techniques. In Scenario 2, the model adopted the policy of returning farmland to forestland. In Scenario 3, the model converted an appropriate amount of forestland to farmland. In Scenario 4, the model adopted development policies to boost economic growth. In Scenario 5, the model slowed the economic development rate. There is more literature on setting different scenarios [27–33], and Table 2 shows the details used when setting each scenario in this paper.

Table 2. Different scenarios.

Variable	Scenario 1	Scenario 2	Scenario 3	Scenario 4	Scenario 5
Net irrigation quota for paddy fields (m^3/hm^2)	Decrease by 20%	No change	No change	No change	No change
Water quota for industrial use (m^3/hm^2)	Decrease by 20%	No change	No change	No change	No change
Forestland (10,000 hm^2)	No change	Increase by 20%	Decrease by 20%	No change	No change
Annual change rate of farmland area	No change	Decrease by 20%	Increase by 20%	No change	No change
Annual growth rate of gross industrial output value	No change	No change	No change	Maximum over the years	0

4.3.2. Simulations under Each Scenario

This model simulated the evolution of the land and water resource system for an arid climate under different scenarios to predict the change range of variables compared with the status quo in 2030 (Table 3).

In Scenario 1, the model decreased the area's water use for agricultural and industrial use, given advanced water-saving technologies, and thus increased the water supply–demand ratio, reduced the water shortage risk index, and increased the water use efficiency. This scenario has a more effective impact on the water subsystem but a less effective impact on the land subsystem.

In Scenario 2, the model slowed the development rate of unused land by adopting the policy of returning farmland to forestland and thus reduced the ecological risk index and increased the coefficient of agricultural land and water resource matching. However, the land use diversity and comprehensive land use degree indices decreased. This means that the policy of returning farmland to forestland can reduce the adverse effects of ecological disturbance, but this scenario may decrease the land use type and land use degree. Therefore, this scenario has a more effective impact on the land subsystems but a less effective impact on the water subsystem.

Table 3. Comparing the simulation results with the status quo under an arid climate.

Variable	Scenario 1	Scenario 2	Scenario 3	Scenario 4	Scenario 5
Unused land	-	2.85%	-	-	-
Land use diversity index	-	−2.60%	1.66%	-	-
Land use ecological risk index	-	−9.97%	8.92%	-	-
Comprehensive index of land use degree	-	−3.09%	2.67%	-	-
Water supply–demand ratio	20.10%	-	-	−0.10%	0.5%
Water use efficiency index	17.81%	-	-	−7.81%	3.59%
Water shortage risk index	−4.44%	-	-	33.22%	−3.22%
Coefficient of agricultural land and water resource matching	−2.35%	26.62%	−6.62%c	−6.62%	0.86%

Note: the symbol of "-"means keep invariant.

In Scenario 3, the model converted forestland to farmland and thus increased the land use diversity and degree indices. However, the ecological risk index increased, and the coefficient of agricultural land and water resource matching decreased. Therefore, this scenario has a more effective impact on the land subsystem but a less effective impact on the water subsystem.

In Scenario 4, the model greatly boosted the regional economy. From Table 3, it can be seen that emphasis on economic development can lead to an increase in water demand for economic use, thus decreasing the water supply–demand ratio, increasing the water shortage risk, and decreasing the coefficient of agricultural land and water resource matching. This scenario has a more effective impact on the water subsystem but a less effective impact on the land subsystem. It emphasizes economic growth but neglects the region's sustainable development.

In Scenario 5, the model decreased the economic development rate. Table 3 shows that decreasing the economic development rate can increase the water supply–demand ratio, reduce the water shortage risk, and increase the coefficient of agricultural land and water resource matching. This scenario has a more effective impact on the water subsystem but a less effective impact on the land subsystem. Furthermore, this scenario can greatly increase the environmental and ecological benefits while reducing the economic benefits.

From the above analysis, it can be seen that no single scenario can improve regional land and water resources in all aspects. Each scenario has its merits and demerits. The region needs to balance and complement different measures to properly improve regional land and water resources. Comparing the scenario results, it can be seen that combining Scenarios 1, 2, and 5 can minimize the evolution risk of regional land and water resources. Table 4 shows the result of the combined scenario in 2030.

Table 4. Comparing the status quo with the system dynamics (SD) model results under the combined scenario in 2030 under an arid climate.

Variable	Unit	Status Quo	Combined Scenario	Change Range
Unused land	10,000 hm²	207.4	251.9	21.46%
Land use diversity index	-	1.27	1.24	−2.05%
Land use ecological risk index	-	0.08	0.07	−9.78%
Comprehensive index of land use degree	-	235.9	228.1	−3.31%
Water supply–demand ratio	-	0.56	0.65	17.45%
Water use efficiency index	-	0.34	0.45	29.45%
Water shortage risk index	-	24.08	22.49	−6.60%
Coefficient of agricultural land and water resource matching	-	0.19	0.23	22.18%

As shown in Table 4, the development of land and water resources is acceptable under the combined scenario. The unused land area and water supply–demand ratio both increased by over 20%; this relieved the declining trend of the land and water resources. In addition, the ecological risk index and water shortage risk index decreased by more than 5%. The water use efficiency index and coefficient of agricultural land and water resource matching also increased by more than 20%. Besides the quantitative improvement, the combined scenario also promoted quality improvement in the system's sustainable development. However, a decrease in the ecological risk index usually results in a decrease in the diversity and degree indices. Since ecological damage may be irreversible, reducing ecological risk substantially but reducing land use diversity and degree appropriately within a manageable range may be a favorable approach for the system's evolution. Therefore, the combined scenario can advance water-saving technologies, adopt the policy of returning farmland to forestland, and decrease the economic development rate to ensure the healthy evolution of regional land and water resources. This is the best plan to use the land and water resources in Heilongjiang Province.

5. Discussion

To discuss the reasons why different scenarios produce different simulation results for each variable, the Vensim software was used to analyze the characteristic variable tree (Figure 9) to analyze the details of the effects of each control variable. Because the "Gross regional output" variable has more appearances and it is a key variable of the economic subsystem, and the economic subsystem is closely related to the land and water subsystem, the "Gross regional output" variable tree is drawn separately.

Considering that the "Gross regional output" variable is directly affected by the "Annual growth rate of gross regional output" variable, it is also influenced by the "Agriculture water demand" variable. Because the "Water use efficiency index" and the "Agriculture water demand" variable is affected by the "Net irrigation quota for paddy fields", the tree containing the "Gross regional output" variable also includes the "Annual growth rate of gross regional output" and "Net irrigation quota for paddy fields." Because the "Gross regional output" tree appears in all of the characteristic variable trees in the water subsystem, the characteristic variable tree in the water resource subsystem includes the "Annual growth rate of gross regional output" and "Net irrigation quota for paddy fields" variables.

The "Annual change rate of farmland area" and "Forestland" variables appear in all characteristic variables of the land subsystem tree, so controlling the conversion between farmland and forestland can directly affect the land resource subsystem, but there is little impact on water subsystems. The "Water quota for industrial use" variable appears in the "water use efficiency index" tree, which shows that increasing industrial water-saving technologies can effectively increase the water use efficiency and water-saving potential. The "Net irrigation quota for paddy fields" and "Annual growth rate of gross regional output" variables appear in the characteristic variable tree in the water resource subsystem. These two variables have a direct impact on the water subsystem but do not directly affect the land resources subsystem, so they have little effect on the land resource subsystem. In addition, "Net irrigation quota for paddy fields" has more appearances than the "Annual growth rate of gross regional output", and it can be seen that the direct impact of the "Net irrigation quota for paddy fields" variable on the water subsystem is greater than that of the "Annual growth rate of gross regional output" variable.

Based on comparative literature, Hamid Balali et al. [23] used SD methods to simulate the groundwater dynamics to depletion under different economic policies and climate change and developed a model that can simulate a farmer's economic behavior and groundwater aquifer dynamics, as well as studied area climatology factors and government economic policies related to groundwater. These authors concluded that climate change can affect the water system, and this study confirms that under conditions where climate and economic changes are difficult to predict, in addition to adopting comprehensive management policies, planning and the application of certain methods to control climate factors such as precipitation can achieve groundwater sustainable management effectively and can reduce the vulnerability of water resource systems. Janez et al. [34] used SD methods to analyze policy trends, incorporate socioeconomic factors into the integrated water system, and monitor the possible environmental and socioeconomic impacts by controlling key parameters to assess the potential impact of water shortages and socioeconomic policies in complex hydrological systems and to provide information for the formulation of regional water resources policies, further confirming the applicability of the SD approach for assessing the potential impact of various policies and measures on regional water systems. Tong et al. constructed a complicated coupling simulation system of water resources-land resources-social economy-ecology based on the method of system dynamics [26] and analyzed the regional land and water resources joint allocation plan, pointing out that in order to realize sustainable use of land and water resources, we should increase the intensity of agricultural water conservation and implement land consolidation, which coincides with the conclusions of this study. This study also points out that the construction of a comprehensive SD model of regional land and water should be combined with relevant hydrological and water environment models to realize accurate quantification research. The results of various studies confirm the applicability of SD in land and water systems and show that climatic factors also have a certain impact on regional land and water systems. In addition, advanced water-saving technologies, adoption of the policy of returning farmland to forestland, and decreasing the economic development rate can ensure the healthy evolution of regional land and water resources.

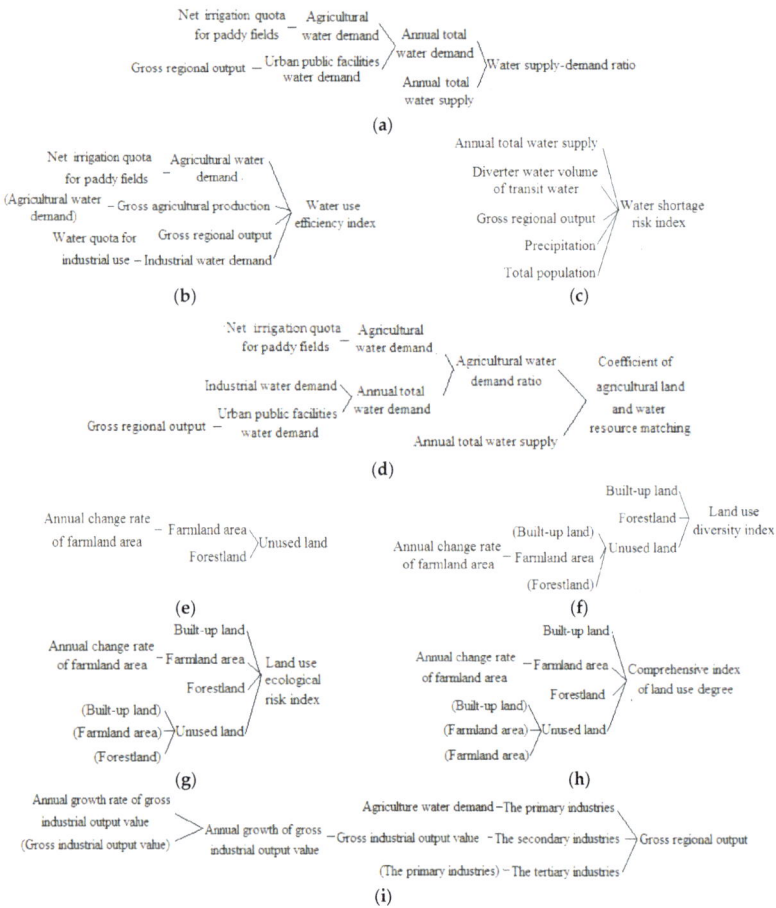

Figure 9. The characteristic variable tree. (a) Water supply-demand ratio variable tree; (b) Water use efficiency index variable tree; (c) Water shortage risk index variable tree; (d) Coefficient of agricultural land and water resource matching variable tree; (e) Unused land variable tree; (f) Land use diversity variable tree; (g) Land use ecological risk index variable tree; (h) Comprehensive index of land use degree variable tree; (i) Gross regional output variable tree.

6. Conclusions and Future Research

1. Through establishing an SD model of the land and water resource system for Heilongjiang Province, this study analyzed the evolution of the system under different climates and concluded that the region faced a risk of supply and demand of land and water resources. This step clarified the problem and the direction of managing regional land and water resources.

2. This paper shows the evolution of regional land and water resources as a result of many driving forces. This study selected and adjusted the variables (i.e., net irrigation quota for paddy fields, water quota for industrial use, forestland area, annual change rate of farmland area, the annual growth rate of industrial gross) to set different scenarios. The model simulated the evolution and analyzed the results and then combined the favorable scenarios to predict the evolution of regional land and water resources in Heilongjiang Province. Through simulation under different scenarios, it can be seen that water-saving technologies can reduce water use, and the policy of returning farmland to forestland can reduce the ecological risk of land use. Decreasing the economic development rate can alleviate the pressure of water shortage. Combining these strategies can promote the healthy evolution of regional land and water resources in the region.

3. The SD model can simulate the dynamic relationships between the variables, as well as predict the response of the land and water resource system to climate change and relevant policies. However, this model has a limitation. Precipitation has a certain impact on the evolution of the system, but it is not accurate and is difficult to predict. Follow-up studies can introduce hydrological considerations into the SD method to reasonably predict precipitation.

This study provides a new perspective for the analysis of the land and water resource system. However, due to the complexity of the land and water resource system, there are still many issues that require further research.

- The model is complicated. The land and water resource system is extremely complex and involves many factors. The research process is still uncertain and lacks mature theoretical research. In view of this, the SD model of land and water resources established by previous scholars can be summarized and analyzed, and the insufficiency of the land and water resource system model of this study can be improved to explore a multi-perspective comprehensive SD model of land and water resources that is suitable for regional conditions so that improved analysis and prediction of land and water resource systems can be achieved.

- The method is simple. The system dynamics analysis method of the land and water resource system in this study is relatively simple, and the analysis result is one-sided. Therefore, in future research, scholars should focus on the complementary expansion of multiple methods and should seek a more reasonable multidisciplinary interactive system to achieve the qualitative and quantitative analysis of land and water resource systems and to obtain more comprehensive research results.

Author Contributions: Y.Z. designed the study, built the model and wrote the paper. Q.J. directed the study. Z.W., Q.F., Z.Z., T.W. and Y.D. provided critical feedback on the manuscript.

funding: The Natural Science Foundation of Heilongjiang Province of China (General Project, grant No. E2016004) and the National Natural Science Foundation of China (grant No. 51679040).

Acknowledgments: The authors wish to thank the reviewers and the editor for their valuable suggestions.

Conflicts of Interest: The authors declare no conflicts of interest.

References

1. Abdul, G.; Manzoor, G.; Ghulam, M. Review Agriculture in the Indus Plains: Sustainability of Land and Water Resources. *Int. J. Agric. Biol.* **2002**, *4*, 429–437.

2. Cuceloglu, G.; Abbaspour, K.; Ozturk, I. Assessing the Water-Resources Potential of Istanbul by Using a Soil and Water Assessment Tool (SWAT) Hydrological Model. *Water* **2017**, *9*, 814. [CrossRef]

3. Wang, Z.L.; Jiang, Q.X.; Fu, Q.; Jiang, X.; Mo, K. Eco-environmental effects of water resources development and utilization in the Sanjiang Plain, Northeast China. *Water Sci. Technol. Water Supply* **2017**. [CrossRef]

4. Forrester, J.W. *Industrial Dynamics*; Productivity Press: Cambridge MA, USA, 1961.

5. Richardson, G.P. System Dynamics: Simulation for Policy Analysis from a Feedback Perspective. In *Qualitative Simulation Modeling and Analysis*; Springer: New York, NY, USA, 1991.

6. Senge, P.M. *The Fifth Discipline: The Art & Practice of the Learning Organization*; Dell Publishing Group Inc.: New York, NY, USA, 1994.

7. Homer J, B. Why we iterate: Scientific modeling in theory and practice. *Syst. Dyn. Rev.* **1996**, *12*, 1–19. [CrossRef]

8. Sehlke, G.; Jacobson, J. System dynamics modeling of transboundary systems: The bear river basin model. *Ground Water* **2005**, *43*, 722–730. [CrossRef] [PubMed]

9. Ren, Y.; Yao, J.; Xu, D.; Wang, Z. A comprehensive evaluation of regional water safety systems based on a similarity cloud model. *Water Sci. Technol. A J. Int. Assoc. Water Pollut. Res.* **2017**, *76*, 594. [CrossRef] [PubMed]

10. Chen, Z.; Wei, S. Application of System Dynamics to Water Security Research. *Water Resour. Manag.* **2014**, *28*, 287–300. [CrossRef]

11. Yin, S.; Dong, J.G.; Wei, C.S.; Wei, J.G. Integrated assessment and scenarios simulation of urban water security system in the southwest of China with system dynamics analysis. *Water Sci. Technol.* **2017**, *76*, 2255–2267. [CrossRef] [PubMed]

12. Yilinuer, A.; Jilili, A.; Long, M.; Alim, S.; Michael, G. System Dynamics Modeling of Water Level Variations of Lake Issyk-Kul, Kyrgyzstan. *Water* **2017**, *9*, 989.

13. Benjamin, L.T.; Hector, M.; Menendez, R.G.; Luis, O.T.; Alberto, S.A. System Dynamics Modeling for Agricultural and Natural Resource Management Issues: Review of Some Past Cases and Forecasting Future Roles. *Water* **2016**, *5*, 40.

14. Liu, D.W. Risk assessment of water shortage in metropolitan area of Beijing, Tianjin and Hebei. *Water Resour. Dev. Res.* **2010**, *10*, 20–24. (In Chinese)

15. Wei, H.X.; Li, Z.D. Ecological Risk Assessment of Water Resources Utilization in Gansu Province. *Water Resour. Plan. Des.* **2016**, *29*, 11–14. (In Chinese)

16. Liu, Y.S.; Gan, H.; Zhang, F.G. Analysis of the Matching Patterns of Land and Water Resources in Northeast China. *Acta Geogr. Sin.* **2006**, *61*, 847–854. (In Chinese)

17. Zhuang, D.F.; Liu, J.Y. Study on the model of regional differentiation of land use degree in China. *J. Nat. Resour.* **1997**, *12*, 105–111. (In Chinese)

18. Wu, W.J.; Shi, P.J.; Hu, W. Study on land ecological risk in oasis city based on LUCC-a case study in the Ganzhou district. *Arid Zone Res.* **2012**, *29*, 122–128.

19. Mo, H.H.; Ren, Z.Y. Study on Changes of Land Ecosystem Value and Ecological Risk: A Case Study in Sand Blowing Region: Over Shenmu County of Shaanxi Province. *J. Desert Res.* **2010**, *30*, 357–362. (In Chinese)

20. National Bureau of Statistics of the People's Republic of China. *China Statistical Yearbook*; China Statistics Press: Beijing, China, 2015.

21. Bureau of Statistics of Heilongjiang Province. *Heilongjiang Statistical Yearbook*; China Statistics Press: Beijing, China, 2015.

22. The People's Government of Heilongjiang Province. *The Thirteenth Five-Year Plan Outline for the National Economy and Social Development in Heilongjiang Province*; The People's Government of Heilongjiang Province: Heilongjiang, China, 2016.

23. Land Resource Bureau. *Heilongjiang Land Use Master Plan (2005–2020)*; Land Resource Bureau: Beijing, China, 2013.

24. Erol, A.; Randhir, T.O. Climatic change impacts on the ecohydrology of Mediterranean watersheds. *Clim. Chang.* **2012**, *114*, 319–341. [CrossRef]

25. Wang, X.J.; Zhang, J.Y.; Wang, J.H.; He, R.M.; Amgad, E.; Liu, J.H.; Wang, X.G.; King, D.; Shahid, S. Climate change and water resources management in Tuwei river basin of Northwest China. *Mitig. Adapt. Strateg. Glob. Chang.* **2014**, *19*, 107–120.

26. Li, X.W.; Gao, X.Z.; Chang, Y.T.; Mu, D.P.; Liu, H.L.; Guo, J.Y. Water storage variations and their relation to climate factors over Central Asia and surrounding areas over 30 years. *Water Sci. Technol. Water Supply* **2017**, ws2017206. [CrossRef]

27. Balali, H.; Viaggi, D. Applying a System Dynamics Approach for Modeling Groundwater Dynamics to Depletion under Different Economical and Climate Change Scenarios. *Water* **2015**, *7*, 5258–5271. [CrossRef]

28. Liu, J.Y.; Hu, Y.C.; Jin, X.Z. Sustainable utilization of water resources in Tianjin based on system dynamics. *J. Huazhong Normal Univ. (Nat. Sci.)* **2014**, *48*, 106–111. (In Chinese)

29. Sun, X.X.; Shen, B.; Yu, J.L.; Liu, X.J.; Mo, S.H. Carrying capacity of water resources of Baoji city based on the system dynamics model. *Xi'an Univ. Arch. Technol. (Nat. Sci. Ed.)* **2007**, *39*, 72–77. (In Chinese) [CrossRef]

30. Tong, F.; Jin, J.L.; Dong, Z.C. Study on regional joint allocation of water and land resources based on improved system dynamics model. *Yangtze River* **2017**, *48*, 43–48. (In Chinese)

31. Nguyen, T.P.L.; Mula, L.; Cortignani, R.; Seddaiu, G.; Dono, G.; Virdis, S.G.P.; Pasqui, M.; Roggero, P.P. Perceptions of Present and Future Climate Change Impacts on Water Availability for Agricultural Systems in the Western Mediterranean Region. *Water* **2016**, *8*, 523. [CrossRef]

32. Huang, R.; Liu, J.M.; Li, L.K. Water resources carrying capacity of Xianyang City based on system dynamics. *J. Drain. Irrig. Mach. Eng.* **2012**, *30*, 57–63. (In Chinese)

33. Jiang, Q.; Fu, Q.; Wang, Z. Dynamic Simulation of Water Resources Sustainable Utilization of Kiamusze Based on System Dynamics. In Proceedings of the Sixth International Conference on Computer and Computing Technologies in Agriculture, Zhangjiajie, China, 19–21 October 2013; pp. 367–375.

34. Sušnik, J.; Vamvakeridou-Lyroudia, L.S.; Savić, D.A.; Kapelan, Z. Integrated System Dynamics Modelling for water scarcity assessment: Case study of the Kairouan region. *Sci. Total Environ.* **2012**, *440*, 290–306. [CrossRef] [PubMed]

![water logo] *water*

MDPI

Article

Optimal Operation of Hydropower Reservoirs under Climate Change: The Case of Tekeze Reservoir, Eastern Nile

Fikru Fentaw Abera [1,2,*], **Dereje Hailu Asfaw** [1], **Agizew Nigussie Engida** [1] and **Assefa M. Melesse** [2]

[1]   School of Civil and Environmental Engineering, Addis Ababa Institute of Technology (AAiT),
      Addis Ababa 1000, Ethiopia; dereje_hasfaw@yahoo.com (D.H.A.); agiz70@yahoo.com (A.N.E.)
[2]   Department of Earth and Environment, Florida International University, Miami, FL 33199, USA;
      melessea@fiu.edu
*   Correspondence: ffentawa@fiu.edu; Tel.: +1-973-424-3654

Received: 3 January 2018; Accepted: 27 February 2018; Published: 5 March 2018

Abstract: Optimal operation of reservoirs is very essential for water resource planning and management, but it is very challenging and complicated when dealing with climate change impacts. The objective of this paper was to assess existing and future hydropower operation at the Tekeze reservoir in the face of climate change. In this study, a calibrated and validated Soil and Water Assessment Tool (SWAT) was used to model runoff inflow into the Tekeze hydropower reservoir under present and future climate scenarios. Inflow to the reservoir was simulated using hydro-climatic data from an ensemble of downscaled climate data based on the Coordinated Regional climate Downscaling Experiment over African domain (CORDEX-Africa) with Coupled Intercomparison Project Phase 5 (CMIP5) simulations under Representative Concentration Pathway (RCP)4.5 and RCP8.5 climate scenarios. Observed and projected inflows to Tekeze hydropower reservoir were used as input to the US Army Corps of Engineer's Reservoir Evaluation System Perspective Reservoir Model (HEC-ResPRM), a reservoir operation model, to optimize hydropower reservoir release, storage and pool level. Results indicated that climate change has a clear impact on reservoir inflow and showed increase in annual and monthly inflow into the reservoir except in dry months from May to June under RCP4.5 and RCP8.5 climate scenarios. HEC-ResPRM optimal operation results showed an increase in Tekeze reservoir power storage potential up to 25% and 30% under RCP4.5 and RCP8.5 climate scenarios, respectively. This implies that Tekeze hydropower production will be affected by climate change. This analysis can be used by water resources planners and mangers to develop reservoir operation techniques considering climate change impact to increase power production.

Keywords: reservoir operation; optimization; SWAT; HEC-ResPRM; climate change; CORDEX-Africa; Tekeze basin

---

1. Introduction

Water resources reservoirs are important tools for integrated water resources development and management [1,2], but nowadays their operation and management is challenging due to various factors [3,4]. The reservoir operates to supply water for municipal consumption, hydropower production, irrigation and industrial needs, flood control, recreation, navigation or ecological requirements. Currently, due to water crisis the global freshwater supply to meet the needs of the different sectors is falling short [5–7]. Factors that contribute to this include population growth, urbanization, climate change, land use change, land degradation and poor water resources management [8,9]. Hence, to alleviate these problems and meet the freshwater and energy demand of communities, it will necessitate optimal operation of water resources reservoirs [10,11].

Various researchers studied the reservoir operation throughout the world to get optimum level of release and optimal volumes of storage considering inflows and needs [12–15]. Most research conducted in the reservoir operations have specific objectives like hydropower [12,14,15], flood control [13], irrigation [3] and environmental [16].

Water resources infrastructures have been designed and managed historically but these designs gave little attention to the effect of climate change and non-stationarity in hydrologic variables [17]. Evidence suggests that these hydrologic variables used for water resources planning and management previously assumed stationarity in time have changed by anthropogenic activities [18]. The increase in temperature, changes in precipitation and evapotranspiration rates due to climate change alters global hydrologic cycle [19]. The streamflow affected by intensity and frequency of precipitation leads to increase the intensity of floods and droughts. These changes affect water resources at local and regional levels [20]. The hydrological processes and water availability affected by a change in the patterns of precipitation and temperature impacts agriculture, industry, communities, hydropower and aquatic life [21]. Climate change impact on fresh water resources may change the mean annual streamflow, shift seasonal flows, increases floods and droughts and changes in sediment fluxes which affects reservoir operation [22–24].

Many researchers in different parts of the world have studied the impacts of climate variability and change on shifts in hydrological regimes and water resources (e.g., [25–28]). These studies assessed the current and future water resources availability and rainfall variability across the globe to support appropriate water resources planning and management. Different studies showed that Africa is highly vulnerable to climate change [29]. Climate change studies showed temperature increased and precipitation pattern changed throughout arid and semi-arid regions of Africa [30–32] and affected the hydrological processes that impacts reservoir operation. Most studies showed the impacts of climate change on African hydropower reservoirs [23,33–36]. Kim and Kaluarachchi [33] and Beyene et al. [34] projected that precipitation and temperature will be increased in the Nile River basin and have positive effect on hydropower production, but Yamba et al. [35] and Hamududu and Killingtveit [36] investigated that in the next 60 years hydropower production show a gradual reduction with large variability in the Zambezi River basin. In the Nile River basin, the rapidly growing hydropower based energy need, population growth, food insecurity and finite water resources will lead to competitions for water in the riparian countries and this will be aggravated by the climate change. Several studies have been conducted on the variability of precipitation and streamflow in the Nile River basin [37–41] that affects reservoir planning and management [3].

Most hydropower reservoir operators concern is existing hydrological variability without foreseeing climate change as a particular serious threat [23]. Hence reservoir operation need to incorporate plans to address hydrologic non-stationarity and uncertainty caused by climate change [20,26,42,43]. Due to this, ensembles of Global Circulation Models (GCMs), scenarios and regional climate models (RCMs) used as input to hydrological model to generate future streamflow [44,45] that can be used as an input for reservoir operations.

Sedimentation may cause serious impacts on reservoir operation and functionality by reducing reservoir storage capacity and shortening reservoir useful life for human benefits. Studies showed that Northern part of the Tekeze basin watersheds are vulnerable to sedimentation and/or soil erosion problems for the sustainable use of small reservoirs developed for irrigated agriculture and Tekeze reservoir [46,47]. This reservoir sedimentation problem may lead serious reduction in reservoir storage capacity, causing future hydropower generation problems. However, rate of sedimentation of Tekeze reservoir still remains unpredicted. More and wide knowledge is still needed to better understand and solve the sediment problem, and hence may improve future reservoir operation. But the focus of this research is to study potential climate change impact on hydropower reservoir operation and management by not varying sedimentation level.

Nowadays, reservoir operation techniques become increasingly important and researchers still searching the best technique. Many authors proposed and reviewed various reservoir operation models and methods [4,48–51]. Labadie [50] extensively reviewed and evaluated various optimization methods and reported that no universally approved algorithm for all reservoir operations. Rani and Moreira [4] investigated that optimization models usually require simulation models for verifying and testing planned operating policies. Dam managers use simulation models more relaxed than optimization models as simulation models are easier to interpret, apply and present to non-professionals [49,50]. But Optimization models give reliable results. In recent years, to overcome these problems, a combination of simulation and optimization models applied in reservoir operation. In this research, US Army Corps of Engineer's Reservoir Evaluation System Perspective Reservoir Models (HEC-ResPRM), a combination of simulation and optimization model is used.

In this study, Tekeze hydropower reservoir was chosen due to: (1) Tekeze basin shows high rainfall variability [37,52] which affects reservoir inflow; (2) Tekeze hydropower reservoir not designed by considering hydrological non-stationarity and climate change; (3) the reservoir has not been optimally operated and sometimes not fully functional during dry periods. Therefore, the objective of this research are to (1) assess impact of climate change on reservoir inflow using Soil and Water Assessment Tool (SWAT) and recent Coordinated Regional climate Downscaling Experiment over African domain (CORDEX-Africa) RCMs under Representative Concentration Pathway (RCP)4.5 and RCP8.5 climate scenarios, and (2) apply HEC-ResPRM optimization model to get optimal release, reservoir level and storage for optimal power production including in the face of climate change.

2. Material and Methods

2.1. Study Area

Tekeze basin, a tributary of Tekeze-Setit-Atbera river part of Eastern Nile (Figure 1) is geographically located from 11°40' to 15°12' N and 36°30' to 39°50' E. The surface area of the Tekeze reservoir watershed is 29,404 km^2. This basin has high mountainous areas in its sources in the central Ethiopian highlands up to 4517 m and low land areas near Ethio-Sudan border as low as 800 m with varying climate depending on altitude change. The rainfall increases with altitude from 600 mm to 1200 mm but it is a reverse for temperature which decreases from 26 °C to 10 °C. This basin has a mean annual inflow of 4.4 Billion cubic meters at Embamadre gauging station and annual potential evapotranspiration of 1778 mm.

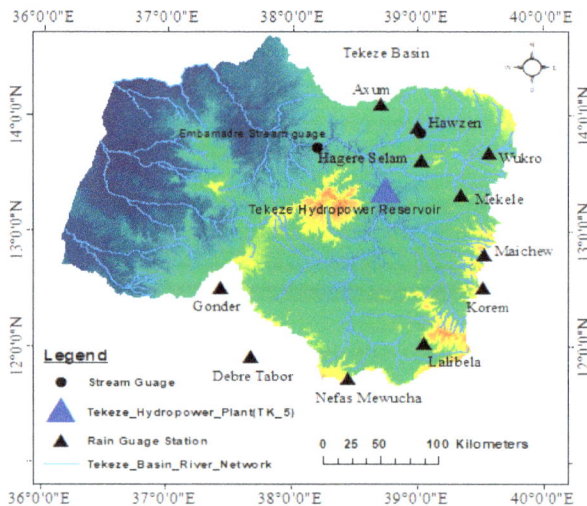

Figure 1. Location of Tekeze hydropower reservoir and weather stations.

Tekeze basin has a large elevation drop from its sources to low land areas near Ethio-Sudan border and offers a significant hydropower potentials of 5960 GWh/year. Tekeze single purpose hydropower reservoir located at 13°21′ N and 38°45′ E (Figure 1) is the second tallest double concrete arch dam in Africa next to Katse arch dam in Lesotho. The purpose of this reservoir is for hydropower production with total installed capacity of 300 MW in four 75 MW Francis turbines at underground power house. The reservoir has a total storage capacity of 9293 million cubic meters (MCM) of which 5293 MCM live storage at 1140 m above sea level (masl) and 4000 MCM below dead storage level (1096 masl). The reservoir also has 147 km^2 surface areas at full supply level with mean annual inflow of 3750 MCM.

2.2. Datasets Used

2.2.1. Historical and Future Hydrology

In this research, four hydrological data periods were analyzed. These were the reservoir inflow data of: (1) observed and RCP scenarios historical records (1994–2008); (2) the near future period (2011–2040), middle future period (2041–2070) and the far future periods (2071–2100). SWAT simulates historical (past) and all future reservoir inflows using precipitation and temperature projections from ensemble outputs of CORDEX-Africa RCMs downscaled from different GCMs from Coupled Intercomparison Project Phase 5 (CMIP5) simulations available in 0.44° resolution for Ethiopian domain under two recent representative concentration pathways (RCP4.5 and RCP8.5) climate scenarios. There are numerous weather stations in Tekeze basin. For this study, stations recording precipitation and temperature data that have long period of records with small data gaps were used. There are more than 20 streamflow gauged stations in the Tekeze basin but most of the stations are found in the small tributaries of Tekeze River which covers only small watershed areas. These stations, except Embamadre station, have large data gaps, short record periods and high amount of missing data. Observed streamflow at Embamadre station was collected from Ethiopian Ministry of Water, Irrigation and Electricity.

2.2.2. CORDEX-Africa

Currently, CORDEX-Africa initiated by World Climate Research Program (WCRP) provides an opportunity for the generation of high resolution regional climate projections over Africa that is used to assess future impacts of climate change at regional and local scales. In this study, results of CORDEX-Africa ensemble RCMs simulations for the past (1951–2005) and future (2006–2100) climate projections downscaled from different GCMs under RCP4.5 and RCP8.5 with spatial resolution of 0.44° is used. CORDEX-Africa RCMs generate an ensemble of high resolution historical and future climate projections at regional scale by downscaling different GCMs forced by RCPs based on the Coupled Intercomparison Project Phase 5 (CMIP5) [32,53]. CORDEX-Africa climate projections use RCP4.5 and RCP8.5 climate scenarios.

RCPs are new climate change scenarios established by CMIP5 [54,55], which can depict a wide variety of possible future climate scenarios. The fifth Assessment Report (AR5) scientific literature selects one mitigation scenario (RCP2.6), two medium stabilization scenarios (RCP4.5 and RCP6.0) and one high emission scenario. RCP2.6 scenario sees emissions peak early, then fall shown to be technically feasible. But one of RCP2.6 scenario key assumptions is the full participation of all developed and developing countries in the world in the short run to reduce all the main emitters, which is not possible in actual cases. Due to this, we decided to choose one medium scenario (RCP4.5) and high scenario (RCP8.5) covering entire range of radiative forcing. RCPs represent pathways of radiative forcing, not linked with exclusive socio-economic assumption in contrary to Special Report on Emission Scenarios (SRES). Any single radiative forcing pathway can result from a diverse range of socio-economic and technological development scenarios. RCP4.5 is a mid-range scenario that stabilizes radiative forcing at 4.5 W/m^2 (approximately 650 ppm CO_2-equivalent) in the year 2100 without exceeding this value, but this does not imply the climate system are stable [53,56]. Whereas RCP8.5 is upper bound of all RCP scenarios that stabilizes radiative forcing at 8.5 W/m^2 (greater than 1370 ppm CO_2-equivalent) in the year 2100 [53,57].

2.2.3. Reservoir Data

HEC-ResPRM, a reservoir operation model, requires data like back ground map of the watershed, reservoir outlet capacities, elevation-area-storage curve, current outflow-energy relationship, power production and flow time series to perform optimal operations. These physical data were used to develop model constraints and allow the model to calculate penalties. All these data were collected from Ethiopian Electric Power Corporation and Ministry of Water, Irrigation and Electricity.

2.3. Methods

2.3.1. Overview of SWAT

Soil and Water Assessment Tool (SWAT) was used to produce inflow projections and assess climate change impact on the streamflow used as an input for reservoir operation. The details of SWAT shown in Neitsch et al. [58]. It is a semi-distributed continuous widely used hydrological model in the Eastern Nile basins [59,60]. Since the objective of the study was to examine streamflow response to climate change, the land phase of the hydrologic cycle simulated by SWAT is based on the water balance equation:

$$SW_t = SW_o + \sum_{i=1}^{t} \left(R_{day} - Q_{surf} - E_a - W_{seep} - Q_{gw} \right) \tag{1}$$

in which SW_t is the final soil water content (mm), SW_o is the initial soil water content on day i (mm), t is the time (days), R_{day} is the amount of precipitation on day i (mm), Q_{surf} is the amount of surface runoff on day i (mm), E_a is the amount of evapotranspiration on day i (mm), W_{seep} is the amount of water entering the vadose zone from the soil profile on day i (mm), and Q_{gw} is the amount of return flow on day i (mm). Surface runoff volume was estimated using modified Soil Conservation Serves-Curve Number (SCS-CN) method.

In this study, SWAT was used together with the ArcSWAT interface, a geographic information system (GIS) based graphical user interface used to facilitate watershed delineation and initial parameterization. The SWAT model requires digital elevation model (DEM), land cover/land use information, soils and basic climate data. The land use/land cover data of the Tekeze basin includes agricultural land, shrub land (range grasses), mixed forests and pasture/grazing lands. SWAT subdivides the watershed in to hydrological response units (HRUs) with a homogeneous land use and soil properties based on topography and quantifies the relative impacts of soil, land use and climate change within each HRU. The 30 m × 30 m DEM, soil properties, land use/land cover and streamflow data were all collected from Ethiopian Ministry of Water, Irrigation and Electricity. Meteorological data of precipitation and temperature (1994–2008) were collected from National Meteorological Service Agency. Sensitivity analysis was done using Latin Hypercube sampling based on One Factor at a Time (LH-OAT) inbuilt in SWAT to identify sensitive parameters that influence model simulations. The sensitive parameters of this study mainly affecting model calibration were curve number (CN-2), soil available water capacity (Sol-AWC), alpha base flow recession constant (ALPHA-BF), soil evaporation compensation factor (ESCO), threshold water depth required for return flow to occur (GWQMN) and saturated hydraulic conductivity (SOL_K). Model calibration adjusts such high sensitive parameters to optimize the agreement between observed and simulated streamflow values at Embamadre station in Tekeze basin. Model performance was assessed using Nash-Sutcliffe efficiency (E_{NS}), coefficient of determination (R^2) and percent of bias (PBIAS). Finally, the historical and future RCP4.5 and RCP8.5 projections of precipitation and temperature were used as input into the calibrated and validated SWAT model to assess the impact of climate change on reservoir inflow.

2.3.2. HEC-ResPRM Optimization Model

In this study, the US Army Corps of Engineer's Reservoir Evaluation System Perspective Reservoir Model (HEC-ResPRM), a reservoir system operations optimizations software package developed to assist planners, operators and managers with reservoir operation plan and decision making, was used. It addresses a reservoir system operation problem of optimal long-term allocation of available water. HEC-ResPRM is a combination of simulation and optimization model when Perspective Reservoir Model (PRM) is integrated in to HEC-RES modeling platform. It is an implementation of HEC-PRM shared with HEC-ResSim in a graphical user interface for creating, running, sorting and analyzing optimization runs. HEC-ResPRM uses HEC's data storage system (HEC-DSS) to store and retrieve of input and output time series data.

It is a monthly network flow programming model and gives optimal values of release and storage by minimizing penalty functions [61,62]. Network flow programming is computationally efficient form of linear programming. A network solver finds optimal flow for the entire network simultaneously based on the unit cost associated with flow along each arc. Optimization problem represented by the network with cost associated with flow as follows:

$$Minimize : \sum_{t}^{n} C_t Q_t \text{ (For all nodes)} \tag{2}$$

$$Subject\ to : \sum Q_t - \sum a_t Q_t = 0 \text{ (For all nodes)} \tag{3}$$

$$L_t \leq Q_t \leq U_t \text{ (For all arcs)} \tag{4}$$

in which n is total number of network arcs; C_t is unit cost, weighting factor for flow along arc t; Q_t is flow along arc t; a_t is multiplier (gain) for arc t; L_t is lower bound on flow along arc t; and U_t is upper bound on flow along arc t. In this case, node represents a reservoir and river or channel junctions. Arcs represent inflow and outflow links in the reservoir system. Each arc has a minimum and maximum flow that it must carry in the reservoir system. The arcs (inflow and outflow links) may transfer water between two points in space (transferring water in channels) or in time (changing pool elevations in the reservoir). Also, flow is conserved in the reservoir (node). Equations (2) through (4) are special forms of linear programming problems solved using primal simplex method.

3. Results and Discussions

3.1. Climate Projections

Projected annual temperature and precipitation showed an increasing trend in 2020s, 2050s and 2080s over Tekeze basin under RCP4.5 and RCP8.5 climate scenarios. Projected mean annual temperature may increase up to 1.1 °C and 3.38 °C under RCP4.5 and RCP8.5 scenarios, respectively in all future time periods. Similarly, mean annual precipitation may increase up to 45% under both scenarios for all future time periods. Figure 2 shows future change rates of temperature in both scenarios for all future time periods. Mean monthly temperature will increase under both scenarios in all time periods except the months of January and February which showed a slightly decreasing trend in 2020s. Figure 3 shows future percentage changes of monthly precipitation amounts for different projected periods under RCP4.5 and RCP8.5 climate change scenarios. For RCP8.5 scenarios, the months of March, April and May would exhibit a decrease in precipitation amount compared to the baseline period whereas RCP4.5 scenario presented an increasing trend. The months of October through February would show an increase in precipitation compared to reference period for bothscenarios and projected periods considered.

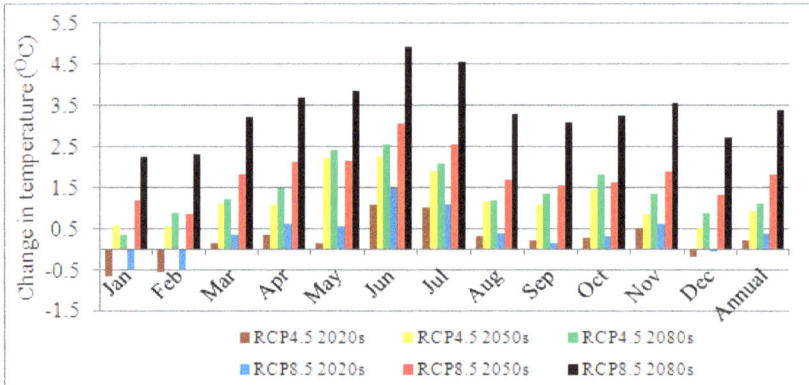

Figure 2. Rates of change of monthly mean temperature for the selected scenarios and projected periods.

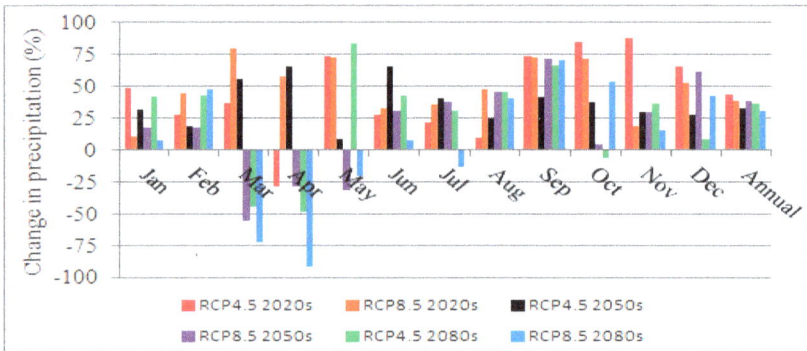

Figure 3. Change of monthly precipitation amount for the selected scenarios and projected periods.

3.2. *Impacts of Climate Change on Streamflow*

Impact of climate change on the streamflow at Embamadre station downstream of the reservoir was analyzed. Observed streamflow data from a period 1994–2002 was used for model calibration and from 2003–2008 was used for validation. Results in Figure 4 show that SWAT successfully simulated annual and monthly streamflow with a reasonable accuracy. The monthly result showed a good performance of SWAT, which indicated by the value of Nash-Sutcliffe efficiency (E_{NS}) 0.70 for calibration and 0.79 for validation and the coefficient of determination (R^2) 0.73 for calibration and 0.80 for validation. Similarly, percent of bias (PBIAS) value of 0.53% during calibration and 0.45% in validation periods showed good fit between observed and simulated streamflow.

Hence, the calibrated and validated SWAT forced to run for historical and future climate scenarios to generate future streamflow for both RCP4.5 and RCP8.5 climate scenarios. The effect of climate change on annual and monthly streamflow was also investigated as a percentage change with respect to the baseline period (1994–2008) under the two scenarios in three time periods 2020s (2011–2040), 2050s (2041–2070) and 2080s (2071–2100).

Figure 5 showed the percentage change of annual and monthly streamflow for both climate scenarios and the three time periods. Mean annual streamflow showed an increasing trend for both RCP4.5 and RCP8.5 for all time periods. Under RCP4.5, the mean annual percentage change of streamflow will increase by 49%, 39% and 47% in the 2020s, 2050s and 2080s, respectively. Similarly, for

RCP8.5, the mean annual percentage change of streamflow increases to 22%, 19% and 2% in the 2020s, 2050s and 2080s, respectively.

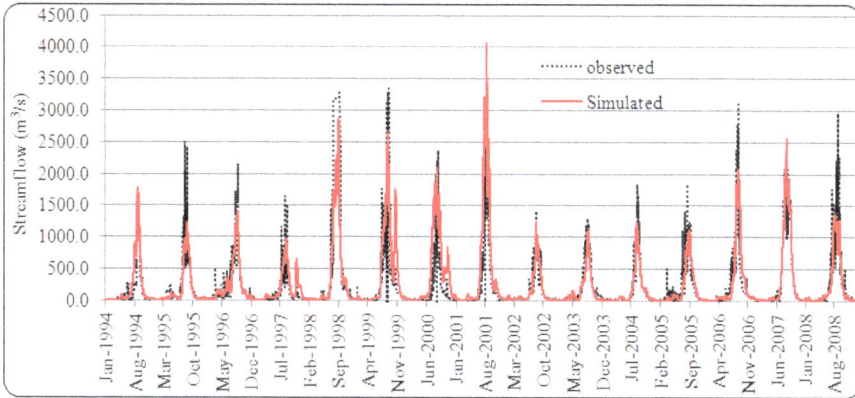

Figure 4. Observed and SWAT simulated monthly stream flow at Embamadre station.

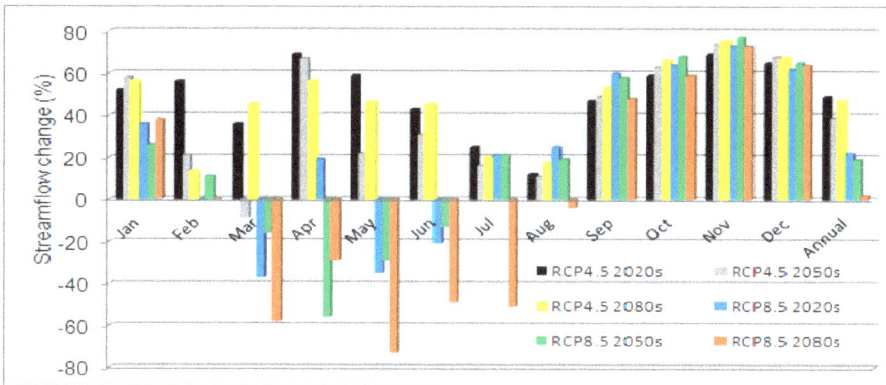

Figure 5. Mean monthly and annual streamflow changes under RCP4.5 and RCP8.5 climate scenarios.

The monthly streamflow change shows a mix of positive and negative trends. Under RCP4.5 climate scenario, change in average monthly streamflow ranges from 12 to 69% in 2020s and 13 to 67% in 2080s but in 2050s streamflow change shows mixed trend that decreases in March to May by up to 9% and increases on other months up to 39%. Mean monthly percentage change of streamflow under RCP8.5 climate scenario showed mixed trends in all time periods. Under RCP8.5, the mean monthly streamflow changes from −37 to 64%, −29 to 68% and −49 to 64% in 2020s, 2050s and 2080s, respectively. Individual month's trend showed that there was an increasing trend from August to February and a decreasing trend from March to July. Therefore, climate change will have a clear impact on the future streamflow an input of reservoir power production in Tekeze basin.

The changes and variability of monthly (inter-annual) streamflow will be much greater than the annual streamflow changes in both scenarios in all time periods. This result showed that it is important for the hydropower reservoir planners and managers to consider, the monthly streamflow variability and changes for future planning and operation of reservoirs.

The total mean annual historical (past) and future Tekeze hydropower reservoir inflow (m^3/s) trends under RCP4.5 and RCP8.5 climate scenarios for all time periods are shown in Figure 6.

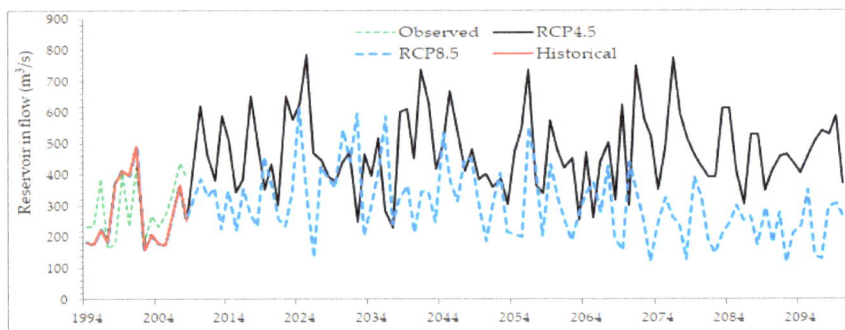

Figure 6. Annual Tekeze hydropower reservoir inflow trend for future time periods.

3.3. Current Reservoir Operation

HEC-ResPRM optimization model run under current baseline condition (2009–2017). This model optimized the current Tekeze hydropower reservoir operation. The current optimized value (Table 1) of HEC-ResPRM optimization model showed an increase in reservoir storage compared to current actual hydropower reservoir operation status. It is also indicated that the mean annual reservoir pool level increased up to 7.87 m (Table 2) that will store more water to produce power throughout the year. It contradicts the current actual Tekeze hydropower reservoir operation which produces insufficient power even very little or no power production during dry months. This implies that Tekeze hydropower reservoir was not optimally operated till now. The reservoir storage dropped to the minimum operating level and sometimes dries in the non-rainy months. Therefore, the current actual reservoir operation is not effective and should consider different well tested reservoir operation techniques under a changing climate.

Table 1. Mean annual optimized power storage under climate change scenarios.

Periods	Optimized Reservoir Storage (Mm³)		Change in Optimized Reservoir Storage (%)	
	RCP4.5	RCP8.5	RCP4.5	RCP8.5
Current optimized	6639		24.0	
2020s	6688	6880	25.0	28.5
2050s	6669	6903	24.6	29.0
2080s	6665	6958	24.5	30.0

Table 2. Mean annual optimized pool level variation under RCP4.5 and RCP8.5 climate scenarios in three future time periods.

Time Periods	Optimized Pool Level (masl)		Pool Level Change (m)	
	RCP4.5	RCP8.5	RCP4.5	RCP8.5
Current optimized	1120.48		7.87	
2020s	1121.06	1123.27	8.45	10.66
2050s	1120.87	1123.37	8.26	10.76
2080s	1120.89	1123.85	8.28	11.24

3.4. Reservoir Operation under Climate Change

The future reservoir inflows generated by SWAT under RCP4.5 and RCP8.5 climate scenarios in three time periods 2020s (2011–2040), 2050s (2041–2070) and 2080s (2071–2100) with other reservoir data were used to run optimization model to get projected optimal reservoir outflow (release), storage and pool level results.

3.4.1. Projected Reservoir Inflow and Outflow

Climate change impacted inflow and outflow (release) hydrograph of Tekeze hydropower reservoir considered in this study are shown in Figure 7a,b. According to the inflow projections based on ensembles of CORDEX-Africa RCM climate model simulations, total inflows to Tekeze hydropower reservoir expected to increase under RCP4.5 and RCP8.5 climate scenarios for all future time periods. Figure 7a,b show that total monthly inflow under RCP4.5 is greater than the total monthly inflow projected under RCP8.5 climate scenarios. However, under RCP4.5 climate scenario, the reservoir inflow projections exhibit high fluctuations inter-annually as compared to RCP8.5 climate scenario and observed historical values. The highest inflow volumes under RCP4.5 were concentrated in the rainy months that spilled easily and affect the dry period reservoir storage level and or release.

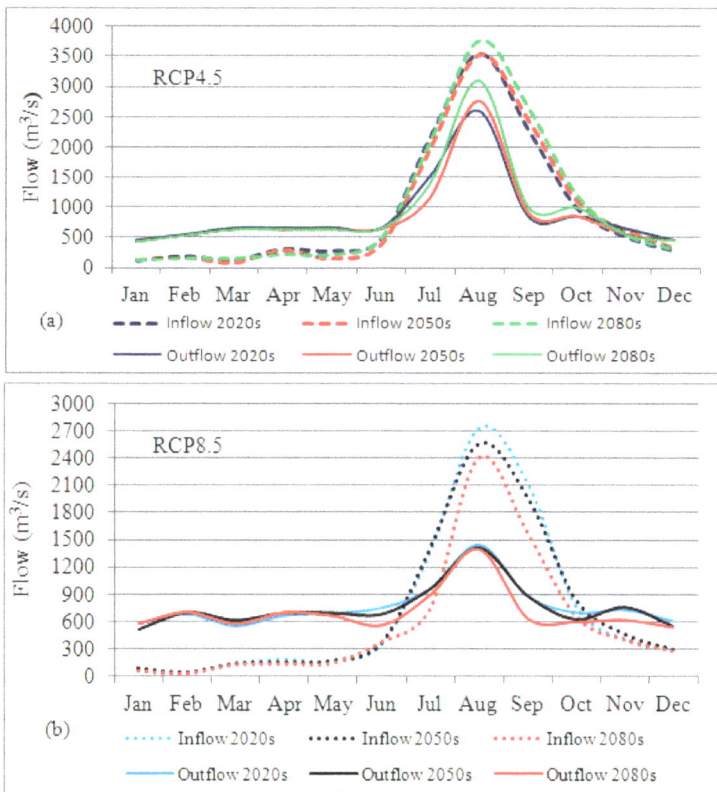

Figure 7. Mean monthly reservoir inflow and optimized outflow (release) for future time periods under: (**a**) RCP4.5 climate scenario; (**b**) RCP8.5 climate scenario.

There would be an increase in excess reservoir inflow during the rainy months of August through October under both RCP4.5 and RCP8.5 climate scenarios in all time periods. This increased spillage of available water inflow occurs because of the effect of climate change that increased the hydropower reservoir inflow under RCP4.5 and RCP8.5 future climate scenarios. According to the latest climate simulations, the overall inflow volume is predicted to be higher during rainy months and provided that the reservoir lack sufficient storage capacity to accommodate these high flows. As a result, Tekeze hydropower reservoir forced to spill water without generating hydropower. This indicates that the increased in overall reservoir inflow volume does not necessarily be advantageous to produce more power. Therefore, decision need to be taken on the amount of water to be released and or stored now and retained for future considering the variations in inflow and demands.

In this study, the reservoir outflow (release) was obtained by HEC-ResPRM optimization model under RCP4.5 and RCP8.5 climate scenarios for 2020s, 2050s and 2080s time periods. In all future time periods (Figure 7a,b) under the two climate scenarios, the reservoir release will be increased to produce more power due to an increased future reservoir inflow and optimum water storage using optimization model. Under RCP4.5 climate scenario average monthly reservoir outflow varies from 353 to 2590 m^3/s in 2020s, 435 to 2757 m^3/s in 2050s and 442 to 3090 m^3/s in 2080s. Similarly, average monthly reservoir outflow varies from 538 to 1445 m^3/s in 2020s, 514 to 1412 m^3/s in 2050s and 577 to 1396 m^3/s in 2080s under RCP8.5 scenario. In both scenarios, the minimum and maximum outflow value occurred during dry and wet periods, respectively. In all time periods, the optimum reservoir outflows (releases) under RCP8.5 climate scenario for the dry months of November through February were greater than the optimum releases under RCP4.5 climate scenario. These changes show that under RCP8.5, the optimized reservoir stored more water in wet months for dry period release and projected higher storage level compared to RCP4.5 climate scenario.

3.4.2. Optimum Reservoir Power Storage under Climate Change

HEC-ResPRM optimized result showed an increase in projected mean annual Tekeze hydropower reservoir storage under RCP4.5 and RCP8.5 climate scenarios. This increase was projected for three future time periods (Table 1) and the projected optimum stored water varies from 24 to 25% (RCP4.5) and 28.5 to 30% (RCP8.5).

HEC-ResPRM model result under both scenarios in current and all future time periods showed a minimum and maximum reservoir storage periods (Figure 8). Tekeze reservoir reached at maximum storage (reservoir filled) in September and stayed somewhat constant optimum storage up to November. During August to September, main rainy months, the reservoir is filled and optimization model keeps the maximum storage up to November. The reservoir storage tends to slightly be decreased starting from end of November until the beginning of February. After February, the reservoir storage decreased down to the optimization model capacity to store energy at a minimum flow and reached a minimum storage level in June to prepare and capture inflows in the wet main rainy months. In all future months, there will be a stored water to produce power which is always greater than the current optimized value.

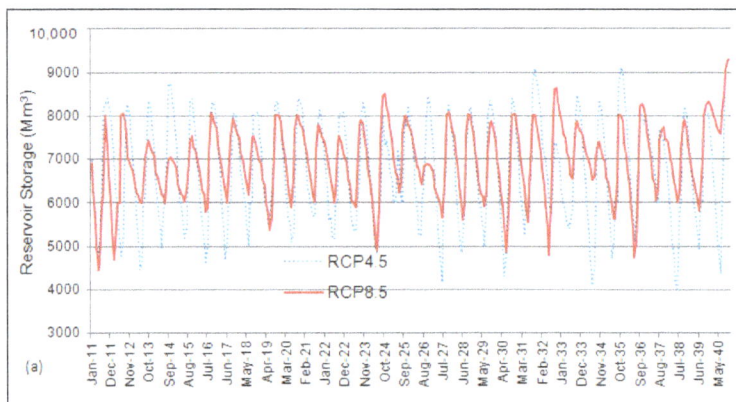

Figure 8. *Cont.*

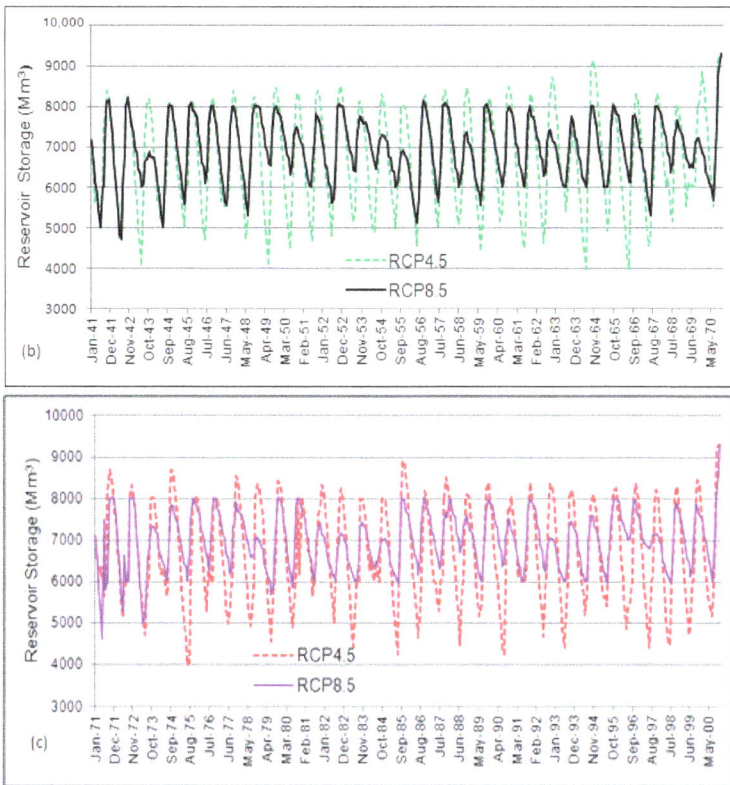

Figure 8. Monthly optimized reservoir storage variations under RCP4.5 and RCP8.5 climate scenarios for: (**a**) 2020s; (**b**) 2050s; (**c**) 2080s.

Figure 8 shows that more power is stored in both scenarios for the next 90 years as compared to current actual and current optimized value. The change in maximum optimum storage increased will be occurred in January and varies from 1693.4 to 1800.8 Mm^3 under RCP4.5 scenario and from 1731.9 to 1851.1 Mm^3 under RCP8.5 scenario in all time periods. The minimum optimal storage change increased will occur in July and varies from 392.2 to 424.7 Mm^3 under RCP4.5 scenario and 803.6 to 956.6 Mm^3 under RCP8.5 scenario in all time periods. This is due to climate change impact on the reservoir inflow and the capacity of the optimization model to operate the reservoir optimally. HEC-ResPRM optimization of future projections tends to make much greater seasonal use of reservoir storage than the current actual operations.

The monthly optimum stored water increases in all months for future time periods under both RCP scenarios as compared to the baseline period (base line varies from 4400 to 6500 Mm^3). Optimized monthly reservoir storage variations are shown in Figure 8. The mean monthly optimum reservoir storage in the future time periods varies for RCP4.5 from 5100 to 8300 Mm^3 in 2020s, 4700 to 8050 Mm^3 in 2050s and 5000 to 8100 Mm^3 in 2080s. It also varies for RCP8.5 from 4900 to 8100 Mm^3 in 2020s, 4850 to 8020 Mm^3 in 2080s and 750 to 7900 Mm^3 in 2080s.

3.4.3. Optimum Reservoir Pool Level (Elevation) under Climate Change

HEC-ResPRM optimization result indicates that Tekeze hydropower reservoir pool level will be increased under RCP4.5 and RCP8.5 scenarios in all time periods. This comparison made with the current mean annual reservoir operation pool level of 1112.61 masl from 2009–2017. In the three projected time periods, optimal pool level (Table 2) change varies from 8.26 to 8.45 m under RCP4.5 and 10.66–11.24 m under RCP8.5 climate scenarios. This is large elevation difference that will store more water in the rainy months for the dry season power production. The optimized pool levels under both scenarios in all time periods are larger than the current optimized pool level. This is due to the impact of climate change and hydrological non-stationarity on reservoir operation. The reservoir storage pool level change in RCP4.5 scenario is lower than RCP8.5 scenario due to increase in each year individual month's fluctuations in RCP4.5 scenarios because of future inflow variability that reduced the mean annual reservoir water storage level.

The Tekeze hydropower reservoir operational level is changing continually due to inflows occurred and releases are made to produce power. The start of dead storage level at 1096 masl (minimum live storage level) has been assumed for power production. Figure 9 shows the optimal reservoir pool level of Tekeze hydropower reservoir generated by HEC-ResPRM optimization model. These optimal pool level results have a similar pattern with the optimal reservoir storage variations and may be considered as rule curves for optimal operation of Tekeze hydropower reservoir under a given scenario and time period. The reservoir pool level stayed at high level every year from August to November when reservoirs filled during the rainy months of August through September. The drop of pool level in June caused due to optimization model constraint reservoir not emptied and a transition zone when the drawdown ends and reservoir refill start.

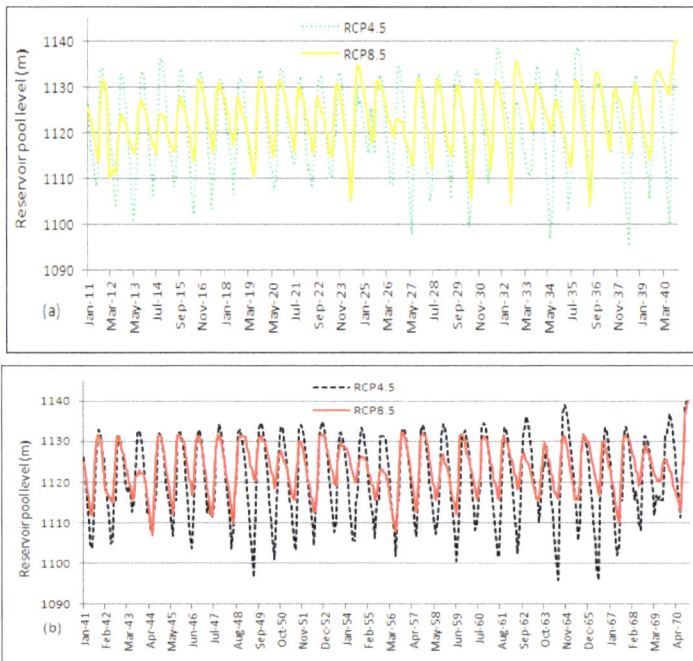

Figure 9. *Cont.*

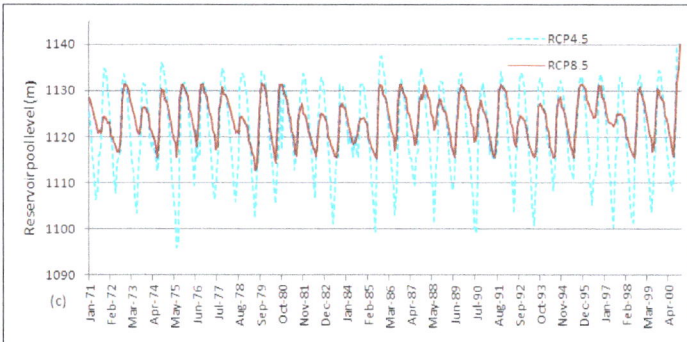

Figure 9. Monthly optimum pool level variations under RCP4.5 and RCP8.5 climate scenarios in three time periods: (**a**) 2020s; (**b**) 2050s; (**c**) 2080s.

There is no doubt that the hydropower system of Tekeze basin will be affected by climate change. With over all predicted increases in precipitation and streamflow, inflow to the reservoir anticipated to increase. Therefore, even though both RCP4.5 and RCP8.5 climate scenarios exhibited higher water inflow volume to Tekeze hydropower reservoir, this did not necessarily result in significantly more hydropower generation. Optimal operation of the reservoir using HEC-ResPRM considerably increases the power production by storing the higher inflow volume to inflow deficiency periods. Based on this research, the potential for hydropower production in the Tekeze hydropower reservoir is predicted to increase if appropriate reservoir operation techniques are used by operators and water managers to store and use the wet months flow to the dry months. Even if many optimization models and techniques have been developed in several fields of water resources system analysis such as hydropower reservoir operation around the world, the adaptation of such techniques and tools by water managers is slow. Researchers and scientists must accept the fact that the gap still exists between research studies and applications in practice. There need to be research on how to translate science to improve management operations of reservoirs for optimal results.

4. Conclusions

This study used a semi-distributed hydrological model (SWAT) and a reservoir optimization model (HEC-ResPRM) to evaluate the hydrological impacts of climate change on Tekeze hydropower reservoir operation in Tekeze basin part of Eastern Nile. We have evaluated climatic data (historical and future period) from the ensemble outputs of CORDEX-Africa RCMs under RCP4.5 and RCP8.5 climate scenarios for the periods of 2020s, 2050s, and 2080s. Calibrated SWAT model was used to generate climate change induced streamflow that was used as an input for optimal reservoir operation modeling. Analysis conducted on Tekeze hydropower reservoir inflows and outflow, reservoir storage volume and reservoir pool levels revealed the following:

1. This study found that the impact of climate change would increase in precipitation, temperature and streamflow in Tekeze basin under RCP4.5 and RCP8.5 climate scenarios over future periods which have an impact on current and future Tekeze hydropower reservoir operation.
2. Projected annual and inter-annual reservoir inflow showed increasing trend under both RCP4.5 and RCP8.5 climate scenarios.
3. HEC-ResPRM incorporates water storage, water surface elevation, release and power generation would provide better understanding of current and future conditions of Tekeze hydropower reservoir operation.

4. Current optimized power storage and pool level show more optimal results than the current actual operation, so it is recommended to change the current operating policy to produce more power throughout the year.

5. The projected increase of reservoir inflow under an ensemble of CORDEX-Africa RCP4.5 and RCP8.5 future climate scenarios lead to optimized reservoir power storage, pool level (head) and release that greatly exceed those historically observed, indicating a shift in current water system behavior.

6. The study showed that climate change clearly affects future reservoir planning and management in Tekeze basin. Therefore, water resources planners, managers and operators should consider climate change impacts in the design, planning and management of reservoir systems.

7. In practice, many reservoir system operators and water managers feel more comfortable to use pre-defined rule curves and simulation results which are easy to understand and operate as most optimal operating rules developed by scientists using sophisticated optimization models and algorithms are mathematically more complex. The use of a combination of simulation and optimization models may solve this problem.

8. This study has not considered the changes in land use/land cover due to socio-economic development in the future. Coupling climate models with projected changes in land use associated with climate change impacts and effect of climate change adaptation on erosion and sediment yield which is necessary to evaluate projected changes in runoff associated with future Tekeze River basin development. Hence, further studies are recommended to quantify future change in streamflow and sedimentation load in Tekeze hydropower reservoir as well as its implication on future hydropower generation.

Acknowledgments: The research presented has been financially supported by the Ethiopian Government via Addis Ababa University (PhD studentship for the first author). The first author would like to give his thanks to Florida International University in assisting for the two months research visit. We acknowledge the data support of Ethiopian Ministry of Water, Irrigation and Electricity, National Meteorological Service Agency and Electric Power Corporation.

Author Contributions: Fikru Fentaw was responsible for this current research article in the framework of his PhD program and initially wrote the manuscript. Dereje Hailu, Agizew Nigussie, and Assefa M. Melesse directed the study by helping to interpret the results, particularly Assefa M. Melesse who contributed more on editing and organizing of this research article.

Conflicts of Interest: The authors declare no conflict of interest.

References

1. Zhou, Y.; Guo, S.; Duan, W.; Chen, H.; Liu, P. Dynamic control of flood limited water level for cascade reservoirs. *J. Hydroelectr. Eng.* **2015**, *34*, 23–30.
2. Yazdi, J.; Moridi, A. Interactive reservoir-watershed modeling framework for integrated water quality management. *Water Resour. Manag.* **2017**, *31*, 2105–2125. [CrossRef]
3. Birhanu, K.; Alamirew, T.; Dinka, M.O.; Ayalew, S.; Aklog, D. Optimizing reservoir operation policy using chance constraint nonlinear programming for Koga Irrigation Dam, Ethiopia. *Water Resour. Manag.* **2014**, *28*, 4957–4970. [CrossRef]
4. Rani, D.; Moreira, M.M. Simulation–optimization modeling: A survey and potential application in reservoir systems operation. *Water Resour. Manag.* **2010**, *24*, 1107–1138. [CrossRef]
5. David, S.; Randolph, B.; Upali, A. Water scarcity in the Twenty-first century. *Int. J. Water Resour. Dev.* **1999**, *15*, 29–42. [CrossRef]
6. Jury, W.A.; Henry, J.V.J. The emerging global water crisis: Managing scarcity and conflict between water users. *Adv. Agron.* **2007**, *95*, 1–76.
7. Rijsberman, G. Water scarcity: Fact or fiction? *Agric. Water Manag.* **2006**, *80*, 5–22. [CrossRef]
8. Larson, K.L.; Polsky, C.; Gober, P.; Chang, H.; Shandas, V. Vulnerability of water systems to the effects of climate change and urbanization: A comparison of Phoenix, Arizona and Portland, Oregon (USA). *Environ. Manag.* **2013**, *52*, 179–195. [CrossRef] [PubMed]

9. Ghashghaie, M.; Marofi, S.; Marofi, H. Using system dynamics method to determine the effect of water demand priorities on downstream flow. *Water Resour. Manag.* **2014**, *28*. [CrossRef]

10. Guo, S.; Chen, J.; Li, Y.; Liu, P.; Li, T. Joint operation of the multi-reservoir system of the Three Gorges and the Qingjiang Cascade Reservoirs. *Energies* **2011**, *4*, 1036–1050. [CrossRef]

11. Zhou, Y.; Guo, S.; Liu, P.; Xu, C. Joint operation and dynamic control of flood limiting water levels for mixed cascade reservoir systems. *J. Hydrol.* **2014**, *519*, 248–257. [CrossRef]

12. Azizipour, M.; Ghalenoei, V.; Afshar, M.H.; Solis, S.S. Optimal operation of hydropower reservoir systems using weed optimization algorithm. *Water Resour. Manag.* **2016**, *30*, 3995–4009. [CrossRef]

13. He, Y.; Xu, Q.; Yang, S.; Liao, L. Reservoir flood control operation based on chaotic particle swarm optimization algorithm. *Appl. Math. Model.* **2014**, *38*, 4480–4492. [CrossRef]

14. Lu, B.; Li, K.; Zhang, H.; Wang, W.; Gu, H. Study on the optimal hydropower generation of Zhelin reservoir. *J. Hydro-Environ. Res.* **2013**, *7*, 270–278. [CrossRef]

15. Cheng, C.-T.; Wang, W.-C.; Xu, D.-M.; Chau, K.W. Optimizing hydropower reservoir operation using hybrid genetic algorithm and chaos. *Water Resour. Manag.* **2008**, *22*, 895–909. [CrossRef]

16. Yin, X.A.; Yang, Z.F. Development of a coupled reservoir operation and water diversion model: Balancing human and environmental flow requirements. *Ecol. Model.* **2011**, *222*, 224–231. [CrossRef]

17. Milly, P.; Betancourt, J.; Falkenmark, M.; Hirsch, R.M.; Kundzewicz, W.Z.; Lettenmaier, P.D.; Stouffer, R.J. Stationarity is dead: Whither water management? *Science* **2008**, *319*, 573–574. [CrossRef] [PubMed]

18. Stocker, D.; Qin, G.-K.; Plattner, L.V.; Allen, S.K. Technical summary. In *Climate Change 2013: The Physical Science Basis. Contribution of Working Group I to the Fifth Assessment Report of the Intergovernmental Panel on Climate Change*; Cambridge University Press: Cambridge, UK, 2013; pp. 31–116, ISBN 978-1-107-41532-4.

19. Huntington, T.G. Climate warming-induced intensification of the hydrologic cycle: An assessment of the published record and potential impacts on agriculture. In *Advances in Agronomy*; Sparks, D.L., Ed.; Academic Press: Cambridge, MA, USA, 2010; Volume 109, pp. 1–53.

20. Vicuña, S.; Garreaud, R.D.; McPhee, J. Climate change impacts on the hydrology of a snowmelt driven basin in semiarid Chile. *Clim. Chang.* **2011**, *105*, 469–488. [CrossRef]

21. Alazzy, A.; Lü, H.; Zhu, Y. Impact of climate change on evaluation of future water demand in the Euphrates and Aleppo basin, Syria. *Proc. Int. Assoc. Hydrol. Sci.* **2014**, *364*, 307–312. [CrossRef]

22. Zhang, C.; Zhu, X.; Fu, G.; Zhou, H.; Wang, H. The impacts of climate change on water diversion strategies for a water deficit reservoir. *J. Hydroinform.* **2014**, *16*, 872–889. [CrossRef]

23. Lumbroso, D.M.; Woolhouse, G.; Jones, L. A review of the consideration of climate change in the planning of hydropower schemes in sub-Saharan Africa. *Clim. Chang.* **2015**, *133*, 621–633. [CrossRef]

24. Lee, S.-Y.; Hamlet, A.F.; Grossman, E.E. Impacts of climate change on regulated streamflow, hydrologic extremes, hydropower production, and sediment discharge in the Skagit River Basin. *Northwest Sci.* **2016**, *90*, 23–43. [CrossRef]

25. Setegn, S.; Melesse, A.; Haiduk, A.; Webber, D.; Wang, X.; McClain, M. Modeling hydrological variability of fresh water resources in the Rio Cobre watershed, Jamaica. *Catena* **2014**, *120*, 81–90. [CrossRef]

26. Ehsani, N.; Vörösmarty, C.J.; Fekete, B.M.; Stakhiv, E.Z. Reservoir operations under climate change: Storage Capacity options to mitigate risk. *J. Hydrol.* **2017**, *555*, 435–446. [CrossRef]

27. Haile, A.T.; Akawka, A.L.; Berhanu, B.; Rientjes, T. Changes in water availability in the Upper Blue Nile Basin under the representative concentration pathways scenario. *Hydrol. Sci. J.* **2017**, *62*, 2139–2149. [CrossRef]

28. Zhu, X.; Zhang, C.; Qi, W.; Cai, W.; Zhao, X.; Wang, X. Multiple climate change scenarios and runoff response in Biliu River. *Water* **2018**, *10*, 126. [CrossRef]

29. Serdeczny, O.; Adams, S.; Baarsch, F.; Coumou, D.; Robinson, A.; Hare, W.; Schaeffer, M.; Perrette, M.; Reinhardt, J. Climate change impacts in Sub-Saharan Africa: From physical changes to their social repercussions. *Reg. Environ. Chang.* **2017**, *17*, 1585–1600. [CrossRef]

30. Conway, D.; Persechino, A.; Ardoin-Bardin, S.; Hamandawana, H.; Dieulin, C.; Mahé, G. Rainfall and water resources variability in Sub-Saharan Africa during the twentieth century. *J. Hydrometeorol.* **2009**, *10*, 41–59. [CrossRef]

31. Hales, S. *Contribution of Working Group II to the Fourth Assessment Report of the Intergovernmental Panel on Climate Change*; Climate Change 2007: Impacts, Adaptation and Vulnerability; Parry, M.L., Canziani, O.F., Palutikof, J.P., van der Linden, P.J., Hanson, C.E., Eds.; Cambridge University Press: Cambridge, UK, 2007.

32. Nikulin, G.; Jones, C.; Giorgi, F.; Asrar, G.; Büchner, M.; Cerezo-Mota, R.; Christensen, O.B.; Déqué, M.; Fernandez, J.; Hänsler, A.; et al. Precipitation climatology in an ensemble of CORDEX-Africa regional climate simulations. *J. Clim.* **2012**, *25*, 6057–6078. [CrossRef]

33. Kim, U.; Kaluarachchi, J.J. Climate change impacts on water resources in the Upper Blue Nile River Basin, Ethiopia. *JAWRA J. Am. Water Resour. Assoc.* **2009**, *45*, 1361–1378. [CrossRef]

34. Beyene, T.; Lettenmaier, D.P.; Kabat, P. Hydrologic impacts of climate change on the Nile River Basin: implications of the 2007 IPCC scenarios. *Clim. Chang.* **2010**, *100*, 433–461. [CrossRef]

35. Yamba, F.D.; Walimwipi, H.; Jain, S.; Zhou, P.; Cuamba, B.; Mzezewa, C. Climate change/variability implications on hydroelectricity generation in the Zambezi River Basin. *Mitig. Adapt. Strateg. Glob. Chang.* **2011**, *16*, 617–628. [CrossRef]

36. Hamududu, B.H.; Killingtveit, Å. Hydropower production in future climate scenarios; the case for the Zambezi River. *Energies* **2016**, *9*, 502. [CrossRef]

37. Abtew, W.; Melesse, A.M.; Dessalegne, T. Spatial, inter and intra-annual variability of the Upper Blue Nile Basin rainfall. *Hydrol. Process.* **2009**, *23*, 3075–3082. [CrossRef]

38. Mengistu, D.; Bewket, W.; Lal, R. Recent spatiotemporal temperature and rainfall variability and trends over the Upper Blue Nile River Basin, Ethiopia. *Int. J. Climatol.* **2014**, *34*, 2278–2292. [CrossRef]

39. Tarekegn, D.; Tadege, A. *Assessing the Impact of Climate Change on the Water Resources of the Lake Tana Basin Using the WATBAL Model*; CEEPA (the Centre for Environmental Economics and Policy in Africa): Pretoria, South Africa, 2006.

40. Melesse, A.M.; Loukas, A.G.; Senay, G.; Yitayew, M. Climate change, land-cover dynamics and ecohydrology of the Nile River Basin. *Hydrol. Process.* **2009**, *23*, 3651–3652. [CrossRef]

41. Setegn, S.G.; Rayner, D.; Melesse, A.M.; Dargahi, B.; Srinivasan, R. Impact of climate change on the hydroclimatology of Lake Tana Basin, Ethiopia. *Water Resour. Res.* **2011**, *47*, W04511. [CrossRef]

42. Jamali, S.; Abrishamchi, A.; Madani, K. Climate change and hydropower planning in the Middle East: implications for Iran's Karkheh hydropower systems. *J. Energy Eng.* **2013**, *139*, 153–160. [CrossRef]

43. Vonk, E.; Xu, Y.; Booij, M.; Zhang, X.; Augustijn, D.C.M. Adapting multireservoir operation to shifting patterns of water supply and demand. *Water Resour. Manag.* **2014**, *28*, 625–643. [CrossRef]

44. Wilby, R.L.; Harris, I. A framework for assessing uncertainties in climate change impacts: Low-flow scenarios for the River Thames, UK. *Water Resour. Res.* **2006**, *42*, W02419. [CrossRef]

45. Raje, D.; Mujumdar, P.P. Reservoir performance under uncertainty in hydrologic impacts of climate change. *Adv. Water Resour.* **2010**, *33*, 312–326. [CrossRef]

46. Haregeweyn, N.; Poesen, J.; Nyssen, J.; Gerared, G.; Verstraeten, G.; de Vente, J.; Deckers, L.; Moeversons, J.; Haile, M. Sediment yield variability in Northern Ethiopia: A quantitative analysis of its controlling factors. *Catena* **2008**, *75*, 65–76. [CrossRef]

47. Wolde, E. Identification and prioritization of sub watersheds for land and water management in Tekeze Dam watershed, Northern Ethiopia. *Int. Soil Water Conserv. Res.* **2016**, *4*, 30–38. [CrossRef]

48. Wurbs, R.A. Reservoir-system simulation and optimization models. *J. Water Resour. Plan. Manag.* **1993**, *119*, 455–472. [CrossRef]

49. Oliveira, R.; Loucks, D.P. Operating rules for multireservoir systems. *Water Resour. Res.* **1997**, *33*, 839–852. [CrossRef]

50. Labadie, J.W. Optimal operation of multireservoir systems: State-of-the-art review. *J. Water Resour. Plan. Manag.* **2004**, *130*, 93–111. [CrossRef]

51. Yeh, W.W.-G. Reservoir management and operations models: A state-of-the-art review. *Water Resour. Res.* **1985**, *21*, 1797–1818. [CrossRef]

52. Ayalew, L. The effect of seasonal rainfall on landslides in the highlands of Ethiopia. *Bull. Eng. Geol. Environ.* **1999**, *58*, 9–19. [CrossRef]

53. Van Vuuren, D.P.; Edmonds, J.; Kainuma, M.; Riahi, K.; Thomson, A.; Hibbard, K.; Hurtt, G.C.; Kram, T.; Krey, V.; Lamarque, J.-F.; et al. The representative concentration pathways: An overview. *Clim. Chang.* **2011**, *109*. [CrossRef]

54. Meinshausen, M.; Smith, S.J.; Calvin, K.; Daniel, J.S.; Kainuma, M.L.T.; Lamarque, J.-F.; Matsumoto, K.; Montzka, S.A.; Raper, S.C.B.; Riahi, K.; et al. The RCP greenhouse gas concentrations and their extensions from 1765 to 2300. *Clim. Chang.* **2011**, *109*, 213. [CrossRef]

55. Taylor, K.E.; Stouffer, R.J.; Meehl, G.A. An overview of CMIP5 and the Experiment design. *Bull. Am. Meteorol. Soc.* **2011**, *93*, 485–498. [CrossRef]

56. Thomson, A.M.; Calvin, K.V.; Smith, S.J.; Kyle, P.G.; Volke, A.; Patel, P.; Delgado-Arias, S.; Bond-Lamberty, B.A.; Wise, M.A.; Clarke, L.E.; et al. RCP4.5: A pathway for stabilization of radiative forcing by 2100. *Clim. Chang.* **2011**, *109*, 77–94. [CrossRef]

57. Riahi, K.; Rao, S.; Krey, V.; Cho, C.; Chirkov, V.; Fischer, G.; Kindermann, G.; Nakicenovic, N.; Rafaj, P. RCP8.5—A scenario of comparatively high greenhouse gas emissions. *Clim. Chang.* **2011**, *109*, 33. [CrossRef]

58. Neitsch, S.; Arnold, J.; Kiniry, J.R.; Williams, J.R.; King, K. SWAT theoretical documentation. *Grassland* **2005**, *494*, 234–235.

59. Van Griensven, A.; Ndomba, P.; Yalew, S.; Kilonzo, F. Critical review of SWAT applications in the Upper Nile Basin countries. *Hydrol. Earth Syst. Sci.* **2012**, *16*, 3371–3381. [CrossRef]

60. Dessie, M.; Verhoest, N.E.C.; Admasu, T.; Pauwels, V.R.N.; Poesen, J.; Adgo, E.; Deckers, J.; Nyssen, J. Effects of the floodplain on river discharge into Lake Tana (Ethiopia). *J. Hydrol.* **2014**, *519*, 699–710. [CrossRef]

61. Faber, B.A.; Harou, J.J. Multiobjective optimization with HEC Res-PRM? Application to the Upper Mississippi Reservoir system. *Oper. Reserv. Chang. Cond.* **2006**. [CrossRef]

62. Ostadrahimi, L.; Mariño, M.A.; Afshar, A. Multi-reservoir operation rules: Multi-swarm PSO-based optimization approach. *Water Resour. Manag.* **2012**, *26*, 407–427. [CrossRef]

Article

An Integrated Method for Accounting for Water Environmental Capacity of the River–Reservoir Combination System

Fen Zhao [1], Chunhui Li [1,*], Libin Chen [2] and Yuan Zhang [3,*]

[1] Ministry of Education Key Lab of Water and Sand Science, School of Environment,
    Beijing Normal University, Beijing 100875, China; zhaofen@mail.bnu.edu.cn
[2] College of Resources and Environment, Yangtze University, Wuhan 430100, China; lbchen@yangtzeu.edu.cn
[3] Sate Key Laboratory of Environmental Criteria and Risk Assessment, Chinese Research Academy of
    Environmental Sciences, Beijing 100012, China
* Correspondence: chunhuili@bnu.edu.cn (C.L.); zhangyuan@craes.org.cn (Y.Z.); Tel.: +86-10-5880-2928 (C.L.)

Received: 31 January 2018; Accepted: 11 April 2018; Published: 14 April 2018

Abstract: The security of drinking water is a serious issue in China and worldwide. As the backup source of drinking water for the Changde City in China, the Huangshi Reservoir suffers from the threat of eutrophication due to the water quality of the reservoir ecosystem being affected by the tributaries that carry Non-Point Source (NPS) pollutants. The calculation of the water environmental capacity (WEC) can provide a scientific basis for water pollution control, which refers to the maximum amount of pollutants that the water can accommodate. In this paper, according to the hydrological characteristics of the river–reservoir combination system, a one-dimensional (1-D) water quality model and the Environmental Fluid Dynamics Code (EFDC) model were chosen to calculate the water environmental capacity of each functional zone in this basin. The quantity control of pollution from the tributaries was conducted based on the combined results of the water environmental capacity calculation from the EFDC model and a one-dimensional (1-D) river water quality model. The results show that total water environmental capacity of the tributaries included a chemical oxygen demand (COD) of 421.97 tons; ammonia nitrogen (NH_3-N) of 40.99 tons; total nitrogen (TN) of 35.94 tons; and total phosphorus (TP) of 9.54 tons. The water environmental capacity of the Huangshi Reservoir region accounts for more than 93% of the total capacity. The reduction targets of the major pollutants in the Huangshi Reservoir and its four major input rivers, which are, namely, the Bamao River, the Longtan River, the Fanjiafang River, and the Dongtan River, have been determined to achieve the water quality objectives for the reservoir in 2020 and 2025. The results will be helpful for the local water quality management and will provide a valuable example for other similar water source reservoirs.

Keywords: drinking water resources; water environmental capacity (WEC); Environmental Fluid Dynamics Code (EFDC) model; the Huangshi Reservoir

1. Introduction

In recent years, instead of flood control and irrigation, drinking water supplementation has become the primary purpose of reservoirs in China [1]. The issue of drinking water security plays a decisive role in the national economy and social wellbeing in China [2,3]. With the enormous development of aquaculture, water pollution in reservoirs has become a serious problem with the water quality of drinking water source directly influencing people's health, which is related to the economic development and stability of the general social situation.

The water environmental capacity (WEC) refers to the maximum amount of pollutants that the water can accommodate under the designed hydrological conditions and the specified environmental objectives without destroying its own function [4,5]. Generally, the calculation of WEC can provide a scientific basis for water pollution control as it provides a baseline for the total pollutant reduction needed. Previous studies mainly focused on various methods of calculating WEC [6–18], including analytical methods, dynamic mechanism model methods, and the calculation of the water environment capacity in a specific area [11,14,16,19–23]. Most case studies mainly use a river [12,15,16,19,23], lake, and reservoir [6,14,20,22] as objects in China, but there has been less research on the WEC of a certain river–reservoir combination system.

In addition, there are many models and equations available to calculate the WEC of rivers, lakes, and reservoirs. For this study, it is important to select the proper water quality simulation model. In calculating WEC, the use of three-dimensional modeling for simulating the water ecosystems has been used extensively and successfully by many studies [24–26]. They can simulate and predict the overall water quality of the water bodies for evaluating the WEC. In this sense, they are superior to the traditional field monitoring methods, which are mainly based on the field data monitored at limited locations [27–31].

Many studies have indicated that the inflow discharges will affect the variations of pollutants or nutrients in reservoirs [3–5,32–34]. As the backup drinking water source reservoir for Changde City in China, the Huangshi Reservoir, with multiple tributaries, is currently suffering from the threat of eutrophication, with the influence of pollutants or nutrients from the tributaries of Huangshi Reservoir having not been taken into consideration. It is necessary to calculate the WEC of the river–reservoir combination system. The purpose of this study is to simulate the WEC and control the amount of pollution in a drinking water reservoir with multiple tributaries using a widely used model, the environmental fluid dynamic model (EFDC), coupled with a 1-D convection equation.

This research paper is arranged as follows: firstly, the EFDC model is applied to analyze the changing trends of the water quality in the Huangshi Reservoir and its tributaries. Secondly, the 1-D water quality model and the EFDC model were chosen to calculate the WEC of each functional zone of the river–reservoir combination system. Finally, the quantity control of pollution of the multiple tributaries is analyzed based on the combined results of the WEC calculation.

2. Study Area

2.1. Study Area

The Huangshi Reservoir is a large-scale water conservancy project, which mainly provides the benefits of irrigation, flood control, power generation, and a habitat for fish, among others. It is located in the Huangshi town, Hunan Province, China. The reservoir is mainly fed by the convergences of the Bamao River, the Longtan River, the Fanjiafang River, the Liujiaxi River, and the Dongtan River. The boundary of the reservoir and the tributaries is shown in Figure 1, with the main characteristics of the tributaries shown in Table 1. The weather is relatively moderate with an annual average temperature of 16.8 °C. The annual average precipitation is about 1465.2 mm, the annual mean evaporation is about 1284.2 mm, and the relative humidity is 77%. The total storage capacity of the reservoir is 6.02×10^9 m^3, and the effective storage capacity is 3.38×10^9 m^3. The watershed has a total area of 494.34 km^2, with forests being the major land use type that accounts for 85.12% of the total basin area.

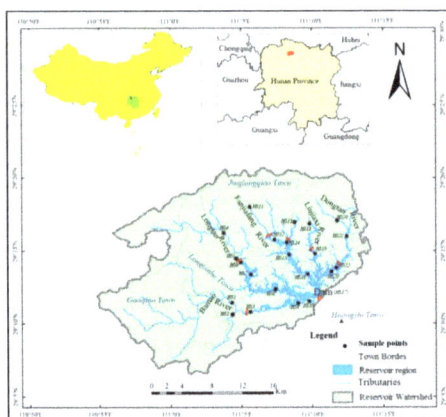

Figure 1. The location of the Huangshi Reservoir and the sample points of water quality.

Table 1. The features of the tributaries.

Rivers	Control Area (km^2)	Length of Reaches (km)	Water Quality Objectives (River Water Quality Standards) *	Flow (m^3/s)
Bamao River	102.9	10	II	1.429
Longtan River	154.4	14.2	II	1.2769
Fanjiafang River	61.2	4	II	0.6726
Dongtan River	40.3	5.8	II	0.4089

* Note: for the values of Water quality objectives, see also Table 2.

Table 2. The surface water environmental quality standard of China [35] (units: mg/L).

Indicators	II	III	IV	V
COD	15	20	30	40
NH$_3$-N	0.5	1	1.5	2
TP	0.1 (lake and reservoir 0.025)	0.2 (lake and reservoir 0.05)	0.3 (lake and reservoir 0.1)	0.4 (lake and reservoir 0.2)
TN	0.5	1	1.5	2

2.2. Water Quality Assessment in the Huangshi Reservoir Basin

A total of 23 representative sampling points was set up in the Huangshi Reservoir area for the assessment of the water quality and hydrobiology (shown in Figure 1), from which two points were monitored in the Bamao River (HS1, HS2), two points were monitored in the Longtan River (HS4, HS5), two points were monitored in the Dongtan River (HS20, HS21), and only one point was monitored in both the Fanjiafang River (HS13) and the Liujiaxi River (HS18). The rest of the 15 points were monitored in the reservoir region. According to the monitored water quality data of the sampling points in the years of 2013–2015 for the surface water environmental quality standard of China (GB3838–2002), the water quality of the Huangshi Reservoir is in accordance with the requirements of drinking water quality. Namely, the water quality of Huangshi Reservoir is classified as degree II and III (shown in Table 2). However, the main water quality indexes, including the chemical oxygen demand (COD), the ammonia nitrogen (NH$_3$-N), the total nitrogen (TN), and the total phosphorus (TP), exceed the environmental quality standards (shown in Tables 3 and 4), with the Longtan River exceeding the standards most compared to with the other four tributaries.

Table 3. The water quality evaluation result of the Huangshi Reservoir.

Sections	Standard	Water Quality	Exceed Factors and Multiples *
HS7	II	III	TN (0.92)
HS8	II	V	TN (3.96)
HS9	II	IV	TN (1.32)
HS16	II	IV	TN (1.18), TP (0.88)
HS17	II	IV	TN (1.70)
HS19	II	IV	TN (0.28), TP (2.36)
HS3	II	V	TN (2.86), TP (0.04)
HS6	II	IV	TN (0.88)
HS10	II	V	TN (2.38)
HS12	II	V	TN (2.06), TP (0.76)
HS14	II	IV	TN (1.54), TP (0.48)
HS15	II	V	TN (1.46), TP (4.68)
HS22	II	IV	TN (1.42), TP (0.92)
HS23	II	IV	TN (1.08), TP (0.08)

* Note: exceed multiples = (measured value − standard value)/standard value.

Table 4. The water quality evaluation result of the tributaries.

Tributaries	Sections	Standard	Water Quality	Exceed Factors and Multiples *
Bamao River	HS1		V	TN (0.90)
	HS2		IV	TN (0.49)
Longtan River	HS4		V	TN (0.90)
	HS5		V	TN (0.88), TP (1.54)
Fanjiafang River	HS11	III	V	TN (0.70)
	HS13		V	TN (0.71), TP (0.60)
Liujiaxi River	HS18		V	TN (0.86)
Dongtan River	HS20		V	TN (0.79)
	HS21		V	TN (0.53)

* Note: exceed multiples = (measured value − standard value)/standard value.

The monitored data of the sampling points set up in the Huangshi Reservoir (Figure 1), the water quality based on the assessment of the hydrobiology with the species, and the quantity measures are shown in Figure 2.

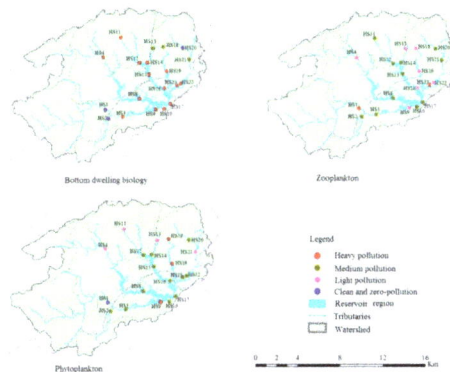

Figure 2. The assessment results of water quality.

The tributaries were seriously affected by agricultural non-point source (NPS) pollution and living pollution, while the water quality in the vicinity of the reservoir was worse. The water quality of the reservoir is increasingly worsening, with the levels of nitrogen and phosphorus nutrient salts exceeding the standard.

3. Methods

3.1. Water Environmental Capacity of the River-Reservoir Combination System

For fulfilling the purpose of water pollutant control in a drinking water reservoir with multiple tributaries, the paper selected 2014 as the status quo year, and 2020 and 2025 as the forecast years.

In order to simulate the WEC and control the amount of pollution of the Huangshi Reservoir with multiple tributaries in 2020 and 2025, we developed an integrated method to evaluate the WEC of the river–reservoir combination system, which consisted of the coupling of the 1-D model and the EFDC model. The technical route is as follows (Figure 3).

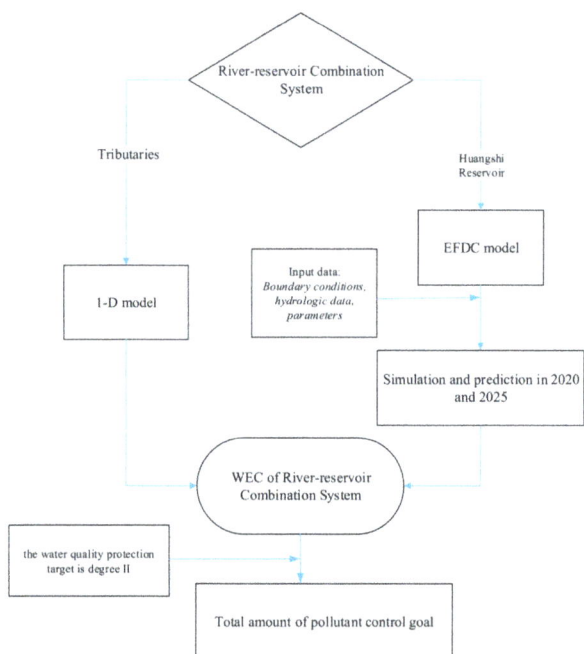

Figure 3. The technical route.

3.2. Water Environmental Capacity Model of the Tributaries

The equation for calculating the WEC was derived based on a 1-D convection-diffusion equation, with the adopted model expressed as follows [5]:

$$M = (C_S - C_0 exp(-kL/u))exp(kL/2u)Q_r \tag{1}$$

where M is the water environmental capacity per unit time (t); C_s is the pollutant concentration of the tail section in mg/L; C_0 is the pollutant concentration of the initial section in mg/L; k is the pollutant degradation coefficient in 1/s; and L is the length of the reach. A plurality of sewage outlets in the reach can be generalized as a centralized outfall and the outfall was located at the middle point of the reach. Therefore, it is equivalent to a concentrated point source, of which the self-purification length is half the length of the river reach. This means that the self-purification length is $L/2$ when the length of the river reach is L. In addition, u is the average flow velocity under the design flow in m/s; and Q_r is the design flow.

In this paper, the climatic characteristics have been considered [36] and the degradation coefficients of COD, NH$_3$-N, TN, and TP are 0.5, 0.25, 0.03, and 0.011/day, respectively.

3.3. Water Environmental Capacity Model of the Huangshi Reservoir

In this study, the EFDC model developed by Hamrick [37] was applied to the Huangshi Reservoir. This model is an integrated modeling system that has been successfully applied to many surface water systems, including rivers, estuaries, lakes, and coastal bays [12,38–40]. The EFDC model includes four major sub-models: (1) a hydrodynamic model; (2) a sediment transport model; (3) a toxic model; and (4) a water quality model. This simulation is mainly based on the EFDC hydrodynamic, convective diffusion, and first-order degradation modules of the water quality model.

The water quality module not only considers the wind direction, wind speed, and evaporation effect on the flow field and the pollutant transport but also the influence of the distribution characteristics of different types of aquatic plants and waves on the bottom stress. Its numerical simulation can estimate C, N, P, and other forms of nutrients, as well as a variety of algae. It is a strong model of water quality and movement and can accurately reflect the level of pollution in water.

In the water quality module, the governing mass-balance equation for each water quality state variable and the temperature of the water was expressed as follows [41]:

$$\frac{\partial C}{\partial t} + \frac{\partial (uC)}{\partial x} + \frac{\partial (vC)}{\partial y} + \frac{\partial (wC)}{\partial z} = \frac{\partial}{\partial x}\left(K_x \frac{\partial C}{\partial x}\right) + \frac{\partial}{\partial y}\left(K_y \frac{\partial C}{\partial y}\right) + \frac{\partial}{\partial z}\left(K_z \frac{\partial C}{\partial z}\right) + S_c \qquad (2)$$

where C is the concentration of a water quality state variable or C is the temperature of the water; u, v and w are the velocity components in the x, y, and z directions, respectively; K_x, K_y and K_z are the turbulent diffusion coefficients in the x, y, and z directions, respectively; and SC is the internal and external sources and sinks per unit volume.

By combining the water quality simulation results simulated by the EFDC model with the environmental target of the Huangshi Reservoir, an inverse method was selected to calculate the WEC, which is able to deduce the concentration of pollutants corresponding to water quality targets.

3.3.1. Hydrological Parameters

The protection of the water quality of the Huangshi Reservoir is strict. The lowest water monthly average flow rate and the lowest water level condition in 2013 and 2014 were chosen to calculate the environmental capacity of the reservoir.

3.3.2. Boundary Conditions

Flow boundary conditions: There are six boundary conditions in the simulated water area, which are formed by the inflow of the five rivers using the flow boundary, and the dam using the water surface elevation.

Meteorological boundary conditions: The data were obtained from the Meteorological Data Sharing Service Network of China, including rainfall, evaporation, solar radiation, cloud coverage, atmospheric pressure, air temperature, air humidity, wind speed, wind direction, and so on.

Water quality boundary conditions: The concentrations of COD, NH$_3$-N, TN, and TP in the five main input rivers were used.

3.3.3. Parameters of Model

Due to the lack of longtime serial hydrological data and water quality monitoring data of inflows and outflows of the Huangshi Reservoir, the parameters of the hydrodynamic model of the Huangshi Reservoir referenced some classical cases of the hydrodynamic model in South China. The degradation coefficients of COD, NH$_3$-N, TN, and TP of the Huangshi Reservoir referenced the research report of comprehensive improvement of the water environment and ecological protection plan in the Shanxi Reservoir area. The degradation coefficients of COD, NH$_3$-N, TN, and TP are 0.03 /day, 0.03/day, 0.03/day, and 0.01/day, respectively. The parameters of the water quality model are shown in Table 5.

Table 5. The parameters of the water quality model.

Parameters	Roughness	Eddy Viscosity Coefficient	Diffusion Coefficient	COD	NH$_3$-N	TN	TP
Coefficient/Degradation coefficient	0.02 m	1 m^2/s	0.2	0.03/day	0.03/day	0.03/day	0.01/day

3.3.4. Model Verification

In this study, four main water quality variables were simulated, namely COD, NH$_3$-N, TN, and TP. The observed data from 2013 to 2014 were used for the verification of the water quality model.

The error statistics of the model verification are listed in Table 6. It is clear that the simulated values agree well with the measured values, which demonstrates that the established water quality model is able to simulate the process of water quality in the Huangshi Reservoir.

Table 6. The error statistics of the model verification (units: mg/L).

Characteristic Values	COD	NH$_3$-N	TN	TP
Compare numbers	12	12	12	12
Average observed	7.78	0.12	1.20	0.15
Average simulated	6.92	0.10	1.15	0.10
Average Error	−0.85	0.01	−0.04	0.02
Relative Error	20.29	28.10	12.99	157.42
Absolute Error	1.68	0.02	0.14	0.03
Root Mean Square Error	2.18	0.03	0.18	0.03
Relative Root Mean Square Error	47.28	70.38	61.56	139.98
Nash-Sutcliffe coefficient	−2.05	−11.89	−4.05	−50.35

4. Results

4.1. Water Quality Simulation and Prediction

The Huangshi Reservoir is the backup source of drinking water for Changde City. There are no industrial districts in the area and water quality is affected mainly by agriculture NPS pollutants, domestic sewage, and aquaculture pollutants.

Based on the pollutant loads discharged into the river in the status quo year of 2014 (shown in Table 7), the increase of the population and the development of urbanization, and the scale of eco-agriculture and the increase of aquaculture, the pollutant loads in the tributaries of the Huangshi Reservoir in 2020 and 2025 have been estimated according to the Report of Huangshi Reservoir protection planning from the Chinese Academy of Environmental Sciences (shown in Table 8). The values have also been converted into concentrations of pollutants in the tributaries of the Huangshi Reservoir, which is shown in Table 9.

Table 7. The pollutant quantity inlets into the river in 2014 (units: tons/year).

Area	COD Inflow	NH$_3$-N Inflow	TN Inflow	TP Inflow
Bamao River	148.66	17.96	27.04	4.859
Longtan River	108.67	13.87	19.65	2.678
Fanjiafang River	116.8	14.6	21.36	3.34
Dongtan River	89.292	12.23	16.23	1.573
Reservoir region	197.64	97.54	179.03	44.375
Total amount	661.062	156.2	263.31	56.825

Table 8. The predicted values of the pollutant loads in the tributaries of the Huangshi Reservoir in 2020 and 2025 (units: tons/year).

Year	Pollutants	Bamao River	Longtan River	Fanjiafang River	Dongtan River
2020	COD	764.73	482.54	428.13	239.27
	NH$_3$-N	70.56	45.65	39.14	22.43
	TN	136.2	86.16	75.46	41.72
	TP	59.98	38.576	32.24	16.446
2025	COD	981.21	604.68	549.45	298.1
	NH$_3$-N	86.48	54.60	48.21	27.08
	TN	172.16	106.32	95.67	51.42
	TP	72.42	44.424	39.21	19.544

Table 9. The predicted values of the pollutant concentrations in the tributaries of the Huangshi Reservoir in 2020 and 2025 (units: tons/years).

Year	Pollutants	Bamao River	Longtan River	Fanjiafang River	Dongtan River
2020	COD	16.97	11.98	20.18	18.55
	NH$_3$-N	1.57	1.13	1.85	1.74
	TN	3.02	2.14	3.56	3.24
	TP	1.33	0.96	1.52	1.28
2025	COD	21.77	15.02	25.90	23.12
	NH$_3$-N	1.92	1.36	2.27	2.10
	TN	3.82	2.64	4.51	3.99
	TP	1.61	1.10	1.85	1.52

The hydraulic conditions adopted the average level of the status year and have no prediction for the future hydraulic conditions, that is, the hydraulic conditions almost stay constant in the model. The annual average wind speed used in Huangshi Reservoir was 2 m/s, the dominant wind direction was northeast, and the annual average evaporation was 1160 mm. The hydrological calculation of WEC was based on the driest monthly flow in 2013 and 2014. The pollutant loads estimated in 2020 and 2025 are allocated on a monthly basis.

The water quality of the Huangshi Reservoir without reduction measures simulated the predicted values of the inflow for COD, NH$_3$-N, TN, and TP in 2020 and 2025. The results are displayed in Figures 4 and 5, which respectively show the following:

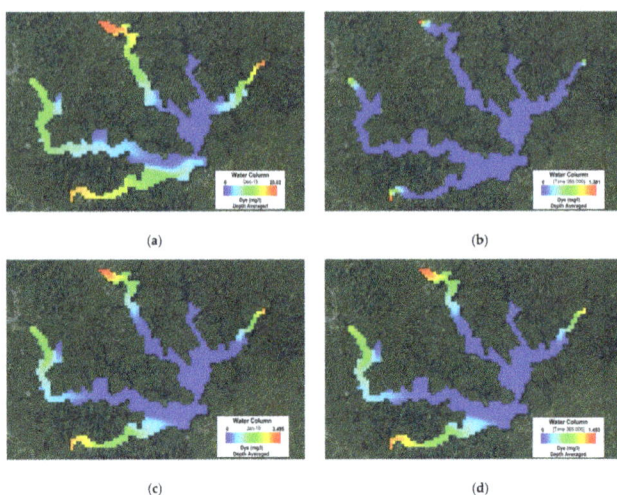

Figure 4. The simulated concentration distribution of COD, NH₃-N, TN, and TP in the Huangshi Reservoir in 2020. ((**a**) COD, (**b**) NH₃-N, (**c**) TN and (**d**) TP).

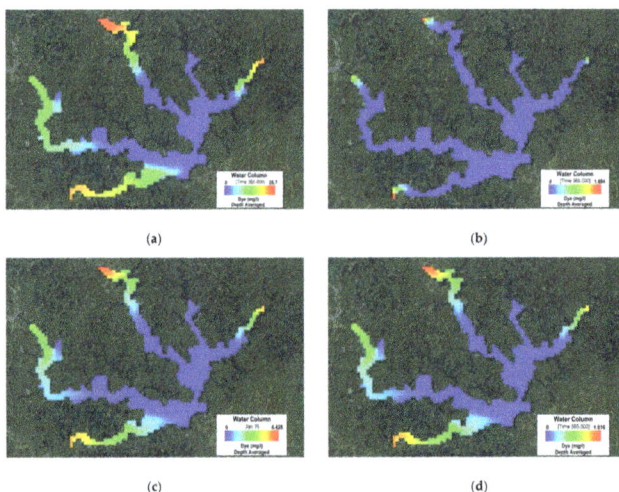

Figure 5. The simulated concentration distribution of COD, NH₃-N, TN, and TP in Huangshi Reservoir in 2025. ((**a**) COD, (**b**) NH₃-N, (**c**) TN and (**d**) TP).

(a) In 2020, the affected areas are the incoming parts of the tributaries, based on the concentration distribution of COD and NH₃-N. The main reservoir area is less affected. TN has a bigger area of the effect. In addition, the influence of TP from the tributaries on the water quality of the reservoir is confined to the inner part of the tributaries in the reservoir.

(b) In 2025, the influences of the tributaries on the water quality of the reservoir are confined to the inner parts of the tributaries in the reservoir, but they have bigger areas of effect on the concentrations of COD, NH₃-N, TN, and TP compared to those in 2020.

4.2. Water Environmental Capacity of the River–Reservoir Combination System

As the Huangshi Reservoir is a water source protection zone, the water quality protection target is degree II. Assuming that the water quality degree of the initial section is degree I when the water quality degree of the reservoir control point is degree II, the results of the WEC for the tributaries of each section are shown in Table 10.

Table 10. The water environmental capacities of the Huangshi Reservoir and tributaries (units: tons/year).

Sections	Water Quality Target	Water Environmental Capacity			
		COD	NH$_3$-N	TN	TP
Bamao River	II	185.46	15.37	13.57	3.60
Longtan River	II	131.32	13.87	12.11	3.22
Fanjiafang River	II	58.56	7.33	6.38	1.69
Dongtan River	II	46.63	4.42	3.88	1.03
Sum of rivers		421.97	40.99	35.94	9.54
Reservoir region (inner part)	II	4850	200	160	38
Total amount		5271.97	240.99	195.94	47.54

The highest WEC values of the COD, NH$_3$-N, TN, and TP were in the Bamao River, while the lowest WEC values of the COD, NH$_3$-N, TN, and TP were in the Dongtan River. The total WEC of the tributaries included a COD of 421.97 tons; an NH$_3$-N of 40.99 tons; a TN of 35.94 tons; and a TP of 9.54 tons. Thus, it is obvious that the WEC of the Huangshi Reservoir region mainly accounts for more than 93% of the total capacity.

4.3. Total Amount of Pollutant Control Goal Based on the Water Environmental Capacity

Based on the pollutant loads discharged into the river in 2014, the utilized peculiarities of the reservoir and tributaries, as well as the water environmental capabilities, were estimated and shown in Table 11.

The utilization rates of COD, NH$_3$-N, TN, and TP in most tributaries were over 100%. The highest utilization rate of COD and TP occurred in the Fanjiafang River, exceeding 100% by 2 times and 1.98 times, respectively. The highest utilization rate of NH$_3$-N and TN occurred in the Dongtan River, exceeding 100% by 2.77 times and 4.18 times, respectively.

The utilization rates of COD and NH$_3$-N in the reservoir region were 13% and 65%, respectively, while there was still a certain WEC. However, the maximum utilization rate of TN and TP exceeded 100% by 1.34 times and 1.2 times, respectively. These were mainly caused by the fish cage culture and the internal release of pollutants. Furthermore, this rate had a significant relationship with the high degree of eutrophication in the reservoir region.

Table 11. The water environmental capacity utilizations of the Huangshi Reservoir in 2014 (units: tons/year).

Area	COD			NH₃-N			TN			TP		
	Inflow	Capacity	Utilization Ratio (%)	Inflow	Capacity	Utilization Ratio (%)	Inflow	Capacity	Utilization Ratio (%)	Inflow	Capacity	Utilization Ratio (%)
Bamao River	148.66	185.46	80	17.96	15.37	117	27.04	13.57	199	4.859	3.6	135
Longtan River	108.67	131.32	83	13.87	13.87	100	19.65	12.11	162	2.678	3.22	83
Fanjiafang River	116.8	58.56	199	14.6	7.33	199	21.36	6.38	335	3.34	1.69	198
Dongtan River	89.292	46.63	191	12.23	4.42	277	16.23	3.88	418	1.573	1.03	153
Reservoir region	197.64	4850	4	97.54	200	49	179.03	160	112	44.375	38	117
Total amount	661.062	5271.97	13	156.2	240.99	65	263.31	195.94	134	56.825	47.54	120

According to the pollutant inflows into the reservoir and the WEC results, the pollutant control target can be determined based on the WEC. The water pollutant reductions of the Huangshi Reservoir region as well as the Bamao River, Longtan River, Fanjiafang River, and Dongtan River—which are the 4 major tributaries into the reservoir—have been determined for 2020 and 2025 (Tables 12 and 13).

Table 12. The pollutants reduction targets of the Huangshi Reservoir in 2020 (units: tons/year).

Area	COD	NH$_3$-N	TN	TP
	Reductions	Reductions	Reductions	Reductions
Bamao River	10.39	6.59	20.93	3.358
Longtan River	7.18	2.73	12.16	0.457
Fanjiafang River	92.74	9.89	20.55	3.062
Dongtan River	63.55	9.93	15.61	1.055
Reservoir region	−4560.48 *	−58.02 *	84.81	23.233
Total amount	−4386.62	−28.88	154.06	31.165

* Note: the symbol of "−"delegates' available capacity without reduction.

Table 13. The pollutants reduction targets of the Huangshi Reservoir in 2025 (units: tons/year).

Area	COD	NH$_3$-N	TN	TP
	Reductions	Reductions	Reductions	Reductions
Bamao River	86.64	13.78	33.19	6.17
Longtan River	61.22	8.19	20.68	1.898
Fanjiafang River	150.44	16.06	29.96	4.961
Dongtan River	103.48	14.23	21.89	1.827
Reservoir region	−4445.73 *	−2.79 *	175.45	44.122
Total amount	−4043.95	49.47	281.17	58.978

* Note: the symbol of "−" delegates' available capacity without reduction.

The pollutants in the tributaries of the Huangshi Reservoir will need a large number of cuts both in 2020 and 2025, with respect to the degree II national surface water quality standards. The largest COD reduction will occur in the Fanjiafang River, with approximately 61.3% and 71.9% of the inflow needing to be cut to meet the water quality objectives, respectively. Tributaries will need a total reduction of 173.8 to 401.8 tons per year, with the residual WEC for COD being approximately 4560.5 tons per year in the reservoir region.

The largest NH$_3$-N reduction will occur in the Dongtan River and the Fanjiafang River in 2020 and 2025. More than 69.2% of the pollutant inflow will need to be cut to meet the water quality objectives in 2020, which corresponds to an increase of 7.1% in 2025. Tributaries will need a total reduction of 29.14 and 52.2 tons per year, respectively, with the residual WEC for NH$_3$-N needing to be approximately 4560.5 and 2.79 tons per year in the reservoir region, respectively.

The largest TN reduction will occur in the Dongtan River in 2020 as well as in the Dongtan River and the Fanjiafang River in 2025. More than 80.1% and 74.59% of the inflow will need to be cut to meet the water quality objectives, respectively. The TN of the entire basin will need to be cut by 44.02% and 58.9%, respectively, which corresponds to an increase of 14.88% in 2025.

The largest TP reduction will occur in the Dongtan River and the Fanjiafang River in 2020 and 2025. More than 64.4% of the inflow will need to be cut to meet the water quality objectives, which corresponds to an increase of 10.19% in 2025. The inflow into the reservoir will be significantly larger than the WEC of the reservoir. The TP of the reservoir area will need to be reduced by 37.9% and 53.73%, respectively, with the entire basin needing to be cut by 39.6% and 55.37%, respectively.

Water pollution control is still an arduous task in all four tributaries and the reservoir region. The pollutants significantly exceed the state standards for drinking water sources in China, especially in the reservoir region, which will induce or aggravate the risk of water eutrophication.

5. Conclusions

This study constructed an integrated method to simulate the WEC of the Huangshi river–reservoir combination system, which coupled a 1-D convection-dispersion model with the EFDC model. It was further applied to study the quantity control of pollution of the input rivers based on the results of the WEC calculation and the degree II national surface water quality standards.

The highest WEC is in the Bamao River, while the lowest WEC is in the Dongtan River. Summarizing the WEC of the tributaries, the COD is 421.97 tons, NH_3-N is 40.99 tons, TN is 35.94 tons, and TP is 9.54 tons.

The utilization rates of COD, NH_3-N, TN, and TP in most of the tributaries are over 100%. The utilization rates of COD and NH_3-N in the reservoir region are 13% and 65%, respectively, and there is still a certain WEC. However, the loads of TN and TP exceed the maximum utilization rates by 1.34 times and 1.2 times, respectively. This is mainly caused by the fish cage culture and the internal release of pollutants. Furthermore, this has a significant relationship with the high degree of eutrophication in the reservoir region.

The pollutants in the tributaries of the Huangshi Reservoir will need a large number of inflow cuts to meet the water quality objectives in 2020 and 2025. The pollutants significantly exceed the state standard, especially in the reservoir region. Therefore, the reduction and control of pollutants are the main objectives.

To exploit the WEC completely, rationally, and continuously, a balance point needs to be found between environmental concerns and the social and economic development. As the study area is the backup source of drinking water for Changde City, the calculation of WEC can enable decision makers to determine load reductions and allocations. Various measures of controlling the total amount of pollutants can be applied to realize the improvement of the water quality. This perspective can be referred as a beneficial supplement for a single study of rivers or lake.

Additionally, in calculating WEC, the uncertainty analysis should be taken into consideration. In a practical river—reservoir combination system there are many uncertainties in terms of variable information while calculating WEC. More variable information will result in a more flexible WEC.

Acknowledgments: This research was supported by the National Water Pollution Control and Treatment Science and Technology Major Project (2017ZX07301-001), National Key Research and Development Program (2017YFC0404505). We would like to extend special thanks to the editor and the anonymous reviewers for their valuable comments in greatly improving the quality of this paper.

Author Contributions: F.Z. conceived and designed the experiments; C.L. performed the experiments; L.C. and F.Z. analyzed the data; L.C. and Y.Z. contributed reagents/materials/analysis tools; F.Z. wrote the paper.

Conflicts of Interest: The authors declare no conflict of interest.

References

1. Liu, W.; Chen, W.; Hsu, M.H. Using a three-dimensional particle tracking model to estimate the residence time and age of water in a tidal estuary. *Comput. Geosci.* **2011**, *37*, 1148–1161. [CrossRef]
2. Su, Q.; Qin, H.; Fu, G. Environmental and ecological impacts of water supplement schemes in a heavily polluted estuary. *Sci. Total Environ.* **2014**, *472*, 704–711. [CrossRef] [PubMed]
3. Shen, Y.; Wang, J.; Zheng, B.; Zhen, H.; Feng, Y.; Wang, Z.; Yang, X. Modeling study of residence time and water age in Dahuofang Reservoir in China. *Sci. China Phys. Mech. Astron.* **2011**, *54*, 127–142. [CrossRef]
4. Zhang, Y. The Development of Basic Concept of Water Environmental Capacity. *Res. Environ. Sci.* **1992**, *5*, 59–61. (In Chinese)
5. Yang, Z.; Cheng, C.; Tan, X.; Cheng, R.; Zhong, M.A. Analysis of water environmental capacity of Guanting Reservoir and its upstream basin. *J. Arid N.A. Res. Environ.* **2015**, *29*, 164–166. (In Chinese)
6. Gao, G.Q. Water environmental carrying capacity calculation and protection measures on Hefang Reservoir. *Appl. Mech. Mater.* **2011**, *1446*, 2537–2540. [CrossRef]
7. Han, L.X.; Yan, F.F.; Peng, H.; Gao, J.J.; Pan, M.M. Methods for calculation of water environment capacity of small and medium river channels. *Adv. Mater. Res.* **2013**, *2115*, 2745–2750. [CrossRef]

8. Gang, Z.; Kun, L.; Guo, F.U.; Guang-Jun, M. Calculation method of river water environmental capacity. *J. Hydraul. Eng. ACSE* **2014**, *45*, 227–234.

9. Li, Y.; Zhao, L.; Li, Y. Index system of water environment capacity promotion in Jilin province in China through ways of structure, engineering and supervision. *Adv. Mater. Res.* **2015**, *3848*, 1155–1159. [CrossRef]

10. Yang, J.F.; Lei, K.; Khu, S.; Meng, W.; Qiao, F. Assessment of water environmental carrying capacity for sustainable development using a coupled system dynamics approach applied to the Tieling of the Liao River Basin, China. *Environ. Earth Sci.* **2015**, *73*, 5173–5183. [CrossRef]

11. Li, K.; Zhang, L.; Li, Y.; Zhang, L.J.; Wang, X.L. A three-dimensional water quality model to evaluate the environmental capacity of nitrogen and phosphorus in Jiaozhou Bay, China. *Mar. Pollut. Bull.* **2015**, *91*, 306–316. [CrossRef] [PubMed]

12. Tao, Y.; Chen, Y.X.; Zhao, X.L.; Yang, L.B.; Lei, K.; Zhou, G. Analysis of seasonality difference on water environmental capacity of Ashi River based on EFDC. *Environ. Eng.* **2017**, *35*, 66–68. (In Chinese)

13. Zhou, X.Y.; Lei, K.; Meng, W.; Khu, S.; Zhao, J.; Wang, M.N.; Yang, J.F. Space–time approach to water environment carrying capacity calculation. *J. Clean. Prod.* **2017**, *149*, 302–312. [CrossRef]

14. Zhang, X.; Han, L.X. A water environmental capacity calculation method of the shallow lake based on water quality response coefficient. *Sichuan Environ.* **2017**, *36*, 47–49. (In Chinese)

15. Wu, J.F.; Gao, Y.; Xie, F.; Chen, L.N. Research on water environment capacity by two-dimensional hydrodynamic and quality model. *Environ. Sci. Manag.* **2017**, *42*, 53–55. (In Chinese)

16. Gu, J.; Li, Z.Y.; Mao, X.D.; Hu, C.F.; Kuang, C.P.; Zhang, W.L. Study of cod environmental capacity in coastal waters of Beidaihe in summer. *Mar. Environ. Sci.* **2017**, *36*, 683–685. (In Chinese)

17. Wu, H.; Yang, X. Calculation of water environmental capacity based on control unit of Shuangtaizi River. *Environ. Sci. Manag.* **2017**, *42*, 61–63. (In Chinese)

18. Chinh, L.V.; Hiramatsu, K.; Harada, M.; Lan, T.T. Estimation of water environment capacity in the cau river basin, vietnam using the streeter-phelps model. *J.- Fac. Agric. Kyushu Univ.* **2017**, *62*, 163–169.

19. Huang, S.L.; Zhang, Y.; Li, Q.; Xu, D.S. Research on water environmental capacity of urban river: A case study of Tuohe River in Suzhou city, northern Anhui province. *Adv. Mater. Res.* **2012**, *1479*, 867–870. [CrossRef]

20. Fan, L.L.; Sha, H.F.; Pang, Y. Water environmental capacity of Lake Taihu. *J. Lake Sci.* **2012**, *24*, 693–697.

21. Xia, Y.; Ma, C.M.; Liu, X.J.; Liu, X.Y.; Deng, Q.L. Comprehensive carrying capacity assessment of water environment in Zhengzhou city. *Appl. Mech. Mater.* **2013**, *2574*, 390–395. [CrossRef]

22. Yan, B.Y.; Xing, J.S.; Tan, H.R.; Deng, S.P.; Tan, Y.N. Analysis on water environment capacity of the Poyang Lake. *Procedia Environ. Sci.* **2011**, *10*, 2754–2759.

23. Liu, Q.; Jiang, J.; Jing, C.; Qi, J. Spatial and seasonal dynamics of water environmental capacity in mountainous rivers of the Southeastern Coast, China. *Int. J. Environ. Res. Public Health* **2018**, *15*, 99. [CrossRef] [PubMed]

24. Missaghi, S.; Hondzo, M. Evaluation and application of a three-dimensional water quality model in a shallow lake with complex morphometry. *Ecol. Model.* **2010**, *221*, 1512–1525. [CrossRef]

25. Missaghi, S.; Hondzo, M.; Melching, C. Three-dimensional lake water quality modeling: Sensitivity and uncertainty analyses. *J. Environ. Qual.* **2013**, *42*, 1684–1698. [CrossRef] [PubMed]

26. Missaghi, S. Three Dimensional Water Quality Modeling in a Shallow Lake with Complex Morphometry; Implications for Cool Water Fish Habitat under Changing Climate. Ph.D. Thesis, University of Minnesota, Minneapolis, MN, USA, July 2014.

27. Heung, W.; Hu, B.Q. Application of interval clustering approach to water quality evaluation. *J. Hydrol.* **2013**, *491*, 1–12.

28. Heung, W.; Hu, B.Q. Application of improved extension evaluation method to water quality evaluation. *J. Hydrol.* **2014**, *509*, 539–548.

29. Peche, R.; Esther, R.E. Development of environmental quality indexes based on fuzzy logic. A case study. *Ecol. Indic.* **2012**, *23*, 555–565. [CrossRef]

30. Ju, H.C.; Yoo, S.H. Using the fuzzy set theory to developing an environmental impact assessment index for a thermal power plant. *Qual. Quant.* **2014**, *48*, 673–680. [CrossRef]

31. Chen, Q.W.; Wu, W.Q.; Blanckaert, K.; Ma, J.F.; Huang, G.X. Optimization of water quality monitoring network in a large river by combining measurements, a numerical model and matter-element analyses. *J. Environ. Manag.* **2012**, *110*, 116–124. [CrossRef] [PubMed]

32. Jordan, T.E.; Weller, D.E.; Correll, D.L. Sources of nutrient inputs to the Patuxent River estuary. *Estuaries* **2003**, *26*, 226–243. [CrossRef]
33. Shen, J.; Haas, L. Calculating age and residence time in the tidal York River using three-dimensional model experiments. *Estuarine Coast. Shelf Sci.* **2004**, *61*, 449–461. [CrossRef]
34. Li, Y.; Acharya, K.; Yu, Z. Modeling impacts of Yangtze River water transfer on water ages in Lake Taihu, China. *Ecol. Eng.* **2011**, *37*, 325–334. [CrossRef]
35. Ministry of Environmental Protection of the people's Republic of China. The surface water environmental quality standard of China. GB3838-2002.
36. Chen, L.B.; Yang, Z.F.; Liu, H.F. Numerical simulations of spread characteristics of toxic cyanide in the Danjiangkou Reservoir in China under the effects of dam cooperation. *Math. Probl. Eng.* **2014**, *2014*. [CrossRef]
37. Hzamrick, H.M. *A Three-Dimensional Environmental Fluid Dynamics Computer Code: Theoretical and Computational Aspects*; The College of William and Mary, Virginia Institute of Marine Science: Williamsburg, VA, USA, 1992.
38. Caliskan, A.; Elci, S. Effects of selective withdrawal on hydrodynamics of a stratified reservoir. *Water Resour. Manag.* **2009**, *23*, 1257–1273. [CrossRef]
39. Wang, C.; Shen, C.; Wang, P.F.; Qian, J.; Hou, J.; Liu, J.J. Modeling of sediment and heavy metal transport in Taihu Lake, China. *J. Hydrodyn. Ser. B* **2013**, *25*, 379–387. [CrossRef]
40. Zhang, C.X.; You, X.Y.; Zhao, S.M. Application of the EFDC model to waterfront planning: A case study in the Tianjin Harbor Economic Zone, China. *Eng. Appl. Comp. Fluid Mech.* **2014**, *8*, 1–13. [CrossRef]
41. Park, K.; Kuo, A.Y. *A Three-Dimensional Hydrodynamic Eutrophication Model: Description of Water Quality and Sediment Process Sub-Models*; The College of William and Mary, Virginia Institute of Marine Science: Williamsburg, VA, USA, 1995.

water

MDPI

Article

Sensitivity Analysis for the Inverted Siphon in a Long Distance Water Transfer Project: An Integrated System Modeling Perspective

Sifan Jin [1], Haixing Liu [1], Wei Ding [1,*], Hua Shang [2] and Guoli Wang [1]

[1] School of Hydraulic Engineering, Dalian University of Technology, Dalian 116024, China; sfjin@mail.dlut.edu.cn (S.J.); hliu@dlut.edu.cn (H.L.); wanggl@dlut.edu.cn (G.W.)

[2] Faculty of Management and Economics, Dalian University of Technology, Dalian 116024, China; shanghua@dlut.edu.cn

* Correspondence: weiding@dlut.edu.cn; Tel.: +86-0411-8470-7904

Received: 3 February 2018; Accepted: 6 March 2018; Published: 8 March 2018

Abstract: Long distance water diversion projects are developed to alleviate the conflicts between supply and demand of water resources across different watersheds. However, the significant scale water diversion projects bring new challenges for the water supply security. This paper presents the flood risk of inverted siphon structure which is used for crossing transversally in the water diversion project through sensitivity analysis. Soboĺ and regionalized sensitivity analysis are used to investigate the sensitive parameters of the integrated model and the sensitive range of the parameters, respectively. The integrated system model consists of the hydrologic model, the sediment transport model and the siphon hydraulic model to determine the flood overtopping duration and volume, which are used to quantify flood risk in this study. The flood overtopping duration and volume indicators are used to quantify flood risk in the sensitivity analysis. The South to North Water Diversion Project in China is used as a case study. The results show the mean rainfall and roughness coefficient of the pipe are the most sensitive parameters in the integrated models, while the sensitive range of these two parameters are distinct. The sensitivity analysis of the inverted siphon provides an insight into the significant contributions to the flood risk. The analysis can provide the guidance for the system operation security.

Keywords: long distance water diversion; inverted siphon; sensitivity analysis; integrated supply system modeling

1. Introduction

Due to urbanization and the uneven distribution of water resources in time and space, long-distance water transfer projects are constructed to alleviate the shortage of water resource and meet the increasing demands [1–3]. Water transfer projects often involve huge capital investment and pose complex security problems [4–6]. For example, canals usually cross hundreds of kilometers of complicated terrain, which leads to water quality deterioration, temperature variation, and long operation response period. Moreover, critical hydraulic structures are essential for ensuring the operation security of the projects, e.g., gate/valve, pump station, and intersection structures. Here, we focus on the inverted siphon structure, which is linked to the river to drain water by gravity that is collected from a relatively small hydrographic basin, so that the river can cross transversally a large artificial open channel carrying fresh water for supply.

Inverted siphon structures are prone to be impacted by the hydraulic transient process and structure stability. The potential failures of inverted siphons can be divided into two categories: (1) structure failure; and (2) operation failure. Structure failure indicates the structure identity is destroyed or collapsed due to structure aging and external forces. Operation failure refers to the flows exceeding the design standard of inverted siphon. That is because the designed flood and flood design standard that are used to guide the structure design are subject to the hydrologic parameter uncertainties (e.g., rainfall, soil moisture, and land use) and the hydraulic uncertainties (e.g., sediment deposition). The extra upstream flow of inverted siphon is retained at the entry and a large flood can potentially enter the canal and contaminate the quality of source water in the canal. Therefore, there is a need to investigate which uncertain parameters (e.g., rainfall, sediments, and pipe roughness coefficient) can significantly contribute to the flooding incidents. This paper utilizes an inverted siphon structure that passes underneath a long distance transfer project to illustrate the flood overtopping risk. The sensitivity analysis with respect to inverted siphons is of vital importance to guarantee the safe operation of the long-distance transfer project.

Sensitivity Analysis (SA) has gathered plenty of attention for describing the sensitivity of parameters in terms of the contributions to the model output [7]. Global SA method, in contrast to local SA, is capable of accounting for the whole range of input parameter variation to avoid the subjective judgment and case-specific characteristics on the parameter range. The global SA can deal with the non-linear and non-monotonic models [8–10]. Another type of SA is regression-related, e.g., Multivariate Adaptive Regression Splines (MARS), Gaussian Process (GP) and Radial Basis Function [11]. They use linear or non-linear models to refit the original model and investigate which parameters can give the relative large reduction in goodness-of-fit. The large reduction represents the sensitive parameters [12]. The regression sensitivity methods have the advantage of less computational effort. Moreover, Soboĺ SA [13] is typically a variance based method. It decomposes the response variances for the specific order SA indices. The Soboĺ SA method can calculate the interactions among input parameters, but it should be noted that the higher order Soboĺ analysis could significantly increase the computational burden with the increase in the input number [14–16]. Regional Sensitivity Analysis (RSA) was developed by Spear and Hornberger [17] and improved by Beven and Binley [18], which is also a global SA method. RSA elaborates the sensitivity variation over the full range for a given parameter. The cumulative possibility of behavioral sets is investigated in the RSA to reflect the parameter interaction implicitly. Both Soboĺ and RSA methods are employed in this paper to ascertain the sensitive parameters in the inverted siphon flooding model [19].

This paper aims to address the sensitivities of any value over the domain of the parameters and further identifies the sensitive parameters in the inverted siphon models. The sensitive parameter screening and sensitive range identification of the parameters are implemented by Soboĺ SA and RSA methods, respectively. The upstream runoff of the siphon is simulated by a local hydrologic model, and the sediment transportation and pipe transmission model are used to calculate the inverted siphon flows. The evaluation indexes (flood overtopping duration and volume) are set up to demonstrate the flood risk of the inverted siphon. The sensitivity analysis results are demonstrated based on an inverted siphon structure across the South to North Water Diversion project.

2. Methods

The sensitivity analysis methods including Soboĺ SA and RSA method are introduced to evaluate the sensitivity of parameters in the integrated system modeling. The integrated system model is set up by integrating hydrologic model, sediment transport model and inverted siphon hydraulic model. Then, the two evaluation indicators, i.e., flood overtopping duration and volume, are used to quantify flood risk.

2.1. Sensitivity Analysis Methods

2.1.1. General

Two global sensitivity analysis methods are employed in this paper for investigating parameter sensitivity in consideration with the interaction of variables and the sensitivity variation over the range of the variables. The model can be represented by a numerical function,

$$Y = f(X) = f(x_1, \ldots, x_n) \tag{1}$$

where Y is the model output (or objective function) and X is the variable set, $x_1, \ldots, x_n$.

2.1.2. Sobol Sensitivity Analysis

The Sobol sensitivity analysis [13,20] is a variance-based method, which uses variance decomposition to derive a variance ratio. It can provide a quantitative description of how individual variables and their interactions affect model performance [21]. An individual model parameter and its interaction with other parameters contribute to the total output variance, and the function is shown as follows:

$$V_S = \sum_{i=1}^{n} V_i + \sum_{i=1}^{n}\sum_{j<i}^{n} V_{ij} + \cdots + V_{1,2,\ldots,n} \tag{2}$$

where V_S is the total variance of the output variable Y; V_i is given by the variance of the conditional expectation $V_i = V[E(Y|x_i)]$ and $V_{ij}(V_{ij} = V[E(Y|x_i, y_i)] - V_i - V_j)$ to $V_{1\ldots k}$ the interactions among k parameters. To assess the role of each variable or interaction between variables, sensitivity measures are needed. The chosen measures are known as Sobol indices. The indices represent the bias in the variance of the output, which is attributed to a variable or a combination of variables. The first-order index (S_i) is

$$S_i = \frac{V_i}{V(Y)} \tag{3}$$

and the second-order index (S_{ij}) is

$$S_{ij} = \frac{V_{ij}}{V(Y)} \tag{4}$$

The total order sensitivity index of a single parameter and this parameter's interaction with other parameters, at least one index being $j \neq i$ from 1 to k, is as follows:

$$S_{Ti} = \sum S_i + \sum_{j \neq i} S_{ij} + \cdots + S_{1\ldots k} \tag{5}$$

The first-order sensitivity index only represents the individual contribution of variable x_i to the model output. The second-order index indicates the interaction effect of two variables $(x_i, x_j, i \neq j)$ on the model output. The total-order index (S_{Ti}) measures the main effect of parameter x_i and its interactions with all the other variables. The Sobol indices are obtained by a sampling process, e.g., Latin Hypercube.

2.1.3. Regionalized Sensitivity Analysis

Regionalized sensitivity analysis (RSA) is proposed by Spear and Hornberger [17] and further extended by Beven and Binley [18]. RSA method is broadly applied in hydrology and environmental system analysis [19,22,23]. The approach is based on the Monte Carlo simulation considering possible combination of uncertain parameters with the given possibility density function. The parameters sampling process can cover the whole distribution range, so RSA also belongs to the global sensitivity analysis category. The sampled parameter sets are divided into behavioral or non-behavioral. If the computational result of a parameter set (objective function evaluations) satisfies the prescribed condition (e.g., less than a threshold), the parameter set is behavioral, vice versa.

RSA results are expressed by the cumulative distribution. The difference between the behavioral and non-behavioral cumulative distributions is larger, and then the parameter is more sensitive.

Kolmogorov-Smirnoff (K-S) test is used to show the maximum vertical distance ($d_{m,n}$) between the behavioral and non-behavioral cumulative distributions. The K-S test function is given as

$$d_{m,n} = \sup_x |S_B(x) - S_{NB}(x)| \tag{6}$$

where $S_B(x)$, $S_{NB}(x)$ are the behavioral and non-behavioral cumulative distributions, respectively.

2.2. Definition of Evaluation Indicators

If the water level at the inlet of the inverted siphon exceeds the embankment crest elevation of the canal (denoted by Z_s), then the flooding water will flow into the main trunk canal, i.e., flood overtopping happens. The event occurrence represents the inverted siphon hydraulic failure. This failure will bring the risks that are the embankment erosion and the water quality pollution for example. The longer duration of flood overtopping leads to more severe hazards and exaggerating the impacted extent on the canal. Therefore, we adopt the flood overtopping duration and flood overtopping volume as the indicators to evaluate the risk of flood overtopping for the inverted siphon.

It is assumed that the water level at the water inlet of the inverted siphon can rise, even if it exceeds the crest elevation of the canal Z_s (i.e., there is a virtual water pond with unlimited crest elevation). The time periods, when water level exceeds Z_s, are defined as the duration of flood overtopping. The flood overtopping volume can be calculated by the difference of maximum flood volume at the inlet and the volume that corresponds to the embankment crest elevation. The volumes are calculated by the water level–storage relationship at the inlet of the inverted siphon.

2.3. System Modeling

2.3.1. General

The input parameters that need to be tested in the sensitivity analysis are assigned by a set of random values ($p_1, p_2, p_3, \cdots, p_n$). The hydrologic model simulates the inflow of the inverted siphon in a given watershed. Sediment transport model is introduced to model the sand movement and deposition. Siphon hydraulic model is used to calculate the flow of the siphon. These three models are integrated and convey the parameter values, as shown in Figure 1. The outputs of the models are runoff, flow, sand content and water level. Two evaluation indexes are formulated by these outputs. The whole flow chart of the methodology is shown in Figure 1.

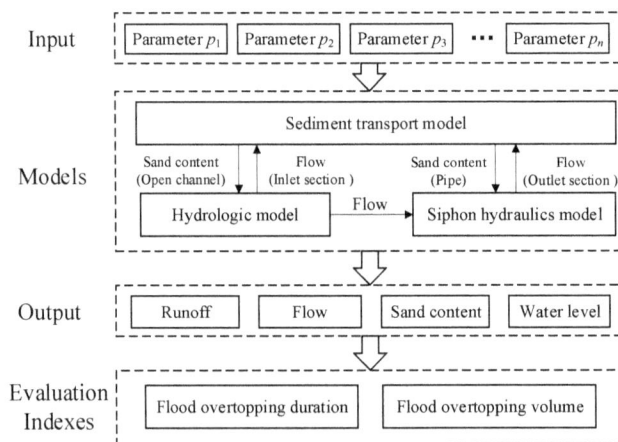

Figure 1. Flow chart of the models.

2.3.2. Hydrologic Model

Due to the lack of monitoring data in small watersheds, the Inferential Formula Method is often used to determine the design floods [24]. This study uses the Inferential Formula Method specified in China's design flood regulations and proposed by Chen [25]. The flood is derived from design rainfall in the assumption that rainfall frequency is the same with flood frequency. The flood calculation includes the following steps:

(1) The maximum rainfall

The maximum rainfall during *t* h (for small watersheds, *t* is set to be 24 h) can be calculated by

$$x_{t,P} = \overline{x_t}\left(1 + C_v \cdot \Phi_p\right) \tag{7}$$

where $x_{t,P}$ represents the amount of rainfall with the frequency P for t h (mm); $\overline{x_t}$ is the average of maximum rainfall for t h (mm); Φ_p is a frequency factor; and C_v is the coefficient of variation of maximum rainfall for t h.

(2) The peak discharge

Rational method [26] is one of the earliest methods to estimate the peak discharge according to the rainfall data. When the runoff generation time (denoted by t_c) is larger than the runoff concentration time (denoted by τ), i.e., $t_c \geq \tau$, the peak discharge is consisted of the runoff from the whole watershed. Assuming the runoff intensity is evenly distributed both in spatial and temporal, and the concentration is irrespective to the channels and slops, then the peak discharge is calculated by

$$Q_{mP} = 0.278\left(\frac{h_\tau}{\tau}\right)F \tag{8}$$

in which, F denotes the area of the watershed (km^2) and τ is the runoff concentration time (h), calculated based on the terrain characteristic for the given area,

$$\tau = 0.278\frac{L}{mJ^{1/3}Q_{mP}^{1/4}} \tag{9}$$

h_τ is the amount of runoff generation during τ hours (mm), given by the following equation under the assumption of excess infiltration, that is, there is runoff generation only when the rainfall intensity is larger than the infiltration rate.

$$h_\tau = x_{t,P}t^{\gamma-1}\tau^{1-\gamma} - \mu\tau \tag{10}$$

In Equations (9) and (10), γ is the rainfall diminishing exponent; μ is the average infiltration rate during τ h (mm/h); L is the longest length from the outlet of main river to watershed outline (km); J is the average slope for the longest path of runoff; m is the runoff parameter; and Q_{mP} is the peak discharge for the flood with the frequency P (m^3/s).

When the runoff generation time is smaller than the runoff concentration time, i.e., $t_c < \tau$, the peak discharge is consisted of the runoff from partial of the watershed, and the peak discharge is approximately calculated by the following equation [25]

$$Q_{mP} = 0.278\left(\frac{h_R}{\tau}\right)F \tag{11}$$

where the runoff generation h_R (mm) is calculated by

$$h_R = x_{t,P}t^{\gamma-1}t_c^{1-\gamma} - \mu t_c \tag{12}$$

and the runoff generation time can be calculated by the following equation:

$$t_c = \left[\frac{(1-\gamma)x_{t,P}t^{\gamma-1}}{\mu} \right]^{\frac{1}{\gamma}} \tag{13}$$

(3) The flood hydrograph

The generalized triangle hydrograph method [25], including the following steps, is applied to obtain the flood hydrograph.

Firstly, the total runoff during t hours is allocated to T periods in terms of the designed runoff hydrograph and the runoff concentration time (τ). The rainfall duration for each period (denoted by t_i and $I = 1, 2, \ldots T$) is larger than the runoff concentration time, that is $t_i \geq \tau$.

Secondly, assume the runoff generation time equals the rainfall time, and the concentration time in each period equals that of the peak discharge for the whole hydrograph (calculated by Equation (9)). Then, the peak discharge resulting from rainfall in ith period can be calculated by the following equation with the generalized triangle hydrograph.

$$Q_{ti} = 0.556 \left(\frac{h_{ti}}{t_i + \tau} \right) F \tag{14}$$

where h_{ti} is the runoff of rainfall in ith period (mm); Q_{ti} is the peak flow at t_i (m³/s). It is worth mentioning that, when $t_i = \tau$ (h), Equation (14) is the same with Equation (8).

Thirdly, the flood hydrograph resulting from rainfall in each period can be obtained with the peak flow, and the time for the flood hydrograph is the sum of rainfall duration and runoff concentration time, i.e., $(t_i + \tau)$. When $t_i = \tau$, the hydrograph resulting from runoff ith periods is symmetric, i.e., equilateral triangle hydrograph. Otherwise, when $t_i > \tau$, the resulting hydrograph is asymmetric. Finally, the flood hydrograph for rainfall during t h can be convoluted.

2.3.3. Sediment Transport Model

The sensitivity analysis is conducted under the worst condition that the water with the given velocity contains the maximum amount of sands. If the velocity in the pipe shows down, the deposition can occur. It is therefore assumed that the flood can carry the maximum amount of sediments for the given flow velocity. Since the fluid regime is complicated and diversified inside the structure, the regression analysis to formulate an empirical formula is more practical. Here the capacity of sediment transport is fitted based on the field data. The Guojunke formula [27], which uses the logarithmic function to fit the field data and performs well for the Yellow River, is adopted here. The Guojunke formula for the open channel is formulated as,

$$S_* = \frac{\frac{1}{20} \left(\frac{V_p^3}{gR\omega} \right)^{1.5}}{1 + \left(\frac{1}{45} \frac{V_p^3}{gR\omega} \right)^{1.15}} \tag{15}$$

where S_* is the sand content, i.e., the mass of sands in a unit volume (kg/m³); V_p is the average velocity of the section (m/s); g is the gravity acceleration (m/s²); R is the hydraulic radius; and ω is the sand deposition rate which is calculated by Zhu-Cheng formula [28] (m/s),

$$\frac{\omega d}{\nu} = \frac{-24\cos^3\alpha + \sqrt{576\cos^6\alpha + \left(18\cos^3\alpha + 3.6\sin^2\alpha\right)d_*^3}}{9\cos^3\alpha + 1.8\sin^2\alpha} \tag{16}$$

$$d_* = \left(\frac{(\rho_s - \rho)g}{\nu^2 \rho} \right)^{\frac{1}{3}} d \tag{17}$$

$$\alpha = \begin{cases} 0 & , d_* \leq 1 \\ \pi / \left(2 + 2.5(\log d_*)^{-3} \right) & , d_* > 1 \end{cases} \tag{18}$$

where d is the diameter of sand, $d = 0.033$ mm for the silty-fine sand; ν is the fluid viscosity (m^2/s); ρ_s is sand density (kg/m^3), $\rho_s = 2650$ kg/m^3; and ρ is water density (kg/m^3), $\rho = 1000$ kg/m^3.

When the floods go into the inverted siphon, the capacity of sediment transport changes. The capacity that the pipe transports sand is determined by the non-deposition velocity. The critical non-deposition velocity (V_c) is given by Wasp equation [29],

$$V_c = 3.28 S_v^{0.243} \left[2gD \left(\frac{\rho_s - \rho}{\rho} \right) \right]^{1/2} \left(\frac{d}{D} \right)^{1/6} \tag{19}$$

$$S_v = \frac{S_*}{\rho_s} \tag{20}$$

where D is the diameter of the pipe (m) and S_v is the ratio of water to sand in the unit volume. When the velocity is larger than V_c, no sediment is deposited, otherwise sediment deposition happens. The diameter parameter of the circular pipe in the original Wasp equation is transformed into the equivalent hydraulic radius of the square pipe.

2.3.4. Siphon Hydraulics Model

The siphon is calculated in terms of the quasi steady flow (i.e., steady flow is calculated in each time interval, and the all simulation snapshots comprise the simulation process over time). The inverted siphon flow computation includes the downstream and upstream open channel flows and pressurized pipe flow. The calculation of the conveyance capacity of the inverted siphon includes flood regulation and hydraulic routing. The flood regulation is based on the water balance equation, given as,

$$Q_A - Q_B = \frac{\Delta W}{\Delta t} \tag{21}$$

where Q_A is the flow at the upstream channel of the inverted siphon (m^3/s), which is known according to the hydrologic model results; Q_B is the flow at downstream channel of the inverted siphon (m^3/s); and ΔW is the volume of retained water at upstream channel during the time of Δt (m^3). The water level at the upstream and downstream channels and flow in the siphon pipe are unknown, i.e., Q_B and ΔW are both unknown, but they meet the hydraulics conditions. Then, the iterative method is applied, and a water level at downstream is given before each iteration and thus the iterative process is implemented as follows,

Step 1. Given a water level Z_B at downstream channel, the flow rate at downstream channel can be calculated with the Chezy equations

$$Q_B = A_c C_c \sqrt{R_c J} \tag{22}$$

$$C_c = \frac{1}{n_c} R_c^{1/6} \tag{23}$$

where Q_B is the flow at downstream channel (m^3/s); C_c is Chezy coefficient of the channel (m$^{1/2}$/s); A_c is the section area of the flow (m^2); n_c is the roughness coefficient at downstream channel which is set to be 0.035 in this study (s/m$^{1/3}$); and R_c is the wetted perimeter of the channel; J is the hydraulic slope.

The iterative initial water level at downstream is assumed to be zero. With the flow rate Q_B, the retaining volume at the inlet ΔW can be calculated with Equation (12). Then, with the relationship between water level and water volume at upstream channel (denoted by $F(Z)$) and the initial water level at the inlet (denoted by $Z_{A,t-1}$, equaling zero at the first time period, otherwise, equaling the ending water level of the last period), the ending water level ($Z_{A,t}$) at upstream channel can be derived by $Z_{A,t} = F^{-1}[\Delta W + F(Z_{A,t-1})]$. This water level minus the presumed water level at downstream can obtain the water level difference ΔZ_1.

Step 2. The difference between downstream and upstream channel water levels (ΔZ_2) is given as,

$$\Delta Z_2 = h_f + h_j + \frac{V_A{}^2 - V_B{}^2}{2g} \tag{24}$$

where V_A and V_B are the velocity at upstream and downstream channel (m/s), and calculated by the ratio of flow to flow area. h_f and h_j are pipe friction headloss and local headloss, respectively (m), and given by

$$h_f = \frac{L_p V_p{}^2}{C_p{}^2 R_p} \tag{25}$$

$$h_j = (K_1 + K_2 + K_3)\frac{V_p{}^2}{2g} \tag{26}$$

where L_p is the length of the inverted siphon (m); C_p is Chezy coefficient of the inverted siphon ($m^{1/2}/s$); R_p is the wetted perimeter of the inverted siphon; K_1, K_2, and K_3 are the local loss coefficients at inlet, outlet and inside of the inverted siphon, respectively; V_p is the velocity of inverted siphon (m/s), calculated by the ratio of flow (Q_p) to flow area (A_p) of the inverted siphon; and A_p can be calculated by

$$A_p = (H_p - z_p) \times W_p \tag{27}$$

where H_p and W_p are the height and width of inverted siphon, respectively (m); z_p is the deposition height (m).

Step 3. In comparison of ΔZ_1 (obtained from water balance) and ΔZ_2 (obtained from energy balance), if the difference between them is less than a threshold, then the calculation process terminates. Otherwise, a new water level at downstream is given following: if $\Delta Z_1 > \Delta Z_2$, the water level at downstream should be increased by a step; if $\Delta Z_1 < \Delta Z_2$, the water level should be decreased by a step. The step is determined according to the search method. Then, the Steps from 1 to 3 are repeated until the stop criteria.

3. Case Description

3.1. Overview of the Study Area

The Central Route of the South-to-North Water Diversion Project, shown in Figure 2, transfers water from Danjiangkou Reservoir on the Han River (a tributary of Yangtze River) to Beijing and Tianjin Cities. This project links up four major basins, including Yangtze River, Huai River, Yellow River and Hai River, and crosses Hebei, Henan, and Hubei Provinces. The main trunk canal is a total length of 1277 km, and crosses 205 rivers with cross-river buildings. The buildings are called river-canal crossing structures.

This study is targeted to the inverted siphon—A typical river-canal crossing structure. The inverted siphon, located on the intersection between main trunk canal of Central Route of the South-to-North Water Diversion Project and Meihe tributary, is taken as an illustrated case study. The drainage area of Meihe tributary is 10.80 km², the longest length from the outlet of the main river to watershed outline is 5 km, and the average slope for the longest path is 0.017. The inverted siphon consists of pipe section, upstream channel and downstream channel sections, as shown in Figure 3. The upstream channel and downstream channel sections are the trapezoidal open channels and the lengths are 55 m and 68 m, respectively. The pipe section includes four 3 × 3 m² square barrels, which have equal heights at the entrance. The horizontal projection length of each pipeline is 111.6 m, and the slopes of the rising and descending legs are 1:5 and 1:4, respectively. Since there is no gate control of the pipelines, the four pipelines operate simultaneously. The peak discharges for a 50-year return period of flood design criterion and 200-year flood check criterion are 209 m^3/s and 294 m^3/s, respectively.

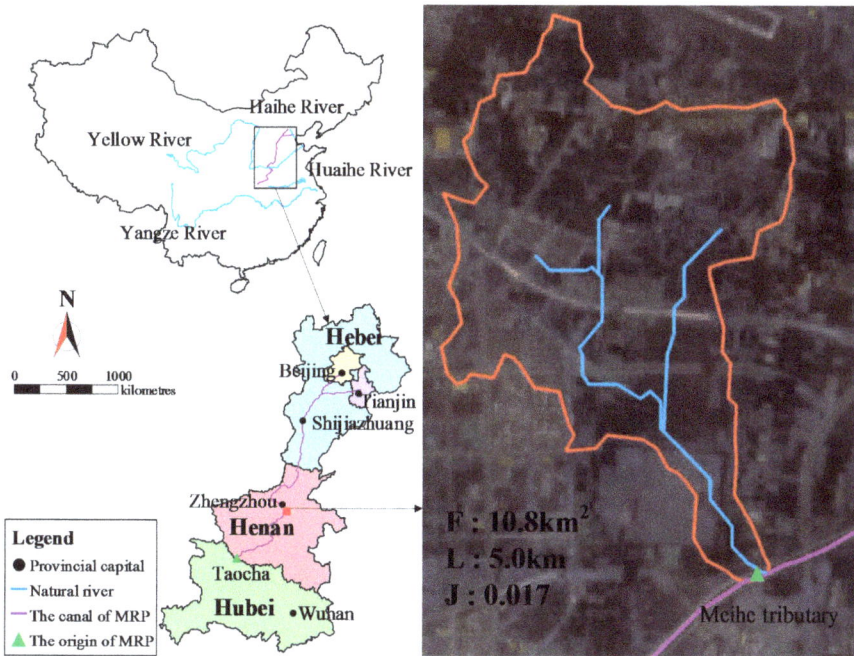

Figure 2. The sketch map of the study area.

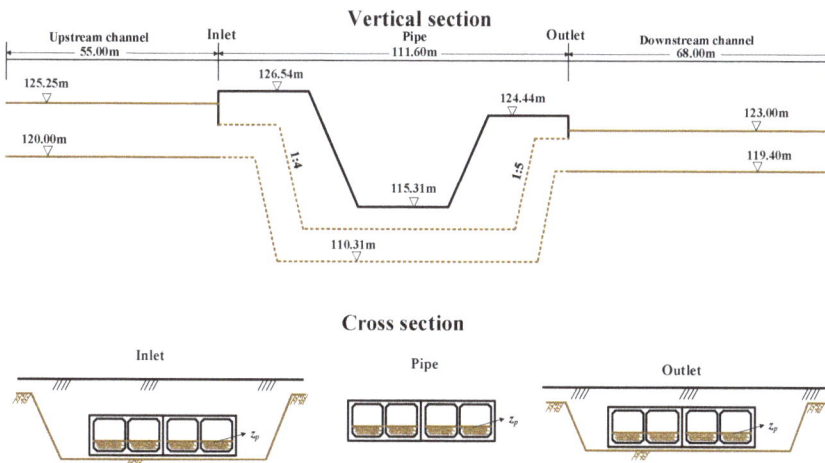

Figure 3. The schematic diagram of the inverted siphon.

3.2. Parameter Uncertainty Description

In the flood risk analysis of the inverted siphon, the parameter uncertainties in the integrated model, including rainfall module ($\bar{x}_{24}$, C_v, and γ), runoff generation module (m), runoff concentration module (u), and hydraulic module (n_c, z_p), are considered. The distribution and feasible range of the parameters are listed in Table 1.

Table 1. Parameters studied for the sensitivity analysis.

Model Response	Parameter	Description	Range	Distribution	Unit
	$\overline{x_{24}}$	Mean annual maximum rainfall in 24 h	90–110	uniform	mm
Rainfall	C_v	Variation coefficient of annual maximum rainfall in 24 h	0.5–0.6	uniform	–
	γ	Rainfall diminishing exponent	0.75–0.80	uniform	–
Runoff	u	Mean infiltration rates	2–5	uniform	mm/h
	m	Confluence coefficient	0.95–1.05	uniform	–
Conduit flow	n_c	Roughness coefficient of the pipe	0.014–0.020	uniform	–
	z_p	Initial deposition height	0–3	truncated normal	m

3.2.1. Mean Value ($\overline{x_{24}}$) and Variation Coefficient (C_v) for the Maximum 24-h Rainfall

According to the "Atlas of the Design Storm and Flood for the Medium and Small-Sized Basins in Henan Province" edited in 1984, the mean value and variation coefficient for the maximum 24-h are 100 and 0.55, respectively. However, they are both closely related to the length of the rainfall data. At the beginning of the design stage for South to North Water Diversion Project, the statistical rainfall parameters for 24 rain gauge stations, which are distributed on different rivers but along the main trunk canal, are validated with the rainfall data prolonged to 2000. The results show that the mean value $\overline{x_{24}}$ ranges within -10%–10% of the designed value and C_v ranges within -0.05–0.05. Regardless of the impact of human activities, it is assumed that the statistical parameters change within the above ranges. That is, the ranges of $\overline{x_{24}}$ and C_v are 90–110 and 0.5–0.6, respectively, and $\overline{x_{24}}$ and C_v are assumed to be uniformly distributed.

3.2.2. The Rainfall Diminishing Exponent (γ), Runoff Concentration Parameter (m) and the Mean Filtration Rate (u)

Statistic parameters of γ, m, and u are obtained from the corresponding contour map in "Atlas of the Design Storm and Flood for the Medium and Small-Sized Basins in Henan Province". The uncertainties in these parameters are caused by observation. The ranges for γ, m and u are 0.75–0.80, 2–5, and 0.95–1.05, respectively, by the upper and lower contour curve evaluation. γ, m and u are assumed to follow the uniform distribution.

3.2.3. The Roughness Coefficient (n_c)

The roughness coefficient of the pipe for the inverted siphon (n_c) changes with sediment deposition. The more sediments, the greater roughness coefficient. The roughness coefficient can reach as large as 0.020 according to the empirical data [24]. However, the inverted siphon was designed according to a fixed value, i.e., 0.014. Therefore, we consider the uncertainty of n_c within range 0.014–0.020 obeying a uniform distribution.

3.2.4. The Initial Deposition Height of Sediment (z_p)

The initial deposition height of sediment in the model is set at the beginning of the flood process. It can be as large as the pipe width, 3 m in this study. z_p is typically small with the larger probability, while small probability corresponds to a large z_p. Thus, the truncated normal distribution, which is widely used when there is little information about the distribution, is assumed for z_p within range 0–3.

4. Results and Discussion

4.1. RSA Results

The parameters in the integrated model, listed in Table 1, are assumed to be independent, and Latin Hypercube sampling method [30] is used. It is noted that the parameters of the hydraulic model (roughness coefficient and initial deposition height) for each conduit (the culvert consists of four squared conduits) are set to be identical. The sample size is 100,000, and one sample includes all parameters values that are randomly assigned. The distributions of runoff concentration time and peak discharge are shown in Figure 4. As can be seen, the peak discharge considering the uncertainties of parameters are all larger than the original designed value, 294 m^3/s.

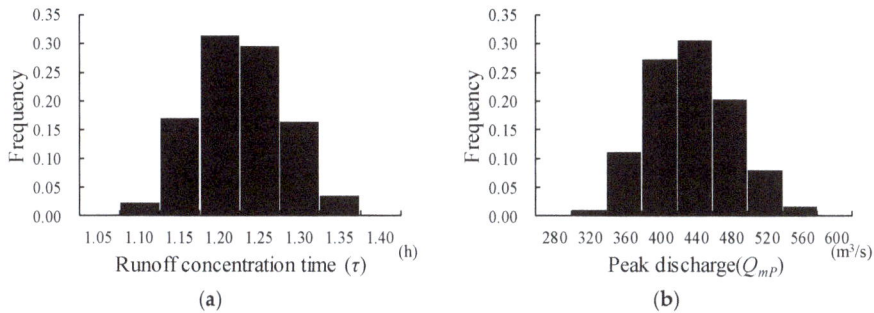

Figure 4. The distribution of runoff concentration time and peak discharge. (**a**) The distribution of runoff concentration time; (**b**) The distribution of peak discharge.

The parameters sets are divided into two subsets in terms of the inverted siphon failure (i.e., non-behavioral) or operation (behavioral), with cumulative distributions shown in Figure 3. The behavioral sample (S_B) is that no water flows into the canal, i.e., the water level at the water intake is smaller than the embankment crest elevation. In contrast, the non-behavioral sample (S_{NB}) is that water flows into the canal, i.e., the water level at the water intake is higher than the crest elevation. In Figure 3, the diagonal line (D-line) represents the parameter has a uniform distribution and the model is not sensitive to this parameter in terms of the chosen likelihood measure. Any deviation from the "D-line" shows a non-uniform distribution and the model is sensitive to this parameter. The larger distance between the S_B and S_{NB} indicates more sensitive range of this parameter.

As shown in Figure 5, all parameters exhibit an obvious shift. In addition, the S_{NB} curve for all the parameters except z_p is close to "D-line", indicating that the effects of parameters within the whole range on the failure of the inverted siphon are almost identical. The cumulative distributions for the mean rainfall ($\overline{x_{24}}$), variation coefficient (C_v), rainfall diminishing exponent (γ), runoff concentration parameter (m) and the roughness coefficient (n_c), show that the values at the lower end of the tested ranges contribute to the greatest number of behaviors, i.e., lower value of these parameters lead to lower flood risk of the inverted siphon. Conversely, the greatest number of behaviors occurs at the higher end of the range for mean filtration rate (u). For initial deposition height (z_p), the greatest number of behaviors and non-behaviors comes from values at the lower end of the range. Meanwhile, the initial deposition height (z_p) has the least impact on flood risk of the inverted siphon, with the smallest shift from the straight line in comparison with the others.

Each parameter sensitivity is tested by the two-sample Kolmogorov–Smirnov (K-S) test method with the confidential level of 95%. Results show that the mean rainfall ($\overline{x_{24}}$) is most sensitive, followed by the roughness coefficient (n_c). Furthermore, the other parameters related with the rainfall, i.e., variation coefficient (C_v), rainfall diminishing exponent (γ), are more sensitive, and thus, the rainfall is the most important factor for the flood risk of the inverted siphon. The roughness coefficient (n_c) is related with the conveyance capacity of the pipe, thus the conveyance capacity of the pipe also is the most important factor for the flood risk of the inverted siphon.

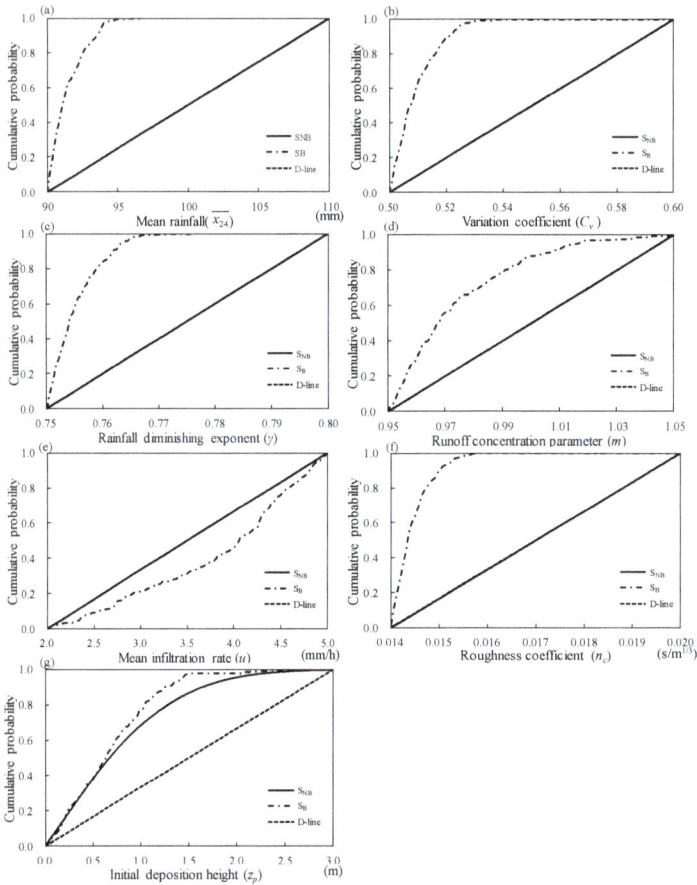

Figure 5. Cumulative distribution of the seven parameters with regard to failure of the inverted siphon. S_B and S_{NB} represent the behavioral and non-behavioral groups, respectively. (**a**) Cumulative distribution of mean rainfall; (**b**) Cumulative distribution of variation coefficient ; (**c**) Cumulative distribution of rainfall diminishing exponent; (**d**) Cumulative distribution of runoff concentration parameter; (**e**) Cumulative distribution of mean filtration rate; (**f**) Cumulative distribution of roughness coefficient; (**g**) Cumulative distribution of initial deposition height.

4.2. Soboĺ Sensitivity Analysis Results

The first-order and total-order sensitivity indices of seven parameters are shown in Figure 4. The black bars represent the first-order index values, which measure the individual parameter contributions to the duration and volume of flood overtopping. The white bars represent the interactive indices, which demonstrate the total interactive contribution of one parameter with all other parameters. The parameter is identified as sensitive when total-order sensitivity index is larger than 0.1.

As can be seen in Figure 6, the total-order index for roughness coefficient (n_c), mean rainfall ($\overline{x_{24}}$), variation coefficient (C_v) and rainfall diminishing exponent (γ), are all larger than 0.1, i.e., sensitive parameters for flood overtopping duration and volume. Parameters m and u, which are related with runoff generation and concentration, are not sensitive. The reason is that flood hydrograph, flood volume, and the conveyance capacity of the pipe are the most important factors for the flood overtopping. A larger flood volume or smaller conveyance capacity leads to more water retained at the inlet, and thus leads to a larger risk of flood overtopping. The magnitude of rainfall is the most important factor of flood volume and flood hydrograph. Thus, the parameters related with the magnitude of rainfall and conveyance capacity are sensitive. The initial deposition height (z_p) is sensitive for the flood overtopping duration, but it is not sensitive for the flood overtopping volume.

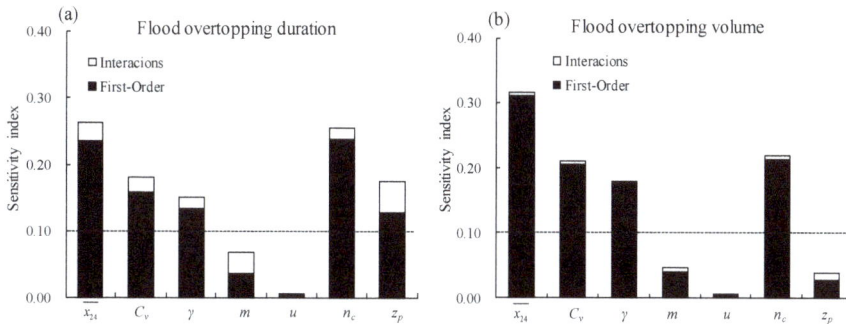

Figure 6. The first-order sensitivity and total-order sensitivity of parameters. (**a**) The first-order sensitivity and total-order sensitivity of parameters for flood overtopping duration; (**b**) The first-order sensitivity and total-order sensitivity of parameters for flood overtopping volume.

The results of the overtopping duration show that $\overline{x_{24}}$ obtains the largest value of total-order index, followed by n_c. These two parameters determine the magnitude of rainfall and the water conveyance capacity of the pipe, respectively. Therefore, the result indicates that rainfall and the water conveyance capacity of the pipe both have a great impact on the overtopping duration in the inverted siphon. The results of the flood overtopping volume show the total-order index for $\overline{x_{24}}$, C_v, γ as well as n_c are sensitive parameters. $\overline{x_{24}}$, C_v and γ determine the runoff volume, which is the input of the inverted siphon, while n_c determines the flood volume transported.

Figure 6 show the interaction between z_p and other parameters is strong, and the interaction for the flood overtopping volume is weaker than the flood overtopping duration. The interaction between any two parameters is shown in Figure 7. As can be seen in Figure 7a, the sum of the second-order index between z_p and other parameters is the largest, and the interaction between z_p and m is strongest for the flood overtopping duration. As for flood overtopping volume, the sum of second-order index between n_c and other parameters is largest. However, Figure 6b shows that the total interactions of z_p is larger than that of n_c, which indicates that higher order interactions (third-order, four-order, etc.) exist between z_p and other parameters. The second-order index value for n_c and γ is strongest, followed by the value for n_c and $\overline{x_{24}}$, indicating that the interaction between n_c and γ, n_c and $\overline{x_{24}}$ are strong. This is because $\overline{x_{24}}$ is the most important factor that determines the flood volume, and a larger $\overline{x_{24}}$ result in a larger flood. A larger γ results in a larger runoff generation under the same magnitude of rainfall, and a larger $\overline{x_{24}}$ results in a larger and thus a larger peak discharge and flood volume, while a larger n_c results in a smaller conveyance capacity.

In conclusion, results of RSA and Soboĺ sensitivity analysis both indicate that the mean rainfall and roughness coefficient of the pipe are two important parameters, which determine the rainfall and water conveyance capacity of the siphon pipe, respectively. These results demonstrate that lower value of the mean rainfall or roughness coefficient of the pipe lead to lower flood risk of the inverted siphon. Thus, the effective measures for reducing the flood risk of the inverted siphon to clean the inverted siphon periodically, i.e., reduce the roughness coefficient of pipe.

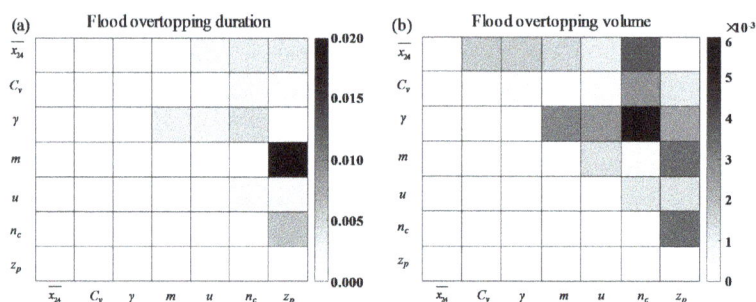

Figure 7. The interaction between each two parameters. (**a**) The interaction between each two parameters for flood overtopping duration; (**b**) The interaction between each two parameters for flood overtopping volume.

5. Conclusions

This paper investigated, by using an integrated system coupling the hydrology and hydraulic models, the impact of rainfall, sediments and pipe roughness coefficient on the failure of inverted siphon through RSA and Soboĺ sensitivity analysis. The flood risk of the inverted siphon is evaluated by the flood overtopping duration and volume. The conclusions are summarized as follows:

(1) RSA and Soboĺ sensitivity analyses both indicate the mean rainfall and the roughness coefficient of the pipe are most sensitive for the flood risk of the inverted siphon. These results imply that lower value of the mean rainfall or roughness coefficient of the pipe lead to lower flood risk of the inverted siphon. Thus, periodically cleaning the inverted siphon is an effective measure for reducing the flood risk of the inverted siphon, i.e., reduce the roughness coefficient of pipe.

(2) The RSA identifies the sensitivity range of safety and failure for the inverted siphon. Effects of all parameters except initial deposition height throughout the feasible range on the failure of the inverted siphon are almost identical. For the safety of the inverted siphon, the smaller values of variation coefficient, rainfall diminishing exponent, runoff concentration parameter, and roughness coefficient of pipe, the more safety of the inverted siphon. For the mean filtration, a higher value leads to the more security of the inverted siphon.

(3) Soboĺ sensitivity analysis reveals the individual and interactive effects of the parameters. The effects of all parameters on flow overtopping duration and volume all parameters are all dominated by the individual effects. For the flood overtopping duration, the interactions between the initial deposition height and other parameters are high, with the largest interaction between initial deposition height and confluence coefficient. For flood overtopping volume, the interaction between rainfall model parameters and hydraulic model parameters are both significant.

Acknowledgments: This research was supported by the National Key Research and Development Program of China (Grant No. 2016YFC0402203) and partly funded by the National Science and Technology Major Project (Grant No. 2014ZX03005001) and National Science Foundation of China (Grant No. 91547116).

Author Contributions: Wei Ding and Haixing Liu conceived and designed the experiments; Sifan Jin performed the experiments; SiFan Jin, Haixing Liu and Wei Ding analyzed the data; Hua Shang and Guoli Wang contributed reagents/materials/analysis tools; Sifan Jin wrote the paper.

Conflicts of Interest: The authors declare no conflict of interest.

References

1. Liu, C.; Zheng, H. South-to-north Water Transfer Schemes for China. *Int. J. Water Resour. Dev.* **2002**, *18*, 453–471. [CrossRef]
2. Koutsoyiannis, D. Scale of water resources development and sustainability: Small is beautiful, large is great. *Hydrol. Sci. J.* **2011**, *56*, 553–575. [CrossRef]

3. Tyralis, H.; Aristoteles, T.; Delichatsiou, A.; Mamassis, N.; Koutsoyiannis, D. A perpetually interrupted interbasin water transfer as a modern Greek drama: Assessing the Acheloos to Pinios interbasin water transfer in the context of integrated water resources management. *Open Water J.* **2017**, *4*, 11.

4. Zhang, Q. The South-to-North Water Transfer Project of China: Environmental Implications and Monitoring Strategy. *JAWRA J. Am. Water Resour. Assoc.* **2009**, *45*, 1238–1247. [CrossRef]

5. Tang, C.; Yi, Y.; Yang, Z.; Cheng, X. Water pollution risk simulation and prediction in the main canal of the South-to-North Water Transfer Project. *J. Hydrol.* **2014**, *519*, 2111–2120. [CrossRef]

6. Wei, G.; Zhang, C.; Li, Y.; Liu, H.; Zhou, H. Source identification of sudden contamination based on the parameter uncertainty analysis. *J. Hydroinform.* **2016**, *18*, 919–927.

7. Pappenberger, F.; Beven, K.J.; Ratto, M.; Matgen, P. Multi-method global sensitivity analysis of flood inundation models. *Adv. Water Resour.* **2008**, *31*, 1–14. [CrossRef]

8. Van Griensven, A.; Meixner, T.; Grunwald, S.; Bishop, T.; Diluzio, M.; Srinivasan, R. A global sensitivity analysis tool for the parameters of multi-variable catchment models. *J. Hydrol.* **2006**, *324*, 10–23. [CrossRef]

9. Chu, J.; Zhang, C.; Fu, G.; Li, Y.; Zhou, H. Improving multi-objective reservoir operation optimization with sensitivity-informed dimension reduction. *Hydrol. Earth Syst. Sci.* **2015**, *19*, 3557–3570. [CrossRef]

10. Song, X.; Zhang, J.; Zhan, C.; Xuan, Y.; Ye, M.; Xu, C. Global sensitivity analysis in hydrological modeling: Review of concepts, methods, theoretical framework, and applications. *J. Hydrol.* **2015**, *523*, 739–757. [CrossRef]

11. Ruppert, D.; Shoemaker, C.A.; Wang, Y.; Li, Y.; Bliznyuk, N. Uncertainty Analysis for Computationally Expensive Models with Multiple Outputs. *J. Agric. Biol. Environ. Stat.* **2012**, *17*, 623–640. [CrossRef]

12. Gan, Y.; Duan, Q.; Gong, W.; Tong, C.; Sun, Y.; Chu, W.; Ye, A.; Miao, C.; Di, Z. A comprehensive evaluation of various sensitivity analysis methods: A case study with a hydrological model. *Environ. Model. Softw.* **2014**, *51*, 269–285. [CrossRef]

13. Soboĺ, I. Quasi-Monte Carlo methods. *Prog. Nucl. Energy* **1990**, *24*, 55–61. [CrossRef]

14. Saltelli, A.; Tarantola, S. On the Relative Importance of Input Factors in Mathematical Models. *J. Am. Stat. Assoc.* **2002**, *97*, 702–709. [CrossRef]

15. Fu, G.; Kapelan, Z.; Reed, P. Reducing the complexity of multiobjective water distribution system optimization through global sensitivity analysis. *J. Water Resour. Plan. Manag.* **2012**, *138*, 196–207. [CrossRef]

16. Zhang, C.; Chu, J.; Fu, G. Soboĺ sensitivity analysis for a distributed hydrological model of Yichun River Basin, China. *J. Hydrol.* **2013**, *480*, 58–68. [CrossRef]

17. Spear, R.C.; Hornberger, G.M. Eutrophication in peel inlet—II. Identification of critical uncertainties via generalized sensitivity analysis. *Water Res.* **1980**, *14*, 43–49. [CrossRef]

18. Beven, K.; Binley, A. The future of distributed models: Model calibration and uncertainty prediction. *Hydrol. Process.* **1992**, *6*, 279–298. [CrossRef]

19. Pappenberger, F.; Iorgulescu, I.; Beven, K.J. Sensitivity analysis based on regional splits and regression trees (SARS-RT). *Environ. Model. Softw.* **2006**, *21*, 976–990. [CrossRef]

20. Saltelli, A.; Tarantola, S.; Chan, K.-S. A quantitative model-independent method for global sensitivity analysis of model output. *Technometrics* **1999**, *41*, 39–56. [CrossRef]

21. Tang, Y.; Reed, P.; Van Werkhoven, K.; Wagener, T. Advancing the identification and evaluation of distributed rainfall-runoff models using global sensitivity analysis. *Water Resour. Res.* **2007**, *43*. [CrossRef]

22. Yang, J. Convergence and uncertainty analyses in Monte-Carlo based sensitivity analysis. *Environ. Model. Softw.* **2011**, *26*, 444–457. [CrossRef]

23. Massmann, C.; Holzmann, H. Analysis of the behavior of a rainfall–runoff model using three global sensitivity analysis methods evaluated at different temporal scales. *J. Hydrol.* **2012**, *475*, 97–110. [CrossRef]

24. Wu, C. *Hydraulics*, 4th ed.; Higher Education Press: Beijing, China, 2008. (In Chinese)

25. Chen, J.; Zhang, G. *Storm-Flood Computation of Small Watershed*; China Conservancy and Hydropower Press: Beijing, China, 1985. (In Chinese)

26. Mulvaney, T.J. *On the Use of Self-Registering Rain and Flood Gauges in Making Observations of the Relations of Rain Falland Flood Discharges in a Given Catchment, Proceedings of the Institute of Civil Engineers of Ireland, Session 1850-1*; Transactions of the Institution of Civil Engineers of Ireland: Dublin, Ireland, 1851; Volume 4, pp. 18–33.

27. Guo, J. Logarithmic matching and its applications in computational hydraulics and sediment transport. *J. Hydraul. Res.* **2002**, *40*, 555–565. [CrossRef]

28. Zhu, L.; Cheng, N. *Settlement of Sediment Particles*; Nanjing Hydraulic Research Institute: Nanjing, China, 1993. (In Chinese)
29. Wasp, E.J.; Kenny, J.P.; Gandhi, R.L. *Solid-liquid Flow Slurry Pipeline Transportation*; Ser. Bulk Materials Handling; Trans Tech Publications: Zurich, Switzerland, 1977; Volume 1, p. 4.
30. McKay, M.D.; Beckman, R.J.; Conover, W.J. A comparison of three methods for selecting values of input variables in the analysis of output from a computer code. *Technometrics* **2000**, *42*, 55–61. [CrossRef]

Article

Joint Operation of Surface Water and Groundwater Reservoirs to Address Water Conflicts in Arid Regions: An Integrated Modeling Study

Yong Tian [1,2], Jianzhi Xiong [1], Xin He [1,2,3], Xuehui Pi [1], Shijie Jiang [1,4], Feng Han [1,2] and Yi Zheng [1,2,5,*]

[1] School of Environmental Science and Engineering, Southern University of Science and Technology, Shenzhen 518055, China; tiany@sustc.edu.cn (Y.T.); xiongjz@mail.sustc.edu.cn (J.X.); hex@sustc.edu.cn (X.H.); pixh@mail.sustc.edu.cn (X.P.); jiangsj@mail.sustc.edu.cn (S.J.); hanf@sustc.edu.cn (F.H.)
[2] Shenzhen Municipal Engineering Lab of Environmental IoT Technologies, Southern University of Science and Technology, Shenzhen 518055, China
[3] Department of Hydrology, Geological Survey of Denmark and Greenland, Copenhagen 1350, Denmark
[4] Department of Civil and Environmental Engineering, National University of Singapore, Singapore 117576, Singapore
[5] Guangdong Provincial Key Laboratory of Soil and Groundwater Pollution Control, Southern University of Science and Technology, Shenzhen 518055, China
* Correspondence: zhengy@sustc.edu.cn; Tel.: +86-755-8801-8030

Received: 31 July 2018; Accepted: 17 August 2018; Published: 19 August 2018

Abstract: At the basin scale, the operation of surface water reservoirs rarely takes groundwater aquifers into consideration, which can also be regarded as reservoirs underground. This study investigates the impact of reservoir operation on the water cycle and evaluates the effect of the joint operation of surface water and groundwater reservoirs on the water conflict in arid regions through an integrated modeling approach. The Heihe River Basin (HRB) in northwestern China is selected as the study area. Our results show that the ecological operational strategies of a reservoir under construction in the upper HRB have a direct impact on the agricultural water uses and consequently affect other hydrological processes. The ecological operation strategy with a smaller water release and a longer duration is beneficial to securing the environmental flow towards the downstream area and to replenishing aquifers. With the joint operation of surface water and groundwater reservoirs, a balance among the agriculture water need, the groundwater sustainability in the Middle HRB and the ecological water need in the Lower HRB can be flexibly achieved. However, the joint operation can hardly improve the three aspects simultaneously. To resolve the water conflict in HRB, additional engineering and/or policy measures are desired.

Keywords: reservoir operation; integrated surface water-groundwater model; Heihe River Basin; environmental flow; irrigation

---

1. Introduction

Irrigation plays an indispensable role in agricultural water supply when rainfall is not sufficient to sustain crop growth. Currently, the irrigated cropland produces more than 40% of the total cereal yield worldwide [1,2]. Globally, about 54% of the total irrigated area is dependent on surface water (SW) such as rivers, lakes, artificial canals and reservoirs [3]. Meanwhile, in arid and semi-arid regions of the world where surface water is not abundantly available, groundwater (GW) abstraction is often required for irrigation purposes. According to Siebert et al. [4], the total area irrigated by groundwater is nearly 114 million hectares globally. In most arid and semi-arid regions, irrigation with groundwater has experienced a considerable increase over the past few decades mainly due to the growing demand for agricultural products and the changing climate [5]. In the North China Plain, one of China's most important agricultural regions, approximately 73% of the groundwater extraction is used for irrigation [6]. In the United States, around 60% of the irrigation relies on groundwater, especially in California's Central Valley and the High Plains [7]. Surface water reservoirs are enlarged natural or artificial lakes, storage ponds or impoundments created using a dam or lock to store water. Operating several surface water reservoirs next to each other collectively in order to serve a defined single purpose has become a common practice. A groundwater aquifer, although buried and unseen, can also act as a lumped water storage unit with a known capacity, which resembles a surface water reservoir [8]. Therefore, using groundwater pumping for irrigation can, in principle, be conceptualized as operating surface water reservoirs. Kundzewicz and Döll [9] analyzed the potential of using the natural storage capacity and the buffering effect provided by groundwater reservoirs to cope with highly variable surface water supply in different years. Scanlon et al. [10] further discussed the possibility of utilizing groundwater reservoirs to better adapt to climate extremes in California's Central Valley and central Arizona.

Taking groundwater as a part of the reservoir chain and considering the joint effect of surface water and groundwater reservoirs simultaneously is a relatively new research topic [11]. For instance, Castle et al. [12] assessed the influence of conjunctive surface water and groundwater use on water availability in the Colorado River Basin by estimating the water storage change of both surface water and groundwater reservoirs from satellite images. Fuchs et al. [13] quantified the resilience of the agricultural system that depends on the conjunctive use of surface and groundwater in the Rincon Valley. Nikoo et al. [14] established optimal operation scheduling rules for a reservoir-river-groundwater joint system through data mining. To our knowledge, however, it is rarely reported that both surface water and groundwater reservoirs are operated jointly for water resources conservation via integrated surface water-groundwater modeling.

Increasing demand for irrigation has created human-nature water conflicts and therefore posed many challenges in water resources management. Such a problem is especially serious in arid and semi-arid regions. As an example, the Heihe River Basin (HRB), our study area, has witnessed this conflict gradually developing in the past 30 years. HRB is the second largest endorheic river basin in China and is agriculturally intensive. In the late 1990s, the growing irrigation in the Middle HRB (MHRB) significantly decreased the streamflow available for the Lower HRB (LHRB) and therefore deteriorated the ecological conditions [15–18]. To protect the fragile ecosystems in the Lower HRB, the central government restricted the surface water diversion since 2000 in the Middle HRB [19]. However, the restriction of surface water usage led to a significant increase of groundwater pumping in the Middle HRB, since it was not strictly regulated [20]. Recently, the construction of a new reservoir in the upper HRB basin has been approved, which aims to enhance the water resources management in the basin. However, how the new reservoir may change the water cycle in the middle and Lower HRB has not been systematically investigated by fully accounting for the groundwater reservoir effect, which motivated this study. Therefore, the main objectives of this study are: (1) to quantify the impact of the new reservoir in the upper HRB on the water cycle in the middle and Lower HRB; and (2) to evaluate the effect of the joint operation of surface water and groundwater reservoirs on the water conflict in HRB, based on integrated surface water-groundwater modeling. Overall, this study provides insights into the water resources management in arid regions.

2. Data and Methods

2.1. Study Area

The Heihe River Basin (HRB) is a typical endorheic river basin in the arid region in the northwest of China. It is located between 97.1° E–102.0° E and 37.7° N–42.7° N. The entire basin can be divided into the upper, middle and lower stream areas, with very different landscapes. The upper stream basin is characterized by a mountainous area on the northern margin of the Qinghai-Tibetan Plateau. It is densely vegetated with forests and grasslands. The Middle HRB is dominated by the Gobi Desert, while many irrigated farmlands exist inside the oases. The Lower HRB is mainly a desert with very little vegetation. More than 30 rivers originated in the upstream area bring approximately 3651 million m^3 of water per year to the midstream area. The largest stream, namely the Heihe River, has a length of 928 km. It originates from the Qilian Mountains, flows towards the north and ends in a terminal lake, the East Juyan Lake (EJL), in the Gobi Desert. Yingluoxia (YLX) and Zhengyixia (ZYX) are usually regarded as the starting points of the midstream and lower stream of the Heihe River, respectively (see Figure 1). In the upstream, the Heihe River has a major tributary called Babaohe River, which flows into the main stream within the vicinity of the municipality Huangzangsi where the new reservoir is under construction. Daily streamflow measurements are carried out at four gaging stations as indicated by the blue points in Figure 1. Qilan, YLX and ZYX gaging stations are located at the main stream of the Heihe River, and Zhamashike gaging station is located at the Babaohe River. Annual streamflow through Qilian, Zhamashike, YLX and ZYX is 8,204,801,770 and 1010 million m^3, respectively.

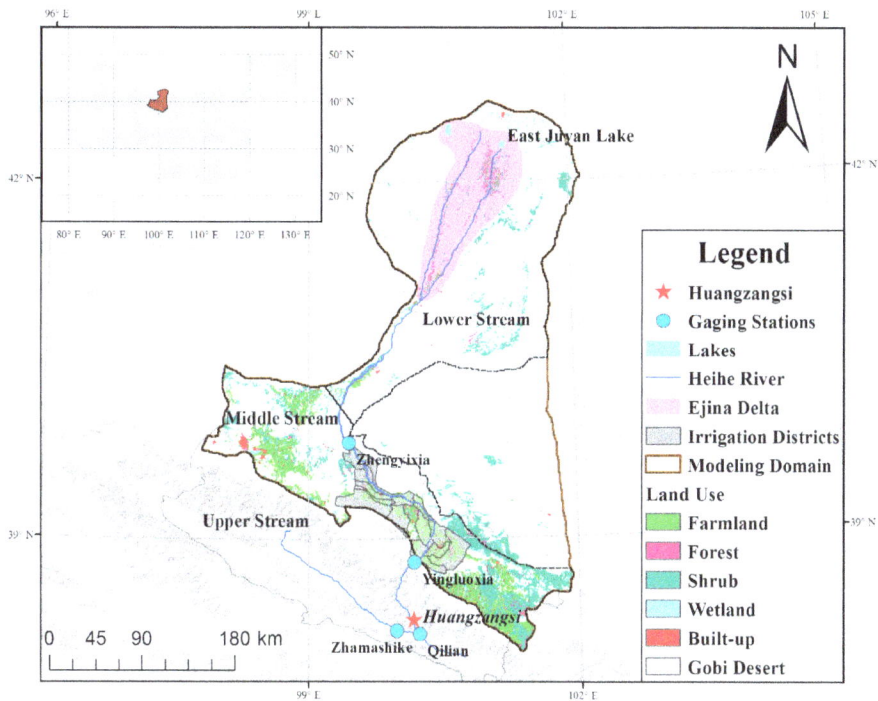

Figure 1. The study area.

In the midstream area, the most common crops are corn and winter wheat. A complicated aqueduct system has been developed in this area. The basic unit for water resources management is called an irrigation district. There are 17 irrigation districts to which stream water from the middle stream of the Heihe River is diverted. Only water consumption from the agricultural sector is considered in these irrigation districts in this study. According to the statistics by the local water resources authority, in 2000–2012, agriculture in the 17 irrigation districts consumed approximately 1860 million m^3 of water per year, 80% of which was diverted from the main Heihe River, and the rest was pumped from the local groundwater aquifer. To secure the environmental flow towards the lower stream under different hydrological conditions, a water allocation plan has been implemented since 2000. For example, in a normal year (i.e., the annual flow from Yingluoxia reaches 1580 million m^3), the flow through Zhengyixia should be no less than 950 million m^3. Since 2000, diversion of the stream flow has been restricted in the midstream area, which has resulted in a substantial increase of groundwater pumping. In recent years, the groundwater pumping has also been regulated. Thus, the agriculture in the midstream area is now facing a great risk of irrigation water shortage [21]. The water resources authorities have therefore been resorting to hydraulic engineering measures, such as reservoirs, in order to mitigate the water shortage problem in the irrigation districts.

In fact, in the past decades, many surface water reservoirs have been built in the natural low-lying lands along the Heihe River in the midstream area. As of 2012, there were 20 reservoirs with a total water storage of 48.65 million m^3. This amount is able to irrigate more than 20,000 ha of farmland. All these reservoirs are replenished by diverting water from the Heihe River, except the one named Shuangquanhu, which is replenished by natural springs. However, the storage capacity of the existing reservoirs is quite small, compared to the irrigation demand. In addition, the loss of water through evaporation is non-negligible, which is approximately 17 million m^3 per year, accounting for more than 1/3 of the total reservoir storage. Therefore, as planned, all the plain reservoirs will stop operating in the near future, except the Shuangquanhu Reservoir. Instead, a large reservoir named Huangzangsi is now being built in the upper mountainous area (Figure 1). The total storage of the Huangzangsi Reservoir is approximately eight-times larger than the sum of all the existing plain reservoirs. More information about this reservoir is provided in Section 2.3. It is expected that, with this new reservoir, the irrigation demand in the midstream area can be better met, and the environmental flow towards the lower stream area can be more secured. This provides a unique opportunity to study the joint operation of the surface water and groundwater reservoirs in the HRB in order to alleviate the human-nature water conflict.

2.2. Hydrological Model

GSFLOW (Coupled Groundwater and Surface Water Flow Model) (Version 1.1.6, U.S. Geological Survey, Reston, VA, U.S.) is the model applied in the present study. It is an integrated surface water-groundwater model developed by the United States Geological Survey (USGS), which can simulate all the major hydrological processes of the terrestrial water cycle [22]. GSFLOW couples the Precipitation-Runoff Modeling System (PRMS) (Version 3.0.5, U.S. Geological Survey, Reston, VA, U.S.) with the Modular Groundwater Flow Model (MODFLOW-2005) (Version 1.9.1, U.S. Geological Survey, Reston, VA, U.S.) to simulate both the surface water hydrology (top of the plant canopy to the root zone) and the three-dimensional groundwater (the base of the root zone to the base of the aquifers) movement. In the surface water domain, hydrologic response units (HRUs) are the basic computing units, which can be either regular grids or irregular polygons. GSFLOW uses a cascade method to route the overland flow and interflow between HRUs and from HRUs to streams and lakes. The subsurface domain is discretized with finite difference grids. To simulate the two-way interactions between surface water and groundwater, GSFLOW defines a "gravity reservoir" as a storage in which an HRU exchanges water with the MODFLOW grid(s) at places where they intersect. The unsaturated zone is defined as the space between the root zone and the top of the groundwater table, which is computed using the Unsaturated Zone Flow package (UZF1) [23]. Streams and lakes are simulated using the Streamflow Routing package (SFR2) [24] and Lake package [25], respectively. In reaches where the stream water is connected to the groundwater, stream-aquifer exchange is calculated based on the head difference using Darcy's law. More details on GSFLOW can be found in Markstrom et al [22].

Tian et al. [20] improved the capacity of GSFLOW by adding two new irrigation modules, so it simulates agricultural water management activities. One of the modules is to distribute the diverted water from streams to farmlands through an aqueduct system, and the other is to distribute the abstracted groundwater to the farmlands close to the pumping wells. These two modules require daily surface water diversion rate and pumping rate as input. Since it is difficult to obtain these data, we developed a third module, the Water Resources Allocation (WRA) module, for GSFLOW recently. Besides integrating functionalities of the two irrigation modules, the WRA module enables one to simulate demand-based diversion and pumping rates. HRUs with farmlands are the basic computing units. The crop type in each HRU and the irrigation quota for each crop need to be predefined. The irrigation quota could be changed from year to year or kept unchanged throughout the entire simulation period. Based on the crop type and irrigation quota, water demand at each HRU is calculated. The demand is supplied by precipitation at first if it rains, then the remaining part is supplied by irrigation. Irrigation in each HRU is assumed to be supplied by either surface water diversion or groundwater pumping, or the combination of the two. The proportions of diversion and pumping in the total irrigation are predefined for each HRU, based on which expected diversion and pumping are first calculated. Then, actual diversion is computed by considering stream routing, and actual pumping is computed by considering the available groundwater in the aquifer. Finally, the actual diversion and pumping is distributed to HRUs by using the method in our previous study [20]. It is noted that the demand for irrigation may not be met 100%, which can be caused by insufficient stream flow or drying of the groundwater wells.

Based on the data availability, the updated GSFLOW model has been applied to the middle and lower streams of HRB over the period 2000–2012 [26]. This period includes wet, dry and normal years and is therefore representative. The model requires daily streamflow at YLX as the boundary conditions. After construction of the Huangzangsi Reservoir, the streamflow at YLX will be greatly changed based on the reservoir's operation, which may lead to significant changes of the water cycle in the middle and lower streams of HRB. Thus, we developed a program to simulate operation of the Huangzangsi Reservoir, which is described in the following section.

2.3. The Huangzangsi Reservoir

The Huangzangsi Reservoir under construction is on the upstream of the Heihe River (Figure 2), approximately 70 km from the Yingluoxia gaging station. It receives the main stream of the Heihe River and the Babaohe River. Table 1 summarizes the physical and hydrological characteristics of the reservoir. The normal storage capacity of the reservoir ranges from 61 million m^3 (the dead storage) to 356 million m^3 (the normal storage). The water surface covers an area of 11.01 km^2 at its normal water level. The Huangzangsi Reservoir is designed to be a multipurpose reservoir, where two main purposes are meeting the agriculture water demand by the Middle HRB and meeting the environmental flow demand by the Lower HRB. It is important to note that the reservoir is not intended to control flooding, thus the normal water level and flood control water level have the same value of 2628 m. The construction of the reservoir is expected to be finished in 2022. All of the reservoir's information was obtained from the environmental impact assessment report provided by the Heihe River Basin Authority [27].

Our study assumes that the Huangzangsi Reservoir is operated based on its rule curves, where the time of decision, the present water level and the overall water demand are the three constraining factors. The overall water demand consists of irrigation water demand in the midstream area, the environmental flow demand in the lower stream area and the minimum base flow demand. Table 2 shows the monthly water demand for the reservoir operation. These values were derived based on statistics and hydrological observations during the period 2000–2012. The irrigation water demand was estimated based on annual statistics by the Water Bureau of the Zhangye City [26]. As Table 2 indicates, irrigation occurs mainly from March–November. The environmental flow demand in the lower stream area was estimated based on the report provided by the Heihe River Basin Authority [27]. The minimum base flow demand occurs from December in the previous year to March, and it is assumed to be 25% of the reservoir's inflow during these months.

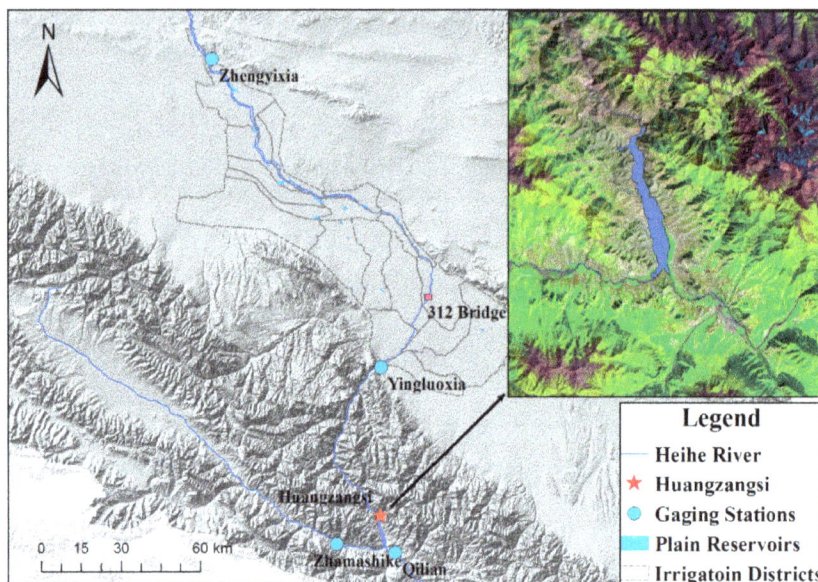

Figure 2. The Huangzangsi Reservoir in the upstream mountainous area and the existing small reservoirs in the plain midstream area.

Table 1. Hydrological and physical characteristics of the Huangzangsi Reservoir.

Hydrological Characteristics	Value
Basin area above dam site	7648 km^2
Annual inflow	1285×10^6 m^3
Averaged annual discharge	40.7 m^3/s
Maximum measured discharge	603 m^3/s
Physical Characteristics	
Maximum water level	2629 m (elevation)
Normal water level	2628 m (elevation)
Flood control water level	2628 m (elevation)
Dead storage level	2580 m (elevation)
Water surface area at normal water level	11.01 km^2
Length of reservoir at normal level	13.5 km
Total storage	406×10^6 m^3
Dead storage	61×10^6 m^3
Normal storage	356×10^6 m^3
Maximum discharge capability	2775 m^3/s

Table 2. Monthly demands considered for the operation of the Huangzangsi Reservoir.

Month	Water Demands		
	Surface Water for Irrigation (10^6 m^3)	Environmental Flow (10^6 m^3)	Base Flow (10^6 m^3)
January	0	0	6
February	0	0	6
March	22	0	9
April	72	140	0
May	183	0	0
June	379	0	0
July	305	82	0
August	344	82	0
September	101	202	0
October	101	0	0
November	219	0	0
December	0	0	9
Total	1727	506	30

For the Huangzangsi Reservoir, there are only two basic rule curves, the Normal Curve (NC) and Critical Curve (CC). The values of NC and CC are determined by the normal water level and dead storage level, respectively, and they are kept unchanged throughout a year. The reservoir operates to maintain the water level below NC (2628 m) and above CC (2580 m). Based on the rule curves, there are three distinct operational strategies, which are described in Table 3. To meet the environmental flow demand, a special category of reservoir operation called environmental flow regulation is designed for the study area. Under this operation, the reservoir discharges water within a short period, and meanwhile, the irrigation districts in the Middle HRB are not allowed to divert water from the main Heihe River. This operation is to ensure that enough flow can reach the Lower HRB and replenish the terminal lake. The short period can range from 3 days–20 days, and it is scheduled in the middle of April, July, August and September.

The water balance of the Huangzangsi Reservoir under the abovementioned operation can be written as:

$$S_{t+1} = S_t + Q_t - R_t - A_t e_t, \quad \forall t \tag{1}$$

where S_t is storage at the beginning of the period t; Q_t is inflow during the period t; R_t is the water release in the period t; A_t is the water surface area at the beginning of the period t; e_t is evaporation rate during the period t; S_t, Q_t, R_t are in units of cubic meters; A_t is in units of square meters; and e_t is in units of meters.

The rule curves and operation strategies are used as constraints during the simulation. First, reservoir storage in any period should not exceed its normal storage and also not be lower than its dead storage:

$$S_{dead} \leq S_t \leq S_{nomal}, \quad \forall t \tag{2}$$

Second, water released in any period should not exceed the reservoir's maximum discharge capability:

$$0 \leq R_t \leq R_{max}, \quad \forall t \tag{3}$$

The reservoir's operation is simulated at a daily time step over the period from 2000–2012. The reservoir's daily inflow is obtained by summing daily streamflow measured at Qilian and Zhamashike gaging stations (Figure 2). The evaporation rate at the water surface is estimated by applying a factor of 0.7 to the daily Pan evaporation measured at Qilian station.

When considering the reservoir operation, the streamflow at YLX is calculated as follows:

$$Q_t^{YLX} = R_t + L_t, \quad \forall t \tag{4}$$

where Q_t^{YLX} is streamflow at YLX; L_t is lateral flow between the Huangzangsi Reservoir and YLX, and is calculated by subtracting the sum of streamflow at Qilian and Zhamashike from the original streamflow at YLX.

Table 3. The basic operational strategies of the Huangzangsi Reservoir based on the rule curves.

Condition	Operational Decision
If storage level > Normal Curve (NC)	Increase water release to keep storage level = NC.
If Critical Curve (CC) < storage level ≤ NC	Perform environmental flow regulation if required. Regulate release to meet agriculture water and base flow demands.
If storage level ≤ CC	Stops release to keep storage level = CC.

2.4. Numerical Modeling Experiments

The model is run from 2000–2012, in total 13 years, at the daily time scale. The actual simulation during that period of time without the new reservoir is considered as the baseline model. Two series of numerical simulations are designed, namely Series A and Series B, in order to reflect the situation if the new reservoir had existed. Series A is aimed at investigating the impact of the different reservoir operational strategies on the hydrological processes in the middle and Lower HRB, where demand of groundwater pumping is fixed. Series B is used to investigate the relationship between the reservoir's operational strategy and groundwater exploitation practices in the study region. Series A specifies three environmental flow conservation schemes, which are shown in Table 4 as Experiments A1, A2 and A3. The total volume of the water released from the reservoir remains the same for the ecological flow purposes in these three experiments, while the water release for non-ecological flow purposes can still be different from one experiment to another. The duration and discharge rate of each operational strategy varies: Experiment A1 has the shortest duration and the largest discharge rate among the three experiments, while Experiment A3 has the longest duration and the smallest discharge rate. In general, a higher discharge in a single release may lead to a larger chance for the streamflow to replenish the terminal lake. Hereafter, the strategies in Experiments A1, A2 and A3 are referred to as the ecological operational strategies. For comparison, an additional experiment, Experiment A0, is also performed, in which only the basic operational strategy is applied without considering the ecological flows in the downstream.

Table 4. Operational strategies of the four experiments in Series A.

A0		A1		A2		A3	
Period	Discharge (m³/s)	Period	Discharge (m³/s)	Period	Discharge (m³/s)	Period	Discharge (m³/s)
N/A	N/A	April 1–5	324	April 1–10	162	April 1–15	108
		July 10–12	318	July 10–15	159	July 10–18	106
		August 10–12	315	August 10–15	158	August 10–18	105
		September 10–15	390	September 10–21	195	September 10–27	130
N/A	N/A	Duration	17 days	Duration	34 days	Duration	51 days
		Av. Discharge	344.5 m³/s	Av. Discharge	172.2 m³/s	Av. Discharge	114.8 m³/s
		Total volume	506 × 10⁶ m³	Total volume	506 × 10⁶ m³	Total volume	506 × 10⁶ m³

Note: Av. is the abbreviation of Average.

Since groundwater recharge plays a critical role in sustaining a healthy ecosystem in the lower stream, it is important to consider the impact of groundwater abstraction together with the reservoir operational strategies. As our previous study [28] revealed, for the Middle HRB, if more groundwater is used in the irrigation districts that are closer to the river and more river water is diverted to the districts that are further away from the river, the water use efficiency for irrigation may be increased. Experiment B is designed to represent this spatial operation of the groundwater reservoir, in which the changes in pumping ratio at different irrigation districts follow the suggestion by Wu et al. [28]. The pumping ratio is defined as the percentage of groundwater in the total irrigation water supply.

Experiment B considers the ecological operational strategy in Experiment A3 (longest release duration with smallest rate), and it assumes that the irrigation water demand at all the irrigation districts remains unchanged. Essentially, Experiment B represents a joint operation of surface water and groundwater reservoirs in this region.

3. Results and Discussion

3.1. Water Balance of the Reservoir

The annual averaged water balances for years 2000–2012 at the Huangzangsi Reservoir for Experiments A0–A3 were calculated and are presented in Table 5. The calculated annual evaporation ranged from 2.9×10^6–4.8×10^6 m^3. It is seen that the ecological operational strategies have an obvious impact on the reservoir's water balance. Compared to Experiment A0, Experiments A1–A3 have much higher annual water release and much lower evaporation and storage changes.

Figure 3a demonstrates the intra-annual fluctuation of the annual average water level. All the water levels fluctuated between the normal water level (2628 m) and the dead storage level (2580 m). As we can see, without the environmental flow consideration (i.e., in Experiment A0), the effective storage of the reservoir was used twice per year: the first time from December–March in the next year, when no irrigation and environmental flow was requested; therefore, most of the inflow was stored in the reservoir, and the water level gradually increased; and the second time from July–October, when the flood season comes, and the inflow to the reservoir significantly exceeded the water demands for irrigation and ecological flow. The reservoir loses water also twice per year: the first time from April–June, when the water level declines rapidly due to the irrigation demands; and the second time in November, where the water level declines once again due to the irrigation demand for keeping enough soil moisture through winter. The fluctuation of water level in the reservoir was more complicated, since the release of environmental flow could reduce the water level in a very short time period of time. Figure 3b illustrates the intra-annual variation of daily evaporation. As can be seen, the ecological operational strategies can substantially reduce the evaporation from the reservoir, especially from April–June and in October.

Table 5. Annual average water balances of the Huangzangsi Reservoir under different operations.

Experiment	Inflow Q (10^6 m^3)	Water Release R (10^6 m^3)	Evaporation E (10^6 m^3)	Storage Change ΔS_r (10^6 m^3)	Average Water Level (Elevation) (m)
A0	1319.1	1297.0	4.8	17.3	2604.69
A1	1319.1	1311.9	2.9	4.3	2593.09
A2	1319.1	1311.0	3.0	5.1	2593.89
A3	1319.1	1308.9	3.2	7.0	2594.45

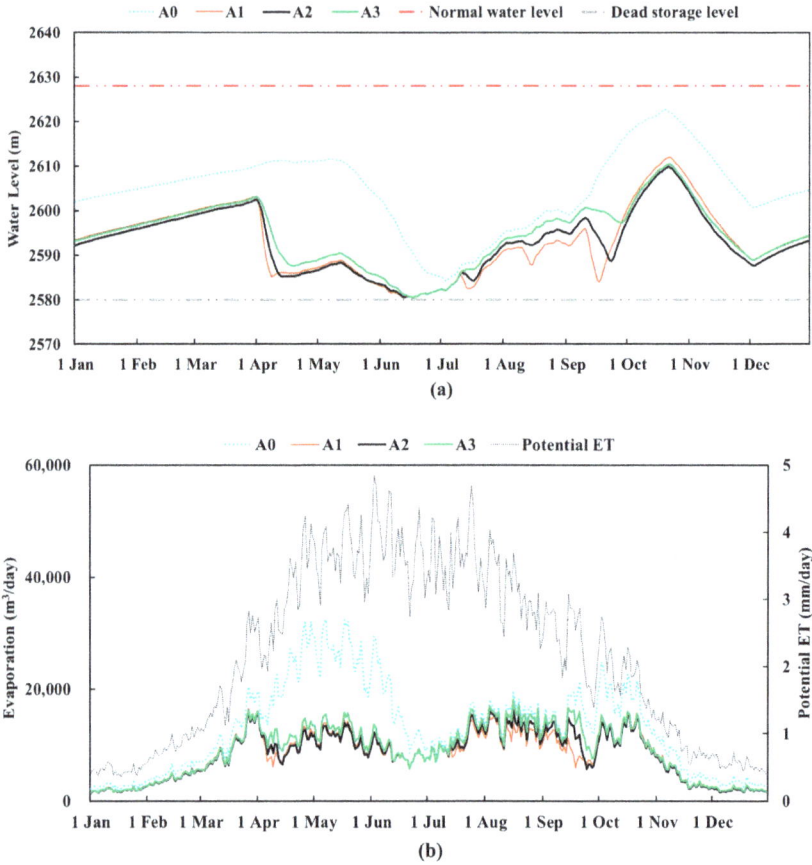

Figure 3. Impacts of the ecological operations on the water balance of the reservoir on a multiyear average basis. (**a**) The daily water level; and (**b**) the daily evaporation from the reservoir.

3.2. Water Balance at the Middle and Lower HRB

Inflow of stream water to our model domain at Yingluoxia (YLX) was derived, following the approach introduced in Section 2.3. A set of variables was examined to evaluate the impact of the ecological operational strategies on the hydrological processes in the middle and Lower HRB systematically. Table 6 summarizes the simulation results of selected variables, which represent five key processes including agricultural water uses, streamflow, stream-aquifer interaction, groundwater flow and evapotranspiration.

Table 6. Key hydrological variables simulated by the Groundwater and Surface Water Flow (GSFLOW) model under different ecological operational strategies. GW, groundwater; MHRB, Middle HRB; ZYX, Zhengyixia; EJL, East Juyan Lake; YLX, Yingluoxia; LHRB, Lower HRB.

Hydrological Process	Variable	Baseline Scenario	Ecological Operations (Experiment ID)			
			A0	A1	A2	A3
Agricultural water uses	SW diversion in MHRB (10^6 m^3)	1456	1672	1483	1402	1336
	GW pumping in MHRB (10^6 m^3)	403	405	405	405	405
	Total supply in MHRB (10^6 m^3)	1859	2078	1888	1807	1741
	Degree of demand fulfillment (%)	87.39	97.67	88.76	84.97	81.85
Streamflow	Streamflow through ZYX (10^6 m^3)	994	807	972	1030	1079
	Streamflow entering EJL (10^6 m^3)	74	33	61	68	74
Stream-aquifer interaction	Stream leakage in YLX-312B (10^6 m^3)	466	480	459	466	476
	GW discharge in 312B-ZYX (10^6 m^3)	−462	−484	−468	−461	−455
	Stream leakage in ZYX-EJL (10^6 m^3)	543	518	553	569	583
Groundwater flow	Areal recharge in MHRB (10^6 m^3)	462	461	462	400	389
	Areal recharge in LHRB (10^6 m^3)	6	6	6	9	10
	ΔS in MHRB (10^6 m^3)	−86	−64	−81	−84	−86
	ΔS in LHRB (10^6 m^3)	15	6	11	14	16
Evapotranspiration	ET in MHRB (10^6 m^3)	1473	1566	1564	1458	1433
	ET in LHRB (10^6 m^3)	1038	859	858	1055	1090

Note: ΔS is storage change in the saturated zone; a negative value indicates loss of water with respect to initial storage, and a positive value indicates gain of water with respect to initial storage.

The ecological operational strategies have a direct impact on the agricultural water uses, which consequently affect other hydrological processes. Under the baseline conditions (i.e., the baseline scenario), 87.39% of the total irrigation water demand in the Middle HRB (MHRB) is fulfilled. This percentage could be increased to as high as 97.67% by employing the function of the new reservoir if the environmental flow is not considered (i.e., Experiment A0). If the environmental flow is taken into consideration (i.e., Experiments A1–A3), more water is needed at the Lower HRB (LHRB), but the degree of demand fulfillment in MHRB would be reduced. For example, in Experiment A3, the annual streamflow through Zhengyixia (ZYX) reaches the maximum value (1079 million m^3), but the degree of demand fulfillment (81.85%) is the lowest. Figure 4 shows the spatial patterns of the degree of demand fulfillment in the four experiments. It can be seen that the degree of fulfillment decreases in all districts from Experiment A0–Experiment A3.

The streamflow through ZYX is the environmental flow for the Lower HRB, which largely determines how much water can eventually enter the East Juyan Lake (EJL). According to the water allocation plan, the streamflow through ZYX should be 1150 million m^3 per year on average in 2000–2012. However, the simulated actual flow (i.e., in the baseline scenario) did not exceed 1000 million m^3 per year, and the goal of the water allocation plan was not achieved. As Table 6 indicates, the operations in Experiments A1–A3 push the environmental flow towards the goal, and the flow in Experiment A3 is very close to the goal. This suggests that a smaller water release with a longer duration is beneficial to achieve the water allocation goal.

As for stream-aquifer interaction, based on our previous study [26], three distinctive segments along the main river can be identified as either gaining or losing streams. YLX to the bridge named 312 Bridge (312B) is a losing segment, where a large amount of streamflow percolates through a thick vadose zone and recharges the aquifer. 312B to ZYX is a gaining segment as a whole, where the aquifer discharges water to the river. ZYX to the terminal lake (EJL) is another losing segment. Based on our prior understanding, the flux of water in each of the three segments due to stream-aquifer interaction are calculated. It is noted that even though small spatial-temporal variations can happen so that a sub-segment of the gain branch can turn to a losing branch for a short period of time, the multi-annual statistics show that the separations of the three segments do not change from year to year. That is, a smaller water release rate with a longer duration in the ecological operations is beneficial to the replenishment of the aquifer in the mid- and lower stream basins, especially in the lower stream segment. In addition, the amount of groundwater discharging to the gaining segment from 312B to ZYX decreases from Experiment A0–Experiment A3.

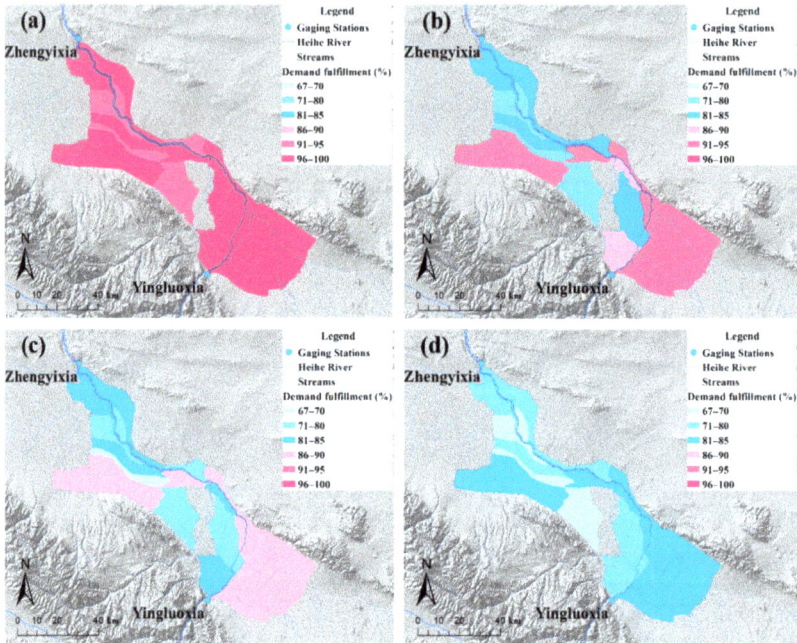

Figure 4. The simulated degree of demand fulfillment under different operations of the Huangzangsi Reservoir. (**a–d**) correspond to Experiments A0–A3.

In the Middle HRB, groundwater recharge takes place mainly due to the percolation of irrigated water. Thus, when the irrigation is reduced (as seen from Experiment A1–Experiment A3), the groundwater recharge decreases accordingly. As seen in Table 6, even though there is still a negative change in the groundwater storage, the speed of decline is slower from Experiment A3–Experiment A1. This is because of the reduced groundwater recharge. Furthermore, our results suggest that the groundwater storage in the Lower HRB is recovered due to increased stream leakage in the ZYX-EJL segment from Experiments A1–A3.

Evapotranspiration (ET) in the study area largely depends on the availability of surface water. From Experiment A1–A3, when the water supply for irrigation decreases, the ET in the Middle HRB decreases accordingly. On the contrary, the simulated ET in the Lower HRB is higher in Experiment A3 than the other two experiments because more environmental flow is available in the Lower HRB for A3, and thus more surface water is available. Figure 5 demonstrates the spatial pattern of annual average ET in the lower HRB.

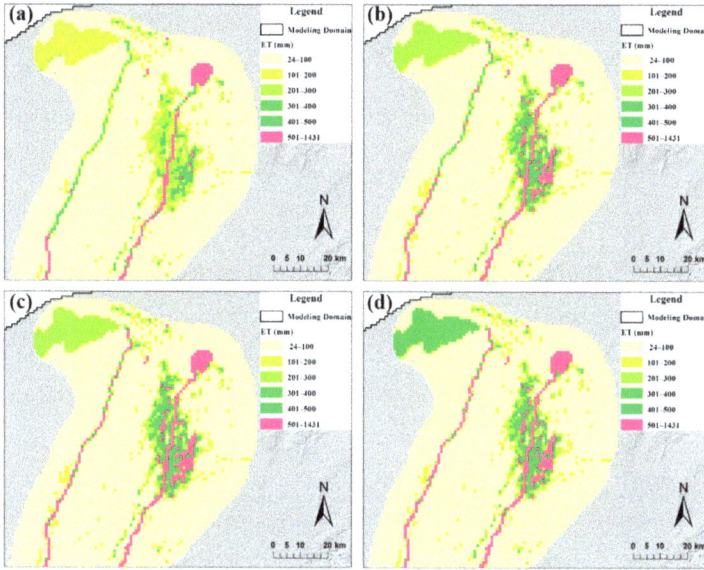

Figure 5. Spatial patterns of annual average Evapotranspiration (ET) in the lower stream under (**a**) Experiment A0, (**b**) Experiment A1, (**c**) Experiment A2 and (**d**) Experiment A3.

3.3. Impacts of the Joint Operation

Several key hydrological variables are selected to investigate the impact of the joint operation in Experiment B. While the overall degree of demand fulfillment is similar in Experiments A3 and B (see Table 7), the spatial pattern of the degree varies. Figure 6a,b show the spatial pattern of the pumping ratio in the 17 irrigation districts in Experiments A3 and B, respectively. Figure 6c and 6d show the change in pumping ratio and the changes in the degree of demand fulfillment. It can be seen that, in the irrigation districts with IDs of 14, 15, 16 and 25 (red areas), the irrigation demand is better met in Experiment B. In contrast, in the irrigation districts with IDs of 23, 30 and 31, the irrigation demand is less met in Experiment B.

Table 7. Key hydrological variables simulated by the GSFLOW model under Experiments A3 and B.

Variable	A3	B	Difference (Percentage Change)
SW diversion in MHRB (10^6 m^3)	1336	1307	−0.29 (−2.14%)
GW pumping in MHRB (10^6 m^3)	405	409	0.04 (0.79%)
Total supply in MHRB (10^6 m^3)	1741	1716	−0.25 (−1.45%)
Degree of demand fulfillment (%)	81.85	80.66	−1.19 (−1.45%)
Streamflow through ZYX (10^6 m^3)	1079	1031	−0.48 (−4.49%)
Stream leakage in YLX-312B (10^6 m^3)	476	458	−0.18 (−3.82%)
GW discharge in 312B-ZYX (10^6 m^3)	−455	−384	−0.71 (−15.59%)
Areal recharge in MHRB (10^6 m^3)	389	388	−0.01 (−0.39%)
Areal recharge in LHRB (10^6 m^3)	10	9	−0.01 (−6.97%)
ΔS in MHRB (10^6 m^3)	−86	−39	0.47 (54.40%)
ΔS in LHRB (10^6 m^3)	16	13	−0.03 (−22.31%)

Note: ΔS is storage change in the saturated zone; a negative value indicates loss of water with respect to initial storage, and a positive value indicates gain of water with respect to initial storage.

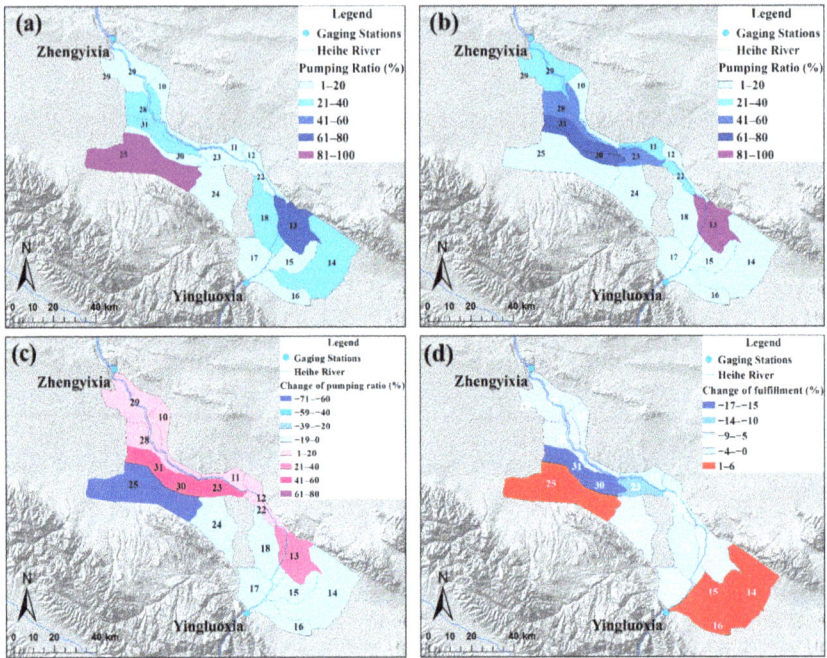

Figure 6. Spatial patterns of the pumping ratios in the 17 irrigation districts. (**a**) The actual pumping ratios over the period 2000–2012 in Experiment A3; (**b**) the assumed pumping ratios in Experiment B; (**c**) changes in the ratios from Experiment A3–B; and (**d**) changes in the degree of irrigation demand fulfillment. The pumping ratio is defined as the percentage of groundwater in the total irrigation water supply.

As Table 7 indicates, the change in pumping ratio changes the surface water-groundwater interactions, which in turn alters the spatial pattern of the degree of fulfillment. Figure 7 compares spatial patterns of annual average Groundwater (GW) recharge at the 17 irrigation districts. It can be seen that the GW recharge decreases in the districts that are near the river and increases in the districts that are far from the river, while the change of the total areal GW recharge is small. Overall, the decline in groundwater storage in HRB can be significantly slowed down (see ΔS in MHRB in Table 7). However, this recovery is at the cost of reduced environmental flow through ZYX (see streamflow through ZYX in Table 7).

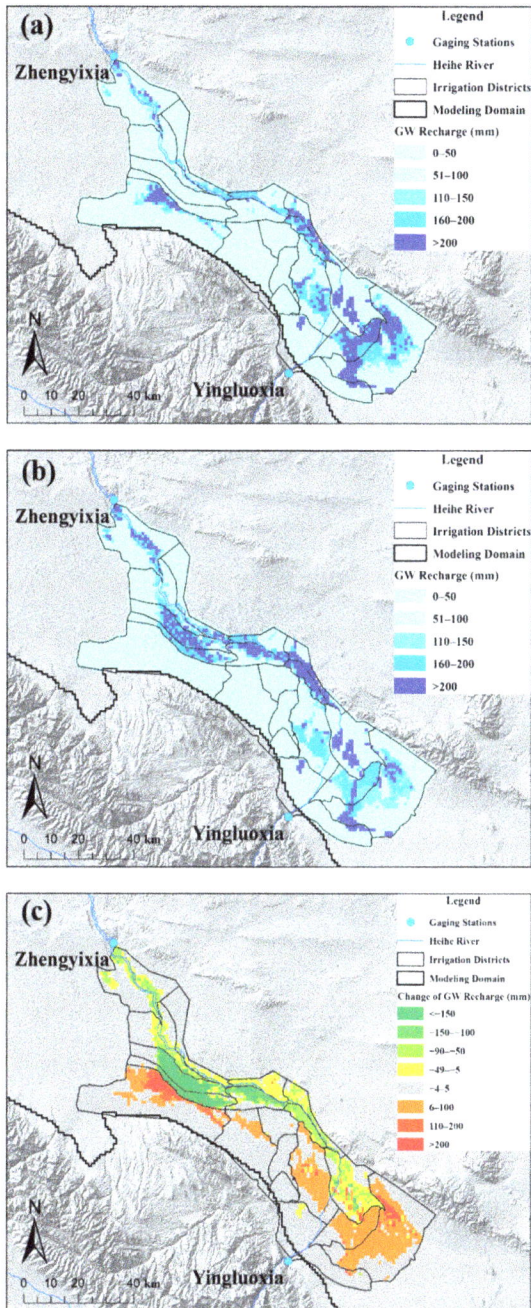

Figure 7. Spatial patterns of annual average Groundwater (GW) recharge at the 17 irrigation districts. (**a**) The annual average GW recharge in Experiment A3; (**b**) the annual average GW recharge in Experiment B; and (**c**) the change in GW recharge from Experiment B–Experiment A3.

Overall, as suggested by Table 7 and Figure 7, the groundwater operation can enhance the flexibility of the water resources management. With the joint operation of surface water and groundwater reservoir, a balance among the agriculture water need, the groundwater sustainability in the Middle HRB and the ecological water need in the Lower HRB can be easily achieved. For example, if the primary goal of the water resources management is to maintain the sustainability of groundwater in the Middle HRB, the groundwater pumping may consider the spatial pattern in Experiment B (Figure 6b). If the primary goal is to meet the ecological water need of the Lower HRB, the groundwater pumping may consider the spatial pattern in Experiment A3 (Figure 6a). However, the joint operation can hardly improve the three aspects simultaneously. To resolve the water conflict in HRB, further engineering and/or policy measures are desired, besides the reservoir development and groundwater regulation. Water-saving irrigation technologies and reducing the weight of agriculture in the regional economy are potential solutions.

4. Conclusions

This study investigates the hydrological impacts of joint operation of surface water and groundwater reservoirs in Heihe River Basin (HRB), using an integrated surface water-groundwater modeling coupled with a reservoir operation simulation model. The integrated model can simulate demand-based diversion and pumping rates, which is specifically designed for arid regions with significant agricultural irrigation. The reservoir operation model evaluates basic and ecological operational strategies. Through a set of numerical experiments, this study further addresses whether and how the joint operation could alleviate the human-water conflict in HRB.

The major findings are summarized as follows. First, based on simulation results of the reservoir operation model, the effective storage of the Huangzangsi reservoir is used twice per year, and the ecological operational strategies can substantially reduce the evaporation from the reservoir. Second, the ecological operational strategies have a direct impact on the agricultural water uses in the Middle HRB, and consequently affect other hydrological processes in the middle and Lower HRB. The ecological operation strategy that has a smaller water release with a longer duration is beneficial to achieve the water allocation goal and to the replenishment of the aquifer in the middle and Lower HRB, but such a strategy may reduce the chance to meet the agriculture water demand of the Middle HRB. Finally, with the joint operation of the surface water and groundwater reservoir, a balance among the agriculture water need, the groundwater sustainability in the Middle HRB and the ecological water need in the Lower HRB can be easily achieved. However, the joint operation can hardly improve the three aspects simultaneously. To resolve the water conflict in HRB, further engineering and/or policy measures are desired, besides the reservoir development and groundwater regulation.

Overall, our study provides insights into the water resources management in arid regions. The study results imply that reservoir operation alone, even considering both surface water and groundwater, may not be sufficient to resolve the typical human-water conflict. Future studies can investigate more management and policy measures, such as using water-saving irrigation technologies and reducing the weight of agriculture in the regional economy.

Author Contributions: Conceptualization, Y.Z. and Y.T. Methodology, Y.T., J.X., X.P. and F.H. Formal analysis, Y.T. Data curation, Y.T. and J.X. Writing, original draft preparation, Y.T. and S.J. Writing, review and editing, Y.Z. and X.H. Supervision, Y.Z. Project administration, Y.Z. Funding acquisition, Y.Z. and Y.T.

funding: This work was funded by the National Natural Science Foundation of China (No. 41501024; No. 91647201; No. 41622111). Additional support was provided by the Southern University of Science and Technology (No. G01296001) and the Guangdong Provincial Key Laboratory of Soil and Groundwater Pollution Control (No. 2017B030301012).

Acknowledgments: The data used in this study, if not collected by the authors or acknowledged in the text, were provided by the Heihe Program Data Management Center (http://www.heihedata.org).

Conflicts of Interest: The authors declare no conflict of interest.

References

1. Doll, P.; Siebert, S. Global modeling of irrigation water requirements. *Water Resour. Res.* **2002**, *38*, 8-1–8-10. [CrossRef]
2. Jagermeyr, J.; Gerten, D.; Heinke, J.; Schaphoff, S.; Kummu, M.; Lucht, W. Water savings potentials of irrigation systems: Global simulation of processes and linkages. *Hydrol. Earth Syst. Sc.* **2015**, *19*, 3073–3091. [CrossRef]
3. Thenkabail, P.S.; Biradar, C.M.; Noojipady, P.; Dheeravath, V.; Li, Y.J.; Velpuri, M.; Gumma, M.; Gangalakunta, O.R.P.; Turral, H.; Cai, X.L.; et al. Global irrigated area map (GIAM), derived from remote sensing, for the end of the last millennium. *Int. J. Remote Sens.* **2009**, *30*, 3679–3733. [CrossRef]
4. Siebert, S.; Burke, J.; Faures, J.M.; Frenken, K.; Hoogeveen, J.; Doll, P.; Portmann, F.T. Groundwater use for irrigation—A global inventory. *Hydrol. Earth Syst. Sci.* **2010**, *14*, 1863–1880. [CrossRef]
5. Garrido, A.; Martinez-Santos, P.; Llamas, M.R. Groundwater irrigation and its implications for water policy in semiarid countries: The spanish experience. *Hydrogeol. J.* **2006**, *14*, 340–349. [CrossRef]
6. Hu, X.L.; Shi, L.S.; Zeng, J.C.; Yang, J.Z.; Zha, Y.Y.; Yao, Y.J.; Cao, G.L. Estimation of actual irrigation amount and its impact on groundwater depletion: A case study in the Hebei plain, China. *J. Hydrol.* **2016**, *543*, 433–449. [CrossRef]
7. Leng, G.Y.; Huang, M.Y.; Tang, Q.H.; Gao, H.L.; Leung, L.R. Modeling the effects of groundwater-fed irrigation on terrestrial hydrology over the conterminous united states. *J. Hydrometeorol.* **2014**, *15*, 957–972. [CrossRef]
8. Velazquez, M.P.; Jenkins, M.W.; Lund, J.R. Economic values for conjunctive use and water banking in southern California. *Water Resour. Res.* **2004**, *40*. [CrossRef]
9. Kundzewicz, Z.W.; Doll, P. Will groundwater ease freshwater stress under climate change? *Hydrolog. Sci. J.* **2009**, *54*, 665–675. [CrossRef]
10. Scanlon, B.R.; Reedy, R.C.; Faunt, C.C.; Pool, D.; Uhlman, K. Enhancing drought resilience with conjunctive use and managed aquifer recharge in California and Arizona. *Environ. Res. Lett.* **2016**, *11*. [CrossRef]
11. Murray-Darling Basin Commission (MDBC). Groundwater: A Resource for the Future. Available online: https://www.mdba.gov.au/sites/default/files/archived/mdbc-GWreports/2173_GW_a_resource_for_the_future.pdf (accessed on 12 August 2018).
12. Castle, S.L.; Thomas, B.F.; Reager, J.T.; Rodell, M.; Swenson, S.C.; Famiglietti, J.S. Groundwater depletion during drought threatens future water security of the Colorado River Basin. *Geophys. Res. Lett.* **2014**, *41*, 5904–5911. [CrossRef] [PubMed]
13. Fuchs, E.H.; Carroll, K.C.; King, J.P. Quantifying groundwater resilience through conjunctive use for irrigated agriculture in a constrained aquifer system. *J. Hydrol.* **2018**. [CrossRef]
14. Nikoo, M.R.; Karimi, A.; Kerachian, R.; Poorsepahy-Samian, H.; Daneshmand, F. Rules for optimal operation of reservoir-river-groundwater systems considering water quality targets: Application of M5P model. *Water Resour. Manag.* **2013**, *27*, 2771–2784. [CrossRef]
15. Li, X.; Cheng, G.; Ge, Y.; Li, H.; Han, F.; Hu, X.; Tian, W.; Tian, Y.; Pan, X.; Nian, Y.; et al. Hydrological cycle in the heihe river basin and its implication for water resource management in endorheic basins. *J. Geophys. Res. Atmos.* **2018**, *123*, 890–914. [CrossRef]
16. Gao, G.; Shen, Q.; Zhang, Y.; Pan, N.; Ma, Y.; Jiang, X.; Fu, B. Determining spatio-temporal variations of ecological water consumption by natural oases for sustainable water resources allocation in a hyper-arid endorheic basin. *J. Clean Prod.* **2018**, *185*, 1–13. [CrossRef]
17. Li, Z.; Liu, W.Z.; Zhang, X.C.; Zheng, F.L. Impacts of land use change and climate variability on hydrology in an agricultural catchment on the loess plateau of China. *J. Hydrol.* **2009**, *377*, 35–42. [CrossRef]
18. Chen, Y.; Zhang, D.; Sun, Y.; Liu, X.; Wang, N.; Savenije, H.H.G. Water demand management: A case study of the Heihe River Basin in China. *Phys. Chem. Earth Parts A/B/C* **2005**, *30*, 408–419. [CrossRef]
19. Wu, X.; Zheng, Y.; Wu, B.; Tian, Y.; Han, F.; Zheng, C.M. Optimizing conjunctive use of surface water and groundwater for irrigation to address human-nature water conflicts: A surrogate modeling approach. *Agric. Water Manag.* **2016**, *163*, 380–392. [CrossRef]
20. Tian, Y.; Zheng, Y.; Wu, B.; Wu, X.; Liu, J.; Zheng, C.M. Modeling surface water-groundwater interaction in arid and semi-arid regions with intensive agriculture. *Environ. Modell. Softw.* **2015**, *63*, 170–184. [CrossRef]

21. Wang, Y.; Xiao, H.L.; Lu, M.F. Analysis of water consumption using a regional input–output model: Model development and application to Zhangye City, Northwestern China. *J. Arid Environ.* **2009**, *73*, 894–900. [CrossRef]

22. Markstrom, S.L.; Niswonger, R.G.; Regan, R.S.; Prudic, D.E.; Barlow, P.M. *Gsflow—Coupled Groundewater and Surfaceewater Flow Model Based on the Integration of the Precipitation-Runoff Modeling System (prms) and the Modular Groundewater Flow Model (modflow-2005)*; USGS: Reston, VA, USA, 2008; p. 240.

23. Niswonger, R.G.; Prudic, D.E.; Regan, R.S. *Documentation of the Unsaturated-Zone Flow (UZF1) Package for Modeling Unsaturated flow Between the Land Surface and the Water Table with Modflow-2005*; USGS: Reston, VA, USA, 2006; p. 62.

24. Niswonger, R.G.; Prudic, D.E. *Documentation of the Streamflow-Routing (SFR2) Package to Include Unsaturated Flow Beneath Streams—A Modification to SFR1*; USGS: Reston, VA, USA, 2005.

25. Merritt, M.L.; Konikow, L.F. *Documentation of a Computer Program to Simulate Lake-Aquifer Interaction Using the Modflow Ground Water Flow Model and the MOC3D Solute-Transport Model*; USGS: Reston, VA, USA, 2000.

26. Tian, Y.; Zheng, Y.; Zheng, C.M.; Xiao, H.L.; Fan, W.J.; Zou, S.B.; Wu, B.; Yao, Y.Y.; Zhang, A.J.; Liu, J. Exploring scale-dependent ecohydrological responses in a large endorheic river basin through integrated surface water-groundwater modeling. *Water Resour. Res.* **2015**, *51*, 4065–4085. [CrossRef]

27. Heihe River Basin Authority. Environmental impact assessment report for the Huangzangsi Reservoir on the Heihe River. Available online: http://www.zhb.gov.cn/gkml/hbb/spwj1/201507/t20150717_306844.htm (accessed on 15 July 2018).

28. Wu, B.; Zheng, Y.; Wu, X.; Tian, Y.; Han, F.; Liu, J.; Zheng, C.M. Optimizing water resources management in large river basins with integrated surface water-groundwater modeling: A surrogate-based approach. *Water Resour. Res.* **2015**, *51*, 2153–2173. [CrossRef]

water

Article

Relationships Among Animal Communities, Lentic Habitats, and Channel Characteristics for Ecological Sediment Management

Mikyoung Choi [1],*, Yasuhiro Takemon [2], Kinko Ikeda [3] and Kwansue Jung [4]

[1] International Water Resources Research Institute, Chungnam National University, Daejeon 34134, Korea; choi.mk1981@gmail.com

[2] Disaster Prevention Research Institute, Kyoto University, Kyoto 611-0011, Japan; takemon.yasuhiro.5e@kyoto-u.ac.jp

[3] ASIA AIR SURVEY CO., LTD., Osaka 530-6029, Japan; kom.ikeda@ajiko.co.jp

[4] Department of Civil Engineering, Chungnam National University, Daejeon 34134, Korea; ksjung@cnu.ac.kr

* Correspondence: choi.mk1981@gmail.com; Tel.: +82-42-821-7745

Received: 29 August 2018; Accepted: 16 October 2018; Published: 19 October 2018

Abstract: This study used a multiscale analysis of relationships among the bitterling and mussel communities, lentic habitat structures with conditions and flooding frequency, and channel characteristics for application in ecological sediment management. From the Kizu River in Japan, 120 lentic habitats were sampled in 2007 and 2010. The floodplain vertical shape index (FVSI), which indicates the degree of convexity or concavity of the vertical shape of a floodplain, was used as channel characteristics using historical cross-section profiles obtained from 1960 to 2012. For examining the relationships between bitterlings/mussels and each habitat condition or structure, abundance values of bitterlings and mussels were transformed into habitat suitability index (HSI). Furthermore, the relationships between the number of habitat structures and FVSI were analyzed. The results indicated that bitterlings and mussels are more abundant in terrace ponds than in active ponds, especially so in terrace ponds located in the lower area of bars with a flooding frequency of 8–16 days/year (bitterlings), those located in the lower area of bars with a flooding frequency of 8 days/year, and those located in the upper area of bars with a flooding frequency of 16–22 days/year (mussels). These ponds tended to have less than 1 cm mud depth that was negatively related to abundance of mussels. These suitable habitat types tended to be located in channels with a floodplain vertical shape index between −0.35 and 0.05. We established countermeasures to prevent channel types with floodplain vertical shape index exceeding 0.05 instead of restoring the previous channel conditions.

Keywords: lentic habitats; bitterling; mussel; floodplain vertical shape index; sediment management

1. Introduction

The interruption of the natural flow and sediment transfer by dams causes changes in the flow, river structure, and ecosystem in the reservoirs and regions downstream of the dam. When reservoirs are filled with sediments, their storage capacity is reduced, affecting the water supply and hydroelectric power [1]. Downstream of dams, the reduced sediment supply results in degradation of the stability of channel structure, and the quality of aquatic habitats [2]. In order to remove sediments accumulated in reservoirs and to transport sediment downstream, several sediment management strategies such as replenishment, sluicing, and bypassing have been tested worldwide [3–5]. Actually, artificial sediment supplies downstream have contributed to enhancing the available spawning habitat for chinook salmon [2] and the lotic habitat quality for invertebrates and fishes [6]. In order to predict the geomorphic and environmental impacts of sediment management and for successful ecological restoration, it is essential to understand river characteristics and conditions under natural and artificial disturbances. The reach-scale channel configuration, such as braided, wandering, and straight channel, can be helpful for linkage between hydraulic conditions and aquatic ecosystems. This can be examined using hydraulic and geomorphic parameters, including discharge and slope [7], sediment load and lateral stability [8], and slope and bed materials [9]. The variations in channel configuration are in response to the flow regime and sediment transport, which further influence the geomorphic units such as ponds, bars, wetlands, backwaters, and pool-riffle sequences. Further, the assemblages of geomorphic units at the reach-scale are related to the habitat diversity and animal communities [10]. The high complexity of landforms generally equates to the high diversity of hydraulics and thus the high biodiversity [11,12]. If research related to the relationship between the channel configuration and ecosystem are efficiently conducted, it is possible to predict the impact of natural or artificial disturbances on the ecosystem [13]. Therefore, integration of ecological and geomorphological perspectives with the acknowledges the multiscale abiotic and biotic structures of a stream system is required, but is an unresolved challenge [14]. Wyrick and Rasternack [12] systematically studied the relationship between geomorphic units, hydraulic parameters, and their relevance for various habitats. Other previous studies on multiscale relationships between abiotic and biotic structures have focused on lotic habitat conditions; the relationships between lotic habitat conditions and channel configurations [15,16], landscape diversity with channel characteristics [17], or habitat diversity in specific channel types [18,19].

This study focused on multiscale relationships among animal communities, lentic habitats, and channel characteristics in the Kizu River. The Kizu River provides suitable lentic habitats such as ponds or backwaters for bitterlings and mussels on its floodplain [20]. However, the diversities of bitterlings and mussels decreased, and the protected bitterling *Acheilognathus longipinnis* disappeared. Therefore, the target animal communities of this study are considered to be bitterlings and mussels. Further, because bitterlings and mussels live only in lentic habitats, lentic habitat structures and conditions was considered as habitat scale. In the case of lentic habitat on floodplains, the environmental conditions are characterized by hydrological connectivity with the main river channels [21]. Flooding provides hydrological connectivity with the exchange of nutrients, sediments and organisms. Therefore, we considered not only structures but also the flooding frequency. First, this study investigates the relationships between the abundance of bitterling/mussel and lentic habitat structures/conditions with the flooding frequency. Second, the relationships between habitat structures and the reach-scale channel characteristics were investigated. The floodplain vertical shape index (FVSI) was used as the reach-scale channel characteristic. As the channel types are not numerical values, FVSI was developed in Reference [22] and represented the concave and convex floodplain shapes. Additionally, the historical changes of FVSI were calculated using the historic cross-section data. Finally, we discussed previous environmental conditions and the application of these characteristics to ecological sediment management.

2. Materials and Methods

2.1. Study Site

The study area was comprised of the lower reaches (0–26 km) of the Kizu River in Japan (Figure 1). The riverbed of the Kizu River degraded, and the vegetation in the floodplain expanded after dam construction over a 65-year period. A total of five dams, namely, Takayama Dam, Syourenji Dam, Murou Dam, Nunome Dam, and Hinachi Dam, which were constructed in 1969, 1970, 1974, 1992, and 1999, respectively, are located in the Kizu River basin. The approximately 6000 m^3/s peak floods occurred before the construction of dams; however, the peak of floods decreased to approximately 3000 m^3/s after construction [23]. The estimated amount of bed-load transportation downstream of the dam was approximately 183,000 m^3/y in the 1960s; further, it decreased to 23,000 m^3/y in the

2000s (Figure 2) [13]. The reduction in the peak discharge and sediment supply upon dam construction influenced the alternation of channels from braided to single or alternating channels. These river channel changes caused the degradation of lentic habitat conditions on the floodplain owing to the reduction of inundation [20].

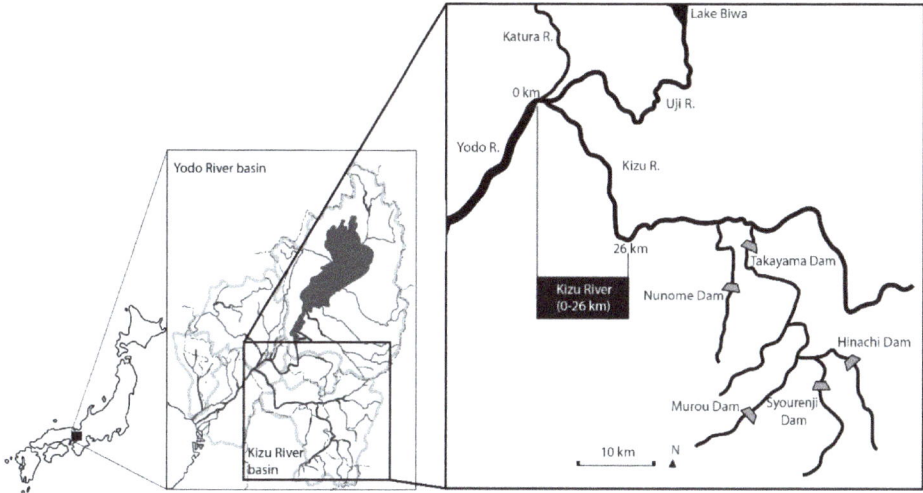

Figure 1. Study site in the Kizu River.

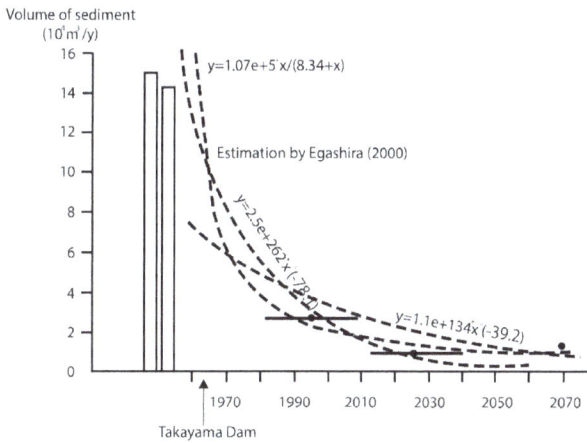

Figure 2. Historic changes in annual transport sediment volume in the Kizu River were estimated in Reference [23].

2.2. Data Acquisition of Bitterling and Mussel Communities

In the case of the Kizu River, because bitterlings and mussels live only in lentic habitats such as ponds and backwaters, the habitat types of pond and backwater were selected for sampling (Figure 3d,e). Prior to conducting the field survey, the size and location of the habitat were identified using aerial photos. In 2006, 190 ponds were detected in aerial photos, whereas 178 ponds were detected in 2010. Among the detected ponds, bitterlings and mussels were sampled in 47 ponds among the 190 ponds in 2007 and in 73 among the 178 ponds in 2010 by the Yodogawa River Bureau. The sampling was conducted in the summer season. Bitterlings were observed in 63 ponds and mussels were observed in 47 ponds among 120 ponds. There are five taxa of bitterlings (*Acheilognathus rhombeus, Acheilognathus tabira tabira, Acheilognathus cyanostigma, Rhodeus ocellatus ocellatus, Tanakia lanceolate*) and five taxa of mussels (*Anodonta calipygos Kobelt, Anodonta woodiana, Lanceolaria grayana cuspidate, Lanceolaria oxyrhyncha (Martens), Unio douglasiae nipponensis*) in sampling ponds. The total number (population) of bitterlings and mussels were divided by the surveyed time (h) and number of people who participated in the sampling (*n*), and this value was used as the abundance of bitterlings and mussels. The abundance value was used to calculate the best model of lentic habitat conditions for bitterlings and mussels. For examining the relationships between bitterlings/mussels and each habitat condition or structure, abundance values transformed into habitat suitability index (HSI) [19–21]. Habitat suitability index was derived by Equation (1).

$$w_i = \frac{u_i / \sum_{i=1}^{n} u_i}{a_i / \sum_{i=1}^{n} a_i} \tag{1}$$

w_i is the ratio for the ith of n habitat categories, u_i is the total abundance in category I, $\sum u_i$ is the total abundance for all habitat categories, a_i is the number of samples from categories i, and $\sum a_i$ is the total number of samples [24]. The indices range from 0 to 1 for each variable, with 0 indicating the least suitable habitat conditions and 1 indicating the optimum habitat condition [25].

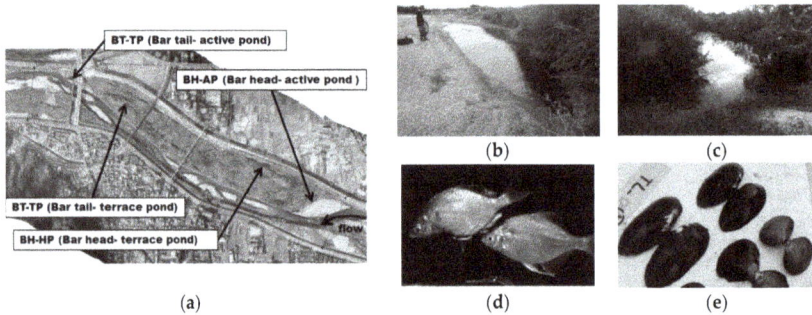

Figure 3. Lentic habitats and target species. (**a**) Image showing classification of lentic habitats; (**b**) an active pond (AP); (**c**) a terrace pond (TP); (**d**) bitterling; and (**e**) mussel.

2.3. Data Acquisition of Lentic Habitat Structure and Conditions

Lentic habitat structures were classified into two types, active pond (AP) and terrace pond (TP), depending on their locations. The aerial photos obtained in 2006 and 2010 were used for this classification. AP is defined as a pond and backwater that is located on active channels such as sand-bars or sanded islands (Figure 3b), and TP is defined as a pond located on a terrace with vegetation (Figure 3c). These ponds were further classified based on the location of upper and lower sites on a particular bar. The bar was divided into two halves based on the direction of stream flow. The ponds located in the upper area were defined as bar head-ponds, whereas those located in the lower area were defined as bar tail-ponds. This classification was based on the assumption that even if the same habitat structure type was located on the same bar, the habitat conditions or the abundance of species differed depending on whether the habitats were located on the upper or lower part of the bar. Thus, four types of lentic habitats were considered: Bar head-active pond (BH-AP), bar head-terrace pond (BH-TP), bar tail-active pond (BT-AP), and bar tail-terrace pond (BT-TP) (Figure 3). A total of 190 ponds were detected in aerial photos in 2006, whereas 178 ponds were detected in 2010. Although only a total of 120 ponds were surveyed, all the detected ponds in aerial photos

(n = 368) had values of the geomorphic parameters of flooding frequency. The term flooding frequency refers to the inundation frequency in a pond per year. The water level data obtained during the period from 1989 to 2008 were converted into discharge using the HQ formula. Using the converted discharge, 1 day discharge (909.1 m^3, at Inooka observatory station), 8 day discharge (249.5 m^3), 16 day discharge (149.0 m^3), 22 day discharge (118.3 m^3), 45 day discharges (72.9 m^3), 71 day discharge (52.9 m^3), 185 day discharge (26.2 m^3) and 365 day discharge were calculated. The water level of the cross section (200 m intervals) according to each discharge was calculated using HEC-RAS software. Further, the values of water levels were overlapped with the digital elevation model (DEM) data using Geomedia 6.1 software. Thus, if the pond would be inundated based on the 1 day discharge, the pond would have a flooding frequency of 1 day/year. If the pond would be inundated based on the 8 day discharge, the pond would have a flooding frequency of 8 days/year. Therefore, the ponds have one of the following flooding frequencies: 1 day, 8 days, 16 days, 22 days, 45 days, 71 days, 185 days, and 365 days per year [26–28]. In this study, the flooding frequency parameter was used as the representative environmental parameter because flooding frequency is an important external condition that determines the internal habitat conditions [21]. This is especially true for lentic habitats such as wetlands or ponds. We classified the structure of the habitats by reflecting the visible geomorphic characteristics and location of the habitat to find visible targets. However, in order to compensate for the limitations of types of habitat structure identified only from aerial photos, the values of flood frequency obtained by numerical simulation were used.

In all, 7 habitat condition parameters were surveyed in 120 ponds. The parameters were as follows: Area, water depth, mud depth, mean grain size (D50), dissolved oxygen (DO), chlorophyll, and wood coverage. The previous studies examined the relationships between mussel and habitat conditions; velocity, substrate size and compaction, water depth [29]; sediment softness, velocity and sediment types [30]; water temperature and quality with the flood pulse frequency [21]. As these studies considered hydraulic habitat conditions and water quality, we determined the habitat shape (area, water depth), the substrate of habitat (mud depth and D50), and the water quality (DO and chlorophyll) to a considerable extent. The method of the survey parameters can be referenced to in the reports of References [26–28]. The wood coverage indicates the shaded shoreline ratio by wood, i.e., 100% wood coverage indicates that the shoreline of the pond was entirely occupied by vegetation such as that depicted in Figure 3c.

2.4. Reach-Scale Channel Characteristics

The reach-scale channel configuration in the Kizu River is categorized into single, semi-wandering-straight and sinuous, wandering-straight and sinuous, bifurcated-straight and sinuous, and braided sinuous using parameters of number of channels and sinuosity per 2 km [13]. Aerial photos obtained between 1948 and 2010 were used for performing this classification. The channel configuration of the Kizu River changed from braided channel (characterized by multi-channels and braided sand bars) to single or slightly wandering channel (characterized by less channels and wide vegetation area on the floodplain) between 1948 and 2010. However, because the channel types are not numerical values, we used floodplain vertical shape index (FVSI). This index indicates the degree of convexity or concavity of the vertical shape of a floodplain [22]. The FVSI value was calculated using cross-section data of 200 m intervals. The relative elevation of the riverbed from the water level was arranged in an ascending order. Further, we determined the area between the arranged shape (B) and the triangle connecting the following three points (A): The riverbed tangent to the water surface, bottom of the bank, and top elevation of the bank (Figure 4). The value is positive if B−A is greater than A, and negative if it is less than A. As one channel type of 2 km has 10 cross sections, 10 values are averaged in 2 km. A positive value means a convex floodplain vertical shape (Figure 4a), and a negative value indicates a concave vertical shape (Figure 4b).

The FVSI was calculated using cross-sectional data obtained between 1960 and 2010, as done in the classification of the reach-scale channel configuration in [13]. The classified reach-scale channel configurations were single, semi-wandering, wandering, bifurcated, braided channels in the Kizu River. Figure 5 depicts the average the FVSI values of the classified channel types by Tukey's tests. The majority of the channel configurations in the Kizu River exhibited a concave floodplain vertical shape with negative values of FVSI. Although a significant difference of FVSI was not observed between single/semi-wandering and wandering/bifurcated, and the difference of wandering/bifurcated and braided, we could identify the difference between single/semi-wandering and braided channels. The single and slightly wandering channels in the Kizu River tended to have slightly convex floodplain vertical shape ($-0.3 <$ values < 0.1), and wandering and bifurcated channels tended to have FVSI values between -0.5 and -0.1. Braided channels have a significantly concave floodplain vertical shape (values < -0.6) (Figure 5).

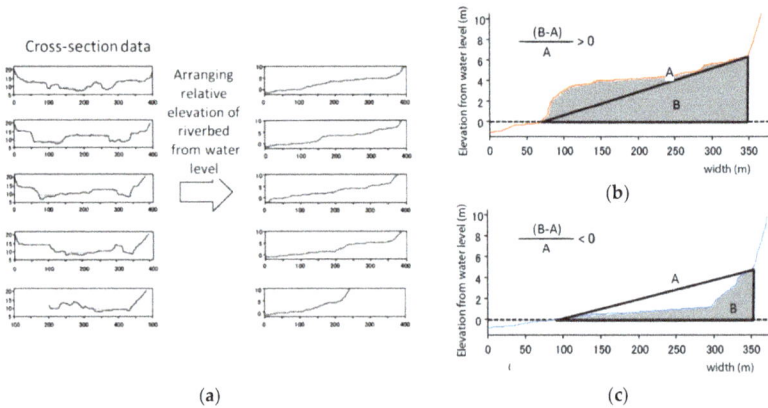

Figure 4. (**a**) Calculation of the floodplain vertical shape index (FVSI); (**b**) a positive value indicates a convex floodplain vertical shape; and (**c**) a negative value indicates a concave vertical shape.

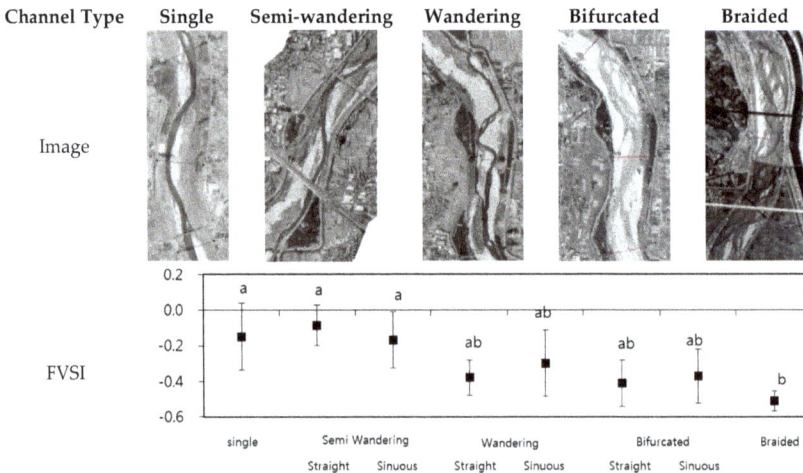

Figure 5. Classification of the channel configurations in the Kizu River [13] and their relationship with FVSI. The squares of FVSI indicate the average FVSI value per channel type (mean $\pm$ SE). Similar letters above histograms indicate significant differences based on Tukey's tests.

2.5. Statistical Analyses

In order to understand the relationships among species, habitat structures and conditions, and channel characteristics, we considered the following items in the analysis:

- A multiple regression model using the parameters of habitat conditions for the abundance of bitterlings and mussels
- Comparison of the habitat suitability index (HSI) of bitterlings/mussels and habitat conditions
- Comparison of the HSI of bitterlings/mussels and habitat structures
- Comparison of habitat structures and habitat conditions
- Comparison of habitat structures and FVSI

We generated multiple linear regression models with a stepwise selection method to examine the effect of habitat conditions on the abundance of bitterlings/mussels. The best model was selected based on the correlation coefficient and significance. An α value of 0.05 indicates statistical significance in the test. Further, only the parameters of habitat conditions that are related to the best models are compared with the HSI of bitterlings/mussels. Additionally, the relationship between the HSI of bitterlings/mussels and habitat conditions were analyzed. Furthermore, the relationships between habitat structures and habitat conditions were analyzed. An HSI exceeding 0.7 indicates excellent suitability. After analyzing the relationships between bitterlings/mussels and habitat structures/conditions, the relationships between the number of habitats and FVSI values were examined.

3. Results

3.1. Relationships among the Bitterlings/Mussels, Habitat Structures, and Habitat Conditions

In order to understand the relationships between the abundance of bitterlings/mussels and habitat conditions, the best models were selected based on surveyed ponds ($n = 120$) by using a multiple regression model. The multiple regression models were generated to test the influences of habitat conditions (area, water depth, D50, DO, Chlorophyll and wood coverage) on abundance of bitterlings and mussels. Only the parameters of habitat conditions related to the best models are listed in Table 1. The abundance of bitterlings was best described by a model consisting of DO and chlorophyll (correlation coefficient: 0.238, $p < 0.05$), and that of mussels was best described by a model consisting of mud depth and wood coverage (correlation coefficient: 0.272, $p < 0.05$). Abundance of bitterlings tended to have a negative relationship with DO (regression coefficient of -1.08), whereas it tended to have positive relationships with chlorophyll (regression coefficient of 2.52). Abundance of mussels tended to have a negative relationship with mud depth (regression coefficient of -1.29) and a positive relationship with wood coverage (regression coefficient of 2.67).

Table 1. Multiple regression analysis to examine the best models of habitat conditions for the abundance of bitterlings and mussels.

Abundance	Area (cm^2)	Water Depth (cm)	Mud Depth (cm)	D50 (mm)	DO (mg/L)	Chlorophyll (µg/L)	Wood Coverage (%)	Best Model R	Best Model p
Bitterling					−1.08 *	2.52 *		0.238	<0.05
Mussel			−1.29				2.67 **	0.272	<0.05

The best models were selected on the basis of the correlation coefficient (R) and significance (p). Values of independent variables indicate the regression coefficient (t). Significance: * $p < 0.05$, ** $p < 0.01$, *** $p < 0.001$.

The HSI values of bitterlings and mussels with the habitat conditions that are selected in the best model are shown in Figure 6. If there are no sampling points in the class or no detection of bitterlings or mussels, the result will exhibit a zero value. Although the HSI values of bitterlings decreased in 15–20 mg/L DO, this exhibited a high habitat suitability in DO case of more than 25 mg/L DO

(Figure 6a). The HSI values of bitterlings tended to increase in chlorophyll between 0 and 500 µg/L and then significantly decreased after 500 µg/L (Figure 6b). The HSI of mussels exhibited the maximum score in mud depth of 0–1 cm (Figure 6c). The HSI values of mussels showed excellent suitability in 100% wood coverage (Figure 6d). There was no sampling pond with wood coverage of 60% and 80%, and number of mussels in ponds with wood coverage of 50, 70 and 90% was very low. Therefore, HSI of mussels showed about zero values in these groups.

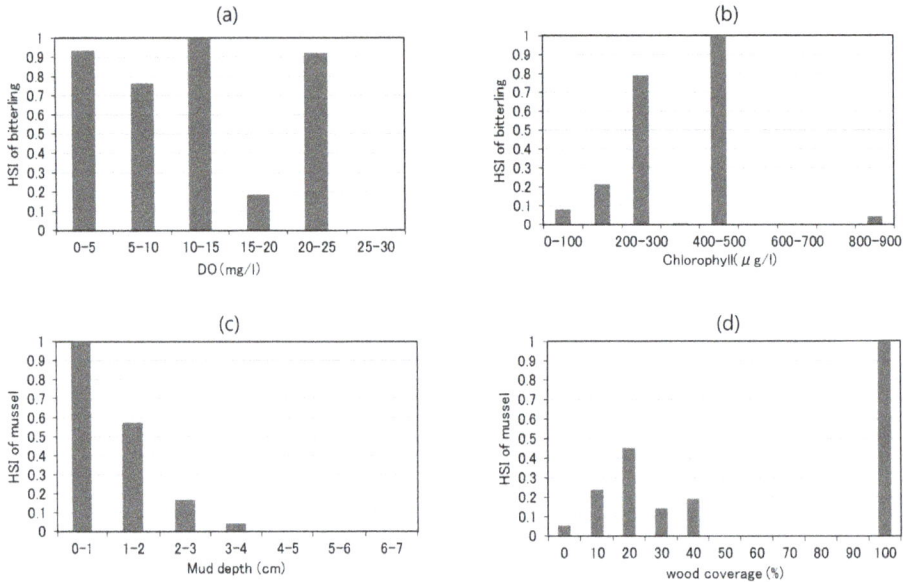

Figure 6. Habitat Suitability Index (HSI) values with habitat conditions were selected for the best models. The HSI of bitterlings and habitat conditions of (**a**) dissolved oxygen (DO) (mg/L) and (**b**) Chlorophyll (µg/L). The HSI of mussels and habitat condition of (**c**) mud depth (cm) and (**d**) wood coverage (%).

The HSI values of bitterlings and mussels are explained by considering the habitat structures with the flooding frequency observed for 120 ponds (Figure 7). Among terrace ponds, BT-TP with a flooding frequency between 8 and 16 days/year showed excellent suitability for bitterlings (Figure 7a). BT-TP and BH-TP with flooding frequencies of 8 days/year and 16–22 days/year exhibited excellent suitability for mussels, respectively (Figure 7b). That is, the terrace ponds (BH-TP and BT-TP) with a flooding frequency between 8 and 22 days/year tended to have high HSI values for bitterlings and mussels.

Figure 8 illustrates the relationships between habitat environmental conditions and habitat structures with flooding frequency. Almost all the habitat structures, except those with a flooding frequency of 45 days of BT-TP, exhibited an average of between 7 and 13 mg/L of DO (Figure 8a). Values of chlorophyll in BT-TP with flooding frequency of 1, 8, 16, and 22 days were higher than other habitat types in the same flooding frequency (Figure 8b), and the values of mud depth in BT-TP with flooding frequencies of 16, 22, and 45 days were highest as compared to other habitat structures (Figure 8c). Further, the relationship between wood coverage and habitat structures with flooding frequency could not detect obvious difference (Figure 8d).

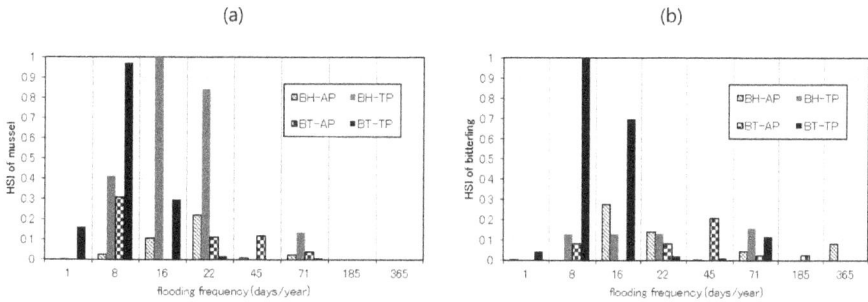

Figure 7. HSI of (**a**) bitterlings and (**b**) mussels, and habitat structures with the flooding frequency. Habitat structures are: BH-AP (bar head-active pond), BH-TP (bar head-terrace pond), BT-AP (bar tail-active pond), and BT-TP (bar tail-terrace pond).

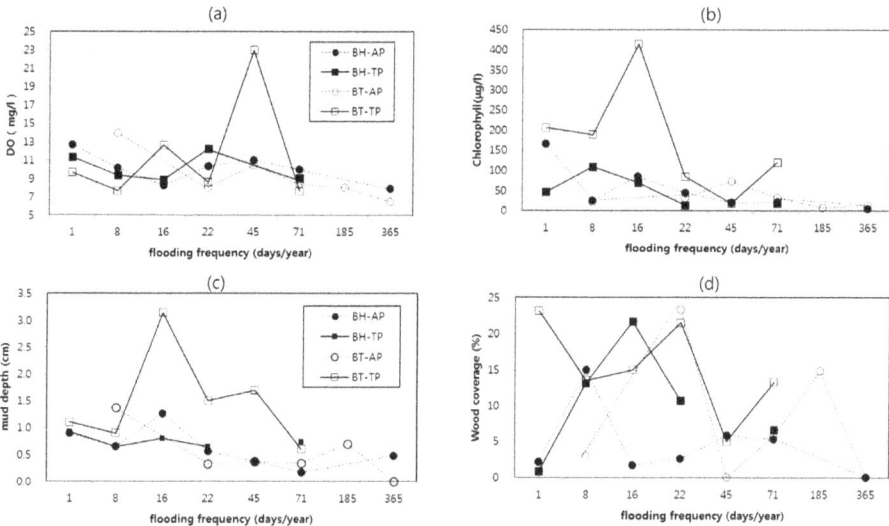

Figure 8. Relationships between the average habitat conditions: (**a**) DO; (**b**) chlorophyll; (**c**) mud depth; and (**d**) wood coverage and habitat types with flooding frequency. Habitat structures are: BH-AP (bar head-active pond), BH-TP (bar head-terrace pond), BT-AP (bar tail-active pond), and BT-TP (bar tail-terrace pond).

3.2. Historical Changes in FVSI

The values of FVSI per year were shown in Figure 9. As the study area extends to a length of 26 km, 12 FVSI values were obtained per year. Average FVSI increased from −0.44 in 1960 to −0.17 in 2010. The cross-sectional floodplain shape of the Kizu River tended to show convex shapes during the 60 years. Since 1990, only values of one or two sections had an FVSI > 0.

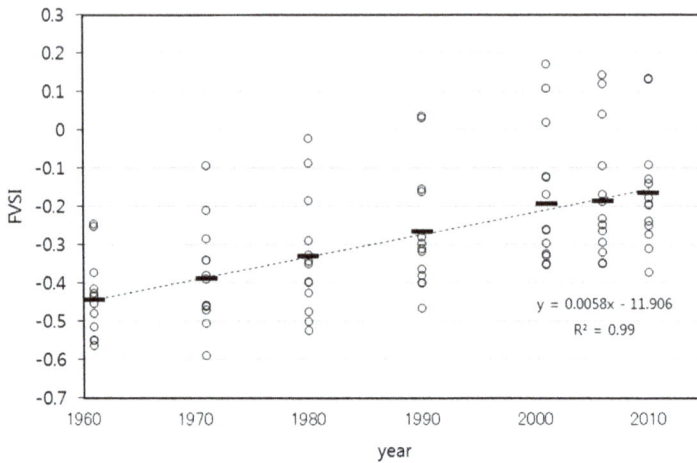

Figure 9. Historical changes in FVSI. Since the study area extends to a length of 26 km, 12 FVSI values per year obtained. The squares indicate the annual average.

3.3. *Relationship between the Habitat Structures and FVSI*

FVSI was classified into six groups (Figure 10). The number of ponds was counted by considering the categorized values of FVSI (Figure 9a), and ratio of ponds with different flooding frequencies was measured by FVSI (Figure 9b). All the ponds detected in aerial photos (n = 386), including 120 surveyed ponds, were used in this relationship. Reaches with FVSI values of less than −0.35 had a high number of APs (Figure 10a), adversely those of more than 0.05 had high TPs (Figure 10b). Based on the increase in FVSI values, the number of APs decreased, whereas the number of TPs increased. The ratio of ponds with flooding frequencies is depicted in Figure 11. The ratio of ponds with a flooding frequency of 1 day/year significantly increased in reaches with FVSI exceeding 0.05. However, ponds with flooding frequencies of less than 16 days/year were not observed in the case of an FVSI of less −0.35. BH-TP and BH-TP with a flooding frequency of 8 days/year and 16–22 days/year exhibited that excellent suitability tended to exist for bitterlings and mussels on a reach with an FVSI that was between −0.35 and 0.05.

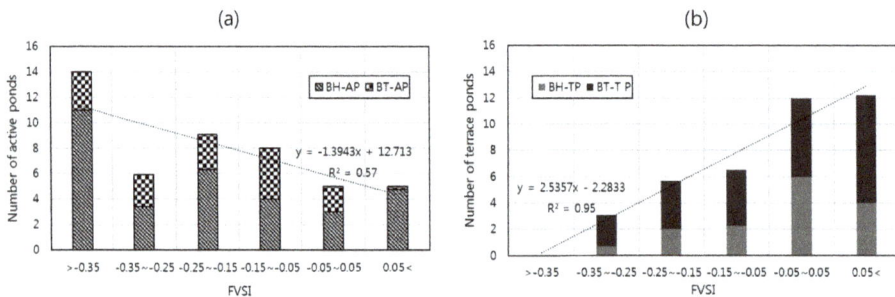

Figure 10. Relationships between the number of habitat types with (**a**) active ponds and (**b**) terrace ponds and FVSI. Habitat structures are: BH-AP (bar head-active pond), BH-TP (bar head-terrace pond), BT-AP (bar tail-active pond), and BT-TP (bar tail-terrace pond).

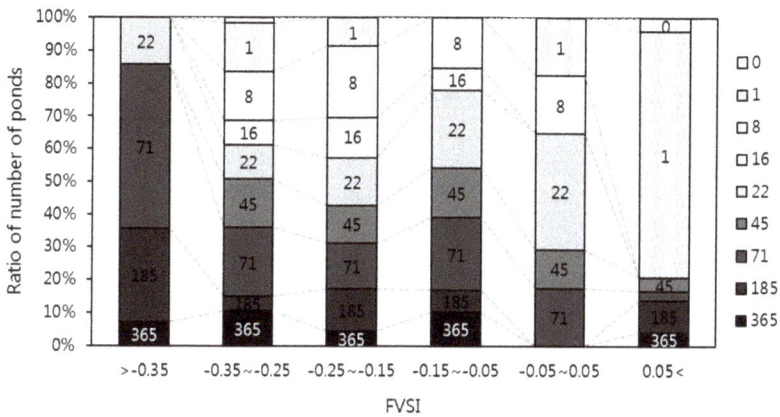

Figure 11. Relationships between the ratio of flooding frequency of habitat and FVSI.

4. Discussion

4.1. Relationships among Species Abundance and Habitat

We attempted to understand the relationships between bitterlings/mussels and habitat structures with flooding frequency. Further, to explain the effect of habitat structures and flooding frequency on the habitat conditions, and relationships between habitat structures and habitat conditions were analyzed. Relationships between bitterling and chlorophyll tended be positive, even though the suitability of the bitterling in chlorophyll over 500 was very low (Table 1; Figure 6b). According to relationship between chlorophyll and habitat structures with flooding frequency, terrace pond with flooding frequency of 1–16 days/year showed relatively high chlorophyll (Figure 8b). Among these ponds, terrace ponds with flooding frequency of 8–16 days/year had high a HSI of bitterlings (Figure 7a). That is, suitability of bitterling had the high values in terrace ponds with a flooding frequency of 8–16 days/year, because they had a high value of chlorophyll that were positively related with bitterling. As the flooding frequency increases, the values of chlorophyll decreases. Maybe this is because frequent flooding reduces the residence time of water [21]. The infrequent flooding of ponds tended to maintain concentration of chlorophyll, and thus tended to influence bitterling habitat. A previous study [31] on CCA analysis between bitterlings/mussels and habitat conditions also showed that these species and chlorophyll have positive relationships.

The bitterling *Acheilognathus longipinnis* is listed on the IUCN red species and is a domestic species with national monument in Japan. However, this bitterling has not been found in the Yodo River system since 2006 [32]. The Kizu River joins the Yodo River. Therefore, the Yodogawa River Bureau and Osaka-fu prefecture implemented the re-introduction of the bitterling *Acheilognathus longipinnis* in the autumn of 2009 as an experimental trial. Mature adults (five hundred individuals) were released into embayments of the Yodo River in autumn 2009 during the spawning season. A total of 133 juveniles were observed in the embayments, but no juveniles were found in the spring of 2011. The bitterling restoration project continues, and this paper could be support of this restoration project.

Mussels were negatively related to the mud depth (Table 1; Figure 6b). Mussels also were more clearly observed in TPs than APs, especially the TPs located in the lower area of bars with a flooding frequency of 8 days/year and those located in the upper area of bars with flooding frequencies of 16–22 days/year than in APs (Figure 7b). Further, these TPs had an average mud depth of 1 cm (Figure 8c). Mussels require an appropriate depth of sediments under stable conditions because they live buried in the riverbed [21], and they filter the surrounding water to obtain food and oxygen [33]. However, deep mud may influence negative relationships with mussels, as shown in Figure 6c. Here, the pond is generated by erosion during large flood events, and is sustained by frequent intermediate flood events. In the case of TPs, the depth of the mud is likely to deepen by the deposition of suspended load by frequent intermediate flood events. Therefore, in the case of BT-TP, a flooding frequency of 16–22 days exhibits considerable mud depth as compared to that obtained at a flooding frequency of 1–8 days. If the mud depth increases, mussels buried in the mud will find it difficult to breathe owing to the reduction in DO. The authors of Reference [21] explained that the survival rates of mussels were low and that the growth was approximately zero in infrequently inundated water bodies (backwater, wetlands) by conducting field-earing experiments in the case of the Kiso River, because infrequently inundated water bodies exhibited hypoxic conditions in substratum-water. Further, other studies [29,30] indicated that the sand size of substrate is more suitable than that of silt (fine size) or gravel (larger size). However, fine substrate sediments have been reported to increase the mortality rate in some freshwater mussel species [34]. We assumed that the mud depth until 1 cm could maintain suitable DO for mussel breathing; if the depth is increased, it could become difficult for mussels to live in the Kizu River.

Further, we detected that, although the same TP with the same flooding frequency, they showed different habitat conditions and abundance of mussels. The mud depth of BH-TP was lower than that observed in BT-TP by considering the same flooding frequency. Thus, the HSI of mussels belonging to BH-TP exhibited higher values as compared to that exhibited by BT-TP with a flooding frequency of 16–22 days. However, we were unable to explain the reason for this difference. We inferred that ponds located on the upper area in the bar have much more powerful river strength than ponds located in the lower area in the flood event. Further, the accumulation of chlorophyll and mud in ponds is prevented by tractive force. In future studies, we should perform various other comparative analyses and include hydraulic aspects.

4.2. Relationships between Habitats Structures and FVSI

FVSI reflects the channel configuration in the Kizu River. The concave reaches with an FVSI < −0.35 had a lot of ponds with frequent flooding frequency more than 22 days/year (Figure 11) and only active ponds (Figure 10b). This indicates the significantly concave reaches with FVSI < −0.35 had low habitat suitability for bitterlings and mussels, because they have few suitable habitat structures (terrace ponds) with flooding frequency (8–22 days/year). These reaches were related with wandering, bifurcated, and braided channels in the Kizu River (Figure 5). Their floodplain is frequently inundated, and thus active channel was wide and area of vegetation was small. Hence, active ponds or backwaters on active channels are mainly located on these reaches. Although these channel types may have low potential of bitterlings and mussels, they may have high potential for other fish or aspects of lotic habitats. For example, Sukhodolov et al. and, Payne and Lapointe [15,16] noted that braided channels supply suitable conditions for fish shelters in terms of depth and velocity. Choi et al. [13] also examined that the bifurcated channel exhibits higher habitat diversity for both lentic and lotic habitats as compared to that exhibited by the single or slightly wandering channels. Since this study focused on lentic habitat with bitterlings and mussels, channel types with FVSI < −0.35 tended to have lower potential than others.

On the other hand, the convex channel reaches with an FVSI > 0.05 had a lot of terrace ponds with infrequent flooding frequency (1 day/year). Although habitat with infrequent flooding was suitable for bitterlings and mussels in the Kizu River, less flooding about 1 day/year showed the low suitability for bitterlings and mussels (Figure 7). However, unfortunately, there was no clear difference of environmental conditions between flooding frequency of 1 day and others (8, 16, 22, 45, 71, 185, and 365 days), and between active ponds and terrace ponds in the same flooding frequency 1 day.

Therefore, the intermediate channel types tended to be suitable excluding reaches of significant concave floodplain vertical shape (FVSI < −0.35) and convex shape (FVSI > 0.05). These suitable reaches were related to single or slightly wandering channels in the Kizu River (Figure 5), and these reaches were detected more after the 1990s (Figure 9). Similarly, studies related to the potential of recent channel conditions of the Kizu River were also noted by Reference [35]. They classified riffle structures into four types (traverse, converge, diverge I and II types), and surveyed the biomass and taxonomic richness of invertebrates per these riffle types. Further, the areas of each riffle type were measured using aerial photos that were obtained over 65 years. Their results exhibited that traverse and converge types of riffle had more biomass and taxonomic richness of invertebrates as compared to the diverge types; further, the areas of these riffle types increased after the 1970s. They discussed that river conditions from the 1980s to the 2010s may have had more biomass and richness than before. Thus, we recommend establishing countermeasures to prevent an increase in single or slightly wandering channel types to an FVSI of more than 0.05 to include many ponds with a flooding frequency of 1 day/year rather than restoring the previous channel conditions.

4.3. Application of FVSI to Sediment Management

Dam constructions cause the regular flow and the reduction of sediment transport. Stabilized flow and low sediment volume results in channel changes from braided (mobile) channel to single or single-thread (stabilized) channels [36]. In the Kizu River, these channel changes occurred over 60 years [13] with a reduction of sediment transport (Figure 2). A stabilized channel shows channel narrowing and incision with vegetation encroachment [2]. The encroachment of riparian vegetation into high level of floodplain by reduction in flood scour and sediment deposition is related to the high value of floodplain vertical shape index. As sediment deposition and vegetation expansion lead to an increase in the relative elevation of floodplains, FVSI was calculated by difference between elevation of floodplain and water level in main channel increases. Thus, positive values (convex shape) of FVSI meant stabilized channel such as single or semi wandering channels, and negative values (concave shape) meant more mobile channels such as wandering, bifurcated and braided channels in the Kizu River (Figure 5). This means that a greater reduction of sediment transport could further increase the FVSI. Therefore, sediment supplies need to maintain the current channel condition, not the more stabilized channel, for bitterlings and mussels. If we want to maintain single or slightly wandering channels with FVSI less than 0.05, we could consider a suitable period as between 1990s and 2000s (Figure 9), and suggest the volume of sediment transport at the time (20,000–50,000 m^3/y) of sediment management. However, this is an indirect and long-term measurement. The consideration of various external and internal conditions should precede before the management.

On the other hand, the sediment management such as replenishment directly leads to improved habitat quality. The sediment excavated at the upstream of dam was moved to downstream by dump truck or other transportation [5]. Then, the riverbed was reformed by magnitude or duration of flushing flow or natural flood. Kondolf [2] stated that sediment supply contributed to enhancing the available spawning habitat for chinook salmon. Wood and Armitage [6] said that replenishment of coarse sediment increased the lotic habitat quality for invertebrates and fishes. Additionally, it is possible to convey sediment from convex floodplain vertical shape (FVSI greater than 0.05 having low potential of bitterlings and mussels) to significant concave floodplain vertical shape (FVSI less than −0.35 having low potential of bitterlings and mussels). The area of convex floodplain vertical shape requires a reduction of sediment deposition on the floodplain for changing the movable channels, and significant concave floodplain vertical shape requires sediment deposition for changing the stabilized channel. These processes could be predicted through simulations or experiments. The experiments and simulations of the sediment replenishment had been undertaken in tributary of the Kizu River [5]. However, ecological evaluation of the sediment management has not yet been properly performed. This study tried to support ecological connectivity between species abundance and sediment management based on relationships among animal communities, habitat structures with flooding frequency, and channel characteristics.

5. Conclusions

The reach-scale channel characteristics helped to determine a suitable target image for ecological sediment management on a catchment scale based on the multiscale relationships. The bitterlings and mussels were more abundant in terrace ponds than in active ponds, especially terrace ponds with a flooding frequency of 8–22 days/year. These suitable pond types tended to be located on the reaches with floodplain vertical shape index between −0.35 and 0.05 under the current channel conditions in the Kizu River. Thus, we recommend establishing long-term countermeasures of required sediment volume to maintain current conditions at sediment supply downstream. In terms of short-term countermeasures, we could suggest the target conditions at experiments and simulation of sediment replenishment. Although this study is based on a simple relationship analysis and requires considerably detailed correlation analysis, it could indicate an eco-geomorphic framework for river management based on the hierarchical interrelationships. In future, we will supplement the biological characteristics of bitterlings and mussels and various habitat conditions based on the hydraulic interpretation.

Author Contributions: Conceptualization, M.C. and Y.T.; Methodology, Y.T.; Software, K.I.; Formal Analysis, M.C.; Investigation, K.I.; Resources, K.I.; Writing-Original Draft Preparation, M.C.; Writing-Review & Editing, M.C.; Supervision, K.J.; Project Administration, K.J.; Funding Acquisition, K.J.

funding: This study was partially supported by "Development of river-bed management techniques for river environments", which is supported by the Japanese Ministry of Land, Infrastructure, Transport and Tourism and by Grants in Aid for Scientific Research (Nos. 25241024) from the Ministry of Education, Culture, Sports, and Technology, Japan and by Korea Ministry of Environment (MOE) as "Water Management Research Program (18AWMP-B079625-05)".

Acknowledgments: We would like to thank Ikeda and Nishii in ASIA AIR Survey CO. Ltd. and the Yodogawa River Bureau for providing historical aerial photos, cross-section data, and historical hydraulic data of the Kizu River.

Conflicts of Interest: The authors declare no conflict of interest.

References

1. Kondolf, G.M.; Gao, Y.; Annandale, G.W.; Morris, G.L.; Jiang, E.; Zhang, J.; Cao, Y.; Carling, P.; Fu, K.; Guo, Q.; et al. Sustainable sediment management in reservoirs and regulated rivers: Experiences from five continents. *Earth's Future* **2014**, *2*, 256–280. [CrossRef]
2. Kondolf, G.M. Hungry Water: Effects of Dams and Gravel Mining on River Channels. *Environ. Manag.* **1997**, *21*, 533–551. [CrossRef]
3. Auel, C.; Kobayashi, T.; Takemon, Y. Effects of sediment bypass tunnels on sediment grain size distribution and benthic habitats. *Int. J. River Basin Manag.* **2017**, *15*, 433–444. [CrossRef]
4. Kantoush, S.A.; Sumi, T.; Kubota, A. Geomorphic response of rivers below dams by sediment replenishment technique. In *River Flow 2010*; Dittrich, A., Koll, K., Aberle, J., Geisenhainer, P., Eds.; Bundesanstalt für Wasserbau: Karlsruhe, Germany, 2010; pp. 1155–1163.
5. Ock, G.; Sumi, T.; Takemon, Y. Sediment replenishment to downstream reaches below dams: Implementation perspectives. *Hydrol. Res. Lett.* **2013**, *7*, 54–59. [CrossRef]
6. Wood, P.J.; Armitage, P.D. Biological effects of fine sediment in the lotic environment. *Environ. Manag.* **1997**, *21*, 203–217. [CrossRef]
7. Leopold, L.B.; Wolman, M.G. *River Channel Patterns: Braided, Meandering and Straight*; US Geological Survey Professional Paper; U.S. Government Printing Office: Washington, DC, USA, 1957; Volume 282B, pp. 39–85.
8. Schumm, S.A. Patterns of alluvial rivers. *Annu. Rev. Earth Planet. Sci.* **1985**, *13*, 5–27. [CrossRef]
9. Rosgen, D.L. A classification of natural rivers. *Catena* **1994**, *22*, 169–199. [CrossRef]
10. Brierley, G.; Fryirs, K. *Geomorphology and River Management: Applications of the River Styles Framework*; Blackwell: Oxford, UK, 2005.
11. Wheaton, J.M.; Brasington, J.; Darby, S.E.; Merz, J.; Pasternack, G.B.; Sear, D.; Vericat, D. Linking geomorphic changes to salmonid habitat at a scale relevant to fish. *River Res. Appl.* **2010**, *26*, 469–486. [CrossRef]

12. Wyrick, J.R.; Pasternack, G.B. Geospatial organization of fluvial landforms in a gravel-cobble river: Beyond the riffle-pool couplet. *Geomorphology* **2014**, *213*, 48–65. [CrossRef]

13. Choi, M.; Takemon, Y.; Yu, W.; Jung, K. Ecological evaluation of reach scale channel configuration based on habitat structures for river management. *J. Hydroinform.* **2018**, *20*, 622–632. [CrossRef]

14. Frothingham, K.M.; Rhoads, B.L.; Herricks, E.E. A multiscale conceptual framework for integrated eco-geomorphological research to support stream naturalization in the agricultural Midwest. *Environ. Manag.* **2002**, *29*, 16–23. [CrossRef]

15. Sukhodolov, A.; Bertoldi, W.; Wolter, C.; Surian, N.; Tubino, M. Implication of channel processes for juvenile fish habitats in Alpine rivers. *Aquat. Sci.* **2009**, *71*, 338–349. [CrossRef]

16. Payne, B.A.; Lapointe, M.F. Channel morphology and lateral stability: Effects on distribution of spawning and rearing habitat for Atlantic salmon in a wandering cobble-bed river. *Can. J. Fish. Aquat. Sci.* **1997**, *54*, 2627–2636. [CrossRef]

17. Beechie, T.J.; Liermann, M.; Pollock, M.M.; Baker, S.; Davies, J. Channel pattern and river-floodplain dynamics in forested mountain river systems. *Geomorphology* **2006**, *78*, 124–141. [CrossRef]

18. Tockner, K.; Ward, J.V.; Arscott, D.B.; Edwards, P.J.; Kollmann, J.; Gurnell, A.M.; Petts, G.E.; Maiolini, B. The Tagliamento River: A model ecosystem of European importance. *Aquat. Sci.* **2003**, *65*, 239–253. [CrossRef]

19. Arscott, D.B.; Tockner, K.; van der Nat, D.; Ward, J.V. Aquatic Habitat Dynamics along a Braided Alpine River Ecosystem (Tagliamento River, Northeast Italy). *Ecosystems* **2002**, *5*, 802–814. [CrossRef]

20. Kizu River Research Group. *Integrated Research of the Kizu River II*; Kizu River Research Group: Kyoto, Japan, 2003. (In Japanese)

21. Negishi, J.N.; Sagawa, S.; Kayaba, Y.; Sanada, S.; Kume, M.; Miyashita, T. Mussel responses to flood pulse frequency: The important of local habitat. *Freshw. Biol.* **2012**, *57*, 1500–1511. [CrossRef]

22. Takemon, Y.; Kobayashi, S.; Choi, M.; Terada, M.; Takebayashi, H.; Sumi, T. River Habitat Evaluation based on Cross-sectional Bed Profile and Frequency Distribution of Relative Elevation. *Adv. River Eng.* **2013**, *19*, 519–524. (In Japanese)

23. Egashira, S.; Jin, H.; Takebayashi, H.; Nida, B.; Nagata, T. Bed variation and sediment budget in the downstream reach of Kizu River. *Annu. J. Hydraul. Eng.* **2002**, *44*, 777–782. (In Japanese) [CrossRef]

24. Jowett, I.G.; Davey, A.J.H. A Comparison of Composite Habitat Suitability Indices and Generalized Additive Models of Invertebrate Abundance and Fish Presence-Habitat Availability. *Trans. Am. Fish. Soc.* **2007**, *136*, 428–444. [CrossRef]

25. Bovee, K.D. *Development and Evaluation of Habitat Suitability Criteria for Use in the Instream Flow Incremental Methodology*; National Ecology Center, U.S. Fish and Wildlife Service: Washington, DC, USA, 1986.

26. *Survey Report on the Kizu River*; Yodogawa River Bureau, ASIA AIR SURVEY CO., LTD.: Osaka, Japan, 2007. (In Japanese)

27. *Survey Report on the Kizu River*; Yodogawa River Bureau, ASIA AIR SURVEY CO., LTD.: Osaka, Japan, 2009. (In Japanese)

28. *Survey Report on the Kizu River*; Yodogawa River Bureau, ASIA AIR SURVEY CO., LTD.: Osaka, Japan, 2010. (In Japanese)

29. Johnson, P.D.; Brown, K.M. The importance of microhabitat factors and habitat stability to the threatened Louisiana pearl shell, Mrgaritifera hembeli (Conrad). *Can. J. Zool.* **2000**, *78*, 271–277. [CrossRef]

30. Yoshihiro, B.A.; Takashi, M. Habitat Characteristics Influencing Distribution of the Freshwater Mussel Pronodularia japanensis and Potential Impact on the Tokyo Bitterling, Tanakia tanago. *Zool. Sci.* **2010**, *27*, 912–916.

31. Terada, M.; Takemon, Y.; Sumi, T. A study on habitat evaluation of tamari for bitterling and mussel. In Proceedings of the JSCE-Kansai, Uji, Japan, 26 August 2011; Volume II-12. (In Japanese)

32. *Global Re-introduction Perspectives: 2011*; IUCN/SSC Re-Introduction Specialist Group: Calgary, AB, Canada, 2011.

33. Reichard, M.; Liu, H.; Smith, C. The co-evolutionary relationship between bitterling fishes and freshwater mussels: Insights from interspecific comparison. *Evolut. Ecol. Res.* **2007**, *9*, 239–259.

34. Ellis, M.M. Erosion silt as a factor in aquatic environments. *Ecology* **1936**, *17*, 29–42. [CrossRef]

35. Kobayashi, S.; Takemon, Y. Historical Changes of Riffle Morphology for Benthic Invertebrate Habitats in the Kizu River. *Annu. Disaster Prev. Res. Inst. Kyoto Univ.* **2013**, *56B*, 681–689. (In Japanese)
36. Surian, N.; Rinaldi, M. Morphological response to river engineering and management in alluvial channels in Italy. *Geomorphology* **2003**, *50*, 307–326. [CrossRef]

Article

water

MDPI

Water Resource Optimal Allocation Based on Multi-Agent Game Theory of HanJiang River Basin

Qi Han, Guangming Tan, Xiang Fu *, Yadong Mei and Zhenyu Yang

State Key Laboratory of Water Resources and Hydropower Engineering Science, Wuhan University,
Wuhan 430072, China; hqwhu@whu.edu.cn (Q.H.); tangm@vip.163.com (G.T.); ydmei@whu.edu.cn (Y.M.);
yangzy95@whu.edu.cn (Z.Y.)
* Correspondence: xfu@whu.edu.cn; Tel.: +86-27-6877-2209; Fax: +86-27-6877-2310

Received: 27 July 2018; Accepted: 28 August 2018; Published: 4 September 2018

Abstract: Water scarcity is an important issue in many countries, and it is therefore necessary to improve the efficiency and equality of water resource allocation for decision makers. Based on game theory (GT), a bi-level optimization model is developed from the perspective of a leader-follower relationship among agents (stakeholders) of a river basin in this study, which consists of a single-agent GT-based optimization model of common interest and a multi-agent cooperative GT-based model. The Hanjiang River Basin is chosen as a case study, where there are conflicts among different interest agents in this basin. The results show that the proposed bi-level model could attain the same improvement of common interest by 8%, with the conventional optimal model. However, different from the conventional optimal model, since the individual interests have been considered in the bi-level optimization model, the willingness of cooperation of individuals has risen from 20% to 80%. With a slight decrease by 3% of only one agent, the increases of interest of other agents are 14%, 18%, 7%, and 14%, respectively, when using the bi-level optimization model. The conclusion could be drawn that the proposed model is superior to the conventional optimal model. Moreover, this study provides scientific support for the large spatial scale water resource allocation model.

Keywords: multi-agent of river basin; game theory; water resources allocation

1. Introduction

Water is essential for human well-being and all activities [1]. Owing to the impact of climate change and human activities, water scarcity has become a common problem in many countries, especially in developing countries [2–4]. Since the imbalance between the supply and demand of water resources is getting more and more prominent, it is urgent for decision makers to solve the conflicts arisen from ineffective and unfair water resource allocation [5].

To enhance the effectiveness and benefits of water resource allocation schemes, a large group of scholars have suggested the use of optimization models. Optimization techniques, such as linear programming, mixed-integer linear programing, dynamic programming, evolutionary computation, artificial neural networks, and so on [5–8], have been trying to find the optimal schemes of water resource allocation. However, these conventional optimization methods usually convert the multi-decision-maker problems of the whole system into a single-decision-maker problem, with a single composite objective [5]. Consequently, based on perfect cooperation, the ideal top-down schemes attained by conventional optimization methods, only emphasize the common interest of the system and ignore individual interests. Nevertheless, in fact, the ideal optimal scheme can't be realized without the willingness to cooperation of individuals [9–11].

Taking individual willingness into consideration, game theory was introduced as a solution to the conflicts caused by ineffective and unfair water resource allocation among multi decision makers. Game theory (GT) has been applied to different water or cost/benefit allocation situations, among users in water resources. Both non-cooperative and cooperative game theory methods have been used to solve water conflicts [12,13]. While non-cooperative game theory is useful in providing strategic insights into conflicts, cooperative game theory is normally helpful in providing an alternative framework for fair and efficient allocation of the incremental benefits of cooperation, among multi decision makers [14]. Using the idea of game theory, Adams et al. [15] advanced a new framework for noncooperative, multilateral bargaining, which can be used to conceptualize negotiation processes. The cooperative water allocation model was designed by Wang et al. [2], aiming at modeling equitable and efficient water allocation among competing users. A multi-objective game-theory model, which could balance economic and environmental concerns in reservoir watershed management, was developed by Lee [3]. Furthermore, Madani et al. [14] proposed a new framework for resolving conflicts over transboundary rivers using bankruptcy methods. However, there are few studies about the conflicts among multi-agent (i.e., water user) at the river basin level, and no strict systems are available to guide practical problems.

The aim of this study is to develop a bi-level optimization model, which consists of the optimization model of water resources allocation among the superiors and the optimization model of water resources allocation among the subordinates. In the bi-level optimization model, not only the maximization of the common interest is realized, but also the individual interest is taken into consideration. In other words, our model is featured by top-down coordination and bottom-up feedback, aiming at proposing an optimal scheme that could be accepted by superiors and subordinates. The Hanjiang River Basin is used as a case study to prove the equality and effectiveness of this bi-level optimization model. Thereby, our model could provide a fundamental basis for water resource allocation on a large scale.

2. Model Description

The bi-level optimization model begins with the framework of the model (Section 2.1), followed by the mathematical formulae of the model, which are presented in Section 2.2.

2.1. Model Framework

As shown in Figure 1, the bi-level optimization model (i.e., multi-agent cooperative game-based optimization model of water resources allocation) consists of the optimization model of water resource allocation among the superiors (in brief, the superiors model), and the optimization model of water resource allocation among the subordinates (in brief, the subordinates model). The former is a single-agent GT-based optimization model of common interest, which is designed to solve the multi-objective optimization problems of water resource management agents. The latter is a multi-agent cooperative GT-based model, which intends to solve multi- agent cooperation problems in water usage. Due to the existence of the cooperation agreement between the superiors and subordinates, the common interest of the superiors could be dealt as an equality constraint on the subordinate model. Both models have their separate objective functions and constrains, and any solution to the subordinates model depends on the corresponding solution to the superiors model.

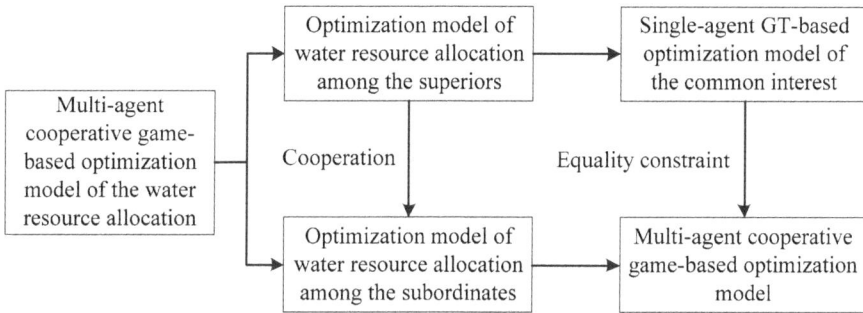

Figure 1. Optimization model frame of water resource allocation.

2.2. Model Formulation

The generalization of multi-agents in a river basin must be the first step. As shown in Figure 2 for demonstration, the whole basin is divided into M zones on the basis of meteorological and hydrological features and natural geographical conditions. Zone 1 is located in the upper and middle reaches of the river's main stream. Zone 2 is located in a tributary of the river basin. Zone M is the most downstream area of the basin, where the tributaries merge with the main stream. According to the spatial distribution, there are mainly three types among these interest agents, including consumptive water users outside the river, users with ecological water needs inside the river, and water energy users inside the river. Water transfer to other basins is also considered in this model. It is assumed that there are L water users outside the river, and the total number of interest agents in zone M is denoted as N_M. R represents the outflow of each zone, and these interest agents are correlated with each other through the runoff.

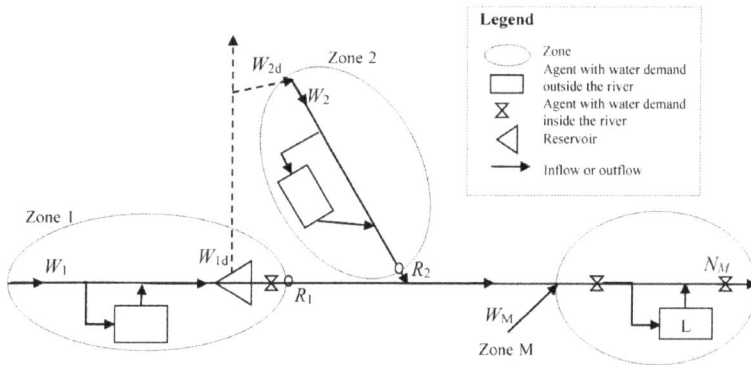

Figure 2. Generalization of multi-agents in a river basin.

2.2.1. The Superiors Model

The optimization of the common interest of water resource management agents in the superior model is described by the following equation:

$$Max\ B^u = \frac{1}{\sum\limits_{z=1}^{M} N_z} \sum_{z=1}^{M} \sum_{i=1}^{N_z} B_{zi}^u = \frac{1}{\sum\limits_{z=1}^{M} N_z} \sum_{z=1}^{M} \sum_{i=1}^{N_z} \left[1 - \left(\frac{D_{zi} - x_{zi}}{D_{zi}} \right)^2 \right] \tag{1}$$

In Equation (1), the objective function of each agent B_{zi}^u represents the water requirement satisfaction index of agents i in Zone z $(z = 1, \cdots, M; i = 1, \cdots, N_z)$. M represents the number of zones in the basin, and N_z represents the number of agents in Zone z $(z = 1, \cdots, M)$. $\sum_{z=1}^{M} N_z$ represents the number of agents in the whole river basin. x_{zi} represents the water allocated to agent i in Zone z, which is also a decisive variable of this model. D_{zi} represents the water demand of each agent. B_{zi}^u represents the standard benefit value (i.e., $[0, 1]$) of agent i in Zone z, which is also the goal of individual interests. B^u represents the total benefits of interest agents, which is also the optimization goal of the common interest in the superior model.

The constraints are as follows:

(1) Water balance equation

$$R_z = W_z + \sum_{k=1}^{K} R_{z-k} - (1 - \varphi_{zi}) \sum_{l=1}^{L} x_{zl} - W_{zd} - \Delta S_z \tag{2}$$

In Equation (2), R_z represents the outflow from Zone z; W_z represents the inflow of Sub-basin z; R_{z-k} represents the outflow from Zone $z - k$; k represents the upstream zones who are related to the flow in Zone z $(k = 1, \cdots, K)$; φ_{zi} represents the water receding coefficient of the user l $(l = 1, \cdots, L)$ outside the river in Zone z; x_{zl} represents the water allocated to user l outside the river in Zone; W_{zd} represents the water transfer from reservoirs in Zone z (the positive value represents the outflow, whereas the negative value represents the inflow); and ΔS_z represents the variation in the poundage of reservoir in Zone z at the specified period.

(2) Constraints on the water available to each agent

$$x_{zi} \leq W_z + \sum_{k=1}^{K} R_{z-k} - W_{zd} - \Delta S_z \tag{3}$$

(3) Water demand constraints

$$0 \leq x_{zi} \leq D_{zi} \tag{4}$$

(4) Constraints on the total amount of water available

$$\sum_{z=1}^{M} \sum_{l=1}^{L} x_{zl} \leq W_k \tag{5}$$

In these equations, W_k represents the total amount of water available to users outside the river basin.

2.2.2. The Subordinates Model

The objective function of the lower multi-agent cooperative game-based optimization model [7] of water resource allocation, based on the Nash bargaining model [8,16,17], is obtained as follows:

$$Max \ B^d = \prod_{z=1}^{M} \prod_{i=1}^{N_z} \left(B_{zi}^d - N_{zi} \right) \tag{6}$$

The constraints are as follows:

$$\frac{1}{\sum\limits_{z=1}^{M} N_z} \sum_{z=1}^{M} \sum_{i=1}^{N_z} B_{zi}^d = B^u \tag{7}$$

$$N_{zi} = Max \ S_{zi} \tag{8}$$

$$B_{zi}^d \geq N_{zi} \geq 0 \tag{9}$$

In this model, x_{zi}, i.e., the water allocated to agent i in Zone z, is the decisive variable. B_{zi}^u, R_z, R_{z-k}, x_{zl}, and B_{zi}^d are also variables in this model, while others are parameters. With a cooperative alliance of all agents, the optimal solution is solved on the condition that the Nash product ($\prod_{z=1}^{M} \prod_{i=1}^{N_z} (B_{zi}^d - N_{zi})$) reaches the maximum. B_{zi}^d and N_{zi} are the agent's benefit in the cooperation mode, and the agent's benefit in the individual (non-cooperative) mode, respectively. $B_{zi}^d - N_{zi}$ is the gain from cooperation. By using the Nash bargaining model, the bargaining process can be simulated among the agents. A compromise among the conflicting objectives could be found, in which agents have considered their own benefits and know that they can get more benefits in the coalition on the whole [17,18].

Constraint (7) denotes the solution of the superior model as the equality constraint of the model.

In Equation (8), N_{zi} represents the benefit of each agent in the non-cooperative mode. The optimization model of individual interests covers objective function (8) and constraints (2)–(4). When the benefit of the most upstream agent with first-mover advantage reaches the maximum $N_{zi} = 1$ in the non-cooperative mode, $N_{zi} = B_{zi}^u$ will facilitate cooperation.

Equation (9) indicates a rational requirement for individuals to participate in cooperation, which is the "bottom line". Failure to reach the bottom line may cause the agent's unwillingness to cooperate and the collapse of their cooperative alliance. The optimization solution of the lower multi-agent cooperative game-based decision-making model for water resource allocation involves identifying appropriate allocation strategies to maintain the cooperative alliance, where the benefit of each individual is not less than that in the non-cooperative mode (or when optimizing the superior model).

Equations (1)–(9) provide the GT-based optimization model that combines the superior and subordinate models of water resource allocation among interest agents, and the optimization model can be solved by the dynamic programming algorithm [6,19]. This optimization describes the carrying capacity of spatially-related water resources using the water available to zones in the same watershed and the pressure load on water resources in each zone using the requirements of interest agents. The model also proposes to achieve the load balance of the entire basin on a large scale through the downward constraints in the optimization of the common interest of the superiors and the fair allocation of the common interest of the subordinates.

3. Case Study

3.1. Study Area

The Hanjiang River basin, with an area of 159,000 km^2, is the largest and most developed tributary of the Yangtze River Basin [20]. Being the ninth largest river in China, with a total length of 1570 km, the Hanjiang River flows through the south of Shanxi Province, the northwest and central region of Hubei Province, and runs into Yangtze River at Wuhan city [21]. After the height of the Danjiangkou Dam being increased from 162.0 m to 176.7 m, Hanjiang River has served as an important water source for the middle route of the South-to-North Water Diversion Project in China, and has formed a connection between north and south China [22–24]. Meanwhile, Hanjiang River is an important channel, which connects the inland area in the northwest with the marine coastal area in the east, and nurtures the natural and human environment of the basin [21].

Because of the rapid growth of population and economy, water demand is expanding and conflicts over water use among interest agents in the Hanjiang River Baisn are intensifying [25]. The bi-level optimization model is established and applied to the Hanjiang River Basin, trying to alleviate conflicts among different individual interests.

3.2. Data and Primary Analyses

The Hanjiang River Basin is divided into three zones ($M = 3$), according to the three-tier zoning of nationwide water resources. Zone 1 is located above the Danjiangkou reservoir, as Danjiangkou

Reservoir is the boundary of the upper and mid-lower reaches of the Hanjiang River Basin. In Zone 1, the interest agents could be conceptualized into one type ($N_1 = 1$), namely, the consumptive water user outside of the river. What needs to be explained is that the water transfer from Danjiangkou reservoir is considered, and the total volume of the transferred water is assumed to be 9.5 billion m^3 per year, through the middle route of South to the North Water Transfer Project [23,26]. Zone 2 is located in the Tangbai river, which is the main tributary of the Hanjiang river. Similarly, the interest agents could be conceptualized into one type ($N_2 = 1$), namely, the consumptive water user outside of the river. Zone 3 is located below the Danjiangkou reservoir, in which the interest agents should be conceptualized into three types ($N_3 = 3$). The first is the users with ecological water needs inside the river, since the flow in Xiangyang is both the spawning ground of the Asian carp and control node of the ecological condition of the main stream of the Hanjiang River. The second is the consumptive water user outside of the Tangbai river. The third is the users with ecological water needs inside the river, because the flow in Xiantao is the critical flow whether the phytoplankton blooms would occur in the river below Qianjiang, and another control node of the ecological condition of the main stream of the Hanjiang River.

As shown in Table 1, the amount of water resources in different zone is resented.

Table 1. The available water resources in different zones.

Content	Unit (Billion m^3)
Total of the Hanjiang River Basin	14.7
Annual inflow of Zone 1 based on a guaranteed rate of 95%	16.7
Annual inflow of Zone 2 based on a guaranteed rate of 95%	1.5
Local inflow of Zone 3	5.5

In Table 2, according to the flow, the annual water demands and water receding coefficients of the three consumptive water users outside the river (i.e., Agents 1, 2 and 4) are presented.

Table 2. The annual water demands and water receding coefficients of Agents 1, 2, and 4.

Agent 1		Agent 2		Agent 4	
annual water demand (million m^3/s)	water receding coefficient	annual water demand (million m^3/s)	water receding coefficient	annual water demand (million m^3/s)	water receding coefficient
510	0.45	560	0.44	3150	0.59

In Table 3, the ecological flow requirement of Agents 3 and 5 are presented, which is the minimum flow of the river to avoid the occurrence of the phytoplankton blooms.

Table 3. The ecological flow requirement of Agents 3 and 5 (m^3/s).

Xiangyang Section		Xiantao Section		
May–October	November–Next April	May–October	November–Next April	February–March
632.3	379.4	623.7	374.3	500

4. Results and Discussion

As mentioned in Section 3.2, interest agents are correlated with each other through the runoff. It has been mentioned above that there are five agents in the Hanjiang River Basin. When the basics of the basin are put into the model, and are solved by the bi-level optimization model with dynamic programming, we see the results as shown in Figure 3.

In Figure 3, the bar in different colors represent different models. Compared to the single-objective individual optimization model (in brief, Scheme 1), the total benefits of the upper model (in brief, Scheme 2) increased by 8%. Since Agent 1 is located in the most upstream river, which could allow for it to take advantage of its spatial location, assuming its individual interests reach the maximum in having enough water resources. Except for Agent 2, whose individual interests increased by 49%, the individual interests of Agents 1, 3, 4, and 5 are decreased by 8%, 10%, 12%, and 2%, respectively, when comparing Scheme 1 with Scheme 2. Although the common interest of the upper optimization model is improved, it is at the sacrifice of some individual interests. In conclusion, only Agent 2 is willing to cooperate, while the rest of the agents are not.

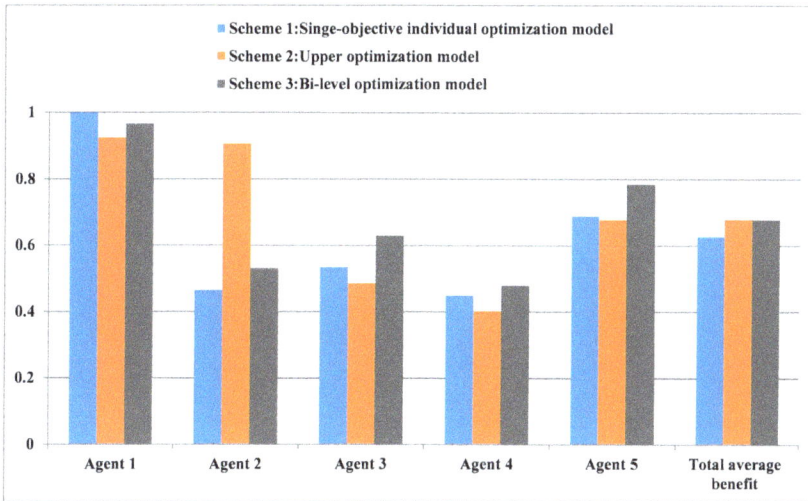

Figure 3. Multi-agent benefits of different optimization models.

Compared to Scheme 1, the total benefits of the bi-level optimization model (in brief, Scheme 3) also increased by 8%. From the perspective of individual interests, except for Agent 1, whose individual interests decreased by 3%, the individual interests of Agents 2, 3, 4, and 5 are increased by 14%, 18%, 7%, and 14%, respectively, comparing Scheme 1 with Scheme 3. Evidently, Scheme 3, which not only emphasizes the common interest but also considers individual interests, is in favor of the cooperation of the agents in the basin, because a slight decline in individual interests can contribute greatly to great increases of other individual and common interests.

When comparing Scheme 2 with Scheme 3, it can be found that the total benefits are the same. However, in Scheme 2, 4 out of 5 of the individuals are unwilling to cooperate due to their damaged interests. Whereas, 4 out of 5 of the individuals tend to cooperate in Scheme 3. Scheme 2 emphasizes the common interest but neglects individual interests, thereby causing in efficiency and inequity. On the contrary, Scheme 3 realized the optimization of the common interest, considering individual interests simultaneously, through bottom-up feedback with optimal individual interests optimized and top-down coordination with the optimal common interest.

5. Conclusions

In this study, the bi-level optimization model is proposed for water resource allocation. Based on the cooperative game-based theory, the bi-level model consists of the superiors model and subordinates model, which could realize a win-win cooperation of common and individual interests in water resource allocation. The following conclusions are drawn, based on the results of the case study of the Hanjiang River Basin.

The results of the case study indicate the effectiveness and fairness of the bi-level optimization model. Compared to the single-objective individual optimization model, our model can bring more common interest, by 8% in total. Four out of five of the interest agents can get more benefits, by 7%–18%, although one out of five of the interest agents' benefit decreases, only by 3%. Compared to the upper optimization model, from the perspective of the common interest, there is no improvement. However, it must be realized that in the upper model, only one out of five of the interest agents can get more benefits, so that this model is only ideal, since most of the interest agents will not choose to cooperate. In a word, both common and individual interests are improved by our model, and these agents are more likely to cooperate to get more benefits.

From the case study, there are some recommendations for decision and policy makers. Before making decisions, policy makers are advised to realize the needs and acceptable values of agents adequately, and adjust the decisions according to the feedback from agents. Afterwards, the allocation could be efficient and equitable.

The bi-level optimization model can be easily applied into other basins, and is possible to solve more problems related to water resources. However, there are still some areas to improve, which require further research.

Author Contributions: Conceptualization, Q.H. and X.F.; Methodology, Q.H. and X.F.; Software, Q.H. and X.F.; Validation, G.T., Y.M. and Z.Y.; Formal Analysis, Z.Y.; Investigation, G.T.; Resources, Q.H. and X.F.; Data Curation, Q.H. and X.F.; Writing-Original Draft Preparation, Q.H.; Writing-Review & Editing, X.F. and G.T.; Visualization, Z.Y.; Supervision, Y.M.; Project Administration, Y.M.; Funding Acquisition, X.F.

funding: This study is financially supported by the National Key Research and Development Project of China (Grant No. 2016YFC0401306).

Conflicts of Interest: The authors declare no conflict of interest.

References

1. Oki, T.; Kanae, S. Global hydrological cycles and world water resources. *Science* **2006**, *313*, 1068–1072. [CrossRef] [PubMed]
2. Wang, L.; Fang, L.; Hipel, K.W. Basin-wide cooperative water resources allocation. *Eur. J. Oper. Res.* **2008**, *190*, 798–817. [CrossRef]
3. Lee, C.S. Multi-objective game-theory models for conflict analysis in reservoir watershed management. *Chemosphere* **2012**, *87*, 608–613. [CrossRef] [PubMed]
4. Ming, B.; Liu, P.; Guo, S.; Zhang, X.; Feng, M.; Wang, X. Optimizing utility-scale photovoltaic power generation for integration into a hydropower reservoir by incorporating long- and short-term operational decisions. *Appl. Energy* **2017**, *204*, 432–445. [CrossRef]
5. Madani, K. Game theory and water resources. *J. Hydrol.* **2010**, *381*, 225–238. [CrossRef]
6. Fayaed, S.S.; El-Shafie, A.; Jaafar, O. Reservoir-system simulation and optimization techniques. *Stoch. Environ. Res. Risk Assess.* **2013**, *27*, 1751–1772. [CrossRef]
7. Madani, K.; Hooshyar, M. A game theory–reinforcement learning (GT–RL) method to develop optimal operation policies for multi-operator reservoir systems. *J. Hydrol.* **2014**, *519*, 732–742. [CrossRef]
8. Nafarzadegan, A.R.; Vagharfard, H.; Nikoo, M.R.; Nohegar, A. Socially-Optimal and Nash Pareto-Based Alternatives for Water Allocation under Uncertainty: An Approach and Application. *Water Res. Manag.* **2018**, *32*, 2985–3000. [CrossRef]
9. Sadegh, M.; Mahjouri, N.; Kerachian, R. Optimal Inter-Basin Water Allocation Using Crisp and Fuzzy Shapley Games. *Water Res. Manag.* **2009**, *24*, 2291–2310. [CrossRef]
10. Ambec, S.; Ehlers, L. Sharing a river among satiable agents. *Games Econ. Behav.* **2008**, *64*, 35–50.
11. Cavalcanti, C.; Schläpfer, F.; Schmid, B. Public participation and willingness to cooperate in common-pool resource management: A field experiment with fishing communities in Brazil. *Ecol. Econ.* **2010**, *69*, 613–622. [CrossRef]

12. Madani, K.; Dinar, A. Non-cooperative institutions for sustainable common pool resource management: Application to groundwater. *Ecol. Econ.* **2012**, *74*, 34–45. [CrossRef]
13. Madani, K.; Dinar, A. Cooperative institutions for sustainable common pool resource management: Application to groundwater. *Water Resour. Res.* **2012**, *48*. [CrossRef]
14. Madani, K.; Zarezadeh, M.; Morid, S. A new framework for resolving conflicts over transboundary rivers using bankruptcy methods. *Hydrol. Earth Syst. Sci.* **2014**, *18*, 3055–3068. [CrossRef]
15. Adams, G.; Rausser, G.; Simon, L. Modelling multilateral negotiations an application to California Water Policy. **1996**, *30*, 97–111. [CrossRef]
16. Nash, J. Two-person cooperative games. *Econometrica* **1953**, *21*, 128–140. [CrossRef]
17. Farhadi, S.; Nikoo, M.R.; Rakhshandehroo, G.R.; Akhbari, M; Alizadeh, M.R. An agent-based-nash modeling framework for sustainable groundwater management: A case study. *Agric. Water Manag.* **2016**, *177*, 348–358. [CrossRef]
18. Madani, K. Hydropower licensing and climate change: Insights from cooperative game theory. *Adv. Water Resour.* **2011**, *34*, 174–183. [CrossRef]
19. Yakowitz, S. Dynamic Programming Applications in Water Resources. *Water Resour. Res.* **1982**, *18*, 673–696. [CrossRef]
20. Yang, G.; Guo, S.; Li, L.; Hong, X.; Wang, L. Multi-Objective Operating Rules for Danjiangkou Reservoir Under Climate Change. *Water Resour. Manag.* **2015**, *30*, 1183–1202. [CrossRef]
21. Ma, T.; Zhou, W.; Chen, L.; Wu, L.; Peter, C.; Dai, G. Concerns about the future of Chinese fisheries based on illegal, unreported and unregulated fishing on the Hanjiang river. *Fish. Res.* **2018**, *199*, 212–217. [CrossRef]
22. Li, W.; Qin, Y.; Sun, Y.; Huang, H.; Ling, F.; Tian, L.; Ding, Y. Estimating the relationship between dam water level and surface water area for the Danjiangkou Reservoir using Landsat remote sensing images. *Remote Sens. Lett.* **2015**, *7*, 121–130. [CrossRef]
23. Xie, A.; Liu, P.; Guo, S.; Zhang, X.; Jiang, H.; Yang, G. Optimal Design of Seasonal Flood Limited Water Levels by Jointing Operation of the Reservoir and Floodplains. *Water Resour. Manag.* **2017**, *32*, 179–193. [CrossRef]
24. Lian, J.; Sun, X.; Ma, C. Multi-year optimal operation strategy of Danjiangkou Reservoir after dam heightening for the middle route of the south-north water transfer project. *Water Sci. Tech. Water Supply* **2016**, *16*, 961–970. [CrossRef]
25. Liu, D.; Guo, S.; Shao, Q.; Liu, P.; Xiong, L.; Wang, L.; Hong, X.; Xu, Y.; Wang, Z. Assessing the effects of adaptation measures on optimal water resources allocation under varied water availability conditions. *J. Hydrol.* **2018**, *556*, 759–774. [CrossRef]
26. Liu, H.; Yin, J.; Feng, L. The Dynamic Changes in the Storage of the Danjiangkou Reservoir and the Influence of the South-North Water Transfer Project. *Sci. Rep.* **2018**, *8*. [CrossRef] [PubMed]

MDPI

St. Alban-Anlage 66

4052 Basel

Switzerland

Tel. +41 61 683 77 34

Fax +41 61 302 89 18

www.mdpi.com

Water Editorial Office

E-mail: water@mdpi.com

www.mdpi.com/journal/water

www.ingramcontent.com/pod-product-compliance
Lightning Source LLC
Chambersburg PA
CBHW051701210326
41597CB00032B/5333